The Geological Society of America
Memoir 164

Blueschists and Eclogites

Edited by

Bernard W. Evans
Department of Geological Sciences
University of Washington
Seattle, Washington 98195

Edwin H. Brown
Department of Geology
Western Washington University
Bellingham, Washington 98225

1986

Published by The Geological Society of America, Inc.
3300 Penrose Place, P.O. Box 9140, Boulder, Colorado 80301

Printed in U.S.A.

GSA Books Science Editor Campbell Craddock

Library of Congress Cataloging-in-Publication Data

Blueschists and eclogites.

   (Memoir / Geological Society of America ; 164)
   Bibliography: P.
   1. Schists—Congresses.   2. Eclogite—Congresses.
I. Evans, Bernard W., 1934–   .   II. Brown, Edwin H.
III. Series: Memoir (Geological Society of America ;
164)
QE475.S3B55   1986   552'.4   86-3082
ISBN 0-8137-1164-9

Contents

Contents

Foreword

A Penrose Conference on "Blueschists and Related Eclogites" was held in Bellingham, Washington, in September 1983. Participants at the conference responded enthusiastically at the time to the suggestion of a volume that would accommodate the results of their ongoing research and be a permanent reminder of their experiences at the meeting. Thus, this volume does not represent the proceedings of the Blueschist Conference, but rather is a collection of papers on that theme submitted for a deadline a year later in response to a general invitation to participants and non-participants active in the blueschist field. We like to think that the sharing of facts and ideas that took place at the conference helped authors to sharpen their observations and refine their conclusions, and generally advanced our understanding of the geological problems presented by high-pressure metamorphic terranes. Rather than an all-encompassing review of the state of knowledge of the subject, the reader will find here a series of papers by authors, not all of whom are in agreement on the various issues, covering aspects of high-pressure metamorphic minerals, phase equilibria, regional petrology, tectonics, structure, and radiometric dating. After three introductory papers on high-pressure phase equilibria, the bulk of the volume contains regional contributions arranged geographically: ten papers from south to north through the western North American Cordillera, eight from north to south through Europe (from Spitzbergen to the Mediterranean), six from west to east through Asia (from Turkey to Japan), and one from New Caledonia. Although the geographic coverage is extensive, reflecting the strong international flavor of the conference, it is not a complete review of the high-pressure belts of the world. Nevertheless, it is hoped that the enthusiast for high-pressure rocks will find much new material of interest here from most of the blueschist terranes of the world.

For decades, blueschists and related rocks—including eclogites—have fascinated petrologists of a tectonic bent because their mineralogy, requiring a history of rapid deep burial and then rapid uplift, seriously challenges our concepts of earth dynamics. The mechanisms of plate tectonics have fit well the requirements for blueschist formation, and the presence of such rocks is today commonly held as primary evidence of an ancient subduction event. To the extent that this connection is valid, a history of the subduction processes can be read from the preserved fabrics and minerals of blueschists.

Interpretation of blueschist facies rocks in light of a tectonic event requires a diversity of geologic tools. Starting with characterization of the protolith, a blueschist researcher may draw upon paleontology, stratigraphy, radiometric dating, and sedimentary and igneous petrology. The metamorphism is understood in terms of field mapping, structural analysis, petrographic study, experimental petrology, thermodynamics, and radiometric dating. The papers in this volume report both on refinements of these tools and on their regional applications. Important constraints on the tectonic processes of blueschist formation emerge.

Uncertainties and problems in blueschist genesis remain and will provide rewarding topics of future research. Calibration of the petrogenetic grid is only approximate, especially with regard to pressure. New minerals of petrologic importance almost certainly remain to be discovered. Kinematic analysis of blueschist metamorphism is in its infancy and promises to provide important connections to tectonics. The rapid uplift and emplacement process by which blueschists are rescued from subduction or collision is obscure, but it is amenable to elucidation by imaginative research. As our calibration of mineral equilibria and our understanding of the controls of mineral zoning improve, blueschist history will increasingly be unravelled in time, pressure, temperature, and strain.

B. W. Evans
E. H. Brown
May, 1985

Geological Society of America
Memoir 164
1986

Experimental investigations of blueschist-greenschist transition equilibria: Pressure dependence of Al₂O₃ contents in sodic amphiboles—A new geobarometer

Shigenori Maruyama*
Moonsup Cho
J. G. Liou
Department of Geology
Stanford University
Stanford, California 94305

ABSTRACT

The blueschist-greenschist facies transition for a model basaltic system Na_2O-CaO-MgO-Al_2O_3-SiO_2-H_2O is defined by a univariant reaction: 6 clinozoisite + 25 glaucophane + 7 quartz + 14 H_2O = 6 tremolite + 9 chlorite + 50 albite; for the Fe_2O_3-saturated basaltic system, by a discontinuous one: 4 epidote + 5 Mg-riebeckite + chlorite + 7 quartz = 7 hematite (magnetite) + 4 tremolite + 10 albite + 7 H_2O. These two reactions were experimentally investigated to determine the nature of the blueschist-greenschist transition. The results have located the first reaction at 350 ± 10°C, 7.8 ± 0.2 Kb and 450 ± 10°C, 8.2 ± 0.4 Kb. Reconnaissance experiments for the second reaction indicate that the minimum pressure for the occurrence of epidote + Mg-riebeckite + chlorite + quartz is about 4 Kb at 300°C for f_{O_2} defined by the hematite-magnetite buffer.

The presently determined P-T location for the blueschist-greenschist transition in the Fe-free basaltic system is about 3 Kb lower than the minimum pressure limit of glaucophane of Carman and Gilbert (1983), but is compatible with the revised stability field of jadeite + quartz determined by Holland (1980). Introduction of Fe^{3+} into the model basaltic system significantly lowers the minimum pressure limit for occurrence of the buffered assemblage sodic amphibole + epidote + actinolite + chlorite + albite + quartz, and the participating phases gradually increase their Fe^{3+}/Al ratio with decreasing pressure. Isopleths of sodic amphibole composition in the buffered assemblage in terms of X_{Gl} are delineated and the effect of Fe^{2+} and temperature on the isopleths are discussed.

The Al_2O_3 content of sodic amphibole coexisting with epidote + actinolite + chlorite + albite + quartz decreases systematically with decreasing pressure and hence can be used as a geobarometer. Pressure estimates for metabasites at Ward Creek of the Franciscan terrane, the Mikabu greenstones of the Sanbagawa belt, the Otago schists of Lake Wakitipu, New Zealand, and the blueschists at Ouegoa, New Caledonia, based on the proposed glaucophane geobarometry, are in agreement with those derived from sodic pyroxene geobarometry.

---

*Present address: Department of Earth Sciences, Toyama University, Toyama, Japan.

INTRODUCTION

The stability of glaucophane with end-member composition $oNa_2Mg_3Al_2Si_8O_{22}$ $(OH)_2$ has been extensively investigated, but the results are not consistent (for reviews, see Maresch, 1977; Carman and Gilbert, 1983). The experimental studies by Ernst (1961, 1963) received a lot of criticism regarding synthetic "glaucophane phases" and proposed polymorphic transition of glaucophane. Carman (1974) and Gilbert and Popp (1973) experimentally delineate and locate the terminal reactions governing the high- and low-pressure limits of glaucophane; their data were recently published (Carman and Gilbert, 1983). Maresch (1977) systematically reviews previous studies and concludes that laboratory synthesis of the end-member glaucophane has not been accomplished. He presents additional experimental data employing natural glaucophane as starting material. Koons (1982) further explores glaucophane stability and concludes that stoichiometric glaucophane does not exist in the NMASH system, and a glaucophane-like amphibole was stable only in an H_2O-undersaturated environment.

In spite of difficulty in laboratory synthesis of end-member glaucophane, differences in terminal reactions governing glaucophane stability, and the differences of 4 to 5 Kb in the minimum pressure value for glaucophane stability, all previous experimental data nearby indicate that glaucophane requires high pressure and low temperature for its formation. This conclusion is compatible with the results deduced from investigations of natural parageneses in many blueschist terranes (e.g., Eskola, 1939; Miyashiro, 1961, 1973; Ernst, 1963, 1981). However, the large discrepancies among the previous glaucophane experiments allow neither precise constraints on P-T conditions for blueschist facies metamorphism nor retrieval calculations of the thermodynamic properties of glaucophane. Moreover, application of the stability of glaucophane in its own bulk composition to natural blueschists is significantly limited by the compositional complexity of sodic amphibole, by the variation in stability of glaucophane with different compositions, and by the variation in coexisting phases in natural blueschists.

Eskola (1939) first proposed the glaucophane-schist facies for metabasites containing glaucophane as a major constituent. Other authors prefer the name of blueschist facies (e.g., Ernst, 1981). Miyashiro (1973, p. 69) expands the definition of blueschist facies to include both glaucophane-bearing metabasites and associated isophysical non-glaucophanitic metamorphic rocks, and concludes that the blueschist facies "may really be intermediate between the more typical glaucophane schist facies and some other facies (e.g., greenschist facies)." Transitional assemblages from a typical blueschist assemblage of glaucophane + lawsonite + chlorite + sodic pyroxene + sphene, to a greenschist assemblage of actinolite + epidote + chlorite + albite + quartz + sphene are common in major subduction zone complexes.

In order to define the phase relations and physical-chemical conditions for the formation of blueschist facies assemblages, basaltic compositions should be considered. In detailed arguments

described elsewhere for low-grade metamorphism of basaltic rocks, we selected the system Na_2O-CaO-MgO-Al_2O_3-SiO_2 H_2O (NCMASH) to model phase equilibria and mineral assemblages for basaltic systems (Liou and others, 1985). For the basaltic composition in this model system, the glaucophane-bearing assemblages are bounded by (1) Pm + Chl + Ab = Cz + Gl + H_2O, marking the transition between the pumpellyite-actinolite and blueschist facies, (2) Cz + Gl + Qz + H_2O = Tr + Chl + Ab, marking the transition between the blueschist and greenschist facies, and (3) Cz + Gl + Qz + H_2O = Hb + Chl + Ab, marking the transition between the blueschist and epidote amphibolite facies (for abbreviations of phases and metamorphic facies see Table 1). Reaction (2) was selected for detailed study in order to establish the nature of the blueschist-greenschist facies transition in both model and natural basaltic systems. The effect of introduction of Fe_2O_3 into the model system was also experimentally and thermodynamically evaluated. Potential use of the Fe^{3+}/Al ratio of sodic amphibole in buffered assemblages as a geobarometer is proposed. The new geobarometer is employed to obtain pressure estimates for blueschist facies metamorphism in the Franciscan of California, New Caledonia, New Zealand, and Sanbagawa terranes of Japan.

EXPERIMENTAL METHODS

Two reactions were experimentally investigated: (1) 6 clinozoisite + 25 glaucophane + 7 quartz + 14 H_2O = 6 tremolite + 9 chlorite + 50 albite for the model system, and (2) 4 epidote + 5 Mg-riebeckite + chlorite + 7 quartz = 7 hematite (magnetite) + 4 tremolite + 10 albite + 7 H_2O for the Fe^{3+}-saturated system. These reactions were studied by employing conventional hydrothermal apparatus and procedures as previously described (e.g., Liou and others, 1983a). Argon was used as the pressure medium for experiments above 2 Kb and H_2O for those at and

TABLE 1. ABBREVIATIONS OF PHASES* AND METAMORPHIC FACIES
USED IN THIS PAPER

Gl = Glaucophane	Qz = Quartz
MRi = Magnesioriebeckite	Ab = Albite
Tr = Tremolite	An = Anorthite
Act = Actinolite	Jd = Jadeite
Hb = Hornblende	Ac = Acmite
Amp = Amphibole	
	HM = Ht-Mt buffer
Bar = Barroisite	Gr = Grossular
Ph = Phengite	Tc = Talc
Chl = Chlorite	Na-Ph = Na-phlogopite
Lw = Lawsonite	Py = Pyrope
Pm = Pumpellyite	
Gar = Garnet	Sph = Sphene
Cz = Clinozoisite	Ht = Hematite
Ep = Epidote	GS = Greenschist facies
Ps = Pistacite	
BS = Blueschist facies	
PA = Pumpellyite-actinolite facies	

*Compositions of most of these phases are the same as those in Liou and others, 1985.

TABLE 2. COMPOSITIONS OF MINERALS USED AS STARTING MATERIALS
FOR THE PRESENT EXPERIMENTAL STUDY

Mineral	Clinozoisite	Epidote	Glaucophane	Mg-riebeckite	Tremolite
SiO_2	40.3	36.8	57.8	53.5	56.5
TiO_2	-	-	-	0.00	0.12
Al_2O_3	30.7	21.4	12.0	0.45	0.98
Fe_2O_3*	2.56	15.5	-	24.5	-
FeO**	-	-	6.45	-	0.33
MnO	-	-	-	0.05	0.03
MgO	0.74	-	13.0	12.5	23.2
CaO	22.6	23.0	1.04	0.27	13.3
Na_2O	-	-	7.00	5.96	0.74
K_2O	-	-	0.67	0.44	0.32
Total	97.10	96.70	97.96	97.67	95.85
	O=12.5	O=12.5	O=23	O=23	O=23
Si	3.090	2.996	7.823	7.649	7.872
Ti	-	-	-	0.000	0.012
Al	2.773	2.050	1.915	0.076	0.161
Fe^{3+}	0.148	0.948	-	2.636	-
Fe^{2+}	-	-	0.730	-	0.034
Mn	-	-	-	0.060	0.004
Mg	0.085	-	2.624	2.664	4.815
Ca	1.856	2.004	0.150	0.041	1.993
Na	-	-	1.837	1.653	0.119
K	-	-	0.115	0.080	0.056
X_{Fe}^{3+}	0.05	0.32			
Σ			15.194	14.858	15.146

*Total Fe as Fe_2O_3
**Total Fe as FeO

below 2 Kb P_{fluid}. The pressure fluctuations were within ±1 percent and temperatures within ±5°C for all experiments. The hematite-magnetite solid oxygen buffer was used for controlling the oxygen fugacity for the second reaction.

Synthetic and natural phases were employed for preparation of the starting materials for the present experiments. These phases include (1) synthetic α-quartz obtained at 2 Kb and 600°C for 18 days, (2) synthetic chlorite of composition $Mg_5Al_2Si_3O_{10}$ $(OH)_8$ obtained from oxide mixtures at 400°C and 2-5 Kb for more than 50 days, (3) natural Amelia albite with X_{An} of 0.01, and (4) natural clinozoisite, epidote, glaucophane, Mg-riebeckite and tremolite. The analyzed compositions of these natural phases are listed in Table 2. It should be pointed out that the natural glaucophane from the Sesia-Lanzo zone in Italy contains 6.45 wt% FeO as total Fe and a minor amount of CaO; the effect of Fe_2O_3 on reaction (1) will be discussed later. These natural phases were carefully separated, purified, and examined by both X-ray diffraction (XRD) and Scanning Electron Microscopy (SEM). The starting materials consisted of reactants and products in subequal proportion together with excess H_2O.

Experimental run products were routinely examined by the petrographic microscope, and by XRD and SEM with Energy Dispersive Analysis (EDAX), and the phase assemblages deter-

mined. Even after 3 months' run at T = 300°-400°C, all the experimental charges contained both reactants and products. The presence of trace amounts of additional phases was searched for by the SEM, but they were not detected. Stability relations were determined by observing which phases grew at the expense of others and the SEM observations were crucial for determination of reaction directions.

EXPERIMENTAL RESULTS

Reaction (1): 6 Cz + 25 Gl + 7 Qz + 14 H_2O = 6 Tr + 9 Chl + 50 Ab

Eighteen experiments were completed for this reaction at temperatures of about 300, 350 and 450°C. The P - T conditions, run durations and growth of glaucophane or tremolite are listed in Table 3 and graphically summarized in Fig. 1. All the product phases are very fine-grained and quantitative analysis by microprobe was not possible. Albitic plagioclases have consistently negative relief; however, their compositions were not determined.

For experiments at pressures less than 5 Kb, growth of the greenschist assemblage at the expense of the blueschist assemblage is obvious by comparison of X-ray diffractograms of run

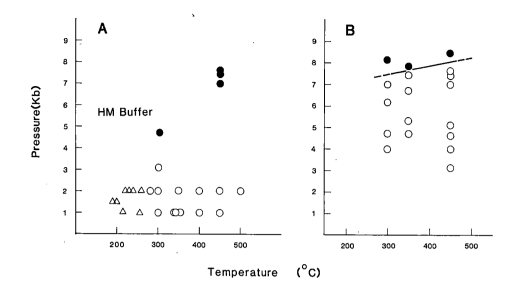

Figure 1. P_{fluid}-T diagrams showing experimental results for (A): reaction (2) $4 Ep + 5 MRi + Chl + 7 Qz = 4 Tr + 10 Ab + 7 Ht + 7 H_2O$ at f_{O_2} defined by the HM buffer; and (B): reaction (1) $6 Cz + 25 Gl + 7 Qz + 14 H_2O = 6 Tr + 9 Chl + 50 Ab$. Closed circles: growth of the high-P assemblage; open circles: growth of the low-P assemblages and open triangles: experiments showing no apparent reaction (see run data in Tables 3 and 4).

products and starting materials. However, such a task is not easy for runs above 7 Kb simply because the reaction rate for the experiments near the equilibrium boundary is extremely sluggish. In such experiments, detailed observations of run products by the SEM with EDAX were extremely useful and the reaction direc-

TABLE 3. EXPERIMENTAL RUN DATA FOR REACTION (1):
Gl + Cz + Qz + H₂O = Tr + Chl + Ab
FOR THE MODEL BASALTIC SYSTEM
(All run products contain
both reactants and products)

Run #	Pressure (Kb)	Temperature (OC)	Run duration (days)	Gl	Tr*
1	4.0	447	63	o	●
2	4.6	450	63	o	●
3	5.1	450	63	o	●
4	8.4	450	63	●	o
5	7.6	450	56	o	●
6	7.4	450	56	o	●
7	3.1	451	56	o	●
8	7.0	450	56	o	●
9	6.7	350	87	o	●
10	4.7	303	56	o	●
11	7.4	350	87	o	●
12	4.6	350	87	o	●
13	5.3	350	87	o	●
14	7.8	350	87	●	o
15	4.0	300	222	o	●
16	6.2	300	65	o	●
17	7.0	300	68	o	●
18	8.1	300	222	●	o

*Reaction direction showing growth of glaucophane ● or tremolite ● (determined by XRD and SEM methods).

tion was deduced by observing overgrowth or dissociation of glaucophane or tremolite. Several examples are illustrated in Fig. 2. Glaucophane is not stable at 350°C, 7.4 Kb as indicated by the occurrence of rounded edges and corners of glaucophane crystals together with the growth of prismatic fine-grained tremolite (Fig. 2-A, -B, -C). The sharp difference in EDAX intensities of Na, Ca and Al peaks provided positive identification of tremolite versus glaucophane. Therefore, at this P and T, the blueschist facies assemblage is not stable.

On the other hand, the growth of the blueschist facies assemblage in the 222-day experiment at 8.1 Kb and 300°C is indicated by the overgrowth of fine prisms of glaucophane on a coarse-grained tremolite (Fig. 2-D, -E, -F). Some fine-grained aggregates of quartz crystals were also found in this and other runs together with clinozoisite at 7.8 Kb and 350°C (Fig. 2-G). The growth of quartz and glaucophane in these experiments suggests that the blueschist facies assemblage is stable at these P-T conditions and the reaction boundary is located at pressures less than 8.1 Kb at 300°C and 7.8 Kb at 350°C. From the available data shown in Table 3 and Fig. 1, we conclude that the boundary between the blueschist and greenschist facies assemblages for the model basaltic system occurs at 350 ± 10°C, 7.8 ± 0.2 Kb and 450 ± 10°C, 8.2 ± 0.4 Kb and the reaction possesses a very gentle P-T slope which is consistent with that deduced by Brown (1974).

Reaction (2): 4 Ep + 5 MRi + Chl + 7 Qz = 4 Tr + 10 Ab + 7 Ht + 7 H₂O

This reaction was investigated at the f_{O_2} defined by the HM buffer. Mg-riebeckite (MRi) is used in this paper as an end-member component. Reaction (2) is a discontinuous reaction and has only one degree of freedom. Thus at a given f_{O_2} defined by the HM buffer, the composition of sodic amphibole will be fixed. However, it should be emphasized that sodic amphibole of the reaction (2) is not always an end-member MRi. As explained in the later section, at higher f_{O_2} than that defined by the HM buffer, sodic amphibole may contain more MRi component than that of reaction (2) fixed at f_{O_2} of the HM buffer. The purposes of the present experiments on this reaction are to delimit the low-pressure limit of Mg-riebeckitic amphibole for the basaltic composition, and to model the nature of the blueschist-greenschist transition involving sodic amphibole in a sliding equilibrium as a function of P, T and bulk composition.

Twenty-five experiments were completed for this reaction; run data are listed in Table 4 and diagrammatically shown in Fig. 1. Most experiments were run for one to two months. Those runs at 7.0 to 7.6 Kb and 450°C yield unambiguous results with regard to the reaction direction. One SEM photo on the run product for 7Kb, 450°C is shown in Fig. 2-H which illustrates the growth of Mg-riebeckitic amphibole. At 400°-500°C, 2 Kb, the growth of the tremolite-bearing assemblage can be detected by XRD only. Tremolites in these run products have a very low Fe peak on the EDAX. Apparently, at such a high oxidation state, tremolite may not be able to accommodate much Fe. However, those runs at temperatures below 250°C are inconclusive even after 2-months' duration. Preliminary data indicate the minimum pressure for the occurrence of epidote + Mg-riebeckitic amphibole + chlorite + quartz is about 4 Kb at 300°C for f_{O_2} values defined by the hematite-magnetite buffer.

Comparison of Glaucophane Stability in the Model System with Those by Previous Studies

Fig. 3 summarizes the glaucophane stabilities from the present and previous experimental and field studies (Maresch, 1973, 1977; Ernst, 1979; Koons, 1982; and Carman and Gilbert, 1983). It is apparent from Fig. 3 that the low-pressure stability limit of glaucophane of its Mg-end member composition extrapolated to 300° and 400°C from Carman and Gilbert is 4 to 5 Kb higher than that determined by Maresch. Such a large discrepancy has not been discussed before but is believed to be due to the difference in starting materials (synthetic glaucophane vs. natural Fe-bearing glaucophane), and in experimental facilities (piston-cylinder vs. low-T cold-seal) and other factors which are not known at present. The results of Carman and Gilbert (1983) are difficult to apply to natural parageneses and the low-P terminal reaction of glaucophane + H_2O to Na-phlogopite + talc solid solution + albite has not been verified from blueschist assemblages. On the other hand, Maresch (1977) compared his experimental data with those P-T estimates of natural blueschists by De Roever and Beunk (1976), Winkler (1979), and others. Hence, Maresch's maximum stability field of glaucophane has

TABLE 4. EXPERIMENTAL RUN DATA FOR REACTION (2):
Ep + MRi + Chl + Qz = Tr + Ab + HM + F
FOR THE Fe_2O_3-SATURATED SYSTEM
(All run products contain
both reactants and products)

Run #	Pressure (Kb)	Temperature (°C)	Run duration (days)	MRi	Tr*
21	2.0	301	65	o	●
22	1.0	400	65	o	●
23	1.0	450	65	o	●
24	1.0	340	65	o	●
25	1.0	355	65	o	●
26	2.0	221	59	o	o
27	2.0	230	59	o	o
28	2.0	240	59	o	o
29	2.0	260	59	o	o
30	2.0	280	59	o	●
31	1.5	190	59	o	o
32	1.5	200	59	o	o
33	2.0	500	24	o	●
34	2.0	450	24	o	●
35	2.0	400	24	o	●
36	2.0	350	33	o	●
37	1.0	346	33	o	●
38	1.0	303	33	o	●
39	1.0	255	33	o	o
40	1.0	215	33	o	o
41	7.6	450	56	●	o
42	7.4	450	56	●	o
43	3.1	451	56	o	●
44	4.7	303	56	●	o
45	7.0	450	56	●	o

*Reaction direction showing growth of magnesiorie-beckite ● or tremolite ● (determined by the XRD and SEM methods).

been widely used and it is consistent with the general scheme of deduced low-T stability fields of various amphiboles suggested by Ernst (1979).

Such a large discrepancy in glaucophane stability as shown in Fig. 3 emphasizes that the stability of glaucophane is highly dependent on starting materials and bulk composition (hence mineral assemblage); its stability in the blueschist-greenschist transition is governed by a continuous reaction. Moreover, the stability of glaucophane on its own composition cannot be used to decipher the phase relations and mineral parageneses of glaucophane-bearing metabasites. The present experimentally determined reaction (1) was suggested by Miyashiro and Banno (1958) and Brown (1974) from geological and petrological observations, and delimits the maximum stability of glaucophane in the model basaltic system. As shown in Fig. 3, reaction (1) occurs at pressures intermediate between those of Carman and Gilbert (1983) on synthetic Fe-free glaucophane and that of Maresch (1973) on natural Fe-bearing glaucophane. The use of natural glaucophane with a substantial Fe content (see Table 2 for composition) as starting material in the present study suggests that the reaction (1) in the Fe-free model system should occur at slightly higher pressures (by about 0.5 Kb from our calculation) than those determined in the present study. Introduction of Fe into the model system significantly lowers the pressure limit of glaucophane and causes the reaction (1) to be multivariant. Both blue-

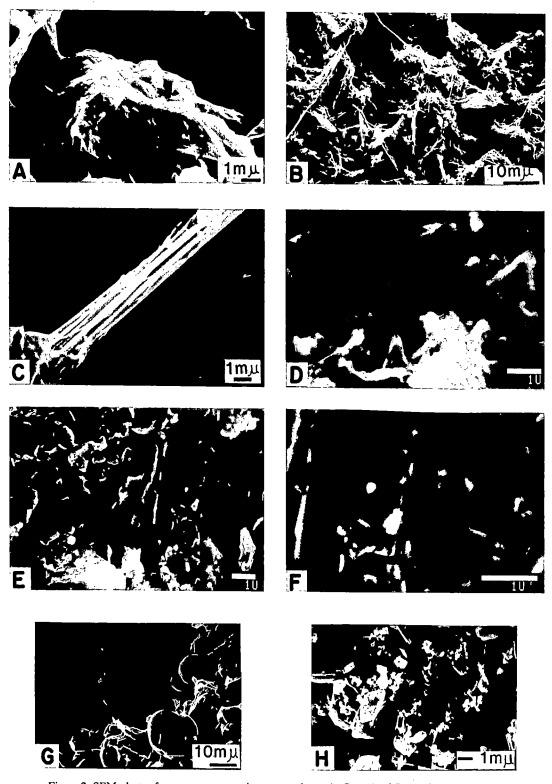

Figure 2. SEM photos for some representative run products: A. Growth of fine-grained tremolite on dissolved surface of glaucophane (Run #11, 7.4 Kb, 350°C, 87 days); B. Growth of fibrous tremolites (Run #11); C. Tremolite needles with chlorite tablets (Run #11); D. Growth of very fine-grained quartz spherules (Run #8, 8.1 Kb, 300°C, 222 days). E. Growth of fine glaucophane prisms mantled around relict tremolite crystal (Run #18). F. Enlarged view of E. G. Newly grown quartz (?) spherules (Run #14, 7.8 Kb, 350°C, 87 days). H. Growth of idiomorphic Mg-riebeckite amphibole (Run #25, 7.0 Kb, 450°C, 56 days).

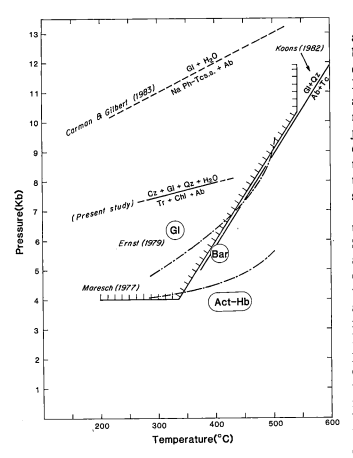

Figure 3. P_{fluid}-T diagram comparing the stabilities of glaucophane and glaucophane-bearing assemblages from present and previous investigations.

schist and greenschist facies assemblages occur together over a large P-T zone.

PHASE RELATIONSHIP FOR BLUESCHIST FACIES ASSEMBLAGES

For high-pressure metamorphic facies series where glaucophane-crossite is ubiquitous for basaltic assemblages, a projection different from those described in Liou and others (1985) was used. As shown in Fig. 4, compositions of lawsonite, clinozoisite, pumpellyite, grossular-pyrope garnet, diopside, glaucophane, tremolite and chlorite are plotted in a CaO-Al_2O_3-MgO diagram. The compositions of these phases are projected from chlorite of clinochlore composition onto two projection lines: one along the Al_2O_3-CaO join used by Brown (1977a) and the other on a *projection line* which is nearly normal to the Al_2O_3-CaO join. The former projection used by Brown has a very compact compositional relation for most blueschist facies minerals, and plots the composition of glaucophane on the negative side. The latter projection, however, spreads the compositions of most phases and plots both tremolite and glaucophane on the same positive side of the diagram.

In order to illustrate the phase relations for blueschist facies assemblages, the Fe^{3+}-component should be taken into account because most Ca-Al silicates as well as sodic amphiboles show extensive Al - Fe^{3+} substitutions. The projection relations in the Fe-bearing model system, Na_2O-CaO-Al_2O_3-Fe_2O_3-FeO-MgO, are illustrated in Fig. 5. A projection plane adopted in this study is normal to the CaO-Al_2O_3-FeO/MgO base and contains the projection line (2) of Fig. 4. Also shown in this diagram is the bulk composition of the average basalt which lies between those of tremolite and epidote. In order to simplify the present discussion, the composition of clinopyroxene in the acmite-jadeite-diopside system is not included.

Fig. 6 uses the projection of Fig. 5 to compare and contrast the phase and compositional relations of metabasites from the Sanbagawa belt and the Franciscan Complex. According to Nakajima and others (1977), well foliated Sanbagawa metabasites contain progressive mineral assemblages from the blueschist through the pumpellyite-actinolite to the greenschist facies. Parageneses of minerals for the blueschist and pumpellyite-actinolite facies rocks for the Sanbagawa belt have been summarized by Banno and others (1978). For the metabasites of the lower pumpellyite-actinolite facies (Maruyama and Liou, 1985), epidote + pumpellyite + hematite, pumpellyite + actinolite + hematite and hematite + actinolite + riebeckitic amphibole are identified. The immiscibility field of two amphiboles for the Sanbagawa metabasites has also been determined by Toriumi (1974). With increasing grade to upper pumpellyite-actinolite facies, a discontinuous reaction: $Pm + Ht + Qz = Act + Ep + Chl + H_2O$ was delineated and its approximate P-T position is shown in

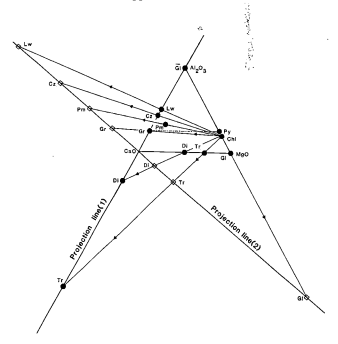

Figure 4. Compositions of common blueschist-greenschist minerals in a ternary diagram CaO-Al_2O_3-MgO (-SiO_2-H_2O) and their projections from chlorite onto a projection line (2) (present study) and a projection line (1) used by Brown (1977a).

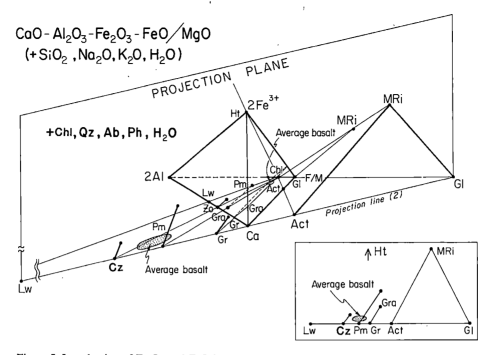

CaO- Al$_2$O$_3$-Fe$_2$O$_3$- FeO/MgO
(+SiO$_2$,Na$_2$O,K$_2$O, H$_2$O)

Figure 5. Introduction of Fe$_2$O$_3$ and FeO into the model system to show dispositions of blueschist-greenschist minerals from the chlorite projection onto a plane normal to the Ca-2Al-F/M base. Also shown are composition of average basalt and compositions of amphiboles in terms of actinolite-glaucophane-Mg-riebeckite components.

Fig. 7. Hence, the 3-phase assemblages Pm + Act + Ep, Ep + Ht + Act, and Ht + Act + riebeckitic amphibole may occur. In greenschist facies rocks, pumpellyite becomes unstable and the assemblages Ep+Act +Ht, and Ep+Act (+Ab+Chl+Qz+Sph) are most characteristic.

Coleman and Lee (1963) classified Franciscan metabasites into 4 types. Type I metabasites were subjected to incipient zeolite facies recrystallization. Type II unfoliated metabasites are characterized by the occurrence of lawsonite, pumpellyite and glaucophane in addition to jadeitic pyroxene, chlorite, albite and aragonite; they have been recrystallized at about 170°C (Taylor and Coleman, 1968). Type III schistose blueschists contain Lw + Cz + Pm, Ep + Pm + Gl, and Lw + Pm + Gl. For iron-rich Type III metasediments, Ep + Ht + riebeckite has also been identified. The Type III rocks have been recrystallized at 270°-305°C. The Type IV tectonic blocks are characterized by coarse-grained eclogite and garnet-epidote-glaucophane-omphacite schists recrystallized at 410°-535°C; neither lawsonite nor pumpellyite are stable.

The differences between the blueschist assemblages shown in Fig. 6 have been emphasized by many investigators, among them Ernst and others (1970); these include the common occurrence of the actinolite-sodic amphibole pair in Sanbagawa metabasites and of jadeitic pyroxene in the Franciscan rocks. They attributed such difference to higher pressure and lower temperature recrystallization for the Franciscan metabasites.

It should be pointed out that the phase relations for the Franciscan metabasites described above are mainly based on the petrological data of Coleman and Lee (1963). Our recent study

of metabasites from the Ward Creek area has significantly modified the relations described above (Liou and others, 1983b, Maruyama and Liou, in prep.). Specifically, we have delineated the coexistence of actinolite + sodic amphibole in the Type III Franciscan metabasites. The blueschist-greenschist transition assemblages of actinolite + sodic amphibole + epidote + albite + chlorite + quartz + sphene (+lawsonite) are ubiquitous in Type III Franciscan metabasites. They have been identified in Ward Creek, Laytonville, Tiburon and Black Butte areas of northern California (see Brown and Ghent, 1983). Apparently, the blueschist-greenschist transitional assemblages are common in the Franciscan Complex and their recrystallization may have occurred at much shallower depths (6-7 Kb) than that previously suggested (8-9 Kb; see below for further explanation).

SLIDING EQUILIBRIA FOR THE BLUESCHIST-GREENSCHIST TRANSITION

In the basaltic model system Na$_2$O-CaO-MgO-Al$_2$O$_3$-SiO$_2$-H$_2$O described by Liou and others (1985), there are 5 univariant lines radiating from each invariant point where 5 phases (+albite + quartz + fluid) coexist. If the additional component Fe$_2$O$_3$ or FeO is introduced into the model system, the univariant lines become continuous reactions and the invariant point of the model system defines a discontinuous reaction. Such a discontinuous reaction may terminate at another invariant point where 4 other discontinuous reactions radiate (Fig. 7).

The boundary between the greenschist and blueschist facies

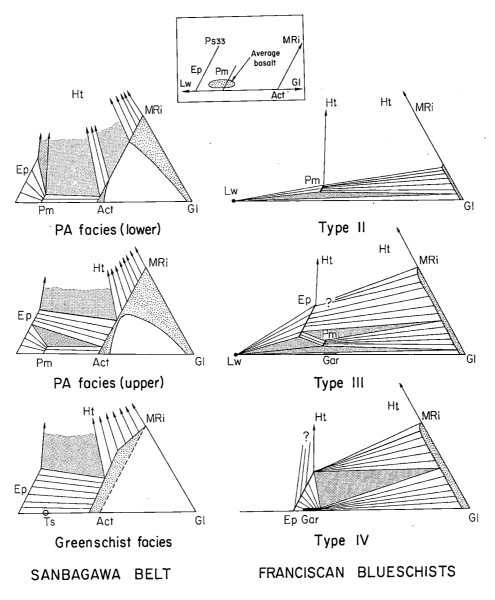

Figure 6. Parageneses of high P/T minerals in the pumpellyite-actinolite facies metabasites from Sanbagawa belt, Japan (Toriumi, 1974; Nakajima and others, 1977; Maruyama and Liou, 1985) and those in Types II, III and IV metabasites of the Franciscan Complex (Coleman and Lee, 1963).

was first proposed by Miyashiro and Banno (1958) and is defined by a reaction: Ep + Gl + Qz + H_2O = Ab + Chl + Act. This reaction is trivariant for natural metabasites; hence greenschist and blueschist assemblages could be interlayered depending mainly on the bulk rock Fe^{+3}/Al ratio and to a lesser extent on the Fe^{+2}/Mg ratio. Brown (1974) proposed a more realistic reaction: Crossite + Ep = Ab + Act + iron-oxide + H_2O for the facies boundary based on a comparative study of minerals in the Otago schist of New Zealand and Shuksan blueschist belt; the composition of each phase is assumed to be fixed because of the continuous nature of the reaction. Later, Laird (1980) suggested a similar reaction without iron-oxide. Many carbonate-bearing reactions related to the blueschist/greenschist facies transition have also

been proposed (e.g., Ernst, 1963; Chatterjee, 1971; Brown, 1977a), but they will not be discussed in this paper.

As shown in Fig. 7, phase relations for the blueschist, greenschist and pumpellyite-actinolite facies assemblages in the model system are illustrated by the CaO - Al_2O_3 - MgO diagram (in the presence of Ab, Qz and H_2O) and related by five univariant lines. The assemblages in the Fe^{3+}-saturated system at high f_{O_2} condition are shown by the chlorite-projection diagrams (Fig. 5) and their phase relations are bounded by 5 discontinuous reaction lines shown with hatch marks. Therefore, the blueschist-greenschist transition equilibria are bounded by the experimentally investigated reactions (1) and (2). Their P - T locations have been qualitatively estimated by Brown (1974, 1977a) based on

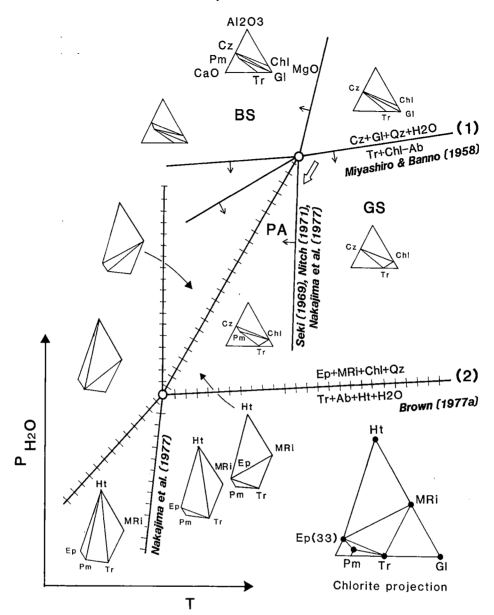

Figure 7. P_{fluid}-T diagram showing phase relations among blueschist, greenschist and pumpellyite-actinolite facies assemblages in the model basaltic system (as shown in simplified ACF plots) and in the Fe_2O_3-saturated system (lines with hatch marks and assemblages shown in the chlorite-projection plots). Arrows refer to the shifts of univariant lines and invariant point due to the introduction of Fe_2O_3 into the model system. The present studied reactions are labelled.

natural parageneses and were experimentally determined in the present study. Both reactions probably possess similar P - T slopes but reaction (1) occurs at a much higher pressure than reaction (2) for the Fe^{3+}-saturated system. Therefore, in the P - T region between these two reactions occurs a greenschist/blueschist transitional assemblage sodic amphibole (glaucophane-Mg-riebeckite) + actinolite + epidote + (+ Ab + Chl + Qz). Compositions and proportions of these phases for a given basaltic bulk composition change systematically as a function of P and T. Hence, this assemblage is termed a buffered assemblage and the composition of

blue amphibole can be used to estimate the pressure of metamorphism.

A schematic P - $X_{Fe}3+$ diagram of Fig. 8 is constructed to illustrate the inferred compositional variation of sodic amphibole of the buffered assemblage as a function of pressure at constant temperature (e.g., 300°C) in a pseudobinary diagram. At f_{O_2} conditions higher than that defined by the HM buffer, sodic amphiboles may vary their compositions along the join Gl - MRi, tremolite may have a very limited FeO substitution, and the stable iron oxide is hematite. P_3 of Fig. 8 refers to the equilibrium

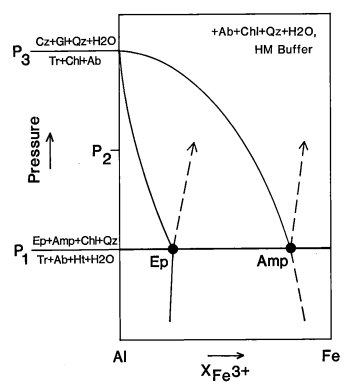

Figure 8. P-$X_{Fe^{+3}}$ plot at constant temperature and f_{O_2} of the HM buffer showing qualitative change in composition in sodic amphiboles for buffered assemblages (+ Chl + Ab + Qz + H_2O): (1) the solid line for epidote-tremolite-sodic amphibole and (2) the dashed line for epidote-hematite-sodic amphibole. P_1 is the equilibrium pressure for the reaction (2) and P_3 is the equilibrium pressure for the reaction (1).

pressure at 300°C for reaction (1) and is about 7.5 Kb from the present experimental data (Fig. 1). With the gradual introduction of Fe_2O_3 into the model system, this reaction is displaced continuously toward lower pressure, and both epidote and sodic amphibole increase their Fe^{3+} contents (Fig. 8). In the Fe^{3+}-saturated system, the discontinuous reaction occurs at P of about 4 Kb (Fig. 1). So long as f_{O_2} is defined by HM, the discontinuous reaction remains at fixed pressure isothermally and has fixed mineral compositions for a given bulk composition. Hence, the P - $X_{Fe^{3+}}$ relation for the discontinuous reaction appears as a horizontal line in Fig. 8. If the P - $X_{Fe^{3+}}$ relations were calibrated, the composition of sodic amphibole from the buffered assemblage could be used as a geobarometer.

Phase assemblages and approximate compositions of sodic amphiboles at P_1, P_2, and P_3 and temperature of 300°C are illustrated in chlorite-projection diagrams in Fig. 9. Tie-lines were schematically drawn for the coexisting phases. Complete solid solution is assumed for the join G1 - MRi and for Cz-Ep (Ps 33) join. These three diagrams illustrate the paragenetic and compositional variations of blueschist-greenschist buffered assemblages as a function of pressure described above. They also display differences in mineral assemblages as a function of bulk composition. For example, basaltic rocks as shown in Fig. 5 have compositions between those of Ep and Tr, whereas ironstones may be very

oxidized and contain abundant hematite. At P = P_3 where reaction (1) occurs, basaltic rocks contain the typical blueschist assemblage epidote + glaucophane (+Ab+Chl+Qz+Sph) whereas ironstones may contain Ep + MRi + Ht. At intermediate pressures (e.g., P = P_2), metabasites contain the buffered assemblage Ep + Tr + sodic amphibole of fixed composition. If pressure continuously decreases, both epidote and sodic amphibole may systematically vary their Fe^{3+}/Al ratios and the compositional change of sodic amphibole is most evident. When pressure is lowered to P_1, the discontinuous reaction (2) occurs. At pressures lower than P_1, basaltic rocks with high Fe^{3+}/Al ratios contain the greenschist assemblage of Ep + Tr + Ht, whereas ironstones may have Ep + Tr + Ht or Tr + MRi + Ht depending on their bulk rock compositions. Compositions of sodic amphiboles from basaltic 3-phase buffered assemblage Ep + Tr + sodic amphibole may differ significantly from those of ironstone amphiboles. The former is shown as solid line in Fig. 8 for the basaltic blue amphiboles and the latter as dashed line in Fig. 8 for those in metamorphosed ironstones.

DISCUSSION AND PETROLOGIC APPLICATIONS

Computation of Isopleths for the Buffered Assemblages

The effect of Fe^{3+}-Al substitutions in glaucophane and epidote solid solutions on the Al-end member reaction (1),

$$\ln K_{P_2} - \ln K_{P_1} = -\Delta V° (P_2 - P_1)/RT \qquad (3)$$

where K_{P_1} and K_{P_2} stand for the equilibrium constants for the reaction at pressures P_1 and P_2, respectively, $\Delta V°$ denotes the standard volume change for the reaction, and R and T represent gas constant and temperature (in Kelvins) respectively. The standard molar volumes of Al-end member phases are from Helgeson and others (1978) except that of glaucophane, which is from Koons (1982). Since reaction (1) is a dehydration reaction, $\Delta V°$ is not constant due to the volume of water (V_{H_2O}) changing with pressure and temperature. However, as a first approximation, ΔV_{H_2O} is assumed to be constant within a limited range of pressure (3-4 kb) at a given temperature. This assumption results in error up to about 0.5 kb in pressure estimates at 300-400°C, but may lie within the uncertainties of the present experimental studies.

The intercrystalline partitioning of Fe^{3+}-Al cations between epidote and glaucophane of the buffered assemblage has to be maintained under equilibrium conditions. Hence, an additional constraint can be introduced by considering the exchange reaction such as:

$$½ Na_2Mg_3Al_2Si_8O_{22} (OH)_2 + ½ Ca_2Fe_2AlSi_3O_{12}(OH) =$$
glaucophane   epidote

$$½ Na_2Mg_3Fe_2Si_8O_{22}(OH)_2 + ½ Ca_2Al_3Si_3O_{12}(OH) \qquad (4)$$
Mg-riebeckite   clinozoisite

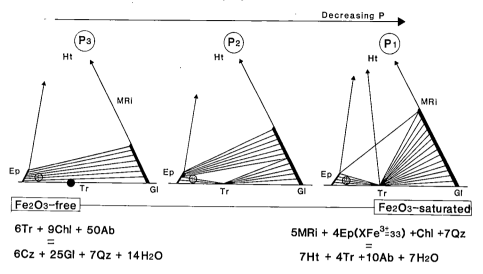

T=300°C
Chlorite projection

Decreasing P

Fe2O3-free

6Tr + 9Chl + 50Ab
=
6Cz + 25Gl + 7Qz + 14H2O

Fe2O3-saturated

$5MRi + 4Ep(XFe^{3+}_{=33}) + Chl + 7Qz$
=
$7Ht + 4Tr + 10Ab + 7H2O$

Figure 9. Three schematic isobaric chlorite-projection diagrams at T = 300°C showing variation of mineral assemblage and composition of sodic amphibole as a function of pressure. Bulk basaltic compositions are shown and the discontinuous reactions for both the Fe_2O_3-free model system and for the Fe_2O_3-saturated system are indicated.

The equilibrium constant for this reaction, K_4 can be expressed as:

$$K_4 = \frac{(a_{MRi})^{\frac{1}{2}} \cdot (a_{Cz})^{\frac{1}{2}}}{(a_{Gl})^{\frac{1}{2}} \cdot (a_{Ep})^{\frac{1}{2}}} \qquad (5)$$

where a_i refers to the activity of the ith phase. The activities of these components are estimated by assuming ideal substitution of Fe^{3+} and Al cations in M(2) sites for glaucophane solid solution and in both M(1) and M(3) sites for epidote solid solutions. Justification for such an assumption for epidote has been described in detail by Nakajima and others (1977). Thus, equation (5) can be rewritten as

$$K_4 = \frac{(X^{Gl}_{Fe, M2}) \cdot (X^{Ep}_{Al, M1, M3})}{(X^{Gl}_{Al, M2}) \cdot (X^{Ep}_{Fe, M1, M3})} = (\frac{Fe^{3+}}{Al})^{Gl}_{M2} / (\frac{Fe^{3+}}{Al})^{Ep}_{M1, M3} \qquad (6)$$

where $X^j_{j, s}$ represents the mole fraction of jth cation in the sth sites for the ith phase.

The isopleths illustrated in Fig. 10 are calculated by solving simultaneously the two equations (3) and (6) assuming constant log K_4 value of 0.6 (Brown, 1974, 1977b; Makanjuola and Howie, 1972). They are drawn for constant compositions for both epidote and sodic amphibole in the buffered assemblage Gl + Ep + Act (+ Ab + Chl + Qz). The results indicate a rapid decrease in pressure with increasing Fe-content, especially at lower pressures. These calculated isopleths of constant X_{Gl} and X_{Ps} do not give estimates of pressure consistent with those obtained from natural parageneses. This apparent inconsistency may be due to the ideal activity-composition relations adopted in this study as well as to other assumptions in the calculation (see

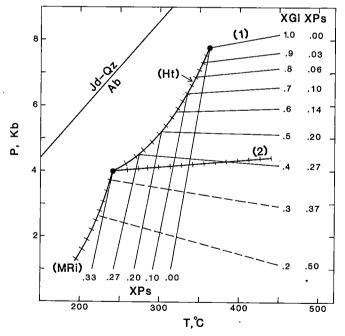

Figure 10. A P-T diagram showing (i) a univariant line for reaction (1), (ii) discontinuous reactions (with hatch marks) for the Fe_2O_3-saturated system designated as (Ht), (MRi), and (2), (iii) calculated isopleths of constant compositions of sodic amphibole (X_{Gl}) and epidote (X_{Ps}) for the blueschist and greenschist transition assemblage (+Ab+Qz+Chl) and (iv) isopleths of epidote for the pumpellyite-actinolite and greenschist transition assemblage from Nakajima and others (1977). Equilibrium line for Jd + Qz = Ab is from Holland (1983). (See Fig. 7 for comparison).

further discussions in later section). The isopleths have very gentle negative slopes at low pressures. Such a relation is very similar to those of the impure jadeite-albite-quartz equilibria described by Essene and Fyfe (1967) and Holland (1983). The nature of the gentle slope for these isopleths confirms their suitability for use as a geobarometer for the blueschist-greenschist facies transition. Both sodic amphibole and epidote in the buffered assemblage systematically increase in Al content with increasing pressure.

Also shown in Fig. 10 are those discontinuous reactions discussed in Fig. 7 and the epidote isopleths for the buffered assemblage Ep + Act + Pm (Ab + Chl + Qz) from Nakajima and others (1977). In contrast to the blueschist-greenschist transition equilibria, the isopleths of Nakajima and others for the pumpellyite-actinolite and greenschist facies transition are very sensitive to temperature change; hence are good for geothermom-etry. Combination of these two sets of isopleths shown in Fig. 10 provides better constraints for P-T relations in high P/T metamorphic facies series.

The Effect of FeO on the Transition Equilibria

Introduction of Fe_2O_3 into the model system as discussed in the previous section significantly displaces the P-T conditions of the equilibrium. The effect of FeO, on the other hand, is difficult to evaluate; it certainly creates an additional degree of freedom for the buffered assemblage. For low-grade metabasites, the compositional change of chlorite expressed as $X_{Fe}{}^*$ (Fe as total iron in chlorite) has been considered to be most sensitive to the change in FeO/MgO ratio of whole rock composition (for discussion see Maruyama and Liou, 1985). Therefore, for routine exercise and as a first approximation, the $X_{Fe}{}^*$ of chlorite can be used to monitor the rock FeO/MgO ratio.

To evaluate the effect of FeO on the transition equilibria, we used compositions of minerals in pumpellyite-actinolite facies metabasites and their associated rocks from the Sanbagawa belt. A greenstone mass in central Shikoku contains rocks with variable composition ranging from basalt to low K_2O rhyolite and with the "buffered" assemblage sodic amphibole + sodic pyroxene ($Jd_{25}Ac_{67}Aug_8$) + epidote (Ps 31-36) + actinolite (+ Chl + Ab + Qz ± Pm). The sodic amphiboles are close to glaucophane-riebeckite solid solution (Maruyama and Liou, 1985). The Al_2O_3 wt% of sodic amphibole is plotted against $X_{Fe}{}^*$ of the coexisting chlorite as shown in Fig. 11A. The results show that the Al_2O_3 content (or the glaucophane component) of sodic amphibole increases very slightly with increasing $X_{Fe}{}^*$ in chlorite. For instance, a change in Al_2O_3 content in sodic amphibole from 2 to 3 wt% corresponds to a change in $X_{Fe}{}^*$ in chlorite from 0.45 to 0.71. Although a similar relation has not been established for other higher pressure blueschist terrane, data available from the Franciscan and New Caledonian metabasites suggest that the transition equilibria for common metabasites, which contain chlorites with $X_{Fe}{}^*$ in the range of 0.4 to 0.5, may not vary significantly within this range.

In order to evaluate the P-X relations of sodic amphibole of

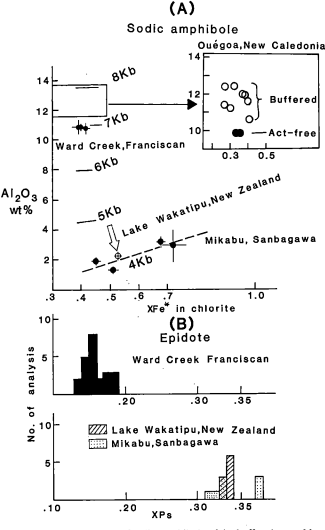

Figure 11. A. Al_2O_3 wt% of sodic amphibole of the buffered assemblage plotted against the $X_{Fe}{}^*$ of chlorite from Sanbagawa, Franciscan, New Zealand and New Caledonia. Note the gentle slope of dashed line for the Sanbagawa samples. See text for details. B. Compositions of epidote (as expressed in X_{Ps} values) from three blueschist terranes showing the pressure dependence of the epidote composition.

the buffered assemblages for natural metabasites at low pressures (e.g., less than 5 Kb), natural parageneses of the Sanbagawa metabasites described above were used. Applying the sodic pyroxene equilibria of Holland (1983) and the epidote geothermometry of Nakajima and others (1976), we obtained 4 Kb and 240°C for the Sanbagawa metamorphism in the Mikabu area. This P-T estimate is consistent with those from other investigations (e.g., Maruyama and Liou, 1985).

Using the 4 Kb value for sodic amphibole composition (X_{G1} = 0.1; Al_2O_3 = 2 - 2.4 wt%) as the low pressure end and 7 Kb for the Franciscan Ward Creek metabasite (X_{G1} = 0.9; Al_2O_3 = 11 wt%), we have drawn five approximate isobaric lines of 4, 5, 6, 7 and 8 Kb as shown in Fig. 11A. This empirical diagram illustrates the proposed geobarometry using the Al_2O_3 content of sodic

Maruyama and Others

TABLE 5. PRESSURE ESTIMATES FOR 4 BLUESCHIST TERRANES IN FRANCISCAN, SANBAGAWA, NEW ZEALAND, AND NEW CALEDONIA

Terrane	Sodic Amphibole Geobarometry (1)*	Maruyama and Liou (1985) (2)*	Brown and Bradshaw (1979) (3)	Brown (1977b) (4)	Brothers and Yokoyama (1982) (5)	Yokoyama and others (in press) (6)*
Ward Creek Franciscan	7-7.5 kb (250-320°C)	6.5 kb	10-11 kb	8 kb		
Mikabu	4 kb (250-300°C)	4 kb		7 kb	5-12 kb (250-600°C)	
Lake Wakatipu	4 kb (300°C)			5.5 kb		
Ouegoa District New Caledonia	7-8kb (410-430°C)			8 kb	4-15 kb (150-550°C)	11-12 kb (410-430°C)

Explanation:
(1) Present study, pressure estimates are discussed in the text and temperatures are from literature.
(2) Pressure estimates using sodic pyroxene equilibria of Holland (1983).
(3) Pressure estimate using the sodic pyroxene equilibria of Newton and Smith (1967).
(4) Pressure estimates using Na-M4 in Ca-amphibole geobarometry of Brown (1977b).
(5) and (6) Pressure estimates using sodic pyroxene equilibria of Holland (1983) and temperature estimates by oxygen isotope data of Black (1974).
*For metabasites with buffered assemblage sodic amphibole + actinolite + epidote (+ Chl + Qz + Ab + sodic pyroxene).

amphibole in the buffered assemblage for common metabasites with X_{Fe}* in chlorite = 0.4–0.5. Sodic amphibole composition is expressed as wt% Al_2O_3 in order to avoid the uncertainty of Fe^{3+} estimates from microprobe analysis. Because of the scarcity of available geobarometry in blueschist facies metamorphism, the proposed glaucophane geobarometer may offer an important tool.

Al_2O_3 Content of Sodic Amphibole as a Geobarometer

The isopleths of sodic amphibole as a function of pressure for buffered assemblages and the proposed empirical isobaric lines shown in Fig. 11A are used to estimate the metamorphic pressures for 4 classic metamorphic terranes (Franciscan, New Caledonia, New Zealand and Sanbagawa). The results are listed in Table 5 together with other pressure estimates made by previous investigators. The Ward Creek metabasites of the Franciscan Complex contain glaucophane together with epidote-clinozoisite, pumpellyite, actinolite, chlorite, albite, and quartz. The sodic amphiboles are high in Al_2O_3 content (Al_2O_3 = 10.5–11.6 wt%) which gives metamorphic pressures of 7.0–7.5 Kb for its formation. Brown and Bradshaw (1979) reported pressures of 10 to 11 Kb for the Type III metabasites from this area based on the Ab-Qz-Jd_{82} curve of Newton and Smith (1967) and temperature estimates of Taylor and Coleman (1968). Using the revised stability of jadeitic pyroxene of Holland (1980) and the systematic study of mineral paragenesis of the Ward Creek Types II and III metabasites, we obtain 5.5 to 8.5 Kb and 200°–350°C, which is consistent with the data based on the glaucophane geobarometry.

In New Caledonia, 8 metabasites with the buffered assemblage have been analyzed by Black (1974) in the Ouegoa district. The sodic amphiboles of these buffered assemblages contain 10.6 to 12.4 wt% Al_2O_3 which yields 7-8 Kb according to the proposed geobarometry of Fig. 11A, whereas those in actinolite-free assemblages have a lower Al_2O_3 content (9.9 wt%). This relationship is consistent with the compositional variation of sodic amphibole shown in Fig. 9. This pressure estimate is similar to the 8 Kb of Brown (1977b) but about 3 Kb lower than that of Yokoyama et al. (this volume). Such a large discrepancy is difficult to explain. One possible reason could be the temperature estimate by Black (1974) of 410–430°C for these rocks. Applying the revised quartz-muscovite fractionation curve (Clayton and others, 1972) may yield a 50°C lower temperature estimate, hence lower pressure value.

The pressure estimates at low pressures by the calculated isopleths of Fig. 10 are consistently lower than the accepted values. This discrepancy is in part due to the effect of FeO described in the previous sections, in part due to the isopleths built upon the equilibrium line of the model system. With increasing Fe^{3+}/Al ratio in sodic amphibole, there is greater uncertainty due to the steeper composition line of X_{G1} (Fig. 8), the oversimplification of mixing model, the constant K_D value and unknown partitioning of Fe^{2+} and Mg among phases in the buffered assemblage. Therefore, the Al_2O_3 content of sodic amphibole as a function of pressure shown in Fig. 11A yields better pressure estimates than the X_{G1} of Fig. 10.

Using the geobarometry of this study and the geothermometry of Nakajima and others (1977), the sample (e.g., NZ32812) from Lake Wakatipu, New Zealand with assemblage sodic amphibole (Al_2O_3 = 2.1 wt%) + epidote (Ps = 33) + chlorite (X_{Fe}^* = 0.53) + actinolite + albite + quartz + hematite may have been recrystallized at about 260°C and 4 Kb pressure. In fact, the New Zealand metabasites contain assemblage Ep + pumpellyite + Act + Chl + Ab + Qz (Kawachi, 1975), which is stable at 3 to 5 Kb and 250°-350°C according to our unpublished experimental data (see Fig. 3 of Liou and others, 1985, for preliminary results). On the other hand, the metabasites in Mikabu, Sanbagawa, have a similar assemblage except for hematite, higher Ps value (31-34) for epidote and similar Al_2O_3 content for sodic amphibole. The mineral assemblage and compositions suggest that the Mikabu metabasites may have occurred at 4 Kb but lower temperature (about 240°C) than those in New Zealand.

It should be noted that epidote minerals are common in blueschist facies rocks and may occur together with sodic amphibole and actinolite (+ Ab + Chl + Qz). In principle, the composition of epidote in this buffered assemblage increases its Ps content with decreasing pressure and can be used as a geobarometer. For example, these minerals in Ward Creek metabasites have Ps values of 13 to 19, and those in Sanbagawa and New Zealand are 31-38 as shown in Fig. 11B. This relation supports the conclusion described in the previous sections that higher pressure is necessary to stabilize glaucophane and clinozoisite in Franciscan and New Caledonia metabasites than crossitic amphibole and epidote in Sanbagawa and New Zealand metabasites.

CONCLUSIONS

Our experimental study together with that of Holland (1980), yields the following conclusions: (1) The blueschist facies metamorphism may occur at shallower depth than previously suggested. For example, jadeite-bearing glaucophane schist may

be stable at 6 Kb, 300°C instead of 8 Kb; (2) Depending on the bulk composition, both blueschist and greenschist facies assemblages may be stable over a considerable P-T range and may form during one-stage metamorphic recrystallization. Coexistence of both facies assemblages in the same outcrop or in a single specimen has been documented in many blueschist terranes in addition to those described above. Two stages of recrystallization with greenschist facies after blueschist facies metamorphism or vice versa are common in many blueschist terranes and have been used to explain such observations. Distinction of one-stage versus two-stage metamorphism can only be done by textural and structural investigation of greenschist-blueschist rocks; and (3) For a given rock composition, the composition of sodic amphibole of the buffered assemblage sodic amphibole + actinolite + epidote (+ Ab + Chl + Qz) varies systematically with pressure; hence it can be used as a geobarometer. Pressure estimates for metabasites in Ward Creek of Franciscan, Mikabu of Sanbagawa, Lake Wakitipu of New Zealand and Ouegoa of New Caledonia using the proposed geobarometry are in agreement with those derived from sodic pyroxene geobarometry.

ACKNOWLEDGMENTS

This research was supported through NSF grant EAR82-04298. We are grateful to Dr. Kawachi for supplying New Zealand samples and to Dr. Philippa Black for compositions of minerals from the New Caledonian blueschist belt. The manuscript was critically reviewed and materially improved by Drs. Edward Ghent, Philippa Black, W. G. Ernst, Ron Frost and Bernard Evans. We thank the foundation and these individuals for their help and support.

REFERENCES CITED

Banno, S., Higashino, T., Otsuki, M., Itaya, T., and Nakajima, T., 1978. Thermal structure of the Sanbagawa metamorphic belt in central Shikoku: J. Phys. Earth, Suppl., 345–356.

Black, P. M., 1974, Oxygen isotope study of metamorphic rocks from the Ouegoa district, New Caledonia: Contr. Mineralogy and Petrology, v. 15, p. 197–206.

Brothers, R. N., and Yokoyama, K., 1982, Comparison of the high-pressure schist belts of New Caledonia and Sanbagawa, Japan: Contr. Mineralogy and Petrology, v. 79, p. 219–229.

Brown, E. H., 1974, Comparison of the mineralogy and phase relations of blueschists from the North Cascades, Washington and greenschists from Otago, New Zealand: Bull. Geol. Soc. Amer., v. 85, p. 333–344.

—— 1977a, Phase equilibria among pumpellyite, lawsonite, epidote and associated minerals in low-grade metamorphic rocks: Contr. Mineral. and Petrology, v. 64, p. 123–136.

—— 1977b, The crossite content of Ca-amphibole as a guide to pressure of metamorphism: Jour. Petrol., v. 18, p. 53–72.

Brown, E. H. and Bradshaw, J. Y., 1979, Phase relations of pyroxene and amphibole in greenstone, blueschist and eclogite of the Franciscan complex, California: Contr. Mineralogy and Petrology, v. 71, p. 67–83.

Brown, E. H. and Ghent, E. D., 1983, Mineralogy and phase relations in the blueschist facies of the Black Butte and Ball Rock areas, northern California Coast Ranges: Am. Mineralogist, v. 68, p. 365–372.

Carman, J. H., 1974, Preliminary data on the P-T stability of synthetic glaucophane (Abstract): Am. Geophy. Union Trans., v. 55, p. 481.

Carman, J. H. and Gilbert, M. C., 1983, Experimental studies on glaucophane stability: Am. Jour. Sci., v. 283-A, p. 414–437.

Chatterjee, N. D., 1971, Phase equilibria in the Alpine metamorphic rocks of the environs of the Dora-Maira Massif, western Italian Alps: Neues Jahrb. Mineralogie Abh., v. 114, p. 211–245.

Clayton, R. N., O'Neil, J. R., and Mayeda, T. K., 1972, Oxygen isotope exchange between quartz and water: Jour. Geophys. Research, v. 77, p. 3057–3067.

Coleman, R. G. and Lee, D. E., 1963, Glaucophane-bearing metamorphic rock types of the Cazadero area, California: Jour. Petrology, v. 4, p. 260–301.

De Roever, E.W.F. and Beunk, F. F., 1976, Blue amphibole-albite-chlorite assemblages from Fuscaldo (S. Italy) and the role of glaucophane in metamorphism: Contr. Mineralogy and Petrology, v. 58, p. 221–234.

Ernst, W. G., 1961, Stability relations of glaucophane: Am. Jour. Sci., v. 259, p. 735–765.

—— 1963, Polymorphism in alkali amphiboles: Am. Mineralogist, v. 48, p. 241–260.

—— 1979, Coexisting sodic and calcic amphiboles from high-pressure metamorphic belts and the stability of barroisitic amphibole: Mineralogy Mag, v. 43, p. 269–278.

——1981, Petrotectonic settings of glaucophane schist belts and some implications for Taiwan: Geol. Soc. China, Memoir v. 5, p. 229–268.

Ernst, W. G., Seki, Y., Onuki, H., and Gilbert, M. C., 1970, Comparative study of low-grade metamorphism in the California Coast Ranges and the outer metamorphic belt of Japan: Geol. Soc. America Mem., v. 124, 276 p.

Eskola, P., 1939, *in* Barth, T.F.W., Correns, C. W., and Eskola, P., *Editors,* Die Entstehung der Gesteine: J. Springer, Berlin, 422 p.

Essene, E. J. and Fyfe, W. S., 1967, Omphacite in California metamorphic rocks: Contr. Mineralogy and Petrology, v. 15, p. 1–23.

Gilbert, M. C. and Popp, R. K., 1973, Properties and stability of glaucophane at high pressure (Abstract): Am. Geophys. Union Trans., v. 54, p. 1223.

Helgeson, H. C., Delany, J. M., Nesbitt, and Bird, D. K., 1978, Summary and critique of the thermodynamic properties of rock-forming minerals: Am. Jour. Sci., v. 278, p. 1–229.

Holland, T.J.B., 1980, The reaction albite = jadeite + quartz determined experimentally in the range 600–1200°C: Am. Mineralogist, v. 65, p. 129–134.

——1983, The experimental determination of activities in disordered and short-range ordered jadeitic pyroxenes: Contr. Mineralogy and Petrology, v. 82, p. 214–220.

Kawachi, Y., 1975, Pumpellyite-actinolite and contiguous facies metamorphism in part of Upper Wakatipu district, South Island, New Zealand: New Zealand Jour. of Geology and Geophysics, v. 18, p. 401–441.

Koons, P. O., 1982, An experimental investigation of the behavior of amphibole in the system Na_2O-MgO-Al_2O_3-SiO_2-H_2O at high pressures: Contr. Mineralogy and Petrology, v. 79, p. 258–267.

Laird, J., 1980, Phase equilibria in mafic schist from Vermont: Jour. Petrology, v. 21, p. 1–37.

Liou, J. G., Kim, H. S., and Maruyama, S., 1983a, Prehnite-epidote equilibria and their petrologic applications: Jour. Petrology, v. 24, p. 321–342.

Liou, J. G., Maruyama, S., and Sasakura, S., 1983b, Paragenesis and compositions of minerals from Ward Creek Franciscan metabasites, Cazadero, California (Abstract): Am. Geophys. Union Trans., v. 64, p. 877.

Liou, J. G., Maruyama, S., and Cho, M., 1985, Phase equilibria and mineral parageneses of metabasites in low-grade metamorphism: Mineral. Mag., v. 49, p. 321–333.

Makanjuola, A. A. and Howie, R. A., 1972, The mineralogy of the glaucophane schists and associated rocks from Ile de Groix, Brittany, France: Contr. Mineralogy and Petrology, v. 35, p. 83–118.

Maresch, W. V., 1973, New data on the synthesis and stability relations of glaucophane: Earth Planetary Sci. Letters, v. 20, p. 385–390.

——1977, Experimental studies on glaucophane: an analysis of present knowledge: Tectonophysics, v. 43, p. 109–125.

Maruyama, S. and Liou, J. G., 1985, The stability of Ca-Na pyroxene in low-grade metabasites of high-pressure intermediate facies series: Am. Mineralogist, v. 70, p. 16–29.

Miyashiro, A., 1957, The chemistry, optics and genesis of alkali amphiboles: Jour. Fac. Sci. Univ. Tokyo, v. 11, p. 57–83.

——1961, Evolution of metamorphic belts: Jour. Petrology, v. 2, p. 277–311.

——1973, Metamorphism and metamorphic belts: John Wiley and Sons, New York, 492 pp.

Miyashiro, A. and Banno, S., 1958, Nature of glaucophanitic metamorphism: Am. Jour. Sci, v. 256, p. 97–110.

Nakajima, T., Banno, S., and Suzuki, T., 1977, Reactions leading to the disappearance of pumpellyite in low-grade metamorphic rocks of the Sanbagawa metamorphic belt in Central Shikoku, Japan: Jour. Petrology, v. 18, p. 263–284.

Newton, R. C. and Smith, J. V., 1967, Investigations concerning the breakdown of albite at depth in the earth: Jour. Geol., v. 75, p. 268–286.

Nitsch, K. H., 1971, Stabilitätsbeziehungen von Prehnit-und Pumpellyit-haltigen Paragenesen: Contr. Mineralogy and Petrology, v. 30, p. 240–260.

Seki, Y., 1969, Facies series in low-grade metamorphism: Jour. Geol. Soc. Japan, v. 75, p. 255–266.

Taylor, H. P. and Coleman, R. G., 1968, O^{18}/O^{16} ratios of coexisting minerals in glaucophane-bearing metamorphic rocks: Geol. Soc. Am. Bull., v. 79, p. 1727–1756.

Toriumi, M., 1974, Actinolite-alkali amphibole miscibility gap in an amphibole composite-grain in glaucophane schist facies, Kanto Mountains, Japan: Jour. Geol. Soc. Japan, v. 80, p. 75–80.

Yokoyama, K., Brothers, R. N., and Black, P. M., Regional eclogite facies in the high-pressure metamorphic belt of New Caledonia. This volume.

Winkler, H.G.F., 1979, Petrogenesis of metamorphic rocks: Springer, Berlin, 5th Edition, p. 348.

MANUSCRIPT ACCEPTED BY THE SOCIETY JULY 29, 1985

Geological Society of America
Memoir 164
1986

Metamorphic temperatures and pressures of
Group B and C eclogites

Robert C. Newton
Department of the Geophysical Sciences
University of Chicago
Chicago, Illinois 60637

ABSTRACT

The Ellis and Green (1979) experimental calibration of the garnet-clinopyroxene K_D (Fe, Mg) geothermometer generally agrees with information from experimental work in simple systems on major and minor phases of Class B and C eclogites and with simple thermodynamic extension of the phase equilibrium work. Correction for Fe^{3+} must be made in microprobe analyses of natural assemblages in applying the garnet-clinopyroxene K_D scale.

Geobarometry of eclogites is more difficult. The absence of plagioclase sets quantitative lower-pressure limits on quartz eclogites. Additional information on equilibration pressures is given by experimental work on the stabilities of the minor eclogite phases lawsonite, zoisite, talc, paragonite, albite and tremolite. Many well-described natural occurrences have garnet and clinopyroxene compositions that demand minimum pressures of 12-13 kbar, based on thermodynamic analysis. Earlier estimates of 6-9 kbar based on long extrapolations of experimental work in dry mafic compositions at high temperatures are not valid. Eclogite could have formed at very high H_2O pressures; P_{H_2O} much lower than P_{total} may have occurred but was not generally necessary.

Crustal eclogites can be roughly grouped in three categories: a low-temperature group (450°-500°C) with pseudomorphs after lawsonite, a high-temperature group (600°-750°C) with talc and kyanite or from terranes with this assemblage in co-metamorphic pelites, and an intermediate-temperature group (500°–600°C) which is neither lawsonite-related nor talc-kyanite-related. Pressures deduced for all of these groups vary from 12 to 18 kbar, and pressures may have been 20 kbar or more during the formation of coarse garnet and omphacite in some occurrences. Such very high pressures must have been the result of subduction at continental margins.

An increasing T and P prograde subduction path is suggested but not proved by differences in composition between rims and cores of garnets and pyroxenes from Tasmania and southern Norway, and possibly, in the spread of eclogite mineral compositions from the Raspas Formation of Ecuador. Secondary greenschist assemblages in many eclogites may indicate uplift at relatively high-temperature conditions. Alternatively, isobaric cooling may have been followed by a discrete period of metamorphism at shallower depth and elevated temperatures prior to uplift to the surface.

The lowest T/P-equilibration conditions were found for the younger (Late Cretaceous-Tertiary) occurrences, which include all of the lawsonite-related examples. Apparent average subduction gradients for these were about 11°C/km of burial or lower. The highest apparent geotherms, of about 16°/km, were found for the oldest (Late Precambrian-Early Cambrian) group, which includes the whiteschist-related samples. This spectrum of apparent subduction geotherms, although based on relatively few

R. C. Newton

> examples and subject to large errors in calculation, may indicate that conditions favorable for the formation of lower-temperature crustal eclogites were more widespread in later geologic history.

INTRODUCTION

Eclogite is a high-pressure modification of gabbro, in many cases consisting almost entirely of the two minerals garnet and omphacitic clinopyroxene. The two major occurrences of eclogite are exotic samples of probable mantle origin ejected in explosive volcanic processes (Group A of Coleman et al., 1965), and layers and tectonic inclusions associated with high-grade and lower-grade metamorphic terranes (Groups B and C of Coleman et al., 1965, respectively).

There is considerable evidence from field relations that some conformable eclogite bodies were co-metamorphosed with the enclosing schists and gneisses (Krogh, 1980); however, some coarse-grained mafic inclusions and detached blocks in blueschist terranes show a somewhat higher-grade metamorphism than the terranes in which they reside (Brown et al., 1981; Moore, 1984, and many others) and the same is probably true of many of the eclogite tectonic blocks and detached boulders. Whether a given eclogite body in a given terrane is co-metamorphic with its enclosing rocks or whether it is an exotic captured product of a different metamorphic regime may be controversial. This paper will concentrate on well-described eclogite occurrences which show evidence of metamorphic compatibility with host rocks; these are most useful in analysis of temperature-pressure conditions in the tectonic terranes of active continental margins.

Many or most Group B and C eclogites are polymetamorphic. A greenschist assemblage of amphibole, epidote and other minerals replacing garnet and omphacite in variable degrees has been reported many times (for example, by Maresch and Abraham, 1981). This polymetamorphism may result from continuous reaction during uplift from deep burial or from one or more discrete later episodes of lower-pressure metamorphism; these contrasting processes are often hard to distinguish in their effects, and it may indeed be hard to tell by textures whether some of the minor minerals in an eclogite were compatible with garnet and omphacite during eclogite-facies conditions or were formed in the uplift career. Finally, it is possible that even earlier assemblages, rarely or never preserved in some eclogites, may have resulted from transient reactions in the burial cycle of subduction. In view of the continuum of T, P states to which eclogites have been subjected, it is perhaps surprising that many eclogites are coarse-grained quasi-equilibrium assemblages with only weak mineral zonation and incipient alteration. This paper concentrates on the seemingly discrete and emphatic major metamorphic signature exhibited by many eclogites in their mineralogies and mineral chemistries.

The distribution of Fe^{2+} and Mg between coexisting garnet and clinopyroxene has been calibrated experimentally as a geothermometer at high temperatures by Ellis and Green (1979). Their temperature scale has been applied with some success to

granulite facies metabasites by Harris et al. (1982) and Johnson et al. (1983), and by Ghent et al. (1983) to upper amphibolite facies rocks. However, the applicability of the Ellis-Green thermometer to the low-temperature range of Group C eclogites and to the ordered primitive-cell omphacites of many eclogites (e.g. Holland, 1979a), is unknown. The substantial quantities of trivalent iron of some eclogite omphacites is another uncalibrated factor. A continuous geobarometer scale for eclogites has not yet been calibrated. The absence of plagioclase from most eclogites prevents use of the temperature-independent barometer scales useful for garnet-clinopyroxene granulites (Newton and Perkins, 1982), except to provide lower-pressure limits on pressures of crystallization.

In view of the intrinsic limitations to continuous geothermometry-geobarometry of eclogites, it is fortunate that many eclogites contain certain amounts of other minerals for which a body of experimental pressure-temperature stability work exists. This fact makes possible a considerable application of the old-fashioned method of the petrogenetic grid, a method which many investigators have found useful for eclogites. Group B and C eclogites commonly contain relatively small amounts of amphibole, kyanite, quartz, talc, rutile, zoisite or clinozoisite, white mica, and occasionally, pseudomorphs of the last two minerals after lawsonite. While some of these accessory minerals are often demonstrably retrogressive, as for example, kelyphitic amphibole after garnet, they are often considered primary: that is, crystallized in equilibrium with garnet and clinopyroxene. The available experimental phase-equilibrium data for many of the accessory minerals makes possible, with thermodynamic extension thereof, a substantial petrogenetic grid with which to pigeonhole many common eclogitic parageneses within relatively narrow P-T sectors.

Diagnostic parageneses in other lithologies, where they can be shown to be co-metamorphic with associated eclogites, can be of considerable help in constraining the metamorphic P, T conditions. These associated lithologies may include quartzofeldspathic rocks, amphibolites and metapelites. The plagioclase of these rocks allows, in some cases, pressure estimates based on the breakdown reactions of albite and anorthite.

THE ACCESSORY PHASES

Figure 1 shows the distribution of accompanying phases reported for several rocks containing major garnet and omphacite, excluding, for the most part, volcanic exotics. The Allalin coronal metagabbros of western Switzerland (Chinner and Dixon, 1973) are not, strictly speaking, eclogites, but some, nevertheless, contain an eclogitic garnet-omphacite assemblage

	Lufilian Arc A	Lufilian Arc B	Hohe Tauern A	Hohe Tauern B	Hohe Tauern C	Margarita Is.	Raspas Fm.	Naustdal	Allalin Gabbro A	Allalin Gabbro B	Muenchberg	Lyell-Colling.	Tillotson Pk.	Guajira Pen.	Garnet Ridge	Breuil-St. Jacq.	Sifnos
QUARTZ	X		X	X	X	X	X	X					X	X		X	X
KYANITE	X	X	X	X	X						X	X	X				
ZOISITE	X		X					E	C		X	X	E	C		E	C
PARAGONITE			X			X	X	X								X	X
GLAUCOPHANE		A	A	X	X	A	A	A			X		A	A	X/A	X	X
TREMOLITE	X																
TALC		X		X?							X	X					
ALBITE	P												X				
LAWSONITE														*	X	*	*
CHLORITOID											X	X					
CHLORITE											X				X		
PHENGITE													X	X		X	X

Figure 1. Common accessory phases, interpreted as primary, of Group B and C eclogites, and some transitional eclogites. Symbols: X = present; E = epidote; C = clinozoisite; A = amphibole (winchite-barroisite to hornblende); P = plagioclase; asterisk denotes pseudomorphs of zoisite or clinozoisite and white mica after lawsonite. Lufilian Arc (Vrana et al., 1975); A = spec #3, B = spec #12; Hohe Tauern (Holland, 1979a): A = TH 418B, B = TH GW2, C = TH 49B; Margarita Is. (Maresch and Abraham, 1981): W 176; Raspas Fm (Feininger, 1980): TF 1369; Naustdal (Krogh, 1980): IB-5$_2$; Allalin Gabbro (Chinner and Dixon, 1973): A = 53748, B = 93648; Muenchberg Gneiss (Matthes et al., 1975): KS-1750/6; Lyell-Collingwood Area (Råheim and Green, 1975): 71-329; Tillotson Peak (Laird and Albee, 1976): A-BM-100; Guajira Peninsula (Green et al., 1964): JPL J-291A; Garnet Ridge (Watson and Morton, 1968): GR-P2-17; Breuil-St. Jacques (Ernst and Dal Piaz, 1978): DBL-379; Sifnos (Okrusch et al., 1978): Si 73-90.

among the grain-boundary reaction products of olivine, in addition to major amounts of other minerals, listed in Figure 1. Some of the Tillotson Peak metabasites of the Green Mountains, Vermont (Laird and Albee, 1981), contain garnet and omphacite of typical eclogitic compositions. Eclogite exotics in the Garnet Ridge, Arizona, and other kimberlite pipes of the Colorado Plateau have parageneses strikingly similar to those of Groups B and C eclogites (Watson and Morton, 1968). The Garnet Ridge eclogites, included in Figure 1 only for comparison, are the only known eclogites with actual lawsonite, which may mean that extraordinarily fast uplift is required to preserve lawsonite outside its field of stability.

In compiling Figure 1, an effort was made to include only those phases which the various authors thought were primary and compatible with garnet-omphacite, rather than alteration products. For example, Ernst and Dal Piaz (1978) reported two amphiboles in some of their eclogites from the Western Italian Alps, but consider that only glaucophane was formed during the period of major eclogite crystallization. Hence, only glaucophane is listed for this occurrence in Figure 1. Where an author expressed some doubt about the primary nature of a phase, a question mark appears. Pseudomorphic evidence of the prior existence of lawsonite has been reported several times in eclogites (Green et al., 1964; Ernst and Dal Piaz, 1978); this is indicated by an asterisk in Figure 1. These pseudomorphs usually consist of a fine-grained aggregate of zoisite or clinozoisite and paragonite.

Plagioclase is often reported in eclogites as a secondary phase, usually thought to indicate retrogressive reaction upon release of pressure. However, sodic plagioclase occurs as an apparently compatible member of some assemblages transitional to eclogites, at Tillotson Peak and the Lufilian Arc of Zambia (Vrana et al., 1975). The presence of sodic plagioclase in co-

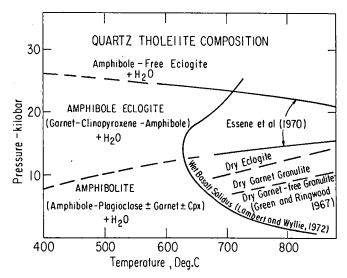

Figure 2. Experimental phase fields of quartz tholeiite composition in wet and dry systems.

metamorphic acid gneisses, as at Kvineset, Norway (Krogh, 1980), and amphibolites, as in western Tasmania (Råheim, 1976), can be of importance in providing upper-pressure limits to the metamorphism.

Most Group B and C eclogites contain some kind of amphibole apparently compatible with garnet and omphacite. The amphibole may be a member of the glaucophane-riebeckite series, a somewhat calcic amphibole of the winchite-barroisite series (Leake, 1978), or hornblende. Colorless to pale-green amphibole, presumably tremolitic, was reported in some of the Zambian eclogites (Vrana et al., 1975).

In addition to quartz and kyanite, other frequently-reported phases are paragonite, zoisite or epidote, and phengite. Talc is less common, but is of importance in association with kyanite, both in eclogites and in co-metamorphosed schists, in defining the P-T field of eclogites of whiteschist terranes (Schreyer, 1973). Chlorite and chloritoid are reported infrequently in eclogites. Calcite and/or dolomite are sometimes reported (Lappin and Smith, 1981; Holland, 1979a). It may be difficult to be sure of the primary nature of carbonates in eclogites.

FOUNDATIONS OF THE PETROGENETIC GRID

Experimental Work

Relevant experimental work includes experiments on basaltic compositions and determination of high pressure, relatively low-temperature univariant equilibrium curves in simple systems. According to the work of Essene et al. (1970), garnet-clinopyroxene assemblages free of amphibole can coexist with an aqueous fluid if pressures are above about 25 kbar (Figure 2). The wet melting curve of quartz tholeiite puts upper-temperature limits of about 650°C for eclogites at $P_{H_2O} = P_{total}$. At pressures below 20-25 kbar, amphibole-garnet-clinopyroxene assemblages

are stable, and at pressures below 10-14 kbar, plagioclase-bearing assemblages are stable. If H_2O pressure is less than the total pressure, the field of anhydrous eclogite expands to lower pressures. Under completely dry conditions, plagioclase- and amphibole-free eclogite may be stable to pressures below 10 kbar, according to a long extrapolation of the experimental work at much higher temperatures and pressures of Green and Ringwood (1967). Very low H_2O activity has been appealed to for the genesis of some crustal eclogites (Fry and Fyfe, 1969; Ahrens and Schubert, 1975), but recent geobarometric work, to be discussed, shows that very high pressures actually did prevail in the formation of many Group B and C eclogites, so that $P_{H_2O} < P_{total}$ is not a necessary boundary condition, in contrast to granulite-facies metamorphism. The indication of Figure 2 is that amphibole-bearing eclogites crystallized in the water-pressure range 10-25 kbar. Additional characterization of the pressures and temperatures is provided by experimental work in simple model systems.

The system $Na_2O-Al_2O_3-SiO_2-H_2O$ has considerable bearing on eclogite petrogenesis. Figure 3 shows the precisely determined upper-pressure stability limits of paragonite, $NaAl_3Si_3O_{10}(OH)_2$, and albite, $NaAlSi_3O_8$ (Holland, 1979b; 1980). The former equilibrium sets maximum pressures of about 25 kbar for eclogites bearing primary paragonite, and direct application of the latter equilibrium may be made to eclogites co-metamorphosed with acid quartzofeldspathic rocks. For more general applications, consideration must be given to the solid solution of jadeite, $NaAlSi_2O_6$, in omphacitic pyroxene, as discussed below. Additional constraints on temperature are set by the experimental high-temperature limit of paragonite and quartz (Chatterjee, 1974).

The stability of tremolite, $Ca_2Mg_5Si_8O_{22}(OH)_2$, provides boundary conditions for blueschist and greenschist assemblages. The high-pressure vapor-absent reaction of tremolite to talc + diopside, recently bracketed experimentally by T.J.B. Holland (personal communication, 1984), together with the albite to jadeite + quartz equilibrium, allows construction of an upper-pressure boundary reaction for the greenschist assemblage tremolite + albite.

The stabilities of lawsonite, $CaAl_2Si_2O_7(OH)_2 \cdot H_2O$, and margarite, $CaAl_4Si_2O_{10}(OH)_2$, were determined experimentally by Newton and Kennedy (1963), Nitsch (1974) and Jenkins (1983), as shown in Figure 3. Although margarite has not been reported in eclogites, its stability limits relative to zoisite + kyanite provide a lower-pressure boundary of 9-11 kbar for some eclogites. The lawsonite dehydration to zoisite, kyanite and quartz sets rather stringent upper-temperature limits on lawsonite eclogites.

Experimental data in the system $Na_2O-CaO-Al_2O_3-SiO_2-H_2O$ on the reactions:

lawsonite + albite = zoisite + paragonite + quartz + H_2O
lawsonite + jadeite = zoisite + paragonite + quartz + H_2O

(Heinrich and Althaus, 1980) relate paragonite and lawsonite

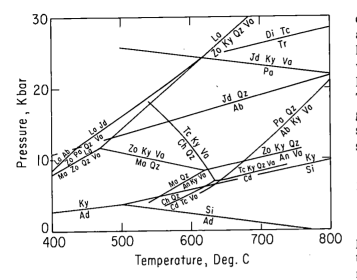

Figure 3. Experimental univariant equilibria in simple systems. *Lawson-ite (La)* + *albite (Ab)* = *zoisite (Zo)* + *paragonite (Pa)* + *quartz (Qz)* + H_2O *(Va)*: Heinrich and Althaus (1980); *La +jadeite (Jd)* = *Zo+P-a+Qz+Va*: Heinrich and Althaus (1980); *La* = *margarite (Ma)* + *Zo+Qz+Va*: Nitsch (1974); *La* = *Zo* + *kyanite (Ky)* + *Qz+Va*: Newton and Kennedy (1963); *Ma+Qz* = *Zo+Ky+Va*: Jenkins (1983); *Ma+Qz* = *anorthite (An)* + *Ky+Va*: Nitsch et al. (1981); *Zo+Ky+Qz=An+Va*: Goldsmith (1981); *Clinochlore (Ch)* +*Qz* = *talc (Tc)* + *Ky+Va*: Massonne et al. (1981); *Ch* +*Qz* = *cordierite (Cd)* + *Tc+Va*: Chernosky (1978); *Cd* = *Tc+Ky+Qz+Va*: Newton (1972); *Jd+Qz* = *Ab*: Holland (1979b); *paragonite (Pa)* = *Jd+Ky+Va*: Holland (1979b); *Pa+Qz* = *Ab+Ky+Va*: Chatterjee (1974); *tremolite (Tr)* = *diopside (Di)* + *Tc*: T.J.B. Holland (unpublished data); *relations among andalusite (Ad), sillimanite (Si) and Ky*: Holdaway (1971).

stability relations (Figure 3) and thus provide a framework for thermodynamic extension of the equilibrium work. In construction of Figure 3, minor adjustment was necessary to interface the determinations of Henirich and Althaus (1980), Nitsch (1974), and Newton and Kennedy (1963).

The end-member composition of glaucophane, $Na_2Mg_3Al_2Si_8O_{22}(OH)_2$, is closely approached in the high-pressure eclogites of the Tauern Window terrane of Austria (Holland, 1979a). Experimental work on this composition should be of particular relevance to this and other eclogites, but several studies have revealed the intrinsic difficulties of non-stoichiometry and appearance of Na-Mg layer silicates, probably metastable, in syntheses (Maresch, 1977; Koons, 1982; Carman and Gilbert, 1983). The upper-pressure breakdown of glaucophane to jadeite and talc has very small volume change, and hence can be perturbed greatly by relatively small departures from the ideal formula. For these reasons, glaucophane equilibria are not currently as useful in petrogenetic analysis of eclogites as other simple system minerals.

The whiteschist assemblage talc, $Mg_3Si_4O_{10}(OH)_2$, and kyanite, Al_2SiO_5, is limited to pressures above 7-9 kbar and temperatures above 540°-630°C (Figure 3), according to the experimental work of Massonne et al. (1981). At lower pressures,

cordierite, $Mg_2Al_4Si_5O_{18} \cdot nH_2O$, appears, and at lower temperatures, clinochlore, $Mg_5Al_2Si_3O_{10}(OH)_8$ plus quartz. The talc-kyanite equilibrium is important in setting apart those eclogites which are apparently co-metamorphic in whiteschist terranes. Figure 3 summarizes the experimental relations discussed above. The univariant equilibria constitute a fairly dense petrogenetic grid, which serves to restrict the possible temperatures and pressures of eclogite metamorphism, especially when extended by simple thermodynamic analysis.

Thermodynamic Extension

The usefulness of the experimental stability relations in simple systems can be considerably extended by equilibrium calculations taking into account compositional departures of the minerals, especially clinopyroxene and plagioclase, from the experimental systems. Activity-composition relations in clinopyroxene are not well known for the composition and temperature ranges characteristic of eclogites. However, approximations based on existing experimental and thermochemical data can provide consistent models. Several similar activity-composition models exist for high-structural-state plagioclase (Orville, 1972; Saxena and Ribbe, 1972; Newton et al., 1980). Although plagioclase is not present in most rocks called eclogites, its absence puts lower-pressure limits on the assemblages garnet-clinopyroxene-quartz and garnet-kyanite-quartz. For this purpose, it is not relevant whether low-state plagioclase or high-stage plagioclase is considered. Calculations based on low-state plagioclase would give somewhat more stringent lower-pressure limits, but activity-composition relations are still unknown for low plagioclase.

The typical clinopyroxene of Group B and Group C eclogites is an omphacite close to the composition $(NaAlSi_2O_6)_{.5}$ $(CaMgSi_2O_6)_{.5}$. Some samples have been shown to be cation-ordered, with alternation of Al and Mg and, to a lesser extent, of Ca and Na in octahedral and 8-coordinated sites parallel to the c-axis (Clark et al., 1969). This ordering tendency probably accounts for the restricted compositional range. Heating experiments on natural ordered omphacite show that disordering takes place rather abruptly at temperatures above 750° (Fleet et al., 1978; Carpenter, 1981). Therefore, it is probable that many or most omphacites of Group B and C eclogites are ordered.

Ganguly (1973) suggested that $NaAlSi_2O_6$ and $CaMgSi_2O_6$ components in omphacite obey the ideal solution relationship of activities equal to mol fractions. Holland (1983) found experimentally that solid solution of $CaMgSi_2O_6$ in jadeite stabilizes clinopyroxene plus quartz relative to albite at 600°C by pressure increments closely predicted by the ideal solution law, for the intermediate composition range. He found evidence that the synthetic intermediate omphacites of his study were ordered. Additional components, involving principally Fe^{2+} (hedenbergite) and Fe^{3+} (acmite) can be allowed for by an ideal molecular-substitution model. The ideality of Al and Fe^{3+} substitution in the jadeite-acmite series was shown experimentally by Newton and

Smith (1967) and Popp and Gilbert (1972). The following molecular activity models result from the foregoing considerations:

$$^a NaAlSi_2O_6 = {}^X NaAlSi_2O_6 = {}^X jd$$
$$= Na\text{-}Fe^{3+}\text{-}Cr \text{ per } 6(0)$$
$$^a CaMgSi_2O_6 = {}^X CaMgSi_2O_6 = {}^X di$$
$$= Mg \text{ per } 6(0)$$

where X is the mol fraction.

Figure 4 shows four additional equilibria calculated from those of Figure 3 for an ordered omphacite of composition $jd_{.5}di_{.5}$. The equilibria of paragonite, albite and lawsonite with omphacite are isopleth lines, while the reaction of tremolite + albite to omphacite + talc + quartz is a true univariant equilibrium. The assumed omphacite composition is slightly metastable for this P-T range, but this fact does not detract from the usefulness of the curve. In making these calculations, the relation $P^\circ - P = nRTln \cdot 5/\Delta V^\circ$ was used for the nondehydration equilibria, where P° is the equilibrium pressure at T for the corresponding reaction with pure jadeite or diopside, ΔV° is the volume change, and n is the coefficient of jadeite or diopside in the balanced reactions. For dehydration reactions, the above formula was modified to accommodate the free energy change of H_2O over the pressure change $P^\circ - P$, using the tables of Sharp (1962) and a method of successive approximations. Molar volumes were taken from Helgeson et al. (1978) and the ΔVs were assumed independent of temperature and pressure.

Geothermometry

Fe^{2+}-Mg exchange between garnet and clinopyroxene is potentially the most useful temperature scale for eclogites. An experimental calibration at high pressures and high temperatures on basaltic compositions was made by Råheim and Green (1974). A correction for ferric iron content of natural clinopyroxenes based on microprobe analyses was suggested by Ryburn et al. (1976), as follows: $Fe^{3+} = 4 - 2Si - Al + Na$, neglecting insignificant amounts of K, Cr and Ti. This amount of ferric iron is subtracted from the total iron of the microprobe analysis, usually recorded as FeO, in applying the K_D thermometer. Additional experimental calibration by Ellis and Green (1979) considered more carefully the effect of Ca in garnet on the exchange. Their expression is:

$$T(K) = \frac{3104 X_{Ca}^{Gt} + 3030 + 10.86P \text{ (kb)}}{lnK_D + 1.9034}$$

where X_{Ca}^{Gt} is the mole fraction of grossular in the garnet and K_D is the distribution ratio $(Fe/Mg)^{Gt}/(Fe/Mg)^{Cpx}$. Subsequent theoretical calibrations were given by Ganguly (1979) and Saxena (1979). A critique of the calibrations of this thermometer is given by Lappin and Smith (1981).

The validities of the various calibrations in the low-temperature range and for ordered omphacite are not known from direct evidence. Harris et al. (1982) and Johnson et al. (1983) preferred the Ellis-Green experimental calibration over

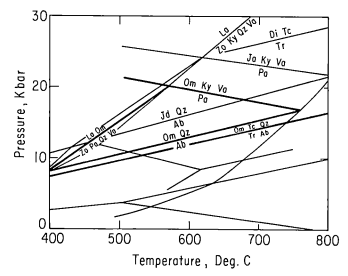

Figure 4. Equilibria calculated from univariant curves in simple systems using ideal solution model of diopside and jadeite (see text). Calculated isopleth lines for omphacite $(Jd_{.5}Di_{.5})$-Pa-Ky-Va, Om-La-Zo-Pa-Qz-Va, and Om-Ab-Qz are shown in bold lines. Univariant equilibrium Om+Tc+Qz = Tr+Ab also shown in bold. Symbols and unlabelled curves same as in Figure 2.

the theoretical calibrations for granulite facies pairs. Ghent and Stout (1981) found that Ganguly's (1979) calibration gave substantially higher temperatures than the experimental scale. In the present study, the Ellis-Green scale is used with the Ryburn et al. (1976) correction for Fe^{3+}, where iron is given only as FeO from microprobe analyses. The main justification of this choice is consistency with other information on the metamorphic temperatures. Desmons and Ghent (1977) found garnet-clinopyroxene temperatures based on Råheim and Green's (1974) calibration, with Fe^{3+} correction for eclogites from the Western Alps, to be in agreement with temperatures calculated for the same specimens from oxygen isotope fractionation by Desmons and O'Neil (1978), and such consistency with isotopic and other data will be shown in the present study.

Geobarometry

Minimum pressures of recrystallization of quartz eclogites lacking plagioclase may be calculated by the method of Perkins and Newton (1981), based on the reaction:

$$CaMgSi_2O_6 + CaAl_2Si_2O_8 = 1/3Mg_3Al_2Si_3O_{12} + 2/3Ca_3Al_2Si_3O_{12} + SiO_2$$

in	in	in	
clinopyroxene	plagioclase	garnet	quartz

The equilibrium P-T condition for this reaction, which is divariant in the system $CaO\text{-}MgO\text{-}Al_2O_3\text{-}SiO_2$ and of higher variance in more complex systems, neglecting differential compressibilities is:

$$\Delta G = 0 \cong \Delta G^\circ + (P-1)\Delta V^\circ + RT\ln \left(\frac{a_{py} \cdot a_{gr}^2}{a_{di} \cdot a_{an}} \right)$$

where ΔG° is the standard free-energy change of the reaction, the a's are the activities of pyrope ($MgAl_{2/3}SiO_4$) and grossular ($CaAl_{2/3}SiO_4$) in garnet solid solution, of diopside in clinopyroxene solid solution, and of anorthite in plagioclase solid solution. Quartz is considered to be pure SiO_2. Perkins and Newton (1981) gave the following parameters, used here, determined from solution calorimetry, heat-capacity measurements, and X-ray diffraction measurements:

$$\Delta G^\circ = 787.4 + 9.907T(K), \text{ in calories,}$$
$$\Delta V^\circ = -23.12 \text{ cm}^3,$$

$$a_{py} = X_{py}\exp \left\{ \frac{(X_{gr}^2 + X_{gr}X_{al})}{RT}(3300 - 1.5T) \right\},$$

$$a_{gr} = X_{gr}\exp \left\{ \frac{(X_{py}^2 + X_{py}X_{al})}{RT}(3300 - 1.5T) \right\},$$

$$a_{an} = \frac{X_{an}(1+X_{an})^2}{4}\exp \left\{ \frac{X_{ab}^2}{RT}(2050 + 9392X_{an}) \right\},$$

where the X's are the mol fractions in the solid solutions, including X_{ab}, the mol fraction of $NaAlSi_3O_8$ (albite) in plagioclase. The activity of diopside in clinopyroxene is evaluated by the ideal-solution ionic model discussed earlier.

The above reaction may be combined with the albite breakdown reaction:

$$NaAlSi_3O_8 = NaAlSi_2O_6 + SiO_2$$
$$\textit{in} \qquad \textit{in} \qquad \text{quartz}$$
$$\text{plagioclase} \quad \text{clinopyroxene}$$

This reaction has the equilibrium condition:

$$\Delta G = 0 \cong (P-P^\circ)\Delta V^\circ + RT\ln \left\{ \frac{a_{jd}}{a_{ab}} \right\},$$

where P° is the experimentally-determined equilibrium pressure of the end-member reaction at T, given by Holland (1980) as $P^\circ = 0.0265T(^\circ C) + 0.35$ kbar, a_{jd} is the $NaAlSi_2O_6$ activity in clinopyroxene evaluated by the ideal ionic-solution model described earlier, and a_{ab} is given from solution calorimetry by Perkins and Newton (1981) as:

$$a_{ab} = X_{ab}^2(2-X_{ab})\exp \left\{ \frac{X_{an}^2}{RT}(6746 - 9392X_{ab}) \right\},$$

At a given temperature, and for specified compositions, hence activities, of the garnet and clinopyroxene components, the two unknowns in the simultaneous solution of the thermodynamic equations of the above two reactions are pressure and the composition of the (fictive) plagioclase, which therefore can be found. Since plagioclase is absent, the pressure determined is a minimum pressure.

An analogous method may be used to find *maximum* pressures of quartz-bearing amphibolites containing garnet and plagi-

oclase, which are sometimes found in close association with eclogites. The basis is the anorthite breakdown reaction:

$$3CaAl_2Si_2O_8 = Ca_3Al_2Si_3O_{12} + 2Al_2SiO_5 + SiO_2$$
$$\textit{in} \qquad\qquad \textit{in} \qquad\quad \text{kyanite} \quad \text{quartz}$$
$$\text{plagioclase} \qquad \text{garnet}$$

Since kyanite is generally lacking, pressures calculated at a given temperature by the thermodynamic method of Newton and Haselton (1981) or the semi-empirical method of Ghent et al. (1979) are upper-pressure limits.

The sources of error in these expressions are several. The errors from calorimetry alone in the garnet-clinopyroxene-plagioclase-quartz method are of the order of ±1.7 kbar (Newton, 1983). Additional possible errors arise from the activity expressions used. The expressions for activities of plagioclase components are appropriate for high structural-stage plagioclase, whereas the plagioclase which would be stable over much of the temperature range considered here would be low plagioclase. This is not a fundamental limitation, however, as the method merely gives minimum pressures, which are still valid in the absence of plagioclase (either high or low state). If activity expressions appropriate for low plagioclase were inserted in the above expressions, the minimum pressures obtained would be somewhat more stringent (that is, higher) than is possible using expressions for high plagioclase. It is likely that the absolute uncertainties in the calculated minimum pressures are of the order of ±2 kbar. The magnitude of this uncertainty discourages overly-quantitative deductions. However, the general pressure ranges deduced below are qualitatively different from, and superior to, some earlier estimates for the same rocks, which justifies the attempt at quantification.

TEMPERATURE-PRESSURE INDICATIONS OF VARIOUS ECLOGITES

Classification of Terranes

A number of eclogite-bearing terranes exist for which good descriptions and mineral analyses are available. Geothermometry and geobarometry are possible for these based on experimental univariant equilibria in simple systems extended by thermodynamic analysis, and by continuous solid solution temperature and pressure indicators. The terranes can be classified into subgroups based on minerals present or inferred to have been present, either in the eclogites themselves or in co-metamorphic host rocks. A simple classification may be based on the actual or inferred former presence of lawsonite (low-temperature group), association with talc-kyanite rocks or whiteschists (high-temperature group) and an intermediate-temperature group with neither lawsonite nor whiteschist association.

Low-Temperature Group. One of the best-described eclogite terranes is *Sifnos*, one of a number of blueschistic terranes in the

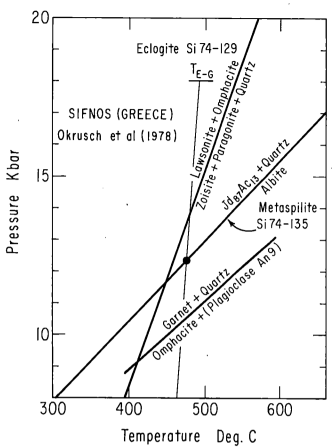

Figure 5. Approximate equilibration conditions of Sifnos (Greece) rocks. T_{E-G} denotes garnet-clinopyroxene K_D (Fe, Mg) temperature from Ellis and Green (1979).

Greek Islands. A variety of blueschist metasediments and co-metamorphic eclogitic metavolcanic rocks were described by Okrusch et al. (1978). The blueschist/eclogite metamorphism at 42 million years ago was followed by a greenschist overprint 21-24 million years ago (Altherr et al., 1979). Eclogite Si 73-90 contains pseudomorphs of clinozoisite and white mica after lawsonite, as well as the apparently primary minerals shown in Figure 1. Geothermometry of mineral cores of Si 74-129 gives temperatures of 470°-490°C, depending on assumed pressures (Figure 5). Closely-associated metapelite Si 74-135 has a pyroxene close to $jd_{87}ac_{13}$, albite and quartz, for which assemblage the calculated P-T line in Figure 5 is given. The indicated equilibration conditions are near 12.4 kbar and 480°C, which agree with the earlier estimates of Okrusch et al. (1978) of P ~14 kbar, T = 450°-500°C. A recent careful study of the oxygen isotopes of Sifnos blueschists gives 455° ± 25°C based on five different mineral pairs (Matthews and Schliestedt, 1984). The near agreement with the Ellis-Green thermometer lends credence to its use for low-temperature eclogites. The P, T conditions given by intersection in Fig. 5 are compatible with a slight temperature transgression of the lawsonite-omphacite stability line and with minimum pressures calculated from the garnet and omphacite compositions of eclogite Si 74-129.

The eclogites of the *Breuil-St. Jacques* area of the western Italian Alps (Ernst and Dal Piaz, 1978) have geologic setting and parageneses similar to Sifnos. However, the eclogites are more mafic and generally lack quartz. Minerals show little chemical zonation. Pseudomorphs of zoisite-epidote-white mica after lawsonite show that the P-T path passed through the lawsonite field. The Ellis-Green temperatures of DBL 379 and MRO 857 are near 500°C. The apparent absence of margarite during any stage of the recrystallization, together with the lawsonite pseudomorphs, fixes the pressure between 11 and 15 kbar. The deduced-temperature range agrees with the 470° ± 50°C preferred by Ernst and Dal Piaz (1978), but the present pressures are higher than their estimated 10 ± 2 kbar. Pseudomorphs after lawsonite have also been found in the eclogite from the *Guajira Peninsula*, Colombia, described by Green et al. (1964) and in some eclogites from the southern Austrian Alps (Holland, 1979a, quoting G. Droop).

Intermediate-Temperature Group

This is a large category, with many well-described representatives. The *Margarita Island* eclogite of northern Venezuela shows a primary assemblage (listed in Figure 1) and a retrogressive greenschist assemblage similar to that displayed by many Alpine eclogites (Maresch and Abraham, 1981). This occurrence is closely related in age and geologic setting to the Guajira Peninsula eclogite. The garnet is practically unzoned, and clinopyroxene is slightly zoned, with more jadeitic rims. Minimum pressures of 11-12 kbar are given by the garnet and clinopyroxene rims with a compatible virtual plagioclase of An 9-11 (Figure 6). The presence of epidote and paragonite and absence of lawsonite suggest upper-pressure limits of about 15-16 kbar at the average Ellis-Green temperature of 515°C. The deduced P-T conditions embrace the 12.5 ± 1 kbar and 450°-525°C preferred by Maresch and Abraham (1981), and are compatible with the absence of talc-kyanite in the enclosing schists (Figure 6).

Eclogite lenses of the Bavarian *Muenchberg Gneiss* of Hercynian age were described by Matthes et al. (1975). Some of the eclogites are kyanite-bearing. The virtual plagioclase An 19 compatible with the garnet and pyroxene of the average light eclogite puts lower pressure limits of about 12 kbar at the average Ellis-Green temperature of 570°C (Figure 7). An upper-pressure limit of 15 kbar may be set by the absence of talc + kyanite in the enclosing schists and gneisses (Figure 7). The temperature range agrees with the 550° ± 50°C preferred by Matthes et al. (1975), but the present pressures are much higher than their suggested 5-6 kbar.

The *Raspas Formation* of Andean Ecuador is an eclogite-bearing horizon of high-pressure schists and amphibolites described by Feininger (1980) in considerable detail. The metamorphic age is not well-established, but known to be pre-early Cretaceous and post-Triassic. Eclogitic layers show a wide range of equilibrium conditions in their variable mineral chemistries, with K_D^{Fe-Mg} (garnet-clinopyroxene) covering the range of at

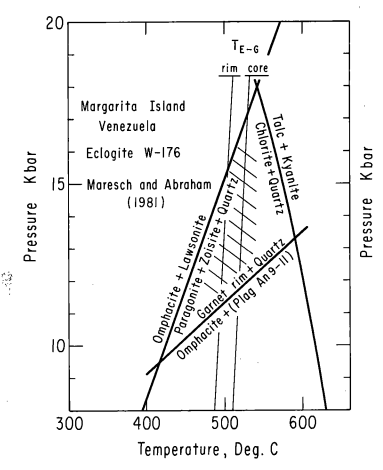

Figure 6. Approximate equilibration conditions for the Margarita Island (Venezuela) eclogite. Plagioclase An 9-11 in parentheses indicates a compatible but absent plagioclase (see text).

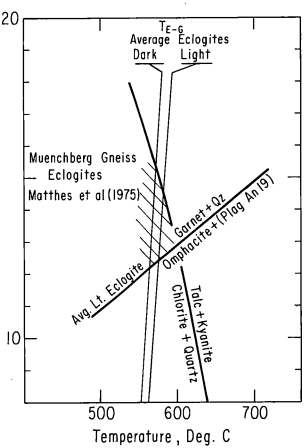

Figure 7. Approximate equilibration conditions for the Muenchberg (Germany) eclogites.

least 21 to 10, indicating temperatures from 540–680°C (Figure 8). The lower temperatures are compatible with the apparent absence of talc-kyanite in enclosing schists. Broad pressure limits of two extreme samples are set approximately by the garnet and pyroxene compositions and by the absence of lawsonite in the lower-temperature sample. The equilibration pressure of the higher-temperature sample appears to be at least 15.5 kbar. Upper-pressure limits are not established for the higher-temperature sample.

The variation of apparent P-T conditions of the Raspas samples may represent either a prograde P-T path (as suggested by Råheim and Green (1975) for several eclogite terranes), variable back-reaction during uplift, or some kind of disequilibrium. The higher temperature samples apparently equilibrated in the talc-kyanite field of stability, and might be placed in the white-schist group of the present classification. Feininger (1980) does not report talc in the eclogites or enclosing metasediments.

Numerous eclogite localities occur in the Basal Gneisses of Norway (Krogh, 1977). Some of these are glaucophane-bearing. On the basis of Sm-Nd age-dates (Griffin and Brueckner, 1980) the eclogite metamorphism is believed to be early Caledonian (~425 m.y.). The eclogites are enclosed in schists and gneisses,

some of which contain plagioclase, rather than jadeite and quartz, which places an upper-pressure limit on the metamorphism. The eclogite at *Naustdal* is limited in pressure at 15.5 kbar constrained at the Ellis-Green core temperature of 540°C by the presence of plagioclase An 19 in enclosing tonalite gneisses. The zoned garnet gives evidence of a prograde increasing P-T path in that the rims are more pyropic than the cores (Figure 9), but it is difficult to be sure of the clinopyroxene composition that was in equilibrium with the garnet cores and rims.

Some of the glaucophanitic mafic schists of Ordovician age at *Tillotson Peak, Vermont* (Laird and Albee, 1976) contain garnet and omphacite which, while not present in sufficient amounts to warrant the term eclogite, are of typical eclogitic compositions. These schists are of interest in that both a blue amphibole and a green amphibole are present in some parageneses, as well as plagioclase. The assemblage tremolite-albite suggests an upper-pressure limit near 10 kbar for sample A-BM-100 at its Ellis-Green temperature of 490°C. The P-T conditions preferred by Laird and Albee are 9 kbar and 450°C.

High-Temperature Group

The reaction of clinochlore and quartz to talc and kyanite

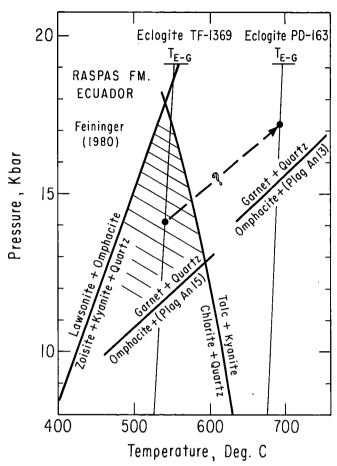

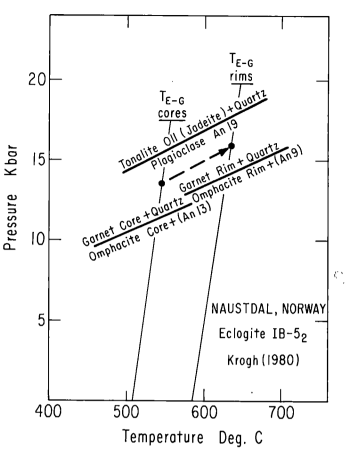

Figure 8. Approximate equilibrium conditions for two eclogites from the Raspas Formation (Ecuador). The dashed arrow suggests a possible P-T subduction path.

Figure 9. Approximate equilibration conditions for the Naustdal, Norway, eclogite, and a possible subduction path based on core and rim analyses of garnet and clinopyroxene. Jadeite in parentheses signifies an absent phase in the enclosing tonalitic gneiss, so that the higher equilibrium line is an upper-pressure bound.

serves to define a high-temperature group of eclogites, the whiteschist group. Temperatures above 550°C for eclogites actually containing talc-kyanite or enclosed in co-metamorphic talc-kyanite schists are indicated by the experimental work of Massonne et al. (1981) (Figure 3).

Some of the very high-pressure eclogites of the *Hohe Tauern Window,* Austrian Alps, contain both talc and kyanite, in addition to the other minerals shown in Figure 1. Holland (1979a) was not sure if the talc is strictly primary or if it belongs to the retrogressive greenschist overprint seen in most specimens. Ellis-Green thermometry is quite uncertain because of the extremely magnesian clinopyroxenes. Holland deduces metamorphic conditions of 620°C and 19.5 kbar and water activity at least 0.9 by an analysis of equilibrium relations in the assemblage omphacite-paragonite-dolomite-quartz, which he thought to be a buffered primary assemblage. He found abundant aqueous fluid inclusions in some minerals to support the deduction of high water pressures. However, Luckscheiter and Morteani (1980) studied fluid inclusions in eclogites from a number of Tauern Window localities, including the Gross Venediger locality of Holland. They found evidence that early fluid inclusions, trapped during garnet growth, were CO_2-rich and H_2O-poor.

The garnet of Holland's (1979a) TH 49B specimen is rich in pyrope and grossular, and the clinopyroxene is ordered omphacite of very nearly the ideal composition, which determine, at the temperature of 620°C, a minimum pressure of 16 kbar and a virtual plagioclase An 11. The assemblage paragonite-omphacite-kyanite may be considered a buffered assemblage, if P_{H_2O} was nearly equal to total pressure during the metamorphism, in which case pressures at 620°C were above 19 kbar. Conversely, this assemblage could be considered to indicate an upper limit if P_{H_2O} was less than the total pressure. In either case, a pressure of 17.5 kbar is conservative, in agreement with Holland's (1979a) analysis.

The high-pressure whiteschist terrane of *western Tasmania,* described by Råheim (1976), is late Precambrian to early Cambrian in age. Eclogites are interlayered with a variety of metasediments, including talc-kyanite-garnet pelitic schists. Also interlayered in the Lyell-Collingwood section are garnet amphibolites which, for reasons of locally higher P_{H_2O} or subtle chemical characteristics, contain plagioclase. The garnets in both eclogites and amphibolites are strongly zoned with more Mg-rich and less Fe-rich rims than cores. From this, Råheim and Green (1975) deduced an increasing-temperature sequence of crystallization which they ascribed to a prograde increasing-pressure subduction path. They showed that this type of garnet zonation is characteristic of several other eclogite occurrences.

Figure 10 shows temperature and pressure estimates based on zoned garnets in an eclogite and an amphibolite from the Lyell-Collingwood area. The garnet core of the eclogite gives, with the clinopyroxene, an Ellis-Green temperature of 620°-640°C. In the absence of a compatible plagioclase An 18, the eclogite gives a lower-pressure limit of 12.6 kbar at this temperature and, in the absence of kyanite, the amphibolite gives an upper-pressure limit of 13.0. These P-T conditions are compatible with the regional presence of whiteschists. Similarly, garnet and omphacite rims of the eclogite give a temperature of 730-750°C, and the corresponding pressure lies between 15.4 and 16.9 kbar. The indicated P-T path, shown by the dashed arrow in Figure 10, is similar to that inferred by Råheim and Green (1975), though their pressure scale is much lower (7-11 kbar). The high temperature indicated by the mineral rims is consistent with the migmatized condition of some of the eclogites (Råheim, 1976).

A somewhat similar occurrence is the *Lufilian Arc of Zambia*, where late Precambrian eclogites are interlayered with metasediments, including whiteschists. Some of the Lufilian eclogites actually contain the whiteschist assemblage. The descriptions of Vrana et al. (1975) indicate that some eclogites contain plagioclase (compositions not given) and at least one contains a pale green amphibole (Vrana et al., 1975, p. 143). Garnets are zoned with more magnesian rims and K_D^{Fe-Mg} (gt-cpx) covers a considerable range. The Ellis-Green temperature for a typical eclogite (their number 3) is about 640°C. The minimum pressure at this temperature is 13.1 kbar for a plagioclase An 15 compatible with the garnet and pyroxene. This plagioclase, with garnet, kyanite and quartz, yields a minimum pressure of 13.4 kbar at 640°C, using the barometer of Newton and Haselton (1981). The inferred P-T conditions are mostly within the talc-kyanite field of stability and close to the equilibrium of albite and tremolite with omphacite, talc and quartz (Fig. 4). The coherent pressures suggest that plagioclase, talc, and tremolite are equilibrium accessories of the eclogite.

The *Allalin* eclogitic coronal metagabbros of western Switzerland have some talc-kyanite associations (Chinner and Dixon, 1973) and thus belong with the high-temperature category. The assemblages chloritoid-kyanite and talc-kyanite may fix the pressure above 15 kbar, based on the experimental work on Fe-chloritoid by Ganguly (1972) combined with the talc-kyanite stability boundary of Massonne et al. (1981), but this is quite uncertain because of the high Mg-content of the Allalin chloritoid.

SUMMARY AND INTERPRETATIONS

Temperature and Pressure Ranges of Crustal Eclogites

Many of the present estimates of recrystallization tempera-

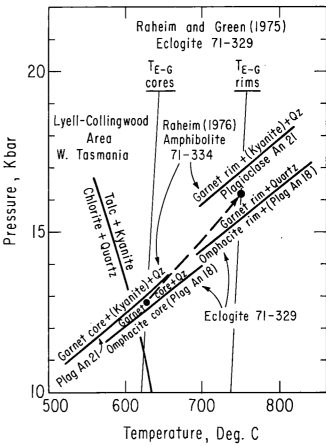

Figure 10. Approximate equilibration conditions of eclogites and garnet amphibolites from the Lyell-Collingwood area, Tasmania, based on rim and core analyses. Compatible plagioclase An 18 (the same for both rim and core compositions of the eclogite) is not present, which gives lower-pressure bounds. Kyanite in parentheses signifies that the phase is absent in the garnet amphibolite, which gives an upper-pressure bound.

tures and pressures of Group B and C eclogites are in substantial accord with values preferred by authors of the original descriptions, including the Sifnos, Margarita, Raspas and Naustdal occurrences.

The Ellis-Green garnet-clinopyroxene thermometer scale appears to be a reasonable working scale; it nearly coincides with the oxygen-isotope thermometer of enclosing blueschistic metasediments and metavolcanics at Sifnos (Matthews and Schliestedt, 1984) and with various phase-equilibrium constraints. The presence of pseudomorphs of zoisite and paragonite after lawsonite implies equilibration near to, but at higher temperatures than, the P-T curve for the reaction of lawsonite and omphacite, and this condition is satisfied for the Sifnos and Breuil-St. Jacques occurrences. Eclogites with talc-kyanite or from whiteschist terranes generally plot within the experimental talc-kyanite field of stability (Figure 11). An apparent exception is the highest temperature group of eclogites from the Raspas Formation, Ecuador. It may be that talc is present but not yet recognized in some of the associated metapelites.

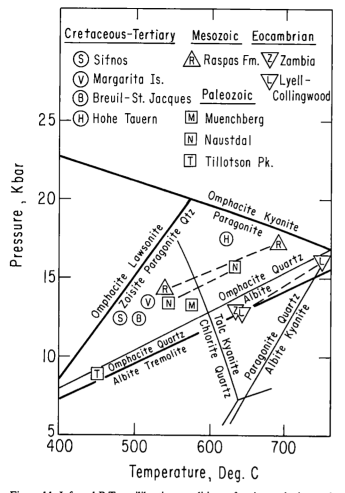

Figure 11. Inferred P-T equilibration conditions of various eclogites and transitional eclogites from the present study. Dashed lines show spread of P-T conditions of different specimens or as indicated by core and rim compositions. The lowest T/P ratios are for the Cretaceous-Tertiary rocks, and the highest T/P ratios are for the late Precambrian-Cambrian examples.

Calculated pressures are generally consistent with experimental petrology. Most of the eclogites plot above the curve for the stability of ideal ordered omphacite plus quartz relative to albite (Figure 11). The rocks of Tillotson Peak, Lufilian Arc and Lyell-Collingwood lie close to, but slightly below, this line. Some of the Tillotson and Lufilian rocks have plagioclase, along with garnet and omphacite, and some closely-associated metabasites from Lyell-Collingwood have plagioclase. The univariant curve tremolite + albite = omphacite + talc + quartz serves as a model for the transition of greenschist and amphibolite facies to eclogite facies, and the Tillotson and Lufilian rocks with tremolite-albite lie close to this line. The pressure threshold of eclogites in the crustal temperature range of 500°-700°C is thus 10-14 kbar. It seems clear that most Group B and C eclogites formed at pressures corresponding to depths at the base of, or below, the normal continental crust. Recent reports of coesite as inclusions in garnet from a quartzite from western Italy (Chopin, 1984) and in clino-

pyroxene from a Norwegian eclogite (Smith, 1983) may demand pressures in excess of 28 kbar. The full petrologic significance of these occurrences has not yet been fully explored. However, it may prove true that some eclogites have gone through an early very-high-pressure stage at temperatures less than those represented by the dominant coarse-grained assemblages, and that records of the early very-high-pressure stage are rarely preserved.

Earlier estimates of lower pressures for some eclogites were perhaps influenced by straight-line extrapolations of the high-temperature, high-pressure experimental work of Green and Ringwood (1967), which could imply eclogite stability relative to garnet granulites in dry rocks at pressures below 10 kbar in the temperature range below 700°C (Figure 2). This approach has suggested to some that crustal eclogites may have sometimes formed in amphibolite-facies and greenschist-facies metamorphism at much reduced H_2O activities as, for example, if pore solutions were saturated brines. However, recent thermodynamic calculations by Wood (1984) on stability of plagioclase in mafic compositions imply that straight-line extrapolations of the high-temperature experimental curves are not warranted: the plagioclase-out-P-T line in mafic compositions has considerable curvature convex to the temperature axis, and pressures above 10 kbar are required for eclogite stability in the low-temperature range even in perfectly dry systems.

The controversy about a possible role of reduced H_2O activity in genesis of some eclogites continues. It may well be that concentrated pore brines and/or dilution of H_2O with CO_2 or CH_4 were effective in certain cases, but it now appears that such mechanisms are not universally necessary. High pressures alone can produce eclogite mineralogy in mafic compositions in a low temperature, hydrous regime.

Evidence For Prograde P-T Paths

The present results for the Lyell-Collingwood eclogites are compatible with, but do not prove, the contention of Råheim and Green (1975) that the rocks record an increasing temperature-pressure trajectory in the zonations of minerals. The apparent descent geotherm was about 17°C/km, if the proper P-T interpretation of Figure 10 is a prograde path. Similar results were obtained by Råheim and Green for the Guajira Peninsula and some Norwegian eclogites. It may be that hydrous melting limited the record in the Lyell-Collingwood rocks because of the temperatures over 700°C given by rim compositions and the presence of migmatized eclogites. The Raspas Formation and Naustdal eclogites may record a similar trajectory in the spread of K_D^{Fe-Mg} values of garnet and clinopyroxene encountered.

Decrease in Subduction Geotherms Over Time

Figure 11 may give an indication of secular change in apparent subduction geotherms, as suggested by deRoever (1956) and Ernst (1972). The highest P/T examples are late Cretaceous-Tertiary, and the lowest P/T examples are the late Precambrian-early Paleozoic occurrences. If this trend is real, the rarity of older eclogites may be explained by the fact that subduction geotherms in the Precambrian generally exceeded the 16°C/km or so which is approximately the lower-pressure boundary of the eclogite field, as modeled by the tremolite-albite breakdown curve. This hypothesis can only be evaluated by investigation of many more eclogite occurrences of different ages, since the relationship will be at best a statistical one, and few occurrences are currently available for comparison. In view of the large uncertainties in temperature and pressure determinations, a significant secular change cannot be regarded as established.

Uplift Paths

Eclogites containing a greenschist overprint suggest a relatively high-temperature uplift path, examples being the Hohe Tauern Window eclogites (Holland, 1979a) and the Margarita Island eclogites (Maresch and Abraham, 1981). The latter authors proposed a nearly isothermal decompression to 5 kbar after eclogite formation. The retrogressive overprints may result either from continuous re-equilibration to lower pressures or one or more discrete periods of re-metamorphism, resulting from extensive sojourns at intermediate depths en route to the surface. The uplift paths convex to the temperature axis are indicative of relatively fast uplift and erosion (England and Richardson, 1977).

ACKNOWLEDGMENTS

National Science Foundation grant #EAR 84-11192 supported the author's research. He acknowledges stimulating and informative conversations with Timothy Holland, Walter Maresch, Alan Matthews, Martin Okrusch, Eberhard Seidel, and David Smith during the Penrose Blueschist Conference and Field Trips, September, 1983. The opinions of Dugald Carmichael, Eric Essene and Edward Ghent on related matters influenced the author greatly.

REFERENCES CITED

Ahrens, T. J. and Schubert, G., 1975, Gabbro-eclogite reaction rate and its geophysical significance: Rev. Geophys. Space Phys. *13*, 383–400.

Altherr, R., Schliestedt, M., Okrusch, M., Seidel, E., Kreuzer, H., Harre, W., Lenz, H., Wendt, I. and Wagner, G. A., 1979, Geochronology of high-pressure rocks on Sifnos (Cyclades), Greece: Contr. Mineral. Petrol. *70*, 245–255.

Brown, E. H., Bernardi, M. L., Christenson, B. W., Cruver, J. R., Haugerud, R. A., Rady, P. M. and Sondergaard, J. N., 1981, Metamorphic facies and tectonics in part of the Cascade Range and Puget Lowland of northeastern Washington: Geol. Soc. Amer. Bull. *92*, 170–178.

Carman, J. H. and Gilbert, M. C., 1983, Experimental studies on glaucophane stability: Am. J. Sci. *283-A* (Orville volume), 414–437.

Carpenter, M. A., 1981, Time-temperature-transformation (TTT) analysis of cation disordering in omphacite: Contr. Mineral. Petrol. *78*, 433–440.

Chatterjee, N. D., 1974, Crystal-liquid vapour equilibria involving paragonite in the system $NaAlSi_2O_8$-Al_2O_3-SiO_2-H_2O: Indian J. of Earth Sci. *1*, 3–11.

Chernosky, J. V., Jr., 1978, The stability of clinochlore + quartz at low pressure: Am. Mineral *63*, 73–82.

Chinner, G. A. and Dixon, J. E., 1973, Some high-pressure parageneses of the Allalin Gabbro, Valais, Switzerland: J. Petrol. *14*, 185–202.

Chopin, C., 1984, Coesite and pure pyrope in high-grade blueschists of the Western Alps: A first record and some consequences: Contr. Mineral Petrol. *86*, 107–118.

Clark, J. R., Appleman, D. E. and Papike, J. J., 1969, Crystal-chemical characterization of clinopyroxenes based on eight new structure refinements: Mineral Soc. Amer. Spec. Pap. *2*, 31–50.

Coleman, R. G., Lee, D. E., Beatty, L. B. and Brannock, W. W., 1965, Eclogites and eclogites: Their differences and similarities: Geol. Soc. Amer. Bull. *76*, 483–508.

de Roever, W. P., 1956), Some differences between post-Paleozoic and older regional metamorphism: Geol. en. Mijnb., new ser. *18*, 123–127.

Desmons, J. and Ghent, E. D., 1977, Chemistry, zonation and distribution coefficients of elements in eclogitic minerals from the Eastern Sesia Unit, Italian Western Alps: Schweiz. mineral. petrog. Mitt. *57*, 397–411.

Desmons, J. and O'Neil, J. R., 1978), Oxygen and hydrogen isotope compositions of eclogites and associated rocks from the Eastern Sesia Zone (Western Alps, Italy): Contr. Mineral Petrol. *67*, 79–85.

Ellis, D. J. and Green, D. H., 1979, An experimental study of the effect of Ca upon garnet-clinopyroxene Fe-Mg exchange equilibria: Contr. Mineral. Petrol. *71*, 13–22.

England, P. C. and Richardson, S. W., 1977, The influence of erosion upon the mineral facies of rocks from different metamorphic environments: J. Geol. Soc. *134*, 201–213.

Ernst, W. G., 1972, Occurrence and mineralogic evolution of blueschist belts with time: Amer. J. Sci. *272*, 657–668.

Ernst, W. G. and Dal Piaz, G. V., 1978, Mineral parageneses of eclogitic rocks and related mafic schists of the Piemonte ophiolite nappe, Breuil-St. Jacques area, Italian Western Alps: Amer. Mineral. *63*, 621–640.

Essene, E. J., Hensen, B. J. and Green, D. H., 1970, Experimental study of amphibolite and eclogite stability: Phys. Earth and Plan. Int. *3*, 378–384.

Feininger, T., 1980, Eclogite and related high-pressure regional metamorphic rocks from the Andes of Ecuador: J. Petrol. *21*, 107–140.

Fleet, M. E., Herzberg, C. T., Bancroft, G. M. and Aldridge, L. P., 1978, Omphacite studies, I. The P2/n-C2/c transformation: Amer. Mineral. *63*, 1100–1106.

Fry, N. and Fyfe, W. S., 1969, Eclogites and water pressure: Contr. Mineral. Petrol. *24*, 1–6.

Ganguly, J., 1973, Activity-composition relation of jadeite in omphacite pyroxene: Theoretical deductions: Earth, Plan. Sci. Lett. *19*, 145–153.

Ganguly, J., 1979, Garnet and clinopyroxene solid solutions, and geothermometry based on Fe-Mg distribution coefficient: Geochim. et Cosmochim. Acta *43*, 1021–1029.

Ghent, E. D., Robbins, D. B. and Stout, M. Z., 1979, Geothermometry, geobarometry, and fluid compositions of metamorphosed calc-silicates and pelites, Mica Creek, British Columbia: Amer. Mineral. *64*, 874–885.

Ghent, E. D. and Stout, M. Z., 1981, Metamorphism at the base of the Samail Ophiolite, southeastern Oman Mountains: J. Geophys. Res. *86*, 2557–2572.

Ghent, E. D., Stout, M. Z. and Raeside, R. P., 1983, Plagioclase-clinopyroxene-garnet-quartz equilibria and the geobarometry and geothermometry of garnet amphibolites from Mica Creek, British Columbia: Canad. J. Earth Sci. *20*, 699–706.

Goldsmith, J. R., 1981, The join $CaAl_2Si_2O_8$-H_2O (anorthite-water) at elevated pressures and temperatures: Amer. Mineral. *66*, 1183–1188.

Green, D. H., Lockwood, J. P. and Kiss, E., 1964, Eclogite and almandine-jadeite-quartz rock from the Guajira Peninsula, Colombia, South America: Amer. Mineral. *53*, 1320–1335.

Green, D. H. and Ringwood, A. E., 1967, An experimental investigation of the gabbro to eclogite transformation and its petrological applications: Geochim. et Cosmochim. Acta *31*, 767–833.

Griffin, W. L. and Brueckner, H. K., 1980, Caledonian Sm-Nd ages and a crustal origin for Norwegian eclogites: Nature *285*, 319–321.

Harris, N.B.W., Holt, R. W. and Drury, S. A., 1982, Geobarometry, geothermometry, and late Archean geotherms from the granulite facies terrain of South India: J. Geol. *90*, 509–528.

Heinrich, W. and Althaus, E., 1980, Die obere Stabilitätsgrenze von Lawsonit plus Albit bzw. Jadeit: Fort. Mineral. *58*, 49–50.

Helgeson, H. C., Delany, J. M., Nesbitt, H. W. and Bird, D. K., 1978, Summary and critique of the thermodynamic properties of rock-forming minerals: Amer. J. Sci. *278-A*, 1–229.

Holdaway, M. J., 1971, Stability of andalusite and the aluminum silicate phase diagram: Amer. J. Sci. *271*, 97–131.

Holland, T.J.B., 1979a, High water activities in the generation of high pressure kyanite eclogites of the Tauern Window, Austria: J. Geol. *87*, 1–28.

Holland, T.J.B., 1979b, Experimental determination of the reaction paragonite = jadeite + kyanite + H_2O, and internally consistent thermodynamic data for part of the system Na_2O-Al_2O_3-SiO_2-H_2O, with applications to eclogites and blueschists: Contr. Mineral. Petrol. *68*, 293–301.

Holland, T.J.B., 1980, The reaction albite = jadeite + quartz determined experimentally in the range 600-1200°C: Amer. Mineral. *65*, 129–134.

Holland, T.J.B., 1983, The experimental determination of activities in disordered and short-range ordered jadeitic pyroxenes: Contr. Mineral. Petrol. *82*, 214–220.

Jenkins, D. M., 1983, Upper stability of synthetic margarite plus quartz: Geol. Soc. Amer. Abstr. with Prog. *15* 604.

Johnson, C. A., Bohlen, S. R. and Essene, E. J., 1983, An evaluation of garnet-clinopyroxene geothermometry in granulites: Contr. Mineral. Petrol. *84*, 191–198.

Koons, P. O., 1982, An experimental investigation of the behavior of amphibole in the system Na_2O-MgO-Al_2O_3-SiO_2-H_2O at high pressures: Contr. Mineral. Petrol. *79*, 258–267.

Krogh, E. J., 1977, Crustal and *in situ* origin of Norwegian eclogites: Nature *269*, 730.

Krogh, E. J., 1980, Compatible P-T conditions for eclogites and surrounding gneisses in the Kristiansund Area, Western Norway: Contr. Mineral. Petrol. *75*, 387–393.

Laird, J. and Albee, A. L., 1981, High-pressure metamorphism in mafic schist from northern Vermont: Amer. J. Sci. *281*, 97–126.

Lambert, I. B. and Wyllie, P. J., 1972, Melting of gabbro (quartz eclogite) with excess water to 35 kilobars, with geological applications: J. Geol. *80*, 693–708.

Lambert, R.St.J., 1983, Metamorphism and thermal gradients in the Proterozoic continental crust: Geol. Soc. Amer. Mem. *161*, 155–165.

Lappin, M. A. and Smith, D. C. 1981, Carbonate, silicate and fluid relationships in eclogites, Selje district and environs, SW Norway. Trans. Roy. Soc. Edinburgh: Earth Sci. *72*, 171–193.

Leake, B. E., 1978, Nomenclature of amphiboles: Mineral. Mag. *42*, 533–563.

Luckscheiter, B. and Morteani, G., 1981, Microthermometrical and chemical studies of fluid inclusions in minerals from Alpine veins from the penninic rocks of the central and western Tauern Window (Austria/Italy): Lithos *13*, 61–77.

Maresch, W. V., 1977, Experimental studies on glaucophane: An analysis of present knowledge: Tectonophyps. *43*, 109–125.

Maresch, W. V. and Abraham, J., 1980, Petrography, mineralogy, and metamorphic evolution of an eclogite from the Island of Margarita, Venezuela: J. Petrol. *22*, 337–362.

Massonne, H. J., Mirwald, P. W. and Schreyer, W., 1981, Experimentelle Überprüfung der Reaktionskurve Chlorit + Quarz = Talk + Disthen im System MgO-Al_2O_3-SiO_2-H_2O: Fort. Mineral. *59*, 122–123.

Matthes, S., Richter, P. and Schmidt, K., 1975, Petrography, geochemistry, and petrogenesis of the eclogites of the Muenchberg gneiss area: N. Jb. Miner. Abh. *126*, 45–86.

Matthews, A. and Schliestedt, M., 1984, Evolution of the blueschist and greenschist facies rocks of Sifnos, Cyclades, Greece: Contr. Mineral. Petrol. *88*, 150–163.

Moore, D. E., 1984, Metamorphic history of a high-grade blueschist exotic block from the Franciscan Complex, California: J. Petrol. *25*, 126–150.

Newton, R. C., 1972, An experimental determination of the high pressure stability limits of magnesian cordierite under wet and dry conditions: J. Geol. *80*, 398–420.

Newton, R. C., 1983, Geobarometry of high-grade metamorphic rocks: Amer. J. Sci. *283-A*, 1–28.

Newton, R. C., Charlu, T. V. and Kleppa, O. J., 1980, Thermochemistry of the high structural state plagioclases: Geochim. et Cosmochim. Acta *44*, 933–941.

Newton, R. C. and Haselton, H. T., 1981, Thermodynamics of the garnet-plagioclase-Al_2SiO_5-quartz geobarometer, *in* Newton, R. C., Navrotsky, A., and Wood, B. J., eds., Thermodynamics of Minerals and Melts: New York, Springer, 129–145.

Newton, R. C. and Kennedy, G. C., 1963, Some equilibrium reactions in the join $CaAl_2Si_2O_8$-H_2O: J. Geophys. Res. *68*, 2967–2983.

Newton, R. C. and Perkins, D., 1982, Thermodynamic calibration of geobarometers based on the assemblages garnet-plagioclase-orthopyroxene (clinopyroxene)-quartz: Amer. Mineral. *67*, 203–222.

Newton, R. C. and Smith, J. V., 1967, Investigations concerning the breakdown of albite at depth in the earth: J. Geol. *75*, 268–286.

Nitsch, K-H., 1974, Neue Erkenntnisse zur Stabilität von Lawsonit: Fort. Mineral. *51*, 34–35.

Nitsch, K.-H, Storre, B. and Töpfer, U., 1981, Experimentelle Bestimung der Gleichgewichtsdaten der Reaktion 1 Margarit + 1 Quarz = 1 Anorthit + 1 Andalusit/Disthen + 1 H_2O: Fort. Mineral. *59*, 139–140.

Okrusch, M., Seidel, E. and Davis, E. N., 1978, The assemblage jadeite-quartz in the glaucophane rocks of Sifnos (Cyclades Archipelago, Greece): N. Jb. Miner. Abh. *132*, 284–308.

Orville, P. M., 1972, Plagioclase cation exchange equilibria with aqueous chloride solution: Results at 700°C and 2000 bars in the presence of quartz: Amer. J. Sci. *272*, 234–272.

Perkins, D. and Newton, R. C., 1981, Charnockite geobarometers based on coexisting garnet-pyroxene-plagioclase-quartz: Nature *292*, 144–146.

Popp, R. K. and Gilbert, M. C., 1972, Stability of acmite-jadeite pyroxenes at low pressure: Amer. Mineral. *57*, 1210–1231.

Råheim, A., 1976, Petrology of eclogites and surrounding schists from the Lyell Highway-Collingwood River area: J. Geol. Soc. Australia *23*, pt. 3, 313–327.

Råheim, A. and Green, D. H., 1974, Experimental determination of the temperature and pressure dependence of the Fe-Mg partition coefficient for coexisting garnet and clinopyroxene: Contr. Mineral. Petrol. *48*, 179–203.

Råheim, A. and Green, D. H., 1975, P, T paths of natural eclogites during metamorphism: A record of subduction: Lithos *8*, 317–328.

Ryburn, R. J., Råheim, A. and Green, D. H., 1976, Determination of the P, T paths of natural eclogites during metamorphism, a record of subduction. A correction: Lithos *9*, 161–165.

Saxena, S. K., 1979, Garnet-clinopyroxene geothermometer: Contr. Mineral. Petrol. *70*, 229–235.

Saxena, S. K. and Ribbe, P. H., 1972, Activity-composition relations in feldspars: Contr. Mineral. Petrol. *37*, 131–138.

Schreyer, W., 1973, Whiteschist: A high-pressure rock and its geologic significance: J. Geol. *81*, 735–739.

Sharp, W. E., 1962, The thermodynamic functions for water in the range -10 to 1000°C and 1 to 250,000 bars: Univ. Calif. Lawrence Rad. Lab. Pap. #UCRL-7118, 1–51.

Smith, D. C., 1983, cited in "Late News": Terra Cognita *3*, 334.

Tarney, J. and Windley, B. F., 1977, Chemistry, thermal gradients and evolution of the lower continental crust: J. Geol. Soc. Lond. *134*, 153–172.

Vrana, S., Prasad, R. and Fediukova, E., 1975, Metamorphic kyanite eclogites in the Lufilian Arc of Zambia: Contr. Mineral. Petrol. *51*, 139–160.

Watson, K. D. and Morton, D. M., 1968, Eclogite inclusions in kimberlite pipes at Garnet Ridge, NE Arizona: Amer. Mineral. *54*, 267–285.

Wood, B. J., 1984, Mineralogic constitution of granulites under lower crustal conditions: EOS *65*, 288.

MANUSCRIPT ACCEPTED BY THE SOCIETY JULY 29, 1985

Printed in U.S.A.

Geological Society of America
Memoir 164
1986

Phase relationships of ellenbergerite, a new high-pressure Mg-Al-Ti-silicate in pyrope-coesite-quartzite from the Western Alps

Christian Chopin
ER 224, Laboratoire de Géologie
Ecole Normale Supérieure
46 rue d'Ulm
75005 Paris, France

ABSTRACT

The new Mg-Al-Ti-silicate ellenbergerite occurs among abundant inclusions within decimetre-size garnets in the pyrope-quartzite layer of the Dora Maira massif, Western Alps, from which coesite relics have been reported. It is associated with pyrope (92 to 98 mole percent end-member), kyanite, talc, chlorite, rutile, zircon, and minor sodic amphibole, which all formed an apparently stable assemblage now exclusively preserved within large garnets.

A petrologic analysis shows that the new mineral, which has the formula $(Mg_{1/3},$ $Ti_{1/3},$ $\square_{1/3})_2Mg_6Al_6Si_8O_{28}(OH)_{10}$ with extensive $Ti \rightleftharpoons Zr$ substitution, is a high-pressure phase with a lower pressure stability limit above 20 kbar and an upper temperature limit near 800°C or less. Its stability field extends from the Mg-carpholite field to the pyrope field and broadly overlaps that of Mg-chloritoid. The high-pressure phase relationships in the magnesian pelitic system are thus considerably modified, in particular through the reactions chlorite + talc + kyanite + rutile + H_2O = ellenbergerite and chlorite + kyanite + rutile + H_2O = ellenbergerite + Mg-chloritoid, which provide new upper pressure limits for two important chlorite-bearing assemblages.

The different assemblages preserved within the garnets and in the matrix record variations of pressure, temperature, and water activity along a prograde metamorphic path passing near 25 kbar, 700°C, and reaching the coesite field. This finding confirms the unusual depth reached by the enclosing continental unit along a low-temperature metamorphic gradient.

INTRODUCTION

A phengite quartzite layer reported by Vialon (1966) in the crystalline basement of the Dora Maira massif, Western Alps, is remarkable for the composition and size of the garnet crystals it contains. The garnet is indeed nearly pure pyrope (90 to 98 mole percent end member) and individual crystals may reach 25 cm in diameter. Recently, the mineral assemblage of the matrix was reinvestigated and the typically centimetre-size garnets of the quartzose matrix were found to preserve coesite relics as inclusions (Chopin, 1984). Considering the far-reaching consequences of this finding, especially in a blueschist and eclogite-bearing crustal terrane, the present study of the larger garnet porphyroblasts and their inclusions has been initiated in order to shed more light on the evolution of these fascinating rocks.

The large garnets are crowded with mineral inclusions. Among them are a conspicuous purple mineral, a sodic amphibole and chlorite that are characteristically absent from the matrix, and from the smaller matrix garnets. The purple pleochroic mineral proved to be a new hydrated, Ti-bearing Mg-Al-silicate with hexagonal symmetry; it has been described as ellenbergerite (Chopin et al., 1985). The present work is an outline of its phase

C. Chopin

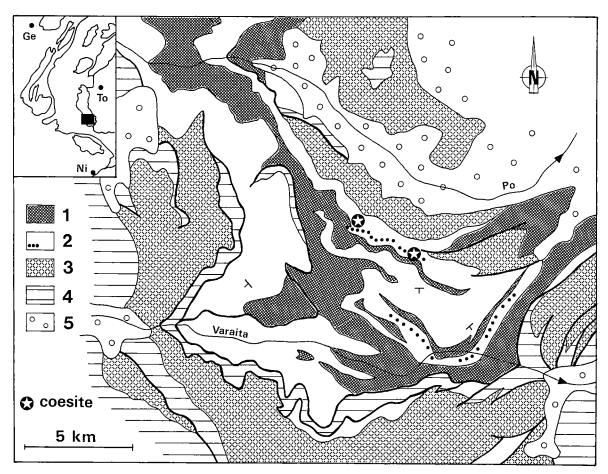

Figure 1. Geological map of the southern Dora Maira massif (after Vialon, 1966). (1) - Paleozoic augen gneisses and metagranites. (2) - Paleozoic fine-grained gneisses and schists with eclogite lenses and marble layers. The outcrops of the pyrope-quartzite are denoted by full circles. Coesite relics occur throughout the pyrope-quartzite layer; the two stars drawn limit the area in which, so far, coesite relics have been found in the country-rock. (3) - Dronero and Sampeyre formations: Upper Paleozoic (?) schists, metaarkoses and metarhyolites. (4) - Mesozoic: Carbonated cover series and Schistes lustrés undifferentiated. (5) - Alluvium. The inset locates the area in the Western Alps, showing the external crystalline massifs, the Penninic Front, the internal crystalline massifs, the Sesia zone and its Dent Blanche outlier. Ge: Geneva, Ni: Nice, To: Torino.

relationships and relevance to the high-pressure metamorphism of pelitic rocks.

GEOLOGICAL AND PETROLOGICAL SETTING

The reader is referred to Vialon (1966) for a description of the regional geology and to Chopin (1984, Fig. 1) for an outline of it. The quartzite layer is a few metres thick but has been mapped by Vialon (1966) over 15 km within a polymetamorphic, Early (?) Paleozoic series of fine-grained felsic gneisses and of schists with some marble and eclogite lenses (Fig. 1). They were intruded by granite bodies and sills, and later underwent Alpine high-pressure, low-temperature metamorphism. The samples to be discussed here have been collected near Parigi, Martiniana Po, Italy, in the same quartzite outcrop as the coesite-bearing samples described by Chopin (1984) and for which P-T estimates

of about 28 kbar and 700°C have been proposed. Recent field work by Chopin and Monié shows that the quartzite layer is closely associated with garnet-kyanite schists, thin marble layers, and amphibolite or eclogite lenses. The whole is a few hundred metres thick and represents, with the metagranites, the only lithologic variations in the monotonous gneiss series. Typical K_D values for Fe-Mg partitioning between garnet and omphacite in eclogites are between 8 and 10. Quartz - kyanite - phengite - eclogite is rather common and a typical pseudomorph of quartz after coesite has been observed in an omphacite grain. Furthermore, centimetre-size garnet porphyroblasts of the garnet-kyanite schists contain in their inner part numerous tiny inclusions of staurolite and magnesian chloritoid and, on their rim, silica inclusions with coesite relics which also occur in kyanite porphyroblasts (Chopin, unpub. data, see Fig. 1). This provides an important record of the prograde metamorphic evolution and leaves little doubt that the whole series enclosing the quartzite layer also experienced unusually high pressures.

PETROGRAPHY

The quartzite layer is rather homogeneous in appearance, usually with evenly distributed centimetre-size garnets. Locally, however, fist-size garnets or even larger ones, mostly euhedral, are clustered within a more micaceous matrix, these clusters defining a very rough layering concordant with the regional foliation. Their matrix is often devoid of quartz and consists of kyanite and centimetre-large phengite flakes. The kyanite grains and sometimes abundant rutile are often concentrated along the garnet rims, as if they had been pushed aside during garnet growth. Kyanite and rutile are nevertheless by far the most abundant mineral inclusions within the garnets. Among the included species, talc and zircon are also always present while chlorite, ellenbergerite and, to a lesser extent, a sodic amphibole are common (up to a few modal percent) but may be missing in some samples. It is noteworthy that the three latter minerals are absent in the immediately surrounding micaceous matrix, as well as in the quartzose matrix and its smaller garnets. Phengite, which is ubiquitous in the matrix, has not been observed as an inclusion, either in the large garnets or in the smaller ones. The only two other minerals observed in thin section are corroded grains of Cl-rich apatite in the cores of two ellenbergerite crystals, and a unique grain of arsenopyrite in the core of the large garnet specimen 4-18/19.

Large, isolated garnets also sporadically occur directly within the quartzose matrix. They often show an outer, 1 to 2-cm-thick shell nearly devoid of inclusions, thus sharply contrasting with the inner part of the porphyroblasts, which contains the same included minerals as the other large garnets. The inner part of such a euhedral 22-cm-diameter crystal (sample 4-18/19) shows a distinct orientation of inclusions. The lineation is defined by abundant elongated prisms of ellenbergerite which have a constant orientation throughout the inner part of the crystal. The outer shell contains only a few oriented kyanite inclusions which tend to parallel the garnet rim during the very last stages of garnet growth. Thus it seems, as will also be shown by the chemical study, that such garnet porphyroblasts have grown first very quickly, without rotation and including numerous minerals, then more slowly under somewhat changing physical conditions. However, it has proved impossible so far to establish a correlation between the nature of the matrix, the existence or lack of an outer shell or an oriented texture, and the presence or absence of ellenbergerite. For instance, a fist-size garnet found in a quartz-bearing matrix is devoid of outer shell and of ellenbergerite inclusions but contains centimetre-size kyanite blades without any preferred orientation. Conversely, large ellenbergerite-rich garnet crystals in a quartz-free matrix may also lack any oriented texture. A further confusing point presently being investigated is that, within a cluster of large garnets, some of the porphyroblasts are obviously zoned in their colour and inclusion distribution, while others are apparently not.

Apart from orientation, the textural relationships among the different minerals within the garnets show little variation from one sample to another. Monomineralic inclusions consist nearly exclusively of kyanite or rutile. Ellenbergerite, chlorite and amphibole occur consistently within polymineralic inclusions containing any combination of these minerals with talc, kyanite, rutile and zircon. The typical included assemblage is kyanite - talc - chlorite ± ellenbergerite ± rutile ± zircon (Figs. 2-5). With perhaps some restriction for zircon, the minerals mentioned may be found in contact with each other and the textural evidence suggests equilibrium among these phases. Ellenbergerite may show euhedral faces, particularly toward the phyllosilicates (Figs. 2, 3, 5) and might have some crystallographic relationship to garnet (Fig. 2). It may also be completely anhedral (Fig. 4). Thus, petrographic evidence suggests that most ellenbergerite is coeval with, or predates, garnet growth. However, in the smallest (2 cm) garnet found to contain it, ellenbergerite is interstitial to pyrope and seems to postdate garnet growth.

The main retrogressive textures observed within garnet are the replacement of ellenbergerite either by a sericitic aggregate or, along rims and cracks, by a nearly opaque aggregate (Fig. 2 and 3). In addition, of the often abundant cracks existing in the garnet, a few broader ones have acted as channels for the penetration of external fluids during retrogression and are now filled by secondary chlorite. Where such fractures reach a polymineralic inclusion containing talc and kyanite, the local breakdown of talc + kyanite to form secondary chlorite and minute (up to 20 μm) quartz grains was observed in a few instances.

The possible presence of silica inclusions and of coesite relics in the large garnets is of course of particular interest to this study, and attention was paid to every silica grain. In fact, individual silica inclusions such as those occurring in the smaller matrix garnets (see Chopin, 1984) are completely absent in the larger ones, even in those residing in a quartz-bearing matrix. The few tiny quartz grains mentioned above are obviously secondary and, in fact, were never seen in contact with pyrope or ellenbergerite. A unique quartz grain, 400-100 μm in size, was found in a talc kyanite - chlorite inclusion in the core of the large garnet 4-18/19. Although some secondary chlorite is also present, the amount of quartz compared to chlorite seems too large to be a breakdown product of talc + kyanite, so that this grain might be primary. If this holds true, it may have been trapped as quartz or as coesite. In the former case, it should have remained quartz throughout the whole metamorphic evolution; in the latter, coesite would have been completely converted to quartz. Both behaviors are easily explained by considering that the silica grain is isolated from the host garnet by much more compressible talc and chlorite (compare Gillet et al., 1984), but the textural evidence does not favour either possibility.

MINERAL CHEMISTRY

From 12 thin sections cut through large garnets, six, repre-

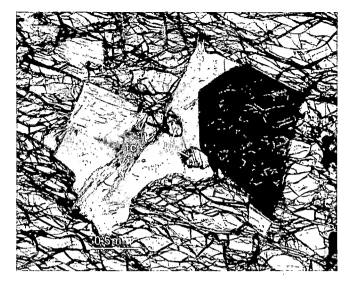

Figure 2. The assemblage ellenbergerite (dark)-chlorite(chl)-talc(tc)-kyanite (ky) in garnet porphyroblast 3-64. Note the absence of preferred orientation, the association of ellenbergerite and pyrope suggesting topotactic relationships, and the opaque rim of ellenbergerite which is absent on the garnet side. Plane polarised light.

Figure 3. Polymineralic inclusion in garnet porphyroblast 3-64. A zoned ellenbergerite crystal is associated with talc, kyanite and minor chlorite (left hand side of the photograph). Plane polarised light.

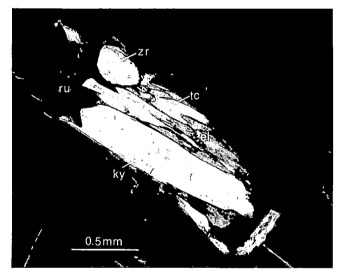

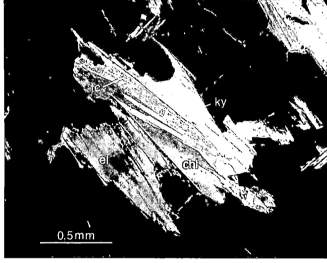

Figure 4. The assemblage ellenbergerite(el)-chlorite-talc-kyanite-rutile(ru)-zircon(zr) in pyrope megacryst 4-18/19. Note the anhedral shape of ellenbergerite and the mineral orientation parallel to the main fracture system of the garnet. Zircon is surrounded by a thin selvage of both fresh and sericitised ellenbergerite. Nearly crossed nicols.

Figure 5. Polymineralic inclusion in garnet megacryst 4-18/19, with the typical assemblage ellenbergerite-chlorite-talc-kyanite-pyrope. Crossed nicols.

senting four garnet crystals (3-64, 3-65, 4-18/19, 4-55), have been selected for microprobe work.

Analytical Technique. The analyses were obtained with a Cameca Microbeam electron probe (40° take-off angle) using wavelength dispersive techniques and ZAF correction with the X-ray absorption coefficients of Theisen and Vollath (1967) instead of Heinrich's (1966) coefficients used in the normal Cameca programme (see Chopin et al., 1985). Typical operating condi-

tions were 15 kV and 15 nA beam current. Standards used were forsterite (Mg, Si), anorthite (Ca, Al), K-feldspar and albite (K, Na), Fe_2O_3, $MnTiO_3$ (Mn, Ti), apatite (P) and zircon (Zr). With this combination of standards the Al/Si ratio tends to be overestimated in garnet, which is particularly obvious for one batch of analyses (Table 1, analyses 8 and 9 for garnet, 4 and 5 for ellenbergerite). Since no particular care was taken for the analysis of trace elements, the routinely obtained values of less than 0.04 wt. percent reported in Table 1 must be regarded with caution. In the absence of any mineralogic evidence for trivalent iron, total

TABLE 1. MINERAL ANALYSES

	Ellenbergerite					Garnet				Talc		Chlorite		Amphibole		
	3-64	4-18/19				rim	4-18/19		core	4-18	3-64	4-18	3-64	4-18	3-64	4-55
	1	2	3	4	5	6	7	8	9	10	11	12	13	14	15	16
SiO2	39.09	38.97	38.26	38.52	38.61	44.68	44.89	44.32	44.62	61.06	61.47	29.73	30.06	55.62	58.28	56.41
P2O5	0.44	0.15	0.19	0.05	0.00	0.10	0.09	0.03	0.05	0.06	--	0.02	--	0.03	--	0.01
TiO2	4.01	3.43	1.17	2.23	2.67	0.06	0.00	0.02	0.06	0.00	0.00	0.03	0.04	0.06	0.08	0.03
ZrO2	0.00	0.81	3.91	2.69	1.92	0.00	0.00	0.02	0.00	0.00	--	0.00	--	0.03	--	0.04
Al2O3	25.12	24.92	24.53	24.83	24.79	25.41	25.42	25.43	25.61	0.73	0.80	20.68	20.40	14.64	14.37	13.45
MgO	22.19	21.92	21.74	21.76	21.80	29.02	28.96	28.53	28.73	30.54	30.37	33.93	33.13	17.71	16.26	15.79
FeO	0.20	0.21	0.27	0.29	0.40	0.54	0.95	1.79	1.50	0.19	0.32	0.21	0.68	1.11	0.75	1.04
MnO	--	0.00	0.02	0.02	0.00	0.02	0.01	0.02	0.00	0.00	0.00	0.03	0.00	0.01	0.03	0.00
CaO	--	0.02	0.02	0.03	0.00	0.41	0.14	0.20	0.15	0.02	0.02	0.04	0.02	1.87	0.75	1.15
Na2O	--	0.00	0.00	0.00	0.00	0.03	0.00	0.07	0.03	0.02	0.03	0.01	0.04	4.31	5.60	6.32
K2O	--	0.00	0.00	0.02	0.00	0.00	0.00	0.00	0.00	0.00	0.02	0.00	0.06	0.21	0.12	---
Total	91.05	90.43	90.11	90.95	90.19	100.27	100.46	100.43	100.75	92.62	93.03	84.65	84.43	95.60	96.24	94.24
	33 oxygens					24 oxygens				11 oxygens		14 oxygens		23 oxygens		
Si	7.92	7.98	7.97	7.96	7.98	6.00	6.02	5.98	5.98	3.97	3.98	2.84	2.88	7.49	7.74	7.70
P	0.08	0.03	0.03	0.01	-	0.01	0.01	-	0.01	-	-	-	-	-	-	-
Ti	0.61	0.53	0.18	0.35	0.42	0.01	-	-	-	-	-	-	-	-	-	-
Zr	-	0.08	0.40	0.27	0.19	-	-	-	-	-	-	-	-	-	-	-
Al	6.00	6.02	6.02	6.05	6.03	4.02	4.01	4.04	4.05	0.05	0.06	2.32	2.30	2.32	2.25	2.16
Mg	6.71	6.69	6.74	6.70	6.71	5.81	5.78	5.73	5.74	2.96	2.93	4.82	4.73	3.56	3.22	3.21
Fe	0.03	0.04	0.05	0.05	0.07	0.06	0.11	0.20	0.17	0.01	0.02	0.02	0.05	0.12	0.08	0.12
Ca	-	-	-	-	-	0.06	0.02	0.03	0.02	-	-	-	-	0.27	0.11	0.17
Na + K	-	-	-	-	-	-	0.01	0.02	-	-	-	-	0.02	1.16	1.46	1.67
Σ cat	21.35	21.37	21.39	21.39	21.40	15.97	15.96	16.00	15.97	6.99	6.99	10.00	9.98	14.92	14.86	15.03
M	0.995	0.995	0.993	0.993	0.990	0.989	0.982	0.968	0.972	0.997	0.994	0.996	0.989	0.966	0.975	0.965

iron is reported as FeO. This view is supported by the near-absence of iron in kyanite and by the colour of ellenbergerite which is most probably due to Fe^{2+}-Ti^{4+} charge transfer (Chopin et al., 1985).

Ellenbergerite. The crystal chemistry of ellenbergerite has been discussed by Chopin et al. (1985) and only a few points essential to the petrologic discussion are addressed here. Ellenbergerite has the general structural formula

$$(Mg, Ti, Zr, \square)_2 Mg_6 (Al, Mg)_6 (Si, P)_2 Si_6 O_{28} (OH)_{10}$$

in which about one percent of the Mg atoms may be replaced by Fe. Accordingly, the formulae given in Table 1 have been calculated on a 33 oxygens anhydrous basis. Of importance is the high water content of about 7.5 wt. percent, whereas fluorine has not been detected. Ellenbergerite composition is entirely defined by three independent parameters, namely Ti/Zr ratio, P content and, accessorily, Mg/Fe ratio. The Ti = Zr substitution may be nearly complete and is responsible for the colour zoning. The most intensely coloured ellenbergerite contains about 4 wt. percent TiO_2, no zirconium and little or no phosphorus. It forms the outer part of zoned grains and also a few homogeneous, smaller grains (analysis 1). The less coloured, most often core zones are Ti-poor and correlatively Zr-rich. They are usually P-poor, rarely P-free, but may contain up to 1.4 P atoms p.f.u. as observed in a strongly zoned crystal containing an apatite inclusion.

Compositional variations, mainly of the Ti/Zr ratio, are more important within a single crystal than between crystals from different parts of a large garnet. This fact is documented in Table 1, in which analyses 2 and 3 are from a single grain in the outer part of the inclusion-rich zone of the garnet crystal 4-18/19, and analyses 4 and 5 from a single grain in the core of this zone. The same holds true for crystals from different garnets, although some garnets (3-64, 3-65) seem to have on the whole more Ti-rich ellenbergerite than others. Also, the P content of ellenbergerite, which is in general rather uniform within a single garnet, may show some variation from garnet to garnet (P_2O_5 between 0 and 0.6 wt. percent in 4-18/19, between 0.4 and 2.9 wt. percent in 4-55).

Garnet. The analyses of the large garnets confirm their extreme pyrope contents (Vialon, 1966) and show a very limited compositional range within and between crystals. The pyrope content varies only between 95 and 98 mole percent in 3-64, 3-65 and 4-18/19. These garnets are free of Cr and Zr; Mn, Ti and Na remain close to the detection limit and the P_2O_5 content never exceeds 0.16 wt. percent (Table 1). Their FeO and CaO contents range between 0.5 and 2.5, and 0 and 0.8 wt. percent, respectively. In a given garnet, the range of CaO variations is in fact much smaller, about 0.2 wt. percent. The same holds true for FeO which varies somewhat erratically throughout the whole garnet, local variations often exceeding those from core to rim, if any. This is in particular the case throughout the inclusion-rich part of garnet 4-18/19 (analyses 8 and 9) whereas FeO decreases and CaO increases towards the rim in the inclusion-poor and ellenbergerite-free outer zone (analyses 6 and 7). This pattern and this compositional range compare well with those of the smaller,

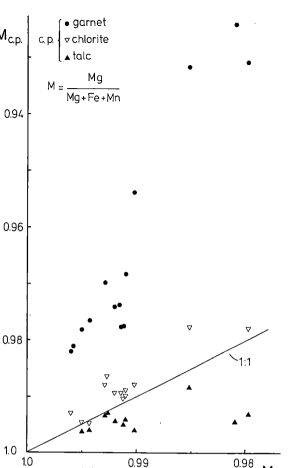

Figure 6. Fe-Mg partitioning between ellenbergerite and the coexisting phases talc, chlorite and garnet, in the four garnet porphyroblasts 3-64, 3-65, 4-18/19 and 4-55.

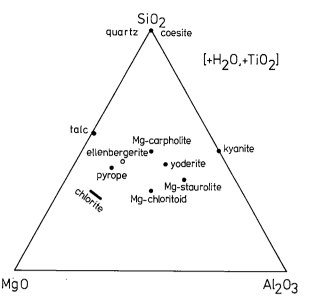

Figure 7. The system TiO_2-MgO-Al_2O_3-SiO_2-H_2O and phases of interest projected from the H_2O and TiO_2 components.

millimetre- to centimetre-size matrix garnets, which have pyrope contents from 91 to 97 mol. percent (Chopin, 1984).

Talc is close to the ideal end-member composition, with small amounts of Al, which seems to be partitioned between octahedral and tetrahedral sites (Table 1).

Chlorite shows little compositional variations and is close to ideal clinochlore, with typical Si contents of 2.85 to 2.90 p.f.u. (Table 1). No systematic variation was observed with the mineral in contact with chlorite nor with its location in the garnet.

Amphibole. Analyses 14 and 16 in Table 1 represent extreme compositions among the six inclusions analysed in three different garnets. These amphiboles referred to as glaucophane by Chopin (1984) on the basis of optics are definitely sodic, but actually deviate from ideal glaucophane by a Na-deficiency and the presence of tetrahedral Al, which are compensated by an excess of Mg over the ideal value of 3 and by the presence of some Ca. In addition, the number of cations may show a small deficiency (analysis 15) and indirect evidence suggests that this may be due to Li incorporation. Indeed, the rare jadeite-rich pyr-

oxene of the quartzose matrix and that of a jadeite-kyanite - almandine quartzite from the same locality both show a slight but consistent excess of Al over Na (Chopin, 1984) relative to a jadeite standard. One of the possible explanations for this is the presence of a few percent of spodumene component in solid solution. The definite identification of ephesite, $NaLiAl_4Si_2O_{10}$ $(OH)_2$, among the breakdown products of jadeite + kyanite in the almandine quartzite substantiates the latter hypothesis and, consequently, the presence of Li at least in the matrix of the pyrope quartzite.

Rutile. Small amounts of Nb have been detected in rutile grains which, in a few instances, show clearer, Nb-free rims.

Kyanite is pure $SiAl_2O_5$, with iron near the detection limit (0 to 0.05 wt. percent Fe_2O_3).

Fe-Mg Partitioning. In spite of the extreme Mg richness of the studied rocks and the very tight clustering of Fe/Mg ratios, a quite consistent picture arises from the iron partitioning among coexisting phases, with iron preference decreasing in the order garnet > amphibole > chlorite > ellenbergerite > talc (Fig. 6). The least Mg-rich and most Mg-rich mineral assemblages in garnet 4-18/19 (Fig. 6) were analyzed a few millimetres apart in the outermost part of the inclusion-rich zone. All the assemblages analysed in the core of this garnet fall between these two, showing that a systematic variation, if any, of Fe/Mg ratio throughout the garnet is of lesser extent than very local variations. This suggests that the growth of the inclusion-rich part of the garnet has proceeded under rather constant physical conditions, or rapidly, possibly after a large overstepping of the equilibrium reaction.

ANALYSIS OF THE PHASE RELATIONSHIPS:

The preceding section has shown that P-poor to P-free ellenbergite compositions are the most common. Thus the role of

TABLE 2. VOLUME CHANGE OF THE REACTIONS

Absent phases	Reaction	Label in text	$\Delta V/V$
(Q, Py)	6 Eb = 5 Tc + 5 Chl + 13 Ky + 4 Ru + 5 V	(1)	+ 6%
(Q, Chl)	18 Eb = 35 Py + 5 Tc + 19 Ky + 12 Ru + 85 V	(7)	+ 0
(Q, Ky)	24 Eb + 6 ·Tc + 19 Chl = 91 Py + 16 Ru + 202 V	(14)	− 4
(Q, Tc)	12 Eb + 5 Chl = 35 Py + 6 Ky + 8 Ru + 80 V	(15)	− 2
(Q, Eb, Ru)	3 Chl + 2 Tc + 4 Ky = 7 Py + 14 V	(3)	− 8
(Q, V)	80 Tc + 35 Chl + 202 Ky + 56 Ru = 84 Eb + 35 Py	(13)	− 6
(Chl, Py)	9 Eb + 35 Q = 20 Tc + 27 Ky + 6 Ru + 25 V	(5)	+ 3
(Chl, Ky)	9 Eb + 7 Tc = 27 Py + 18 Q + 6 Ru + 52 V		− 2
(Chl, Tc)	9 Eb = 20 Py + 7 Ky + 5 Q + 6 Ru + 45 V		− 0
(Chl, Eb, Ru)	Tc + Ky = Py + 2 Q + V	(4)	− 4
(Py, Ky)	27 Chl + 91 Q + 6 Ru = 3 Eb + 25 Tc + 38 V		− 6
(Py, Tc)	4 Chl + 5 Ky + 7 Q + 2 Ru = 3 Eb + V		− 7
(Py, Eb, Ru)	3 Chl + 14 Q = 5 Tc + 3 Ky + 7 V	(2)	− 4

Abbreviations, formulae, and molar volumes (in cm^3)

Chl	: chlorite	$Mg_5Al_2Si_3O_{10}(OH)_8$	211.5*
Ky	: kyanite	$SiAl_2O_5$	44.1[†]
Py	: pyrope	$Mg_3Al_2Si_3O_{12}$	113.3[†]
Ru	: rutile	TiO_2	18.8[†]
Tc	: talc	$Mg_3Si_4O_{10}(OH)_2$	136.2[†]
Q	: quartz	SiO_2	22.7[†]
V	: fluid	H_2O	14.8[§]
Eb	: ellenbergerite	$Mg_{6.67}Ti_{0.66}Al_6Si_8O_{28}(OH)_{10}$	386.0[§]

Data sources:
*Chernosky (1974)
[†]Robie et al. (1978)
[§]see text

phosphorus will not be further considered and the discussion will first focus on the P-, Fe- and Zr-free ellenbergerite, i.e. on the TiO_2-MgO-Al_2O_3-SiO_2-H_2O (Ti-MASH) system.

The TiO_2-MgO-Al_2O_3-SiO_2-H_2O system

The ellenbergerite formula may then be simplified to $(Mg_{2/3}, Ti_{2/3}, \square_{2/3})$ $Mg_6Al_6Si_8O_{28}(OH)_{10}$. This corresponds to the proportions 2:20:9:24:15 in the Ti-MASH system, which are used throughout the following discussion. A projection of the system through H_2O and TiO_2 (Fig. 7) reveals the most important phase relationships of ellenbergerite which, like pyrope and yoderite, projects within the talc-kyanite-chlorite triangle. The most simple ellenbergerite-forming reaction is thus

$$\text{chlorite + talc + kyanite + rutile + } H_2O \text{ = ellenbergerite.} \quad (1)$$

A Schreinemakers analysis of the system has been performed considering the solid phases ellenbergerite (Eb), pyrope (Py), chlorite (Chl), talc (Tc), kyanite (Ky), rutile (Ru) and either quartz or coesite (Q), and taking into account three experimentally determined curves, namely

$$\text{chlorite + quartz = talc + kyanite + } H_2O \quad (2)$$
$$\text{chlorite + talc + kyanite = pyrope + } H_2O \quad (3)$$
$$\text{talc + kyanite = pyrope + } SiO_2 + H_2O \quad (4)$$

(Massonne et al., 1981, for reaction 2; unpublished data of the

author for reactions 3 and 4). The analysis leads to two alternative chemographies which both satisfy the criterion that less hydrated products are on the high-temperature side. In one case, ellenbergerite displays an upper pressure stability limit, in the other a lower pressure stability limit. Additional consideration of the volume criterion (Table 2) unequivocally defines the latter case as the unique solution, which is represented in Fig. 8. The 386 cm^3 molar volume of ellenbergerite used in the calculations is the average of several cell parameter determinations. The volume changes calculated for the reactions (Table 2) must be regarded as only indicative, since a constant value is given to the molar volume of H_2O (corresponding to 25 kbar, 580°C or 30 kbar, 800°C), and the compressibility and thermal expansion of the solids is ignored. However the resulting picture remains unchanged for a larger molar volume of H_2O, or by considering coesite instead of quartz. Thus it may be confidently deduced from Fig. 8 that ellenbergerite is a high-pressure phase.

Interestingly, the ellenbergerite-forming reaction (1) closes the talc - chlorite - kyanite - rutile stability field on the high-pressure side; it proceeds with a significant volume change of about 6 percent (Table 2) and should thus be quite pressure-dependent. Likewise the curve for the reaction

$$\text{ellenbergerite + } SiO_2 \text{ = talc + kyanite + rutile + } H_2O \quad (5)$$

yields an upper pressure limit for the talc - kyanite - rutile stability field.

Among the stable invariant points of Fig. 8, the "quartz"-absent one is of particular relevance to this study since it involves all the phases present in the natural assemblage, and it should actually be stable. It lies on the curve of reaction (3) which has been approximately located by Schreyer (1968) and is presently being refined by the author. This curve extends stably from near 20 kbar, where Mg-staurolite or yoderite may become stable, to about 30 kbar, where it reaches the upper pressure stability limit of chlorite + kyanite, which react to talc + magnesiochloritoid. This creates an invariant point (Chopin, 1984, Fig. 5), which can now be more precisely located between about 30 and 31 kbar, 700 and 720°C, according to the experimental data given by Chopin and Schreyer (1983) and in Chopin and Monié (1984) for the reaction

$$\text{chlorite} + \text{kyanite} = \text{talc} + \text{magnesiochloritoid.} \qquad (6)$$

Since magnesiochloritoid is absent and chlorite + kyanite obviously stable in the natural assemblages studied here, the [Q] invariant point of Fig. 8 must lie at pressures lower than 30 kbar water pressure and be stable.

Moreover, magnesian, talc-kyanite-rutile-bearing rocks metamorphosed at pressures approaching 20 kbar are known from the Eastern Alps (Holland, 1979; cf. Miller, 1977) and, for somewhat lower pressures, from the Western Alps (Chopin and Monié, 1984). In neither case has ellenbergerite been observed, in spite of favourable bulk compositions. This suggests that the lower-pressure part of the ellenbergerite field as limited by the invariant points [Q] and [Py] lies at water pressures in excess of 20 kbar. Considering the positive slope of the curve of the quartz- and pyrope-absent reaction (1), these few constraints allow an approximate location of the invariant points in the P_{H_2O}-T plane. The [Py] invariant point would lie between 20 and 25 kbar at temperatures near 500°C, and the [Q] invariant point at about 25 kbar (the maximum possible value being 30 kbar) at temperatures between 700 and 750°C. The reaction connecting the [Q] and [Chl] invariant points, namely

$$\text{ellenbergerite} = \text{pyrope} + \text{talc} + \text{kyanite} + \text{rutile} + H_2O \qquad (7)$$

proceeds with a very small volume change (Table 2) and should be nearly pressure-independent. Its curve should therefore intersect the steep curve of reaction (4) at quite high-pressure, creating the [Chl] invariant point well within the coesite field, at temperatures near 800°C or less. Reaction (7) thus represents the upper temperature stability limit of ellenbergerite, making it unlikely that this mineral is involved in melting reactions.

This first attempt at deciphering the phase relations of ellenbergerite concerns the phases present in the natural assemblage. In fact, in the P-T range deduced, other high-pressure phases, such as Mg-carpholite, Mg-chloritoid, Mg-staurolite and yoderite, have also to be considered. It is assumed in the following that ellenbergerite displays no stable phase relations with yoderite because this phase is stable only at temperatures in excess of

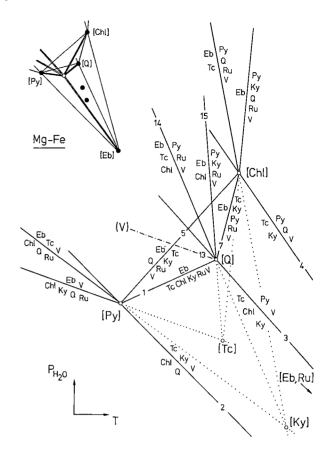

Figure 8. Chemographic analysis of the system TiO_2-MgO-Al_2O_3-SiO_2-H_2O involving the solid phases pyrope, talc, chlorite, kyanite, quartz (or coesite), ellenbergerite and rutile. The construction is based on the volume data of Table 2 and on the assumption that dehydration proceeds with temperature increase. Abbreviations: see Table 2. The inset shows the position of the relevant Fe-Mg-univariant curves (double lines) relative to the main univariant curves (simple lines) of the pure Mg system.

700°C and probably has an upper pressure limit near 25 kbar (Schreyer and Seifert, 1969, p. 423). The same assumption is made with the Al-rich phase Mg-staurolite, although only limited data are available on its lower stability limit (Schreyer and Seifert, 1969). Mg-carpholite and Mg-chloritoid deserve more attention, since the stability field of ellenbergerite as outlined above broadly overlaps that of Mg-chloritoid and, in its lower-temperature part, that of Mg-carpholite.

The curve of reaction (2) extends stably up to 18 kbar, 540°C, where it reaches the stability field of Mg-carpholite (Chopin and Schreyer, 1983). If we correctly assume that the [Py] invariant point, which lies on this curve, is in excess of 20 kbar, it must then be metastable. This implies a new set of low-temperature phase relations involving Mg-carpholite, which are shown in a tentative manner in Fig. 9. The [Q] invariant point of Fig. 8 remains unaffected and lies, as deduced in the preceding section, on the curve of reaction (3) at a lower pressure than the invariant point involving the solid phases pyrope, Mg-chloritoid,

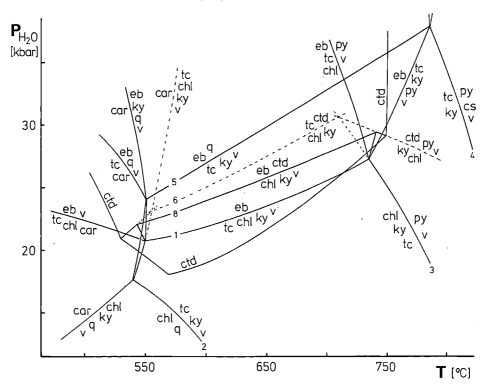

Figure 9. Phase relations of ellenbergerite in the system TiO_2-MgO-Al_2O_3-SiO_2-H_2O *with excess TiO_2*: a tentative outline. The grid used as basis for the MASH subsystem is drawn after Chopin and Schreyer (1983) for the relations of Mg-carpholite (Car) and Mg-chloritoid (Ctd), after Massonne et al. (1981) for reaction (2), and after unpublished experimental results of the author for reactions (3) and (4). The curve of the reaction $Eb+Q = Tc+Ky+V$ is unconstrained on the high-pressure side; it may lie at still higher, geologically unrealistic pressures. For clarity, the curves of reactions (1), (8) and (6) have been drawn to converge toward the lower temperatures, whereas they should converge on the high-temperature side. The stability field of Mg-chloritoid is merely outlined; the assemblages alternative to chloritoid are, with increasing temperature, carpholite+chlorite+diaspore, kyanite+chlorite+diaspore, kyanite+chlorite+corundum, and pyrope+corundum. Mg-staurolite and the quartz-coesite(cs) transition have been ignored.

chlorite, talc and kyanite (Chopin, 1984, Fig. 5). The latter invariant point thus becomes metastable in the presence of excess TiO_2. The resulting phase relations are also depicted in Fig. 9. An interesting feature is that the curve of reaction (6), which is so important for barometry in high-grade pelitic blueschists (Chopin and Schreyer, 1983; Chopin and Monié, 1984), becomes in fact entirely metastable in the presence of rutile. Indeed, the upper pressure limit of chlorite + kyanite + rutile is now given by the reaction

chlorite + kyanite + rutile + H_2O = ellenbergerite + Mg-chloritoid   (8)

which involves little H_2O but proceeds with a significant volume decrease of about 7 percent. Another important point is that the main ellenbergerite-forming reaction remains unchanged; it is still reaction (1), the curve of which extends stably from the Mg-carpholite field to the pyrope field, as that of reaction (8) does.

Thus, this chemographic analysis reveals that the high-pressure phase relations in the Ti-MASH system, hitherto essentially the same as in the MASH system, are considerably modified

by the existence of ellenbergerite. Reactions (1) and (8) represent new upper pressure stability limits for the assemblages chlorite - talc - kyanite - rutile and chlorite - kyanite - rutile, respectively.

Zr-ellenbergerite

The role of Zr-ellenbergerite is conveniently discussed in the Zr-MASH system. Projected through $ZrSiO_4$ and H_2O, ellenbergerite plots within the talc-kyanite-chlorite triangle so that a ZrO_2 phase does not need to be considered in the analysis, which is basically the same as in Ti-MASH. The most important ellenbergerite-forming reaction is

chlorite + talc + kyanite + zircon + H_2O = ellenbergerite   (9)

which is very similar to reaction (1) and proceeds with a volume decrease of about 6 percent. Again, the existence of ellenbergerite makes reaction (6) metastable in the Zr-bearing system and, for the same reasons as those developed for reaction (1), reaction (9) must take place at pressures higher than 20 kbar (on the low-temperature side) and lower than 30 kbar (on the high-temperature side).

One may speculate on the position of reaction (9) relative to its Ti-counterpart reaction (1), by considering the relative stabilities of end-members in isomorphous series, such as the Mg, Fe and Mn end-members of carpholite, chloritoid, cordierite and garnet, among others. The systematics arising is that the lower pressure stability limit is shifted toward higher pressure as the size of the substituting cation decreases. Therefore, one may tentatively conclude that Zr-ellenbergerite becomes stable at lower pressure than Ti-ellenbergerite, an important point for interpreting the zoning pattern.

A new phase of the MgO-Al₂O₃-SiO₂-H₂O system?

Considering the crystal structure of ellenbergerite (Chopin et al., 1985), it is quite tempting to speculate on a Mg Mg = (Ti, Zr) $\Box$ substitution within the Mg(2) site, which would lead to the hypothetical Mg-Al-silicate $Mg_8Al_6Si_8O_{28}(OH)_{10}$, compositionally close to pyrope + H_2O. A chemographic analysis essentially similar to that developed above shows that this end-member would be a high-pressure phase of the MASH system, in which reaction (6) and the invariant point involving the solid phases pyrope, chloritoid, talc, chlorite, and kyanite would become metastable. Again, on the basis of ionic radius alone, one would expect this "Mg-ellenbergerite" to become stable at higher pressure than Ti-ellenbergerite.

However, as long as ellenbergerite has not been found in a TiO_2- and ZrO_2-undersaturated rock, or as a specific experimental study has not been undertaken in the MASH system, this event remains purely conjectural and is thus not further discussed here. It may be merely noticed that the presence of rutile and zircon in the natural assemblages studied here sets the Ti + Zr contents of ellenbergerite at the maximum value.

THE METAMORPHIC EVOLUTION

In the natural assemblages, the phases relevant to the systems discussed in the preceding section are talc, kyanite, chlorite, pyrope, ellenbergerite, rutile and zircon. Obviously, it is hardly possible to find in thin section a small area in which seven minerals show contacts one to each other, so that actual equilibrium among the seven phases is difficult to demonstrate. Whereas garnet inclusions containing all the phases but zircon are rather common, less inclusions contain zircon in addition to the six phases. In such cases, zircon is usually included within ellenbergerite and probably does not participate any more in the equilibrium.

The role of iron must also be considered (see Fig. 6), this component providing one more degree of freedom. An assemblage represented by an invariant point in the pure magnesian system becomes univariant. The locus of its representative points in the P_{H_2O}-T plane is a curve emanating from the invariant point of the magnesian system, in every point of which the mineral

compositions are uniquely defined. The orientation of this curve relative to the univariant curves of the magnesian system (Fig. 8, inset) is deduced from the data on Fe-Mg partitioning (Fig. 6), considering the shift of the Mg-curves after iron incorporation. The univariant curves of the Fe-Mg-system converge into an invariant point of higher order which has been qualitatively located on the inset of Fig. 8. Nevertheless, even if the role of iron is important with regard to the phase rule, the actual mineral compositions are so Mg-rich that the iron contents do not materially affect P-T estimates that could be derived from the magnesian system. Indeed, calculations made to approximate the shift of the curve of reaction (3) due to Fe incorporation show it to be negligible as compared to the experimental uncertainty on the location of the curve.

The mineral assemblages preserved in the large garnet porphyroblasts, even in the core zones, do not record a history any earlier than that of the incoming of ellenbergerite. Only the *possible* presence of relic quartz can be an indication that the chlorite-quartz pair was formerly present in these rocks. Nevertheless the main stages of the mineral evolution which led to the actual state can be reconstructed. With increasing pressure (and temperature), the compositional field of chlorite coexisting with quartz narrows to form either talc + garnet on the Fe-rich side (cf. Chopin, 1981; Chopin and Schreyer, 1983), or talc + kyanite on the Mg-rich side, until the univariant reaction

$$\text{chlorite + quartz = talc + kyanite + garnet + } H_2O \qquad (10)$$

is reached. This reaction which is fundamental to the understanding of the few talc - kyanite - garnet assemblages reported so far (Råheim and Green, 1974; Udovkina et al., 1978) is actually by-passed here because of the highly magnesian bulk-rock composition. Indeed, talc and kyanite are formed and quartz is exhausted before reaction (10) is reached. The incoming of pyropic garnet occurs at a higher pressure from the talc - kyanite - chlorite assemblage through reaction (3'), the divariant equivalent of reaction (3). Upon further pressure increase, more pyrope is produced by the divariant breakdown of chlorite + talc + kyanite, all the mafic phases becoming increasingly Mg-rich until the conditions of ellenbergerite formation are reached. If the assumption that Zr-ellenbergerite has the lower stability limit holds true, the zoning pattern observed in ellenbergerite is prograde. The univariant ellenbergerite forming reaction is then

$$\text{talc + kyanite + chlorite + pyrope + zircon + } H_2O = \text{Zr-ellenbergerite.} \quad (11)$$

As soon as Zr-ellenbergerite is formed, rutile participates in the reaction instead of zircon and ellenbergerite composition continuously evolves toward Ti-ellenbergerite. These large variations of Ti/Zr ratio, as compared to those of Mg/Fe ratio in ellenbergerite and garnet, suggest that the stability limits of Ti- and Zr-ellenbergerite do not significantly differ.

The coexistence in garnet of the six solid phases talc, kyanite, chlorite, pyrope, rutile and, as discrete grains and rims, Ti-ellenbergerite, apparently implies that the conditions of ellenbergerite formation have not been overstepped, i.e. that this relic paragenesis records exactly the univariant formation of ellenbergerite, according to the reaction

$$\text{talc} + \text{kyanite} + \text{chlorite} + \text{pyrope} + \text{rutile} + H_2O = \text{Ti-ellenbergerite.} \quad (12)$$

This may not be as unlikely as it seems. First, the curve of reaction (12), which emanates from the [Q] invariant point of Fig. 8 (inset), has a gentle positive slope, forming a small angle with metamorphic gradients, thus making the preservation of nearly univariant conditions by a metamorphic rock more likely. Second, the development of the hydration reaction (12) may result in a lowering of the water activity, which would allow the coexistence of the six solid phases over a divariant field. The Mg end-member of the water-conserving reaction

$$\text{talc} + \text{kyanite} + \text{chlorite} + \text{rutile} = \text{ellenbergerite} + \text{pyrope} \quad (13)$$

is shown in Fig. 8 as reaction (V). Very importantly, this reaction is the only one which produces pyrope *together with* ellenbergerite, and, as suggested by textural evidence, the considerable growth of pyrope must be accounted for along with the appearance and growth of ellenbergerite. Thus, this may be evidence for a decrease of water activity after the incoming of ellenbergerite.

The absence of ellenbergerite in the outer zone of the large garnets, and of ellenbergerite and chlorite in the surrounding matrix shows that these more hydrated phases have disappeared during the later metamorphic evolution, while garnet has grown further. This should mainly reflect a temperature increase, since the divariant ellenbergerite- and chlorite-consuming reactions

$$\text{ellenbergerite} + \text{chlorite} + \text{talc} = \text{garnet} + H_2O \quad (14')$$
$$\text{ellenbergerite} + \text{chlorite} = \text{garnet} + \text{kyanite} + H_2O \quad (15')$$

and especially reaction (7'), which leads to the matrix assemblage talc - kyanite - pyrope, have a rather steep slope (Fig. 8). These reactions, too, conveniently account for the Mg enrichment in the outer zone of large garnets (Table 1, 4-18/19). As shown by the assemblage found in the quartzose matrix, the conditions of the equilibrium

$$\text{talc} + \text{kyanite} = \text{garnet} + SiO_2 + H_2O \quad (4')$$

are finally reached within the coesite field (Chopin, 1984). As compared to the conditions of equilibrium (3') preserved within the large garnets, equilibrium (4') implies higher temperatures and/or pressures, even for very different water activities. Calculations made using unpublished experimental data of the author on reactions (3) and (4) (see Fig. 9) and the water data of Halbach and Chatterjee (1982) show that, if a water activity of 1.0 prevails for reaction (3) and of 0.2 for reaction (4), the two curves still differ in temperature by about 50°C at pressures near the quartz-coesite transition. This result leads to a reevaluation of the temperature estimate made for the matrix assemblage, which, unless water activity was extremely low, is probably closer to 750°C than to 700°C as originally proposed in Chopin (1984), for the same minimum pressure of 28 kbar.

An estimate of the conditions recorded by the assemblage preserved in garnet can only be tentative, yet is constrained to be close to the conditions of equilibrium (3), for an initially high water activity. The latter point is suggested by the presence of the hydrated phases ellenbergerite, chlorite and sodic amphibole among the inclusions, whereas talc - kyanite - pyrope and sodic pyroxene occur instead in the matrix. In addition to the occurrence of ellenbergerite, the composition of chlorite and amphibole suggests unusually high pressures of formation. Inasmuch as the decrease of tetrahedral Al through the AlAl = SiMg substitution is likely to be favoured by a pressure increase, the unusual Si content of chlorite (2.9 p.f.u.) in this kyanite-bearing assemblage suggests higher pressures than the 2.7 to 2.8 values known in talc - chloritoid - quartz ± kyanite, high-pressure assemblages from the Gran Paradiso (Chopin, 1981) and Monte Rosa massif (Chopin and Monié, 1984). Likewise, the departure of amphibole composition from ideal glaucophane suggests that the upper stability limit of glaucophane was approached as the large garnets crystallised (compare Carman and Gilbert, 1983; Koons, 1982), bearing in mind that the highest-pressure talc - kyanite-bearing assemblages known until recently contain pure glaucophane (Holland, 1979). These are reasons why pressures near 25 kbar or more, and thus temperatures near 700°C, seem likely for the formation of the ellenbergerite-bearing assemblage.

These conditions are rather close to those of the supposed metamorphic maximum recorded in the matrix and, therefore, much of the prograde history of these rocks still remains conjectural. Nevertheless, this study strengthens the results obtained during the study of the matrix and further enhances the importance of these rocks, which indicate that continental material may be subducted to depths of about 100 km along very low temperature gradients.

CONCLUSION

Since titanium and zirconium occur in every pelitic rock, the present analysis applies to one of the most common rock types. It provides at least a general framework for interpreting the petrographic observations and for a forthcoming experimental study, even if a few puzzling points remain, such as the absence of phengite inclusions in garnet, or the juxtaposition of silica-saturated and undersaturated systems.

In more general terms, it is noteworthy that the pelitic system which has been so extensively studied in nature and in experiment may still yield new rock-forming silicates. This reflects of course very unusual metamorphic conditions but is in keeping with the trend of metamorphic petrology during the last decade(s). Indeed, the wealth of new minerals or mineral assemblages recently reported from common, pelite-like compositions—ferro- and magnesiocarpholite (De Roever, 1951, and

Goffé et al., 1973, respectively), yoderite and talc-kyanite (McKie, 1959), magnesiochloritoid and talc-chloritoid (Bearth, 1963; Miller, 1977), talc-phengite (Abraham and Schreyer, 1976; Chopin, 1981), sudoite (Fransolet and Bourguignon, 1978), pure pyrope and coesite (Chopin, 1984), very Mg-rich staurolite (Schreyer et al., 1984; Ward, 1984), and finally ellenbergerite— shows that progress is being made primarily in the area of high-pressure assemblages. Interestingly, several of these discoveries were anticipated by the experimental reconnaissance work of Schreyer (1968), covering P-T conditions which, at that time, were not clearly known to have existed in nature.

ACKNOWLEDGMENTS

Helpful reviews and comments by B. W. Evans, W. Maresch, R. C. Newton and W. Schreyer on an earlier version of the manuscript are gratefully acknowledged. O. Medenbach, Bochum, contributed the photomicrographs.

NOTE ADDED IN PROOF

Contrary to the expectations laid down in an abstract by Schreyer, Baller and Chopin (Terra Cognita, v. 5, p. 327, 1985), it proved impossible to ascertain the synthesis of Ti-free ellenbergerite in the MASH system, whereas ellenbergerite synthesis could be readily achieved in the Ti-MASH system at pressures in excess of 35 kbar and temperatures of 700 to 750°C (Schreyer, 1985, Fortschritte der Mineralogie, v. 63, Heft 2, in press).

REFERENCES CITED

Abraham, K., and Schreyer, W., 1976, A talc-phengite assemblage in piemontite schist from Brezovica, Serbia, Yugoslavia: Journal of Petrology, v. 17, p. 421–439.

Bearth, P., 1963, Chloritoid und Paragonit aus der Ophiolith-Zone von Zermatt-Saas Fee: Schweizerische mineralogische und petrographische Mitteilungen, v. 43, p. 269–286.

Carman, J. H., and Gilbert, M. C., 1983, Experimental studies on glaucophane stability: American Journal of Science, v. 283-A, p. 414–437.

Chernosky, J. V., 1974, The upper stability of clinochlore at low pressure and the free energy of formation of Mg-cordierite: American Mineralogist, v. 59, p. 496–507.

Chopin, C., 1981, Talc-phengite: a widespread assemblage in high-grade pelitic blueschists of the Western Alps: Journal of Petrology, v. 22, p. 628–650.

Chopin, C., 1984, Coesite and pure pyrope in high-grade blueschists of the Western Alps: a first record and some consequences: Contributions to Mineralogy and Petrology, v. 86, p. 107–118.

Chopin, C., and Monié, P., 1984, A unique magnesiochloritoid-bearing high-pressure assemblage from the Monte Rosa, Western Alps: petrologic and $^{40}Ar/^{39}Ar$ radiometric study: Contributions to Mineralogy and Petrology, v. 87, p. 388–398.

Chopin, C., and Schreyer, W., 1983, Magnesiocarpholite and magnesiochloritoid: two index minerals of pelitic blueschists and their preliminary phase relations in the model system $MgO-Al_2O_3-SiO_2-H_2O$: American Journal of Science, v. 283-A, p. 72–96.

Chopin, C., Klaska, R., Medenbach, O., and Dron, D., 1985, Ellenbergerite, a new high-pressure Mg-Al-(Ti, Zr)-silicate with a novel structure based on face-sharing octahedra. Contributions to Mineralogy and Petrology, in press.

De Roever, W. P., 1951, Ferrocarpholite, the hitherto unknown ferrous iron analogue of carpholite proper: American Mineralogist, v. 36, p. 736–745.

Fransolet, A. M., and Bourguignon, P., 1978, Di/trioctahedral chlorite in quartz veins from the Ardenne, Belgium: Canadian Mineralogist, v. 16, p 365–373.

Gillet, Ph., Ingrin, J., and Chopin, C., 1984, Coesite in subducted continental crust: P-T path deduced from an elastic model: Earth and Planetary Science Letters, v. 70, p. 426–436.

Goffé, B., Goffé-Urbano, G., and Saliot, P., 1973, Sur la présence d'une variété magnésienne de la ferrocarpholite en Vanoise (Alpes francaises)-Sa signification probable dans le métamorphisme alpin. Académie des Sciences Paris, Comptes rendus, v. 277, serie D, p. 1965–1968.

Halbach, H., and Chatterjee, N. D., 1982, An empirical Redlich-Kwong type equation of state for water to 1000°C and 200 kbar: Contributions to Mineralogy and Petrology, v. 79, p. 337–345.

Heinrich, K.F.J., 1966, X-ray absorption uncertainty. In The electron microprobe. New York, John Wiley and sons, p. 296–377.

Holland, T.J.B., 1979, High water activities in the generation of high-pressure kyanite eclogites of the Tauern Window, Austria: Journal of Geology, v. 87, p. 1–27.

Koons, P. O., 1982, An experimental investigation of the behavior of amphibole in the system $Na_2O-MgO-Al_2O_3-SiO_2-H_2O$ at high pressure: Contributions to Mineralogy and Petrology, v. 79, p. 258–267.

Massonne, H.-J., Mirwald, P. W., and Schreyer, W., 1981, Experimentelle Überprüfung der Reaktionskurve Chlorit+Quarz = Talk+Disthen im System $MgO-Al_2O_3-SiO_2-H_2O$: Fortschritte der Mineralogie, v. 59, p. 122–123.

McKie, D., 1959, A new hydrous magnesium iron aluminosilicate from Mautia Hill, Tanganyika: Mineralogical Magazine, v. 32, p. 282–307.

Miller, C., 1977, Chemismus und phasenpetrologische Untersuchungen der Gesteine aus der Eklogitzone des Tauernfensterns, Österreich: Tschermaks mineralogische und petrographische Mitteilungen, v. 24, p. 221–277.

Råheim, A., and Green, D. H., 1974, Talc-garnet-kyanite-quartz schist from an eclogite-bearing terrane, Western Tasmania. Contributions to Mineralogy and Petrology, v. 43, p. 223–231.

Robie, R. A., Hemingway, B. S., and Fisher, J. R., 1978, Thermodynamic properties of minerals and related substances at 298.15 K and 1 bar pressure and at higher temperature: Geological Survey Bulletin 1452.

Schreyer, W., 1968, A reconnaissance study of the system $MgO-Al_2O_3-SiO_2-H_2O$ at pressures between 10 and 25 kbar: Carnegie Institution of Washington, Yearbook, v. 66, p. 380–392.

Schreyer, W., and Seifert, F., 1969, High-pressure phases in the system $MgO-Al_2O_3-SiO_2-H_2O$: American Journal of Science, v. 267-A, p. 407–443.

Schreyer, W., Horrocks, P. C., and Abraham, K., 1984, High-magnesium staurolite in a sapphirine-garnet rock from the Limpopo Belt, Southern Africa: Contributions to Mineralogy and Petrology, v. 86, p. 200–207.

Theisen, R., and Vollath, D., 1967, Tables of X-ray mass attenuation coefficients. Düsseldorf, Verlag Stahleisen.

Udovkina, N. G., Muravitskaya, G. N., and Laputina, I. P., 1978, Phase equilibria in the talc-garnet-kyanite rocks of the Kokchetav Block, Northern Kazakhstan. Isvestiya Akademia Nauk SSSR, Geological Section, v. 7, p. 55–64 (in Russian).

Vialon, P., 1966, Etude géologique du massif cristallin Dora Maira, Alpes cottiennes internes, Italie [Thèse d'état] Université de Grenoble. 282 p.

Ward, C. M., 1984, Magnesium staurolite and green chromian staurolite from Fjordland, New Zealand. American Mineralogist, v. 69, p. 531–540.

MANUSCRIPT ACCEPTED BY THE SOCIETY JULY 29, 1985

Geological Society of America
Memoir 164
1986

Petrology and tectonic implications of the blueschist-bearing Puerto Nuevo melange complex, Vizcaino Peninsula, Baja California Sur, Mexico

Thomas E. Moore*
Geology Department
Stanford University
Stanford, California 94305

ABSTRACT

A serpentinite-matrix melange complex with blocks of greenschist, blueschist, metagabbro, orthogneiss, amphibolite, and eclogite is present within a small structural window beneath Upper Triassic ophiolite about 2 km north of Puerto Nuevo in the central Vizcaino Peninsula. The melange complex consists of three units: lower (more than 200 m thick) and upper (less than 100 m thick) serpentinite breccia units and the intervening exotic block unit (less than 100 m thick). The tectonic contact of the complex with the serpentinized (mainly chrysotile-bearing) basal harzburgite unit of the ophiolite is gradational and subparallel to the ophiolite stratigraphy; rocks below the melange complex are not exposed.

The serpentinite breccia units are composed of pebble- to boulder-sized fragments of serpentinized harzburgite, dunite, pyroxenite, and minor rodingitized diabase in a friable foliated matrix of sheared chrysotile. The clasts were probably all derived from disruption of the lower part of the overlying ophiolite. The exotic block unit contains abundant metamorphosed blocks, 50 cm to 0.5 km in diameter, in an indurated and scaly antigorite matrix. The exotic blocks include metabasalt, metagabbro, metachert, meta-tuff, metaperidotite, and chromitite of ophiolitic character, and metagraywacke, metato-nalite, and metarhyolite blocks of possible arc affinity. Metabasite blocks of the former group have major and trace element characteristics of ocean-floor basalts and exhibit pumpellyite-actinolite (ab+ep+pm+act±chl±wt mica), greenschist (act+chl+ep+ab+wt mica), blueschist (crossite+ab+ep+wm±cc±lw), and epidote-amphibolite (barroisitic amph+ep+ab+wt mica±grnt) and eclogite (grnt+cpx) facies assemblages. Blueschists contain abundant epidote and are interlayered with greenschists. This moderate temperature/high pressure series is analogous to that displayed by the Sanbagawa belt in Japan. Blocks of arc affinity exhibit greenschist assemblages. Some blocks experienced calcium metasomatism (rodingization) prior to incorporation into the melange, whereas others display tremolitic amphibole+chlorite+fuchsite-bearing rinds which resulted from magnesium metasomatism (blackwall) after incorporation into the melange.

The melange complex probably represents a thrust fault within the basal ultramafic part of the Sierra de San Andres ophiolite. The exotic metamorphic rocks are inferred to have been dragged up from one or more underlying metamorphosed ocean-floor and volcanic-arc terranes. Metamorphism, composition, lithology, tectonic position, and preliminary age data argue that the Puerto Nuevo melange should not be correlated with

*Present address: Branch of Alaskan Geology, U.S. Geological Survey, M.S. 904, 345 Middlefield Road, Menlo Park, California 94025.

the melange complexes on Cedros and the San Benitos Islands nor with the melanges of the Franciscan Complex of Alta California. The Puerto Nuevo melange is instead interpreted as an older, more inboard, blueschist-bearing unit analogous to the blueschist units of the Klamath and Sierra Nevada Mountains. The multiplicity of subduction-zone complexes, ophiolites, and volcanic-arc sequences suggests that the Vizcaino Peninsula region is composed of a collage of tectonostratigraphic terranes that may be allochthonous relative to each other and the North American craton.

INTRODUCTION

The structurally lowest unit exposed in the Vizcaino Peninsula is the blueschist-bearing Puerto Nuevo melange complex found in a 12 km^2 fenster below the Upper Triassic Sierra de San Andres ophiolite (Moore, 1976, 1979, 1983). Other blueschist localities on the Pacific margin of Mexico are on Cedros Island (Klienast and Rangin, 1982), the San Benitos Islands (Cohen and others, 1963), and Santa Margarita Island (Fig. 1). The blueschists on Cedros Island structurally underlie Upper Jurassic ophiolite (Kimbrough, 1982) and have been dated at 148 to 104 m.y. B.P. (Suppe and Armstrong, 1972). All of the Baja California blueschist localities have been widely considered to represent the southern extension of the Franciscan Complex into Mexico (Cohen and others, 1963; Suppe and Armstrong, 1972; Jones and others, 1976, 1978; Kilmer, 1977, 1979; Rangin, 1977, 1978; Gastil and others, 1978; Klienast and Rangin, 1982). The purpose of this paper is to describe the petrology and field relations of the Puerto Nuevo melange and to propose a model for its emplacement. Additionally, I suggest that the tectonic position and metamorphic character of the inclusions of the melange indicate that it represents a southern extension of the older, more inboard, Klamath and Sierra Nevada Mountains belt of blueschists rather than part of the Franciscan Complex.

GEOLOGIC SETTING

The Vizcaino Peninsula is a mountainous region located on the western side of the Baja California peninsula, about 600 km south of the U.S.-Mexico border (Fig. 1). The Vizcaino Peninsula is underlain by Triassic ophiolite, Jurassic island-arc rocks, Mesozoic blueschist, and Cretaceous submarine-fan deposits. These rocks are separated from the extensive Cretaceous volcanic and plutonic core of the Baja California peninsula (the Alisitos arc) by the Vizcaino basin (Fig. 1, inset). This basin is filled with Quaternary sediment which covers the probable fault-bounded eastern margin of the Vizcaino terrane (Coney and others, 1980).

The Puerto Nuevo melange complex, first reported by Jones and others (1976) and described in detail by Moore (1983), is exposed in the core of an antiform under the disrupted Sierra de San Andres ophiolite of the Vizcaino Norte terrane (Moore, 1983, 1985). This ophiolite has been dated at 220 m.y. B.P. (Late Triassic) by Barnes and Mattinson (1981) and Kimbrough (1982), and contains the oldest known rocks in the Vizcaino Peninsula. The ophiolite is depositionally overlain by Upper Triassic tuffaceous chert, tuff, and volcaniclastic sandstone (Fig. 2).

These rocks are structurally overlain in turn by the volcanogenic clastic rocks of the Upper Jurassic to Lower Cretaceous Eugenia Formation. Barnes (1982) interprets the Eugenia Formation as proximal deposits derived from the coeval Cedros-San Andres arc of Barnes and Berry (1979). The Eugenia Formation is overlain by the middle to Upper Cretaceous Valle Formation which comprises a fore-arc basin filled with plutonic and volcanic debris probably shed westward from the Alisitos volcanic-batholithic belt in the interior of the Baja California peninsula (Patterson, 1980).

STRATIGRAPHY OF THE MELANGE COMPLEX

The melange complex at Puerto Nuevo consists of three units, each with a serpentinite matrix (Fig. 2). These are, in ascending order, the lower serpentinite breccia unit, less than 200 m thick; the exotic block unit, also called the exotic block horizon by Moore (1979; 1980), less than 100 m thick; and the upper serpentinite breccia unit, less than 100 m thick. The units are sheet-like, but locally discontinuous, bodies which are delineated by the distinctive texture of the serpentinite matrix and composition of the enclosed blocks of the interior exotic block unit (Fig. 3). The upper and lower serpentinite breccia units consist of a nearly identical suite of endogenous ophiolitic clasts and serpentinite matrix and can only be distinguished by stratigraphic position.

The melange complex is structurally, but gradationally, overlain by serpentinized harzburgite of the Sierra de San Andres ophiolite. The harzburgite is overlain in turn, by fault blocks composed of the gabbro, sheeted dikes, and pillowed volcanic rocks of the upper part of the ophiolite along low-angle thrust faults (Fig. 4). The ophiolite has a thickness of less than 3000 m, but may have been tectonically thinned along numerous low-angle faults (Moore, 1976). The base of the melange complex is not exposed.

A northwest-trending high-angle fault divides the melange complex into eastern and western parts (Fig. 4). The stratigraphy of the larger eastern part defines a broadly crested, doubly-plunging antiform elongate in a northeasterly direction. The limbs of the antiform, estimated by mapping of the exotic block unit and the contact with the overlying ophiolite, dip variably up to about 20°. Foliations of the serpentinite matrix of all three units are random and do not reflect dip on the limbs of the antiform. However, structural measurements of the dominant foliation of

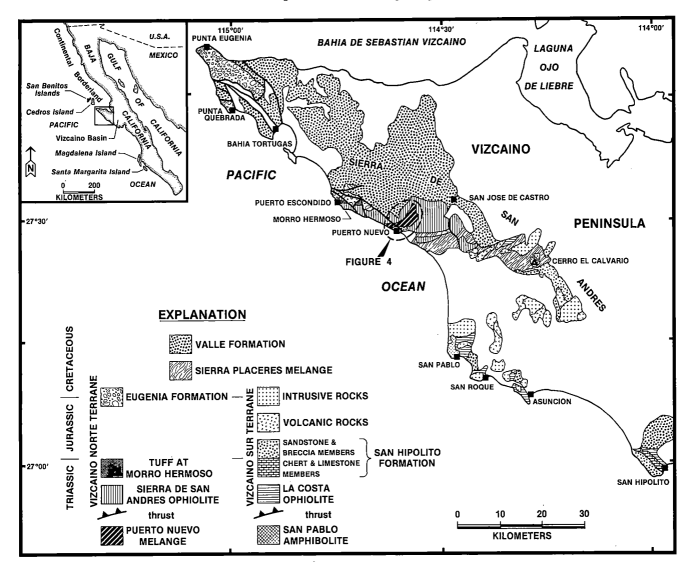

Figure 1. Index map, simplified geologic map and terrane map of the Vizcaino Peninsula.

blocks in the exotic block unit define a broad symmetrical fold about a steeply dipping, northeast-trending axial plane. This orientation of foliated blocks is interpreted to have resulted during emplacement of the melange.

West of Arroyo Casitas, the melange complex forms a northeast-trending belt with the exotic block unit defining a steeply dipping zone in the interior of the melange. This part of the melange appears to be tightly folded into a northeast-trending anticline with a near-vertical axial plane, but may also be a nearly vertically dipping homocline that faces toward the northwest. The unit that lies south of the melange is critical to an understanding of the structure of this area, but is covered by thrust sheets composed of gabbro of the Sierra de San Andres ophiolite and sedimentary rocks of the Valle Formation.

Serpentinite Breccia Units

The lower and upper serpentinite breccia units of the me-

lange complex consist of similar chaotic mixtures of pebble- to boulder-size fragments of massive serpentinite and diabase in a matrix of schistose, friable, fine-grained serpentinite (Fig. 5A). In most exposures, the included fragments are closely packed, but matrix-supported, with matrix/fragment ratio roughly equal to one. Locally, the block-in-matrix character of the unit is less well-defined, and the serpentinite breccia consists of shear polyhedra of fractured and disrupted serpentinite.

The matrix of the serpentinite breccia units is pervasively cut by thin, obliquely intersecting shear surfaces which wrap around the serpentinite and diabase fragments. The shear surfaces are undulating and anastomosing, but are generally subparallel and define an overall matrix foliation. Zones of shearing range from 1 cm to about 20 cm in thickness. They consist of poorly lithified, finely comminuted chrysotile, tremolitic amphibole, and magnesium carbonate.

Diabase, serpentinized dunite and pyroxenite, and massive

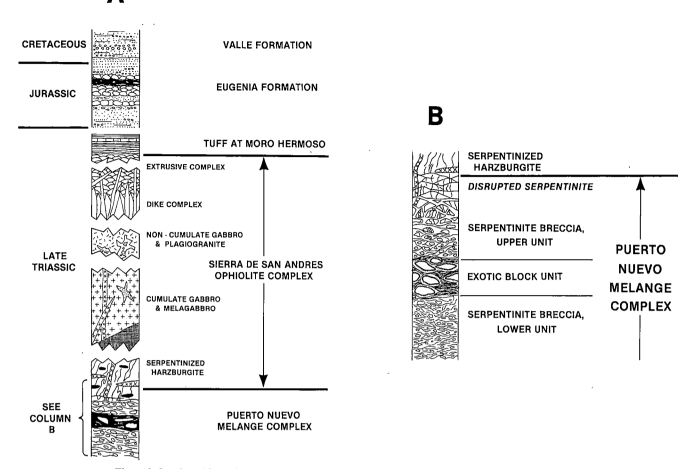

Figure 2. Stratigraphic sections: (A) generalized columnar section of the Vizcaino Norte terrane; (B) the Puerto Nuevo melange complex.

serpentinite clasts are common in the serpentinite breccia units, but the most abundant clast type is serpentinized harzburgite. The serpentinized harzburgite clasts have bastite pseudomorphs after orthopyroxene and consist mostly of chrysotile. The diabase clasts are persistent within the serpentinite breccia units and have equigranular ophitic textures that are replaced by rodingite assemblages consisting of hydrogrossular, diopside, prehnite, tremolite-actinolite, and chlorite. The massive serpentinite clasts, which consist of a fine-grained mosaic of antigorite crystals that locally display fine discontinuous banding, are uncommon and restricted to the lower serpentinite breccia unit.

The clasts contained within the serpentinite breccia units were probably derived mainly from the overlying ultramafic part of the Sierra de San Andres ophiolite. The antigorite clasts are mineralogically and texturally similar to the matrix of the exotic block unit and may represent fragments derived from it. For this reason, the clast assemblage in the serpentinite breccia units is viewed as locally derived.

Exotic Block Unit

The exotic block unit consists of indurated scaly serpentinite which envelops abundant blocks of diverse composition (Fig.

5B). The matrix is composed of massive leathery antigorite which exhibits a moderate to well-developed foliation defined by subparallel shear surfaces. Locally, the serpentinite matrix displays chaotic foliation and porphyroclastic texture.

Numerous blocks of various compositions and metamorphic grades are enclosed in the matrix of the exotic block unit. The blocks range from 1 m to over 500 m in size, and display a variety of compositions and mineral assemblages. All are metamorphosed and appear to be exotic to the Vizcaino Peninsula.

Many of the exotic blocks are surrounded by brightly colored metasomatic halos composed of fine- to coarse-grained tremolite-actinolite, chlorite, talc, and/or sparse chromium white mica. The metasomatic rinds vary in thickness from zero to several meters, commonly even on the same block. The nonsymmetric arrangement of the alteration zones on some blocks suggests that the rinds may have been partly removed by relatively late shearing between the blocks and matrix.

PETROLOGY OF EXOGENOUS TECTONIC INCLUSIONS

The tectonic blocks within the exotic block unit display a wide range of composition, texture, and grade of metamorphism.

Figure 3. Photographs showing stratigraphic relationships: (A) contact between the lower serpentinite breccia (below line) and exotic block (above line) units. Note large block in exotic block unit; (B) exotic block unit (foreground) overlain by upper serpentinite breccia unit (near road in background). Contact lies in arroyo and is hidden from view. Note the well-developed alteration rind on block located at right-center.

T. E. Moore

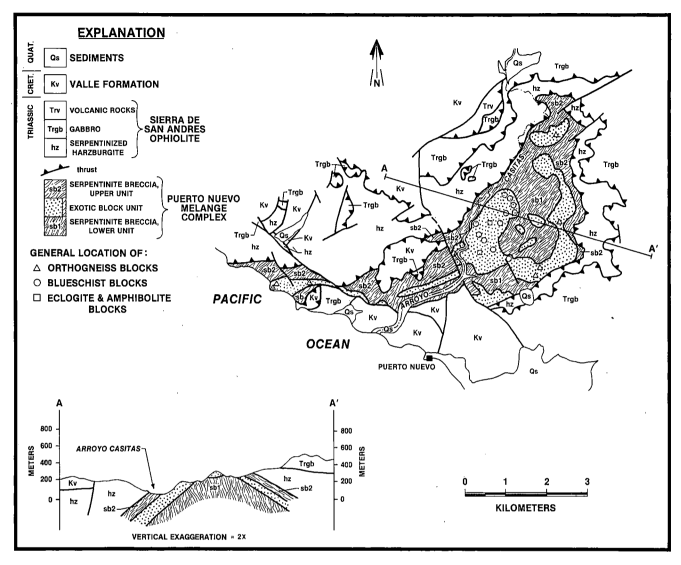

Figure 4. Geologic map and section of the Puerto Nuevo area.

Although the pre-melange relationships of the blocks are difficult to ascertain, thin-section study of more than 80 tectonic blocks allows classification of the blocks into several groups which have similar metamorphic grade and protolith composition. Mafic metavolcanic rocks are the most abundant types, but blocks of metaperidotite, metachert, metasandstone, and tonalitic orthogneiss are common. Other rare, but significant, varieties include metagabbro, metarhyolite, and chromitite. Both foliated and unfoliated blocks are present and some, especially metachert blocks, exhibit complex deformational histories.

Greenschist facies tectonic blocks are predominant within the melange; blue amphibole-bearing blocks are locally very abundant. Type III tectonic blocks (Coleman and Lee, 1963) are the most abundant blueschists, but Types II and IV are well represented. Blocks containing eclogite, epidote-amphibolite, or pumpellyite-actinolite facies mineral assemblages are rare and are spatially associated with the blueschists.

Pumpellyite-Actinolite Facies Tectonic Blocks

Metavolcanic Rocks. One greenstone block contains the assemblage albite + epidote + pumpellyite + actinolite + chlorite + white mica + quartz + sphene which is characteristic of the pumpellyite-actinolite facies (Coombs and others, 1976). The block is fine-grained, massive, unfoliated, and retains a volcanic-breccia texture. Relict lapilli-size volcanic clasts consist of sparsely porphyritic plagioclase metabasalt with abundant plagioclase microlites. These clasts are contained in a fine-grained metahyaloclastite matrix that contains sparse plagioclase crystals.

Greenschist Facies Tectonic Blocks

Mafic Metavolcanic Rocks. Massive, generally fine-grained mafic greenschists are the most abundant blocks in the exotic block unit. The common mineral assemblage of the blocks

Figure 5. Photographs showing matrix textures in the melange complex: (A) pebble- to boulder-sized clasts of serpentinized harzburgite (hz) and rodingitized diabase (db) (note circled hand sledge for scale) in upper serpentinite breccia unit; (B) a tectonic inclusion of metaperidotite enveloped by the massive, scaly, indurated antigorite matrix of the exotic block unit.

T. E. Moore

is albite + epidote + actinolite + chlorite + quartz + sphene ± calcite. Many massive blocks retain igneous features such as volcanic-breccia structures, phenocrysts, and amygdules. Microphenocrysts replaced by albite and epidote attest to the earlier presence of plagioclase, but no relict mafic minerals have been identified. Some blocks display strong foliations and polydeformational histories. These typically lack relict volcanic textures and contain sieved albite porphyroblasts.

Metagabbro. One massive medium-grained greenschist facies block exhibits relict cumulate-gabbro texture. This block contains approximately 40 percent albite, 50 percent green actinolitic amphibole, and a few percent epidote and sphene. The actinolitic amphibole displays colorless actinolite overgrowths and is partially replaced by chlorite and calcite.

Orthogneiss. Coarse-grained, leucocratic orthogneiss blocks, as large as 500 m, are locally present in the exotic block unit. The blocks contain about 65 percent feldspar, 25 percent quartz, and lesser amounts of mafic minerals and display moderate to well-developed stretching foliations. Their mineral assemblages consist of albite + quartz + chlorite + epidote + white mica + calcite ± magnetite + sphene. Relict textures indicate that albite has replaced euhedral plagioclase, quartz is interstitial, and mafic minerals (biotite?) are entirely replaced by chlorite. There is no evidence for the former presence of potassium feldspar and the protolith of the orthogneiss is therefore interpreted as coarse-grained tonalite.

Metarhyolite. One massive, fine-grained, leucocratic metavolcanic block contains the metamorphic mineral assemblage quartz + albite + white mica + epidote + calcite. The rock contains abundant epidote microporphyroblasts, but retains a relict sparsely porphyritic volcanic texture with an equal number of euhedral quartz and feldspar phenocrysts. The groundmass consists of a mosaic of quartz and feldspar with fine-grained white mica. This high-silica metavolcanic rock (PN-47, Table 1) probably had a rhyolite porphyry protolith.

Metasandstone. Metasandstone blocks are uncommon in the exotic block unit. Most are foliated, folded, and range from fine- to coarse-grained. The blocks contain the mineral assemblage quartz + albite + chlorite + stilpnomelane + white mica + sphene + magnetite ± epidote ± garnet ± actinolite. Relict grains of apatite and tourmaline are locally present. Stilpnomelane is post-deformational in at least one block.

The subequal abundance of quartz, albite, and micas in coarser grained blocks indicates that the protolith was a quartz-bearing lithic sandstone. The quartz-poor composition and abundance of sodic, calcic, and mafic minerals in some very fine-grained metasandstone blocks suggests that their protolith may have been compositionally immature volcanic sandstone.

Blueschist Facies Tectonic Blocks

Metavolcanic Rocks. Approximately 10 percent of the tectonic inclusions in the exotic block unit are metavolcanic rocks that contain blueschist facies assemblages. These blocks are commonly massive, but locally display faint relict igneous features such as volcaniclastic and pillow structures. Other relict volcanic textures include pseudomorphs of albite after plagioclase phenocrysts and amygdules. Strongly foliated blocks are coarser grained than massive varieties and commonly exhibit compositional layers and multistage deformational histories. Bulk compositions suggest that many of the foliated blocks may have had a tuffaceous protolith whereas others display apparent volcanic-breccia textures that have undergone a marked stretching or flattening deformation.

At least two types of blueschist facies mineral assemblages have been recognized in the metavolcanic rocks. The most common mineral assemblage is characterized by abundant epidote. This assemblage is commonly found in blocks that contain both blueschist and greenschist assemblages which are interlayered or interdigitated on a scale of 3 to 100 cm. Greenschist layers contain the assemblage albite + actinolite + epidote + white mica + chlorite + quartz + magnetite + sphene ± calcite. In blueschist layers, this assemblage is joined by crossite or glaucophane which locally completely replaces actinolite in the assemblage. The blueschist layers typically have very fine-grained or fibrous blue amphibole and somewhat coarser grained green amphibole and display zoning and overgrowths which indicate recrystallization under disequilibrium conditions.

Two blocks contain the mineral assemblage albite + acmitic clinopyroxene + epidote + quartz + sphene + magnetite ± crossite ± white mica. In one block, a fine-grained, massive, green acmite-rich lens, 60 cm in length, is enclosed within an albite-crossite blueschist. The lens is oriented parallel to compositional layering and retains relict volcanic textures. The second block is also unfoliated and retains a volcanic texture with abundant relict plagioclase microphenocrysts and microlites, 0.5 mm-size amygdules, and a megascopic pillow breccia-like structure. The plagioclase in this block is replaced by albite and lawsonite. The block has numerous veins that contain either acmite ± lawsonite ± albite ± crossite or white mica + chlorite + albite + epidote + oxychlorite ± calcite ± sphene. The former, apparently earlier, assemblage contains coarse clinopyroxene; the other assemblage displays abundant white mica and chlorite as well as coarse, unstrained albite and epidote crystals that project from the walls of the veins.

Metagabbro. One clast of coarse-grained, unfoliated, metagabbro that retains a subophitic gabbroic texture was observed in the exotic block unit. Albite and white mica comprise about 60 percent of the rock and pseudomorph plagioclase. Relict mafic minerals are replaced by an early generation of coarse-grained crossite and abundant sphene accompanied by euhedral epidote. A later assemblage is represented by fine-grained overgrowths and patchy replacement of early crossite by a paler colored crossite and by fine-grained granoblastic albite and white mica that replace coarse-grained feldspar.

Metachert. Siliceous metasedimentary blocks, although a minor fraction of the blocks of the exotic block unit, comprise a significant fraction of the blueschists in the melange. Most blocks are characterized by pronounced layering and strong foliation, but a few blocks display no preferred orientation of minerals. The metachert blocks are generally fine-grained and ferruginous, and contain the mineral assemblage quartz + crossite + white mica ± garnet ± stilpnomelane ± albite ± epidote ± sphene ± hematite. Although most of these blocks have a ferruginous chert protolith, several samples contain a significant amount of albite and epidote which may reflect a tuffaceous component.

Epidote-Amphibolite and Eclogite Facies Tectonic Blocks

Tectonic blocks of metabasite with epidote-amphibolite and eclogite assemblages are 2 m to 5 m in size, but several blocks are more than 50 m in length. The epidote-amphibolite blocks contain the assemblage blue-green (probably barroisitic) hornblende + epidote + white mica + albite + ilmenite + sphene ± quartz. In a few rocks, albite is a minor phase, present interstitially or as poikiloblastic grains. Sphene is typically present as coarse- to fine-grained replacement of rutile and locally as exsolved blebs from the zoned barroisitic amphibole.

Some of the amphibolite blocks contain an eclogitic assemblage consisting of Na-pyroxene + garnet + white mica + epidote. The Na-pyroxene may be partly replaced by barroisitic amphibole in some samples. The garnet is locally abundant and commonly is present as strongly zoned idioblastic porphyroblasts that contain abundant inclusions of epidote. A few amphibolites, however, contain Na-pyroxene, but lack garnet. In these rocks, the clinopyroxene is apparently stable with barroisitic amphibole, epidote, white mica, and albite.

Most of the samples contain retrograde assemblages consisting of chlorite and actinolite. One 500-m-size tectonic block contains the fine-grained nematoblastic greenschist assemblage albite + actinolite + chlorite + epidote + white mica ± quartz ± sphene and abundant coarse rutile. The presence of rutile suggests that the block may have had an epidote-amphibolite or eclogite facies mineral assemblage which was recrystallized to the greenschist assemblage.

Metamorphosed Ultramafic Tectonic Blocks

Tectonic inclusions of metamorphosed ultramafic rock are a persistent rock type in the exotic block unit. Protolith lithologies include peridotite, which is by far the most abundant, clinopyroxenite, and chromitite. The blocks are chiefly 1 m to 5 m in size, massive, and indurated. A few display a foliation, cataclasis, or layering composed of magnetite-rich layers.

The metaperidotite consists of a random interlocking network of radiating laths of antigorite ± chlorite. Locally, aligned antigorite laths comprise blocky mats of serpentine which replace pyroxene and contain curved parallel structures suggestive of the deformed cleavage surfaces of orthopyroxene in harzburgite tectonite. One block retains cumulate textures. In some blocks, carbonate porphyroblasts have partially or completely replaced

pyroxene and one sample exhibits a second generation of carbonate which has overgrown the earlier, blocky grains.

The clinopyroxenite is generally very dense, unfoliated, fine-to medium-grained, and consists mostly of igneous clinopyroxene which displays probable adcumulate textures. Clinopyroxene grains are partially replaced by actinolite, chlorite, and crosscutting veins of antigorite.

The prograde assemblage of the metamorphosed tectonic blocks of ultramafic affinity of the exotic block horizon is antigorite + magnetite + calcite ± chlorite ± actinolite. This assemblage is stable throughout pumpellyite-actinolite, greenschist, blueschist, and epidote-amphibolite facies metamorphic conditions (Evans, 1977).

Metasomatized Tectonic Blocks

About ten percent of the tectonic blocks consist of mineral assemblages which probably developed in response to metasomatic exchange between the tectonic blocks and their surrounding serpentinite matrix. Three varieties of metasomatites have been recognized: 1) epidote-clinozoisite ± sphene ± chlorite ± garnet, 2) tremolite-actinolite ± talc ± sphene ± chlorite ± chromian mica, and 3) chlorite ± magnetite ± sphene. The first assemblage is present in blocks that have sharp contacts with the surrounding serpentinite matrix of the melange and probably reflects rodingitization resulting from calcium metasomatism during serpentinization of peridotite (Coleman, 1966) prior to incorporation into the melange.

The second and third assemblages are 'blackwall' alteration assemblages that are attributed by Lan and Liou (1981) to magnesium metasomatism during alteration of a calcium-depleted serpentinite. Both assemblages are present in zones around the same tectonic blocks. In several places, the alteration selvages display structural features such as foliation and folds along their inner margins. These structures mimic the fabric contained in the enclosed tectonic inclusions and indicate that parts of the tectonic inclusions themselves have been replaced by the alteration minerals. These features indicate that the "blackwall" metasomatic reactions have occurred essentially *in situ.*

Whole-Rock Chemistry

Nine samples of metabasite and one of metarhyolite were selected for major element analysis by wave-length-dispersive X-ray fluorescence techniques at the U.S. Geological Survey analytical laboratory in Denver and trace element analysis by energy-dispersive X-ray fluorescence techniques at the U.S. Geological Survey analytical laboratory in Menlo Park. The results, together with modal compositions, are presented in Table 1.

All analyzed eclogite, epidote-amphibolite, blueschist, and greenschist samples apparently have tholeiitic bulk compositions. The SiO_2 content, calculated on an anhydrous basis, averages about 49.6 weight percent (44.9 to 53.4 wt %) and TiO_2 is moderately low, averaging about 1.5 weight percent. The K_2O

TABLE 1. MAJOR ELEMENT (wt %) AND TRACE ELEMENT (ppm) COMPOSITIONS AND MINERAL ASSEMBLAGES (%)
OF SELECTED TECTONIC INCLUSIONS FROM THE EXOTIC BLOCK UNIT

	Metarhyolite		Greenschist				Blueschist			Eclogitic-amphibolite
	PN-47	PN-70	PN-101E	PN-105	PN-111	PN-99	PN-127A	PN-79	PN-78	PN-94B
SiO_2	72.3	49.2	48.3	51.1	49.0	48.1	41.0	48.5	44.3	46.3
Al_2O_3	15.3	17.0	16.3	13.8	15.5	16.4	16.3	14.2	20.0	13.0
Fe_2O_3*	0.97	10.4	10.7	11.3	11.4	9.01	13.2	13.7	9.74	16.7
MgO	0.41	6.56	3.21	8.01	6.43	7.66	10.6	6.37	3.60	6.59
CaO	1.37	8.38	13.1	7.14	8.91	9.74	3.98	4.49	9.97	11.5
Na_2O	5.88	4.03	4.08	4.15	4.03	3.38	2.31	4.96	3.45	3.21
K_2O	1.52	0.21	0.90	0.20	0.05	0.28	2.36	1.17	1.78	0.13
TiO_2	0.02	1.16	1.44	0.89	1.27	1.67	2.49	0.99	0.70	2.11
P_2O_5	0.07	0.12	0.23	0.08	0.15	0.27	0.20	0.15	0.06	0.15
MnO	0.07	0.19	0.21	0.16	0.18	0.15	0.25	0.25	0.1	0.47
LOI	1.44	2.92	1.11	2.82	2.73	2.81	6.68	5.12	6.66	0.56
Total	99.55	100.17	99.58	99.65	99.65	99.47	99.37	99.9	100.36	100.72
Cr	9	249	217	304	289	208	321	42	348	107
Ni	6	88	74	147	102	114	128	37	102	102
Zn	52	80	93	77	96	75	136	150	78	88
Zr	59	66	88	50	74	152	152	54	44	111
Y	9	22	25	19	21	26	41	25	25	36
V	3	295	358	329	334	261	419	321	227	509
Quartz	40	tr	tr	tr	tr	tr	--	tr	--	--
Albite	45	50	30	30	25	40	20	25	20	2
Actinolite	--	20	30	40	50	15	10	10	--	--
Crossite	--	--	--	--	--	--	15	35	1	--
Blue-Green Horneblende	--	--	--	--	--	--	--	--	--	25
Na Pyroxene	--	--	--	--	--	--	--	--	--	30
Acmite	--	--	--	--	--	--	--	--	40	--
Epidote-Clinozoisite	5	10	30	15	20	20	3	3	1	10
Lawsonite	--	--	--	--	--	--	--	--	10	--
Calcite	2	tr	--	2	tr	--	10	10	10	--
Garnet	--	--	--	--	--	--	--	--	--	25
White Mica	8	--	--	2	--	6	20	5	15	5
Chlorite	--	15	1	10	5	15	20	10	2	--
Oxychlorite	--	--	8	--	--	--	--	--	1	--
Magnetite	--	--	--	1	--	--	--	--	--	--
Sphene	--	5	2	2	2	1	5	3	1	3
Rutile	--	--	--	tr	--	3	--	--	--	1
Hematite	--	--	--	--	tr	--	--	--	--	--

*Total Fe as Fe_2O_3

content in most of the metabasites is low except in rocks which contain abundant white mica. The rocks also contain moderately low P_2O_5 (average 0.16 wt %) and have low Al_2O_3/TiO_2 (average 13.2) and CaO/TiO_2 (average 7.5) ratios.

All major element abundances, except titanium, may have been considered modified by low-grade metamorphism (Cann, 1970; Hart and others, 1974). The tectonic environment of eruption of basaltic rocks is probably more accurately reflected by the relative abundances of trace elements that are immobile or only slightly mobile under low-grade metamorphic conditions (Pearce and Cann, 1973; Pearce and Norry, 1979). Plots using Ti, Cr, Ni, Y, and V have proved to be the most diagnostic of magmatic affinity and are utilized below to determine the tectonic affinity of metabasite samples from the Puerto Nuevo melange. All of the samples contain between 11.5 and 20 weight percent (anhydrous) CaO plus MgO and contain no relict features indicative of a cumulate igneous history.

The metabasite samples contain moderate to high concentrations of Cr (100-400 ppm) and Ni (75-150 ppm), and have

Ti/V ratios of about 30, similar to those of abyssal tholeiites (Shervais, 1982). The samples plot in or near to the field of mid-ocean-ridge basalts (field B) on the Ti-Zr-Y diagram of Pearce and Cann (1973) (Fig. 6A). Diagrams plotting Ti vs Zr and Zr/Y vs Zr (Fig. 6B, C) also show that all but two samples plot within the field of mid-ocean-ridge basalts. Garcia (1979) has shown that relative abundances of Ti and Zr are nearly identical for mid-ocean-ridge and island-arc tholeiites. A plot of Ti vs Cr (Fig. 6D), designed to distinguish between these magma types (Pearce, 1975), shows that all but one sample plot within the field of ocean-floor basalts. Although the protoliths of the metabasite blocks cannot be proved to be comagmatic, the minor- and trace-element geochemistry of the metabasites are similar and suggest that they may have originated at a mid-ocean ridge.

The single analysis of metarhyolite is characterized by high silica, calcium, and alkali and low iron, magnesium, and titanium contents. Because the analysis plots near the junction of the fractionated parts of the calc-alkalic and tholeiitic trends on an AFM

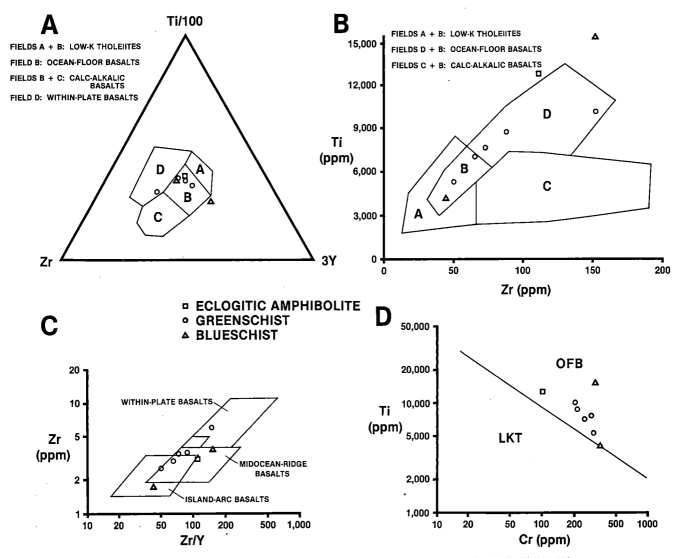

Figure 6. Compositions of selected tectonic inclusions from the exotic block unit: (A) Zr-Ti/100-3Y plot; (B) Ti-Zr plot; (C) Zr/Y-Zr plot; (D) Ti-Cr plot. Compositional fields are from Pearce and Cann (1973), Pearce (1975), and Pearce and Norry (1979). Abbreviations: OFB - ocean-floor basalt; LKT - low-potassium tholeiite.

diagram, its petrologic affinity is unclear. However, its large relict plagioclase and quartz phenocrysts make it texturally more similar to rocks of its composition in the calc-alkalic rather than the tholeiitic series. For these reasons, the metarhyolite is provisionally considered to have a protolith of calc-alkalic affinity.

Metamorphic Conditions

The data presented in the previous sections and summarized in Figure 7 indicate that the tectonic inclusions in the exotic block unit contain blueschist, pumpellyite-actinolite, greenschist, epidote-amphibolite, and eclogite facies mineral assemblages. Metabasites of the exotic block unit are characterized by interlayered blueschists and greenschists, by the ubiquitous presence of abundant epidote and actinolite, and by the local occurrence of

acmitic and omphacitic clinopyroxene. Lawsonite is rare and the $CaCO_3$ phase is calcite. In the blueschist metabasites, blue amphibole is commonly crossite. Jadeitic pyroxene has not been observed in either the metabasites or metagraywackes and the epidote-amphibolite blocks contain barroisitic amphibole.

The exotic blocks may represent a facies series similar to that of the Sanbagawa metamorphic rocks in southeastern Japan. The Sanbagawa belt (Ernst and Seki, 1967; Ernst and others, 1970; Banno and others, 1978) comprises a coherent sequence composed in large part of mafic volcanic rock metamorphosed under conditions transitional between blueschist and greenschist, and epidote amphibolite facies. Mineral assemblages of Sanbagawa metabasites are characterized by ubiquitous actinolite and epidote, crossite rather than glaucophane, and calcite rather than aragonite. Lawsonite and jadeitic pyroxene are rare, but acmitic

Mineral Phase	Grade of Metamorphism	Pumpellyite-Actinolite Facies	Greenschist Facies					Blueschist Facies			Epidote-Amphibolite Facies	Eclogite Facies		Metasomatites	
	Lithology	Metabasite	Metavolcanic Rocks	Metagabbro	Orthogneiss	Metadacite	Meta-sandstone	Metavolcanic Rocks	Metagabbro	Metachert	Metabasite	Metabasite	Ultramafic Rocks	Rodingite	Blackwall
QUARTZ															
ALBITE															
ACTINOLITE															
CROSSITE															
BARROISITIC AMPHIBOLE															
ACMITIC CLINOPYROXENE															
OMPHACITE															
PUMPELLYITE															
EPIDOTE-CLINOZOISITE															
LAWSONITE															
CALCITE															
GARNET															
WHITE MICA															
CHLORITE															
STILPNOMELANE															
OXYCHORITE															
FUCHSITE															
ANTIGORITE															
RUTILE															
SPHENE															
MAGNETITE															
PYRITE															

Figure 7. Summary diagram showing mineral assemblages for tectonic inclusions from the exotic block unit. Dashed line indicates presence of only minor amounts or sporadic occurrence of mineral; dotted line indicates secondary mineral assemblage.

clinopyroxene is common in lower grade rocks and omphacite is locally present in the highest grade rocks.

In contrast to the Sanbagawa belt, the metamorphic rocks of the Franciscan Complex are divided into two groups: those that were metamorphosed after deposition of the oldest (Late Jurassic) Franciscan sedimentary rocks (the so-called in situ metamorphic rocks) and the coarser grained tectonic blocks of high-grade blueschist, amphibolite, and eclogite which generally yield older isotopic metamorphic ages (Coleman and Lanphere, 1971). Glaucophane, lawsonite, aragonite and jadeite pyroxene are relatively common in the Franciscan rocks and albite, epidote, actinolite, and acmitic clinopyroxene are not as common as in the Sanbagawa belt. Both the in situ and high-grade blueschists of the Franciscan Complex probably were metamorphosed under similar geothermal gradients (J. G. Liou, personal communication, 1982).

Studies of progressive metamorphism in the Sanbagawa belt show that the lowest grade rocks were metamorphosed at about 200°C and 3 Kb; higher grade rocks are estimated to have been metamorphosed at about 500°C and 7 Kb (Ernst and Seki, 1967). The in situ metamorphic rocks of the Franciscan Complex, in contrast, are generally estimated to represent considerably higher pressures: lower grade rocks at about 200°C and 6 Kb and higher grade rocks at about 400-500°C and at more than 7 Kb (Ernst and Seki, 1967; Cloos, 1982).

Although accurate estimates of the physical conditions of metamorphism of the tectonic blocks of the exotic block unit in the Puerto Nuevo melange complex must await more detailed petrologic study and determination of mineral chemistries, the metamorphic inclusions of the Puerto Nuevo melange have mineral assemblages more similar to those of the Sanbagawa belt

than the Franciscan Complex. These similarities suggest that prograde metamorphism of the Puerto Nuevo melange blocks occurred at between 200°C and 3 Kb and 500°C and 7 Kb and at higher temperatures and lower pressures than Franciscan metamorphic rocks. Retrograde metamorphism of the blocks is largely greenschist facies and probably represents later metamorphism at somewhat lower pressures or higher temperatures.

Distribution of Exotic Blocks

The generalized distribution of blocks composed of orthogneiss, blueschist-greenschist, and epidote-amphibolite within the exotic block unit is shown on Figure 4. Blue amphibole-bearing tectonic blocks are restricted to a narrow north-trending zone in the central area of the melange along the east side of Arroyo Casitas. The rarer eclogite and epidote-amphibolite blocks are located in a restricted area near the center of the blueschist-bearing zone, where they constitute a significant fraction of the blocks. Orthogneiss blocks are present along the eastern flank of the melange antiform, where several small, widely scattered blocks define a north-trending zone. Near the middle of this zone is the single known metarhyolite block. Several orthogneiss blocks are also present in the westernmost exposure of the exotic block unit. In that area, a single kilometer-sized block is present and probably occupies more volume than the total of all other orthogneiss blocks in the melange.

The blueschist and epidote-amphibolite blocks are spatially associated within the exposed part of the melange complex. This association suggests that they may be genetically related. Likewise, the metarhyolite block is associated with the orthogneiss blocks, suggesting that there is a genetic relationship between the two rock types. The areas occupied by orthogneiss blocks do not overlap with the area that contains the blueschist and epidote-amphibolite blocks. The block lithology of the exotic block unit, therefore, is spatially zoned, an indication that mixing during melange formation was incomplete. The spatial separation of the two groups of blocks may reflect the structural configuration of the source regions that contributed debris to the melange complex.

CRYPTIC TERRANES IN THE EXOTIC BLOCK UNIT

The exotic blocks in the Puerto Nuevo melange may be fragments of two distinct geologic elements: one of ocean-floor and the other of volcanic-arc affinity. Blocks of ocean-floor affinity are far more numerous and are characterized by quartz-poor meta-igneous rocks, including peridotite, clinopyroxenite, chromitite, cumulate and non-cumulate gabbro, diabase, mafic volcanic rocks and ferruginous chert. The metamorphosed mafic volcanic rocks locally retain relict pillow and pillow-breccia textures and small, sparse amygdules, indicating a subaqueous site of extrusion. The mafic volcanic blocks retain tholeiitic bulk compositions and exhibit trace element characteristics of mid-ocean-ridge basalts, possibly modified by post-extrusive hydrothermal metamorphism. This group of metamorphosed ocean-floor rocks may constitute a dismembered metamorphosed ophiolite (Penrose Conference Participants, 1972), although the presence of a cogenetic ophiolitic sequence cannot be firmly established. Its lack of structural coherence, presence in serpentinite-matrix melange, generally poor preservation of igneous structures, and high P/T dynamic, subduction-zone metamorphism distinguish it from the overlying coherent Sierra de San Andres ophiolite which exhibits low P/T temperature ocean-floor metamorphism (Moore, 1983).

Exotic blocks in the Puerto Nuevo melange that may be of volcanic-arc origin include metamorphosed granitic rocks (orthogneiss), metarhyolite, and metagraywacke. The large size of the orthogneiss blocks and their lack of mafic xenoliths, screens, or country rocks are atypical of plagiogranite associated with ophiolites and indicate that they are fragments of a large tonalitic pluton.

Metamorphism of the oceanic-derived blocks within the exotic block unit varies from pumpellyite-actinolite to blueschist-greenschist to epidote-amphibolite and eclogite facies; this facies series is interpreted to result from subduction metamorphism under moderate temperature/high pressure conditions. The volcanic-arc-derived blocks all contain greenschist facies assemblages and bulk compositions which allow the possibility that their metamorphism occurred in either an arc or subduction-zone environment.

The exotic blocks may have been derived from two separate tectonostratigraphic subterranes, possibly introduced into the melange at widely different times from tectonically separate sources. Alternately, the blocks may represent fragments of a single composite terrane.

AGE OF METAMORPHISM AND CORRELATION OF THE MELANGE

Potassium-argon-age dating of coarse-grained crossite from a single metabasite block in the Puerto Nuevo melange yielded an age of 173.3 ± 69.3 m.y. B.P. (D. Krummenacher, written communication, 1980). The uncertainty of the age determination allows, but does not prove, possible temporal correlation with the 148 to 104 m.y. B.P. age of blueschist blocks from the San Benitos Islands and Cedros Island (Suppe and Armstrong, 1972), and the ages of about 160 m.y. B.P. from high-grade blueschist of the Franciscan Complex (Coleman and Lanphere, 1971; Coleman, written communication, 1983). The age determination could also be in agreement, however, with the age of 175 m.y. B.P. to 190 m.y. B.P. blueschists from the Sierra Nevada (Schweickert and others, 1980), the 220 m.y. B.P. age of blueschist in the Stuart Fork Formation in the Klamath Mountains in Oregon (Hotz and others, 1977), and the maximum possible age of 167 m.y. B.P. of the Condrey Mountain Schist of the Klamath Mountains (Coleman, written communication, 1983). Further age refinement is necessary.

Most workers have generally correlated the Puerto Nuevo melange complex with melanges on Cedros Island and the San Benitos Islands (e.g. Klienast and Rangin, 1982). These Baja California melanges have been correlated with the Franciscan Complex because of their location oceanward of the Sierra Nevada/Peninsular Ranges batholithic belt (Jones and others, 1976). The blueschist-bearing melange complex at Puerto Nuevo, however, differs in several ways from the melanges of Cedros Island, the San Benitos Islands and the Franciscan Complex. The Puerto Nuevo melange is overlain by Triassic ophiolite (Barnes and Mattinson, 1981) whereas the Franciscan and Cedros Island melanges are overlain by Jurassic (Coast Range) ophiolite (Hopson and others, 1981; Kimbrough, 1982). The Franciscan complex and offshore Baja California melanges are predominantly pelitic-matrix melanges which contain abundant sedimentary and metasedimentary blocks, whereas the Puerto Nuevo melange lacks both pelitic matrix and abundant sedimentary and metasedimentary blocks. The metamorphosed exotic blocks in the Franciscan and offshore Baja California melanges typically contain glaucophane, lawsonite, jadeite, and aragonite and the mineral assemblages of zeolite to prehnite-pumpellyite, blueschist and eclogite facies whereas exotic blocks at the Puerto Nuevo are characterized by the minerals crossite, actinolite, epidote, acmite, calcite and barroisitic hornblende and the mineral assemblages of pumpellyite-actinolite, blueschist-greenschist, epidote-amphibolite and eclogite facies. Additionally, the Puerto Nuevo melange appears to be a contact zone between the Sierra de San Andres ophiolite and an underlying coherent metamorphosed ocean floor - volcanic arc terrane(s) (see below) whereas the Franciscan, Cedros Island and San Benitos Islands melange are thick, imbricated subduction complexes.

In tectonic position and petrologic character, the Puerto Nuevo melange resembles the older blueschist occurrences in the

Sierra Nevada and the Klamath Mountains rather than the Franciscan Complex. The Sierran blueschists are located in a sliver of serpentinite-matrix melange along the Melones fault zone east of the Franciscan Complex and Jurassic ophiolite and arc terranes and may itself be a metamorphosed ophiolite (Schweickert and others, 1980). The blueschists of the Stuart Fork Formation and the Condrey Mountain Schist in the Klamath Mountains (Hotz, 1973) are present in thrust sheets to the east of, or 'inboard' of, the Franciscan Complex.

Although the metamorphic age of the inclusions in the Puerto Nuevo melange is not well constrained, the above features show that any correlation of the melange with the Franciscan Complex and the melanges of Cedros and the San Benitos Islands should be viewed with caution. Instead, the available evidence indicates that the Puerto Nuevo melange complex may represent an older 'inboard' belt of blueschists analogous to those of the Klamath and Sierra Nevada Mountains.

MODEL FOR EMPLACEMENT OF THE MELANGE COMPLEX

Any model proposed for the origin and emplacement of the melange complex at Puerto Nuevo must take into consideration the following features: 1) The melange complex has an orderly stacking sequence that consists of upper and lower serpentinite breccia units containing only locally-derived clasts and an interior exotic block unit; 2) the exotic block unit contains two spatially distinct groups of inclusions which are: a) ocean-floor rocks metamorphosed in the pumpellyite-actinolite, transitional blueschist-greenschist, and eclogite facies, and b) silicic rocks of possible volcanic-arc affinity metamorphosed in the greenschist facies; 3) the melange complex is structurally overlain by the Sierra de San Andres ophiolite which displays ocean-floor hydrothermal, but not subduction, metamorphism.

Based on studies of the Franciscan Complex in California and modern deep-sea trench deposits, most blueschist-bearing pelite-matrix melanges have been interpreted to have been formed by sedimentary or tectonic processes associated with subduction beneath a thick accretionary prism (Ernst and others, 1970; Ernst, 1973; Cloos, 1982). Although some or all of the exotic blocks of the Puerto Nuevo melange may have participated in subduction metamorphic processes, the melange as a whole was emplaced by processes imposed on, or unrelated to, deposits produced by subduction accretion. This interpretation is supported by the absence of a pelitic matrix, the lack of high-pressure metamorphism of the serpentinite breccia units, and the juxtaposition of the exotic block unit to the Sierra de San Andres ophiolite which was metamorphosed under high-temperature, low-pressure ocean-floor conditions.

Although geologists engaged in early reconnaissance studies suggested a simple accretionary-prism origin (Jones and others, 1976; Rangin, 1977, 1978), other workers argue that the Puerto Nuevo melange is only secondarily related to subduction accretion along the continental margin of Mexico. Based on a limited study of the melange, Moore (1976) suggests that it was em-

placed as a serpentinite diapir derived from a Mesozoic subduction zone lying beneath Baja California. Karig (1980) argues that blueschist-bearing melanges on Cedros Island and the San Benitos Islands were emplaced by vertical displacements along interarc strike-slip faults. More recently, Klienast and Rangin (1982) suggest that the blueschist-bearing melanges on Cedros and the San Benitos Islands and the Vizcaino Peninsula resulted from high-pressure, low-temperature metamorphism by tectonic loading beneath thick nappes emplaced by an interarc collision during the mid-Cretaceous along the continental margin of Baja California.

A multistage tectonic history can explain many observed features of the melange complex at Puerto Nuevo. The petrology of the exotic blocks argues that possibly two metamorphosed tectonostratigraphic terranes, one of ocean-floor and high P/T metamorphic affinity and the other of volcanic-arc affinity, may have provided fragments to the melange. The incompletely mixed character of the various exotic blocks may indicate that the fragments were emplaced close to the site of formation of the melange under its ophiolitic cover. However, the contrasting metamorphic character of the blocks in the melange and the ocean-floor metamorphism of the overlying ophiolite indicate that significant amounts of displacement have occurred subsequent to their metamorphism.

Interaction along a thrust zone crosscutting an inferred thrust contact between the metamorphic terranes (shown as composite) and the Sierra de San Andres ophiolite is suggested to have resulted in the formation of the Puerto Nuevo melange complex (Fig. 8). In Figure 8A, the Sierra San Andres ophiolite complex is thrust onto the composite metamorphosed volcanic-arc and ocean-floor terranes. The symmetry of the stratigraphy of the melange and the gradational contact with the overlying serpentinized harzburgite unit of the ophiolite is interpreted to indicate that serpentinized harzburgite may also underlie the unexposed lower contact of the melange. The second thrust, shown as dashed line in Figure 8A, is postulated to crosscut the primary fault contact between the ophiolite and underlying composite metamorphic terranes. Pieces and fragments of the composite metamorphic terranes were probably dragged up along the second fault (Fig. 8B) to form the exotic block unit. The upper and lower serpentinite breccia units of the melange may have formed by progressive disruption, brecciation, and further serpentinization of the partly serpentinized harzburgite, dunite, pyroxenite and rodingitized diabasic rocks that comprise the upper and lower walls of the thrust. The gradational contact of the melange with the ophiolite was produced by cumulative strain along the margins of the thrust. Along the thrust zone, now represented by the exotic block unit, higher temperatures produced by shearing or the introduction of fluids at depth may have allowed recrystallization of the matrix serpentinite to antigorite. Later, the entire sequence of thrust sheets of ophiolite, melange and underlying terranes were folded and uplifted during post-Miocene deformation (Robinson, 1979), forming the antiform now defined by the stratiform units in the Arroyo Casitas area (Fig. 8C). This uplift,

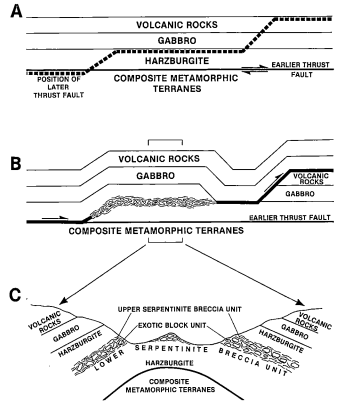

Figure 8. Model of emplacement of the Puerto Nuevo melange complex: (A) ophiolite complex, which structurally overlies a composite metamorphic terrane composed of ophiolitic and volcanic-arc elements, is cut by a later thrust; (B) displacement along thrust drags up fragments of the underlying metamorphic terrane into lowest part of ophiolite and disrupts the harzburgitic walls of the thrust; (C) Miocene regional folding uplifts the igneous and metamorphic rocks, forming an antiform.

coupled with associated serpentinization and mobility of the matrix serpentinite, may have mostly destroyed the tectonic fabric imposed on the serpentinite by the earlier thrusting event.

CONCLUSIONS

The Vizcaino Peninsula and adjacent islands are underlain by the Upper Triassic Sierra de San Andres and La Costa ophio- lites, the Upper Jurassic Choyal ophiolite (on Cedros Island), the Jura-Cretaceous Cedros and San Andres arc complexes, the Cretaceous Alisitos arc, and two distinct blueschist-bearing melange complexes: the Cedros Complex on Cedros Island and the Puerto Nuevo melange on the Vizcaino Peninsula (Rangin, 1978; Gastil and others, 1978; Kilmer, 1979; Barnes and Berry, 1979; Barnes and Mattinson, 1981; Kimbrough, 1982; Moore, 1983). The presence of a number of ophiolitic, volcanic-arc and blueschist assemblages in the Cedros-Vizcaino Peninsula area and the Magdelena Bay area (Blake and others, 1984), and the juxtaposition of contrasting stratigraphic packages in central and eastern Baja California (Gastil and Miller, 1981) suggest that the Vizcaino Peninsula and probably the remainder of Baja California consists of a number of discrete geologic terranes. Tectonostratigraphic terranes of disparate and possibly far-traveled origin have been recognized throughout Alta California, Oregon, Washington, British Columbia, and Alaska by Coney and others (1980). The multiple subduction zone, ophiolite, and arc complexes of the Vizcaino Peninsula and Baja California support the hypothesis that the southern part of the Cordillera, like the northern part, is composed of a collage of geologic terranes that may be allochthonous relative to each other and cratonal North America.

ACKNOWLEDGMENTS

My understanding of the Puerto Nuevo melange complex benefited greatly from discussions with J. G. Liou, R. G. Gastil, D. L. Kimbrough, M. C. Blake, Jr., S. Maruyama, R. G. Coleman, B. M. Page, C. Carlson, and P. Schiffman. I also thank Carter Hull for unpublished X-ray diffraction and petrographic descriptions of the mineralogy of one tectonic block, field assistants Glen La Tourelle and Robin Gallo, and typist Kathy Small. Financial support from the National Science Foundation (Grant EAR 76-22650 and 80-08527 to J. G. Liou), Geological Society of America (Research Grants 2503-79 and 2696-80), Shell Fund of Stanford University, and the Consejo de Minerales of the Republic of Mexico are greatly appreciated. The U.S. Geological Survey analytical laboratory provided major element chemical analyses. Various versions of the manuscript were reviewed by M. C. Blake, Jr., D. S. Cowan, R. G. Coleman, B. W. Evans, J. G. Liou, B. M. Page, and S. S. Sorensen.

REFERENCES CITED

Banno, S., Higashino, T., Otsuki, M., Itaya, T., and Nakajima, T., 1978, Thermal structure of the Sanbagawa metamorphic belt in central Shikoku: Journal of Physics of the Earth, v. 26, Suppl., S 345-356.
Barnes, D. A., 1982, Basin analysis of volcanic arc-derived Jura-Cretaceous sedimentary rocks, Vizcaino Peninsula, Baja California Sur, Mexico [Ph.D. thesis]: University of California, Santa Barbara, California, 249 p.
Barnes, D. A., and Berry, K. D., 1979, Jura-Cretaceous paleogeography: The Eugenia Group, western Vizcaino Peninsula, Baja California Sur, Mexico, *in* Gastil, R. G., and Abbott, P. L., eds., Baja California Geology: Department of Geological Sciences, San Diego State University (fieldtrip guidebook for the 1979 Geological Society of America Annual Meeting, San Diego, Cali- fornia), p. 53-64.
Barnes, D. A., and Mattinson, J. M., 1981, Late Triassic-Early Cretaceous age of eugeoclinal terranes, western Vizcaino Peninsula, Baja California Sur, Mexico [abs.]: Geological Society of America Abstracts with Programs, v. 13, p. 43.
Blake, M. C., Jr., Jayko, A. S., Moore, T. E., Chavez, V., Saleeby, J. B., and Seel, K., 1984, Tectonostratigraphic terranes of Magdalena Island, Baja California Sur, *in* Frizzell, V. A., Jr., Geology of the Baja California peninsula: Pacific Section, Society of Economic Paleontologists and Mineralogists, v. 39, p. 183-192.
Cann, J. R., 1970, Rb, Sr, Y, Zr, and Nb in some ocean-floor basaltic rocks: Earth and Planetary Science Letters, v. 10, p. 1-11.
Cloos, M., 1982, Flow melanges: a numerical modeling and geologic constraints

on their origin in the Franciscan subduction complex, California: Geological Society of America Bulletin, v. 93, p. 330–345.

Cohen, L. H., Condie, K. C., Kuest, L. J., Jr., Mackenzie, G. S., Meister, F. H., Pushkar, P., and Steuber, A. M., 1963, Geology of the San Benitos Islands, Baja California, Mexico: Geological Society of America Bulletin, v. 74, p. 1355–1370.

Coleman, R. G., 1966, New Zealand serpentinites and associated metasomatic rocks: New Zealand Geological Survey Bulletin, v. 76, 102 p.

Coleman, R. G., and Lee, L. E., 1963, Glaucophane-bearing metamorphic rock types of the Cazadero area, California: Journal of Petrology, v. 4, p. 260–301.

Coleman, R. G., and Lanphere, M. A., 1971, Distribution and age of high-grade blueschists, associated eclogites, and amphibolites from Oregon and California: Geological Society of America Bulletin, v. 82, p. 2397–2412.

Coney, P. J., Jones, D. L., and Monger, J.W.H., 1980, Cordilleran suspect terranes: Nature, v. 288, p. 329–333.

Coombs, D. S., Nakamura, Y., and Vuagnat, M., 1976, Pumpellyite-actinolite facies schists of the Taveyanne Formation near Loeche, Valais, Switzerland: Journal of Petrology, v. 17, p. 440–471.

Ernst, W. G., 1973, Blueschist metamorphism and P-T regimes in active subduction zones: Tectonophysics, v. 26, p. 229–246.

Ernst, W. G., and Seki, Y., 1967, Petrologic comparison of the Franciscan and Sanbagawa metamorphic terranes: Tectonophysics, v. 4, p. 463–478.

Ernst, W. G., Seki, Y., Onuki, H., and Gilbert, M. C., 1970, Comparative study of low-grade metamorphism in the California Coast Ranges and outer metamorphic belt of Japan: Geological Society of America, Memoir 124, 276 p.

Evans, B. W., 1977, Metamorphism of alpine peridotite and serpentinite: Annual Review, Earth and Planetary Science, v. 5, p. 397–447.

Garcia, M. O., 1979, Petrology of the Rogue and Galice Formations, Klamath Mountains, Oregon: identification of a Jurassic island arc sequence: Journal of Geology, v. 86, p. 29–41.

Gastil, R. G., Morgan, J., and Krummenacher, D., 1978, Mesozoic history of peninsular California and related areas east of the Gulf of California, *in* Howell, D. G., and McDougall, K. A., eds., Mesozoic paleogeography of the western United States: Pacific Section, Society of Economic Paleontologists and Mineralogists, Pacific Coast Paleogeography Symposium 2, p. 107–116.

Gastil, R. G., and Miller, R. H., 1981, Lower Paleozoic strata on the Pacific plate of North America: Nature, v. 292, p. 828–829.

Hart, S. R., Erlank, A. J., and Kable, E.J.D., 1974, Seafloor basalt alteration: some chemical and Sr isotope effects: Contributions to Mineralogy and Petrology, v. 44, p. 219–240.

Hopson, C. A., Mattinson, J. M., and Pessagno, E. A., Jr., 1981, Coast Range ophiolite, western California, *in* Ernst, W. G., ed., The geotectonic development of California, Rubey volume 1: Prentice-Hall, Englewood Cliffs, New Jersey, p. 418–510.

Hotz, P. E., 1973, Blueschist metamorphism in the Yreka-Fort Jones area, Klamath Mountains, California: U.S. Geological Survey Journal of Research, v. 1, p. 53–61.

Hotz, P. E., Lanphere, M. A., and Swanson, D. A., 1977, Triassic blueschist from northern California and north-central Oregon: Geology, v. 5, p. 659–663.

Jones, D. L., Blake, M. C., Jr., and Rangin, C., 1976, The four Jurassic belts of northern California and their significance to the geology of the southern California borderland, *in* Howell, D. G., ed., Aspects of the geologic history of the California borderland: Pacific Section, American Association of Petroleum Geologists Miscellaneous Publication 24, p. 343–362.

Jones, D. L., Blake, M. C., Jr., Bailey, E. H., and McLaughlin, R. J., 1978, Distribution and character of upper Mesozoic subduction complexes along the west coast of North America, *in* Burns, K. L., and Rutland, R.W.R., eds., Structural characteristics of tectonic zones: Tectonophysics, v. 47, p. 207–222.

Karig, D. E., 1980, Material transport within accretionary prisms and the "knocker" problem: Journal of Geology, v. 88, p. 27–39.

Kilmer, F. J., 1977, Reconnaissance geology of Cedros Island, Baja California, Mexico: Southern California Academy of Science Bulletin, v. 76, p. 91–98.

Kilmer, F. J., 1979, A geological sketch of Cedros Island, Baja California, Mex-

ico, *in* Gastil, R. G., and Abbott, P. L., eds., Baja California Geology: Department of Geological Sciences, San Diego State University (fieldtrip guidebook for the 1979 Geological Society of America Annual Meeting, San Diego, California), p. 11–28.

Kimbrough, D. L., 1982, Structure, petrology, and geochronology of Mesozoic paleooceanic terranes on Cedros Island and the Vizcaino Peninsula, Baja California Sur, Mexico [Ph.D. thesis]: University of California, Santa Barbara, 395 p.

Klienast, J. R., and Rangin, C., 1982, Mesozoic blueschists and melanges of Cedros Island (Baja California, Mexico): a consequence of nappe emplacement or subduction?: Earth Planetary Science Letters, v. 59, p. 119–138.

Lan, C., and Liou, J. G., 1981, Occurrence, petrology, and tectonics of serpentinites and associated rodingites in the Central Range, Taiwan: Memoir Geological Society of China, no. 4, p. 343–387.

Moore, T. E., 1976, Structure and petrology of the Sierra de San Andres ophiolite, Baja California Sur, Mexico [M.S. thesis]: San Diego State University, 83 p.

Moore, T. E., 1979, Geological summary of the Sierra de San Andres ophiolite, *in* Gastil, R. G., and Abbott, P. L., eds., Baja California, geology: Department of Geological Sciences, San Diego State University (fieldtrip guidebook for the 1979 Geological Society of America Annual Meeting, San Diego, California), p. 95–106.

Moore, T. E., 1980, The blueschist-bearing melange complex near Puerto Nuevo, Vizcaino Peninsula, Baja California Sur, Mexico (abs.): EOS, v. 61, no. 46, p. 1155.

Moore, T. E., 1983, Geology, petrology, and tectonic significance of paleooceanic terranes of the Vizcaino Peninsula, Baja California Sur, Mexico [Ph.D. thesis]: Stanford University, Stanford, California, 376 p.

Moore, T. E., 1985, Stratigraphy and tectonic significance of the Mesozoic tectonostratigraphic terranes of the Vizcaino Peninsula, Baja California Sur, Mexico, *in* Howell, D. G., ed., Tectonostratigraphic terranes of the Circum-Pacific region: Circum-Pacific Council for Energy and Earth Science series, number 1, Houston, Texas, p. 315–329.

Pearce, J. A., 1975, Basalt geochemistry used to investigate past tectonic environments on Cyprus: Tectonophysics, v. 25, p. 41–67.

Pearce, J. A., and Cann, J. R., 1973, Tectonic setting of basic volcanic rocks determined using trace element analysis: Earth Planetary Science Letters, v. 19, p. 290–300.

Pearce, J. A., and Norry, M. J., 1979, Petrogenetic implications of Ti, Zr, Y, and Nb variations in volcanic rocks: Contributions to Mineralogy and Petrology, v. 69, p. 33–47.

Penrose Conference Participants, 1972, Penrose field conference: ophiolites: Geotimes, v. 17, p. 24–25.

Rangin, C., 1977, Sur un trait tectonique majeur de la hordure continentale pacifique: le dispositif Franciscain en Basse California (Mexique): C. R. Somm. Soc. Geol. Fr., v. 4, p. 227–230.

Rangin, C., 1978, Speculative model of Mesozoic geodynamics, central Baja California to northeast Sonora (Mexico), *in* Howell, D. G., and McDougall, K. A., eds., Mesozoic paleogeography of the western United States: Pacific Section, Society of Economic Paleontologists and Mineralogists, Pacific Coast Paleogeography Symposium 2, p. 85–106.

Robinson, J. W., 1979, Structure and stratigraphy of the northern Vizcaino Peninsula with a note on a Miocene reconstruction of the peninsula, *in* Gastil, R. G., and Abbott, P. L., eds., Baja California Geology: Department of Geological Sciences, San Diego State University (fieldtrip guidebook for the 1979 Geological Society of America Annual Meeting, San Diego, California), p. 77–82.

Schweikert, R. A., Armstrong, R. L., and Harakel, J. E., 1980, Lawsonite blueschist in the northern Sierra Nevada, California: Geology, v. 8, p. 27–31.

Shervais, J. W., 1982, Ti - V plots and the petrogenesis of modern and ophiolitic lavas: Earth and Planetary Science Letters, v. 59, p. 101–118.

Suppe, J., and Armstrong, R. L., 1972, Potassium-argon dating of Franciscan metamorphic rocks: American Journal of Science, v. 272, p. 217–233.

MANUSCRIPT ACCEPTED BY THE SOCIETY JULY 29, 1985

Geological Society of America
Memoir 164
1986

Petrologic and geochemical comparison of the blueschist and greenschist units of the Catalina Schist terrane, southern California

Sorena Svea Sorensen*
Department of Earth and Space Sciences
University of California at Los Angeles
Los Angeles, California 90024

ABSTRACT

In the Catalina Schist terrane, a nappe of blueschist facies melange is overlain by one of isoclinally-folded greenschist facies rocks. The blueschist unit contains meta-graywacke, metashale, metaconglomerate, greenstone, quartz schist, blueschist, and eclogite blocks in a fine-grained, schistose matrix which probably recrystallized from a mixture of ultramafic and clay-rich, quartzose detritus.

The greenschist unit consists of metabasites, graywacke-composition grayschist, and quartz schist. Some metabasites are glaucophanic greenschists which initially re-crystallized in the blueschist facies; others originally were epidote amphibolites. Both are overprinted by greenschist facies minerals.

Most metabasites from both tectonic units have major, minor, and trace element characteristics of ocean-floor tholeiites. The white micas of the blueschist unit probably recrystallized under substantially higher P_{fluid}/T conditions than those of the greenschist unit. Amphiboles from metabasites of the greenschist unit are crossites, barroisites, and sodic actinolites in glaucophanic greenschists, and hornblendes rimmed by actinolites in epidote amphibolites. Crossites from the glaucophanic greenschists probably formed at somewhat higher T/P conditions than glaucophanes and crossites from the lawsonite-bearing metabasites of the blueschist unit. The mineral assemblages and metamorphic histories of the two units evidently represent two distinct, relatively high P/T trajectories.

INTRODUCTION: METAMORPHIC GEOLOGY OF THE CATALINA BLUESCHIST AND GREENSCHIST UNITS

The Catalina Schist terrane consists of three tectonic units which are exposed on Santa Catalina Island (Platt, 1975, 1976) and are distributed throughout much of the California Continental Borderland (Figure 1). The structurally lowest unit of the Catalina Schist terrane is a blueschist facies unit, which is overlain

*Present address: Department of Mineral Sciences, National Museum of Natural History, Smithsonian Institution, Washington, D.C. 20560.

by a greenschist facies unit along a thrust contact (Figure 2). The Catalina blueschist and greenschist units are overlain in turn by a slab of zoisite + An_{05-15} plagioclase + magnesiohornblende-bearing amphibolite and yet structurally higher ultramafic rocks which contain garnet + hornblende ± diopside-bearing blocks (Figure 2; Platt, 1975, 1976; Sorensen, 1984). The similarity of K-Ar ages determined for amphibolites, greenschists, and blue-

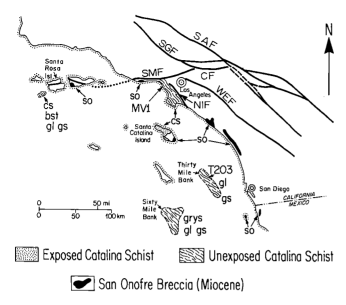

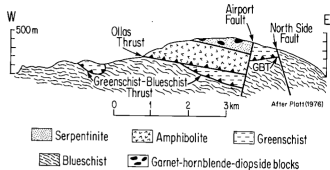

Figure 2. Diagrammatic cross section of the Catalina schist terrane, after Platt (1976).

Figure 1. Index map, showing exposures of the Catalina Schist terrane and the San Onofre Breccia in coastal southern California and the California Continental Borderland, and areas underlain by Catalina Schist. Abbreviations are as follows: cs = Catalina Schist, so = San Onofre Breccia, bst = blueschist, gl gs = glaucophanic greenschist, grys = grayschist, SAF = San Andreas fault, SMF = Santa Monica Fault, CF = Cucamonga fault, WEF = Whittier - Elsinore fault, NIF = Newport-Inglewood fault. Well MV-1 (Mobil Venice 1) and dredge haul T203 are also shown.

schists (Suppe and Armstrong, 1972) led Platt to propose that the Catalina schist metamorphism reflects an inverted thermal gradient developed below hot, hanging-wall peridotite in a newly-formed subduction zone.

The apparent metamorphic zonation displayed by the Catalina Schist terrane on Santa Catalina Island probably transcends the tectonic boundaries between the amphibolite, greenschist, and blueschist units.

Platt (1976, p. 60) noted that in exposures of the greenschist unit between the North Side fault and the Airport fault (Figure 2), blue amphibole is abundant near the lower (blueschist unit—greenschist unit) contact, whereas biotite and garnet occur near the upper (greenschist unit—amphibolite unit) contact. Similar relationships appear in other areas where the upper and/or lower contacts of the greenschist unit are exposed. Mineral assemblages of the higher-grade metabasites exposed near the upper contact indicate epidote amphibolite facies, and consist of An_{00}-An_{15} plagioclase + calcic amphibole + clinozoisite + garnet ± chlorite ± white mica ± quartz ± ilmenite ± rutile ± sphene; interlayered high-grade grayschist consists of An_{00} - An_{05} plagioclase + quartz + white mica + garnet + clinozoisite + calcic amphibole + chlorite + graphite ± biotite ± rutile ± sphene ± hematite (Table 1). The metabasites that are inferred to be structurally lower in the greenschist unit lack garnet, rutile and ilmenite, and contain pistacitic epidote; interlayered grayschists contain garnet relics partially to entirely replaced by epidote + chlorite. Metabasites from the structurally lowest part of the Catalina greenschist unit are glau-

cophanic greenschists with the mineral assemblage albite + pistacitic epidote + sodic-calcic or sodic amphibole ± actinolite + chlorite ± white mica ± quartz + sphene + hematite ± magnetite ± calcite (Table 1). Sodic-calcic and calcic amphibole appear to replace sodic amphibole. Lawsonite inclusions are present in one epidote porphyroblast in one sample of glaucophanic greenschist (9580SC9). The greenschist unit evidently had an early metamorphic zonation from epidote amphibolite to epidote (or lawsonite?) + crossite-bearing assemblages (Sorensen, 1984; Jacobson and Sorensen, 1986). The greenschist unit is thoroughly recrystallized and penetratively deformed. Early isoclinal folds are refolded by later open to isoclinal folds; both have an axial planar schistosity. Still younger open folds, which are sporadically developed throughout the unit, are accompanied by crenulations (Platt, 1976).

The blueschist unit contains a variety of mineral assemblages and displays a higher P/T facies series than the greenschist unit. In some localities, the blueschist unit consists of blocks of metagraywacke, metashale, metaconglomerate, greenstone, and quartz schist in a fine-grained schistose matrix. Rare garnet ± epidote-bearing high-grade blueschist and eclogite blocks, both of which have lower-grade blueschist facies retrograde mineral assemblages, may represent exotic blocks in this blueschist unit 'melange' (Table 1). Structural geometries of these areas are complex, and reflect localized deformation around the blocks, which can be up to tens of meters in size. Large blocks are deformed little internally, but intensely at their margins. The schistose melange matrix is tightly folded with great variability of axial orientations near most tectonic blocks larger than a meter in diameter. Other areas of the blueschist unit appear to be moderately large (km^2) bodies of "coherent" metagraywacke. Glaucophane + lawsonite-bearing mineral assemblages are ubiquitous in metagraywacke, metaconglomerate, metashale, and quartz schist as well as in metabasites (Table 1).

Albite in some metagraywackes and metashales contains feathery sprays of jadeite with anomalous interference colors. The fine-grained jadeite prisms may be tens of μm long, but even the coarsest grains range from 5-10 μm in other dimensions; the phase was identified by its optical properties. Jadeitic pyroxene + glaucophane + lawsonite-bearing mineral assemblages at Panoche

TABLE 1. MINERAL ASSEMBLAGES OF METASEDIMENTS AND METABASITES

Metasediments	qtz	fld	chl	mus	law	epi	Ca-am	Na-am	bio	stilp	gar	jad	sph	gph	hem	ilm	apt	pyr	cal
Blueschist Unit																			
98801 mat	X	a	X	m	m			X					X	X	am			X	X
9128010 msh	X	a		X				X				sa	X	X				X	
714802 msh	X	a	m	X	X			X				sa		X					
716802 mgr	X	a		X	X			X					X	X					
9148012 mgr	X	a	m	X	X			X					X						
713803 mgr	X	a	m	X	X			X				sa	X	X	ia			X	
Greenschist Unit																			
7168012 gry	X	a	X	X			X						rc	X					
815812 gry	X	a	X	X			X		ia		X		rc	X					
714806 gry	X	An5		X			X			ag	ia		rc	X	ap				
0Gar4-1 gry	X	a	X	X									rc	X					
714801 gry	X	a	X	X	X	X							X	X					
716803 gry	X	a	X	X			X						X	X					
99803 qst	X	a	X	X		m			rt		X		ric						

Metabasites	qtz	fld	chl	mus	law	epi	Ca-am	Na-am	bio	stilp	gar	jad	sph	gph	hem	ilm	apt	pyr	cal
Blueschist Unit																			
LHCNGL6 mdi	X	a	X	X	X		rt	X					X		am			X	X
OH-1 gsn	X	a	X	X	X	rt?		X				sa?	X						
716806 gsn	X	a	X		X			X		X			X						
719805 bst	X		X	X	X			X					X		am				
L295 ecl				X				X		ag		O	rc				X		
79817' ecl				X	ag			X		ag		O	rc						
Greenschist Unit																			
814814 eam		a		X		X	X				X		X						
9148011 eam		a	X	X		X	X			m	X		X						
7198012 eam	X	An15				X	X		X		X		ric			X			
911801 eam/gs?		a		X		X	X			rt			rc		m				
91180SC1 gs	X	a	X	X		X	X						X						
7188013 gs	X	a	X	X		X	X				X		X		am				X
9580SC9 ggs	X	a	X	X	ie	X	X	X					X						
9580SC5 ggs	X	a	X	X		X	X	X					X						X
T203B ggs	X	a	X	X		X		X			X		X		am			X	X
MV-1 ggs	X	a	X	X		X		X			X		X					X	

Note: Abbreviations are: mat=melange matrix; msh=metashale; mgr=metagraywacke; gry=grayschist; qst=quartz schist; mdi= metadiorite; gsn=greenstone; bst=blueschist; ecl=eclogite; eam=epidote amphibolite; gs=greenschist; "X"=mineral is present; "m"=present in minor amounts; Anoo=anorthite content of plag; sa=sprays in albite; am=after magnetite; ia= inclusions in albite; rt=relict; ag=after garnet; rc=rutile cores; ic=ilmenite cores; ie=inclusions in epidote; ap=after pyrite; O= omphacite.

Pass and Pacheco Pass are interpreted to represent the highest pressures and temperatures of recrystallization of Franciscan metagraywackes (Ernst, 1965; 1971). Mineral assemblages of greenstones consist of lawsonite + glaucophane along with varying amounts of albite, chlorite, white mica and stilpnomelane (Table 1). Greenstones may also contain epidote, but this mineral does not appear to be part of the stable assemblage. Platt (1976, p. 35), reported occurrences of omphacitic clinopyroxene in some greenstones from the blueschist unit.

Despite the diversity of mineral assemblages within each unit, the blueschist and greenschist units can be delineated by the occurrences of lawsonite, epidote, sodic amphibole, and calcic amphibole (Table 1). Metasediments and metabasites of the blueschist unit characteristically contain lawsonite + sodic amphibole. All metasediments in the greenschist unit lack lawsonite and most contain either calcic amphibole or epidote in addition to albite, quartz, white mica, and chlorite (Table 1). Nearly all metabasites of the greenschist unit contain epidote + calcic amphibole. The glaucophanic greenschists are mineralogically distinguished from blueschists by the occurrence of epidote + sodic amphibole in the former, lawsonite + sodic amphibole in the latter. The rare high-

grade blueschists may indeed contain garnet, epidote, and sodic amphibole, but garnet and epidote are invariably partially to mostly replaced by lawsonite. Glaucophanic greenschists such as MV-1 or T-203 (Table 1) exhibit small modal amounts of sodic amphibole in an albite + epidote + chlorite schist.

The metamorphic zonation of the Catalina Schist terrane is accompanied by changes in the proportions of ultramafic, mafic, and sedimentary lithologies. The amphibolite unit consists of perhaps 70 percent ultramafic (serpentinite and ultramafic schist) + gabbroic rocks, about 20 percent basaltic rocks, and about 10 percent sedimentary rocks (semipelitic schist, as well as spessartine quartzite presumably derived from chert).

In contrast to the mafic/ultramafic association of the amphibolite unit, both the greenschist and blueschist units have a protolith of basaltic rocks, clastic sediments, and cherts. The Catalina greenschist unit consists of about 50 percent metabasite schist, 40 percent graywacke-composition grayschist, and about 10 percent Fe- and Mn-rich quartz schist. Rare ultramafic lithologies are represented by tremolite-fuchsite-chlorite pods and lenses intercalated with grayschist. The rocks are thoroughly recrystallized and lack relict igneous or sedimentary features. All

TABLE 2. MAJOR ELEMENT, WHOLE-ROCK XRF ANALYSES OF METASEDIMENTS AND METABASITES

Sample	98801 mat	916807 mat	716802 mgr	9148012 mgr	713803 mgr		7168012 gry	815812 gry	714801 gry	0Gar4-1 gry
Weight % Oxide										
SiO2	60.8	54.1	68.6	60.4	60.9		55.4	55.6	63.5	69.0
TiO2	0.4	0.2	0.6	0.6	0.9		1.0	0.9	0.6	0.7
Al2O3	5.3	3.8	13.4	15.2	12.6		16.1	23.4	12.2	14.3
FeO*	6.6	6.0	4.9	5.8	8.2		8.2	8.2	7.2	4.7
MgO	19.6	24.1	3.3	5.8	5.9		6.4	4.9	4.8	2.4
CaO	6.1	9.4	4.4	3.7	4.5		2.4	2.9	4.4	0.5
Na2O	1.4	1.3	2.2	1.8	2.6		2.7	2.8	3.2	2.3
K2O	0.6	0.6	1.5	2.5	1.0		2.8	2.4	1.5	3.1
Total	100.8	99.5	98.9	95.8	96.9		95.0	101.1	97.4	97.0

Sample	LHCNGL6 mdi	OH-1 gsn	716806 gsn	719805 bs	L295 ecl	79817' ecl	911801 eam/gs?	7188013 gs	9580SC9 ggs	9580SC5 ggs	MV-1 ggs	T203B ggs
Weight % Oxide												
SiO2	57.0	50.4	45.9	48.3	47.6	47.3	51.8	47.8	52.0	52.7	46.2	45.8
TiO2	0.6	1.2	2.0	0.5	2.2	0.8	1.3	1.5	2.3	1.9	3.2	2.7
Al2O3	15.5	15.7	19.0	16.4	14.0	14.6	14.8	14.3	13.6	14.5	15.6	17.4
FeO*	7.0	9.8	8.1	7.8	12.1	12.0	8.9	11.3	11.6	10.0	15.1	11.8
MgO	4.0	7.7	6.5	7.6	6.4	4.7	6.9	9.1	6.5	8.3	8.5	6.1
CaO	9.7	12.7	11.1	10.4	9.9	11.3	8.2	5.7	8.9	6.8	6.6	10.7
Na2O	2.0	2.2	1.8	3.4	3.8	3.9	3.0	3.3	2.9	3.1	2.0	3.2
K2O	0.6	1.1	0.7	0.6	0.3	0.2	0.4	0.3	0.3	0.8	1.0	1.4
Total	96.4	100.8	95.1	95.0	96.3	94.8	95.3	93.3	98.1	98.1	98.2	99.1

Note: *FeO = total iron as FeO. Abbreviations as in Table 1.

rock types may be interlayered within a single outcrop on a centimeter to several-meter scale.

The disrupted nature of many localities of the Catalina blueschist unit makes assessments of proportions of rock types difficult; however, most "coherent" areas of the blueschist unit are metagraywacke/metashale sequences, and Platt's (1976) estimate of about 75 percent metagraywacke/metashale seems reasonable. The remainder of the blueschist unit consists of subequal amounts of the metabasite lithologies (greenstone, blueschist, eclogite) and ultramafic schist (herein called "melange matrix"), with about 5 percent metachert and Fe/Mn-rich quartz schist.

Subsea and subsurface recovery of Catalina Schist basement rocks indicates that the exposed Catalina Schist is a fairly representative sample of a widespread, regional metamorphic terrane which is predominantly composed of the blueschist and greenschist units. Apart from Santa Catalina Island, rocks identical to those found in the blueschist and greenschist units are common in the subsurface of the western margin of the Los Angeles Basin (Schoellhamer and Woodford, 1951; Yeats, 1973; Sorensen, 1984). Subsea dredge hauls and dart cores from Thirty Mile Bank and Sixty Mile Bank (Figure 1) yielded lithologies of the Catalina greenschist unit; a dredge haul ~16 km SSW of Santa Rosa Island contained rock types found in the blueschist and greenschist units.

Provenance studies of the San Onofre Breccia, which occurs along the coastal areas of southern California (Figure 1; Woodford, 1924; 1925; Stuart, 1976), concluded that the current exposures of Catalina Schist are representative of the source terrane of the aforementioned Miocene conglomerate. The San Onofre Breccia contains about 34 percent of greenschist unit detritus (Stuart, 1976), which suggests that despite its limited areal extent and relative thinness (~200 m) on Santa Catalina Island, on a regional scale the greenschist unit is volumetrically significant in the Catalina Schist terrane.

In the Catalina Schist terrane, then, a blueschist unit and a greenschist unit with similar protoliths but with distinct structural and metamorphic histories are juxtaposed by a low-angle fault contact in an inverted metamorphic sequence. This study employs major, minor and trace element geochemical techniques and examines the mineral chemistry of white micas and amphiboles from both metasediments and metabasites in order to compare some aspects of the geochemical and crystallochemical evolution of the two units.

MAJOR, MINOR, AND TRACE ELEMENT GEOCHEMISTRY OF METASEDIMENTS AND METABASITES

Whole-rock, major element X-ray fluorescence (XRF) analyses of metasediments and metabasites from the Catalina greenschist and blueschist units are reported in Table 2. Minor- and trace-element data were also obtained for the metabasites, using instrumental neutron activation analysis (INAA), and are given in Table 3. Analytical procedures and sample preparation for both techniques are discussed in Appendix I. Metagraywackes from the blueschist unit and grayschists from the greenschist unit

TABLE 3. WHOLE-ROCK, MAJOR, MINOR, AND TRACE ELEMENT INAA ANALYSES OF METABASITES

Sample	LHCNGL6 mdi	OH-1 gsn	716806 gsn	719805 bs	L295 ecl	79817' ecl	91180SCl eam	91180SCl⁺ eam/gs?	7188013 gs	9580SC9 ggs	9580SC5 ggs	MV-1 ggs	T203B ggs	CI**
Na, mg/g	13.0	13.6	13.2	25.6	26.0	34.6	5.90	26.0	31.8	27.9	31.7	22.4	35.0	
(Na$_2$O, wt%)	(1.8)	(1.8)	(1.8)	(3.5)	(3.5)	(4.7)	(0.8)	(3.5)	(4.3)	(3.8)	(4.3)	(3.0)	(4.7)	
K, mg/g	3.2	7.13	4.18	3.95	1.45	..	1.45	1.22	1.61	2.35	5.27	6.21	8.83	
(K$_2$O, wt%)	(0.39)	(0.86)	(0.50)	(0.47)	(0.17)	(..)	(0.27)	(0.15)	(0.29)	(0.46)	(0.63)	(0.74)	(1.06)	
Ca, mg/g	60.2	84.5	..	81.7	72.4	76.5	129	62.1	43.5	56.0	48.3	44.3	61.8	
(CaO, wt%)	(8.4)	(11.8)	..	(11.4)	(10.1)	(10.7)	(18.1)	(8.7)	(6.1)	(7.8)	(6.8)	(5.2)	(8.7)	
Mn, mg/g	0.85	1.36	1.34	1.80	2.37	1.61	1.74	1.19	1.40	1.62	1.55	1.55	1.66	
(MnO, wt%)	(0.11)	(0.17)	(0.17)	(0.23)	(0.31)	(0.21)	(0.22)	(0.15)	(0.18)	(0.21)	(0.20)	(0.30)	(0.21)	
Fe, mg/g	48	75	66	66	115	103	76.9	76.1	94.1	98	84	111	81	
(FeO*, wt%)	(6.2)	(9.7)	(8.5)	(8.5)	(14.8)	(13.3)	(9.9)	(9.8)	(12.1)	(12.6)	(10.8)	(14.3)	(10.4)	
(μg/g)														
Sc	29.9	49.5	47.5	34.4	48.5	48.0	40.7	42.6	45.4	56.2	40.7	53.7	29.9	
Cr	16.0	471	198	53.3	133	38.3	64	338	97	169	241	126	74	
Co	14.0	154	41.8	40.9	41.7	40.4	50.7	3.6	40.3	39.3	46.6	53.2	51.0	
Ni	..	..	..	..	133	..	287	..	..	..	..	220	125	
Zn	..	88.2	95.4	88.2	192	..	..	..	138	..	..	..	..	
Ga	19.7	16.9	22.6	17.0	..	..	20.5	14.1	18.8	18.6	20.0	23.0	19.2	
Rb	24.9	17.9	14.6	13.2	..	..	..	52.7	..	..	19.7	41.7	29.7	
Sr	1110	68	235	380	..	422	405	..	334	219	..	157	420	
Zr	290	..	185	..	307	..	205	225	140	..	168	808	227	
Sb	0.04	0.56	0.69	0.29	0.13	..	0.43	..	..	..	0.22	0.02	0.38	
Cs	0.39	0.89	3.87	0.81	2.26	..	..	1.9	0.51	0.96	0.62	2.80	0.83	
Ba	137	148	888	175	..	281	279	386	70	116	141	345	215	
La	4.51	1.28	9.47	5.05	9.26	2.95	5.71	4.21	2.71	4.62	3.57	6.67	11.4	0.236
Ce	9.99	5.29	23.5	8.39	27.4	7.76	16.8	11.1	9.57	11.9	11.3	17.4	25.3	0.616
Nd	..	4.85	19.9	8.97	24.0	..	14.7	13.5	10.3	11.5	11.9	11.7	18.8	0.457
Sm	2.65	2.18	5.73	2.01	7.02	2.09	3.65	3.19	3.11	4.38	3.60	5.37	4.90	0.149
Eu	0.81	0.86	1.97	0.76	2.43	1.01	1.38	1.38	1.25	1.65	1.27	1.94	2.03	0.056
Tb	0.56	0.63	1.33	0.52	1.71	0.43	0.71	0.75	0.92	1.09	0.86	1.41	0.94	0.0355
Dy	4.9	4.7	8.8	2.23	10.0	..	5.83	6.52	4.52	9.31	..	..	6.23	0.245
Yb	2.25	2.63	4.4	1.74	5.72	1.75	2.29	2.57	3.15	3.86	3.13	4.58	2.32	0.159
Lu	0.42	0.48	0.78	0.28	0.94	0.33	0.48	0.49	0.52	0.70	0.60	0.84	0.49	0.0245
Hf	2.03	1.76	3.76	0.83	5.08	1.61	2.10	2.24	2.53	3.30	3.18	4.60	3.39	
Ta	0.14	0.52	0.42	0.09	0.40	0.17	0.13	0.12	0.16	0.23	0.19	0.30	0.86	
Th	0.52	..	..	..	0.55	0.58	0.38	0.05	..	0.33	0.30	0.37	0.94	
U	0.40	..	..	..	0.32	..	0.23	..	..	0.72	0.80	1.00	0.35	

Note: Abbreviations are as in Table 1. Weight % oxides of Na, K, Ca, Mn, and Fe given in parentheses. **CI values for REE calculated from the data of Ebihara et al. (1982). ⁺Sample 91180SCl has a TiO2 content of 1.4 weight %, determined by XRF.

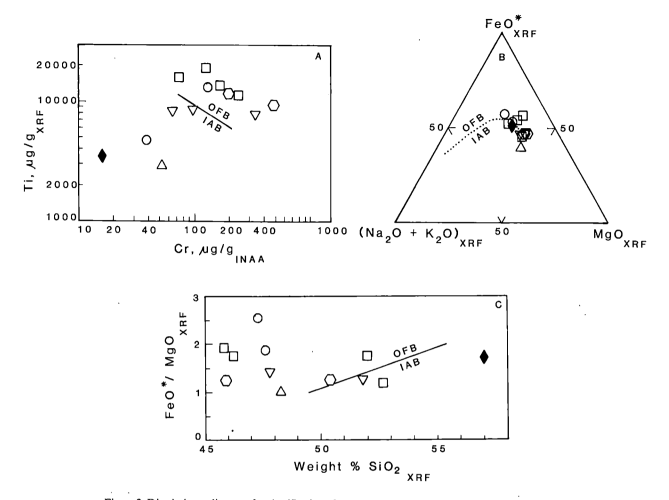

Figure 3. Discriminant diagrams for classification of ocean-floor (OFB) and island-arc (IAB) basalts. Ti versus Cr diagram (3A) from Pearce (1975), A (Na_2O + K_2O) - F (FeO*) - M(MgO) diagram (3B) from Miyashiro (1974), and FeO*/MgO versus SiO_2 diagram (3C) from Irvine and Baragar (1971). FeO* refers to total iron as ferrous iron. Plotted are whole-rock XRF (Ti, Na, K, Mg, Fe, Si) and INAA (Cr) data for metabasites from the Catalina blueschist and greenschist units. μg/g = ppm. Fields for OFB and IAB were empirically determined in the studies cited above. Symbols are as follows: for metabasites of the Catalina blueschist unit, circles = eclogites; hexagons = greenstones, upright triangle = blueschist 719805, filled diamond = metadiorite LHCNGL6, a blueschist facies clast from a metaconglomerate body; for the metabasites of the Catalina greenschist unit, inverted triangles = greenschists, squares = glaucophanic greenschists.

have similar proportions of FeO, MgO, and Al_2O_3, although grayschists are richer in Na_2O and K_2O and poorer in CaO than the blueschist unit metagraywackes (Table 2). The metagraywackes contain volcanic and plutonic clasts; feldspar grains in metagraywackes and in clasts from metaconglomerates typically exhibit lawsonite pseudomorphs after plagioclase. The higher CaO-contents of the metagraywackes relative to the grayschists may be due to a large fraction of such volcanogenic detritus. The melange matrix is relatively poor in aluminum and alkalis and rich in MgO, SiO_2 and CaO (Table 2). Mineral assemblages of the melange matrix are variable, and may include Mg-chlorite ± quartz ± talc ± serpentine (?) ± graphite ± white mica ± sphene ± albite ± calcite ± glaucophane ± fuchsite. The whole-rock com-

positions suggest that the melange matrix consists of a mixture of ultramafic (serpentinite?) and carbonaceous, quartzose sediments.

In general, metabasites from both the greenschist and blueschist units have the major, minor, and trace element characteristics of tholeiitic basalts (Tables 2, 3; Figure 3). Most analyses of these rocks plot within empirically-delineated fields for ocean-floor tholeiites on Ti versus Cr, A-F-M, and FeO* (total iron = ferrous iron) versus SiO_2 discriminant diagrams (Figure 3A, B, C; Irvine and Baragar, 1971; Miyashiro, 1974; Pearce, 1975). However, Garcia (1978) concluded that low-grade metamorphism could cause large, coherent changes in the major-element chemistry of metabasites, and render plots such as A-F-M and FeO* versus MgO invalid for classifying such rocks.

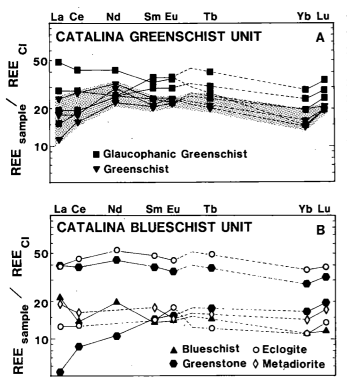

Figure 4. Rare-earth element (REE) contents of metabasites from the Catalina greenschist and blueschist units, normalized to CI (calculated from the data of Ebihara et al., 1982). All REE determined by instrumental neutron activation analysis (INAA). Symbols are indicated on the figure. The shaded portion of Figure 4A represents the range of REE in greenschist.

Three metabasites from the Catalina blueschist unit and one greenschist sample have Ti and Cr contents which plot well within the island-arc basalt field of Figure 3A. Greenschist 7188013 (Tables 1, 2, 3) has the mineral assemblage albite + chlorite + white mica + calcite + hematite + stilpnomelane + iron-poor epidote + sphene. Calcic amphibole occurs only as inclusions in the albite porphyroblasts. The mineral assemblage and the relatively Ca-poor, Fe-rich bulk chemistry of 7188013 suggest that the rock might have had a different mafic protolith than other greenschist and glaucophane greenschist samples. Sample 7188013 also plots in the IAB field in the A-F-M (Figure 3B) and FeO*/MgO versus SiO_2 (Figure 3C) diagrams. Eclogite 79817′, metadiorite conglomerate clast LHCNGL6, and glaucophane + lawsonite-bearing blueschist 719805 (Tables 1, 2, 3) all plot within the IAB field in Figure 3A. The eclogite sample exhibits a low-temperature assemblage of lawsonite + glaucophane + white mica + chlorite + stilpnomelane + sphene which partially replaces the eclogite mineral assemblage of garnet + clinopyroxene; the block from which the sample was taken has a well-developed metasomatic rind. The trace element composition of this eclogite may have changed during the reequilibration and metasomatism which resulted in rind formation. Metadiorite LHCNGL6 has textural and mineralogic evidence for a plutonic origin; its major-, minor-, and trace-element chemistry suggest

that this plutonite was calc-alkaline. Sample LHCNGL6 plots in the island-arc basalt field in both of the major-element discriminant diagrams (Figures 3B, C) as well as in the Ti versus Cr plot, has 1100 μ/g Sr, and 16 μg/g Cr. The metadiorite clast is representative of the numerous calc-alkaline metaplutonic clasts in metaconglomerates of the Catalina blueschist unit. Glaucophane + lawsonite-bearing blueschist 719805 plots within the IAB field in Figure 3A. The major-element analysis of 719805 lies within the IAB field on the A-F-M diagram (Figure 3B) and in the OFB field on the FeO*/MgO versus SiO_2 plot (Figure 3C). The relatively high Na- and Fe-contents of 719805 may reflect a basaltic andesite protolith, or could be the result of pre- or synmetamorphic metasomatism; concentrations of both Na and Fe in basalts can be significantly changed during sea-floor metamorphic processes (Humphris and Thompson, 1978; Hart et al., 1974; Marshall, 1975).

The rare-earth-element (REE) contents of the metabasites suggest that most of these rocks were derived from ocean-floor tholeiites (Table 3, Figure 4). The flat to light rare-earth-element (LREE)-depleted patterns with LREE 5-50X CI (CI chondrite data from Ebihara et al., 1982) and heavy rare-earth elements (HREE) 10-40X CI which are exhibited by the suite are similar to the REE characteristics of ocean-floor basalts (Lofgren et al., 1981). The REE patterns of the Catalina rocks also resemble those of greenschists and blueschists of the Shuksan Schist of Washington state (Dungan et al., 1983), which were interpreted as the signature of an ocean-floor basalt protolith. However, REE patterns similar to the ones obtained for the Catalina metabasites have also been observed in greenstone terranes and interpreted to result from REE mobility accompanying secondary metamorphic/metasomatic processes (Hellman et al., 1979).

The samples from the blueschist unit which appeared to have different major-element and Ti/Cr contents than the two greenstones and the relatively less altered eclogite also have slightly different REE patterns. The LREE-enriched REE pattern of metadiorite LHCHGL6, like other aspects of its chemistry, indicates its calc-alkaline nature. Altered eclogite 79817′ has a flat REE pattern with an apparent positive Eu-anomaly; blueschist 719805 has relatively high Tb/Yb and a rather erratic, but nonetheless LREE-enriched REE pattern. These samples display relatively low amounts of many of the incompatible and highly incompatible elements (Table 3), and it is possible that the REE patterns and trace-element contents of these two samples may reflect alteration.

Metabasites from the greenschist unit display fairly systematic increases of Mn, Sc, Hf, Sm, and Yb with Fe-contents (Figure 5), and of Fe, Mn, Sc, La, Sm, Yb, and Hf with Lu-contents (Figure 6). Greenschists appear to be poorer in Fe and incompatible elements than glaucophanic greenschists, as also observed for comparable lithologies of the Shuksan suite (Dungan et al., 1983). The ranges of Sc, Cr, Co, and Ni (Figures 5, 6; Table 3) in metabasites from both the blueschist and greenschist units are comparable with those obtained by Schilling et al. (1983) for ocean-floor basalts along the mid-Atlantic ridge. Sm,

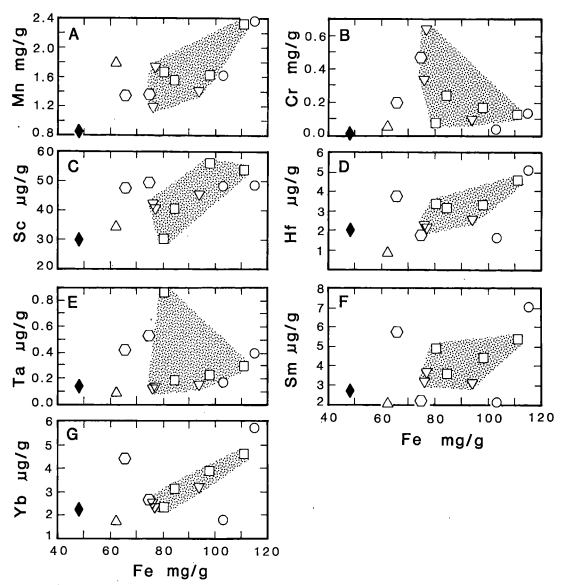

Figure 5. Mn, Cr, Sc, Hf, Ta, Sm, and Yb versus Fe for metabasites from the Catalina greenschist and blueschist units. All elements determined by INAA and reported in S.I. units. Symbols as in Figure 3. The shaded portion of Figure 5 encompasses the range of compositions of metabasites from the greenschist unit.

Yb, and Hf appear to exhibit more "coherent" behavior with Lu than does La (Figure 6E, F, G, and H), however, the magmatically-distinct (calc-alkaline) metadiorite also plots within the trends of progressively better correlations of La, Sm, Yb, and Hf with Lu. These element-enrichment trends probably reflect protolith characteristics (that is, concomitant Fe and incompatible-element enrichment resulting from magmatic fractionation of such elements) but some alteration and disturbance of elemental abundances cannot be ruled out. The major-, minor-, and trace-element contents of metabasites from the blueschist and greenschist units suggest that most of these rocks are variably fractionated and possibly variably-altered ocean-floor basalts. Calc-alkaline plutonites occur as clasts in metaconglomerates of the blueschist unit.

MINERAL CHEMISTRY OF WHITE MICA AND AMPHIBOLE FROM THE CATALINA BLUESCHIST AND GREENSCHIST UNITS

White Mica

Averaged, representative microprobe analyses of white micas in metasediments and metabasites from the blueschist and greenschist units are reported in Table 4 (see Appendix I for analytical procedure). White mica formulae are based on 11 oxygens, and all iron is assumed to be ferrous.

The white micas are celadonitic, with Si ranging from 3.1 -3.8 per 11 oxygen formula. The white micas all exhibit interlayer deficiencies; interlayer site occupancies range from 0.79 - 0.95

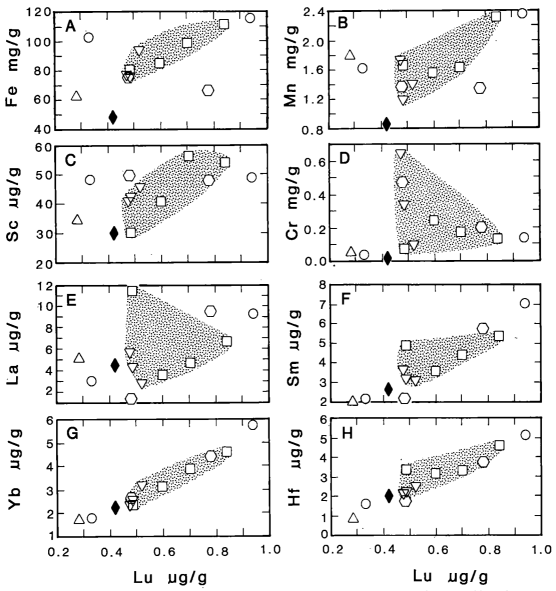

Figure 6. Fe, Mn, Sc, Cr, La, Sm, Yb, and Hf versus Lu in metabasites from the Catalina greenschist and blueschist units. All elements determined by INAA. Symbols as in Figure 3. The shaded area encompasses the range of compositions of metabasites from the greenschist unit.

(Table 4). Small amounts of BaO (0.2 - 0.3 weight percent) were detected in some white micas (as noted in Table 4), but are stoichiometrically insufficient to fill the interlayer site. Formula $(Mg + Fe^{+2})$ contents are equal to or higher than the excess of Si above 3.0, which suggests that the high Si-contents in the formula indeed reflect a celadonite substitution. The celadonite substitution $((Mg, Fe^{+2}) + Si = 2Al)$ in white micas is thought to increase with fluid pressure, and to reflect relatively high-pressure, low-temperature metamorphic conditions (Ernst, 1963; Guidotti and Sassi, 1976). Velde (1965) determined that Si-contents of 3.3 per formula unit in white mica indicated minimum pressures of 5 kb at about 500°C and 3 kb at 400°C. Massonne and Schreyer (1983) obtained results suggesting higher pressures for Si-contents of 3.3, and created isopleths of Si in the muscovite stability field

ranging from 3.1 - 3.7 in the temperature range of 300° - 700°C at pressures up to 20 kb. Their limiting assemblage contains K-feldspar and phlogopite, which are absent in the Catalina rocks. Pressure estimates based on these data are probably minimum pressures, and should be viewed as preliminary conclusions.

In general, white micas from the rocks of the blueschist unit display higher Si contents than greenschist-unit equivalents (Figure 7A, B, C). White micas which occur in the matrix of the glaucophanic greenschists do not appear to contain significantly more Si than those of epidote amphibolites and greenschists (Table 4). The minimum pressures for white micas of the greenschist unit suggested by the geobarometer of Massonne and Schreyer (1983), using an average of 3.3 Si for grayschists and

S. S. Sorensen

TABLE 4. ANALYSES OF WHITE MICAS

Sample	714806 gry	825812 gry	714801 gry	7168013 gry	99803 qst	9148011 eam	814814 eam	911801 eam/gs?	9580SC9 ggs	714802 msh	9128010 msh	716802 mgr	719805 bst	OH-1 gsn
Number of analyses	4	4	4	3	4	3	6	4	4	3	3	4	1	4
SiO_2, wt %	47.7	48.7	49.5	49.6	47.6	50.3	50.2	49.4	48.7	51.8	56.7	55.0	55.1	54.6
TiO_2	0.4	0.2	0.2	0.2	0.4	0.2	0.0	(0.06)	0.1	(0.02)	0.0	0.0	(0.01)	(0.03)
Al_2O_3	28.4	28.9	27.7	28.2	30.2	27.2	24.0	27.5	25.7	24.8	19.8	22.2	25.6	20.8
FeO	2.0	2.4	3.1	3.0	2.1	4.5	4.6	4.4	4.8	3.1	2.1	2.9	3.4	4.4
MgO	2.6	2.8	3.0	3.3	2.3	3.4	5.4	3.7	3.2	3.7	6.6	5.2	5.1	4.8
Na_2O	0.6	0.2	0.2	0.2	0.7	0.2	0.2	(0.06)	0.2	(0.08)	(0.02)	(0.03)	(0.02)	(0.04)
K_2O	9.2	10.4	10.3	9.9	9.4	10.5	10.5	10.9	10.3	9.6	9.4	10.3	9.8	11.1
	90.9	93.6	94.0*	94.4*	92.7*	96.3	94.9	95.9**	93.0	93.0	94.62	95.63	99.0	95.7
Si	3.32	3.31	3.36	3.34	3.25	3.36	3.42	3.32	3.38	3.52	3.76	3.65	3.53	3.66
Al^{IV}	0.68	0.69	0.64	0.66	0.75	0.64	0.58	0.68	0.62	0.48	0.24	0.35	0.47	0.34
Al^{VI}	1.64	1.62	1.58	1.54	1.68	1.50	1.34	1.50	1.48	1.51	1.30	1.38	1.45	1.31
Ti	0.02	0.01	0.01	0.01	0.02	0.01	0.0	0.0	0.0	0.0	0.0	0.0	0.0	0.0
Mg	0.27	0.28	0.31	0.33	0.23	0.34	0.55	0.37	0.33	0.37	0.65	0.51	0.48	0.48
Fe^{+2}	0.12	0.14	0.17	0.17	0.12	0.25	0.26	0.25	0.28	0.18	0.12	0.16	0.18	0.25
Total VI	2.05	2.05	2.07	2.05	2.05	2.10	2.15	2.12	2.09	2.06	2.07	2.05	2.11	2.04
Na	0.08	0.03	0.03	0.03	0.09	0.03	0.03	0.01	0.03	0.01	0.0	0.0	0.0	0.0
K	0.82	0.90	0.89	0.86	0.82	0.89	0.91	0.93	0.91	0.83	0.79	0.87	0.79	0.95
(Na + K)	0.90	0.93	0.92	0.89	0.91	0.92	0.94	0.94	0.94	0.84	0.79	0.87	0.79	0.95

Note: Abbreviations as in Table 1. Cations on the basis of an 11-oxygen formula, all iron ferrous. *BaO analyzed, 0.3 wt %. **BaO analyzed, 0.2 wt %.

TABLE 5. ANALYSES OF AMPHIBOLES FROM THE BLUESCHIST UNIT

Sample	9128010 msh	714802 msh	716802 mgr	LHCNGL6 mdi(c)*	LHCNGL6 mdi(r)**	OH-1 gsn	719805 bst	L21785 ecl**
Number of analyses	9	6	5	1	1	9	8	14
SiO_2, wt %	58.0	56.0	54.5	46.2	55.2	53.6	57.1	57.6
TiO_2	1.3	(0.05)	0.3	1.3	(0.03)	(0.04)	..	(0.01)
Al_2O_3	9.8	11.4	9.2	6.5	9.2	5.2	10.6	11.3
FeO	9.6	14.6	14.0	15.5	12.3	17.7	11.8	10.7
MgO	11.6	7.0	8.4	13.7	9.8	10.3	11.1	10.2
MnO	(0.09)	0.0	0.2	0.2	0.05)	0.2	0.2	(0.02)
CaO	0.5	0.0	0.4	10.4	0.8	3.2	1.2	1.0
Na_2O	7.2	7.5	7.2	1.4	6.8	5.4	6.8	7.4
K_2O	0.0	0.0	(0.03)	(0.09)	0.0	(0.08)	(0.02)	(0.01)
	98.09	96.55	94.23	95.29	94.18	95.72	98.82	98.74
Si	7.88	7.90	7.88	6.86	7.89	7.74	7.74	7.83
Al^{IV}	0.12	0.10	0.12	1.14	0.11	0.26	0.26	0.17
Al^{VI}	1.45	1.80	1.44	0.0	1.44	0.62	1.44	1.72
Ti	0.13	0.0	0.03	0.14	0.0	0.0	0.0	0.0
Fe^{+3} (est)	0.30	0.23	0.47	0.84	0.54	1.12	0.61	0.21
Mg	2.35	1.47	1.81	3.03	2.08	2.22	2.24	2.07
Fe^{+2}	0.79	1.49	1.22	1.09	0.93	1.02	0.73	1.00
Mn	0.01	0.0	0.02	0.03	0.01	0.02	0.02	0.0
Total VI	5.03	5.00	5.00	5.13	5.00	5.00	5.04	5.00
Excess VI	0.03	0.0	0.0	0.13	0.0	0.0	0.04	0.0
Ca	0.07	0.0	0.06	1.66	0.12	0.49	0.17	0.15
NaMg	1.90	2.00	1.94	0.21	1.88	1.51	1.79	1.85
Na_A	0.0	0.05	0.08	0.20	0.01	0.0	0.0	0.10
K	0.0	0.0	0.0	0.02	0.0	0.01	0.0	0.0
A Total	0.0	0.05	0.08	0.22	0.01	0.01	0.0	0.0

Note: Abbreviations are as in Table 1. *(c) = core; (r) = rim analysis. **Sample from same eclogite block as L295. Ferric iron estimate as in the text.

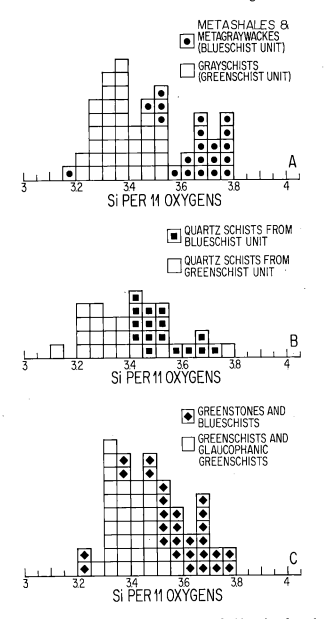

Figure 7. Si-contents, per 11 oxygen formula, of white micas from the blueschist and greenschist units. Each box represents one analysis. Figure 7A depicts white micas from clastic metasediments, 7B, white micas from quartz schists, and 7C, white micas from metabasites. Lithologies of the blueschist unit are indicated by filled symbols, and those of the greenschist unit by empty boxes.

3.4 Si for metabasites, range from about 5 kb at 300°C to about 9 kb at 450°C. The significantly higher average Si contents of white micas from the blueschist unit (3.7 Si for metagraywackes and metashales, 3.6 Si for metabasites) yield minimum-pressure estimates of about 12-15 kb at 300°C. The Si contents of white micas suggest that: 1) metamorphism of the blueschist unit occurred at relatively higher fluid pressures/lower temperatures than that of the greenschist unit, and 2) white micas in the matrix of the glaucophanic greenschists formed under lower P_{fluid}/T than those in glaucophane and crossite-bearing blueschists, green-

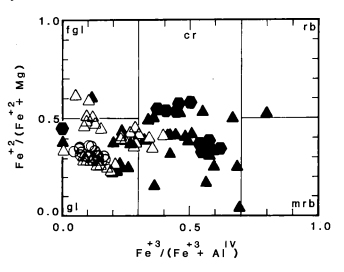

Figure 8. Compositions of sodic amphiboles from metasediments and metabasites of the blueschist unit. Abbreviations are: gl = glaucophane, gfl = ferroglaucophane, cr = crossite, rb = riebeckite, mrb = magnesiorie-beckite. Symbols are as follows: open triangles = metagraywackes and metashales, filled triangles = blueschists, filled hexagons = greenstones, open circles = eclogites.

stones, and eclogites in the blueschist unit, but at fluid pressures comparable to those of greenschists and epidote amphibolites in the greenschist unit.

Amphibole

Sodic Amphiboles From the Blueschist Unit. Representative analyses of amphiboles from greenstones, blueschists, eclogite L21785, metagraywackes, and metashales are reported in Table 5. All amphiboles were normalized to 23 oxygens, and ferric iron estimates were made using the method of Papike et al. (1974). The maximum ferric iron estimate which can solve the charge balance is reported as the formula, and is plotted. The International Mineralogical Association nomenclature (Leake, 1978) is used throughout the discussion. Analytical procedures are described in Appendix 1.

Amphiboles from the blueschist unit are typically sodic, although some greenstones may display sodic-calcic or calcic amphiboles. Ca-, Al-, and Ti-rich relict igneous amphiboles, mantled by sodic amphiboles, are found in metaplutonic clasts from metaconglomerates of the blueschist unit (Tables 1, 5). Ca contents of sodic amphiboles from metashale and metagraywacke resemble those of amphiboles from blueschist and eclogite, and are in general less than 0.18 Ca per 23 oxygens. Amphiboles from metasediments contain little iron, whereas sodic amphiboles from greenstones and blueschists are crossitic (Figure 8). The higher iron contents of the metabasite amphiboles are in part a function of higher whole-rock iron contents, although glaucophanes from eclogite L21785 are compositionally similar to the metasediment glaucophanes.

S. S. Sorensen

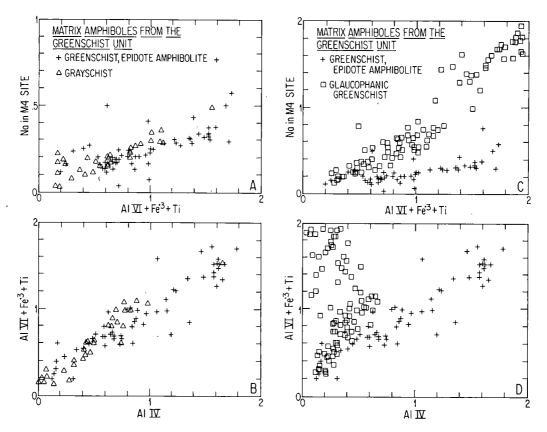

Figure 9. Composition of matrix amphiboles from greenschists and epidote amphibolites (+), from grayschists (triangles) and from glaucopanic greenschists (squares). Figures 9A, B compare metasediments and metabasites lacking sodic amphibole, whereas Figures 9C, D contrast metabasites with sodic amphiboles and those with calcic amphiboles. Figures 9A, C depict NaM4 versus the sum of octahedral Al, Fe^{+3}, and Ti; Figures 9B, D illustrate the sum of octahedral Al, Fe^{+3}, and Ti versus tetrahedral Al.

Calcic Amphiboles from the Greenschist Unit. Calcic amphiboles from grayschists, greenschists, and epidote amphibolites are similar in composition. They range from tremolites to magnesiohornblendes with moderate Ti and estimated Fe^{+3} contents. Amphiboles from the metabasites may have higher Al, Fe^{+3}, and Ti-contents than those from metasediments (Figure 9A, B). Hornblendic amphiboles in both grayschists and metabasites may exhibit actinolitic rims which are richer in Si and poorer in Al and Na than grain cores (Table 5). Such discontinuous zoning implies that hornblende-bearing rocks, which originally were epidote amphibolites, partially reequilibrated under greenschist facies conditions.

Sodic, Sodic-Calcic, and Calcic Amphiboles from the Glaucophanic Greenschists. Glaucophanic greenschists typically carry both sodic and sodic-calcic or calcic amphiboles. Textural occurrences of two amphibole assemblages suggest that sodic amphibole formed before sodic-calcic or calcic amphibole. In most of the two amphibole-bearing glaucophanic greenschists, sodic amphiboles in the matrix are discontinuously zoned to rims which are compositionally indistinguishable from nearby, slightly zoned or unzoned, sodic-calcic or calcic amphiboles. Sodic am-

phiboles may occur exclusively as inclusions in epidote porphyroblasts which are set in a matrix of calcic amphibole (see analyses of such an inclusion/matrix relationship in sample 948010, Table 5).

The mineral chemistry of calcic amphiboles which occur in glaucophanic greenschists is distinct from that of calcic amphiboles which occur in greenschists or epidote amphibolites. The former display high NaM4 at a given $(Al^{VI} + Fe^{+3} + Ti)$ and higher $(Al^{VI} + Fe^{+3} + Ti)$ at $Al^{IV} < 1$ than the latter (Figure 9 C, D). The nearly 1:1 correlation of NaM4 and $(Al^{VI} + Fe^{+3} + Ti)$ for matrix amphiboles from glaucophanic greenschists suggests a glaucophanic substitution (Figure 9 C); the nearly 1:1 correlation of $(Al^{VI} + Fe^{+3} + Ti)$ with Al^{IV} for matrix amphiboles from greenschists and epidote amphibolites (Figure 9D) suggests tschermakitic substitutions characteristic of metamorphic hornblendes (Laird and Albee, 1981 a; b).

Most matrix amphiboles from greenschists and epidote amphibolites have NaM4 < 0.4, whereas those from glaucophanic greenschists have a bimodal distribution of sodic-calcic and calcic amphiboles with NaM4 < 0.8 and sodic amphiboles with NaM4 > 1.2 (Figure 10 A, B). Amphiboles which occur as inclusions in

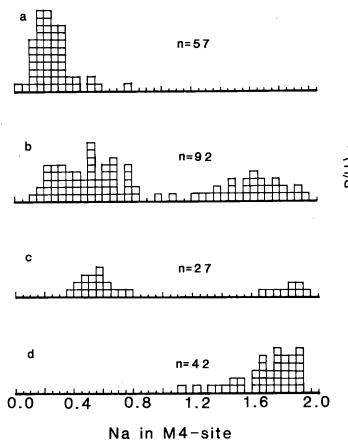

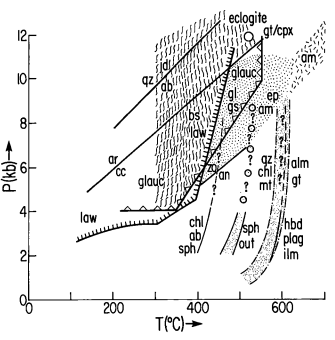

Figure 11. Multivariant phase equilibria and estimated P-T conditions of metamorphism of the Catalina blueschist and greenschist units. References to the experimental studies which produced these data are found in the text. The point labelled "eclogite gt/cpx" is a pressure and temperature estimate for one sample of eclogite from the blueschist unit, as described in the text. Abbreviations are as follows: gt = garnet, cpx = clinopyroxene, jd = jadeite, qz = quartz, ab = albite, ar = aragonite, cc = calcite, glauc = glaucophane, law = lawsonite, zo = zoisite, an = anorthite, chl = chlorite, sph = sphene, mt = magnetite, alm = almandine, hbd = hornblende, plag = plagioclase, ilm = ilmenite, bs = blueschist unit, gl gs = glaucophanic greenschist of the greenschist unit, am = amphibolite unit. Shaded fields show P-T estimates for the blueschist, greenschist, and amphibolite units.

Na in M4-site

Figure 10. Histograms depicting the NaM4 contents of amphiboles from the Catalina greenschist unit metabasites. Figure 10A represents matrix amphiboles from greenschists and epidote amphibolites, 10B shows matrix amphiboles from glaucophanic greenschists, 10C illustrates amphiboles from glaucophanic greenschist which occur as inclusions in albite, and 10D depicts amphiboles from glaucophanic greenschists which occur as inclusions in epidote.

albite also have a bimodal distribution in NaM4, but the calcic amphiboles have NaM4 ranging from 0.4 - 0.8, and the sodic amphiboles have NaM4 > 1.6 (Figure 10C). Amphiboles which occur as inclusions in epidote are richest in NaM4 (Figure 10 A). Matrix amphiboles with calcic compositions probably reequilibrated in the greenschist facies; the aluminous sodic and sodic-calcic amphiboles included in albites did not become as zoned to relatively Al-poor calcic compositions as the former. The sodic amphiboles in glaucophanic greenschists apparently were armored by epidote prior to crystallization of sodic-calcic or calcic amphiboles and, like the lawsonite inclusions in one epidote porphyroblast of sample 9580SC9, are relicts of early blueschist facies metamorphism.

Glaucophanic greenschists evidently developed from a crossite- and epidote-bearing assemblage similar to that of Franciscan "high-grade" blueschists such as those described by Coleman and Lanphere (1971) or from glaucophane-bearing blueschists. Crossites from the glaucophanic greenschists, however, are less aluminous than those from metabasites of the blueschist

unit. CaM4 contents of crossites which occur as inclusions in epidotes are higher than those of crossites from greenstone, blueschist, and eclogite. The latter also contain more Al^{VI} and NaM4 (Table 5). These differences probably result from higher P/T metamorphism of the crossite or glaucophane + lawsonite-bearing rocks of the blueschist unit relative to the crossite or glaucophane + epidote-bearing assemblages of the glaucophanic greenschists, as is also suggested by the high celadonite contents of white micas from the blueschist unit.

The mineral chemistry of amphiboles from metasediments and metabasites of the Catalina greenschist unit suggests that both glaucophanic greenschist and epidote amphibolite mineral assemblages have a greenschist facies overprint.

MINERAL ASSEMBLAGES AND THE RELATIVE P-T CONDITIONS OF METAMORPHISM OF THE CATALINA BLUESCHIST AND GREENSCHIST UNITS

Figure 11 is a compilation of multivariant, experimentally-

TABLE 6. ANALYSES OF AMPHIBOLES FROM THE GREENSCHIST UNIT

Sample	815812 gry	714801 gry(c)*	714801 gry(r)*	7198012 eam	814814 eam(c)*	814814 eam(r)*	911801 eam/gs?	91180SC1 gs	9580SC9 ggs M**	9580SC9 ggsEI**	9580SC9 ggsAI**	948010 ggsM**	948010 ggsEI**	948010 ggsAI**
Number of analyses	6	1	1	8	4	3	3	7	6	3	2	10	13	4
SiO_2, wt %	51.5	51.5	53.5	43.8	47.0	51.8	44.9	51.0	52.1	56.1	52.2	51.4	54.4	48.4
TiO_2	0.1	0.1	..	0.8	0.4	0.2	0.6	(0.08)	0.1	(0.04)	(0.05)	(0.05)	(0.09)	0.1
Al_2O_3	5.8	5.7	0.8	13.3	12.0	7.4	11.6	5.8	4.6	6.6	3.3	3.0	6.7	6.3
FeO	15.4	13.7	16.2	18.4	12.0	11.1	16.4	11.5	18.9	18.9	19.2	18.4	17.9	19.3
MgO	13.9	14.3	14.1	9.2	13.0	15.9	11.4	16.0	11.7	8.7	12.0	12.2	8.5	9.9
MnO	0.2	0.3	0.3	0.2	0.5	0.5	0.2	(0.09)	0.3	0.2	0.3	0.3	0.1	0.2
CaO	11.1	10.7	11.7	10.7	11.2	11.8	11.3	11.8	8.8	2.1	9.1	10.1	2.2	8.5
Na_2O	1.0	1.3	0.3	2.1	1.8	1.0	2.3	1.2	2.4	6.8	2.3	1.4	6.3	3.2
K_2O	0.3	0.2	0.1	0.3	0.3	0.2	0.6	0.2	0.2	(0.06)	0.2	0.2	(0.04)	0.3
	99.3	97.4	97.0	98.8	98.2	99.9	99.3	97.67	99.1	99.5	98.65	97.05	96.23	96.2
Si	7.30	7.31	7.85	6.37	6.69	7.16	6.49	7.25	7.44	7.85	7.51	7.53	7.85	7.24
Al^{IV}	0.70	0.69	0.15	1.63	1.31	0.84	1.51	0.75	0.56	0.15	0.49	0.47	0.15	0.76
Al^{VI}	0.27	0.27	0.0	0.65	0.71	0.36	0.47	0.22	0.21	0.94	0.07	0.05	0.99	0.35
Ti	0.01	0.01	0.0	0.09	0.04	0.02	0.07	0.01	0.01	0.0	0.0	0.0	0.01	0.01
Fe^{+3}(est)[†]	0.55	0.72	0.15	0.81	0.54	0.64	0.65	0.56	0.94	0.71	0.92	0.76	0.69	0.69
Mg	2.93	3.05	3.09	1.99	2.76	3.27	2.46	3.39	2.49	1.82	2.57	2.67	1.83	2.21
Fe^{+2}	1.28	0.92	1.84	1.43	0.89	0.64	1.33	0.82	1.31	1.51	1.39	1.49	1.47	1.72
Mn	0.03	0.03	0.04	0.02	0.06	0.06	0.02	0.01	0.04	0.02	0.04	0.04	0.01	0.02
Excess IV	0.07	0.0	0.12	0.0	0.0	0.0	0.0	0.0	0.0	0.0	0.0	0.01	0.0	0.0
Ca	1.69	1.64	1.64	1.67	1.71	1.75	1.75	1.80	1.35	0.31	1.40	1.59	0.34	1.36
Na_{M4}	0.24	0.36	0.04	0.33	0.29	0.25	0.25	0.20	0.65	1.68	0.60	0.40	1.66	0.64
Na_A	0.04	0.0	0.03	0.26	0.21	0.02	0.40	0.13	0.01	0.17	0.05	0.0	0.10	0.29
K	0.05	0.03	0.01	0.06	0.05	0.03	0.11	0.14	0.04	0.01	0.04	0.04	0.01	0.06

Note: Abbreviations as in Table 1. *(c) = core, (r) = rim. **M = matrix, EI = occur as inclusions in epidote, AI = occur as inclusions in albite, [†] = ferric iron estimate made by procedure described in text.

determined phase equilibria for the metamorphism of the Catalina blueschist and greenschist units, including phase relations for the greenschist to amphibolite facies transition in tholeiites (Moody et al., 1983), the lawsonite stability field (Liou, 1971), the glaucophane stability field of Maresch (1977), the aragonite → calcite transformation (Boettcher and Wyllie, 1968), the reaction jadeite + quartz = albite (Newton and Smith, 1967) and the reaction quartz + Fe-chlorite + magnetite = almandine garnet (Hsu, 1968). Epidote amphibolites and almandine garnet-bearing grayschists from the upper part of the Catalina greenschist unit contain rutile or ilmenite mantled by sphene. In a metamorphosed tholeiitic basalt bulk composition, sphene breaks down to form ilmenite at temperatures >450°C, pressures between 1 and 4 kb, and oxygen fugacities defined by the iron-magnetite, nickel-nickel oxide, and hematite-magnetite buffers (Moody et al., 1983). Ilmenite mantled by sphene occurs in an epidote amphibolite sample which also contains almandine-rich garnet partially replaced by low-iron epidote and chlorite. Meta-basites with almandine-rich garnet and with ilmenite or rutile rimmed with sphene probably initially recrystallized at temperatures between 450° - 575°C, then partially reequilibrated within the sphene stability field.

Mineral assemblages of glaucophanic greenschists represent low-temperature, high-pressure greenschist facies metamorphic conditions. The matrix amphiboles in these metabasites are sodic actinolites, barroisites and crossites. Ernst (1979) proposed that a barroisitic amphibole stability field would range from temperatures of about 350°C at 4-5 kb, to 450°C at 5-7 kb. The absence of lawsonite and the presence of pistacitic epidote + crossitic amphibole in the glaucophanic greenschists are evidence for

temperatures of 350° - 450°C at pressures of 4-7 kb (Figure 11; Brown, 1974; Brown and Ghent, 1983). Glaucophanic greenschists contain calcite, which is stable at pressures lower than 10 kb at 400°C, and sphene, which is stable at temperatures less than 450°C. The mineral assemblages of the greenschist unit probably formed at pressures of 4-9 kb, and temperatures ranging from 350° - 450°C, an estimate compatible with the mineral chemistry of the white micas.

Rare omphacitic clinopyroxene + garnet-bearing eclogite and garnet + epidote-bearing blueschist exotic blocks, apparently from the blueschist unit melange, contain higher-temperature mineral assemblages than metasediments, low-grade blueschists, or greenstones. The exotic blocks exhibit retrograde blueschist facies mineral assemblages with lawsonite + glaucophane + cela-donitic white mica + chlorite ± stilpnomelane. Temperatures of about 510° - 530°C at pressures of 6-14 kb for the last equilibration of garnet and pyroxene from an eclogite with a blueschist overprint are estimated using the garnet + clinopyroxene geothermometer of Ellis and Green (1979). The mole fraction of jadeite in the clinopyroxene is about 48 percent, yielding a minimum-pressure estimate of approximately 12 kb, using the graphical solution of Gasparik and Lindsley (1980) for the effect of diopside-jadeite solid solutions on the jadeite + quartz-forming reaction.

The ubiquitous occurrence of the mineral assemblage glaucophane + lawsonite, the presence of jadeite forming at the expense of albite in some metagraywackes and metashales, and the paucity of chlorite in metagraywackes are consistent with temperatures between 200 - 400°C for crystallization of this blueschist facies assemblage. However, somewhat higher minimum temperatures of ~300°C seem more in keeping with: 1) the suggestion that glaucophane + lawsonite + jadeite-bearing metagraywackes represent relatively high temperatures as well as pressures of recrystallization (Ernst, 1965; 1971), 2) the absence of zeolite facies assemblages, as noted by Platt (1976, p. 38), and 3) the resemblance of many lithologies to the type III Franciscan metabasalts of Ward Creek, which yielded oxygen isotope temperatures of 270 - 315°C (Taylor and Coleman, 1968). The celadonite contents of the white micas suggest minimum pressures of 12-15 kb at 300°C. The lower minimum is probably more reasonable in light of the textures displayed by the occurrences; assemblages lacking jadeite could have formed at pressures as low as the minimum pressures for glaucophane + lawsonite-bearing assemblages (Figure 11).

DISCUSSION

The blueschist and greenschist units of the Catalina Schist terrane represent relatively high P/T metamorphism of a protolith consisting mostly of graywacke and ocean-floor basalts. The units exhibit distinct structural and metamorphic histories.

The greenschist unit initially equilibrated in a temperature gradient which, at apparently constant pressure, formed assemblages ranging from epidote amphibolite to blueschist. These relatively early assemblages are overprinted by greenschist facies assemblages. If the latter represent a single greenschist "event," and the present exposures of the unit represent excavation of a coherent slab, the greenschist assemblages could be the result of partial thermal equilibration at depth; alternatively, the lower P/T assemblages may have developed during uplift.

The blueschist unit exhibits no evidence of retrograde greenschist facies metamorphism; high-grade, exotic eclogites and blueschists display relatively lower-temperature lawsonite-bearing assemblages replacing the calcium-rich phases garnet or epidote. The systematically higher celadonite contents of white micas

from the blueschist unit suggest metamorphism occurred at higher pressures than the overlying greenschist unit.

The differences in mineral assemblages, mineral chemistry, and metamorphic history between these tectonic units suggest that the "inverted metamorphism" of the Catalina Schist terrane may represent a late-stage juxtaposition of rocks formed in widely separated regions of a subduction zone. The metamorphic environments which produced the Catalina blueschist and greenschist units are probably comparable to those which produced the Franciscan Complex of the California Coast Ranges (blueschist unit) and the Sambagawa Schist terrane of Japan (greenschist unit) as described by Ernst and Seki (1967) and Ernst et al. (1970).

ACKNOWLEDGMENTS

This research was financially supported by NSF grant EAR 83-12702/Ernst, as well as by a Penrose grant from the Geological Society of America and a National Science Foundation Graduate Fellowship awarded to the author.

Some of the samples used in this study were provided by J. E. Schoellhamer, J. G. Vedder, and R. F. Yerkes of the United States Geological Survey. The late A. K. Baird of Pomona College generously devoted his time and XRF equipment for the whole-rock, major-element analyses. Neutron-activation experiments and data reduction were carried out in the laboratory of J. T. Wasson, with the able assistance of G. W. Kallemeyn, F. T. Kyte, and J. N. Grossman. R. E. Jones helped obtain the microprobe mineral analyses.

M. D. Barton, W. G. Ernst, J. G. Liou, and W. M. Thomas read an early version of the manuscript, and returned thorough reviews with excellent suggestions for improvement. B. W. Evans, J. Laird, and D. Moore are thanked for their painstaking and constructive reviews of the submitted manuscript.

APPENDIX 1: ANALYTICAL PROCEDURES

X-RAY FLUORESCENCE (XRF) AND INSTRUMENTAL NEUTRON ACTIVATION ANALYSIS (INAA). Whole-rock samples for XRF were broken into chips and powdered in a tungsten carbide disc mill. Rock chips used for INAA were abraded with silicon carbide paper, washed with spectroscopic grade acetone, and pulverized to flour in a stainless steel percussion mortar.

Splits of powders for XRF were fused with $Li_2B_4O_7$ at ~1050°C for 20 minutes, using the method of Welday et al. (1964), then reground in pica (ball) mills and pressed into pellets with a cellulose backing. Pomona College rock standards (Welday et al., 1964) were fused, reground, and pressed along with samples.

The elements Si, K, Ca, Ti, and Fe were determined using an energy dispersive system (EDAX): Na, Mg, and Al were counted using a crystal spectrometer with a gas proportional counter. Both types of equipment were used at Pomona College, Claremont, California. The background-corrected data were reduced using a working curve method of standard counts versus standard concentrations. Almost all unknown

counts were well within the calibration range, but exceptions are noted in the tables. Analyses are believed to be ±2 percent for Ca, Fe, Mg, Na, and Al, and ±5 percent for Si, Ti, and K.

One hundred milligrams of rock flour were weighed into polyvials for INAA. These were irradiated at the U.C.L.A. Engineering Nuclear Reactor, which has a neutron flux of 2×10^{12} neutrons cm^{-2} sec^{-1}. Three detectors were used. Two are Ge (Li) gamma-ray detectors; one has <1.85 keV resolution at 1.33 MeV, with 17 percent efficiency, and the other has <1.75 keV resolution at 1.33 MeV, with 21 percent efficiency. The third is a 31 percent efficiency, high-purity germanium detector with <1.70 resolution at 1.33 MeV. Mixed element standards were made from known concentration solutions evaporated on high purity MgO or SiO_2 powders. Four different counts of sample activity were made, beginning a few hours after irradiation and ending after four weeks. The data were reduced using a version of the SPECTRA gamma-ray analysis program of Baedecker (1976). Sample concentrations of most elements are ±5-10 percent.

ELECTRON PROBE MICROANALYSIS AND
CALCULATION OF MINERAL FORMULAE

Mineral analyses were made using the U.C.L.A. ARL-EMX micro-probe with automated crystal spectrometers. Accelerating voltage was 15 kV, and sample currents were 15 nAmps for plagioclase and white micas, and 18 nAmps for all other minerals. A minimum spot was always used. The standards are a variety of analyzed minerals as well as synthetic silicate glasses. Control samples were analyzed at least once every two hours.

The analyses were corrected for background and time. Inter-element and matrix effects were corrected using the method and alpha factors of Albee and Ray (1970). Major element analyses are reproducible to within 2 percent.

REFERENCES CITED

Albee, A. L., and Ray, L., 1970, Correction factors for electron probe microanalysis of silicates, oxides, carbonates, phosphates, and sulfates: Analytical Chemistry, v. 42, p. 1408–1414.

Baedecker, P. A., 1976, SPECTRA: Computer reduction of gamma-ray spectroscopic data for neutron activation analysis: in Taylor, R. E., ed., Advances in Obsidian Glass Studies, Archaeological and Geochemical Perspectives, Noyes Press, New Jersey, p. 334–349.

Boettcher, A. L., and Wyllie, P. J., 1968, The calcite-aragonite transition measured in the system $CaO-CO_2-H_2O$: Journal of Petrology, v. 76, p. 314–330.

Brown, E. H., 1974, Comparison of the mineralogy and phase relations of blueschists from the North Cascades, Washington, and greenschists from Otago, New Zealand: Geological Society of America Bulletin, v. 85, p. 333–344.

Brown, E. H., and Ghent, E. D., 1983, Mineralogy and phase relations in the blueschist facies of the Black Butte and Ball Rock areas, northern California Coast Ranges: American Mineralogist, v. 68, p. 365–372.

Coleman, R. G., and Lanphere, M. A., 1971, Distribution and age of high-grade blueschists, associated eclogites, and amphibolites from Oregon and California: Geological Society of America Bulletin, v. 82, p. 2397–2411.

Dungan, M. A., Vance, J. A., and Blanchard, D. P., 1983, Geochemistry of the Shuksan greenschists and blueschists, North Cascades, Washington: variably fractionated and altered metabasalts of oceanic affinity: Contributions to Mineralogy and Petrology, v. 82, p. 131–146.

Ebihara, M., Wolf, R., and Anders, E., 1982, Are Cl chondrites chemically fractionated? A trace element study: Geochimica et Cosmochimica Acta, v. 46, p. 1849–1862.

Ellis, D. J., and Green, D. H., 1979, An experimental study of the effect of Ca upon garnet-clinopyroxene Fe-Mg exchange equilibria: Contributions to Mineralogy and Petrology, v. 71, p. 13–22.

Ernst, W. G., 1963, Significance of phengitic micas from low-grade schists: American Mineralogist, v. 48, p. 1357–1373.

Ernst, W. G., 1965, Mineral parageneses in Franciscan metamorphic rocks, Panoche Pass, California: Geological Society of America Bulletin, v. 76, p. 879–914.

Ernst, W. G., 1971, Petrologic reconnaissance of Franciscan metagraywackes from the Diablo Range, central California Coast Ranges: Journal of Petrology, v. 12, p. 413–437.

Ernst, W. G., 1979, Coexisting sodic and calcic amphiboles from high-pressure metamorphic belts and the stability of barroisitic amphibole: Mineralogical Magazine, v. 43, p. 269–278.

Ernst, W. G., and Seki, Y., 1967, Petrologic comparison of the Franciscan and Sanbagawa metamorphic terranes: Tectonophysics, v. 4, p. 463–478.

Ernst, W. G., Seki, Y., Onuki, H., and Gilbert, M. C., 1970, Comparative study of low-grade metamorphism in the California Coast Ranges and the Outer Metamorphic Belt of Japan: Geological Society of America Memoir 124, 276 pp.

Garcia, M. O., 1978, Criteria for the identification of ancient volcanic arcs: Earth-Science Reviews, v. 14, p. 147–165.

Gasparik, T., and Lindsley, D. H., 1980, Phase equilibria at high pressure of pyroxenes containing monovalent and trivalent ions: in Prewitt, C. T., ed., Pyroxenes, Reviews in Mineralogy, v. 7, p. 309–336.

Guidotti, C. V., and Sassi, F. P., 1976, Muscovite as a petrogenetic indicator mineral in pelitic schists: Neues Jahrbuch fur Mineralogie Abhandlungen, v. 127, no. 2, p. 97–142.

Hart, S. R., Erlank, A. J., and Kable, E.J.D., 1974, Sea floor basalt alteration: some chemical and Sr isotopic effects: Contributions to Mineralogy and Petrology, v. 44, p. 219–240.

Hellman, P. L., Smith, R. E., and Henderson, P., 1979, The mobility of the rare earth elements: evidence and implications from selected terranes affected by burial metamorphism: Contributions to Mineralogy and Petrology, v. 71, p. 23–44.

Hsu, L. C., 1968, Selected phase relationships in the system Al-Mn-Fe-Si-O-H: a model for garnet equilibria: Journal of Petrology, v. 9, no. 1, p. 40–83.

Humphris, S. E., and Thompson, G., 1978, Hydrothermal alteration of oceanic basalts by seawater: Geochimica et Cosmochimica Acta, v. 42, p. 107–125.

Irvine, T. N., and Barager, W.R.A., 1971, A guide to the chemical classification of the common volcanic rocks: Canadian Journal of Earth Sciences, v. 8, p. 523–548.

Jacobson, C. E., and Sorensen, S. S., 1986, Metamorphic history and amphibole compositions of the Rand Schist and the greenschist unit of the Catalina Schist: two relatively high P/T metamorphic terranes in southern California: in press, Contributions to Mineralogy and Petrology.

Laird, J., and Albee, A. L., 1981a, High-pressure metamorphism in mafic schists from northern Vermont: American Journal of Science, v. 281, p. 97–126.

Laird, J., and Albee, A. L., 1981b, Pressure, temperature, and time indicators in mafic schist: their application to reconstructing the polymetamorphic history of Vermont: American Journal of Science, v. 281, p. 127–175.

Leake, B. E., 1978, Nomenclature of amphiboles: American Mineralogist, v. 63, p. 1023–1052.

Liou, J. G., 1971, P-T stabilities of laumonite, wairakite, lawsonite, and related minerals in the system $CaAl_2Si_2O_8-SiO_2-H_2O$: Journal of Petrology, v. 12, p. 379–441.

Lofgren, G. E., Bence, A. E., Duke, M. B., Dungan, M. A., Green, J. C., Haggerty, S. E., Haskin, L. A., Irving, A. J., Lipman, P. W., Naldrett, A. J., Papike, J. J., Reid, A. M., Rhodes, J. M., Taylor, S. R., and Vaniman, D. T., 1981, Petrology and chemistry of terrestrial, lunar, and meteoritic basalts: in Kaula, W., ed., Basaltic volcanism of the terrestrial planets, Basaltic Volcanism Study Project, Pergamon Press, p. 132–157.

Maresch, W. V., 1977, Experimental studies on glaucophane: an analysis of present knowledge: Tectonophysics, v. 43, p. 109–125.

Marshall, M. C., 1975, Petrology and chemical composition of basaltic rocks recovered on Leg 32, Deep Sea Drilling Project: in Larson, R. L., et al., eds., Initial Reports of the Deep Sea Drilling Project, v. 32, p. 563–570.

Massonne, H.-J., and Schreyer, W., 1983, A new experimental phengite barometer and its application to a Variscan subduction zone at the southern margin of the Rhenohercynicum: Terra Cognita, v. 3, nos. 2-3, p. 187.

Miyashiro, A., 1974, Volcanic rock series in island arcs and active continental margins: American Journal of Science, v. 274, p. 321–355.

Moody, J. B., Meyer, D., and Jenkins, J. E., 1983, Experimental characterization of the greenschist/amphibolite boundary in mafic systems: American Journal of Science, v. 283, p. 48–92.

Newton, R. C., and Smith, J. V., 1967, Investigations concerning the breakdown of albite at depth in the earth: Journal of Geology, v. 75, p. 268–286.

Papike, J. J., Cameron, K. L., and Baldwin, K., 1974, Amphiboles and pyroxenes, characteristics of *other* than quadrilateral components and estimates of ferric iron from microprobe data: Geological Society of America Abstracts With Programs, v. 6, no. 7, p. 1053–1054.

Pearce, J. A., 1975, Basalt geochemistry used to investigate past tectonic environment on Cyprus: Tectonophysics, v. 25, p. 41–67.

Platt, J. P., 1975, Metamorphic and deformational processes in the Franciscan Complex, California: some insights from the Catalina Schist Terrain: Geological Society of America Bulletin, v. 86, p. 1337–1347.

Platt, J. P., 1976, The petrology, structure, and geologic history of the Catalina Schist terrain, southern California: University of California Publications in the Geological Sciences, v. 112, p. 1–111.

Schilling, J.-G., Zajac, M., Evans, R., Johnston, T., White, W., Devine, J. D., and Kingsley, R., 1983, Petrologic and geochemical variations along the mid-Atlantic ridge from 29°N to 73°N: American Journal of Science, v. 283, p. 510–586.

Schoellhamer, J. E., and Woodford, A. O., 1951, The floor of the Los Angeles Basin, Los Angeles, Orange, and San Bernardino Counties, California: United States Geological Survey Oil and Gas Investigations Map OM-117, 2 pp.

Sorensen, S. S., 1984, Petrology of basement rocks of the California Continental Borderland and the Los Angeles Basin: Unpublished Ph.D. dissertation, University of California, Los Angeles, 423 pp.

Stuart, C. J., 1976, Source terrane of the San Onofre Breccia — preliminary notes: in Howell, D. G., ed., Aspects of the geologic history of the California Continental Borderland, Pacific Section, American Association of Petroleum Geologists Miscellaneous Publication 24, p. 309–325.

Suppe, J., and Armstrong, R. L., 1972, Potassium-argon dating of Franciscan metamorphic rocks: American Journal of Science, v. 272, p. 217–233.

Taylor, H. P., and Coleman, R. G., 1968, O^{18}/O^{16} ratios of coexisting minerals in glaucophane-bearing metamorphic rocks: Geological Society of America Bulletin, v. 79, p. 1727–1756.

Velde, B., 1965, Phengite micas: synthesis, stability, and natural occurrence: American Journal of Science, v. 263, no. 10, p. 886–913.

Welday, E. E., Baird, A. K., McIntyre, D. B., and Madlem, K. W., 1964, Silicate sample preparation for light-element analyses by X-ray spectroscopy: American Mineralogist, v. 49, p. 889–903.

Woodford, A. O., 1924, The Catalina metamorphic facies of the Franciscan Series: University of California Publications in the Geological Sciences, v. 15, p. 49–68.

Woodford, A. O., 1925, The San Onofre Breccia: its nature and origin: University of California Publications in the Geological Sciences, v. 15, p. 159–280.

Yeats, R. S., 1973, Newport-Inglewood fault zone, Los Angeles Basin, California: American Association of Petroleum Geologists Bulletin, v. 57, p. 117–135.

MANUSCRIPT ACCEPTED BY THE SOCIETY JULY 29, 1985

Geological Society of America
Memoir 164
1986

Blueschists in the Franciscan Complex of California: Petrotectonic constraints on uplift mechanisms

Mark Cloos
Department of Geological Sciences
and
Institute for Geophysics
University of Texas at Austin
Austin, Texas 78713

ABSTRACT

High-pressure/low-temperature metamorphic rocks in the Franciscan Complex of western California are primarily found as small blocks in mud-matrix melange and as extensive coherent, bedded sheets or slabs. The highest-pressure rocks, the high-grade sodic and/or calcic amphibole-epidote-garnet-omphacitic pyroxene-bearing blueschists, amphibolites, or eclogites, are found as blocks (typically meters to a few tens of meters across) in the melanges. Low-grade schistose blueschists are found as small blocks in melanges and in extensive coherent belts. Many Franciscan greenstones and metagray-wackes in both the melanges and coherent tracts locally contain sodic pyroxene + quartz, lawsonite, and/or aragonite but are neither blue nor strongly schistose.

Petrotectonic constraints on models for uplift and preservation of Franciscan high-pressure/low-temperature metamorphic rocks include: (1) high-grade blueschist and eclogite blocks found in mud-matrix melange terranes underwent cooling under high-pressure conditions and were once immersed in serpentinite; whereas lower-temperature/high-pressure rocks found both as blocks in melanges and extensive coherent tracts are typically only incipiently recrystallized and show no evidence of former immersion in serpentinite; (2) extensive tracts of schistose blueschists are largely in fault contact with the base of the overriding plate in northern California, whereas both the high-grade blocks and the jadeitic pyroxene + quartz-bearing coherent units in northern California are not juxtaposed against the overriding plate; (3) uplift of both blueschist blocks in melange and extensive coherent tracts typically occurred without thoroughly penetrative strain; (4) high-grade blocks in mud-matrix melanges must have been displaced oceanward and upwards from beneath the base of the overriding plate; (5) synsubduction uplift of Franciscan blueschists to depths less than 10 kilometers or so is indicated by the lack of retrograde greenschist facies alterations, the widespread preservation of aragonite, and the present exposure of blueschists where subduction continues off northernmost California and southern Oregon; and (6) the scarcity of high-pressure metamorphic detritus in Franciscan sediments indicates that large tracts of blueschists were not exposed in California by synsubduction erosion.

INTRODUCTION

The Franciscan Complex of western California (Bailey et al., 1964) is an assemblage of rocks that accumulated, were deformed and were variably metamorphosed during convergence between the North American and paleo-Pacific plates (Hamilton, 1969; Dickinson, 1970; Ernst, 1970; Page, 1970; Berkland et al., 1972). The Complex extends into southwestern Oregon (Dott, 1965; Koch, 1966). The oldest blueschists in the Franciscan are around 150 to 155 Ma old (Coleman and Lanphere, 1971;

Franciscan Complex of California

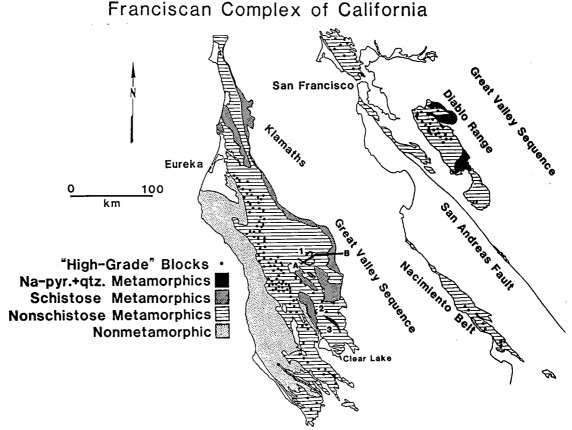

Figure 1. Geologic map of the Franciscan Complex of California. Data from Geologic Map of California (Jennings, 1977), Ernst (1971), and Coleman and Lanphere (1971). The Eastern Belt is largely composed of the South Fork Mountain Schist. The eastern portion of the Central Belt contains mud-matrix melange with greenstone blocks and extensive tracts of bedded sedimentary rocks. The western portion of the Central Belt is primarily composed of mud-matrix melange with relatively abundant high-grade blocks. The Western (Coastal) Belt is labeled nonmetamorphic but locally contains laumonite. High-grade blocks are principally Type IV high-pressure metamorphics of Coleman and Lee (1963). Schistose metamorphics are dominantly textural grade 3 rocks in the classification of Blake et al. (1967). Regions labeled as nonschistose are locally slaty or phyllitic and could be classified as textural grade 2. 1) Taliaferro Complex of Suppe (1973), 2) Elk Mountain Sequence of Etter (1979), 3) Pacific Ridge Complex of Suppe and Foland (1978).

McDowell et al., 1984), indicating that a period of subduction began in the Late Jurassic. Along the California margin, subduction gave way to San Andreas-related strike-slip motion in the mid-Tertiary as a plate triple junction migrated northward from southern California (Atwater, 1970). Active arc volcanism in the Cascade chain (McBirney et al., 1974) and recent folding and faulting of modern trench sediments (detected in marine seismic reflection profiles; Silver, 1972), indicate that slow subduction continues in northernmost California, Oregon and Washington.

Franciscan blueschists are primarily found in two structural settings, both where convergence continues, and further south, where it has stopped. They are found 1) as blocks, meters to tens of meters in diameter, sitting in mud-matrix melange, and 2) as variably deformed sheets or slabs, composed largely of metasediments, that are several kilometers thick and laterally extensive for tens of kilometers (Figure 1). Similar settings are found for

blueschists in many other emergent, inactive subduction complexes around the world (see Ernst, 1975; various authors in this volume). In this report, the petrologic and structural characteristics of Franciscan blueschists and related rocks are reviewed, and fundamental constraints on tectonic models for their generation, uplift and preservation are identified. Details of mineralogical associations can be found in the referenced works.

The Nacimiento Belt is west of the San Andreas fault and has been translated many hundreds of kilometers northwards (see Page, 1982). Numerous post-subduction strike-slip fault truncations are present. The Belt is included in this report because of numerous similarities to the Franciscan rocks further north (Bailey et al., 1964; Ernst, 1980). The tract of Franciscan blueschists on Catalina Island (offshore Los Angeles) is not included because they form a small area of exposure and differ in radiometric age and nature of metamorphism from the Franciscan rocks further north. They are discussed by Platt (1975, 1976) and Sorensen (this volume).

Post-subduction strike-slip related processes obscure geologic relationships. Strike-slip movements are pronounced in the area of San Francisco Bay (Aydin and Page, 1984) and are becoming substantial further north (Kelsey and Hagans, 1982; Kelsey and Cashman, 1983). Many fault-bounded sedimentary basins (McLaughlin and Nilsen, 1982) and small volcanic centers (Johnson and O'Neil, 1984) can be recognized on the Geologic Map of California (Jennings, 1977; scale 1:750,000).

The differentiation of subduction-related deformation and metamorphism from transform-related effects can be difficult (and is the subject of much recent debate). The writer concurs with Ernst (1984), that the dominant effect of transform-related strike-slip faulting appears to be a shuffling of subduction-generated tectonostratigraphic units. The features of Franciscan blueschists focused on herein are those which the writer believes to be most clearly related to their generation and uplift.

BLUESCHISTS IN THE COAST RANGES OF CALIFORNIA

Franciscan rocks display tremendous variations in their sedimentological, structural and metamorphic characteristics. Although numerous distinct terranes can be identified in the Coast Ranges, only the major subdivisions are outlined in this report.

North of the latitude of Clear Lake, three subparallel belts are recognized (Bailey et al., 1964; Blake and Jones, 1974). The Western (Coastal) Belt is largely Tertiary in age and is only incipiently metamorphosed (Evitt and Pierce, 1975; Bachman, 1978; McLaughlin et al., 1982; Underwood, 1984). Albitization of plagioclase, and veins of calcite and laumontite, are present. Only a few blocks of blueschist have been found in narrow strips of melange that extend through portions of the Coastal Belt (McLaughlin et al., 1982). Because of the scarcity of blueschists, this Belt is not discussed further. Blueschists, eclogites and other high-grade rocks having diagnostic high-pressure mineral assemblages are found as blocks in the mud-matrix melanges with scaly cleavage in the western portion of the Central Belt (see Figure 1). Variably recrystallized coherent units and local bodies of mud-matrix melange in which greenstones are the most abundant blocks constitute the eastern portion of this Belt (primarily the Yolla Bolly Terrane of Blake et al., 1984b). In this zone, high-grade blocks are rare. Extensive (tens of kilometers) sheets or slabs of thoroughly recrystallized phyllitic to schistose (textural grade 2, slaty to phyllitic, and textural grade 3, schistose, of Blake et al., 1967) blueschist facies metasediments (primarily the South Fork Mountain Schist; Pickett Peak Terrane of Blake et al., 1984a, b) constitute the Eastern Belt.

In the area around San Francisco and north to about the latitude of Clear Lake, the effects of post-subduction strike-slip faulting are profound. Numerous tectonostratigraphic units have been mapped as Franciscan in this area (Blake et al., 1984a; McLaughlin and Ohlin, 1984). Both coherent tracts of blueschist

and belts of melange containing blocks of blueschist are present. As it is clear that post-subduction deformation has significantly modified the relationships between units, structural relations in this portion of the Franciscan are not described in detail.

South of San Francisco, mud-matrix melange containing blueschist and eclogite blocks, together with extensive tracts of coherent metasediments, constitute the core of the Diablo Range (Ernst, 1970; Page, 1981). Although the diagnostic high-pressure minerals (Ernst, 1971a) lawsonite and aragonite are present throughout the Range (McKee, 1962a, b; Ernst, 1965; 1971b), most of the coherently bedded rocks are at most only slaty or phyllitic (textural grades 1 or 2). The assemblage jadeitic pyroxene + quartz is found in extensive tracts and is more widespread in occurrence here than north of San Francisco (Figure 1). The coherent units of the Diablo Range are markedly less schistose (only locally textural grade 3) than those in the Eastern Belt of northern California despite the fact that they contain the higher-pressure mineral assemblage of jadeitic pyroxene + quartz. The degree of textural reconstitution is more similar to that of coherent units in the eastern portion of the Central Belt (Blake et al., 1984a, b correlate these units as the Yolla Bolly Terrane). Much of the mud-matrix melange in the Diablo Range is similar to the melange of the western portion of the Central Belt in northern California.

Franciscan rocks exposed between the central California coast and the western edge of the Salinian Block are referred to as the Nacimiento Belt, Block or Terrane (Ernst, 1980) or the Sur-Obispo Belt (Page, 1981). Numerous other subdivisions have been made (Vedder et al., 1983). Much of the region is composed of blueschist-bearing, mud-matrix melange (Hsu, 1969), but portions are extensive, coherent, variably schistose (mostly textural grade 2) metasediments along the eastern margin of the Belt (Gilbert, 1972). Although no blocks of eclogite have been reported, gneissic blueschists similar to those in melanges east of the San Andreas Fault are present. Overall, the melange is quite similar to that in the western portion of the Central Belt in northern California and the Diablo Range. The assemblage jadeitic pyroxene + quartz is present but scarce. The overall structural and metamorphic patterns are similar to those in northern California except that truncation and dislocation of units along strike-slip faults are extensive (Ernst, 1980; Page, 1982; Vedder et al., 1983).

FRANCISCAN BLUESCHISTS IN MUD-MATRIX MELANGE

The extensive tracts of mud-matrix melange containing blocks of blueschists and rare eclogites that extends from Oregon to central California is one of the most characteristic features of the Franciscan. The high-pressure/low-temperature metamorphic blocks are minor components compared to the volume of sandstone blocks and black mudstone matrix with scaly cleavage. Substantial amounts of bedded terrigenous sediments and small amounts of bedded oceanic sediments such as pelagic cherts and

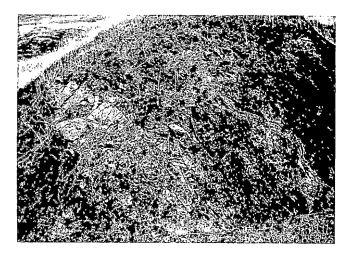

Figure 2. Typical roadcut (about 3-meters high) through mud-matrix melange in northern California. Note the pinch-and-swell of small pieces of greenstone and graywacke.

limestones are locally present. Most of the bedded sandstone and shale sequences were deposited directly upon the trench slope or offscraped after accumulation on the seafloor. The diverse array of materials juxtaposed in the melanges displays a wide range in degree of metamorphism and mechanical behavior during deformation.

Paleomagnetic studies show that some of the pelagic limestones, cherts and associated greenstones were translated northwards long distances from the original latitudes of deposition (Alvarez et al., 1980; Curry et al., 1984; Gromme, 1984). It is probable that many greenstones and associated pelagic sediments are fragments of seamounts, atolls or fracture-zone escarpments that were detached from the incoming plate (Bailey et al., 1964; Hsu, 1971; Wachs and Hein, 1975; MacPherson, 1983).

The small-scale structures of the Franciscan mud-matrix melanges have been discussed by Maxwell (1974), Cowan (1974, 1978, 1985), Page (1978), and Cloos (1982). These chaotic mixtures must be differentiated from bodies which are variably disrupted but still bedded (Underwood, 1984; Cowan, 1985). The term broken formation seems appropriate for describing many of the bedded units (Hsu, 1969). Broken formation that recrystallized under high-pressure/low-temperature conditions is classified as a coherent unit in this report.

General structural aspects of the mud-rich melanges that contain blocks of schistose blueschist are briefly summarized here. The mud-matrix has a distinctive scaly or anastomosing cleavage. All rock types for which contact relations with the mud-matrix can be observed show the development of pinch-and-swell or boudinage structures (Figure 2). Blocks typically have a roughly ellipsoidal shape. Distortion of blueschists and greenstones has been accommodated largely by cataclastic flow which is particularly intense along the margins and necked regions of the blocks. Elongate clasts are oriented with their long axes subparallel to the foliation of the mud-matrix. Small sandstone clasts in exotic-block-bearing melange rarely display evi-

dence of intense cataclasis and recognizable sedimentary structures are uncommon. In northern California, where the melanges in the Central Belt are several tens of kilometers in width, subdivisions (melange units) a few kilometers across have been mapped largely upon the relative proportions of different types of blocks (Maxwell, 1974). Fossils have been recovered from limestone, chert and sandstone blocks and bedded slope-basin deposits atop melange. Most fossils from the Central Belt are Late Jurassic to Cretaceous in age (Pessagno, 1973; Blake and Jones, 1974; Murchey and Jones, 1984; Sliter, 1984).

Nothing typifies the problems in understanding the uplift and preservation of blueschists in the Franciscan more than the knocker problem (Bailey et al., 1964; Karig, 1980). "Knocker" is an informal field term for the large, rounded blocks that create the characteristic knobby topography in grassy, mud-matrix melange terranes (Figure 3; Coleman and Lanphere, 1971). Although it is difficult to estimate size, few high-grade blueschist knockers are more than several tens of meters across, while the rare eclogite blocks are typically smaller. Knockers of greenstone with relict volcanic textures or structures can be hundreds of meters across. High-grade blueschist or eclogite knockers were clearly metamorphosed under higher temperature conditions than the surrounding mud-matrix and most nearby blocks; thus their place of origin and mechanism of emplacement into the less recrystallized melanges are problematic.

The literature on the Franciscan melanges is voluminous and confusing. Explanations for their origin fall into two broad classes. Some workers regard most of them as chaotic sedimentary deposits (olistostromes) that have undergone variable degrees of tectonism (Hsu and Ohrbom,1969; Maxwell, 1974; Gucwa, 1975; Cowan, 1978; Page, 1978; Aalto, 1981), while others interpret most of them as tectonic mixtures resulting from the deformation of variably lithified sedimentary materials (Hsu, 1971; Cowan, 1974; Hamilton, 1978; Cloos, 1982, 1984). It is ex-

Figure 3. View of several knockers of blueschist and eclogite on Tiburon Peninsula, San Francisco Bay. Note the hummocky grass-covered hillsides which are typical of mud-matrix melange terranes in the Franciscan.

tremely difficult to differentiate between these two end-member processes even where exposure is good (Hsu, 1974). Certainly both tectonic and sedimentary processes that result in the development of chaotic mixtures occur in the subduction zone environment (Cowan, 1985). The volumetric importance of these end-member processes is uncertain. Because the primary mechanism of blueschist uplift is unidentified, surficial sedimentary processes cannot be the sole mechanism by which the blueschist-bearing mud-matrix melanges developed. In addition, the fact that only minor amounts of blueschist detritus are found reworked in Franciscan sandstones and conglomerates (discussed below) is evidence against extensive exposure of blueschists during the period(s) of convergence that created the Franciscan. Finally, even if mixing were the result of sedimentary processes, significant tectonism is still required to account for the cataclasis and pinch-and-swell of blueschist and greenstone blocks.

Nevertheless, to some workers structural arguments for a tectonic origin of melange seem no more compelling because the required tectonic strains are enormous and the mechanism of uplift of blueschists seems unclear. Certainly, one major problem is quantifying the magnitude of tectonic strain of the melange. Although primary compositional layering and metamorphic foliation are recognizable in less cataclastic portions of many mafic blocks of blueschist, this only indicates that the strain is not penetrative and that the blocks were relatively rigid compared to the foliated mud-matrix. The overall preferred orientation of rounded, elongate, variably cataclastic blocks, the pinch-and-swell structures, the separation of boudins, and the lack of bedding, lamination, or megafossils in most meter-sized fragments of graywacke encased in exotic-block-bearing melange all indicate that the effects of a very large tectonic shear strain were concentrated in the mud-matrix and soft bodies of sandstone (see Cloos, 1982, 1984). In short, structural features in the melanges provide few direct constraints on the depth of burial of the melange and the mechanism of uplift of included blueschists.

HIGH-GRADE BLOCKS OF NA-AMPHIBOLE BLUESCHISTS, ECLOGITES AND AMPHIBOLITES

The most distinctive and highly-studied knockers are coarse grained, high-grade sodic amphibole + epidote + garnet + rutile gneissic blueschists, omphacitic pyroxene + garnet + rutile eclogites, and calcic amphibole + epidote + garnet + rutile amphibolites (the Type IV rocks of Coleman and Lee, 1963). They are volumetrically minor (<<1%) components of the mud-matrix melanges all the way from Oregon (Ghent and Coleman, 1973) to the southern Diablo Range (Ernst, 1965). Gneissic blueschists are also found west of the San Andreas fault in mud-matrix melange of the Nacimiento Belt (Coleman and Lanphere, 1971). The high-grade blocks were the subject of several of the earliest petrologic studies on Franciscan rocks (Switzer, 1945, 1951; Borg, 1956; Bloxam, 1959). In practice, they are typically differentiated from low-grade blueschists by one or more criteria. High-grade blueschists contain garnet instead of chlorite, rutile

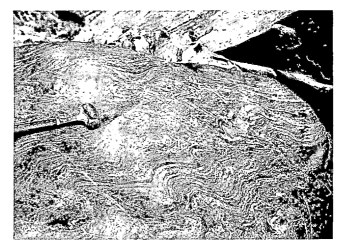

Figure 4. Refolded folds in high-grade blueschist block eroded out of melange near Alder Point in northern California. Such folds are observed in many places but gneissic or massive blocks are also quite common.

instead of sphene, epidote instead of lawsonite, and have an average grain size of about one millimeter or larger.

The high-grade blocks described in this section are concentrated in the western portion of the Central Belt (Figure 1). A few blueschist blocks are also present in strips of mud-matrix melange in the Coastal Belt (McLaughlin et al., 1982) and bodies of serpentinite-matrix melange in the Eastern Belt (Blake and Jones, 1977). The last two occurrences have not been studied in detail. Almost all high-grade knockers have a mafic bulk composition. The various forms of sodic amphibole-bearing blueschist blocks greatly outnumber eclogite and amphibolite blocks. Some have a few quartz-rich interlayers that are probably recrystallized chert. Although coarsely recrystallized blocks of metasediments are extremely rare, one large block exposed in the Laytonvillem Quarry in northern California contains a section of metamorphosed metal-rich sediments which is over ten meters thick. This block has received much attention because of its excellent exposure, accessibility and extraordinary mineralogy (including the minerals deerite, howieite and zussmanite; Chesterman, 1966; Wood, 1982).

Except for compositional layering, primary igneous or depositional structures are not recognizable in high-grade blueschists. Layers range from millimeters to meters in thickness and are commonly discontinuous where they are thin. Folds, in some cases refolded (Figure 4), and boudinage are found, but many blocks are massive. Regeneration of foliation in fold hinges indicates that some recrystallization was syndeformational.

Petrologic studies indicate that the high-grade blueschist and eclogite blocks were recrystallized at temperatures of 400 to 600°C and pressures of 6 to 10 kb (Taylor and Coleman, 1968; Ernst et al., 1970; Ghent and Coleman, 1973; Brown and Bradshaw, 1979). Most radiometric dates range between 140 to 155 Ma (Coleman and Lanphere, 1971; McDowell et al., 1984; Figure 5). These are the highest-pressure, highest-temperature meta-

M. Cloos

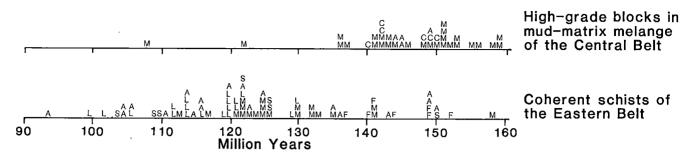

High—grade blocks in
mud—matrix melange
of the Central Belt

Coherent schists of
the Eastern Belt

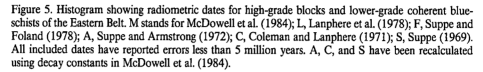

Figure 5. Histogram showing radiometric dates for high-grade blocks and lower-grade coherent blue-schists of the Eastern Belt. M stands for McDowell et al. (1984); L, Lanphere et al. (1978); F, Suppe and Foland (1978); A, Suppe and Armstrong (1972); C, Coleman and Lanphere (1971); S, Suppe (1969). All included dates have reported errors less than 5 million years. A, C, and S have been recalculated using decay constants in McDowell et al. (1984).

morphic rocks in the Franciscan, and record the oldest Ar isotopic ages.

Many of the high-grade blocks have remnants of a coarse grained actinolite ± chlorite ± talc rind up to about one meter thick (Figure 6). The discontinuous rind rarely covers more than a few square meters, and is usually thickest where found in embayments along the margins of the blocks. The actinolite and chlorite are Mg-rich and formed during metasomatic interaction with serpentinizing peridotite (Coleman and Lanphere, 1971; Moore, 1984). In many cases, the gneissic foliation in the outermost portions of the blocks is bent into parallelism with the margin of the blocks (Bailey et al., 1964; Coleman and Lanphere, 1971). Weathering and spalling of slabs accents the wrap-around foliation and creates their elliptical knocker shape. Typically remnants of the rind can be found on all sides of the block—a fact which indicates that the rounded shape developed while they were encased in serpentinite.

Almost all high-grade blocks display evidence that the initial high-pressure/low-temperature assemblages have been partially replaced or overprinted by a high-pressure but lower-temperature, low-grade blueschist assemblage (Ernst, 1965; Coleman et al., 1965; Ernst et al., 1970; Coleman and Lanphere, 1971; Hermes, 1973; Moore, 1984). No evidence of a retrograde greenschist metamorphism has been reported for the high-grade blocks. In many cases, the low-grade blueschist alterations post-date the last penetrative deformation because the replacing minerals are randomly oriented and crosscut small-scale structures. Primary minerals such as garnet are pseudomorphed by sodic amphibole, chlorite, lawsonite, white mica, aragonite, pumpellyite, albite, sphene, and apatite (Figure 7). More remarkably, some blocks contain cavities which are lined by euhedral crystals of the same high-pressure minerals that replace minerals in the host rock (Figure 8; Fyfe et al., 1958; Essene et al., 1965; Essene and Fyfe, 1967). The low-grade blueschistic alterations are primarily manifestations of a hydration which is most intense along the margins of the blocks (Figure 9). These features indicate the blocks underwent cooling and variable degrees of hydration at depths of 15 kilometers or more, following thoroughly penetrative syntectonic

recrystallization at similar or greater depths. The preservation of pseudomorphs and cavities indicates that retrograde alterations and transport to the surface occurred without thoroughly penetrative deformation of the interior of many blocks.

Figure 6. Thick actinolite-chlorite rind along the margin of the high-grade blueschist block at the famous quarry just south of Laytonville on Highway 101. Discontinuous patches of actinolite ± chlorite rind are found in embayments on the margins of most high-grade blueschists, eclogites, and amphibolites.

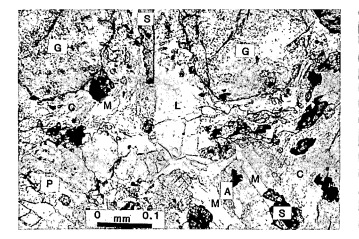

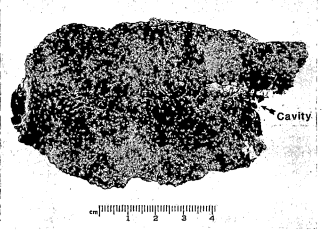

Figure 7. Thin section from an eclogite block along Kincaid Road just west of Mount Hamilton in the Diablo Range. Similar alterations are common and most intense along the margins of most high-grade blocks. Primary minerals are G, garnet and P, sodic pyroxene. Partial replacement by M, white mica, C, chlorite, L, lawsonite, and A, sodic amphibole.

Figure 8. Cavities up to a centimeter across lined by sodic pyroxene, sodic amphibole, aragonite, and sphene in the same eclogite illustrated in Figure 7.

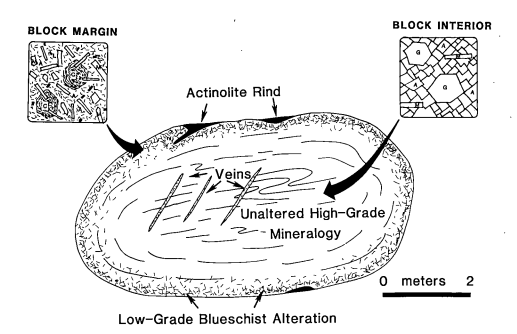

Figure 9. Schematic composite diagram illustrating the pattern of alteration of many high-grade blocks from the mud-matrix melanges of the Central Belt. Cataclastic zones that crosscut all of the illustrated features are omitted. The primary foliation and lineation are folded in the interior of some blocks. Commonly, the foliation is bent to parallel the margins of the blocks. This forms the wrap-around foliation which is the basic reason for the rounded-knocker shape. The interior of the block contains primary high-temperature/high-pressure mineral assemblages which are variably replaced and pseudomorphed by low-grade blueschist assemblages with little, if any, preferred orientation. This alteration is typically most extensive along the margin of the blocks. Veins or cavities in the interior of the block crosscut the primary folds and are commonly lined with the same minerals that pseudomorph the primary minerals in other parts of the block. The actinolite ± chlorite rind is typically present only in patches along the block margins. A stands for sodic amphibole, M, white mica, G, garnet, C, chlorite, and L, lawsonite.

Figure 10. Blueschist block with remnants of actinolite rind encased in mud-matrix melange, near San Simeon, California. Hammer is on the rind near the contact of the black-mud matrix.

Figure 11. Large, low-grade, sodic amphibole + lawsonite blueschist block displaying pinch-and-swell in mud-matrix melange, near San Simeon, California.

Contact relations between the high-grade blocks and the mud-matrix of the melange are rarely observed. Because of the presence of the actinolite rinds, some workers have assumed that all high-grade blocks simply eroded out of serpentinite diapirs, and that most blocks are now found as erosional lags on the surface. Because some high-grade block with actinolite rinds have been found encased in mud-matrix melange (Figure 10), and because many blocks are found in areas with few, if any, serpentinite masses (Bailey et al., 1964), it seems most likely that most high-grade blocks are presently sitting in, or were eroded out of, mud-matrix melange.

Nonetheless, some Franciscan blueschists have been uplifted in serpentinite diapirs which penetrated into sedimentary deposits of the Great Valley forearc basin (e.g. New Idria in central California; Coleman, 1961, 1980). Some serpentinite bodies have spread on the surface as extrusive flows or been reworked into surficial sediments of the forearc basin (Dickinson, 1966; Cowan and Mansfield, 1970; Lockwood, 1971; Phipps, 1984). While a diapiric process could explain the uplift of the volumetrically minor high-grade blocks, their subsequent emplacement and dispersal in Franciscan mud-matrix melange and their scarcity as fragments in conglomerates or other unambiguous sedimentary deposits of the Franciscan or Great Valley remains unexplained. Moreover, serpentine diapirism does not explain the uplift of the other types of high-pressure blocks in the melange, nor the extensive coherent tracts of blueschist discussed below.

LOW-GRADE BLOCKS OF SCHISTOSE BLUESCHIST

Commonly associated with the high-grade blocks are low-grade blueschists that vary from schistose to weakly foliated. Low-grade blueschists are finely crystalline and generally contain chlorite instead of garnet, and pumpellyite and/or lawsonite instead of epidote. Aragonite is common in veins. Low-grade blue-schists are found as blocks in melanges as well as coherent terranes. Low-grade (Type III of Coleman and Lee, 1963) mafic blueschists containing the assemblage sodic amphibole + chlorite + lawsonite are much more abundant than high-grade blocks but even so they are volumetrically minor constituents of the entire Franciscan (Bailey et al., 1964). A few form large (few hundred meters across) knockers, but blocks less than ten meters across are more common. Low-grade, sodic amphibole-bearing blocks generally have a strong foliation and typically display crenulations in which grains in the hinges are broken. Retrograde replacements are rarely as obvious as the pseudomorphing relations found in high-grade blocks, but partial replacement by albite + chlorite + pumpellyite (the same metamorphic assemblage found in greenstones discussed below) is evident in some cataclastic zones. Actinolite rinds are very rarely found on low-grade blueschist blocks. Unlike the high-grade blocks, most low-grade blocks show no evidence of a former association with serpentinite.

The contact relations of low-grade, sodic amphibole-bearing blueschists with the surrounding mud-matrix melange are rarely seen. Pinch-and-swell of these blocks makes them appear ductile on a mesoscopic scale, but more detailed observation indicates that flow has been by cataclasis that tends to be concentrated in zones along the margins and necked portions of elongate blocks (Figure 11). The cataclastic zones are commonly eroded off wave-washed seastacks or knockers on hillsides.

OTHER HIGH-PRESSURE METAMORPHIC ROCKS

The eastern portions of the Central Belt and the entire Eastern Belt in northern California, most of the Diablo Range in central California, and portions of the Nacimiento Belt contain high-pressure metamorphic rocks that are neither blue from the presence of sodic amphibole nor highly schistose. Most metagraywacke and greenstone blocks encased in the mud-matrix of the melanges in the Central Belt are included in this class of rock.

The nature of coherent tracts of this form of blueschist is discussed in the next section. Metamorphism under high-pressure conditions is indicated by the presence of jadeitic pyroxene + quartz, lawsonite, and/or aragonite (see Brace et al., 1970 for a discussion of tectonic overpressures in this class of Franciscan rock). Primary features such as phenocrysts and vesicles in mafic rock types and clastic grains in psammitic units are commonly recognizable. In many cases, partial recrystallization has occurred with little modification of the primary textures or structures and evidence of metamorphism is only detected by petrographic examination.

Bodies of sandstones retaining clastic textures but containing metamorphic albite + chlorite + white mica + pumpellyite are widespread throughout the eastern portions of the Central Belt and the Diablo Range. Sodic pyroxene + quartz, lawsonite, and aragonite are locally developed, particularly in the Diablo Range (McKee, 1962a, b; Bailey et al., 1964; Ernst, 1965, 1971b, 1980; Ernst et al., 1970; Suppe, 1973; Cowan, 1974; Worrall, 1981; Blake et al., 1982; Blake and Jayko, 1983; Blake et al., 1984a, b; Maruyama et al., 1985).

Greenstones have received relatively little detailed study anywhere in the Franciscan. Igneous textures and structures are typically much more evident in hand sample than is evidence of high-pressure metamorphism. Many greenstones encased in mud-matrix melange are variably brecciated. Some of the fragmentation is due to primary igneous or volcanic processes. However,the fact that nearby schistose blueschists are deformed cataclastically during tectonism makes it seem likely that the fragmentation that characterizes many greenstones is also due to tectonism. It is clear that the primary cause of fragmentation of many greenstones may be indeterminate. The most common metamorphic mineral assemblage is albite + chlorite + pumpellyite. Veins of aragonite, sodic pyroxene, lawsonite, and sodic amphibole have been found in greenstones in localities ranging from northern to central California (Essene and Fyfe, 1967; Suppe, 1973; Brown and Bradshaw, 1979; Ernst et al., 1970; Worrall, 1981; MacPherson, 1983). More studies are clearly needed to identify regional trends.

One of the least-studied aspects of the Franciscan metamorphism is the rocks of pelitic bulk composition. The black, mud-matrix of the melange is voluminous but poorly exposed. The matrix contains the assemblage quartz + albite + white mica + chlorite and is so fine-grained that it appears unmetamorphosed to workers in the field (Cloos, 1983). This is the stable mineral assemblage for pelitic bulk compositions over a remarkably wide range of pressure at low temperature. The degree of recrystallization of clays and organics indicates the matrix attained temperatures of 100 to 150°C. At these temperatures, the presence of albite indicates the maximum pressures that could have been attained were 6 to 8 kb. Although not apparent from field observations, the scaly mud-matrix typically found in the melanges could have been subducted to depths of 10 to 20 kilometers if temperatures remained low. Subduction to such depths is clearly indicated for mud-matrix melanges in the Diablo Range (discussed below).

FRANCISCAN BLUESCHISTS IN COHERENT BELTS NORTHERN CALIFORNIA

In northern California, coherent metamorphic rocks outcrop continuously for several hundred kilometers along the eastern edge of the Franciscan (Figure 1). Highly schistose rocks (textural zone 3) that constitute the easternmost portion of the Franciscan in California are known as the South Fork Mountain Schist (Blake et al., 1967; 1969). A similar unit known as the Colebrook Schist is present in southwestern Oregon (Coleman, 1972). In northern California, the association of the Schist (mostly textural zone 3) and nearby slaty or phyllitic rocks (mostly textural zone 2), has been termed the Pickett Peak Terrane by Blake et al. (1984b; see also Jayko et al., this volume).

Extensive tracts of markedly less schistose, but coherently bedded metasediments and lesser amounts of greenstone block-bearing mud-matrix melange comprise the eastern portion of the Central Belt. In most places, this zone separates the Eastern Belt from the zone of mud-matrix melange with relatively abundant high-grade blocks (Suppe, 1973; Blake et al., 1982; see Figure 1) and has been termed the Yolla Bolly Terrane by Blake et al. (1984b). In areas, mud-matrix melange and the coherent metasediments are interfingered, intersliced, or interlayered (Suppe, 1973; Jones et al., 1978; McLaughlin and Ohlin, 1984). Rare fossils present in the less schistose rocks are Late Jurassic to Early Cretaceous in age (Ghent, 1963; Blake and Jones, 1974; Murchey and Jones, 1984). The problem of preservation and uplift of the coherent metamorphic units in the Franciscan is similar to that of uplifting and preserving any extensive tract of blueschist.

The protoliths of the coherent units in the easternmost portion of the Franciscan were largely fine-grained sediments (Blake and Jones, 1974). Interlayers of mafic or cherty materials are generally volumetrically minor, but several large, mappable mafic units are present (Suppe, 1973; Worrall, 1981; Blake and Jayko, 1983). Outcrop-scale or smaller isoclinal folds are widespread (Wood, 1971; Bishop, 1977; Suppe, 1973; Monson and Aalto, 1980; Worrall, 1981) and a few kilometer-scale folds are present (Blake and Jayko, 1983). An upside-down metamorphic zonation is defined by variations in mineralogy and degree of textural reconstitution (Kilmer, 1962; Blake et al., 1967). Some mafic layers contain both sodic amphibole and actinolite, mineral associations which are transitional between typical blueschist and greenschist facies assemblages (Blake et al., 1967; Ghent, 1965; Worrall, 1981).

In the eastern portion of the Central Belt, many of the rocks are coherent and not strongly schistose. Locally, blocks of greenstones are common in mud-matrix melange that seems to lack gneissic-blueschists with actinolite rinds. Lawsonite and/or pumpellyite are typically present instead of epidote. Aragonite seems to be widespread in veins. The assemblage of jadeitic pyroxene + quartz is locally developed in several coherent sheets of bedded metagraywacke that are not in fault contact with the Coast Range ophiolite (Figure 1). The largest tract of jadeitic pyroxene + quartz-bearing metasediments in northern California is the Talia-

A

■ Taliaferro Metamorphic Complex

◩ Franciscan Metasediments

▨ South Fork Mountain Schist

☐ Coast Range Ophiolite

From Suppe (1973)

Figure 12. Cross-section illustrating the slabiform nature of the Taliaferro Metamorphic Complex in the northeastern Franciscan. The schistose slab contains jadeitic pyroxene + quartz which are absent from the less schistose units above and below (see Figure 1 for location).

B.

Figure 13. Faults marked by narrow zones of cataclasis in coherent blueschist-facies metasediments (textural grade 3) in the South Fork Mountain Schist east of the Taliaferro Metamorphic Complex.

ferro Metamorphic Complex (Suppe, 1972, 1973). Smaller tracts of particularly high-pressure metasediments are the Elk Mountain Sequence (Etter, 1979), the Pacific Ridge Complex (Suppe and Foland, 1978), and several small bodies in the San Francisco Bay area (Brothers, 1954; Bloxam, 1956, 1960).

Petrologic studies by Ghent (1965), Suppe (1973), Brown and Ghent (1983) have shown that parts of the Eastern Belt schists were recrystallized at temperatures of 250 to 350°C and pressures between about 5 to 7 kb. Somewhat higher pressures and/or lower temperatures seem to be indicated for the jadeitic pyroxene + quartz-bearing units.

Suppe (1973), Suppe and Foland (1978), Worrall (1981), and Blake and Jayko (1983), among others, have shown that the transitions between mineralogic and textural zones in the eastern portion of the Central Belt and Eastern Belt are commonly abrupt. Major faults are usually inferred which topographic relations indicate vary from steeply to gently dipping. In many places, faults are oriented at low angles with respect to the attitude of bedding or foliation in adjoining units (Figure 12). Exposed fault zones are typically marked by cataclastic zones from a few centimeters to meters across (Figure 13).

Individual coherent units defined by mineralogy or texture are up to several kilometers thick and extend for tens of kilometers. Individual layers of chert or mafic rocks can rarely be traced for more than a few kilometers but some horizons extend for 10 kilometers or so and define large folds (Blake and Jayko, 1983; Jayko et al., this volume). Large folds may be more common than generally recognized because outcrop is discontinuous and suitable marker beds are scarce. Some folds are clearly synrecrystallization because the metamorphic fabric was regenerated in the hinges and/or ductile compositional layers were thinned or thickened in the limbs or around boudins (Wood, 1971; Bishop, 1977; Worrall, 1981). Small folds characterized by kinking (Figure 14) or layer-parallel slippage are widespread. Veins of quartz, aragonite, calcite, and albite commonly crosscut early-formed structures in the schist, and in some places are themselves folded

(Figure 14). Mineralized veins, kink folds and narrow cataclastic zones indicate that much of the early syntectonic recrystallization was followed by one or more stages of less penetrative, more brittle style of deformation.

Lanphere et al. (1978) infer a metamorphic age for the schist of the Eastern Belt of around 120 Ma based primarily upon Ar-isotopic dating of whole rocks. Other radiometric-dating studies on the South Fork Mountain Schist by Suppe and Foland (1978) and McDowell et al. (1984) show similar results but have been interpreted to indicate a prolonged cooling history for the Belt since at least 140 Ma ago (Figure 5).

The outcrop width of the Eastern Belt schists is highly variable (Figure 1), typically a few kilometers, but ranging from a few tens of kilometers to locally absent. Much of the South Fork Mountain Schist Belt is in high-angle-fault contact with the Coast Range ophiolite. Northeast of Clear Lake, mud-matrix melange and slaty coherent units are locally in fault contact with the

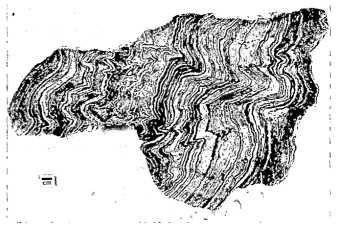

Figure 14. Small kink or chevron folds of foliation and veins that are widespread in the South Fork Mountain Schist of the Eastern Belt.

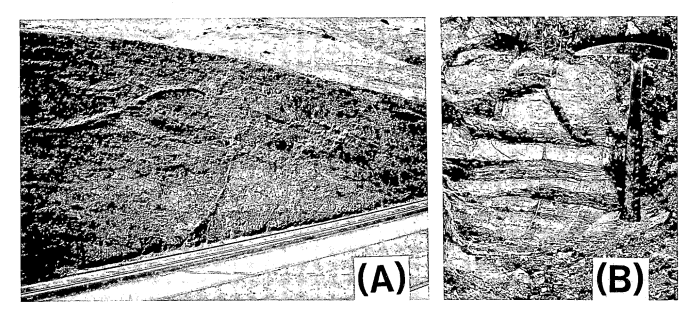

Figure 15. a) Roadcut in coherently bedded, nonschistose jadeitic pyroxene + quartz-bearing sediments near Pacheco Pass in the Diablo Range. A few steeply dipping faults offset the bedding. b) Interbedded sandstone and shale in jadeitic pyroxene-bearing sediments near Pacheco Pass. Note the excellent preservation of bedding despite subduction and uplift from depths of more than 15 kilometers.

ophiolite and even Great Valley forearc basin sediments. Just south of the Klamaths, Eastern Belt schists are in contact with Klamath basement rocks and locally with Great Valley sediments. The master fault separating the Franciscan and the overriding ophiolitic block is generally known as the Coast Range Thrust (Bailey et al., 1970), despite the fact that in most places it is steeply dipping. The discontinuity across the fault in depth of burial (as determined from mineral assemblages developed during metamorphic recrystallization) is at least 5 to 10 kilometers (several kilobars). This indicates that the Franciscan side must have moved upward to attain its present position, probably during the Late Cretaceous or Tertiary (Page, 1966, 1981; Berkland,

1972; Suppe, 1979; Hopson et al., 1981; Korsch, 1983). From this perspective it is unlikely that the Coast Range fault marks the original surface of underthrusting.

DIABLO RANGE

Large areas of the Diablo Range are coherently bedded sequences of sediments (Figure 15). In many areas, clastic grains are incipiently replaced by jadeitic pyroxene + quartz, or lawsonite and aragonite veins are common (McKee, 1962a, b; Ernst, 1965; Ernst et al., 1970, 1971b; Raymond, 1973; Cowan, 1974; Maruyama et al., 1985). Across most of the Range, penetrative strain is slight and sedimentary structures and textures are well preserved (Figure 15). This has made it possible to study paleocurrent directions (Telleen, 1977), petrofacies (Jacobson, 1978) and provenance of conglomerates (Figure 16, Fyfe and Zardini, 1967; Platt et al., 1976; Moore and Liou, 1979). Thus, it is perhaps not surprising that the regional metamorphism was not recognized until the late 1950's (McKee, 1962a). Ernst (1971b) completed a regional petrographic analysis and showed that textural and mineralogic variations in the high-pressure metamorphics show no systematic pattern around the Range. Cowan (1974) and Crawford (in Page, 1981) have shown that some of the coherent bedded slabs of metasediments are interfingered, intersliced, interlayered or engulfed in the melange on a kilometer scale.

Vitrinite-reflectance studies by Bostick (1974), petrologic analysis of conglomerates by Moore and Liou (1979) and studies of crystallinity of bedded metashales (Cloos, 1983), all indicate maximum temperatures of 100 to 200°C for recrystallization in

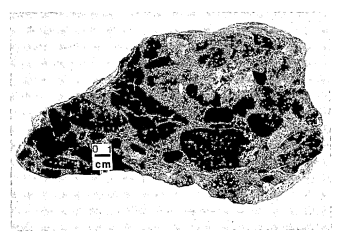

Figure 16. Blueschist facies metaconglomerate from near Pacheco Pass. Note the rounded shape of cobbles despite subduction and uplift from depths of more than 15 kilometers.

much of the Diablo Range. The widespread, but sporadic, distribution of jadeitic pyroxene + quartz indicates minimum pressures between 5 and 7 kb. As elsewhere in the Franciscan, there is no evidence of retrograde greenschist facies alterations.

The sediments of the Great Valley Sequence are upturned along the margins of the Diablo Range and form a giant anticlinorium. Fragments of the Coast Range ophiolite are found along the margins and within the Range (Evarts, 1977; Bauder and Liou, 1979). The present boundaries of the Franciscan core are post-Cretaceous high-angle faults (Page, 1966; Ernst et al., 1970; Raymond, 1970, 1973) along which the Franciscan core of the range rose more than 10 kilometers to juxtapose jadeitic pyroxene + quartz-bearing rocks against Great Valley strata. In central California, it is particularly clear that the Franciscan rocks of the Diablo Range were transported and recrystallized beneath the Great Valley forearc basin and underlying Coast Range ophiolite that comprise the overriding North American plate.

NACIMIENTO BELT

Part of the easternmost portion of the Nacimiento Belt consists of semicoherent metagraywacke that contains lawsonite and very rare jadeitic pyroxene + quartz (Gilbert, 1972; Ernst, 1980). Outcrop is generally quite poor. Some of the rocks have a semischistose appearance (locally approaching textural grade 3), similar to portions of the Eastern Belt schist. Despite extensive strike-slip displacements across the Nacimiento Belt, the overall pattern of mineralogic and textural variations is strikingly similar to that in northern California (Ernst, 1980).

THE TIMING OF UPLIFT

Important constraints on the timing of uplift of Franciscan blueschists come from the study of unmetamorphosed sedimentary sequences atop deformed and metamorphosed basement rocks (many of these units may be remnants of trench-slope basins; see Maxwell, 1974) and regional sedimentological patterns. Franciscan sandstones contain little, if any, diagnostic high-pressure metamorphic detritus. The sands appear to have been derived from a volcanic-plutonic arc complex (Dickinson et al., 1982). High-pressure metamorphic debris is rare in Franciscan conglomerates of Mesozoic age (Seiders and Blome, 1984).

In northern California, small amounts of blueschist-bearing conglomerate have been reported in the Coastal Belt (O'Day and Kramer, 1972). The oldest occurrence of significant blueschist detritus appears to be in the Rice Valley outlier. Berkland (1973) found cobbles of lawsonite-bearing metagraywacke in the Paleocene portions of the basin. Blueschist detritus is also reported in Early Tertiary sediments around Covelo in Round Valley (Clark, 1940). In northern California, it appears that small amounts of blueschists were exposed and subject to sedimentary reworking beginning in the Early Tertiary.

In central California, widespread exposure of the Franciscan rocks occurred in the mid-Tertiary. Significant amounts of Franciscan detritus are found in the Lospe, Vaqueros and Temblor Formations of Oligocene and Miocene age (Page, 1970b; Nilsen and Clarke, 1975; Nilsen, 1984). A particularly large volume of Franciscan debris is found in nonmarine Pliocene and Pleistocene gravels around the Diablo Range, indicating major uplift and exposure at that time (Page, 1981). Most interestingly, several centimeter-sized, well-rounded pebbles of foliated glaucophane schist have been found in conglomerates in Franciscan rocks that core the Diablo Range (Moore and Liou, 1980). These small pebbles of blueschist are volumetrically trivial, but nonetheless significant because they provide direct evidence for some synsubduction sedimentary reworking of blueschists. They may have been derived from older subduction complexes in Oregon and the Klamaths (Hotz et al., 1977), or the Sierras (Schweickert et al., 1980), or the cobbles may be resubducted Franciscan debris.

In the Nacimiento Belt, small amounts of reworked blueschist are found in Franciscan sediments in several places. A small body of sand-matrix sedimentary breccia of Late Cretaceous age along Las Tablas Creek east of Morro Bay contains small clasts of blueschist (Cowan and Page, 1975). A small amount of blueschist debris has also been found in the Cambria Slab, a trench-slope basin of Late Cretaceous age (Smith et al., 1979).

Although there is some evidence for synsubduction uplift and exposure of Franciscan blueschists during the Cretaceous, the volumes of blueschist reworked in surficial sediments are exceedingly minor in California. The first clear evidence for extensive exposure and surficial reworking of blueschists seems to correspond with the beginning of transform motion in the Early to Mid-Tertiary (Atwater, 1970).

OTHER CONSTRAINTS ON THE ORIGIN AND UPLIFT OF FRANCISCAN BLUESCHISTS

At actively convergent margins with a thick accretionary complex, the top of the descending plate typically attains depths in excess of about 15 kilometers only arcward of the forearc high (see Karig, 1983). Greater depths are attained beneath the crystalline overriding plate. Likewise, sufficiently high temperatures to form the high-grade blueschists and eclogites will only be attained during the initial stages of fast, continuous plate convergence in a metamorphic aureole beneath the overriding plate (see, among others, Platt, 1975; Suppe and Foland, 1978; Cloos, 1982, 1984, 1985; Brown et al., 1982). From the perspective, it appears that part of the uplift problem for Franciscan high-grade blocks is moving them from beneath the overriding plate.

Further constraints on the mechanism of uplift of blueschists come from thermal modeling. Calculations show that if post-subduction uplift from depths of 20 to 30 kilometers occurs only at normal erosional rates of denudation, greenschist facies assemblages will develop because a near-normal thermal gradient is reattained faster than the rocks reach shallow levels by erosion (Oxburgh and Turcotte, 1974; Draper and Bone, 1981). Simple post-subduction erosion cannot explain the uplift of Franciscan blueschists because they lack any evidence of a retrograde greenschist facies alteration.

The modeling by Draper and Bone (1981), however, indicates that normal rates of erosion are adequate to unroof blueschists without retrograde greenschist alterations if they somehow attained depths less than approximately 10 kilometers before subduction ended. This requires some mechanism of synsubduction uplift. Synsubduction uplift to the surface is clearly indicated for the Franciscan blueschists exposed in northern California and southern Oregon.

Direct evidence for synsubduction uplift at deep levels in the Franciscan accretionary complex comes from the fact that aragonite is widespread (Brown et al., 1962; Coleman and Lee, 1962). Studies on the rate of replacement of aragonite by calcite (Carlson and Rosenfeld, 1981) indicate that even in the absence of water, the rate of replacement of aragonite by calcite is so fast that no aragonite would be found in Franciscan blueschists if temperatures in excess of about 175°C were attained during uplift to confining pressures less than about 5 kb. This indicates that pressure-temperature trajectories during the first 10 to 20 kilometers of uplift were less than about 10 °C/km. Thermal modeling indicates that the long-term maintenance of such gradients is possible only in active subduction shear zones (cf. Oxburgh and Turcotte, 1974; Bird, 1978; Honda and Uyeda, 1983; Wang and Shi, 1984).

SUMMARY: MAJOR CONSTRAINTS OF THE UPLIFT OF FRANCISCAN BLUESCHISTS

Blueschists in the Franciscan occur in two tectonic setting: as blocks in mud-matrix melange and as extensive coherent sheets or slabs. Major observations which all models for the uplift and preservation of high-pressure/low-temperature rocks must explain include:

(1) The high-grade mafic blocks are concentrated in mud-matrix melange in the western portion of the Central Belt (but also are found sporadically in portions of the Eastern and Coastal Belts). They were heated to higher temperatures and subducted to greater depths than the matrix of the melange. The blocks underwent cooling at high pressures, and were once immersed in serpentinite. Most of the oldest radiometric dates, 140 to 155 Ma, are measured on the high-grade blocks.

(2) Coherent, extensive belts of schists and nonschistose high-pressure metasedimentary sequences are concentrated in the eastern portion of the Franciscan near the base of the overriding North American plate. Most radiometric dates vary from about

90 to 140 Ma for the less coarsely recrystallized, lower-temperature, metasediments. These rocks show no evidence of former association with serpentinite. Less intensely deformed, high-pressure metasedimentary sequences are interfingered with melange throughout the Diablo Range and locally in the eastern portion of the Central Belt. In northern California, the highest-pressure coherent metasedimentary sheets which contain jadeitic pyroxene + quartz occur as kilometer-thick units that are not juxtaposed against the base of the overriding plate.

(3) Uplift of both blueschist blocks in melange and extensive coherent tracts occurred without thoroughly penetrative strain. Cataclasis associated with uplift of blocks encased within mud-matrix melange is most extensive on the margins or necked portions of blocks. In coherent units, cataclasis was concentrated along faults typically spaced from meters to hundreds of meters apart. Many of the faults which separate mapable subdivisions are subparallel to the bedding or foliation of juxtaposed units.

(4) The mineral assemblages of high-grade blocks require minimum pressures of 7 to 10 kb. In actively convergent margins, such pressures are attained only beneath the bottom of the overriding plate. It appears that these blocks were displaced seaward and upwards from the base of the overriding plate.

(5) The lack of greenschist facies alterations, the widespread preservation of aragonite, and the present exposure of blueschists where subduction continues in northern California and southern Oregon all indicate that Franciscan blueschists were uplifted to shallow depths (less than 10 kilometers) by some synsubduction mechanism(s).

(6) The scarcity of high-pressure metamorphic detritus in Franciscan sediments indicates that large tracts of blueschist were not exposed in California by synsubduction erosion.

ACKNOWLEDGMENTS

This paper was significantly improved through reviews from M. C. Blake, William Carlson, R. G. Coleman, Trevor Dumitru, W. G. Ernst, Mark Helper, and Carl Jacobson. No doubt each of them would have placed somewhat different emphasis on portions of this paper. Their comments were greatly appreciated. Recent field studies in the Franciscan have been supported by Grant 14546-G2 from the Petroleum Research Fund of the American Chemical Society, Grant EAR-8512533 from the National Science Foundation, and by a grant from the University Research Institute of the University of Texas at Austin.

REFERENCES CITED

Aalto, K. R., 1981, Multistage melange formation in the Franciscan complex, northernmost California: Geology, v. 9, p. 602–607.

Alvarez, W., Kent, D. V., Premoli Silva, I., Schweickert, R. A., and Larson, R. A., 1980, Franciscan complex limestone deposited at 17° south paleolatitude: Geological Society of America Bulletin, v. 91, p. 476–484.

Atwater, T., 1970, Implications of plate tectonics for the Cenozoic tectonic evolution of western North America: Geological Society of America Bulletin, v. 81, p. 3513–3535.

Aydin, A., and Page, B. M., 1984, Diverse Pliocene-Quaternary tectonics in a transform environment, San Francisco Bay region, California: Geological Society of America Bulletin, v. 95, p. 1303–1317.

Bachman, S. B., 1978, A Cretaceous and early Tertiary subduction complex, Mendocino coast, northern California: *in* Howell, D. G., and McDougall, K. A., eds., Mesozoic Paleogeography of the Western United States, Society of Economic Paleontologists and Mineralogists Pacific Section, Pacific Coast Paleogeography Symposium 2, p. 419–430.

Bailey, E. H., Irwin, W. P., and Jones, D. L., 1964, Franciscan and related rocks and their significance in the geology of western California: California Division of Mines and Geology Bulletin 183, 176 p.

Bailey, E. H., Blake, M. C., and Jones, D. L., 1970, On-land Mesozoic oceanic crust in California Coast Ranges: U.S. Geological Survey Professional Paper 700-C, p. 70–81.

Bauder, J. M., and Liou, J. G., 1979, Tectonic outlier of Great Valley sequence in Franciscan terrain, Diablo Range, California: Geological Society of America Bulletin, v. 90, p. 561–568.

Berkland, J. O., 1972, Paleocene "frozen" subduction zone in the Coast Ranges of northern California: 24th International Geological Congress, Montreal, Section 3, p. 99–105.

Berkland, J. O., 1973, Rice Valley outlier-new sequence of Cretaceous-Paleocene strata in northern Coast Ranges, California: Geological Society of America Bulletin, v. 84, p. 2389–2406.

Berkland, J. O., Raymond, L. A., Kramer, J. C., Moores, E. M., and O'Day, M., 1972, What is Franciscan?: American Association of Petroleum Geologists Bulletin, v. 56, p. 2295–2302.

Bird, P., 1978, Stress and temperature in subduction shear zones: Tonga and Mariana: Royal Astronomical Society Geophysical Journal, v. 55, p. 411–434.

Bishop, D. B., 1977, South Fork Mountain Schist at Black Butte and Cottonwood Creek, northern California: Geology, v. 5, p. 595–599.

Blake, M. C., and Jones, D. L., 1974, Origin of Franciscan melanges in northern California: Society of Economic Paleontologists and Mineralogists Special Publication 19, p. 345–357.

Blake, M. C., and Jones, D. L., 1977, Plate tectonic history of the Yolla Bolly Junction, northern California: Geological Society of America Guidebook for 73rd Cordilleran Section Meeting, Sacramento, California, April 1977, no. 9, 17 p.

Blake, M. C., and Jayko, A. S., 1983, Preliminary geologic map, Yolla Bolly-Middle Eel Wilderness, California: U.S. Geological Survey Miscellaneous Field Studies Map MF-1595-A, 1:62,500.

Blake, M. C., Irwin, W. P., and Coleman, R. G., 1967, Upside down metamorphic zonation, blueschist facies, along a regional thrust in California and Oregon: U.S. Geological Survey Professional Paper 575-C, p. 1–9.

Blake, M. C., Irwin, W. P., and Coleman, R. G., 1969, Blueschist-facies metamorphism related to regional thrust faulting, Tectonophysics, v. 8, p. 237–246.

Blake, M. C., Jayko, A. S., and Howell, D. G., 1982, Sedimentation, metamorphism and tectonic accretion of the Franciscan assemblage of northern California: Geological Society of London Special Publication 10, p. 433–438.

Blake, M. C., Howell, D. G., and Jayko, A. S., 1984a, Tectonostratigraphic terranes of the San Francisco Bay region: in Blake, M. C., ed., Franciscan Geology of Northern California; Society of Economic Paleontologists and Mineralogists Pacific Section, v. 43, p. 5–22.

Blake, M. C., Jayko, A. S., and McLaughlin, R. J., 1984b, Tectonostratigraphic terranes of the northern Coast Ranges, California: in Howell, D. G., ed., Tectonostratigraphic Terranes of the Circumpacific Region; Earth Science Series, Circumpacific Council for Energy and Mineral Resources, v. 1.

Bloxam, T. W., 1956, Jadeite-bearing metagraywackes in California: American Mineralogist, v. 41, p. 488–496.

Bloxam, T. W., 1959, Glaucophane schists and associated rocks near Valley Ford, California: America Journal of Science, v. 257, p. 95–112.

Bloxam, T. W., 1960, Jadeite-rocks and glaucophane schists from Angel Island, San Francisco Bay, California: American Journal of Science, v. 258, p. 555–573.

Borg, I. Y., 1956, Glaucophane schists and eclogites near Healdsburg, California: Geological Society of America Bulletin, v. 67, p. 1563–1584.

Bostick, N. H., 1974, Phytoclasts as indicators of thermal metamorphism; Franciscan assemblage and Great Valley sequence (Upper Mesozoic), California: Geological Society of America Special Paper 153, p. 1–17.

Brace, W. F., Ernst, W. G., and Kallberg, R. W., 1970, An experimental study of

tectonic overpressure in Franciscan rocks: Geological Society of America Bulletin, v. 81, p. 1325–1338.

Brothers, R. N., 1954, Glaucophane schists from the North Berkeley Hills, California: American Journal of Science, v. 252, p. 614–626.

Brown, E. H. and Bradshaw, J. Y., 1979, Phase relations of pyroxene and amphibole in greenstone, blueschist and eclogite of the Franciscan complex, California: Contributions to Mineralogy and Petrology, v. 71, p. 67–83.

Brown, E. H., and Ghent, E. D., 1983, Mineralogy and phase relations in the blueschist facies of the Black Butte and Ball Rock areas, northern California Coast Ranges: American Mineralogist, v. 68, p. 365–372.

Brown, E. H., Wilson, D. L., Armstrong, R. L., and Harakel, J. E., 1982, Petrologic structural and age relations of serpentinite, amphibolite and blueschist in the Shuksan suite of the Iron Mountain-Gee Point area, North Cascades, Washington: Geological Society of America Bulletin, v. 93, p. 1087–1098.

Brown, W. H., Fyfe, W. S., and Turner, F. J., 1962, Aragonite in California glaucophane schists and the kinetics of the aragonite-calcite transformation: Journal of Petrology, v. 3, p. 566–582.

Carlson, W. D., and Rosenfeld, J. L., 1981, Optical determination of topotactic aragonite-calcite growth kinetics: metamorphic implications: Journal of Geology, v. 89, p. 615–638.

Chesterman, C. W., 1966, Mineralogy of the Laytonville Quarry, Mendocino County, California: California Division of Mines and Geology Bulletin, v. 190, p. 503–507.

Clark, S. G., 1940, Geology of the Covelo District Mendocino County, California: University of California Publications in the Geological Sciences, v. 25, p. 119–142.

Cloos, M., 1982, Flow melanges: Numerical modeling and geologic constraints on their origin in the Franciscan subduction complex, California: Geological Society of America Bulletin, v. 93, p. 330–345.

Cloos, M., 1983, Comparative study of melange matrix and metashales from the Franciscan subduction complex with the basal Great Valley sequence, California: Journal of Geology, v. 91, p. 291–306.

Cloos, M., 1984, Flow melanges and the structural evolution of accretionary wedges: Geological Society of America Special Paper 198, p. 71–80.

Cloos, M., 1985, Thermal evolution of convergent plate margins: Thermal modeling and reevaluation of isotopic Ar-ages for blueschists in the Franciscan Complex of California: Tectonics, v. 4, p. 421–433.

Coleman, R. G., 1961, Jadeite deposits of the Clear Creek area, New Idria district, San Benito Co., California: Journal of Petrology, v. 2, p. 207–247.

Coleman, R. G., 1972, The Colebrooke Schist of southwestern Oregon and its relation to the tectonic evolution of the region: U.S. Geological Survey Bulletin, v. 1339, 61 p.

Coleman, R. G., 1980, Tectonic inclusions in serpentinites: Archives Des Sciences, Societe De Physique Et D'Historie Naturelle De Geneve, v. 33, p. 89–102.

Coleman, R. G. and Lee, D. E., 1962, Metamorphic aragonite in the glaucophane schists of Cazadero, California: American Journal of Science, v. 60, p. 577–595.

Coleman, R. G. and Lee, D. E., 1963, Glaucophane bearing metamorphic rock types of the Cazadero area, California: Journal of Petrology, v. 4, p. 260–301.

Coleman, R. G., Lee, D. E., Beatty, L. B., and Brannock, W. W., 1965, Eclogites and eclogites: their differences and similarities: Geological Society of America Bulletin, v. 76, p. 483–508.

Coleman, R. G. and Lanphere, M. A., 1971, Distribution and age of high-grade blueschists, associated eclogites, and amphibolites from Oregon and California: Geological Society of America Bulletin, v. 82, p. 2397–2412.

Cowan, D. S., 1974, Deformation and metamorphism of the Franciscan subduction zone complex northwest of Pacheco Pass, California: Geological Society of America Bulletin, v. 85, p. 1623–1634.

Cowan, D. S., 1978, Origin of blueschist-bearing chaotic rocks in the Franciscan complex, San Simeon, California: Geological Society of America Bulletin, v. 89, p. 1415–1423.

Cowan, D. S., 1985, Structural styles in Mesozoic and Cenozoic melanges in the

Western Cordillera of North America: Geological Society of America Bulletin, v. 96, p. 451–462.

Cowan, D. S., and Mansfield, C. F., 1970, Serpentinite flows on Joaquin Ridge, southern Coast Ranges, California: Geological Society of America Bulletin, v. 81, p. 2615–2628.

Cowan, D. S., and Page, B. M., 1975, Recycled Franciscan material in Franciscan melange west of Paso Robles, California: Geological Society of America Bulletin, v. 86, p. 1089–1095.

Curry, F. B., Cox, A., and Engebretson, D. C., 1984, Paleomagnetism of Franciscan rocks in the Marin Headlands: *in* Blake, M. C., ed., Franciscan Geology of Northern California: Society of Economic Paleontologists and Mineralogists Pacific Section, v. 43, p. 89–98.

Dickinson, W. R., 1966, Table Mountain serpentinite extrusion in California Coast Ranges: Geological Society of America Bulletin, v. 77, p. 451–472.

Dickinson, W. R., 1970, Relations of andesites, granites, and derivative sandstones to arc-trench tectonics: Reviews of Geophysics and Space Physics, v. 8, p. 813–860.

Dickinson, W. R., Ingersoll, R. V., Cowan, D. S., Helmold, K. P., and Suczek, C. A., 1982, Provenance of Franciscan graywackes in coastal California: Geological Society of America Bulletin, v. 93, p. 95–107.

Dott, R. H., 1965, Mesozoic-Cenozoic tectonic history of the southwestern Oregon coast in relation to Cordilleran orogenesis: Journal of Geophysical Research, v. 70, p. 4687–4707.

Draper, G., and Bone, R., 1981, Denudation rates, thermal evolution, and preservation of blueschist terrains: Journal of Geology, v. 89, p. 601–613.

Ernst, W. G., 1965, Mineral parageneses in Franciscan metamorphic rocks, Panoche pass, California: Geological Society of America Bulletin, v. 76, p. 879–914.

Ernst, W. G., 1970, Tectonic contact between the Franciscan melange and the Great Valley sequence, crustal expression of a Late Mesozoic Benioff zone: Journal of Geophysical Research, v. 75, p. 886–901.

Ernst, W. G., 1971a, Do mineral parageneses reflect unusually high pressure conditions of Franciscan metamorphism?: American Journal of Science, v. 271, p. 81–108.

Ernst, W. G., 1971b, Petrologic reconnaissance of Franciscan metagraywackes from the Diablo Range, Central California Coast Ranges: Journal of Petrology, v. 12, p. 413–437.

Ernst, W. G., 1975, Systematics of large-scale tectonics and age progressions in alpine and circum-Pacific blueschist belts; Tectonophysics, v. 26, p. 229–246.

Ernst, W. G., 1980, Mineral paragenesis in Franciscan metagraywackes of the Nacimiento Block, a subduction complex of the southern California Coast Ranges: Journal of Geophysical Research, v. 85, p. 7045–7055.

Ernst, W. G., 1984, California blueschists, subduction, and the significance of tectonostratigraphic terranes: Geology, v. 12, p. 436–440.

Ernst, W. G., Seki, Y., Onuki, H., and Gilbert, M. C., 1970, Comparative study of low-grade metamorphism in the California Coast Ranges and the outer metamorphic belt of Japan: Geological Society of America Memoir 124, 276 p.

Essene, E. J., Fyfe, W. S., and Turner, F. J., 1965, Petrogenesis of Franciscan glaucophane schists and associated metamorphic rocks, California: Contributions to Mineralogy and Petrology, v. 11, p. 695–704.

Essene, E. J. and Fyfe, W. S., 1967, Omphacite in California metamorphic rocks: Contributions to Mineralogy and Petrology, v. 15, p. 1–23.

Etter, S. D., 1979, Geology of the Lake Pillsbury area, northern Coast Ranges, California: Unpublished Ph.D. Dissertation, University of Texas at Austin.

Evarts, R. C., 1977, The geology and petrology of the Del Puerto ophiolite, Diablo Range central California Coast Ranges: *in* Coleman, R. G., and Irwin, W. P., eds., North American Ophiolites: Oregon Department of Geology and Mineral Industries Bulletin 95, p. 121–140.

Evitt, W. R., and Pierce, S. T., 1975, Early Tertiary ages from the Coastal Belt of the Franciscan complex, northern California: Geology, v. 3, p. 433–436.

Fyfe, W. S., and Zardini, R., 1967, Metaconglomerate in the Franciscan formation near Pacheco Pass, California: American Journal of Science, v. 265,

p. 819–830.

Fyfe, W. S., Turner, F. J., and Verhoogen, J., 1958, Metamorphic reactions and metamorphic facies: Geological Society of America Memoir 73, 259 p.

Ghent, E. D., 1963, Fossil evidence for maximum age of metamorphism in part of the Franciscan Formation, northern Coast Ranges: California Division of Mines and Geology Special Report 82, p. 41.

Ghent, E. D., 1965, Glaucophane-schist facies metamorphism in the Black Butte area, northern Coast Ranges, California: American Journal of Science, v. 263, p. 385–400.

Ghent, E. H. and Coleman, R. G., 1973, Eclogites from southwestern Oregon: Geological Society of America Bulletin, v. 84, p. 2471–2488.

Gilbert, W. G., 1972, Franciscan rocks near Sur fault zone, northern Santa Lucia range, California: Geological Society of America Bulletin, v. 84, p. 3317–3328.

Gromme, S., 1984, Paleomagnetism of Franciscan basalt, Marin County, California, revisited: *in* Blake, M. C., ed., Franciscan Geology of Northern California: Society of Economic Paleontologists and Mineralogists Pacific Section, v. 43, p. 113–119.

Gucwa, P. R., 1975, Middle to late Cretaceous sedimentary melange, Franciscan complex, northern California: Geology, v. 3, p. 105–108.

Hamilton, W., 1969, Mesozoic California and the underflow of the Pacific mantle: Geological Society of America Bulletin, v. 80, p. 2409–2430.

Hamilton, W., 1978, Mesozoic tectonics of the western United States: *in* Howell, D. G., and McDougall, K. A., eds., Mesozoic paleogeography of the western United States, Society of Economic Paleontologists and Mineralogists, Pacific Section, Pacific Section Paleogeography Symposium 2, p. 33–70.

Hermes, D. O., 1973, Paragenetic relationships in an amphibolitic tectonic block in the Franciscan terrain, Panoche Pass, California: Journal of Petrology, v. 14, p. 1–32.

Honda, S., and Uyeda, S., 1983, Thermal process in subduction zone-a review and preliminary approach on the origin of arc volcanism: *in* Shimozuru, D., and Yokoyama, I., eds., Arc Volcanism: Physics and Tectonics; Terra Scientific Publishing Company (TERRAPUB), Tokyo, p. 117–140.

Hopson, C. A., Mattinson, J. M., and Pessagno, E. A., 1981, Coast Range ophiolite, western California: *in* Ernst, W. G., ed., The Geotectonic Development of California, Prentice-Hall, Englewood Cliffs, New Jersey, p. 418–510.

Hotz, P. E., Lanphere, M. A., and Swanson, D. A., 1977, Triassic blueschist from northern California and north-central Oregon: Geology, v. 5, p. 659–663.

Hsu, K. J., 1969, Preliminary report and geologic guide to Franciscan melanges of the Morro Bay-San Simeon area, California, California Division of Mines Special Report 35, 46 p.

Hsu, K. J., 1971, Franciscan melanges as a model for eugeosynclinal sedimentation and underthrusting tectonics: Journal of Geophysical Research, v. 76, p. 1162–1170.

Hsu, K. J., 1974, Melanges and their distinction from olistostromes: Society of Economic Paleontologists and Mineralogists Special Publication 19, p. 321–333.

Hsu, K. J. and Ohrbom, R., 1969, Melanges of the San Francisco Peninsula-Geologic reinterpretation of type Franciscan: American Association Petroleum Geologists Bulletin, v. 53, p. 1348–1367.

Jacobson, M. I., 1978, Petrologic variations in Franciscan sandstone from the Diablo Range, California: *in* Howell, D. G., and McDougall, K. A., eds., Mesozoic Paleogeography of the Western United States; Pacific Section, Society of Economic Paleontologists and Mineralogists, Pacific Coast Paleogeography Symposium 2, p. 401–417.

Jennings, C. W., 1977, Geologic map of California: California Division of Mines and Geology, Scale 1:750,000.

Johnson, C. M., and O'Neil, J. R., 1984, Triple junction magmatism: a geochemical study of Neogene volcanic rocks in western California: Earth and Planetary Science Letters, v. 71, p. 241–262.

Jones, D. L., Blake, M. C., Bailey, E. H., Bailey, and McLaughlin, R. J., 1978, Distribution and character of upper Mesozoic subduction complexes along the west coast of North America: Tectonophysics, v. 47, p. 207–222.

Karig, D. E., 1980, Material transport within accretionary prisms and the "knocker" problem: Journal of Geology: v. 88, p. 27–39.

Karig, D. E., 1983, Deformation in the forearc: Implications for mountain belts: in Hsu, K. J., ed., Mountain Building Processes, Academic Press, London, p. 59–72.

Kelsey, H. M., and Hagans, D. K., 1982, Major right-lateral faulting in the Franciscan assemblage of northern California in late Tertiary time: Geology, v. 10, p. 387–391.

Kelsey, H. M., and Cashman, S. M., 1983, Wrench faulting in northern California and its tectonic implications: Tectonics, v. 2, p. 565–576.

Kilmer, F. H., 1962, Anomalous relationship between the Franciscan formation and metamorphic rocks, northern Coast Ranges, California: Geological Society of America Special Paper 68, p. 210.

Koch, J. G., 1966, Late Mesozoic stratigraphy and tectonic history, Port Orford-Gold Beach area, southwestern Oregon coast: American Association of Petroleum Geologists Bulletin, v. 50, p. 25–71.

Korsch, R. J., 1983, Franciscan-Knoxville problem: relationship between an accretionary prism and adjacent forearc basin: American Association of Petroleum Geologists Bulletin, v. 67, p. 29–40.

Lanphere, M. A., Blake, M. C., and Irwin, W. P., 1978, Early Cretaceous metamorphic age of the South Fork Mountain Schist in northern Coast Ranges of California: American Journal of Science, v. 278, p. 798–815.

Lockwood, J. P., 1971, Sedimentary and gravity-slide emplacement of serpentinite: Geological Society of America Bulletin, v. 82, p. 919–936.

MacPherson, G. J., 1983, The Snow Mountain volcanic complex: An on-land seamount in the Franciscan terrain, California: Journal of Geology, v. 91, p. 73–92.

Maruyama, S., Liou, J. G., and Sasakura, Y., 1985, Low-temperature recrystallization of Franciscan greywackes from Pacheco Pass, California: Mineralogical Magazine, v. 49, p. 345–355.

Maxwell, J. C., 1974, Anatomy of an orogen: Geological Society of America Bulletin, v. 85, p. 1195–1204.

McBirney, A. R., Sutter, J. F., Naslund, H. R., Sutton, K. G., and White, C. M., 1974, Episodic volcanism in the central Oregon Cascade Range: Geology, v. 2, p. 585–589.

McDowell, F. W., Lehman, D. H., Gucwa, P. R., Fritz, D., and Maxwell, J. C., 1984, Glaucophane schists and ophiolites of the northern Coast Ranges: Isotopic ages and their tectonic implications: Geological Society of America Bulletin, v. 95, p. 1373–1382.

McKee, B., 1962a, Widespread occurrence of jadeite, lawsonite and glaucophane in central California: American Journal of Science, v. 260, p. 596–610.

McKee, B., 1962b, Aragonite in the Franciscan rocks of the Pacheco Pass area, California: American Mineralogist, v. 47, p. 379–387.

McLaughlin, R. J., and Nilsen, T. H., 1982, Neogene non-marine sedimentation and tectonics in small pull-apart basins of the San Andreas fault system, Sonoma County, California: Sedimentology, v. 29, p. 865–876.

McLaughlin, R. J., and Ohlin, H. N., 1984, Tectonostratigraphic framework of the Geysers-Clear Lake region, California: in Blake, M. C., ed., Franciscan Geology of Northern California: Society of Economic Paleontologists and Mineralogists Pacific Section, v. 43, p. 221–254.

McLaughlin, R. J., Kling, S. A., Poore, R. Z., McDougall, K., and Beutner, E. C., 1982, Post-middle Miocene accretion of Franciscan rocks, northwestern California: Geological Society of America Bulletin, v. 93, p. 595–605.

Monsen, S. A., and Aalto, K. R., 1980, Petrology, structure and regional tectonics: South Fork Mountain Schist of Pine Ridge Summit, northern California: Geological Society of America Bulletin, v. 91, p. 369–373.

Moore, D. E., 1984, Metamorphic history of a high-grade blueschist exotic blocks from the Franciscan complex, California: Journal of Petrology, v. 25, p. 126–150.

Moore, D. E., and Liou, J. G., 1979, Mineral chemistry of some Franciscan blueschist facies metasedimentary rocks from the Diablo Range, California: Geological Society of America Bulletin, v. 90, p. 1737–1781.

Moore, D. E., and Liou, J. G., 1980, Detrital blueschist pebbles from Franciscan metaconglomerates of the northeast Diablo Range, California: America Journal of Science, v. 280, p. 249–264.

Murchey, B. L., and Jones, D. L., 1984, Age and significance of chert in the Franciscan complex, in the San Francisco Bay region: in Blake, M. C., ed., Franciscan Geology of Northern California, Society of Economic Paleontologists and Mineralogists Pacific Section, v. 43, p. 23–30.

Nilsen, T. H., 1984, Oligocene tectonics and sedimentation, California: Sedimentary Geology, v. 38, p. 305–336.

Nilsen, T. H., and Clarke, S. H., 1975, Sedimentation and tectonics in the Early Tertiary continental borderland of central California: U.S. Geological Survey Professional Paper 925, 64 p.

O'Day, M., and Kramer, J. C., 1972, The "Coastal Belt" of northern California Coast Ranges: in Moores, E. M., and Matthews, R. A., eds., Geologic Guide to the northern Coast Ranges: Lake, Mendocino and Sonoma Counties, California: Geological Society of Sacramento Annual Field Trip Guidebook, p. 51–56.

Oxburgh, E. R., and Turcotte, D. L., 1974, Thermal gradients and regional metamorphism in overthrust terrains with special reference to the eastern Alps: Schweizerische Mineralogische und Petrographische Mitteilungen, v. 54, p. 641–662.

Page, B. M., 1966, Geology of the Coast Ranges of California: California Division of Mines and Geology Bulletin, v. 190, p. 255–276.

Page, B. M., 1970a, Sur-Nacimiento fault zone of California: Continental margin tectonics: Geological Society of America Bulletin, v. 81, p. 667–690.

Page, B. M., 1970b, Time of completion of underthrusting of Franciscan beneath Great Valley rocks west of Salinian block, California: Geological Society of America Bulletin, v. 83, p. 957–972.

Page, B. M., 1978, Franciscan melanges compared with olistostromes of Taiwan and Italy: Tectonophysics, v. 47, p. 223–246.

Page, B. M., 1981, The southern Coast Ranges: in Ernst, W. G., ed., The Geotectonic Development of California, Rubey Volume 1, Prentice-Hall, New Jersey, p. 329–417.

Page, B. M., 1982, Migration of the Salinian composite block, California, and disappearance of fragments: American Journal of Science, v. 282, p. 1694–1734.

Pessagno, E. A., 1973, Age and significance of radiolarian cherts in the California Coast Ranges: Geology, v. 1, p. 153–156.

Phipps, S. P., 1984, Ophiolitic olistostromes in the basal Great Valley sequence, Napa County, northern California Coast Ranges: Geological Society of America Special Paper 198, p. 103–125.

Platt, J. B., Liou, J. G., and Page, B. M., 1976, Franciscan blueschist facies metaconglomerates, Diablo Range, California: Geological Society of America Bulletin, v. 87, p. 581–591.

Platt, J. P., 1975, Metamorphic and deformational processes in the Franciscan complex, California: some insights from the Catalina Schist terrane: Geological Society of America Bulletin, v. 86, p. 1337–1347.

Platt, J. P., 1976, The petrology, structure and geologic history of the Catalina Schist terrain, southern California: University of California Publications in the Geological Sciences, v. 112, 111 p.

Raymond, L. A., 1970, Cretaceous sedimentation and regional thrusting northeastern Diablo Range, California: Geological Society of America Bulletin, v. 81, p. 2123–2328.

Raymond, L. A., 1973, Tesla-Ortigalita fault, Coast Range thrust fault and Franciscan metamorphism, northeastern Diablo Range, California: Geological Society of America Bulletin, v. 84, p. 3547–3562.

Schweickert, R. A., Armstrong, R. L., and Harakal, J. E., 1980, Lawsonite blueschist in the northern Sierra Nevada, California: Geology, v. 8, p. 27–31.

Seiders, V. M., and Blome, C. D., 1984, Clast compositions of upper Mesozoic conglomerates of the California Coast Ranges and their tectonic significance: in Blake, M. C., ed., Franciscan Geology of Northern California; Society of Economic Paleontologists and Mineralogists Pacific Section, v. 43, p. 135–148.

Silver, E. A., 1972, Pleistocene tectonic accretion of the continental slope off Washington: Marine Geology, v. 13, p. 239–249.

Sliter, W. V., 1984, Foraminifers from Cretaceous limestone of the Franciscan complex, Northern California: in Blake, M. C., ed., Franciscan Geology of

Northern California; Society of Economic Paleontologists and Mineralogists Pacific Section, v. 43, p. 149–162.

Smith, G. W., Howell, D. G., and Ingersoll, R. V., 1979, Late Cretaceous trench-slope basins of central California: Geology, v. 7, p. 303–306.

Suppe, J., 1969, Times of metamorphism in the Franciscan terrain of the northern Coast Ranges, California: Geological Society of America Bulletin, v. 80, p. 135–142.

Suppe, J., 1972, Interrelationships of high-pressure metamorphism, deformation and sedimentation in Franciscan tectonics, U.S.A.: XXIV International Geological Congress, Montreal, Section 3, p. 552–559.

Suppe, J., 1973, Geology of the Leech Lake-Ball Mountain region, California: University of California Publications in the Geological Sciences, v. 107, 82 p.

Suppe, J., 1979, Structural interpretation of the southern part of the northern Coast Ranges and Sacramento Valley, California: Summary: Geological Society of America Bulletin, Pt. 1, v. 90, p. 327–330.

Suppe, J., and Armstrong, R. L., 1972, Potassium-argon dating of Franciscan metamorphic rocks: American Journal of Science, v. 272, p. 217–233.

Suppe, J. and Foland, K. A., 1978, The Goat Mountain schists and Pacific Ridge complex: a redeformed but still intact Late Mesozoic Franciscan schuppen complex: *in* Howell, D. and McDougall, K., eds., Mesozoic Paleogeography of the Western United States, Society of Economic Paleontologists and Mineralogists Pacific Coast Paleogeography Symposium 2, p. 431–451.

Switzer, G., 1945, Eclogite from the California glaucophane schists: American Journal of Science, v. 243, p. 1–8.

Switzer, G., 1951, Mineralogy of the California glaucophane schists: California Division of Mines and Geology Bulletin, v. 161, p. 51–70.

Taylor, H. P., and Coleman, R. G., 1968, O^{18}/O^{16} ratios of coexisting minerals in glaucophane bearing metamorphic rocks: Geological Society of America Bulletin, v. 79, p. 1727–1756.

Telleen, K. E., 1977, Paleocurrents in part of the Franciscan complex California: Geology, v. 5, p. 49–51.

Underwood, M. B., 1984, A sedimentologic perspective on stratal disruption within sandstone-rich melange terranes: Journal of Geology, v. 92, p. 369–385.

Vedder, J. G., Howell, D. G., and McLean, H., 1983, Stratigraphy, sedimentation and tectonic accretion of exotic terranes, southern Coast Ranges, California: American Association of Petroleum Geologists Memoir 34, p. 471–496.

Wachs, D., and Hein, J. R., 1975, Franciscan limestones and their environments of deposition: Geology, v. 3, p. 29–33.

Wang, C. Y., and Shi, U. L., 1984, On the thermal structure of subduction complexes: A preliminary study: Journal of Geophysical Research, v. 89, p. 7709–7718.

Wood, B. L., 1971, Structure and relationships of late Mesozoic schists of northwest California and southwest Oregon: New Zealand Journal of Geology and Geophysics, v. 14, p. 219–239.

Wood, R. M., 1982, The Laytonville Quarry (Mendocino County California) exotic block: iron-rich blueschist facies subduction zone metamorphism: Mineralogical Magazine, v. 45, p. 87–99.

Worrall, D. M., 1981, Imbricate low-angle faulting in uppermost Franciscan rocks, south Yolla Bolly area, northern California: Geological Society of America Bulletin, v. 92, p. 703–729.

MANUSCRIPT ACCEPTED BY THE SOCIETY JULY 29, 1985

Geological Society of America
Memoir 164
1986

Geochronology of high-pressure–low-temperature Franciscan metabasites: A new approach using the U-Pb system

James M. Mattinson
Department of Geological Sciences
University of California
Santa Barbara, California 93111

ABSTRACT

The U-Pb isochron method is a promising new approach to the geochronology of high-pressure–low-temperature metabasites. In samples with favorable U/Pb ratios, metamorphic minerals such as sphene, apatite, lawsonite, glaucophane, garnet, and hornblende partition U and Pb in such a way as to provide a range of U/Pb ratios suitable for isochron dating. In a manner analogous to Rb-Sr isochron dating, these U-Pb isochrons provide not only ages, but also information on the initial isotopic composition of Pb at the time of metamorphism, a significant petrogenetic tracer. Sphene is the key mineral for dating. Its relatively high-U/Pb ratio results in the evolution of moderately radiogenic Pb, and it is highly resistant to resetting. Some metabasites have U/Pb ratios that are extremely low, perhaps owing to severe U depletion at some stage of metamorphism. These samples are not useful for dating, but still provide valuable data on the initial isotopic composition of Pb.

Analysis of a Type III metabasalt (blueschist) from the Taliaferro complex near Leech Lake Mountain yields a U-Pb isochron age of 162 ± 3 Ma., slightly older than the widely quoted 150-155 Ma. K-Ar ages for high-grade Franciscan tectonic blocks. Two garnet amphibolite blocks from the Catalina Schist terrane yield identical ages, with an isochron for both samples giving an age of 112.5 ± 1.1 Ma. These samples, plus three more Franciscan metabasites, have isotopic compositions of initial Pb ($^{206}Pb/^{204}Pb$ = 18.40 -18.85; $^{207}Pb/^{204}Pb$ = 15.55 - 15.66) that plot distinctly above the field for modern MORB; instead they plot in the fields for some island arcs and granodiorites from the Sierra Nevada.

INTRODUCTION

Blueschists and related high-pressure–low-temperature metamorphic rocks are now widely regarded as having formed in subduction-accretion complexes (Ernst, 1977, provides a valuable summary of the tectonic settings, mineral sequences, and relevant experimental phase-equilibrium studies). The ages of blueschist metamorphism within a terrane are thus critical to deciphering the timing and nature (episodic?, continuous?) of the subduction and accretion processes, and possible relationships with adjacent magmatic belts.

Most published ages of blueschists and related high-P/T metamorphic rocks have been determined by the K-Ar method, plus a very few Rb-Sr isochron determinations. As will be dis-

cussed below, these methods have provided a valuable geochronologic framework for understanding the general timing of high-P/T metamorphism, but have left open some important questions. In particular, a wide range of ages is typically obtained with the K-Ar method for different minerals from the same sample. This clearly indicates loss of radiogenic Ar from some if not all of the minerals. The most retentive phases, usually phengitic mica or hornblende, may or may not have lost some Ar, complicating interpretations of the actual time of metamorphic recrystallization.

This paper describes a new approach to the dating of high-P/T metabasites: application of the U-Pb isochron method. Preliminary work, discussed below, shows that, in favorable cases, accurate and precise ages may be obtained with this method. The key mineral is sphene (titanite), which when endowed with a favorable U/Pb ratio, provides moderately radiogenic points on the isochron. Sphene is highly resistant to resetting, and thus should yield accurate metamorphic crystallization ages for rocks that crystallized below about 500°C. Moreover, the isochron approach, in combination with complete analysis of the isotopic composition of Pb, yields information about the initial ratio of Pb in the sample at the time of metamorphism. The initial Pb isotopic composition is a powerful petrogenetic tracer, with the potential for distinguishing MORB-type, arc, and continental affinities.

BACKGROUND

High-P/T Rocks and Their Tectonic Significance

Blueschists and related high-P/T metamorphic rocks are generally recognized as having formed in subduction or collision zones, where the combination of high pressures (about 3 to more than 13 kb) and low temperatures (150 ± 50°C for low grade blueschists; 500°C or more for higher grade blueschists, eclogites, and amphibolites) can be simultaneously achieved (an excellent summary paper containing extensive references is Ernst, 1977). The Franciscan complex of western North America, with its abundant blueschists, and similar complexes elsewhere in the world, thus have been long regarded as trench assemblages— collections of oceanic off-scrapings and trench deposits—that have been tectonically shuffled and locally carried to great depth (for example, Hamilton, 1969, 1978; Ernst, 1970, 1974; Dickson, 1970, 1972; Page, 1970a, b).

High-P/T metamorphic rocks within the Franciscan complex occur in two tectonic settings: as "knockers" or exotic blocks within much lower grade rocks, usually metagraywackes; and as large tracts of intact Franciscan in which metamorphic zones can be mapped with some consistency. A detailed discussion is beyond the scope of this brief treatment, but in general, the high-grade exotic blocks are thought by most workers to represent the products of an early, possibly pre-Franciscan, period of subduction. The blocks are believed to have returned rapidly to the surface via some sort of reverse flow back up the the subduction zone (Cloos, 1982), major strike-slip faulting in the forearc region (Karig, 1980), or serpentinite diapirism (Carlson, 1981). The "intact" sequences of Franciscan are thought, in contrast, to have been metamorphosed *in situ.* Thus, the intact Franciscan represents the exposed portions of a fossil subduction complex. Ernst (1980) gives a recent treatment for the southern California Coast Ranges, as does Page (1981).

Blueschist belts such as those in the Franciscan complex are thus key elements in the overall tectonic picture of convergent-plate margins. The actual metamorphism clearly requires subduction to great depths of "packets" of light crustal rocks. Such subduction may be continuous or episodic, perhaps related to

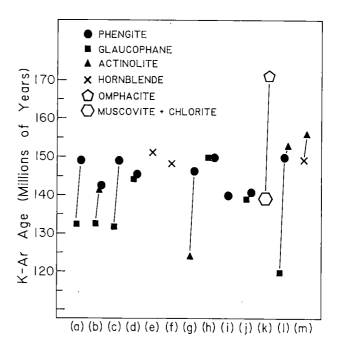

Figure 1. Summary of K-Ar ages on minerals from high-grade Franciscan tectonic blocks after Coleman and Lanphere (1971). Tie-lines (and overlapping symbols) represent different minerals from the same rock. Letters represent sample localities from Coleman and Lanphere (1971). Note the strong clustering of the oldest ages in the range of 150 to 155 Ma, and the marked spread in ages for different minerals from the same sample.

periods of rapid convergence (for example, see Mattinson and Echeverria, 1980). Establishing accurate chronologies for high-P/T metamorphism in complexes like the Franciscan is critical to understanding the nature of the processes of accretion, subduction, and metamorphism. For example, can discrete episodes of metamorphism be documented over large areas, or is there a continuum? If episodes of metamorphism can be distinguished, how do they correlate with pulses of magmatic activity in adjacent magmatic belts and periods of rapid subduction deduced from marine geophysics? The contributions of existing work, and the potential of the U-Pb method are explored in the following sections.

Previous Geochronologic Work

Previous geochronologic work on the blueschists and related metamorphic rocks of the Franciscan complex has been chiefly by the K-Ar method. The major "classic" studies are those of Coleman and Lanphere (1971) and Suppe and Armstrong (1972). Coleman and Lanphere reported a detailed study of mineral ages of high-grade blueschists, eclogites, and amphibolites that occur as exotic blocks in the Franciscan complex. A summary of their results is presented in Figure 1. Two major points emerge. First, there is a tight clustering of the *oldest* ages of the blocks at around 150 to 155 Ma. These ages are predominantly for phengitic white mica and hornblende. This age has been

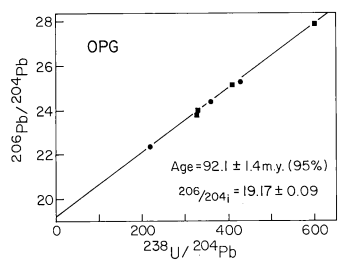

Figure 2. U-Pb isochron diagram for sphene, shown as squares, and "plagioclase" (a mixture of pumpellyite, albite, and quartz); shown as circles) after Mattinson and Echeverria, (1980). The sphene and "plagioclase" are metamorphic minerals from a siliceous differentiate in the Ortigalita Peak Gabbro. See text for details.

widely accepted as representing a major early (pre-Franciscan?) subduction event. Second, ages from coexisting minerals, for example, glaucophane, are typically lower; indicating Ar loss, probably during later, lower-grade events. Coleman and Lanphere (1971) also report one older age of about 170 Ma. on omphacite (Figure 1). They reject this as being too old (excess Ar?) because it is markedly older than the strong cluster of mica and hornblende ages at 150–155 Ma.

Suppe and Armstrong (1972) report a very large body of K-Ar dates on Franciscan metamorphic rocks, including both high-grade exotic blocks (mineral ages) and lower-grade intact Franciscan (chiefly whole-rock ages). The results of Suppe and Armstrong (1972) for the high-grade exotic blocks corroborate those of Coleman and Lanphere (1971), indicating maximum ages of 150 to 155 Ma. on white mica and hornblende, and significantly lower ages for glaucophane. Total-rock ages for the lower-grade, intact Franciscan metamorphic rocks range widely, but are always younger than the ages of the high-grade blocks, as would be expected from their geologic relations. Several periods of subduction, either discrete events or parts of a continuum, may be represented by the age data, but the "smearing out" of the ages precludes more detailed resolution.

Several other studies of age relationships in the Franciscan range from those reporting a few ages (for example, Maxwell, 1974) to a very detailed K-Ar study of the South Fork Mountain Schist by Lanphere and others (1978). Of particular interest, however, are the limited Rb-Sr data on Franciscan metamorphic rocks. Peterman and others (1967) report a Rb-Sr isochron age of 112 ± 16 Ma. on metagraywackes from the Yolla Bolly terrane. This is in general agreement with K-Ar ages from similar metagraywackes to the south. Additional Rb-Sr data are presented by Suppe and Foland (1978). Some samples of the Goat Mountain

Schist yield ages in agreement with K-Ar ages, but two samples yield apparent isochron ages of 180 Ma. These older ages are similar to the K-Ar age on omphacite reported by Coleman and Lanphere (1971), raising the possibility "of geologic significance for these 180 m.y. dates" (Suppe and Foland, 1978).

The only published U-Pb isochron for high-P/T metamorphism I am aware of is for metamorphic sphene (titanite) and "plagioclase" (= pumpellyite, albite, and quartz) by Mattinson and Echeverria (1980). This isochron is presented in Figure 2. The dated sample is a siliceous differentiate from a large gabbroic sill that intruded an intact Franciscan sequence at Ortigalita Peak after the Franciscan had been deformed in an accretionary prism. The sill and surrounding Franciscan rocks were then both metamorphosed to blueschist grade. The sill was emplaced about 95 Ma. ago (based on zircon dates, Mattinson and Echeverria, 1980), and the blueschist metamorphism followed only about 3 Ma. later based on the mineral isochron (Figure 2). Thus the sill was intruded in the accretionary wedge just prior to subduction. The 92 Ma. age of subduction (blueschist metamorphism) agrees well with K-Ar ages from intact graywacke sequences in the Diablo Range (Suppe and Armstrong, 1972), where the Ortigalita Peak gabbro is exposed. The proportion of siliceous igneous rocks in the Franciscan complex is vanishingly small, however. The application of U-Pb methods to metabasites, which are abundant in the Franciscan and similar complexes, is developed in the following section.

U-PB SYSTEMATICS AND GEOCHRONOLOGY

Application of U-Pb isochron methods to high P/T metamorphic rocks depends on a number of factors, each of importance, and exactly analogous to the factors required for successful Rb-Sr isochron work. First, the U-Pb balance of the total rock must be considered. For determination of precise ages, a relatively high U/Pb ratio, together with sufficient quantities of Pb for precise mass spectrometric analysis is most desirable. The actual U/Pb ratio is dependent on two main factors: the initial U and Pb complement of the protolith; and the possible effects of open system behavior during alteration or metamorphism, or both. Basalts, especially those derived from depleted mantle (MORB-type), typically contain variable, but low amounts of both U and Pb (for example, 0.029 to 0.52 ppm U and 0.22 to 2.14 ppm Pb from Juan de Fuca-Gorda Ridge area basalts, Church and Tatsumoto, 1975). However, local additions of large quantities of Pb to ocean floor basalts are possible by the mineralizing action of submarine hot springs ("black smokers"). Open-system behavior during metamorphism could lead to exchange of Pb or U or both with surrounding rocks of different character (for example, metagraywackes), or possibly to the loss of these elements during dehydration reactions accompanying prograde metamorphism. Possible examples of open-system behavior will be presented later.

The second major requirement for U-Pb isochron dating is complete isotopic homogenization of Pb during metamorphism,

J. M. Mattinson

TABLE 1. U-Pb DATA FOR FRANCISCAN METABASITES

Sample*	Pb(tot)†	238U†	208Pb/204Pb §	207Pb/204Pb §	206Pb/204Pb §	238U/204Pb
TC Sa(ID)	0.7723	6.294	..	..	35.70+0.10	659.8+3.3
(IC)	..	..	39.79+0.13	16.29+0.05	32.96+0.10	..
TC Sb(ID)	0.8560	5.761	..	..	32.33+0.11	526.6+2.6
(IC)	..	..	40.26+0.13	16.32+0.05	32.52+0.10	..
TC Sc(ID)	0.7580	6.098	..	..	35.05+0.12	647.5+3.2
(IC)	..	..	39.98+0.40	16.42+0.16	35.32+0.30	..
TC La(ID)	1.908	1.299	..	..	19.98+0.03	44.90+0.22
(IC)	..	..	39.12+0.10	15.72+0.03	19.98+0.03	..
TC Lb(ID)	2.140	1.441	..	..	20.07+0.03	44.70+0.22
(IC)	..	..	39.42+0.08	15.75+0.03	20.05+0.02	..
TC Ga(ID)	0.9420	0.6430	..	..	19.90+0.04	45.25+0.23
(IC)	..	..	39.56+0.08	15.68+0.03	19.86+0.02	..
TC Gb(ID)	1.085	0.7270	..	..	19.96+0.03	44.66+0.22
(IC)	..	..	39.87+0.10	15.73+0.03	19.97+0.03	..
HGB-2 Sa	2.217	8.060	40.83+0.09	15.83+0.02	23.23+0.03	254.4+0.7
HGB-2 Sb	2.187	8.422	40.88+0.09	15.84+0.02	23.55+0.03	272.5+0.8
HGB-2 Sc	2.029	8.110	41.06+0.08	15.87+0.02	23.77+0.02	284.2+0.9
HGB-2 Am	0.5227	0.0973	38.70+0.08	15.65+0.02	19.02+0.02	12.05+0.04
HGB-2 Gt	0.0293	0.0462	39.10+0.20	15.67+0.08	20.71+0.10	104.6+1.0
SC-2 Sa	3.794	7.739	38.77+0.09	15.78+0.02	21.27+0.03	140.1+0.4
SC-2 Sb	3.616	7.466	38.68+0.08	15.74+0.02	21.18+0.02	137.7+0.4
SC-2 Am	1.163	0.3176	38.53+0.08	15.64+0.02	19.07+0.02	17.65+0.05
SC-2 Apa	4.826	1.611	38.50+0.08	15.64+0.02	19.11+0.02	21.57+0.06
SC-2 Apb	5.017	1.616	38.52+0.08	15.64+0.02	19.11+0.02	20.83+0.06
SC-2 Gt	0.0753	0.1358	38.86+0.14	15.74+0.05	20.96+0.06	120.3+0.04
WC-1 S	2.583	0.1809	38.37+0.08	15.65+0.02	18.61+0.02	4.48+0.02
PNP-1 S	2.252	0.0729	38.11+0.08	15.55+0.02	18.46+0.02	2.06+0.01
AN-1 S	1.565	0.1529	38.22+0.08	15.59+0.02	18.58+0.02	6.25+0.03

*S, L, G, Am, Gt, and Ap represent sphene, lawsonite, glaucophane, amphibole (hornblende), garnet, and apatite, respectively. Different fractions of the same mineral are designated a, b, c. Sample TC was analyzed using a 208Pb tracer, necessitating separate digestions for concentrations (ID) and isotopic composition (IC).

†Concentrations of total Pb and 238U are listed in ppm.

§The reported isotopic compositions have been normalized to "absolute" values by applying fractionation corrections of 0.1 to 0.125% per mass unit, depending on the temperature of the mass spectrometer run. These corrections are based on replicate analyses of NBS Pb standards.

so that each new mineral phase contains Pb of the same isotopic composition. The thorough mineralogic reconstitution of almost all well-crystallized blueschists, eclogites, and amphibolites almost assures the necessary isotopic homogenization.

The third requirement is the availability of mineral phases with different relative partition coefficients for U and Pb, such that a wide range of U/Pb ratios is represented in the metamorphic mineral assemblage. In this regard, sphene (titanite) appears to provide the highest U/Pb ratios, whereas the amphiboles, lawsonite, and apatite have relatively low U/Pb ratios. Garnet appears to have intermediate U/Pb ratios, although the absolute amounts of both U and Pb are extremely low, hindering precise isotopic analysis. The availability of sphene is of particular importance, in that studies of the behavior of the U/Pb system in sphene from igneous rocks indicate closed-system behavior below about 500°C (Mattinson, 1978, 1982). For rocks metamorphosed at or below about 500°C then, sphene should record the actual time of metamorphism, rather than a later cooling age.

Assuming that the three requirements discussed above are

met, the plotting on an isochron diagram of $^{206}Pb/^{204}Pb$ versus $^{238}U/^{204}Pb$ for the mineral phases will yield information on the age of metamorphism and the initial composition of Pb at the time of metamorphism. In addition, the data can be subjected to rigorous statistical analysis of the type commonly applied to Rb-Sr isochrons (for example York, 1969), in order to assess the quality of fit of the data points to the isochron.

Taliaferro Complex Type III Metabasalt

Type III metabasalt from the Taliaferro Complex of Suppe (1969) is a typical, well-crystallized blueschist. The sample was kindly provided by Dr. M. C. Blake, Jr. of the U.S. Geological Survey, who notes the possibility that, although considered part of the intact Taliaferro complex by Suppe, this blueschist may actually be a tectonic block brought up in serpentinite. The major minerals in the rock are glaucophane and lawsonite, with lesser amounts of aragonite, white mica, and sphene. U-Pb analytical data for this sample are summarized in Table 1, and a U-Pb

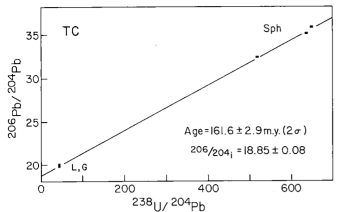

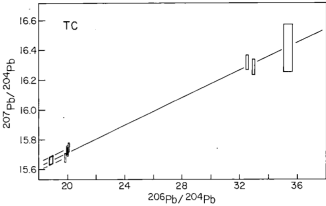

Figure 3. U-Pb isochron diagram for minerals from the Taliaferro complex, Type III metabasalt (TC). Three magnetic fractions of sphene (sph) are shown in the upper-right portion of the diagram, and two analyses each of lawsonite (L) and glaucophane (G) are represented by the small rectangle in the lower-left portion of the diagram. The size of the rectangles represents the uncertainties in the measured ratios. The age, initial ratio, and uncertainties (all at the two-sigma level) have been determined using the regression methods of York (1969) as discussed in Faure (1977).

Figure 4. $^{207}Pb/^{204}Pb$ versus $^{206}Pb/^{204}Pb$ plot for sample TC. The sizes of the rectangles represent the uncertainties in the measured ratios. The lower end of the regression line is fairly well controlled by the cluster of four lawsonite and glaucophane points with $^{206}Pb/^{204}Pb$ ratios of about 20. The initial isotopic composition of Pb for the sample is diagrammatically represented by the heavy parallelogram, the intersection of the initial $^{206}Pb/^{204}Pb$ determination from Figure 3, and the regression through the $^{207}Pb/^{204}Pb$ and $^{206}Pb/^{204}Pb$ ratios for the various minerals. For a $^{206}Pb/^{204}Pb$ initial ratio of 18.85 ± 0.08, regression analysis indicates an initial $^{207}Pb/^{204}Pb$ ratio of 15.66 ± 0.02 (two-sigma confidence level).

isochron derived from the data is shown in Figure 3. Assuming that none of the unanalyzed phases contains significant concentrations of U or Pb, the U-Pb balance of the rock is controlled by glaucophane and lawsonite. The U/Pb ratios and Pb isotopic compositions of these two minerals are essentially identical (Table 1), but lawsonite has higher absolute concentrations of both U and Pb by a factor of about two. As can be seen on the isochron (Figure 3), the U/Pb ratio of these minerals is sufficient to have produced a small but significant increase in the $^{206}Pb/^{204}Pb$ ratio (from about 18.85 to about 20) in the time since metamorphic recrystallization (162 Ma.). Three different fractions of sphene, however, have U/Pb ratios some 10 to 15 times greater than those for glaucophane and lawsonite, producing substantial enrichments of radiogenic ^{206}Pb. The three magnetic fractions of sphene spread out slightly along the isochron (Figure 3), and along with the two duplicate analyses each for glaucophane and lawsonite, define an excellent isochron. The good fit of all the points to the isochron indicates clearly that there was complete isotopic homogenization of Pb during metamorphism, and probably no subsequent isotopic disturbance of the system.

The 162 ± 3 Ma. age for the Taliaferro Complex metabasalt (a preliminary age of 164 ± 2 was reported earlier in Mattinson, 1981) can be compared with previously reported K-Ar ages from the same complex. Suppe (1969) reported maximum ages (recalculated using new decay constants) of about 154 ± 3 Ma. for white mica from metabasalt; and Suppe and Armstrong (1972) reported ages ranging from about 90 to 115 Ma. from total-rock metagraywackes of the Taliaferro Complex. The metagraywacke and metabasalt ages clearly represent distinct younger and older

events. The small but significant difference between the U-Pb isochron age and the K-Ar white mica age suggests that the white mica has lost a small amount of radiogenic Ar during the 90-115-Ma. event recorded in the metagraywackes. On this basis, many of the previously reported ages for white micas from high-grade blueschists and other high-P/T metamorphic rocks with complex thermal histories may be only minimum ages for metamorphic recrystallization.

The isotopic composition of initial Pb for the sample (that is, at the time of metamorphism) can be obtained directly from the isochron diagram in the case of the $^{206}Pb/^{204}Pb$ ratio, and by calculation or regression on a $^{207}Pb/^{204}Pb$ versus $^{206}Pb/^{204}Pb$ plot in the case of the $^{207}Pb/^{204}Pb$ ratio. As noted earlier, the isotopic composition of initial Pb is a powerful tracer of petrogenetic processes, reflecting as it does the long-term, time-integrated ratio of U (and Th) to Pb in the eventual source(s) of, say, basaltic magmas. "Depleted" mantle, undepleted or enriched mantle, and various crustal sources each generate distinctive isotopic signatures as a result of different parent-daughter decay histories over substantial portions of the earth's history. In many cases then, the isotopic composition of initial Pb in a sample can be used to place constraints on the tectonic setting in which the magma was generated.

The initial Pb from the Taliaferro Complex metabasalt, as deduced from the isochron diagram (Figure 3) and a plot of $^{207}Pb/^{204}Pb$ versus $^{206}Pb/^{204}Pb$ (Figure 4) is 206:207:204 = 18.85 ± 0.08 : 15.66 ± 0.02 : 1.0. The significance of these data, together with initial Pb data from a number of other Franciscan metabasites is considered later.

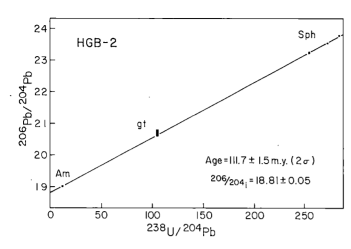

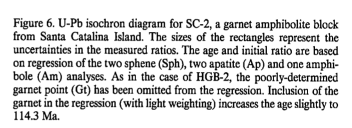

Figure 5. U-Pb isochron diagram for HGB-2, a garnet amphibolite block from Santa Catalina Island. The sizes of the rectangles represent the uncertainties in the measured ratios. The age and initial ratio are based on regression of the three sphene (Sph) and one amphibole (Am) analyses only. The garnet point (Gt) is less well determined because of the very low levels of Pb in the garnet. Inclusion of garnet in the regression lowers the age very slightly to 111.3 Ma. (the poorly-determined garnet point is given only light weighting by the regression program).

Figure 6. U-Pb isochron diagram for SC-2, a garnet amphibolite block from Santa Catalina Island. The sizes of the rectangles represent the uncertainties in the measured ratios. The age and initial ratio are based on regression of the two sphene (Sph), two apatite (Ap) and one amphibole (Am) analyses. As in the case of HGB-2, the poorly-determined garnet point (Gt) has been omitted from the regression. Inclusion of the garnet in the regression (with light weighting) increases the age slightly to 114.3 Ma.

Santa Catalina Island Garnet Amphibolite

High-grade garnet amphibolite occurs as tectonic blocks in the Catalina Schist terrane (Platt, 1975). The garnet amphibolite blocks represent the highest grade of metamorphism in the terrane, with T and P estimated at 600°C and 10 kb, respectively, by Platt (1975), and up to 660°C and 10 kb, respectively by Sorensen (1984a). The garnet amphibolites consist predominantly of garnet, brown hornblende, and clinopyroxene, plus minor sphene, apatite, and zoisite. Mineral separates from two blocks (HGB-2 and SC-2) were kindly provided by Dr. Sorena Sorensen of UCLA (now at the Smithsonian Institution).

U-Pb data for the two blocks are presented in Table 1. For HGB-2, brown hornblende, garnet, and three fractions of sphene were analyzed. Hornblende, the most abundant constituent in the rock, contains a moderate amount of Pb, about 0.5 ppm, and less than 0.1 ppm U. Thus, the measured Pb is relatively non-radiogenic (Table 1). Garnet, the other major constituent of the rock that was analyzed, contains very small amounts of both Pb and U, on the order of 0.03 to 0.05 ppm. However, unlike the hornblende discussed above, the garnet has a relatively high U/Pb ratio, so that the measured Pb contains a significant radiogenic component. Sphene contains the highest Pb and U concentrations of any mineral analyzed from HGB-2, about 2.0 to 2.2 ppm and 8.0 to 8.4 ppm, respectively. The small differences in concentrations among the three analyzed fractions probably reflect heterogeneities (zoning?, impurities?) as a result of the small sample sizes and relatively coarse-grained nature of the mineral separates. The relatively high U/Pb ratio in the sphene fractions, almost three times as great as that observed for garnet, and more

than 20 times as great as that observed for hornblende, has resulted in a highly favorable degree of radiogenic enrichment of the Pb in the sphene. The spread in U/Pb ratios for all the mineral phases analyzed permits the construction of a well-defined, precise isochron (Figure 5) that indicates an age of 111.7 ± 1.5 Ma. (two-sigma confidence level), with an initial $^{206}Pb/^{204}Pb$ ratio of 18.81 ± 0.05.

For the second block, SC-2, brown hornblende, garnet, two fractions of apatite, and two fractions of sphene were analyzed (Table 1). The relative patterns of Pb and U concentrations and U/Pb ratios for the hornblende, garnet, and sphene are similar to those for HGB-2, discussed above, with garnet having the lowest absolute concentrations, hornblende the lowest U/Pb ratio, and sphene the highest U concentration and U/Pb ratio. However, the absolute concentrations of Pb and U for SC-2 are significantly different from those of HGB-2. In particular, the major minerals in the two samples, hornblende and garnet, have Pb concentrations that are about twice as great, and U concentrations that are about three times as great for SC-2 than for HGB-2. Sphene from SC-2 has a much higher Pb content than does sphene from HGB-2, but about the same U content (Table 1). The possible significance of these differences in Pb and U concentrations will be discussed later. Apatite from SC-2 contains very high concentrations of Pb—about five ppm—plus about 1.6 ppm U. Thus the Pb in apatite is only slightly more radiogenic than Pb in the hornblende. The SC-2 mineral data are plotted on a U-Pb isochron diagram in Figure 6. Despite the less radiogenic nature of Pb from the sphene fractions (owing to the lower U/Pb ratios than for the HGB-2 sphenes), a relatively precise isochron is still defined by the data, with an age of 114.1 ± 2.1 Ma. (two-sigma confidence level), and an initial $^{206}Pb/^{204}Pb$ ratio of 18.74 ± 0.03.

The small differences in age and initial ratio for the two samples, HGB-2 and SC-2, are not significant statistically. Regressing the data for both samples together yields an age of 112.5 ± 1.1 Ma. and an initial $^{206}Pb/^{204}Pb$ ratio of 18.76 ± 0.03 (errors at the two-sigma level).

The U-Pb isochron age(s) reported here may be compared with K-Ar ages for Santa Catalina metamorphic rocks, including a garnet amphibolite, reported by Suppe and Armstrong (1972). The K-Ar ages (recalculated using new decay constants) range from about 98 ± 2 Ma. (glaucophane from blueschist) to 112 ± 3 Ma. (hornblende from garnet amphibolite). The K-Ar hornblende age is in agreement with the U-Pb ages from this work, confirming the similarity in "blocking temperatures" (at about 500°C) of the K-Ar system in hornblende and the U-Pb system in sphene proposed by Mattinson (1978, 1981). However, in the case of these high-grade amphibolites, with estimated crystallization temperatures of 600 to 660°C, the ages must still be considered minimum ages for the metamorphism. Preservation of high-P/T mineral assemblages throughout the Catalina Schist terrane, and of the inverted metamorphic gradient in the terrane (Platt, 1975), suggests rapid cooling (and uplift?) after metamorphism, however, so the ages probably are close approximations of the actual time of metamorphism.

Initial $^{206}Pb/^{204}Pb$ ratios derived from the U-Pb isochron diagrams for each amphibolite sample (HGB—2 and SC-2) were discussed above. Complete Pb isotopic data for both samples are plotted in Figure 7. The lower half of Figure 7 shows a plot of $^{207}Pb/^{204}Pb$ versus $^{206}Pb/^{204}Pb$. The scale for the $^{207}Pb/^{204}Pb$ axis is greatly expanded to show the small differences in radiogenic enrichment of the ^{207}Pb from the various minerals (which vary in $^{207}Pb/^{204}Pb$ ratio from about 15.64 to 15.87). A best-fit regression line to the data is very tightly controlled at its lower end by the clustering of the two hornblende and two apatite analyses. This permits the determination of a highly precise initial $^{207}Pb/^{204}Pb$ for the amphibolites. For an initial $^{206}Pb/^{204}Pb$ ratio of 18.76 ± 0.03, the initial $^{207}Pb/^{204}Pb$ ratio is 15.633 ± 0.016.

The upper half of Figure 7 is a plot of $^{208}Pb/^{204}Pb$ versus $^{206}Pb/^{204}Pb$. The mass 208 isotope of Pb is, of course, produced by the decay of ^{232}Th, whereas masses 207 and 206 are produced by the coupled decay of the U isotopes 235 and 238, respectively. Thus the array of data on the $^{208}Pb/^{204}Pb$ versus $^{206}Pb/^{204}Pb$ plot is a function not only of the age of the sample, but also of the Th/U ratio of each mineral analyzed. Strictly speaking, then, the significance of a regression line through data points for the various minerals, in the absence of actual measurements of Th concentrations, is questionable. Nevertheless, two points seem clear from inspection of the plot. First, the minerals from the two samples, HBG-2 and SC-2, define very different trends which converge to a common $^{208}Pb/^{204}Pb$ value for $^{206}Pb/^{204}Pb$ values in the range of the initial values indicated in Figures 6 and

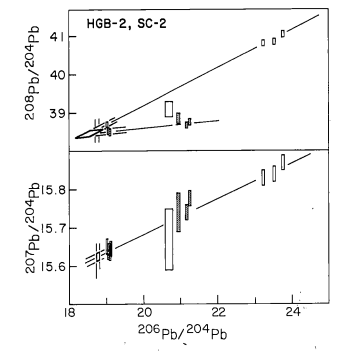

Figure 7. Plot of $^{207}Pb/^{204}Pb$ and $^{208}Pb/^{204}Pb$ versus $^{206}Pb/^{204}Pb$ for HGB-2 (open rectangles) and SC-2 (cross-hatched rectangles). The sizes of the rectangles represent the uncertainties in the isotopic ratios. In the lower half of the diagram, the heavy parallelogram represents the initial $^{206}Pb/^{204}Pb$ and $^{207}Pb/^{204}Pb$ ratios for the samples as indicated by the intersection of the initial $^{206}Pb/^{204}Pb$ ratio determined from the U-Pb isochrons (see Figures 5 and 6, and text) with the regression line through the $^{207}Pb/^{204}Pb$ and $^{206}Pb/^{204}Pb$ ratios (garnet, shown as the two largest rectangles, has been omitted from the regression). For an initial $^{206}Pb/^{204}Pb$ ratio of 18.76 ± 0.03, the initial $^{207}Pb/^{204}Pb$ ratio is 15.633 ± 0.016 (two-sigma uncertainty).

7. Thus both samples had essentially identical Pb isotopic compositions about 112 Ma. ago: ^{208}Pb: ^{207}Pb: ^{206}Pb: ^{204}Pb: = 38.55 ± 0.05 : 15.633 ± 0.016 : 18.76 ± 0.03 : 1.0. The possible significance of these results in terms of protolith petrogenesis is discussed in the next section. Second, the sharp divergence of the two trends indicates clearly that HGB-2 has a much higher Th/U ratio than does SC-2. In fact, using sphene as an indicator, a simple calculation that assumes only that the ^{208}Pb-^{232}Th system closed at the same time as the ^{206}Pb-^{238}U system; and that both samples had a common $^{208}Pb/^{204}Pb$ ratio at the time of metamorphism (see discussion above), suggests that the Th/U ratio of HGB-2 is almost ten times as great as the Th/U ratio of SC-2. Whether this marked difference reflects primary protolith characteristics, or open system behavior at some stage during alteration or metamorphism, is not clear. However, Sorensen (1983, 1984a, b) has documented major metasomatic effects in the rinds that surround these blocks. Thus, significant, metasomatically-produced variations in U and Th concentrations would not be surprising. The major minerals in HGB-2 (hornblende and garnet) have U concentrations that are about a factor of three lower than are the U concentrations in the same minerals in SC-2 (Pb is also

TABLE 2. CALCULATED INITIAL Pb

Sample*		$^{207}Pb/^{204}Pb$	$^{206}Pb/^{204}Pb$
WC-1 S	obs.	15.65$_0$	18.61
	100 Ma	15.647	18.54
	160 Ma	15.64$_4$	18.50
PNP-1 S	obs.	15.55$_0$	18.46
	100 Ma	15.54$_8$	18.43
	160 Ma	15.547	18.41
AN-1 S	obs.	15.59$_0$	18.58
	100 Ma	15.58$_5$	18.48
	160 Ma	15.58$_2$	18.42

*S designates sphene analysis. "obs." is observed data from Table 1. 100 Ma and 160 Ma rows represent initial ratios calculated using these assumed ages and the observed U/Pb data for each sample (Table 1).

lower by a factor of about two); so U loss from HGB-2 (during metamorphism?) is a possible major factor in the difference in Th/U ratios.

Other Franciscan Results

Several other samples from the Franciscan complex and elsewhere have been analyzed. Unfortunately, in all these cases, sphene, the critical mineral, has been characterized by a highly unfavorable U/Pb ratio (on the order of 100 times lower than for the samples discussed above). Data for three such Franciscan samples from Ward Creek, Panoche Pass, and Annapolis are listed in Table 1. Inspection of the analyses reveals that all three sphene separates have Pb concentrations similar to those in the sphenes from the Taliaferro and Catalina samples. However, the U concentrations are extremely low, so that they have lower $^{238}U/^{204}Pb$ ratios than do any of the minerals from the successfully dated samples. Further work is underway on this problem, but it seems likely that the bulk rocks must be severely depleted in U. One possible mechanism for this depletion is the removal of U (in fluids?) during the crystallization of high-grade, anhydrous mineral assemblages. For example, the typical eclogite assemblage of omphacitic pyroxene, garnet, and rutile seems unlikely to provide sites for U substitution. Sphene, crystallizing later as a part of a retrograde blueschist assemblage, will simply find very little U available to incorporate. Thus, some coarsely-crystallized blueschists have proven unsuitable for dating. Such samples at least provide precise determinations of initial Pb, which can be readily calculated from the observed Pb isotope ratios, the U/Pb ratios, and assumed ages. Initial $^{206}Pb/^{204}Pb$ and $^{207}Pb/^{204}Pb$ ratios, calculated for ages of 100 and 160 Ma. are summarized in Table 2, along with the previously discussed initial ratios for the dated samples. All of the initials are plotted in a lead evolution diagram ($^{207}Pb/^{204}Pb$ versus $^{206}Pb/^{204}Pb$) in Figure 8, where they can be compared with data from modern mid-ocean-ridge

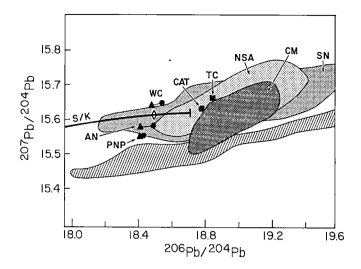

Figure 8. Isotopic composition of initial Pb from Franciscan metabasites. TC = Taliaferro complex Type III metabasalt; CAT = Santa Catalina garnet amphibolite data; WC = Ward Creek blueschist; AN = blueschist-eclogite from Annapolis area, PNP = blueschist from Panoche Pass area. Solid squares represent data inferred from isochron and Pb/Pb plots. Circles and triangles represent data calculated from measured Pb isotopic compositions and U/Pb ratios assuming ages of 100 and 160 Ma., respectively. Cross-hatched area is the field of mid-ocean ridge basalts (MORB) and oceanic-island basalts (OIB) from Bass and others (1973), Church and Tatsumoto (1973), and Sun (1980). Field-labeled CM is the field of volcanic rocks from the High Cascades of Washington, Oregon, and California from Church and Tilton (1973). Field-labeled NSA is the field of volcanic rocks from North and South Atlantic island arcs (Lesser Antilles from Armstrong and Cooper, 1971; and South Sandwich Islands from Barreiro, 1983). Field-labeled SN represents granodiorites from the Sierra Nevada from Chen and Tilton (in Tilton, 1983). The heavy curve labeled S/K is a model Pb-evolution curve approximating crustal evolution from Stacey and Kramers (1975). The vertical bar at the end of the curve represents modern crustal Pb, and the diamond on the curve represents 160 Ma-old crustal Pb, for reference.

and ocean-island basalts, and magmatic-arc rocks. All five Franciscan samples show a moderate spread in both $^{206}Pb/^{204}Pb$ and $^{207}Pb/^{204}Pb$ ratios. Their isotopic compositions are clearly distinct from the field of MORB and OIB (Figure 8), having $^{207}Pb/^{204}Pb$ ratios that are significantly higher than the MORB-OIB trend for given $^{206}Pb/^{204}Pb$ ratios. Thus, the Pb in the Franciscan metabasites cannot have been derived exclusively from an oceanic (depleted mantle) source. Instead, they resemble Pb's from some Mesozoic and Recent magmatic arcs. Nelson and DePaolo (1982) reach a similar conclusion for Franciscan eclogites and blueschists on the basis of initial Nd and Sr ratios, and Nd/Sm ratios. A question that awaits further research concerns the extent to which the observed Pb isotopic compositions accurately reflect primary protolith values, as opposed to reflecting the influence of metasomatic fluids and exchange with surrounding rocks such as metagraywackes. Isotopic analyses of a wider range of Franciscan metamorphic rocks will perhaps provide an answer. Meanwhile, the fact that metabasites from a wide range of settings yield isotopic compositions that are consistently more radiogenic in terms of $^{207}Pb/^{204}Pb$ ratios than is the "depleted mantle array" argues against derivation of Franciscan high-P/T metabasites from MORB-type oceanic crust.

CONCLUSIONS

U-Pb isotopic mineral analysis is a powerful new tool in the study of high-pressure–low-temperature metamorphic rocks. Depending on the U/Pb ratio of the protolith, and the nature of the metamorphic or metasomatic processes, or both, involved in the formation of high-P/T metamorphic rocks, the U-Pb method can provide precise ages, useful petrogenetic information, and perhaps an added understanding of the role of fluids in Pb isotopic exchange and U depletion.

For the determination of metamorphic ages, the key mineral phase is sphene (titanite). Under favorable circumstances, sphene concentrates sufficient U relative to Pb during metamorphic crystallization to permit precise dating of samples older than several tens of Ma. Sphene is a relatively "well behaved" mineral, being highly resistant to weathering, alteration, and to loss of Pb below about 500°C. Thus, U-Pb isochron ages based on sphene appear to be most comparable to K-Ar hornblende ages in terms of temperature response, and on the basis of limited comparison, give more accurate (older) ages than K-Ar on white mica.

Pending the completion of more U-Pb work, it is premature to do more than speculate, but the 162 Ma. age reported here for the Taliaferro(?) metabasalt suggests that the main early period of bleuschist metamorphism took place somewhat earlier than previously thought on the basis of K-Ar work. If this is the case, then this period of metamorphism overlaps almost exactly with the time of formation of the Coast Range ophiolite (Hopson and others, 1981; Mattinson, in prep.), and with a major episode of

arc magmatism in the Klamath and Sierran regions (Wright and Sharp, 1982).

The 112 Ma. age reported here for garnet amphibolites occurring as tectonic blocks in the Catalina Schist terrane confirms the oldest ages (K-Ar on hornblende) previously reported for the Catalina Schist terrane by Suppe and Armstrong (1972). This combination of high grade and relatively young age mark the Catalina Schist terrane as distinct from any other known Franciscan terrane.

The isotopic compositions of initial Pb from the above samples, and from three other Franciscan high-P/T metamorphic rocks (all metabasites), show a moderate range in isotopic composition, with $^{206}Pb/^{204}Pb$ = 18.40 to 18.84, and $^{207}Pb/^{204}Pb$ = 15.55 to 15.66. All of these values plot above the field of oceanic basalts, in terms of their $^{207}Pb/^{204}Pb$ ratios. Instead, the results more closely resemble arc-type values. If the reported initial Pb is simply inherited from the protoliths of the metabasites, then an oceanic crustal origin can be ruled out. The observed values may not accurately reflect the nature of the protoliths of the metamorphic rocks, however. Considering the possibility of metasomatic effects during metamorphism, conclusions based on isotopic compositions are tentative at best, pending further work.

ACKNOWLEDGMENTS

M. C. Blake, Jr., and C. A. Hopson have generously offered advice, encouragement, and samples for this project, sharing willingly their extensive knowledge of the Mesozoic evolution of Western North America. S. S. Sorensen provided samples (pure mineral separates!) from Santa Catalina Island, and W. S. Wise kindly supplied a large blueschist-eclogite sample from a new "knocker" locality near Annapolis, Sonoma County. The research was supported by a National Science Foundation Grant EAR-8205823.

APPENDIX 1. SAMPLE DESCRIPTIONS AND LOCALITIES

TC-1: Type III metabasalt from the Taliaferro complex of Suppe (1969), medium-grained blueschist consisting of glaucophane, lawsonite, aragonite, white mica, and sphene. Collected by M. C. Blake, Jr., just north of Leech Mountain (M. C. Blake, Jr., pers. comm., 1984).

HGB-2: garnet amphibolite from the Catalina Schist terrane of Platt (1975). Consists of garnet, hornblende, clinopyroxene, and minor sphene, apatite, and zoisite. Occurs as a tectonic block in serpentinite. Collected by S. S. Sorensen about 0.5 miles due north of the Santa Catalina airport.

SC-2: garnet amphibolite from the Catalina Schist terrane of Platt

(1975). Consists of garnet, clinopyroxene, hornblende, and minor sphene, apatite, and zoisite. Float block collected by S. S. Sorensen in the creek bed of Buffalo Springs Canyon 0.75 mi. northeast of Rancho Escondito.

WC-1: blueschist from the Ward Creek area. Sphene separate supplied by M. C. Blake, Jr.

PNP46-1: garnet-bearing blueschist. Consists of glaucophane, garnet, lawsonite, white mica, and sphene. Collected by C. A. Hopson from the Panoche Pass area

AN-1: coarse-grained interlayered blueschist and eclogite. Analyzed sphene is a single large grain from a vein. Collected by W. S. Wise near the town of Annapolis in Sonoma County.

APPENDIX 2. ANALYTICAL PROCEDURES

Samples of sphene, hornblende, and garnet weighing from about 10-50 mg were digested in small stainless steel-jacketed teflon "bombs" in 50% HF plus a few drops of concentrated HNO_3 for several days at 210-215°C. Isotopic tracers ("spikes") were added prior to digestion. In the early stages of this work, a mixed ^{208}Pb-^{235}U tracer was used, and a

separate mechanical split of the sample was digested to determine the isotopic composition of Pb. Most of the work reported here used a mixed $^{205}Pb/^{235}U$ tracer, eliminating the need for a separate digestion to determine P isotopic composition. Following the first digestion, the samples were evaporated to dryness, and the digestion repeated. Following the second digestion and evaporation, 0.1 ml of concentrated HNO$_3$ was added, and the samples were again dried. This last step was repeated three times to break down fluorides. The sample was then converted to chloride form by evaporation with 0.2 ml 6N HCl three times. Following the final evaporation to dryness, one ml of mixed acids (one part 6N HCl plus two parts 1N HBr) was added to each sample, and the bombs were heated overnight at 135°C to dissolve the samples. Pb was then separated by HBr chemistry (1N HBr) on a 0.15 ml resin volume teflon column using Dowex 1×8 resin. The procedure was repeated to further purify the Pb prior to mass spectrometry. U was separated using 7N HNO$_3$ and Dowex 1×8, first on a 0.5-ml column, then on a 0.15-ml column. Pb blanks were on the order of 0.2 to 0.4 ng throughout the study—

significant only for the garnet analyses. U blanks were negligible. Separation of Pb and U for apatite was identical to that for the other minerals, but the apatite was digested directly in the HCl-HBr-column solution in small screw-top PFA Teflon capsules at 80°C on a hot plate. Pb blanks for the apatite work were on the order of 0.1 ng Pb. U blanks were negligible. Pb was run on the mass spectrometer using the usual silica-gel and phosphoric-acid method. All work except for the earliest (Taliaferro complex) utilized the UCSB MAT261 multicollector instrument, permitting the simultaneous collection of four Pb isotopes. U was run as the metal by loading with phosphoric acid and powdered graphite. All Pb and U runs were normalized using the results of replicate analyses on NBS Pb and U standards.

REFERENCES CITED

Armstrong, R. L., and Cooper, J. A., 1971, Lead isotopes in island arcs: Bulletin Volcanologic, v. 35, p. 27–63.

Barreiro, Barbara, 1983, Lead isotopic compositions of South Sandwich Islands volcanic rocks and their bearing on magmagenesis in intra-oceanic island arcs: Geochimica et Cosmochimica Acta, v. 47, p. 817–822.

Bass, M. H., Moberly, R., Rhodes, J. M., Shih, C., and Church, S. E., 1973, Volcanic rocks cored in the central Pacific, Leg 17, Deep Sea Drilling Project: in Initial Reports of the Deep Sea Drilling Project, v. 17, Washington, D.C., United States Government Printing Office, p. 429–503.

Carlson, C., 1981, Upwardly mobile melanges, serpentinite protrusions and transport of tectonic blocks in accretionary prisms: Geological Society of America Abstracts with Programs, v. 13, p. 48.

Church, S. E., and Tatsumoto, Mitsunobu, 1975, Lead isotope relations in oceanic ridge basalts from the Juan de Fuca-Gorda Ridge area, N.E. Pacific Ocean: Contributions to Mineralogy and Petrology, v. 53, p. 253–279.

Church, S. E., and Tilton, G. R., 1973, Lead and strontium studies in the Cascade Mountains: bearing on andesite genesis: Geological Society of America Bulletin, v. 84, p. 431–454.

Cloos, M., 1982, Flow melanges: numerical modeling and geological constraints on their origin in the Franciscan subduction complex, California: Geological Society of America Bulletin, v. 93, p. 330–345.

Coleman, R. G., and Lanphere, M., 1971, distribution and age of high-grade blueschists, associated eclogites, and amphibolites from Oregon and California: Geological Society of America Bulletin, v. 82, p. 2397–2412.

Dickinson, W. R., 1970, Relation of andesites, granites, and derived sandstones to arc-trench tectonics: Reviews of Geophysics and Space Physics, v. 8, p. 813–860.

Dickinson, W. R., 1972, Evidence for plate tectonic regimes in the rock record: American Journal of Science, v. 272, p. 551–576.

Ernst, W. G., 1970, Tectonic contact between the Franciscan melange and the Great Valley sequence, crustal expression of a Late Mesozoic Benioff zone: Journal of Geophysical Research, v. 25, p. 886–901.

Ernst, W. G., 1974, Metamorphism and ancient continental margins: in Burke, C. A., and Drake, C. L., eds., The Geology of Continental Margins: New York, Springer-Verlag, p. 907–919.

Ernst, W. G., 1977, Mineral parageneses and plate tectonic settings of relatively high-pressure metamorphic belts: Fortschritte der Mineralogie, v. 54, p. 192–222.

Ernst, W. G., 1980, Mineral paragenesis in Franciscan metagraywackes of the Nacimiento block, a subduction complex of the southern California Coast Ranges: Journal of Geophysical Research, v. 85, p. 7045–7055.

Faure, Gunter, 1977, Principles of Isotope Geology: New York, John Wiley and Sons, Inc., 464 p.

Hamilton, W. B., 1969, Mesozoic California and the underflow of the Pacific mantle: Geological Society of America Bulletin, v. 80, p. 2409–2430.

Hamilton, W. B., 1978, Mesozoic tectonics of the western United States: in Howell, D. G., and McDougall, K. A., eds., Mesozoic Paleogeography of the Western United States: Society of Economic Paleontologists and Mineralo-

gists, Pacific Section, Pacific Coast Paleogeography Symposium 2, p. 33–70.

Hopson, C. A., Mattinson, J. M., and Pessagno, E. A., Jr., 1981, Coast Range ophiolite, western California: in Ernst, W. G., ed., The Geotectonic development of California, W. R. Rubey Vol. 1, New York, Prentice-Hall, p. 418–510.

Karig, D. E., 1980, Material transport within accretionary prisms and the "knocker" problem: Journal of Geology, v. 88, p. 27–39.

Lanphere, M. A., Blake, M. C., Jr., and Irwin, W. P., 1978, Early Cretaceous metamorphic age of the South Fork Mountain Schist in the northern Coast Ranges of California: American Journal of Science, v. 278, p. 798–815.

Mattinson, J. M., 1978, Age, origin, and thermal histories of some plutonic rocks from the Salinian Block of California: Contributions to Mineralogy and Petrology, v. 67, p. 233–246.

Mattinson, J. M., 1981, U-Pb systematics and geochronology of blueschists: preliminary results [abs.]: EOS (American Geophysical Union Transactions), v. 62, p. 1059.

Mattinson, J. M., 1982, U-Pb "blocking temperatures" and Pb-loss characteristics in young zircon, sphene, and apatite: Geological Society of America Abstracts with Programs, v. 14, no. 7, p. 588.

Mattinson, J. M., and Echeverria, L. M., 1980, Ortigalita Peak gabbro, Franciscan complex: U-Pb ages of intrusion and high pressure-low temperature metamorphism: Geology, v. 8, p. 589–593.

Maxwell, J. A., 1974, Anatomy of an orogen: Geological Society of America Bulletin, v. 85, p. 1195–1204.

Nelson, B. K., and DePaolo, D. J., 1982, Sr and Nd composition of Franciscan eclogite and blueschist: a sampling of subducted crust?: EOS (American Geophysical Union Transactions), v. 63, p. 1133.

Page, B. M., 1970a, Sur-Nacimiento fault zone of California: continental margin tectonics: Geological Society of America Bulletin, v. 81, p. 667–690.

Page, B. M., 1970b, Time of completion of underthrusting of Franciscan beneath Great Valley rocks west of the Salinia block, California: Geological Society of America Bulletin, v. 81, p. 2825–2834.

Page, B. M., 1981, The southern Coast Ranges: in Ernst, W. G., ed., The Geotectonic development of California, W. R. Rubey, Vol. 1, New York, Prentice-Hall, p. 329–417.

Peterman, Z. E., Hedge, C. E., Coleman, R. G., and Snavely, P. D., Jr., 1967, ^{87}Sr/^{86}Sr ratios in some eugeosynclinal sedimentary rocks and their bearing on the origin of granitic magma in orogenic belts: Earth and Planetary Science Letters, v. 2, p. 433–439.

Platt, J. P., 1975, Metamorphic and deformational processes in the Franciscan complex, California: some insights from the Catalina Schist terrane: Geological Society of America Bulletin, v. 86, p. 1337–1347.

Sorensen, S. S., 1983, The formation of metasomatic rinds around amphibolite facies tectonic blocks, Catalina Schist terrane, Southern California: Geological Society of America Abstracts with Programs, v. 15, no. 5, p. 436.

Sorensen, S. S., 1984a, Petrology of basement rocks of the California continental borderland and Los Angeles Basin [Ph.D. Thesis]: Los Angeles, University of California, 423 p.

Sorensen, S. S., 1984b, Trace element effects of eclogite/peridotite metasomatism, Catalina Schist terrane, Southern California: Geological Society of America Abstracts with Programs, v. 16 (in press).

Stacey, J. S., and Kramers, J. D., 1975, Approximation of terrestrial lead evolution by a two stage model: Earth and Planetary Science Letters, v. 26, p. 207–221.

Sun, S. S., 1980, Lead isotopic study of young volcanic rocks from mid-ocean ridges, oceanic islands, and island arcs: Philosophical Transactions of the Royal Society of London, v. A297, p. 409–445.

Suppe, John, 1969, Times of metamorphism in the Franciscan terrain of the northern Coast Ranges, California: Geological Society of America Bulletin, v. 80, p. 135–142.

Suppe, John, and Armstrong, R. L., 1972, Potassium-argon dating of Franciscan metamorphic rocks: American Journal of Science, v. 272, p. 217–233.

Suppe, John, and Foland, K. A., 1978, The Goat Mountain schists and Pacific Ridge complex: a redeformed but still-intact late Mesozoic schuppen complex: *in* Howell, D. G., and McDougall, K. A., eds., Mesozoic Paleogeography of the Western United States: Society of Economic Paleontologists and Mineralogists, Pacific Section, Pacific Coast Paleogeography Symposium 2, p. 431–451.

Tilton, G. R., 1983, Evolution of depleted mantle: the lead perspective: Geochimica et Cosmochimica Acta, v. 47, p. 1191–1197.

Wright, J. E., and Sharp, W. D., 1982, Mafic-ultramafic intrusive complexes of the Klamath-Sierran region, California: Remnants of a Middle Jurassic arc complex: Geological Society of America Abstracts with Programs, v. 14, no. 4, p. 245–246.

York, D., 1969, Least squares fitting of a straight line with correlated errors: Earth and Planetary Science Letters, v. 5, p. 320–324.

MANUSCRIPT ACCEPTED BY THE SOCIETY JULY 29, 1985

Geological Society of America
Memoir 164
1986

Blueschist metamorphism of the Eastern Franciscan belt, northern California

A. S. Jayko
M. C. Blake, Jr.
U.S. Geological Survey
345 Middlefield Road
Menlo Park, California 94025

R. N. Brothers
University of Auckland
Auckland, New Zealand

ABSTRACT

Rocks of the Eastern Franciscan belt, northern California, are divided into two tectonostratigraphic terranes metamorphosed to the blueschist facies, both with a distinct lithologic association and deformational history. The easternmost terrane, the Pickett Peak terrane of Early Cretaceous isotopic age, consists of crenulated mica schist and gneissic to schistose metagraywacke, with lesser alkalic mafic metaigneous rocks and scarce metachert. The Pickett Peak terrane retains evidence of three periods of penetrative deformation, the first of which is characterized by segregation layering, and the second and third by crenulation cleavages. Blueschist-facies conditions persisted during the first two deformations.

The Yolla Bolly terrane of Late Jurassic and Early Cretaceous paleontologic age lies structurally below and to the west of the Pickett Peak terrane. It is characterized by voluminous metagraywacke and lesser argillite, coherent interbedded radiolarian chert, and alkalic gabbroic dikes and sills. The Yolla Bolly terrane retains evidence for two phases of penetrative deformation that were coaxial with the second and third phases of deformation in the Pickett Peak terrane. The first phase of deformation (parallel to the second phase in the Pickett Peak terrane) was also accompanied by blueschist-facies metamorphism.

INTRODUCTION

Along the Pacific margin of North America are several linear belts of mélange or broken formation consisting largely of sheared argillite and graywacke (Jones and others 1977). Tectonically interleaved blocks and slabs of mafic volcanic rock, serpentinite, radiolarian chert, and limestone are common but volumetrically insignificant (Bailey and others 1964; Cowan 1974; Jones and others 1976). Regionally these rocks are metamorphosed to blueschist, prehnite-pumpellyite, or zeolite facies, and locally they may contain small blocks of eclogite, amphibolite, or blueschist. Most researchers currently agree that the sedimentary rocks from these belts were originally deposited along the interface between converging plates and were subsequently accreted to a plate margin, and that the blocks of igneous and metamorphic rock were incorporated during convergence. The Franciscan assemblage of the California Coast Ranges, which is probably the best known of these melange and blueschist belts, has been the subject of numerous studies.

The Franciscan assemblage north of San Francisco, California, was originally divided into three north-south-trending belts (Figure 1; Bailey and Irwin 1959; Irwin 1960; Berkland and others 1972); more recently, each belt has been subdivided into one or more tectonostratigraphic terranes (Blake and others,

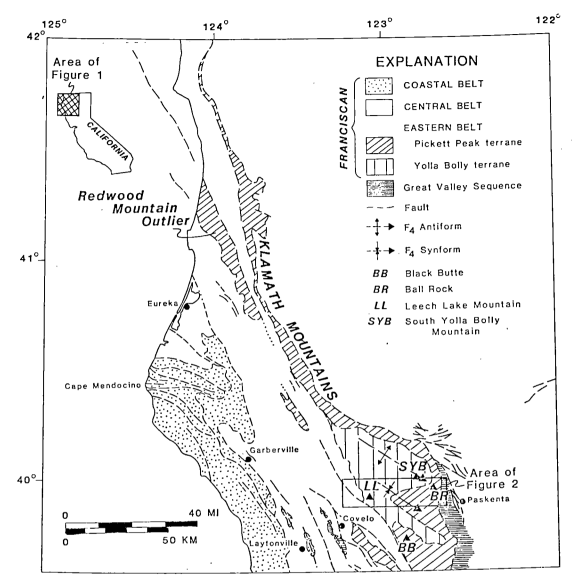

Figure 1. Northern California, location of study area in relation to areas of Franciscan assemblage. Generalized after Blake and others (1982).

1982). The Coastal, Central, and Eastern Franciscan belts, as originally defined, represent the major tectonic units that are distinguished on the basis of lithology, age, structural style, and metamorphism. The subdivision of belts into terranes further refines the different units, and so both designations are used.

The Eastern Franciscan belt consists of several imbricate nappes, regionally metamorphosed to low- and moderate-grade blueschist facies (Irwin, 1960; Berkland and others 1972; Blake and others 1984). It is bounded on the east by the Coast Range ophiolite (Hopson and others 1981), on the northeast by the Klamath Mountains (Irwin 1981), and on the west by the Central Franciscan belt (Blake and Jones 1974).

Regional tectonic and metamorphic studies of the Eastern Franciscan belt by Blake and others (1967, 1981), Suppe (1973), and Worrall (1981) have definitively established the tectonostra-

tigraphic relations and general metamorphic characteristics of these rocks. A west-to-east increase in intensity of metamorphism has been recognized, although the progression is generally discontinuous and interrupted by postmetamorphic imbricate thrust faults (Suppe 1973; Worrall 1981; Blake and Jayko 1983). The Eastern Franciscan belt was recently divided into two lithologically and structurally distinct terranes, the Pickett Peak and the Yolla Bolly (Blake and Jayko 1983).

This study reports on the deformational characteristics and associated metamorphic grade within thrust plates of the Eastern Franciscan belt; in particular, it documents the differences in structural style and metamorphism between the Pickett Peak and Yolla Bolly terranes. A transect across the Eastern Franciscan belt from its east contact with the dismembered Coast Range ophiolite to the Central Franciscan belt (Figure 2) was mapped in

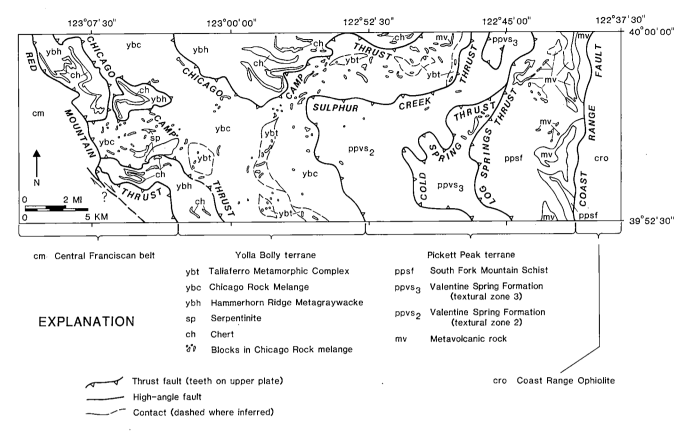

Figure 2. Generalized geologic map of study area.

detail and sampled for petrologic studies (Jayko 1984). The petrologic results and a brief summary of the structural studies are presented here. The Pickett Peak terrane was found to have an earlier phase of deformation and blueschist-facies metamorphism that was not observed in the Yolla Bolly terrane. In addition, much of the Pickett Peak terrane appeared to reach higher temperature conditions than the Yolla Bolly terrane.

GENERAL GEOLOGY

The Eastern Franciscan belt in northern California consists of the Pickett Peak and Yolla Bolly terranes (Figures 1 and 2). The Pickett Peak terrane structurally overlies the Yolla Bolly terrane along an east-dipping low-angle fault and consists of two fault-bound units, the structurally lower, metagraywacke-dominated Valentine Spring Formation of Worrall (1981) (Figure 2) and the overlying quartz-mica schist and Chinquapin Metabasalt Member of the South Fork Mountain Schist (Blake and others 1967). The Pickett Peak terrane has been interpreted as a fragment of a seamount province or oceanic plateau that was scraped off the subducting oceanic lithosphere at the North American continental margin (Jayko 1984) and metamorphosed during the Early Cretaceous, approximately 125 ± 10 m.y. ago. The metamorphic age is based on K-Ar and Ar^{39}-Ar^{40} dating of whole

rock, white mica, and (or) glaucophane (Suppe 1973; Lanphere and others 1978; McDowell and others 1984).

The Yolla Bolly terrane is divided into four tectonic units: Chicago Rock mélange, the metagraywacke of Hammerhorn Ridge, the broken formation of Devils Hole Ridge (Blake and Jayko 1983), and the Taliaferro Metamorphic Complex (Suppe 1973), that contain megafossils and (or) radiolarians of Late Jurassic and Neocomian age (Suppe 1973; Blake and others 1981; Jayko 1984). The Devils Hole Ridge unit does not occur at the latitude of this transect so is not shown on Figure 2. The structurally highest unit, the Chicago Rock melange, consists predominantly of metamorphosed, tectonized argillite and metagraywacke with intercalated thin-bedded radiolarian chert, minor greenstone, scarce pods of serpentinite, and rare blocks of amphibolite and mafic blueschist. The chert and greenstone appear to be indigenous to the unit on the basis of stratigraphic relations (Jayko 1984); however, the serpentinite, amphibolite, and mafic blueschist appear to be tectonically interleaved. Introduction of these "exotic" blocks into the Chicago Rock mélange may have been associated with the large-scale faulting that juxtaposed the Yolla Bolly and Pickett Peak terranes.

Intermixed with the Chicago Rock melange are many slabs and thin thrust sheets of the Taliaferro Metamorphic Complex (Suppe 1973), which are of slightly higher-grade than the sur-

Jayko and Others

TABLE 1. SUMMARY OF STRUCTURES CHARACTERIZING DEFORMATION PHASES, D_1-D_4

	D_1	D_2	D_3	D_4
Pickett Peak terrane				
Folds	Isoclinal to tight	Tight to open	Open to conjugate, chevron, box	Open warping
Axial planes	Refolded	SW dipping and refolded	Steeply SE+NW dipping	Steep NW-SE trend
Fabric	Mineral segregation into bands up to 3 to 4 mm	Strong transposition of S_1 development of crenulation cleavage	Crenulation clevage	----
Lineations	Dispersed, mineral and streaking most abundant	Varies but SE and NW,	NE and SW fairly well grouped	----
Asymmetry	----	NE vergence	Varies	SW limb of regional antiform, overturned locally
Yolla Bolly terrane				
Folds	----	Isoclinal to tight	Tight to open	Open
Axial plane	----	Predominantly SW-dipping	Steeply NW+SE dipping	Steep NW-SE trend
Fabric	----	Weak to strong schistosity, flattening and elongation of clasts	Locally poorly developed crenulation cleavage	----
Lineations	----	Predominantly SE to S plunging	Predominantly SW plunging	----
Asymmetry	----	NE vergence	SW to NE	Same vergence as Pickett Peak

rounding argillaceous matrix but are also considered to be part of the Yolla Bolly terrane because of lithologic similarities (Blake and Jayko 1983). The Taliaferro Metamorphic Complex, as defined in this study, includes the jadeite- and crossite-bearing metagraywacke and chert of Suppe (1973) but excludes the older, approximately 160 m.y. blocks of coarse-grained blueschist and amphibolite. The Hammerhorn Ridge unit, a massive graywacke with a continuous horizon of interbedded radiolarian chert, structurally underlies the Chicago Rock melange. The interbedded chert and graywacke of the Hammerhorn Ridge unit are locally intruded by Ti-rich alkalic gabbro sills and dikes. This distinctive lithologic association of interbedded chert and graywacke intruded by alkalic titaniferous gabbroic rock also characterizes tectonic inclusions of the Taliaferro Metamorphic Complex. The Devils Hole Ridge unit, a deformed graywacke and argillite unit, lies structurally below the Hammerhorn Ridge unit to the north of this study area.

In addition to the alkalic-titaniferous intrusive bodies, the Yolla Bolly terrane contains intrusive and extrusive keratophyres and quartz keratophyres. Chemical analyses of these rocks indicates a bimodal silica content ranging from 44 to 50 percent SiO_2 in the alkalic gabbro and from 68 to 76 percent SiO_2 in the keratophyres and quartz keratophyres (Jayko 1984).

The Yolla Bolly terrane is nearly identical in metamorphic-mineral assemblages and lithologic association to several other units that have been mapped in the eastern part of the Franciscan assemblage, including the Pacific Ridge Complex of Suppe and Foland (1978), the Summit Spring Formation (MacPherson 1983), the Eylar Mountain sequence (Crawford 1975), the graywacke and chert of the Mount Oso (Raymond 1974) and Pacheco Pass areas (Ernst and others 1970).

On the basis of radiolarian microfossils and rare *Buchia* megafossils, the protolith age of the Yolla Bolly terrane is Late Jurassic and Early Cretaceous (Blake 1965; Suppe 1973; Worrall 1981; Lanphere and others 1978; Jayko 1984). The strata were apparently deposited along a continental margin receiving micaceous quartzofeldspathic detritus (Jayko 1984) (possibly undergoing minor extension similar to the California Continental Borderland, as indicated by alkalic intrusive rocks) that later became imbricated and metamorphosed during middle Cretaceous time, approximately 90–105 m.y. ago (Blake and others 1981). This metamorphic age, based on K-Ar dating of whole rock and white mica separates (Suppe 1973) from metasedimentary rocks, probably represents the timing of accretion to the Pickett Peak terrane.

Older blueschist and amphibolite blocks, 150–164 m.y. old

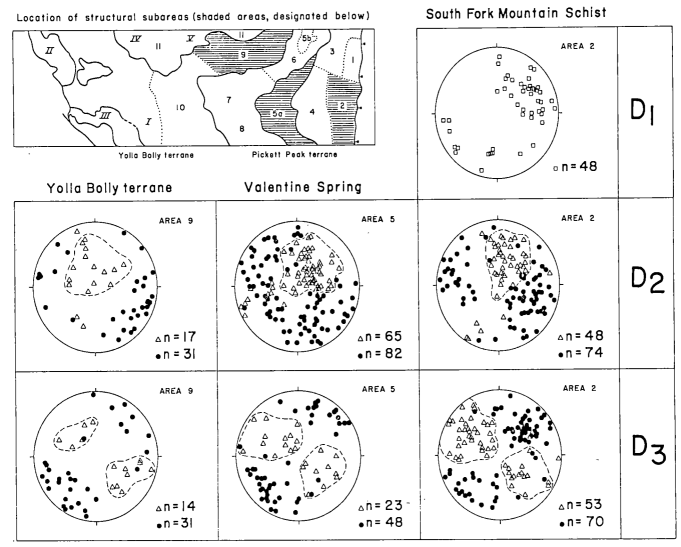

Figure 3. Representative stereonet plots of D_{1-3} structural elements, lower-hemisphere equal-area net. From Jayko (1984).

(Suppe, 1973; Mattinson, 1981), also locally incorporated into the Chicago Rock mélange and previously included by Suppe (1973) in his Taliaferro Metamorphic Complex, are generally Nevadan in age and were likely incorporated during postaccretionary tectonic events possibly during the Late Cretaceous or early Tertiary.

STRUCTURE

Multiple phases of deformation are recognized in both the Pickett Peak and Yolla Bolly terranes. Four phases of deformation are evident in the Pickett Peak terrane: two characterized by well-developed schistosity (D_1 and D_2), one by a weakly developed crenulation cleavage (D_3), and the last by a broad, open regional warping (D_4). The first two phases of deformation (D_1 and D_2) were accompanied by blueschist-facies metamorphism. Three of these phases of deformation (D_2, D_3, and D_4) are present in the Yolla Bolly terrane. Table 1 summarizes the field criteria used to differentiate the deformation phases D_{1-4}.

The various phases of deformation are distinguished by differences in the morphology and orientation of the foliation and folds (Figures 3 and 4). The D_1 foliation, S_1, in the Pickett Peak terrane is characterized by segregation layering in the higher grade metasedimentary rocks and by a pervasive spaced cleavage in lower grade metagraywacke (terminology after Powell 1979). The D_1 folds, F_1, are tight to isoclinal, with metamorphic amphiboles morphologically oriented parallel to the F_1 axial planes in metachert and metavolcanic rock; these orientations are inferred to be a-lineations indicating tectonic transport direction.

The D_2 foliation S_2, in the Pickett Peak terrane is a discrete

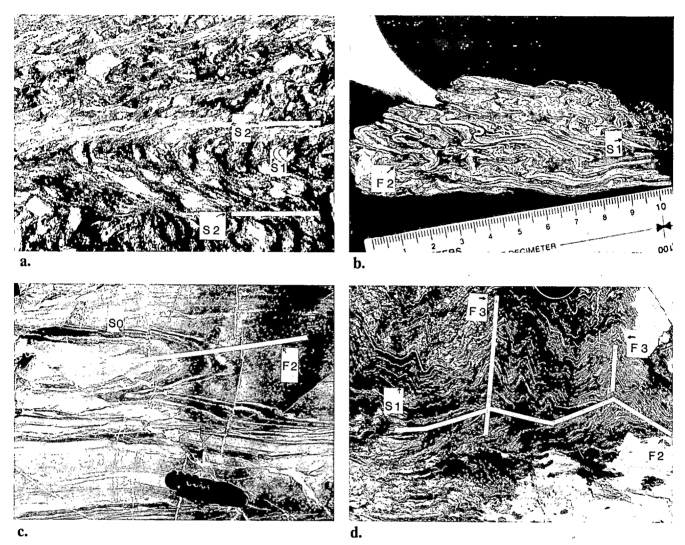

Figure 4. a. S_2 foliation in TZ 3A of the Pickett Peak terrane. b. Asymmetric F_2 fold in mica schist of the Pickett Peak terrane. F_2 fold and S_2-axial-plane foliation in metagraywacke of the Yolla Bolly terrane (D_2 is first deformation in Yolla Bolly rocks). S_0, bedding d. S_1, F_2, and F_3 in metagraywacke of the Valentine Spring Formation.

crenulation cleavage that commonly transposes S_1 (Figure 4a); in mica schist, it is typically the conspicuous foliation observed in outcrop. The D_2 folds, F_2, in the Pickett Peak terrane are tight to open, commonly asymmetric, and overturned to the northeast (Figure 4b), with axial planes dipping southwest and with fold hinges and strikes of axial planes trending northwest-southeast. Folds plunge both northwest and southeast. In the Yolla Bolly terrane, the F_2 folds are generally tight to isoclinal and are oriented parallel to F_2 folds of the Pickett Peak terrane (Figure 4c). In metagraywacke of the Yolla Bolly terrane, S_2 is a micro-styolitic to fairly smooth, disjunctive, spaced cleavage, with random to strong microlithon-fabric realignment.

The D_3 foliation, S_3, in both terranes is also a discrete crenulation cleavage that is less well developed than S_2 in most rock types and generally oriented at a high angle to S_2. The D_3

folds are commonly kink or box shaped (Figure 4d), with axial planes striking northeast and dipping gently to the northwest or southeast, and hingelines plunging northeast or southwest.

The D_4 phase of broad, low-amplitude northwest-trending folding is inferred from the regional map pattern (Figure 1) and the folding of metamorphic isotects (Blake and Jayko 1983). Obvious small-scale structures (outcrop or thin section) associated with this folding are rare, although a few asymmetric chevron folds are oriented subparallel to the southeast plunging regional trend of F_4 structures.

Six major faults bound the Franciscan units (Figure 2, Table 2). From east to west: the Coast Range thrust (Bailey and others 1970), the Log Springs thrust (Suppe 1973), the Cold Springs thrust (Jayko 1984), the Sulphur Creek thrust (Worrall 1981), the Chicago Camp thrust (Blake and others 1984), and the Red

TABLE 2. FAULT CRITERIA USED IN THIS STUDY

Fault	Comments
Coast Range	Linear contact separating sheared serpentinite and mica shist, with intervening unmetamorphosed sheared argillite. Contact can be precisely mapped in the field.
Log Springs	Westward limit of mica schist. This lithologic break coincides with a pronounced fabric break in the northwestern part of the study area. Significant difference in metamorphic grade in metavolcanic rock occus above and below the contact; however, metagraywacke is very similar. The contact can be located in the field within a few meters; however, a fault surface has not been observed.
Cold Spring	A pronounced fabric break between TZ 2B and 3A metagraywacke within a the Valentine Spring Formation. Contact can be located in the field within a few meters; however, a fault surface has not been identified.
Sulphur Creek	TZ 2A-2B boundary, with appearance of numerous blocks of greenstone, chert, serpentinite, and higher grade blueschist in the lower plate, and locally strong break in fabric orientation. Contact can be located within meters in the northeast part of study map area but depends primarily on the textural-zone boundary toward the southwest. In the southwest, the boundary is somewhat obscured by the great abundance of sheared argillite and very thin bedded graywackes, which do not lend themselves to textural determination.
Chicago Camp	Lithologic contact separating predominantly argillite of the Chicago Rock mélange from massive Hammerhorn Ridge graywacke-chert unit; paleontologic ages for both units are the same. This contact can be located within several tens of meters in the field. The contact forms a prominent break in slope locally.
Red Mountain	Lithologic break between more coherent Yolla Bolly terrane and Central belt mélange. The contact forms a prominent break in slope locally.

Mountain thrust (Blake and Jayko 1983). Initial activity on the Cold Springs and Chicago Camp thrust faults postdates D_1 and predates D_2 (Jayko 1984). The Sulphur Creek and Log Springs thrust faults were active after D_2, and the Log Springs may also have been active after D_3 (Jayko 1984). The Red Mountain thrust fault is either synchronous with D_4 or postdates it, and the Coast Range thrust postdates D_4 (Worrall 1981; Hopson and others 1981; Jayko 1984).

Except for the Coast Range thrust, actual surfaces of the major faults described in this report are exceedingly difficult to identify in the field, primarily because of poor exposures. However, the contact between units can generally be located within a few meters or tens of meters on the basis of the distinguishing characteristics of each unit. The thrust nature of the faults is inferred because in most places either higher grade and (or) older rocks occur structurally above lower grade and (or) younger rocks. Table 2 summarizes the structural criteria used to define these faults.

Textural Reconstitution

The criteria for determining textural reconstitution used in this study were first developed by Hutton and Turner (1936) for quartzofeldspathic lithic graywackes in New Zealand and later applied to Franciscan rocks by Blake and others (1967). Textural grades or zones (TZ) are a qualitative description of the degree of penetrative deformation in graywacke. Bishop (1972) coined the term "isotect" to define the boundary surfaces that separate areas of different textural grade.

The textural zones range from TZ 1, which is characterized by no obvious flattening in thin section and no preferred orientation of foliation in outcrop, to TZ 3B, which is characterized by gneissic banding and quartz segregations thicker than 2 mm (Figures 5a-d). Table 3 summarizes the microscopic and mesoscopic criteria for the textural grade designations used in this report.

The surface distribution of textural zones (Figure 6) shows the increasing reconstitution from west to east as documented by earlier workers (Blake and others 1967; Suppe 1973; Worrall 1981). In addition, Figure 6 shows several other important features. 1) In the Yolla Bolly terrane, most of the rocks are TZ 2A, except in the locus of tight D_2 isoclinal folds (not shown in Figure 6, see Figure 2), where TZ 2B fabric was formed parallel to the axial planes, and in tectonic inclusions of TZ 2B and 3A rocks in the Taliaferro Metamorphic Complex. 2) The fault separating TZ 3A and 2B graywacke of the Valentine Spring Formation formed before D_2. 3) TZ 3A graywacke is present in both the South Fork Mountain Schist and the Valentine Spring Formation and occurs on both sides of the Log Springs thrust. 4) the Yolla Bolly terrane is generally lower in textural grade than the Pickett Peak terrane.

Radiolarians from chert interbedded with graywacke in the Yolla Bolly terrane were identified in graywackes up to TZ 2B. They can be observed in chert associated with graywacke up to the TZ 2B-3A transition but cannot be extracted owing to similarities in SiO_2 crystallinity between matrix and fossil.

Metavolcanic rock of a given textural grade is generally less deformed than the nearly metagraywacke. In general, metavol-

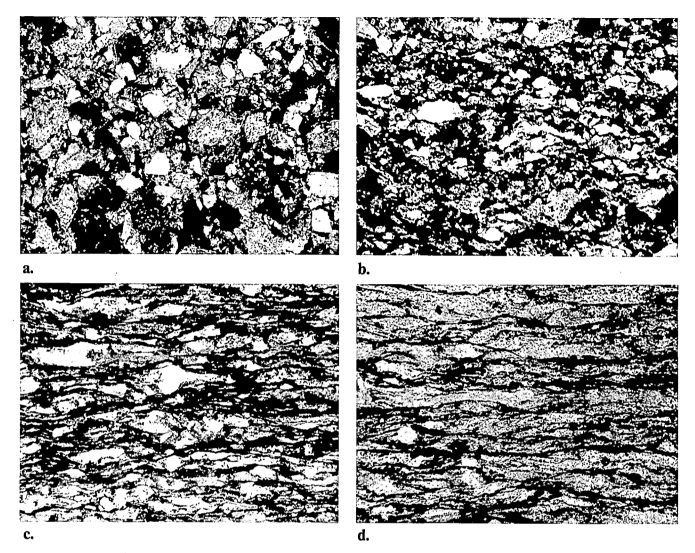

Figure 5. a. TZ 1 metagraywacke of the Yolla Bolly terrane. b. TZ 2A metagraywacke of the Yolla Bolly terrane. c. TZ 2B metagraywacke of the Yolla Bolly terrane. TZ 3A metagraywacke of the Pickett Peak terrane.

canic rock of TZ 1 and 2A have no obvious schistosity in outcrop, and mesoscopic igneous textures and minerals are preserved. In TZ 2B, the metavolcanic rock more commonly has a pronounced schistosity comparable to that of TZ 2A metagraywacke. In TZ 3A, metavolcanic rock commonly shows a strong schistosity and local segregation layering.

METAMORPHISM

Rocks of the study area show a west-to-east progressive increase in metamorphic grade (Blake and others 1967; Suppe 1973; Worrall 1981). Three possible isograds can be recognized within the Eastern Franciscan Belt, all within the Pickett Peak terrane (Figure 7). Here the term "isograd" is used purely in a descriptive sense as the first or last occurrence of a phase rather than as a boundary representing a reaction (reaction isograd of

Winkler 1979), because numerous faults disrupt a true prograde sequence. Except for blocks in melange, clastic metasedimentary rocks show blue-amphibole-in and paragonite-in isograds from west to east, and in situ metavolcanic rocks show an isograd representing lawsonite-out. In addition to the appearance or disappearance of index minerals, the relative abundance of some phases varies considerably. For example, in the Pickett Peak terrane, blue amphibole typically constitutes 30 to 60 percent of the mafic metaigneous rock, whereas in the Yolla Bolly terrane, blue amphibole, if present, commonly constitutes only about 1 percent or less of the in situ metaigneous rocks, even in rocks of comparable bulk composition (Jayko 1984). Although lawsonite occurs throughout the study area in clastic metasedimentary rocks, its grain size and abundance increase considerably toward the east. Pumpellyite is also widespread in metavolcanic rocks throughout the study area but is generally restricted to metagray-

TABLE 3. METAGRAYWACKE TEXTURAL ZONES

Textural Zone	Foliation	Thin Section Description	Outcrop Description	Mineral Phases Observed
1	None	Grains do not appear flattened, absence of tectonic fabric, opaque residue around grains, boundaries between different grains gradational, randomly oriented grains	Massive, no preferred cleavage orientation	*cel pu lw ba jd par
2A	Anastamosing	Incipient flattening of lithic clasts and reorientation of micaceous and argillaceous clasts, cleavage is spaced-disjunctive, rough, microanastamosing; kinked micas and chlorite	Incipient development of broadly spaced ellipsoidal anastamosing cleavages with a preferred orientation	
2B	Planar-parallel	Obvious to extreme flattening of all clasts, strong shape elongation within microlithons, cleavage domains much smoother	Rock breaks into almost perfectly planar slabs, commonly difficult to break across this fabric	
3A	Banded	Incipient to distinctive quartz-mica segregation, increase in size of metamorphic white mica, quartz and albite are polycrystalline, a few detrital clasts may still be preserved. Polygonal quartz and albite, larger grains divided into smaller polycrystalline domains	Extremely strong planar fabric, hand samples appear noticeably banded, with quartz-mica segregation up to 2 mm wide and tens of cm long	
3B	Banded	No detrital clasts preserved	Banding thicker than 2 mm	

*cel = celadonite; pu = pumpellyite; lw = lawsonite; ba = blue amphibole; jd = jadeite; par = paragonite.

EXPLANATION: ○ TZ 2A, ◑ TZ 2B, ● TZ 3A, ⬩ High-angle fault (balls on down-thrown block), ⬩ Thrust fault (teeth on upper plate)

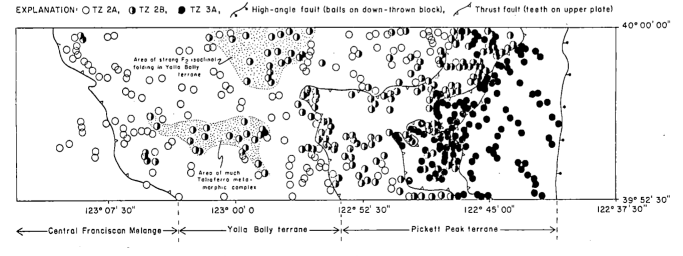

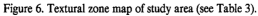

Figure 6. Textural zone map of study area (see Table 3).

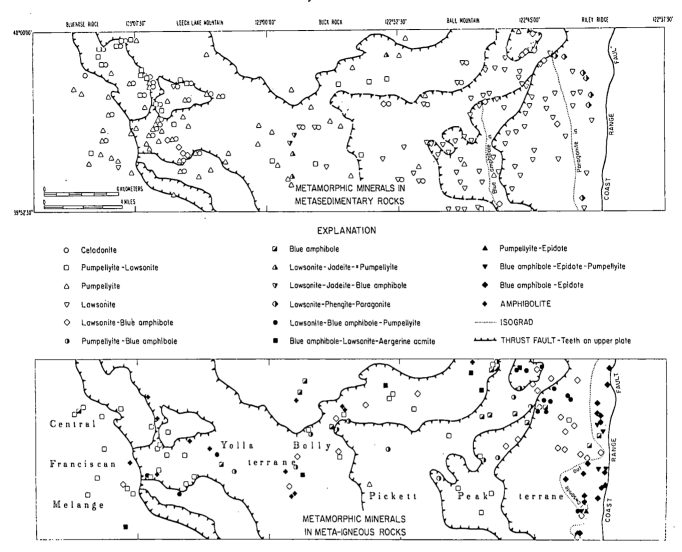

EXPLANATION

O	Celadonite	▫	Blue amphibole	▲	Pumpellyite-Epidote
▫	Pumpellyite-Lawsonite	△	Lawsonite-Jadeite-*Pumpellyite	▼	Blue amphibole-Epidote-Pumpellyite
△	Pumpellyite	▽	Lawsonite-Jadeite-Blue amphibole	◆	Blue amphibole-Epidote
▽	Lawsonite	◇	Lawsonite-Phengite-Paragonite	✦	AMPHIBOLITE
◇	Lawsonite-Blue amphibole	●	Lawsonite-Blue amphibole-Pumpellyite	⌐⌐⌐⌐	ISOGRAD
⊙	Pumpellyite-Blue amphibole	■	Blue amphibole-Lawsonite-Aergerine acmite	▲▲▲	THRUST FAULT-Teeth on upper plate

Figure 7. Schematic maps of study area, showing metamorphic isograds and distribution of metamorphic minerals.

wacke of the Yolla Bolly terrane. Both pumpellyite and celadonite commonly occur in TZ 2A metagraywacke, but not in TZ 2B metagraywacke.

Pickett Peak Terrane

South Fork Mountain Schist. The South Fork Mountain Schist consists of metasedimentary rocks and a metaigneous member, the Chinquapin Metabasalt Member. The Chinquapin Metabasalt Member can be divided into the lawsonite and epidote zones of the blueschist facies (Turner 1981), which occur in two distinct areas. In the metabasalt, the epidote-zone assemblage lie to the east of lawsonite-zone rocks and are adjacent to the Coast Range thrust. Intercalated clastic metasedimentary rocks in the South Fork Mountain Schist lack epidote and contain lawsonite ± blue amphibole in addition to other constituent minerals. Tables 4 through 6 list the characteristic mineral assemblages.

Blue amphibole, in varying amounts, occurs in 99% of the metavolcanic rocks, accompanied by either lawsonite (in the west) or epidote (in the east). Pumpellyite was observed in about 80%, albite in about 90%, and relict igneous clinopyroxene in about 50% of the thin sections. Accessory metamorphic minerals include acmitic-omphacitic pyroxene (Brown and Ghent 1983), which generally rims or replaces relict igneous clinopyroxene, white mica, stilpnomelane, hematite, calcite or aragonite, quartz, opaque minerals, and sphene. Green amphibole rimmed by blue amphibole was observed in one sample, and two amphiboles were recognized on an X-ray diffraction pattern from an additional sample. The green amphibole-bearing sample is also unique in that it contains coarse-grained stilpnomelane and pronounced coarse-grained gneissic albite-amphibole segregation layers.

The metaigneous rocks of the Chinquapin Metabasalt Member are all schistose and have a foliation commonly defined

TABLE 4. MINERAL ASSEMBLAGES OBSERVED IN METAIGNEOUS ROCKS*

	alb	chl	pu	ba	lw	ep	ga	stilp
	alb	chl	pu	ba				
	alb	chl	pu		lw			
	alb	chl	pu		lw			
Chinquapin	alb	chl		ba	lw			
Metabasalt		chl	pu	ba	lw			
Member		chl		ba	lw			
		chl	pu	ba	lw	ep		
	alb	chl		ba		ep		
	alb	chl	pu	ba		ep		
		chl	pu			ep		
	alb			ba			ga	stilp
	alb	chl	pu					
Valentine	alb	chl	pu		lw			
Spring	alb	chl	pu	ba	lw			
Formation	alb	chl		ba	lw			
	alb		pu	ba	lw			
		chl	pu			ba		
Taliaferro		chl	pu			ba		
Metamorphic	alb	chl	pu	lw		ba		
Complex	alb	chl	pu	lw		ba	±ga	
		chl		lw		ba		
	alb	chl		lw		ba		
Metagraywacke of	alb	chl		pu				
Hammerhorn Ridge	alb	chl		pu		ba		
Chicago Rock	alb	chl	lw	pu				
Mélange	alb	chl	lw			ba		
	alb	chl				ba		
	alb			ga	ba		ep	pu
Amphibolite	alb			ga	ba	lw		
blocks	alb	chl		ga	ba	lw		
	alb	chl		ga	ba			

*see Table 6.

TABLE 5. MINERAL ASSEMBLAGES IN METASEDIMENTARY ROCKS*

	qtz	alb	phg/wm	chl	lw	par	ba	pu	jd
South Fork	qtz	alb	phg	chl	lw				
Mountain	qtz	alb	phg	chl	lw	par			
Schist	qtz	alb	phg	chl					
	qtz	alb	phg	chl	lw		ba		
	qtz	alb	wm	chl					
Valentine	qtz	alb	wm	chl	lw				
Spring	qtz	alb	wm	chl	lw		ba		
Formation	qtz	alb	wm	chl	lw			pu	
	qtz	alb	wm	chl				pu	
Taliaferro	qtz	alb	wm	chl	lw				
Metamorphic	qtz	alb	wm	chl	lw		ba		
Complex	qtz	alb	wm	chl	lw		ba	pu	
	qtz	alb	wm	chl	lw		ba	pu	jd
	qtz	alb	wm	chl	lw		ba		jd
	qtz		wm	chl	lw		ba		jd
Metagraywacke of	qtz	alb	wm	chl					
Hammerhorn Ridge	qtz	alb	wm	chl				pu	
and Chicago Rock	qtz	alb	wm	chl	lw			pu	
Mélange	qtz	alb	wm	chl	lw				

*see Table 6.

by intergrown fibrous blue amphibole, pumpellyite, and chlorite. The rocks generally have a poorly developed banding formed by pumpellyite-rich versus blue amphibole-rich layers. Mineral segregation into layers rich in lawsonite and layers rich in blue amphibole + pumpellyite are also present. Rare incipient pods and lenses of quartz, and (or) albite segregations, are present.

The metavolcanic rocks have two distinct fabrics: either schistose with a very fine grained blue amphibole matrix, with or without relict clinopyroxene augen; or gneissic with relatively coarse grained, subhedral to euhedral blue amphibole, and epidote or lawsonite. Relict igneous textures are absent.

The metamorphic mineral assemblages in metasedimentary rocks of the South Fork Mountain Schist commonly include quartz + albite + chlorite + white mica + lawsonite ± calcite ± aragonite ± blue amphibole ± stilpnomelane ± sphene ± hematite ± tourmaline. Fine-grained blue amphibole has been observed only in highly lithic or tuffaceous metagraywacke. Paragonite, detected by X-ray diffraction, is sporadically distributed with phengite in about 20% of the samples. Paragonite has not previously been described from the South Fork Mountain Schist. Jadeitic pyroxene and pumpellyite were not observed in any of the mica schist samples.

Metacherts containing the assemblage crossite + quartz + magnetite ± stipnomelane ± barite ± spessartine ± hematite is a minor constituent associated with the metavolcanic rocks. Spessartine garnet in the metachert is widespread north of the study area (Blake 1965); however, it was observed at only one locality within this transect. A spessartine-in isograd may be present but has not been mapped owing to scarcity of the metachert.

Valentine Spring Formation

Metavolcanic rocks within the Valentine Spring Formation are intercalated with metagraywacke of TZ 2B and 3A. Mineral assemblages in these metavolcanic rocks commonly are chlorite + albite + lawsonite + pumpellyite or chlorite + albite + pumpellyite + blue amphibole with accessory calcite, aragonite, celadonite, white mica, sphene, opaque minerals, and quartz. The assemblage chlorite + albite + pumpellyite is most common in TZ 2B rocks, whereas the assemblage chlorite + albite + lawsonite + pumpellyite + blue amphibole is more common in TZ 3A rocks. Relict igneous textures are found in 25 percent of the samples and include both porphyritic and hypidiomorphic granular. Relict igneous clinopyroxene, in some samples rimmed by acmitic pyroxene, is present in about 35% of the samples. Stilpnomelane and epidote have not been observed in Valentine Spring metavolcanic rocks. About half of the metavolcanic rocks show textures suggesting slightly annealed cataclastic fabrics; the remaining samples are either slightly schistose or nonschistose and lack a relict igneous texture.

Metagraywacke is characterized by the presence of lawsonite, the absence of pumpellyite, and rare blue amphibole in lithic TZ 3A metagraywacke of the northeastern part of the study area. Blue amphibole occurs as incipient fibrous clusters intergrown in albite and chlorite. Although pumpellyite is common in TZ 2A rocks of the underlying Yolla Bolly terrane, it is rare to absent in TZ 2B rocks of the Valentine Spring Formation.

Metachert, which is known from only two small areas in the Valentine Spring Formation, commonly contains the assemblage

TABLE 6. MINERAL ASSEMBLAGES IN METACHERTS*

South Fork	qtz	ba	opq	stlp			bar	alb gt	carb
Mountain	qtz	ba	opq		chl				
Schist	qtz	ba	opq			hem			
	qtz			stlp					
	qtz			stlp					
Valentine	qtz	ba	bar	opq	stlp	hem			
Spring	qtz	ba							
Formation	qtz	ba		opq	stlp				
	qtz				stlp	hem			
Taliaferro	qtz	ba	chl						
Metamorphic	qtz	ba							
Complex	qtz	ba		opq	stlp				
	qtz	ba				aeg			
Metagraywacke of	qt		chl						
Hammerhorn Ridge	qtz			opq					
Chicago Rock	qtz				hem				
Mélange	qtz								

*pu = pumpellyite; lw = lawsonite; ep = epidote; chl = chlorite; ba = Na-amphibole; stilp = stilpnomelane; ga = Ca-amphibole; bar = barite; gt = spessartine; wm = phengite; par = paragonite; hem = hematite; opa = opaque; jd = jadeite; aeg = aegerine-acmite; alb = albite; qtz = quartz

crossite + hematite ± chlorite ± stilpnomelane ± barite in addition to quartz. It generally is recrystallized completely and lacks any remnants of radiolarian tests.

Yolla Bolly Terrane

Taliaferro Metamorphic Complex. Metabasites of the Taliaferro Metamorphic Complex are characterized by the mineral assemblage lawsonite + blue amphibole + chlorite + albite + pumpellyite, with accessory phases of quartz ± calcite ± aragonite ± sphene ± opaque minerals ± white mica. Both lawsonite and blue amphibole are major phases. The lawsonite is coarser grained (0.1–0.3 mm) than in other Yolla Bolly terrane metavolcanic rocks (typically, 0.03–0.07 mm) and in metavolcanic rocks of the Valentine Spring Formation (0.05–0.1 mm). The rocks are either coarse grained, lacking foliation and relict igneous texture, fine grained and schistose with a poorly developed mineral segregation, or brecciated and partially annealed. Epidote is absent, but relict green amphibole cores in blue amphiboles occur in about 30% of the samples. Although pumpellyite is fairly common, it appears to be a retrograde phase overprinting blue amphibole and lawsonite, and also occurs in veins cutting the rock. Relict igneous pyroxene is rare, occurring in only 2 of 26 samples.

Metagraywacke of the Taliaferro Metamorphic Complex is TZ 2B to 3A. The metasedimentary rocks are unique in that they commonly contain jadeitic pyroxene in addition to the assemblage lawsonite ± pumpellyite ± blue amphibole (Suppe 1973), and jadeite was one of the criteria for distinguishing the Taliaferro Metamorphic Complex from surrounding rocks (Suppe 1973). The mineral assemblage is quartz + albite + chlorite ± pumpellyite ± jadeitic pyroxene ± lawsonite ± blue amphibole ± sphene ±

calcite or aragonite. The albite veins that are common in these rocks seem to be a good field criterion for mapping the Taliaferro Metamorphic Complex.

Metagraywacke of Hammerhorn Ridge and Chicago Rock Melange. The mineral assemblages in metabasites of the Chicago Rock melange and metagraywacke of Hammerhorn Ridge are typically either chlorite + albite + pumpellyite (75 percent) or chlorite + albite + pumpellyite + lawsonite (25 percent). Trace amounts of incipient blue amphibole rimming igneous green amphibole and (or) relict pyroxene occur in about 30 percent of the samples and are most common in the coarse-grained gabbroic intrusive rock; 70 percent of the blue amphibole occurs in this rock type. Relict igneous titaniferous clinopyroxene is preserved in 50 percent of the samples, and igneous(?) green and (or) brown amphibole is also present locally. Pseudomorphs by albite, calcite and (or) pumpellyite after plagioclase are common. Plagioclase in a few of gabbroic samples contains intergrown lawsonite. Igneous textures are commonly well preserved and typically include hypidiomorphic granular and porphyritic, with extremely rare cumulate varieties. The metabasites are massive and lack any tectonite fabric.

Metagraywacke in the two units typically contain trace to minor amounts of lawsonite and pumpellyite in addition to the ubiquitous assemblage quartz + albite + white mica + chlorite + lithic fragments, trace amounts of celadonite also occur. Accessory metamorphic minerals include sphene ± hematite ± calcite ± aragonite. Detrital crystals of biotite, white mica and tourmaline (common), epidote (rare), hornblende (rare), and lithic volcanic fragments are preserved in the metagraywacke.

Lawsonite and pumpellyite are typically 0.01 mm wide parallel to (010). They are not abundant enough to identify with whole-rock X-ray diffraction. Detrital textures and clasts are well preserved in these rocks, although flattening is evident locally (TZ 2A to 2B).

Amphibolite and Blueschist Blocks. The amphibolite blocks that occur locally in the Chicago Rock mélange generally consist of pale-green amphibole and albite, with accessory sphene ± opaques ± pumpellyite ± blue amphibole ± epidote ± lawsonite. The blue amphibole typically rims green amphibole, and the lawsonite is porphyroblastic and overprints the amphibolite texture, suggesting that these phases are retrograde with respect to the earlier amphibolite or greenschist assemblage. Blueschist blocks containing the assemblage lawsonite + blue amphibole + white mica ± sphene ± calcite or aragonite ± pumpellyite ± opaques assemblages either show prograde assemblages, or are completely retrograded amphibolites.

Regional Distribution of Metamorphic Minerals

In this section, we describe the general occurrence, geographic distribution, and significance of the key metamorphic minerals. Lawsonite and aragonite occur within metasedimentary rocks, whereas pumpellyite and blue amphibole occur within metaigneous rocks, throughout the Eastern Franciscan Belt in both the Yolla Bolly and Pickett Peak terranes. Pumpellyite and pumpellyite + lawsonite are the common metamorphic index minerals in metagraywacke of the Yolla Bolly terrane, whereas generally only lawsonite appears in metagraywacke of the Pickett Peak terrane. Jadeitic pyroxene occurs only in blocks and slabs of metagraywacke of the Taliaferro Metamorphic Complex within the Chicago Rock melange. Epidote was found only in metavolcanic rocks from the eastern part of the Pickett Peak terrane, and in amphibolite blocks in the Chicago Rock melange.

Sheet Silicates. Celadonite, the Fe-Mg end member of muscovite, is common in metagraywacke of the Yolla Bolly terrane. It occurs commonly in trace amounts, as small (approximately 0.1 mm diameter) pods in the matrix, commonly with only one or two pods per thin section. Like pumpellyite, celadonite generally is absent in TZ 2B metagraywacke of the Pickett Peak terrane and appears to be restricted to TZ 2A metagraywacke of the Yolla Bolly terrane. Its disappearance may be related to a reaction forming phengite and chlorite (Ernst 1965).

Phengite and paragonite coexist locally in the South Fork Mountain Schist. The phengitic composition is only inferred on the basis of the white mica composition in other high-pressure belts (Ernst 1965; Ernst and others 1970; Black 1975). Paragonite was detected by X-ray diffraction.

Paragonite has been reported in low-grade metasedimentary rocks of the Alps (Frey 1970). It appears to be stable in Al-rich rocks, but in less Al-rich pelites it does not form until higher metamorphic grades (Hoffman and Hower 1979). The reaction lawsonite + albite = zoisite + paragonite + H_2O has been experimentally determined at 350° at 6 kb and at 450° at 12 kb (Holland 1979); however, neither zoisite nor epidote were observed in samples that contained paragonite. In addition, lawson-

ite is abundant and occurs in euhedral crystals showing no obvious signs of reaction.

In the Alpine rocks, Frey (1970) suggested that paragonite formed from clays in the sequence: irregular mixed-layer illite to montmorillonite, to regular mixed-layer illite-montmorillonite, to mixed paragonite-phengite, to paragonite. Chatterjee (1973) was able to synthesize, but not reverse, the reaction albite + Na - montmorillonite = paragonite + quartz between 335–315°C and 2–7 kb. The presence of paragonite in the South Fork Mountain Schist may be due to a reaction involving albite plus mixed-layer mica and clays in the temperature regime of Chatterjee's (1973) experiment.

Suppe (1973) reported stilpnomelane in metagraywacke from both the Yolla Bolly and Pickett Peak terranes. However, Kerrick and Cotton (1971), Raymond (1972), and Black (1975), showed that much of what has been called stilpnomelane in low-grade metasedimentary rocks is an oxychlorite or Fe-vermiculite. Stilpnomelane was observed in a few samples of metachert and, more rarely, in metavolcanic rock of the Pickett Peak terrane. Its identification was verified by X-ray diffraction.

Blue Amphibole. Blue amphibole was observed in tectonic blocks from the metagraywacke of the Taliaferro Metamorphic Complex, and from metagraywacke of the Valentine Spring Formation and the South Fork Mountain Schist. The appearance of blue amphibole in in situ metagraywacke approximately coincides with a pronounced increase in the abundance of blue amphibole in metaigneous rocks of the Pickett Peak terrane. The characteristic assemblage lawsonite + pumpellyite + chlorite ± trace amounts of fibrous incipient crossitic blue amphibole in lower grade assemblages in the metavolcanic rocks suggests reactions such as chlorite + albite = blue amphibole (Miyashiro and Banno 1958) or 4 albite + 1 chlorite = 1 glaucophane + 1 paragonite + $2H_2O$ (Goffe 1977).

Blueschist-facies assemblages from the Ankarama area in Turkey (Okay 1980) show textural evidence for the reaction yellow-green Na-pyroxene (after relict igneous pyroxene) + chlorite + quartz = Na amphibole + lawsonite. This reaction is consistent with the assemblages and textures observed in metaigneous rocks of the Valentine Spring Formation and in some metaigneous rocks of the Chinquapin Metabasalt Member.

Lawsonite and Pumpellyite. Lawsonite and pumpellyite are common constituents in metagraywacke and metaigneous rocks of both the Yolla Bolly and Pickett Peak terranes. The assemblages lawsonite + pumpellyite + chlorite + white mica + quartz + albite ± calcite or aragonite is typically found in TZ 2A metagraywacke of the Yolla Bolly terrane. In the metagraywacke, detrital textures commonly are well preserved, although evidence for pressure solution and elongation of clasts and for incipient recrystallization is widespread. Lawsonite and pumpellyite generally are extremely fine grained (approx. 0.01 mm) and irregularly distributed throughout thin sections.

Epidote. The appearance of epidote and the disappearance of lawsonite in metaigneous rocks of the Pickett Peak terrane was investigated by Brown and Ghent (1983) in the Ball Rock area

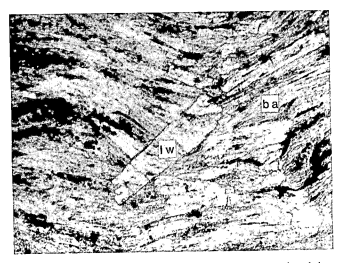

Figure 8. Lawsonite (lw), blue amphibole (ba), and opaque minerals in a sample of the Chinquapin Metabasalt Member showing lawsonite overprinting S_1 and F_2 crenulation fold. Fabric defined by fibrous blue amphibole.

(Figure 8b). Epidote first occurs in metavolcanic rocks west of the lawsonite-out isograd at Ball Mountain (Suppe 1973; Brown and Ghent 1983) and can be found in the same rock with lawsonite. Brown and Ghent (1983) proposed the reaction lawsonite + blue amphibole = epidote + quartz + albite + chlorite + H_2O for these rocks. With increasing temperature and pressure, this reaction is followed by the reaction blue amphibole + pumpellyite + quartz = epidote + calcic amphibole + albite + chlorite + H_2O. An epidote + Ca-amphibole + albite + chlorite assemblage occurs in metabasalt at North Yolla Bolly Mountain (Blake 1965), north of the study area and at Black Butte south of the study area (Ghent 1965; Brown and Ghent 1983). Epidote was not observed in metasedimentary rocks of the South Fork Mountain Schist within the study area.

Paragenesis

Textural relations indicate mineral growth during D_1, and to a much lesser extent D_2 in the Pickett Peak terrane, and during D_2 in the Yolla Bolly terrane (note the absence of D_1 structures in the Yolla Bolly terrane). Blue amphibole, lawsonite, pumpellyite, and chlorite in metavolcanic rocks, and quartz, albite, white mica, chlorite, and lawsonite in mica schist and TZ 3 metagraywacke are generally oriented subparallel to S_1 foliation in the Pickett Peak terrane. These minerals are typically deformed by D_2 and D_3 microstructures. Less common are samples in which lawsonite and blue amphibole overprint S_1 or F_2 folds. In a few thin sections of metavolcanic rock, however coarse-grained, euhedral lawsonite porphyroblasts overprint both S_1 schistosity and F_2 crenulations (Figure 8). In other thin sections, lawsonite is strongly oriented parallel to the S_1 fabric, and earlier lawsonite grain ends are truncated, pinched out, or oriented such that their long axis is normal to the foliation. Lawsonite tablets commonly

contain S_1 inclusion trails that deflect the S_1 fabric, features indicating synkinematic and postkinematic growth. Therefore, the metamorphic assemblage grew during D_1 and, at least in part, after D_2, as indicated by rare overprint minerals.

In a few samples of the Chinquapin Metabasalt Member, sprays of blue amphibole that crosscut the S_1 fabric indicate late or post-D_1 growth. In a few thin sections, blue amphibole is morphologically oriented subparallel to S_2 surfaces. However, the blue amphibole is generally fibrous, and slight undulose extinction suggests that the mineral is probably rotated into the direction of S_2 foliation. Veins of chlorite, and more rarely, pumpellyite that crosscut both S_1 and S_2 fabrics indicate that these phases continued to be stable after the fabrics developed.

In metavolcanic rock associated with TZ 2B metagraywacke of the Valentine Spring Formation, the S_1 fabric generally is poorly developed or absent. The critical crosscutting relations between metamorphic minerals and deformational fabrics were not observed. In the lowest grade metagraywacke and metavolcanic rocks of the Yolla Bolly terrane, lawsonite and blue amphibole generally are very fine grained and randomly oriented.

PRESSURE-TEMPERATURE CONDITIONS

Pressure Conditions

Aragonite and jadeite reaction curves, which have been determined experimentally, provide the best constraints for the pressure conditions of metamorphism. Suppe (1973) first showed the widespread distribution of aragonite in the Leech Lake-Ball Mountain area. Aragonite, identified by X-ray and petrographic methods, occurs in both veins and groundmass from the Yolla Bolly and Pickett Peak terranes (Suppe 1973). Jadeite, though more restricted in occurrence, is useful for establishing pressure estimates.

Experimental data (Boettcher and Wyllie 1967; Johannes and Puhan 1971; Crawford and Hoersch 1972) suggest that aragonite is stable above pressures of 4.5 kb at 100°C and 8.5 kb at 300–320°C, corresponding to depths of 18–30 km. The presence of aragonite also indicates that the rocks were dry during reduction in pressure and temperature conditions. Aragonite reacts to form calcite, essentially instantaneously with respect to geologic time, at almost any pressure and at temperatures above 50°C, when a fluid is present and when the rocks are within the calcite-stability field (Brown and others 1962; Metzger and Barnard 1968; Bischoff 1969). Therefore, apparently either the rocks were dry by the time they reached the calcite-stability field, or they had cooled below 50°C before entering that field (Metezger and Barnard 1968).

The jadeite component of metamorphic pyroxenes has been used to estimate pressure conditions (Newton and Smith 1967; Brown and Ghent 1983). Pyroxene compositions in metavolcanic rocks of the Yolla Bolly terrane fall in the aegirine-jadeite field (Echeverria 1978) and those of the Pickett Peak terrane fall in the chloromelanite, aegirine-jadeite, and aegirine fields (Brown and

Ghent 1983). Overall, the jadeite content in pyroxenes ranges from 10 to 40 percent. Gabbroic rock of the Yolla Bolly terrane contains 30 to 40 mol% jadeite, whereas the average Pickett Peak terrane component is about 20 ± 10%. The apparent variation in the jadeite component by as much as 10% within a thin section indicates that equilibrium was not attained on that scale. Jadeite in metagraywacke of the Taliaferro Metamorphic Complex is optically similar to analyzed jadeite of more than Jd_{80} (Coleman 1961; Kerrick and Cotton 1971).

Pressure minimums obtained from experimental data and thermodynamic calculations by Brown and Ghent (1983) yield 5–6.7 kb on Jd_{20} and 6–7.7 kb on Jd_{30} between 200 and 350°C. These data, as applied to the study area, are consistent with the pressure estimated from aragonite stability.

Temperature Conditions

Temperature conditions can be deduced from oxygen isotope ratios (Taylor and Coleman 1968) and phase relations (e.g. Brown and Ghent 1983) from rocks in the study area. Temperature conditions can also be inferred from vitrinite reflectance (Bostick 1974) on correlative rocks farther south in the Diablo Range, California.

Pickett Peak Terrane. Metavolcanic rocks of the Pickett Peak terrane are similar to the type III metabasalts of Coleman and Lee (1963) that yielded O^{18}/O^{16} temperatures of 270–315°C. Brown and Ghent (1983) applied a different calibration (Bottinga and Javoy 1973) to a quartz-magnetite pair from the South Fork Mountain Schist reported by Taylor and Coleman (1968), and obtained a temperature of 330°C. Preliminary O^{18}/O^{16} data on metachert from the South Fork Mountain Schist (Jim Drotleff, oral communication 1983) yield temperatures of 330–345°C in samples collected within 1.5 to 3 km of the Coast Range thrust. Maximum temperatures were no greater than 400°C, as limited by the maximum-stability field of lawsonite ± albite and the jadeite reaction curve (Figure 9, curves 1 and 3).

Nitsch (1971) determined that the reaction curve pumpellyite + chlorite + quartz goes to epidote + actinolite + H_2O between about 340° to 380°C at about 2.5 to 9 kb respectively, and Schiffman and Liou (1977) experimentally determined the reaction curve for end-member Mg-pumpellyite, which forms zoisite-garnet-chlorite-quartz and H_2O between 300° and 375° at 3 to 7.5 kb. Extrapolation of these curves suggests a maximum temperature of about 380°C up to about 7 to 9 kb pressure.

Yolla Bolly Terrane. Potassium feldspar is estimated to be stable up to about 100°C and, possibly, to 125°C before it reacts with smectite to form chlorite, illite, and quartz (Hower and others 1976; Hoffman and Hower 1979). A total of 79 samples of TZ 2A to 2B graywacke were stained for potassium feldspar; only one sample contained any detectable potassium feldspar (less than 1 percent). Assuming that detrital potassium feldspar was originally present in the graywacke, because it is a common mineral in lower grade clastic sedimentary rocks of the Franciscan assemblage (Bailey and Irwin 1959), then temperatures were probably above 100–125°C.

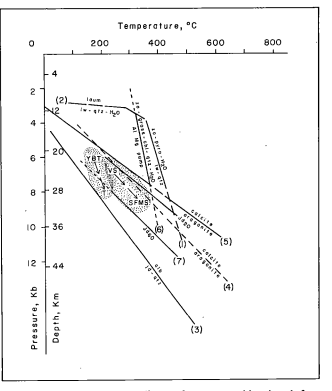

Figure 9. Pressure-temperature diagram for metamorphic minerals from the study area. Reaction curves: (1) Nitsch, 1971; (2) Liou, 1971; (3) Newton and Smith, 1967; (4) Johannes and Puhan, 1971; (5) Crawford and Hoersch, 1972; (6) Schiffman and Liou, 1977; and (7) Brown and Ghent, 1983. YBT = Yolla Bolly terrane: Chicago Rock melange and metagraywacke of Hammerhorn Ridge undivided; T = Taliaferro Metamorphic Complex; VS_2 = Valentine Spring TZ 2B; VS_3 = Valentine Spring TZ 3A; and SFMS = South Fork Mountain Schist.

Jadeite- and aragonite-bearing Franciscan rocks from the Diablo Range, central California, which are correlated with the Taliaferro Metamorphic Complex of the Yolla Bolly terrane (Blake and others 1982), were measured for vitrinite reflectance (Bostick 1974). The rocks are mostly massive graywacke with lesser amounts of interbedded shale, continuous horizons of radiolarian chert, and intrusive diabasic and gabbroic sills (Ernst and others 1970; Echeverria 1978, 1980). They contain the assemblage lawsonite + pumpellyite + quartz + albite + chlorite + white mica ± jadeite ± calcite ± aragonite and are of TZ 2A to 2B. Eight graywacke samples yielded reflectance temperatures of 125–150°C (Bostick 1974); this temperature range may also be comparable for Yolla Bolly terrane rocks.

CONCLUSIONS

The structural, metamorphic, and geochronologic data indicate two periods of high pressure/temperature metamorphism in the Eastern Franciscan Belt of northern California. The earliest event (D_1), which was restricted to the Pickett Peak terrane and accompanied by blueschist metamorphism, is poorly dated at about 125 m.y. This event is inferred to be related to accretion of that terrane to North America during subduction followed by collision of oceanic fragments. The D_1 mineral lineations in the Pickett Peak terrane indicate northeast to southwest movement directions; however, the vergence direction could not be determined. A second blueschist event (D_2), which is recorded in both the Pickett Peak and Yolla Bolly terranes, appears to have occurred about 90–105 m.y. ago, presumably coincident with accretion of the Yolla Bolly terrane. One of the anomalous results of this study is that the pronounced asymmetry of D_2 folds in the Pickett Peak terrane (Jayko 1984) indicates southwest to northeast vergence and suggests that the upper plate (continental basement? Coast Range ophiolite?) moved eastward relative to the lower plate (Jayko 1984).

D_3 was a postmetamorphic event that affected both the Yolla Bolly and Pickett Peak terranes. D_3 structures formed within a northwest-southeast compressional regime (Jayko 1984)

and may be related to the thrusting that resulted in uplift of the blueschist. Although D_3 structures in the study area are not asymmetric, mapping to the north (Blake and Jayko 1983) indicates a consistent northwest to southeast vergence in D_3 folds, and suggests that the upper plate was moving eastward relative to the lower plate. The timing of D_3 is poorly constrained; however, cobbles of lawsonite-bearing metagraywacke and metachert of the Yolla Bolly terrane(?) occur in Paleocene or early Eocene conglomerates to the south in Rice Valley (Berkland 1973) indicating that the terranes were up to the surface by this time. D_4 structures are inferred to be early Tertiary in age and related to pre-San Andreas strike-slip faulting.

ACKNOWLEDGMENTS

We thank David J. Borns, Keith Howard, Dave Miller, Stephen Phipps, and Dan M. Worrall for their constructive reviews.

REFERENCES CITED

Bailey, E. H., and Irwin, W. P., 1959, K-spar content of Jurassic and Cretaceous graywackes of the northern Coast Ranges and Sacramento Valley, California: American Association of Petroleum Geologists Bulletin, v. 42, p. 2797–2807.

Bailey, E. H., Irwin, W. P., and Jones, D. L., 1964, Franciscan and related rocks and their significance in the geology of western California: California Division Mines and Geology Bulletin 183, 177 p.

Bailey, E. H., Blake, M. C., Jr., and Jones, D. L., 1970, On-land Mesozoic oceanic crust in California Coast Ranges, *in* Geological Survey research, 1970: U.S. Geological Survey Professional Paper 700-C, p. C70–C81.

Berkland, J. O., 1973, Rice Valley Outlier—new sequences of Cretaceous-Paleocene strata in northern Coast Range, California: Geological Society of America Bulletin, v. 84, p. 2389–2406.

Berkland, J. O., Raymond, L. A., Kramer, J. C., Moores, E. M., and O'Day, M., 1972, What is Franciscan?: American Association of Petroleum Geologists Bulletin, v. 56, p. 2295–2302.

Bischoff, J. L., 1969, Temperature controls of the aragonite-calcite transformation in aqueous solution: American Mineralogist, v. 54, p. 149–155.

Bishop, D. G., 1972, Progressive metamorphism from prehnite-pumpellyite to greenschist facies in the Dansey-Pass area, New Zealand: Geological Society of America Bulletin, v. 83, p. 3177–3197.

Black, P. M., 1975, Mineralogy of New Caledonian metamorphic rocks. IV. sheet Silicates from the Quegoa district: Contributions to Mineralogy and Petrology, v. 49, p. 269–284.

Blake, M. C., Jr., 1965, Structure and petrology of low grade metamorphic rocks, blueschist facies: Yolla Bolly area, northern California [Ph.D. thesis]: Stanford, California, Stanford University, 91 p.

Blake, M. C., Jr., Harwood, D. S., Helley, E. J., Irwin, W. P., Jayko, A. S., and Jones, D. L., 1984, Preliminary geologic map of the Red Bluff 1:100,000 quadrangle, California: U.S. Geological Survey Open-File Report, OF-84-105, scale 1:100,000.

Blake, M. C., Jr., Howell, D. G., and Jones, D. L., 1982, Tectonostratigraphic terrane map of California: U.S. Geological Survey OPen File Report 82-593, scale 1:750,000.

Blake, M. C., Jr., Irwin, W. P., and Coleman, R. G., 1967, Upside-down metamorphic zonation, blueschist facies, along a regional thrust in California and Oregon: U.S. Geological Survey Professional Paper 575-C, p. C1–C9.

Blake, M. C., Jr., and Jayko, A. S., 1983, Geologic map of the Yolla Bolly-Middle Eeel Wilderness and adjacent roadless areas, northern California: U.S. Geological Survey Miscellaneous Field Studies Map, MF-1565-A, scale

1:62,500.

Blake, M. C., Jr., Jayko, A. S., and Howell, D. G., 1981, Geology of a subduction complex in the Franciscan assemblage of northern California: Oceanologica Acta, 1981, p. 267–272.

Blake, M. C., Jr., and Jones, D. L., 1974, Origin of Franciscan melanges in northern California, *in* Dott, R. H., and Shaver, R. H., eds., Modern and ancient geosynclinal sedimentation: Society Economic Paleontologists and Mineralogists, Special Publications 19, p. 255–263.

Boettcher, A. L., and Wyllie, P. J., 1967, Biaxial calcite inverted from aragonite: American Mineralogist, v. 52, p. 1527–1529.

Bostick, N. H., 1974, Phytoclasts as indicators of thermal metamorphism, Franciscan assemblage and Great Valley sequence (Upper Mesozoic), California, *in* Dutcher, R. R., Hacquebard, P. A., Schopf, J. M., and Simon, J. A., eds., Carbonaceous materials as indicators of metamorphism: Geological Society of America Special Paper 15, p. 1–17.

Bottinga, Y., and Javoy, M., 1973, Comments on oxygen isotope geothermometry: Earth and Planetary Science Letters, v. 20, p. 250–265.

Brown, E. H., and Ghent, E. D., 1983, Mineralogic and phase relations in the blueschist facies of the Black Butte and Ball Rock areas, northern California Coast Ranges: American Mineralogist, v. 68, p. 365–372.

Brown, W. H., Fyfe, W. S., and Turner, F. J., 1962, Aragonite in California glaucophane schists, and the kinetics of the aragonite-calcite Transformation: Journal of Petrology, v. 3, p. 566–582.

Chatterjee, N. D., 1973, Low-temperature compatibility relations of the assemblage quartz-paragonite and the thermodynamic status of the phase rectorite: Contributions to Mineralogy and Petrology, v. 42, p. 259–271.

Coleman, R. G., 1961, Jadeite deposits of the Clear Creek area, New Idria district, San Benito County, California: Journal of Petrology, v. 2, p. 209–247.

Coleman, R. G., and Lee, D. E., 1963, Glaucophane-bearing metamorphic rock types of the Cazadero area, California: Journal of Petrology, v. 4, p. 266–301.

Cowan, D. S., 1974, Deformation and metamorphism of the Franciscan subduction zone complex northwest of Pacheco Pass, California: Geological Society of America Bulletin, v. 85, p. 1623–1634.

Crawford, K. E., 1975, The geology of the Franciscan tectonic assemblage near Mount Hamilton, California [Ph.D. thesis]: University of California, Los Angeles, 137 p.

Crawford, W. A., and Hoersch, A. L., 1972, Calcite-aragonite equilibrium from 50° to 150°C: American Mineralogist, v. 57, p. 995–998.

Echeverria, L. M., 1978, Petrogenesis of metamorphosed intrusive gabbros in the Franciscan complex, California [Ph.D. thesis]: Stanford, California, Stanford University, 187 p.

—— 1980, Oceanic basaltic magmas in accretionary prisms: The Franciscan intrusive gabbros: American Journal of Science, v. 280, p. 697–724.

Ernst, W. G., 1965, Mineral parageneses in Franciscan rocks, Panoche Pass, California: Geological Society of America Bulletin, v. 76, p. 879–914.

Ernst, W. G., Seki, Y., Onuki, H., and Gilbert, M. C., 1970, Comparative study of low-grade metamorphism in the California Coast Ranges and the outer metamorphic belt of Japan: Geological Society of America Memoir no. 124, 276 p.

Frey, Martin, 1970, The step from diagenesis to metamorphism in pelitic rocks during Alpine orogenesis: Sedimentology, v. 15, p. 261–279.

Ghent, E. D., 1965, Glaucophane-schist facies metamorphism in the Black Butte area, northern Coast Ranges, California: American Journal of Science, v. 263, p. 385–400.

Goffe, Bruno, 1977, Succession de subfacies metamorphism en Vanoise meridional (Savoie): Contributions to Mineralogy and Petrology, v. 62, p. 23–41.

Hoffman, J., and Hower, J., 1979, Clay mineral assemblages as low-grade metamorphic geothermometers: application to the thrust faulted disturbed belt of Montana, U.S.A., *in* Scholle, P. A., and Schluger, P. R., eds., Aspects of diagenesis: Society of Economic Paleontologists and Mineralogists Special Publication 26, p. 55–79.

Holland, T. J., 1979, Experimental determination of the reaction paragonite = jadeite + kyanite + H_2O, and internally consistent thermodynamic data for part of the system $Na_2O-Al_2O_3-SiO_2-H_2O$ with application to eclogites and blueschists: Contributions to Mineralogy and Petrology, v. 68, p. 293–301.

Hopson, C. A., Mattinson, J. M., and Pessagno, E. A., Jr., 1981, Coast Range ophiolite, western California, *in* Ernst, W. C., ed., The geotectonic development of California: Englewood Cliffs, N. J., Prentice Hall, p. 418–510.

Hower, J., Eslinger, E., Hower, M. E., and Perry, E. A., 1976, Mechanisms of burial metamorphism of argillaceous sediments: mineralogical and chemical evidence: Geological Society of America Bulletin, v. 87, p. 725–737.

Hutton, C. O., and Turner, F. J., 1936, Metamorphic zones in northwest Otago: Royal Society of New Zealand, Transactions and Proceedings, v. 65, p. 405–406.

Irwin, W. P., 1960, Geologic reconnaissance of the northern Coast Ranges and Klamath Mountains, California, with a summary of the mineral resources: California Division of Mines Bulletin 179, 80 p.

—— 1981, Tectonic accretion of the Klamath Mountains, *in* Ernst, W. C., ed., The geotectonic development of California: Englewood Cliffs, N.J., Prentice Hall, p. 29–49.

Jacobson, M. I., 1978, Petrologic variations in Franciscan sandstone from the Diablo Range, California, *in* Howell, D. G., and McDougall, K. A., eds., Mesozoic Paleogeography of the western United States: Pacific Coast Paleogeography Symposium 2, Pacific Section, Society of Economic Paleontologists and Mineralogists, p. 401–417.

Jayko, A. S., 1984, Structure and deformation in the Eastern Franciscan belt, northern California [Ph.D. thesis]: Santa Cruz, University of California, 204 p.

Johannes, W., and Puhan, D., 1971, The calcite-aragonite transition, reinvestigated: Contributions to Mineralogy and Petrology, v. 31, p. 28–38.

Jones, D. L., Blake, M. C., Jr., Bailey, E. H., and McLaughlin, R. J., 1977, Distribution and character of Upper Mesozoic subduction complexes along the west coast of North America: Tectonophysics, v. 47, p. 207–222.

Jones, D. L., Blake, M. C., Jr., and Rangin, Claude, 1976, The four Jurassic belts of northern California and their significance to the geology of the southern California borderland, *in* Howell, D. G., ed., Aspects of the geologic history of the California Continental Borderland: American Association of Petro-

leum Geologists, Pacific Section Miscellaneous Publication 24, p. 343–362.

Kerrick, D. M., and Cotton, W. R., 1971, Stability relations of jadeite pyroxene in Franciscan metagraywackes near San Jose, California: American Journal of Science, v. 271, p. 350–369.

Lanphere, M. A., Blake, M. C., Jr., and Irwin, W. P., 1978, Early Cretaceous metamorphic age of the South Fork Mountain Schist in the northern California Coast Ranges: American Journal of Sciences, v. 278, p. 798–816.

Liou, J. G., 1971, P-T stabilities of laumontite, wairakite, lawsonite, and related minerals in the system $Ca_2Al_2Si_2O_8-SiO_2-H_2O$: Journal of Petrology, v. 12, p. 379–411.

MacPherson, G. P., 1983, The Snow Mountain Volcanic Complex: An on-land seamount in the Franciscan terrain, California: Journal of Geology, v. 91, p. 73–92.

Mattinson, D. M., 1981, U-Pb systematics and geochronology of blueschist: preliminary results: American Geophysical Union Transactions, v. 62, p. 1059.

McDowell, F. W., Lehman, D. H., Gucwa, P. R., Fritz, Deborah, and Maxwell, J. C., 1984, Glaucophane schists and ophiolites of the northern California Coast Ranges: Isotopic ages and their tectonic implications: Geological Society of America Bulletin, v. 95, p. 1373–1382.

Metzger, W. J., and Barnard, W. M., 1968, Transformation of aragonite to calcite under hydrothermal conditions: American Mineralogist, v. 53, p. 295–299.

Miyashiro, A., and Banno, S., 1958, Nature of glaucophanitic metamorphism: American Journal of Science, v. 256, p. 97–110.

Newton, R. C., and Smith, J. V., 1967, Investigations concerning the breakdown of albite at depth in the earth: Journal of Geology, v. 75, p. 268–286.

Nitsch, K. H., 1971, Stabilitatsbeziehungen von Prehnit-Pumpellyit-haltigen Paragenesen: Contributions to Mineralogy and Petrology, v. 30, p. 240–260.

Okay, A. I., 1980, Lawsonite zone blueschists and a sodic amphibole producing reaction in the Tavsanti region, northwest Turkey: Contributions to Mineralogy and Petrology, v. 75, p. 179–186.

Powell, C. M., 1979, A morphological classification of rock cleavage: Tectonophysics, v. 58, p. 21–35.

Raymond, L. A., 1972, Franciscan geology of the Mt. Oso area, California [Ph.D. thesis]: Davis, University of California, 185 p.

—— 1974, Possible modern analogs for rocks of the Franciscan Complex, Mount Oso area, California: Geology, v. 2, p. 143–146.

Schiffman, P., and Liou, J. G., 1977, Synthesis and stability reactions of Mg-pumpellyite: Journal of Petrology, v. 21, p. 441–474.

Suppe, John, 1973, Geology of the Leech Lake Mountain-Ball Mountain region, California; A cross section of the northeastern Franciscan belt and its tectonic implications: University of California Publications in Geological Sciences, v. 107, 82 pp.

Suppe, John, and Foland, K. A., 1978, The Goat Mountain Schists and Pacific Ridge Complex: A redeformed but still intact Late Mesozoic Schuppen complex, *in* Howell, D. G., and McDougall, K. A., eds., Mesozoic Paleogeography of the western United States: Pacific Coast Paleogeography Symposium 2, Pacific Section, Society of Economic Paleontologists and Mineralogists, p. 431–451.

Taylor, H. P. and Coleman, R. G., 1968, O^{18}/O^{16} ratios of coexisting minerals in glaucophane-bearing metamorphic rocks: Geological Society of America Bulletin, v. 79, p. 1727–1755.

Turner, F. J., 1981, Metamorphic petrology: New York, Hemisphere, 525 p.

Winkler, H. G., 1979, Petrogenesis of metamorphic rocks: New York, Springer-Verlag, 348 p.

Worrall, D. M., 1981, Imbricate low-angle faulting in uppermost Franciscan rocks, South Yolla Bolly area, northern California: Geological Society of America Bulletin, v. 92, p. 703–729.

MANUSCRIPT ACCEPTED BY THE SOCIETY JULY 29, 1985

Printed in U.S.A.

Geological Society of America
Memoir 164
1986

Deformation and high P/T metamorphism in the central part of the Condrey Mountain window, north-central Klamath Mountains, California and Oregon

Mark A. Helper
Department of Geological Sciences
University of Texas at Austin
Austin, Texas 78713

ABSTRACT

The Condrey Mountain Schist occupies a window through Late Triassic(?) amphibolite facies mélange of the Western Paleozoic and Triassic Belt in the north central Klamath Mountains. Along the western margin of the window, the schist comprises a sequence of multiply deformed, greenschist facies metavolcanic and fine-grained metasedimentary rocks, which are in thrust or high-angle fault contact with graphite-quartz-mica schist exposed in the window interior. Transitional blueschist-greenschist facies parageneses are developed in metabasites, metacherts, and metalliferous metasedimentary rocks in the graphitic schist in the central part of the window. All units record progressive, polyphase, deformational and metamorphic histories.

The earliest stages of deformation in the blueschists generated isoclinal intrafolial folds (F_1), a layer-parallel transposition foliation, and a strong crossite lineation. The foliation is folded by two later sets of coaxial, isoclinal to tight folds (F_2 and F_3) that produce kilometer-scale, N-S trending, recumbent folds in the central part of the window. The same deformational sequence is recorded in the graphitic schist by continued regeneration of the transposition foliation through at least the second set of folds, and tight to isoclinal folding of the resulting surfaces. Both lithologies have been further deformed by extension parallel to F_3 axes (boudinage and fracturing) and a late folding (F_4) that produced kink bands, box folds, and chevron folds with E-W trending axes. Fold styles and asymmetries suggest that the early stages of progressive deformation (F_1 folding, transposition) resulted from noncoaxial deformation. Shear strains diminish during F_2 and F_3 folding and are replaced by irrotational flattening strains by the time of boudinage.

The mineral assemblages formed prior to F_2 folding indicate greenschist-blueschist facies conditions during transposition. F_2 and F_3 folding were accompanied by the growth of deerite that lies in the axial planes of minor folds in meta-ironstones, indicating P-T conditions similar to those existent during transposition. Boudinage is concurrent with and followed by the static growth of chlorite, actinolite, albite, stilpnomelane, spessartine, and the Ba-silicate, cymrite. Ferroglaucophane rims on crossite may also have grown at this time. Pressure and temperature estimates, the relative time framework of deformational events, and the noncoaxial geometry of ductile strain are all consistent with, but not restricted to, a subduction zone environment. High shear strains may reflect descent and burial, whereas flattening and static mineral growth occur during uplift. Regional relationships favor an interpretation that relates metamorphism and ductile deformation to a Middle Jurassic subduction event, but do not preclude a Late Jurassic age for deformation and metamorphism.

125

INTRODUCTION

The Condrey Mountain Window in the north-central Klamath Mountains of northern California and southwestern Oregon is one of several areas in the western United States where high-pressure facies-series rocks of Mesozoic age are exposed (Coleman 1971; Misch 1966, 1977; Ernst 1984, and references therein). Based on the extent of lithologic and metamorphic continuity, these high-pressure belts or terranes can be grossly divided into two structurally distinct groups: mélanges and coherent terranes. In the former, blueschist and eclogite parageneses are developed in mafic blocks that are often older than, and may have metamorphic and structural histories different from, an enclosing pelitic or serpentine matrix (Coleman and Lanphere 1971; Mattinson and Echeverria 1980; Moore 1984). In contrast, rocks containing blueschist facies assemblages in coherent terranes are exposed in areally extensive tracts of high-pressure facies-series tectonites that may show an orderly progression of metamorphic facies and have rock units that share common ages and structural histories. The belts are coherent in the sense that the rock units within them have not been tectonically mixed or dismembered. The Condrey Mountain Schist is of this latter group, as are the majority of other Mesozoic blueschists in the western United States (Coleman 1971; Misch 1966, 1977; Wood 1971; Hotz 1973).

The common mineralogic associations of both of the above groups have been fairly well studied and are known to require metamorphism under high P/T conditions. The structural histories of the two groups, specifically the kinematic or strain patterns developed during recrystallization and uplift, are not as well documented. Inasmuch as a thorough understanding of the tectonic evolution of high P/T belts must derive from both structural and petrologic considerations, tectonic interpretation are particularly well-served by studies that integrate these aspects. Coherent blueschist facies terranes are well-suited for such studies in that they often preserve well-developed sequences of superimposed minor structures that can be traced or correlated over large areas. Although markers that record the geometry and magnitude of strain are rarely preserved, changes in the style of deformation and the deformation mechanisms during metamorphism, as reflected by minor structures and mineral textures, can often offer clues to the strain history.

In this paper, I present data from the central part of the Condrey Mountain Window on the structure and petrology of transitional blueschist-greenschist facies rocks of the Condrey Mountain Schist. Minor structures, mineral textures, and the relationships between deformation and metamorphism are used to evaluate metamorphic conditions during deformation and to examine possible tectonic settings for deformation and metamorphism.

REGIONAL SETTING

The Klamath Mountains comprise a series of east-dipping,

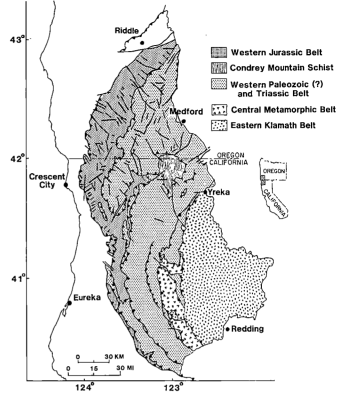

Figure 1. Generalized geologic map of the Klamath Mountains (modified from Irwin 1981), showing major thrust belts and the location of the Condrey Mountain Window.

westward younging, thin, thrust sheets (Irwin 1960, 1972; Hotz 1971; Barnes et al. 1982) (Figure 1). The eugeoclinal nature of the sediments in each of the belts, the relatively widespread occurrences of ophiolitic mélange, ultramafic rocks and blueschists, and the imbricated geometry of the thrust sheets have led many workers (e.g. Hamilton 1969; Irwin 1981; Davis et al. 1978; Wright 1982; Saleeby et al. 1982) to suggest that the Klamath Mountains are the product of protracted convergent-margin accretion. The Condrey Mountain Schist occupies a subcircular window in the north central Klamath Mountains through the largest of the thrust sheets, the so-called western Paleozoic and Triassic Belt (WPTB) (Hotz 1979; Irwin 1960) (Figure 1). The metamorphic grade of the WPTB is greenschist facies away from the window but it increases, in a roughly concentric manner, to amphibolite facies at the window margins (Kays and Ferns 1980; Hotz 1979; Medaris and Welsh 1980). The elevation of the amphibolite facies rocks and the Condrey Mountain Schist to their present levels has been ascribed to a structural dome centered on the Condrey Mountain Window (Coleman and Helper 1983; Mortimer and Coleman 1984).

Within the Condrey Mountain Window the schist can be divided into two main units (Figure 2). Along the western and northeastern margins of the window, the schist comprises a multiply deformed, thoroughly recrystallized sequence of greenschist facies metavolcanics and fine-grained metasediments (Coleman et

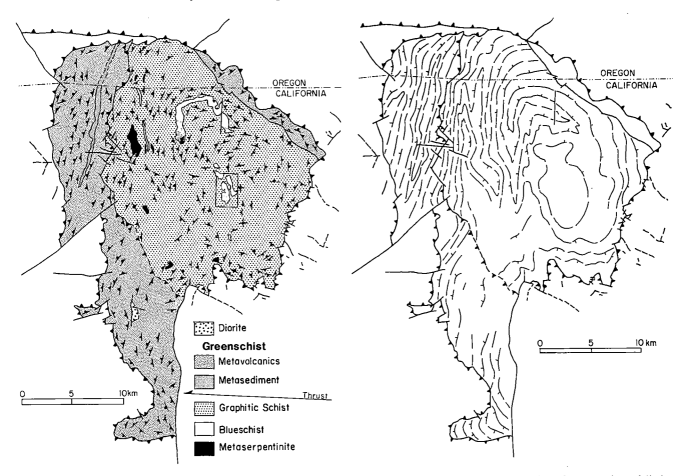

Figure 2. Geologic map of the Condrey Mountain Window, after Coleman et al. (1983), Helper (1985), Hotz (1967), Kays and Ferns (1980), and Barrows (1969). Strike and dip symbols show attitudes of the prominent foliation (S_T); thrust symbol shows teeth on upthrown plate. Area of Figure 4 is indicated by the reference box.

Figure 3. Strike line map of the orientation of the prominent foliation (S_T) in the Condrey Mountain Window. S_T orientations define an asymmetric dome. Data are from same sources given in Figure 2.

al. 1983). Actinolite and white mica from this unit have yielded K-Ar ages of 125 ± 3 Ma and 142 ± 2 Ma (Coleman et al. 1983; Helper 1985), respectively. The greenschists are in thrust or high-angle-fault contact with graphite-quartz-mica schist exposed in the window interior. The graphitic schist contains volumetrically minor amounts of interlayered and infolded metaserpentinite, metavolcanics, metalliferous metasediment, pelitic schist, and metachert which contain greenschist-blueschist facies parageneses (Donato et al. 1980; Helper 1983a, b) (Figure 2). The age of metamorphism in the central part of the window has not yet been resolved. Phengite and crossite from one pelitic schist interlayer within the area described in this report give K-Ar cooling ages of 128 ± 2 Ma and 127 ± 6 Ma, respectively (Helper 1985), whereas Coleman (written communication) has obtained an $^{40}Ar/^{39}Ar$ sodic amphibole age of 167 ± 12 Ma and a K-Ar white mica age of 118 ± 2 Ma from a mafic layer elsewhere in the central window. Suppe and Armstrong (1972) obtained a K-Ar whole rock age of 159 ± 3 Ma and Lanphere and others (1968) reported a K-Ar mica age of 144 Ma (recalculated using constants in Steiger and Jager 1977). A single, concordant U/Pb zircon age of

a deformed tonalite that intrudes greenschist near the southern margin of the window indicates a minimum depositional age of 170 ± 1 Ma (Saleeby et al. 1984).

The strikes of the folded interlayers and the main foliation in the graphitic schist conform to a roughly concentric pattern that mimics the outline of the window margins. Dip attitudes define an asymmetric dome with shallowly dipping eastern, northern and southern flanks and a steeply dipping western flank (Figure 3). The mapped area (Figure 2) is located near the center of the dome and encompasses several north-facing glacial headwalls and ridges along a major drainage divide. In contrast to the extensive vegetative cover so characteristic of the Condrey Mountain Window, exposures in the mapped area are exceptionally good; unit contacts can in places be traced for up to 0.5 kilometers along strike. Outcrop mapping at a scale of about 1:16,000 shows that some of the greenschist-blueschist interlayers define megascopic, recumbent, isoclinal folds (Figure 4). The structural and petrologic relationships in the area of recumbent folding (hereafter referred to as the Dry Lake area) are the focus of this report.

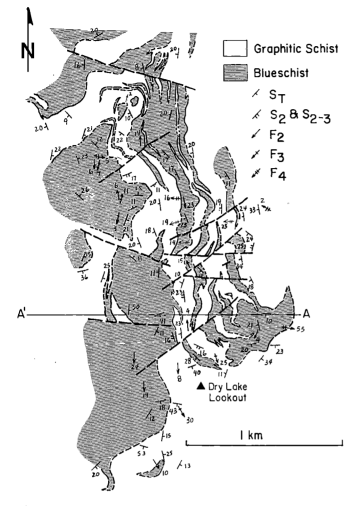

Figure 4. Geologic map of the Dry Lake area showing representative attitudes of linear and planar structures discussed in text. Mapped exposures occupy northeast and north facing headwalls of glacial valleys. The topography in the area is mimicked by the contacts, which are flat-lying or dip gently south.

ROCK UNITS

Two major units were recognized during mapping: graphitic quartz-mica-schist and blueschist. Within the Dry Lake area there are, however, several types of blueschist, and the contact between blueschist and graphitic schist is often marked by a 1–20 meter thick zone containing actinolite-bearing greenschist, chloritic schist, metachert, sulfide-rich spessartine-stilpnomelane schist and/or pelitic blueschist. Contacts between all units are gradational. Although most units in the contact zone are thin and discontinuous (none are persistent enough to serve as mappable marker beds), their consistent relative order of occurrence throughout much of the area defines a unique sequence suggestive of a relict stratigraphy. This sequence is schematically illustrated in Figure 5 and briefly described below. Assemblages in samples from most units are given in Table 1.

Graphitic quartz-mica schist throughout the area shows little variation in appearance and mineralogy. The schist is fine- to medium-grained, gray to steely blue-black, and well foliated. Fine, alternating, discontinuous laminae of quartz plus albite and muscovite plus graphite define the foliation, imparting a weak compositional banding. In some fine-grained, graphite-rich varieties, chlorite pseudomorphs after garnet have been observed.

With proximity to blueschist contacts, the amount of graphite in the schist decreases, accompanied by an increase in chlorite and the appearance of sphene. This assemblage grades into a light green, chlorite-rich schist that is free of amphibole but contains porphyroblastic albite and Fe-Mn garnet. Calcite is also commonly present. This zone is usually less than 1 meter thick and is transitional to interlayered greenschist and blueschist, metachert or metalliferous metasediment.

Metalliferous metasedimentary layers are relatively rare and are generally less than a meter thick. Fine-grained spessartine and extremely fine, disseminated sulfide and calcite make up a groundmass for coarse-grained, randomly oriented blades of stilpnomelane, ferrostilpnomelane and chlorite. Thin spessartine and sulfide-rich laminae give these rocks a faint dark brown and pink compositional banding.

Ferruginous and manganiferous metachert layers are found in both the contact zone and as thin lenses within more massive blueschist. At a few localities within the contact zone, 1-2 meter-thick, nearly pure quartzite layers contain the assemblage quartz + riebeckite + acmite + magnetite. Elsewhere near contacts, and within massive blueschists, the assemblage in thin quartzites is typically free of pyroxene, contains more sodic amphibole, and fine-grained Mn-rich garnet, phengitic mica, chlorite, and stilpnomelane. White mica and sodic amphibole define a foliation parallel to a compositional banding defined by quartz-, garnet-, and mafic mineral-rich layers. Randomly oriented sprays of porphyroblastic, white, lath-shaped cymrite have been found in this assemblage in one rock composed predominantly of crossite, quartz, and fine-grained garnet. A third assemblage found at two localities is quartz + deerite + minnesotaite + magnetite + stilpnomelane. Accessory apatite is common in all metachert lithologies.

Fine-grained greenschist (containing actinolite but no crossite) is restricted in occurrence to layers adjacent to chloritic schist or to thin (<5 meters) metavolcanic layers in graphitic schist. Within thicker metavolcanic packages, actinolitic greenschist is interlayered with crossite blueschist near graphitic schist contacts; away from contacts, green, chlorite-epidote-rich layers within blueschists do not contain actinolite and where actinolite is present, crossite is present as well.

Blueschists are principally of two types: a flaggy, layered, fine-grained, bluish gray-green schist and a more massive, fine- to medium-grained deep blue schist. A third, less common, coarser-grained variety is of pelitic bulk composition and appears to be restricted to thin zones adjacent to metacherts and metalliferous metasedimentary horizons. The flaggy variety is most abundant and is composed of alternating blue, crossite-rich, and green, chlorite- plus epidote-rich laminae. Elongate crossite and white

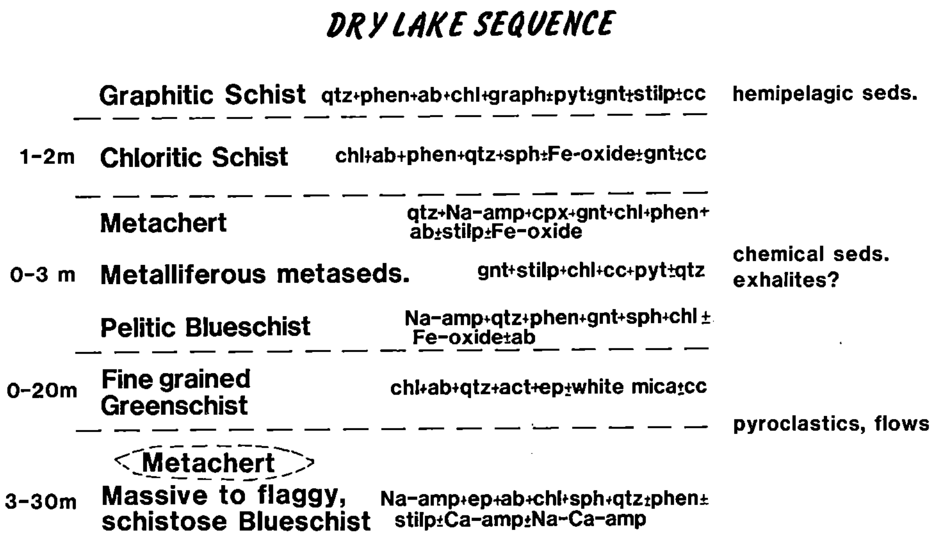

DRY LAKE SEQUENCE

Graphitic Schist qtz+phen+ab+chl+graph±pyt±gnt±stilp±cc   hemipelagic seds.

1-2m   **Chloritic Schist**   chl+ab+phen+qtz+sph±Fe-oxide±gnt±cc

Metachert   qtz+Na-amp+cpx+gnt+chl+phen+ ab±stilp±Fe-oxide

0-3 m   **Metalliferous metaseds.**   gnt+stilp+chl+cc+pyt±qtz   chemical seds. exhalites?

Pelitic Blueschist   Na-amp+qtz+phen+gnt+sph+chl± Fe-oxide±ab

0-20m   **Fine grained Greenschist**   chl+ab+qtz+act+ep±white mica±cc

pyroclastics, flows

<Metachert>

3-30m   **Massive to flaggy, schistose Blueschist**   Na-amp+ep+ab+chl+sph+qtz±phen± stilp±Ca-amp±Na-Ca-amp

Figure 5. Sequence of units observed near blueschist-graphitic schist contacts. Contacts between all units are folded and gradational. The middle three units are referred to as the "contact zone" and are included with blueschist in Figure 4.

mica define a foliation parallel to the compositional layering. Massive blueschists contain the same assemblage as flaggy schists but are composed of up to 80% sodic amphibole and lack a fine compositional layering.

STRUCTURE

At least four generations of superposed folds are present in blueschists within the Dry Lake area. The first three generations are nearly coaxial and define the large-scale structure of the area. Fourth generation folds are of small amplitude and do not affect the distribution of rock types. Deformation during F_1 produced intrafolial folds, a layer-parallel foliation in all units, and a strong linear fabric in blueschists. Mesoscopic folds of this generation are rare. F_2 and F_3 folds deform the layer-parallel foliation and are best developed in the hinges of major folds in the blueschist; in the graphitic schist near-total transposition through F_3 has made minor folds of these generations rare. F_3 folding is followed by boudinage.

In contrast to earlier folds, fourth generation folds are equally well-developed in both units and are generally more symmetric, open and upright. This contrast is also reflected by the orientation of F_4 axes, which trend roughly normal to the earlier structures. Deformational styles from F_1 folding to boudinage are summarized in Figure 6 and discussed below.

TABLE 1. MINERAL ASSEMBLAGES FOR ROCK TYPES FOUND IN THE DRY LAKE AREA

Rock Type* Sample No.	MV 263A	MV 263B	MV DLF3	P 268A	MC 269	MC 268D	MC 284	MC 267	GS 283A	CS 271	MMS 265C
Na-Amphibole	X	X	X	X	X	X	X				
Ca-Amphibole	X										
NA-Ca Amphibole			X								
Sodic pyroxene					X						
Garnet				X	.	X	X			X	X
Epidote	X	X	X			minor	rare				
Chlorite	X	X	X	X		X	X		X	X	X
Calcite											X
Albite	X	X	X	X		X	X	X	X	X	X
Phengite				X		X	minor		X	X	
Stilpnomelane			X			X	X				X
Sphene	X	X	X							X	
Quartz	rare	minor	X	X	X	X	X	X	X	X	
Magnetite					rare	X	X		X		
Hematite				X		X			X	X	
Other**	1					1	1,2	3,4	1,5	1	1

*MV = metavolcanic; P = pelite; MC = metachert; GS = graphitic schist; CS = chloritic schist; MMS = metalliferous metasediment.
**1 - pyrite; 2 - cymrite; 3 - deerite; 4 - minnesotaite; 5 - graphite.

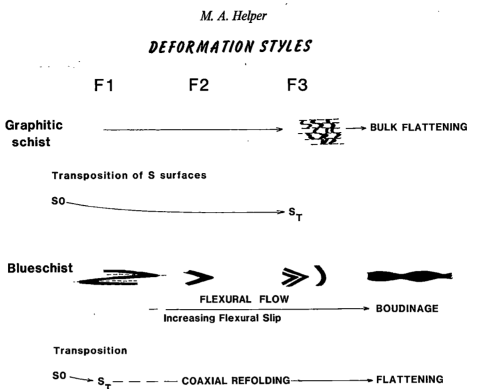

Figure 6. Summary diagram comparing progressive deformational styles and minor structures developed in graphitic schist and blueschist. See text for discussion.

Transposition Foliation (S_T) and F_1 Structures

The most pervasive structural element common to both units is a nearly flat-lying foliation parallel to unit contacts and to the compositional layering in the blueschists. Foliation surfaces are defined by the preferred orientation of mica and sodic amphibole.

In graphitic schist the foliation is not strictly planar but deviates around ubiquitous quartz phacoids and rootless, interfolial isoclinal folds. The latter are defined by thin quartz or mica plus graphite laminae, have amplitudes of less than 10 centimeters and, where sufficiently well exposed, can be seen to be asymmetric. The limbs of the folds are truncated at very low angles by mica- and graphite-rich bands. At several localities the foliation encloses at least two generations of isoclinal folds, in rare cases three, which interfere to produce Type 3 patterns (Ramsay 1967:530–533). In all cases folds have limbs and axial planes subparallel to the foliation. The presence of these folds and the layer-parallel orientation of the foliation indicates that the foliation is the product of repeated isoclinal folding and transposition and as such should not be considered as strictly an S_1 surface. Following Williams and Campagnoni (1983) this surface is designated S_T, the transposition foliation.

In the blueschists, structures synchronous with the formation of the transposition foliation are intrafolial isoclinal folds and a pronounced sodic amphibole lineation. The lineation lies in the plane of the S_T foliation and in most places gives these rocks a strong lineated-schistose (L-S) fabric (Figure 7). F_1 intrafolial folds are defined by thin layers rich in quartz and albite, sodic amphibole, or epidote plus sphene and are similar in style and size to those seen in the graphitic schist. Serial slabs cut perpendicular to intrafolial fold axes show these folds to be highly noncylindrical at the scale of a hand specimen. The sodic amphibole lineation is best developed in coarser units and in thin section can be seen to parallel intrafolial fold axes. Although varying greatly in orientation, on the more planar limbs of large F_2 and F_3 folds the lineation generally trends more NE–SW than later fold axes.

F_2 and F_3 Folds

Two sets of nearly coaxial, N–S trending folds deform the S_T foliation in the blueschists. Both are so similar in style and orientation that, lacking superpositional relationships, they are difficult to distinguish from one another. Where structural overprinting is evident, folds are designated as F_2 or F_3; elsewhere they are grouped as F_{2-3}. Neither has an associated penetrative axial planar surface, although a localized crenulation cleavage is often developed in hinge regions. The resulting superposed crenulations on S_T surfaces often provided the best means of assigning relative ages.

F_2 minor folds vary in style from asymmetric, recumbent isoclinal folds to more open upright structures, depending on structural position and the degree of F_3 overprinting (Figure 8). F_2 minor folds are similar in profile, although not greatly thickened in the hinge regions, and appear to have formed by a mechanism involving both ductile flow and flexural slip. Folds of this

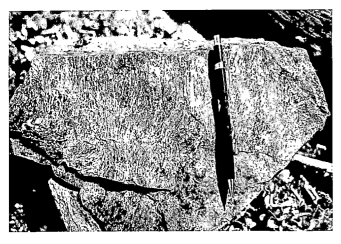

Figure 7. L-S fabric in coarse-grained pelitic blueschist. Sodic amphibole lineation (parallel to pencil) is well- developed in both coarse- and fine-grained blueschist units. Lineation in this photo is gently folded by F_3 folds. Pencil is 13 cm in length.

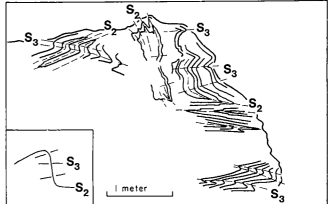

Figure 8. Field sketch of mesoscopic F_2 and F_3 interference pattern in flaggy blueschist. Surface being folded (S_T) is defined by interlayered chlorite-rich and sodic amphibole-rich laminae that are paralleled by a mica and sodic amphibole schistosity. Folded F_2 axial traces (S_2) define a Z-shaped F_3 fold (inset). View is looking down plunge (south) of F_2 and F_3 axes.

generation are not strictly cylindrical but do not depart greatly from this form. Culminations and depressions along fold axes are of small amplitude and relatively long wavelengths. Most large F_2 folds are S-shaped when viewed down plunge.

F_2 fold axes are somewhat dispersed but trend mostly NNW-SSE and plunge at shallow angles to the south (Figure 9, F_2). The 20–30° scatter in orientations along poorly defined small circles is primarily the result of later deformation by nearly coaxial F_3 folds and later boudinage, but is also due, in part, to the noncylindrical nature of the original folds. The north or south variation in plunge direction is the result of F_4 folding. The orientation of the S_2 and S_3 axial surfaces in the blueschists indicates that most F_2 and F_3 folds are recumbent and coplanar with the S_T surface (Figure 9, S_{2-3}, S_T Blue). Poles to S_T surfaces (Figure 9, S_T Blue) define a very broad point maximum, indicating that S_T is roughly planar and dips shallowly west. The point maximum does, however, show the effects of F_2, F_3, and F_4 folding by the distribution of some poles away from the maximum density along crudely defined, steeply dipping ENE–WSW (F_2, F_3) and N–S striking (F_4) girdles.

Like F_2 folds, F_3 folds are similar in style, but show slightly less thickening in hinge regions and a greater component of flexural slip. Folds range from tight to relatively open and recumbent with angular hinges, to nearly concentric, with rounded hinges. This variation in style can often be seen within a single F_3 fold profile (Figure 8). Large F_3 folds are predominantly Z-shaped when viewed down plunge but minor (parasitic) folds of this generation have asymmetries that vary in a manner consistent with their positions on the large F_3 folds in the area.

The orientation of F_3 structures closely resembles that shown by F_2. F_2 and F_3 axial planes are coplanar except on the short limbs of F_3 folds. F_3 fold axes show a distribution similar to F_2 but have trends that are slightly more N–S (Figure 9, F_3). The scatter of F_3 axes in part reflects the formation of these folds on

previously folded surfaces, since it can be seen in the field that F_3 axis orientations are affected by the position of the folds relative to F_2 limbs or hinges. F_3 axes do not, however, lie in a well-defined plane, but more closely approximate small circles, possibly suggesting tightening of F_2 folds during F_3 folding.

Minor F_2 and F_3 folds are uncommon in the graphitic schist, but where present are tighter and less cylindrical than F_2 and F_3 folds in blueschist. Axial surfaces to these folds are subparallel to the transposition foliation. Poles to the transposition foliation in the graphitic schist define a girdle with an axis that falls in the range of blueschist F_2 and F_3 fold axes (Figure 9, S_T Black), indicating that both units share a common F_2-F_3 fold history.

Boudinage and F_4 Structures

The earliest structures that post-date F_3 folds in blueschists are elongate to equant, lozenge-shaped boudins that have been extended in the S_T plane subparallel to F_2 and F_3 axes. Boudins are most common in flaggy blueschist and range in size from large, 10-meter-long boudins with highly fractured neck regions to meter long boudins with highly attenuated ends. F_2 and F_3 crenulations wrap into boudin neck regions and are clearly cross-cut by quartz-filled extension fractures associated with stretching. In blueschist layers in areas where individual boudins are not well developed boudinage is commonly reflected by a gentle, periodic, dome and basin-style warping of the S_T surface. On such surfaces and where sufficiently well-exposed in three dimensions, boudins can be seen to be tabular or lozenge-shaped, indicating formation by shortening at a high angle to the S_T surface. In graphitic schist the only layers with sufficient ductility contrast to record boudinage are quartz veins and quartz-rich layers, which may owe their present phacoidal shape to this period of flattening.

A final deformational event is recorded in both units by F_4 kink bands and chevron and box folds that have steeply inclined

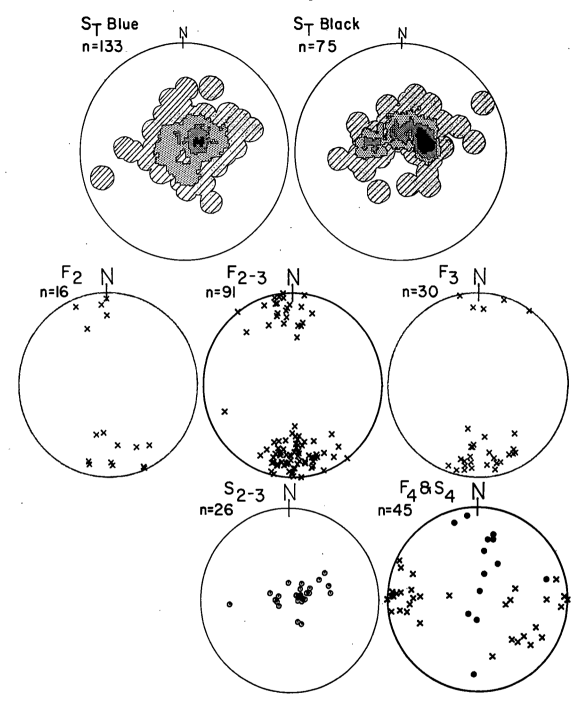

Figure 9. Lower hemisphere, equal-area (contoured) and equal-angular (others) diagrams of structural elements from Dry Lake area discussed in text. S_T poles are contoured at concentrations of 0, 5, 10, and >15%.

axial planes and predominately E–W trending axes (Figure 9, F_4 and S_4). Folds of this generation are generally small in amplitude and pass downward into slip horizons parallel to the transposition foliation.

Large-scale Structure

The senses of asymmetry of minor F_2 and F_3 folds and the type of F_2-F_3 interference patterns that developed in different parts of the map area yield a consistent pattern that can be used to interpret the megascopic structure. The megascopic structure is illustrated in Figure 10 by a series of down-plunge profiles along a single line of section. Unit contacts are everywhere parallel to the S_T foliation so that in all profiles contacts also reflect the orientation of the S_T surface. The upper diagram shows axial

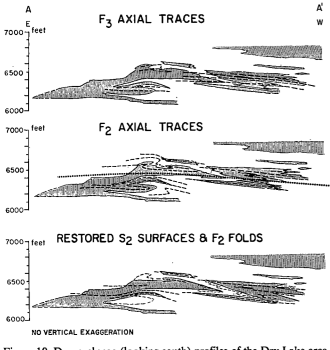

A   A'
F_3 AXIAL TRACES
E   W
7000⌐ feet

6500

6000⌐

7000⌐ feet F_2 AXIAL TRACES

6500

6000⌐

7000⌐ feet   RESTORED S_2 SURFACES & F_2 FOLDS

6500

6000⌐

NO VERTICAL EXAGGERATION

Figure 10. Down-plunge (looking south) profiles of the Dry Lake area along line of section A–A' in Figure 4. Profiles were constructed by first removing the throw on the faults and then projecting all data down plunge into the line of section. F_4 fold axes lie at low angles to the plane of section so that the profiles primarily show the influences of F_2 and F_3 folding. The upper two profiles show the large scale F_3 structure, the lower diagram F_2 folding. See text for discussion.

traces to known F_3 folds. Axial traces to F_2 folds (middle profile) define the major F_3 structure. Asymmetric F_3 folds of the S_2 surface have both clockwise and anticlockwise senses of rotation in the center of the profile but are symmetric near the western end of the profile. The asymmetric folds define a vergence boundary (Figure 10, dotted line; Bell 1981) to a large, symmetric, recumbent F_3 fold that closes to the west.

To examine the asymmetries of F_2 folds it is first necessary to consider the effects of the overprinting F_3 folds. F_2 and F_3 folds are nearly coaxial, so that F_2 folds axes are not appreciably reoriented in the plane of the cross section when the effects of F_3 folding are removed. The lower profile shows the orientation of S_2 surfaces prior to F_3 folding "restored" by removing F_3 folds. F_2 folds are consistently S-shaped, indicating an anticlockwise sense of rotation. If the major eastward closing fold in the lower left corner of the profile is an F_2 structure, then the minor, S-shaped F_2 folds could be interpreted as parasitic on an overturned limb of the large F_2 fold. Alternatively, if the major eastward closure is a large F_3 fold, then no large F_2 fold structures are present in the area. Under either interpretation, the map pattern and minor structures clearly show the presence of large, kilometer-scale, recumbent isoclinal folds, indicating that the style of structures observed in outcrop are also present on a regional scale.

STRUCTURAL PETROLOGY

Introduction

Nearly all of the mineral assemblages present in the Dry Lake area preserve evidence of sequential mineral growth. Compositional zoning in amphiboles and epidote, preservation of reaction textures, and the presence of two-amphibole assemblages in which amphiboles of different composition are aligned with structural elements of different ages (discussed below) all point to a progression in metamorphic conditions rather than a single metamorphic peak or culmination. The most widespread and best preserved assemblages indicate synkinematic epidote-crossite grade metamorphism (as defined by Brown 1974), but nearly all assemblages record evidence of a later static overprint. No changes in grade are detectable across the area, indicating fairly widespread attainment of uniform metamorphic conditions. The relative time of mineral growth with respect to deformation is shown in Figure 11 and discussed below.

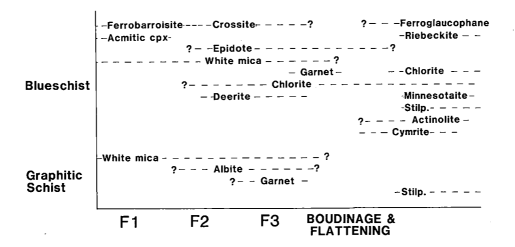

Figure 11. Ages of mineral growth relative to stages of folding and boudinage.

M. A. Helper

TABLE 2. MINERAL COMPOSITIONS[a]

| | DLF3 | | 263A | 269 | | 268D | | 265C |
	Ferrobarr.	Cross.	Act.	Rieb.	Acmite	Garnet	Phengite	Stilp.
SiO_2	49.2	52.4	51.5	51.8	53.6	39.0	49.1	42.8
TiO_2	0.14	0.11	0.00	0.00	0.00	0.16	1.19	0.33
Al_2O_3	8.86	7.16	3.80	2.84	0.04	20.6	26.5	9.78
FeO*	21.9	21.4	18.5	36.0	31.8	16.2	7.01	27.0
MnO	0.27	0.27	0.25	0.41	0.00	18.4	0.00	5.68
MgO	7.03	7.11	11.2	0.00	0.00	0.00	0.73	1.95
CaO	5.55	3.55	10.2	0.52	0.20	8.23	0.03	0.24
Na_2O	4.20	6.05	1.77	6.09	13.9	0.00	0.21	0.46
K_2O	0.55	0.17	0.32	0.00	n.a.	n.a.	8.34	4.35
	97.70	98.22	97.54	97.66	99.54	102.59	93.11	92.59

Cations per formula unit[b]

Si	7.20	7.60	7.57	7.93	2.00	3.07	3.34	8.00
Al<4>	0.80	0.40	0.43	0.07	0.00	0.00	0.66	0.00
Al<6>	0.73	0.83	0.23	0.44	0.00(2)	1.92	1.46	2.14
Fe···	1.00	0.71	0.41	1.44	0.99	0.00	0.00	0.00
Fe··	1.69	1.88	1.86	3.17	0.00	1.07	0.04	4.22
Ti	0.02	0.01	0.00	0.00	0.00	0.01	0.06	0.05
Mn	0.03	0.03	0.03	0.05	0.00	1.23	0.00	0.90
Mg	1.53	1.54	2.46	0.00	0.00	0.00	0.07	0.54
Ca	0.87	0.55	1.61	0.09	0.01	0.70	0.00(2)	0.05
Na	1.19	1.70	0.51	1.81	1.00	0.00	0.03	0.17
K	0.10	0.03	0.06	0.00	--	--	0.72	1.04

Am .36
Sp .41
Gr .23
Py 0

[a]All mineral compositions determined by electron microprobe using combined energy and wavelength dispersive analyses.
[b]Mineral formula and the fractions of ferrous and ferric iron present are calculated as follows: pyroxene, sum of cations = 4, Fe··· = Na - Al; garnet, sum of cations = 8, all Fe as Fe··; muscovite, sum of cations less (K+Na+Ca) = 6, all Fe as Fe··; Stilpnomelane, Si = 8, all Fe as Fe··; Amphibole, see text.
*Total Fe as FeO.
n.a. = not analysed.

Amphiboles and Pyroxene

Sodic, sodic-calcic, and calcic amphiboles are present in the blueschists at Dry Lake. Calcic and sodic-calcic amphiboles are restricted to greenschists and flaggy and massive blueschists; sodic amphiboles are found in all blueschists and most metacherts. Amphibole mineral formulae and the amount of ferric and ferrous iron represented by total FeO in microprobe analyses were calculated by either normalizing total cations less K + Na + Ca to 13 (about ¾ of all analyses) or total cations less K to 15, and assuming a net cationic charge of 46. The normalization procedures, site filling scheme, and crystal-chemical limits used in evaluating normalizations follow those reviewed in Robinson and others (1982).

Sodic amphiboles from 3 metacherts, 4 metabasites, and 2 pelitic blueschists are compared in Figure 12; three of these analyses are given in Table 2. Except for riebeckite in one metachert sample, all are crossite or ferroglaucophane. Sodic amphiboles in all three rock types show fairly narrow ranges of $Fe^{2+}/(Fe^{2+} + Mg)$ and correspondingly wide ranges of $Fe^{3+}/(Fe^{3+} + Al^{VI})$ (Figure 12). The latter is due in part to difficulties inherent in estimating ferric iron contents but also reflects true variations due to zoning.

Zoning patterns are complex but show consistent variations within any given lithology. In most metacherts zoning is expressed optically by deep blue crossite cores and lighter blue ferroglaucophane rims and patches along amphibole cleavage traces. Zoning in all cases is optically continuous, although grain centers often contain greater numbers of inclusions, most of which are quartz. Similar though less pronounced zoning is seen in amphiboles in pelitic blueschists. In both lithologies the decrease in $Fe^{3+}/(Fe^{3+} + Al^{VI})$ from grain centers to edges also marks a net decrease in the total number of trivalent cations. This decrease is mimicked by a decrease in sodium and an increase in calcium in the M4 site, corresponding to the glaucophane substitution: Na^{M4}, $(Fe^{3+}, Al^{VI}) = Ca^{M4}$, (Fe^{2+}, Mg, Mn). A small increase in Al^{IV} accompanies the decrease in $Fe^{3+} + Al^{VI}$ from grain centers to edges, as does an increase in A site occupancy (the edenite substitution: $(Na^A, K), Al^{IV} = \square^A, Si)$.

Sodic amphiboles in metavolcanic lithologies show the same compositional range as that seen in pelites and metacherts (Figure 12; zoning is, however, more irregular and patchy, often to the extent that a clear temporal relationship between compositional domains is difficult to establish. In addition, some samples contain a second calcic or sodic-calcic amphibole phase. In samples where two amphiboles are present, continuous zoning is not opti-

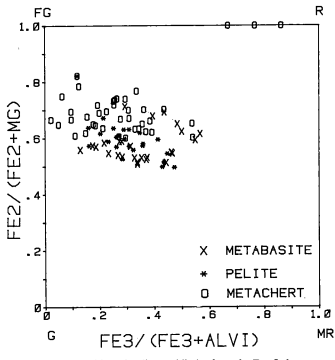

Figure 12. Composition of sodic amphiboles from the Dry Lake area.

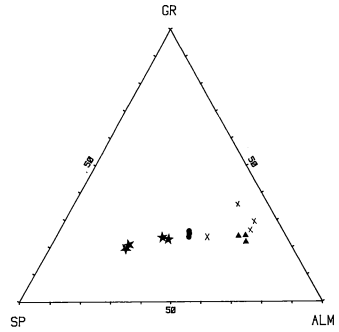

Figure 13. Garnet compositions. Stars-pelite; dots-metachert; X's-chloritic schist; triangles-metalliferous metasediment. All garnets lack a detectable pyrope component.

cally evident. Instead, in some cases earlier amphiboles may be mantled or partially replaced by a later amphibole phase. One sample studied in detail (DLF-3; Table 2) contains pleochroic deep bluish green to blue ferrobarroisite and medium blue to lavender crossite. In sections cut perpendicular to an F_{2-3} fold axis, coexisting medium-grained ferrobarroisite and medium- to fine-grained crossites show an extremely strong preferred orientation of prism faces on fold limbs. Across fold hinges, however, ferrobarroisites are strained, and within hinges one finds basal sections of small, new grains of crossite growing in areas of highly strained ferrobarroisite. These relationships indicate crystallization of sodic-calcic and sodic amphibole prior to F_{2-3} folding (both define a strong lineation on the folded S_T surface) and the continued crystallization of sodic amphibole during F_{2-3} folding.

Coexisting actinolite and crossite-ferroglaucophane are present in some fine-grained flaggy blueschists. The two amphiboles do not constitute true stable coexistence, however, in that sodic amphibole aligned in S_T is being replaced by chlorite, and the actinolite present shows little preferred orientation. Fine-grained actinolite is also present in greenschist layers near the graphitic schist-blueschist contacts. No sodic amphibole is present and the actinolites are apparently unzoned, although the fine grain size makes such determinations difficult. Actinolite in greenschists is only very weakly aligned in the S_T plane and does not define a linear fabric like that displayed by sodic amphiboles in blueschists. Actinolite in both lithologies contains significant amounts of Na in the M4 site (0.3-0.4 cations per formula unit). Actinolite is also found in undeformed veins that crosscut rocks in the contact zone.

Sodic pyroxenes are only found in a few aluminum- and magnesium-poor ferruginous metacherts. Those analysed are nearly pure acmite (Table 2) in the divariant assemblage quartz + acmite + riebeckite + magnetite ($+H_2O$). In rocks containing this assemblage, acmite defines the S_T plane and is reacting with magnetite and quartz to form randomly oriented sprays and needles of riebeckite.

Garnet

Garnets are present in metacherts, metalliferous metasediments, pelitic blueschists, and chloritic schists in the contact zone between graphitic schists and blueschists. In metacherts and metalliferous metasediments they occur in pink, coticule bands of very small, inclusion-free granular idioblasts. In chloritic and pelitic schists they are also porphyroblastic. In all units garnets are grossular-spessartine-almandine solid solutions that contain a fairly constant mole fraction of grossular component of between 20 and 30 percent (Figure 13). Zoning in idioblasts if difficult to detect because of the fine grain size but small grains within a single thin section may show differences in mole fractions of spessartine and almandine components of up to 15 percent. Porphyroblasts within chloritic schist are weakly zoned, with interiors slightly richer in spessartine relative to almandine.

Rare chlorite pseudomorphs after garnet within fine-grained graphitic schist west of the Dry Lake area contain white mica inclusions that retain the orientation of primary foliation outside the pseudomorph, indicating static growth of garnet after the formation of the transposition foliation and subsequent replace-

Figure 14. Deerite (black blades) - minnesotaite (colorless blades) - magnetite (small, equant, black grains) - stilpnomelane (not shown) - metachert. Deerite lies in axial plane of minor F_{2-3} fold. Short dimension of photo is 1.5 mm. Sample is from a 0.3 m thick metachert layer in the contact zone, approximately 1.5 km NNW of Dry Lake Lookout.

ment of garnet by chlorite. Similar garnet pseudomorphs occur in some chloritic schists that do not contain fresh garnet. Fine-grained ilmenite and sphene(?) inclusions in these pseudomorphs define F_{2-3} microfolds. White mica and some of the chlorite in the surrounding matrix is deflected around pseudomorphs and pressure shadows of chlorite and quartz are present in areas adjacent to pseudomorph boundaries at high angles to the foliation. Chlorite within pseudomorphs and most chlorite within the surrounding matrix is randomly oriented. These relationships indicate garnet growth after F_{2-3} folding and during or before flattening, growth of chlorite during flattening, and replacement of garnet by chlorite after flattening.

White Mica and Stilpnomelane

Phengitic white mica is a major constituent of pelitic blueschist, graphitic schist, and chloritic schist but is absent or present only in minor amounts in blueschists and metalliferous metasediments. Micas show a fairly limited range of celadonite substitution in all lithologies, having between 3.2 and 3.5 Si per formula unit (p.f.u.). All are alkali deficient (0.7–0.9 K p.f.u.) and contain only trace amounts of Na and Ca (Table 2). Fe/Mg ratios for all lithologies are near unity, with the exception of metacherts, where micas are highly enriched in iron relative to magnesium.

All rocks containing white mica possess a strong mica fabric developed during the formation of the transposition foliation. Later F_2-F_4 folding did not produce a new mica fabric but instead resulted in the formation of localized crenulation cleavages. Within the hinges of microfolds, micas show a variety of textures. Micas in tight F_2 and F_3 fold hinges are characterized by highly serrate boundaries, indicating grain boundary migration (Etheridge and Hobbs 1974), and by sparse, small, new grains aligned

axial planar to F_2 and F_3 folds. In contrast, mica in the hinges of F_4 folds are kinked and show undulatory extinction.

Stilpnomelane is present as a minor phase in nearly all Fe-rich lithologies (especially metacherts, massive blueschists, and metalifferous metasediments) as medium- to coarse-grained, randomly oriented, porphyroblastic blades and sprays that crosscut the S_T foliation, and in undeformed quartz veins within these units. The total lack of any preferred orientation is taken as evidence of static growth, indicating that stilpnomelane is not in equilibrium with the syntectonic assemblages with which it is found (c.f. Brown 1971).

Deerite and Cymrite

Deerite, present in two iron-rich metachert samples (perhaps more aptly termed ironstones because of the near absence of all oxides except FeO, Fe_2O_3, and SiO_2), occurs as well-formed needles and laths that are oriented with long dimensions subparallel to axial planes of F_{2-3} crenulations (Figure 14). Fine, platey bundles of randomly oriented minnesotaite radiate from most deerite crystal faces.

The barium silicate cymrite is present as randomly oriented, porphyroblastic, bowtie-shaped clusters up to 6 mm in length in one crossite- and garnet-rich metachert layer.

CONDITIONS DURING FOLDING

Without a precise knowledge of the continuous reactions responsible for the observed changes in mineral chemistry, it is not possible to directly evaluate the magnitude or direction of change in metamorphic conditions during the F_1-F_3 folding. In any event, the estimation of pressures and temperatures during deformation in these rocks is hampered by a lack of assemblages or subassemblages for which calibrated mineral equilibria are available. A rough measure of the metamorphic conditions during deformation can, however, be gained by comparison with blueschist belts in which conditions are better known, and by using the meager data available on the stability of a few of the phases present. A petrogenetic grid of equilibria pertinent to the minerals and assemblages present in the Dry Lake area is given in Figure 15. The suggested field for synkinematic mineral growth is based on the following considerations:

1) Epidote-crossite assemblages that grew prior to F_2 and F_3 folding are similar in nearly all aspects to those described by Brown and others (1982), Haugerud and others (1981), and Brown (1974) in the regional blueschists of the Shuksan Suite, for which Brown and O'Neil (1982) obtained oxygen isotope temperatures of 350–410°C. Corresponding pressures are difficult to constrain. Lawsonite, glaucophane, and aragonite, which, along with the jadeite content in pyroxenes, suggest pressures of 7–9 kb for Shuksan crossite-epidote assemblages (Brown and O'Neil 1982; Brown 1983), are not present in the Condrey Mountain Schist. The absence of lawsonite, which is stable relative to paragonite + zoisite above 7 kb at 350°C and above about

9 kb at 410°C (Holland 1979; Heinrich and Althaus 1980), may suggest somewhat higher temperatures and/or lower pressures than those estimated for Shuksan schists, but its breakdown products have not been found either, indicating perhaps an absence of rock of the appropriate bulk composition. Similarly, the presence of calcite instead of aragonite at these temperatures might argue for pressures less than 9 kb (Johannes and Puhan 1971), but might also be explained by inversion during uplift. Minimum pressures at the upper end of this temperature range are roughly constrained by the absence within metavolcanic units of coexisting actinolite, albite, and Fe-oxide, as seen within Otago greenschists (Brown 1974), suggesting pressures above about 5 to 6 kb (Scott 1974; Henley 1975). The presence of early barroisitic amphibole may also suggest a minimum pressure in the 5 kb range at these temperatures (Ernst 1979).

2) Conditions during F_2-F_3 folding are within the range experienced during transposition, as indicated by the growth of small crossite grains parallel to F_{2-3} fold axes and by the alignment of deerite in F_{2-3} axial planes. Wood (1979) estimates that below 350°C, at pressures less than 5 kb, deerite breaks down to minnesotaite and magnetite. Above 5 kb the high temperature limit for deerite stability is poorly constrained.

A static phase of mineral growth is recorded by randomly oriented actinolite, chlorite, cymrite, and stilpnomelane, the replacement of deerite by minnesotaite, and the late growth of Mn-rich garnet. Ferroglaucophane that rims and replaces crossite may have also grown at this time. This high P/T greenschist facies "overprint" appears to have been accompanied by flattening in the earliest stages, as evidenced by flattening of the S_T fabric around garnet pseudomorphs and late-growing albite poikiloblasts. Conditions during flattening cannot be directly constrained but must occupy a position in P-T space between synkinematic conditions during F_2 and F_3 folding and static mineral growth.

No evidence exists that bears directly on the relative ages of different static minerals, thus it is not possible to determine whether all were formed under the same conditions or instead represent conditions at different points in time. Some indication of conditions during static mineral growth is given by the reaction of deerite to minnesotaite, which occurs at 4–5 kb at temperatures of 200–360°C in meta-ironstones (Wood 1982, Figure 16). A maximum temperature of 350–360°C might also be indicated for metatuffs, based on the stability of end-member ferroglaucophane (Hoffman 1972), although the effects of Ca, Mg, and Fe^{3+} substitutions on ferroglaucophane stability are not known. Pressures estimated from cymrite stability (Nitsch 1980) are slightly higher than those indicated by the breakdown of deerite; at 350°C a minimum of 5 kb is required. At 300°C cymrite is stable above about 4.3 kb (Nitsch 1980: Fig. 1). Within the precision of the above estimates it appears that all statically recrystallized minerals grew under similar conditions. Static mineral growth is roughly constrained to below 360°C at pressures less than about 5 kb.

It is noteworthy that static textures are best developed in

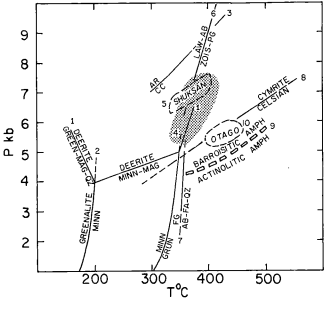

Figure 15. Petrogenetic grid and inferred stability field (shaded) for synkinematic mineral parageneses in the Dry Lake area. Sources of mineral equilibria and other fields: (1), (2), (4) Wood (1982); (3) Johannes and Puhan (1971); (5), (10) Brown and O'Neil (1982); (6) Holland (1979), Heinrich and Althaus (1980); (7) Hoffmann (1972); (8) Nitsch (1980); (9) Ernst (1979).

rock at or near graphitic schist contacts in the contact zone. Re-equilibration down grade is, of course, fundamentally governed by the availability of fluid, of which the graphitic schist was apparently an abundant source. Within this zone, fluid compositions were apparently buffered by graphite, resulting in low partial pressures of oxygen and, in places, relatively high partial pressures of CO_2. Two reactions that stabilize carbonate-bearing or carbonate-free chloritic schist relative to carbonate-free sodic amphibole-bearing rocks at high P_{CO2} and low P_{O2} respectively are:

epidote + crossite + CO_2 = albite + chlorite + $CaCO_3$ + Fe-oxide

(Vance 1957)

Na-amphib. + muscovite + chlorite (Mg-rich) =
    stilpnomelane + albite + chlorite (Fe-rich) + O_2 (Brown 1974)

both of which are compatible with the assemblages present in chloritic schists at graphitic schist contacts. The presence of actinolite in greenschist near these contacts is more difficult to assess. The lack of an L-S actinolite fabric like that displayed by sodic amphiboles in blueschists suggests the two amphiboles are not temporally equivalent, yet there is no indication of any earlier sodic or calcic amphibole growth in greenschists that would correspond to sodic amphibole growth in blueschists. In view of this apparent total reequilibration, actinolite stability could be seen as a response to reducing conditions that prevailed throughout the metamorphic history (e.g. Vance 1957) or as the result of final reequilibration under P/T conditions outside the sodic amphibole stability field (e.g. Brown 1974). Bulk compositional differences might also be a factor. Reaction textures are absent so that it is not possible to directly evaluate the relative effects of these influences. High sodium contents in the M4 site of all actinolites do, however, suggest P/T conditions greater than those normally associated with intermediate- or low-pressure facies-series greenschists (Brown 1977; Laird and Albee 1981), an inference consistent with the pressures and temperatures estimated from other static phases.

DISCUSSION

Significance of Folding

The similarity in styles, orientations, and metamorphic conditions during F_1-F_3 folding suggests that deformation during this period was progressive. The style of intrafolial folding, the constant asymmetry of F_2 folds and the nearly coaxial and coplanar orientations of F_2 and F_3 folds further suggest that folding was largely the product of noncoaxial deformation. No mylonite or zones of localized intense deformation are present, implying that shear strains were fairly uniform throughout. Given these conditions, it is not surprising that the number of overprinting intrafolial fold generations in the graphitic schist varies widely from locality to locality and does not match, in absolute numbers, the number of generations of equivalent folds in the blueschists. Under such conditions, the number of fold generations is a function of lithology and the local shear strain environment, and need not be correlative from area to area. The similarity in style and orientation of the S_T surface in both units and the lack of any features that document decoupling or high shear along the graphitic schist-blueschist contacts suggests that during the generation of the S_T foliation both units were behaving similarly. Such a situation could only occur if ductility contrasts between graphitic schist and blueschists were low. High shear strains are indicated by the pronounced L-S fabric in the blueschists.

The onset of F_2 and F_3 folding in the blueschists marks a change in the conditions under which deformation was occurring. The constant asymmetry of F_2 folds and parallelism of S_2 and S_T surfaces indicate that deformation during F_2 was still largely by shearing, but the steady-state transpositional foliation in the blueschists was no longer being regenerated. This change may be viewed as a response to a decrease in magnitude of shear strains, a decrease in fluid content or temperatures, and/or an increase in superimposed irrotational flattening strains. In blueschists the change from F_1 to F_3 fold styles documents a decrease in the component of intralayer flow during folding and an increase in the amount of flexural slip. Although poorly constrained, the paragenetic sequence does not indicate significant differences in temperature or fluid content from F_1 to F_3, suggesting that the change in deformation style may have been due to a decrease in shear strains. A decrease in the magnitude of shear strains might also be indicated by the incomplete rotation of F_2 and F_3 axes toward parallelism with the sodic amphibole lineation, and per-

haps by the development of parasitic F_3 folds with opposite senses of asymmetry on the limbs of a large, symmetric F_3 structure. An increase in superimposed irrotational flattening strains may have accompanied decreasing shear, as indicated by later boudinage, but this remains largely undocumented for the time period of F_2 and F_3 folding.

Shearing had ceased by the time of boudinage. Boudins are oblate and lozenge-shaped and show no signs of rotation during formation. The shape and orientation of the boudins indicate formation by shortening at a high angle to the foliation. Their slightly oblong form indicates that the direction of maximum extension in the S_T plane was roughly N-S, subparallel to F_3 axes, but an overall lozenge shape implies that the magnitude of extension in this direction was not much greater than that experienced in other directions in the S_T plane. The same irrotational flattening is recorded microstructurally by the deflection of S_T around albite and garnet pseudomorph porphyroblasts in chloritic schists.

F_4 folding appears to have occurred under near-surface conditions. The purely flexural style of folding, an E-W trend or strike of most structures and the absence of an associated metamorphic fabric all contrast sharply with conditions during earlier deformation and indicate that F_4 folding is not part of the earlier ductile deformational continuum. F_4 folds and faults are the only structures common to both the schist and the overriding WPTB thrust plate and are thought to be related to deformation associated with the final stages of domal uplift (Helper 1985).

Tectonic Implications

The presence of high P/T assemblages in rocks that record high shear strains has been widely interpreted as evidence for subduction (e.g. Coleman 1972; Ernst 1975; Dewey and Bird 1970; Miyashiro 1972). The deformational and metamorphic histories described above can be explained in this context. High shear strains would occur during descent whereas irrotational flattening following a decrease in shear strains could reflect uplift. Admittedly, there is very little evidence within the paragenetic sequence to suggest that decreasing shear or flattening was accompanied by decompression, yet within the precision of the P-T estimates this possibility is certainly not precluded. Although megascopic F_2 and F_3 fold asymmetries within the Dry Lake area can not be used to deduce a sense of shear, the gross geometry of the large-scale structure (Figures 9, 10) suggests that transport during F_2 and F_3 folding was roughly E-W.

The wide range of metamorphic ages within the window and the unknown age of the Condrey Mountain Schist protolith limits attempts to place the history of this unit in a regional context. Similarities in lithology and structural position have suggested to some workers (e.g. Klein 1977) that the schist represents an inlier of the 157-153 Ma, back-arc basinal flysch (Harper 1980; Saleeby et al. 1982) of the Western Jurassic Belt (Figure 1). This interpretation is in conflict with the two oldest metamorphic ages from the window and the minimum age of the protolith indicated by the 171 Ma U/Pb age (Saleeby et al. 1984), but is compatible with recent gravity modeling (Barnes et al. 1982) that shows that the intervening amphibolites of the WPTB plate west of the window are underlain at shallow depths by low density rocks. Thrusting of the overlying WPTB plate along the eastern edge of the Western Jurassic Belt is Nevadan in age (150-145 Ma, Harper and Wright 1984) and was westward directed (Gray 1985). This thrusting has an age that is similar to the inferred age of thrusting along the window margins (153-156 Ma; Helper 1985), and has a direction compatible with the E–W transport seen in the Dry Lake area. If the correlation between the two units is valid and the greenschist-blueschist facies metamorphism and F_1-F_3 structures are interpreted as Nevadan features (this is acceptable within the precision of the oldest metamorphic ages) then the parageneses within the window suggest the possibility that Nevadan thrusting and imbrication in the Klamaths was related to subduction. The lack of an accretionary wedge mélange or coeval magmatism of this age might suggest subduction was very short-lived, an idea lent support by the age of Franciscan blueschists (see summary by Cloos, this volume), which indicate that subduction was outboard of the Klamaths before 140 Ma (McDowell et al. 1984; Suppe and Foland 1978). Short-lived subduction is also compatible with the transitional greenschist-blueschist facies observed, inasmuch as the thermal regime established during subduction is a strong function of the amount of convergence (Hsui and Toksöz 1979; Wang and Shi 1984; Cloos 1985).

If the age of metamorphism is more closely reflected by the older isotopic ages (167, 159 Ma), then ductile deformation within the window is pre-Nevadan and cannot be directly related to the thrusting of the overlying WPTB plate. The overlying WPTB plate records an early isoclinal folding and flattening event that is synchronous with a pre-163 Ma, post-190 Ma metamorphism that reached upper amphibolite facies near the window margins (Hotz 1967; Mortimer 1983). Thrusting of the WPTB over the Condrey Mountain Schist clearly postdates this event. The marked contrast in metamorphic facies across the thrust makes correlation of events in the two plates unlikely, and such a correlation would require that thrusting is not expressed by any of the ductile deformation within the central window. It thus appears that if the F_1-F_3 folds are pre-Nevadan, they have no analogs within any of surrounding rocks. Such an interpretation requires that the schists are allochthonous with respect to the surrounding thrust plates and record a Middle Jurassic subduction event not recognized elsewhere in the central Klamaths. Recent work along the western margin of the window (Helper 1985) supports this latter interpretation, but more age data from within the window are needed before either interpretation can be adequately evaluated.

Uplift of the blueschists following boudinage is reflected by a period of relatively high P/T static mineral growth at pressures below those experienced during the earlier deformation, indicat-

ing nonpenetrative deformation or hydrostatic conditions during ascent. The only structures present that might be of this age are undeformed actinolite, quartz, calcite, and quartz + albite veins, some of which contain stilpnomelane. The presence of these minerals in veins is consistent with vein formation during static mineral growth, indicating some deviatoric stresses were active, and that deformation was of a more brittle nature than that experienced during earlier flattening. The final stage of uplift is marked by the formation of a regional dome. Mortimer and Coleman (1984) present arguments for a minimum of 7 km of uplift during dome formation, which they suggest is Neogene in age, although it is not yet clear that all of the uplift has to be this young (Helper 1985).

CONCLUSIONS

The progressive deformation and metamorphism of blueschists and graphitic schist in the central part of the Condrey Mountain Window are consistent with conditions during subduction and uplift. Progressive ductile deformation in both units is synchronous with metamorphism at transitional blueschist-greenschist facies conditions. By analogy with similar assemblages in the Shuksan Suite, conditions during folding are inferred to be in the range of 360–410°C at pressures greater than 6 kb. Deformational styles reflect noncoaxial deformation in which shear strains decrease through time and are succeeded by irrotational flattening. Synkinematic metamorphism is followed by a period of relatively high P/T mineral growth under static conditions of nonpenetrative strain. The breakdown of deerite and growth of cymrite during this event suggest pressures of about 4–5 kb at temperatures less than about 360°C. Regional relationships and isotopic ages favor an interpretation that relates metamorphism and ductile deformation to a Middle Jurassic subduction event, but do not preclude a Late Jurassic (Nevadan) age for deformation and metamorphism.

ACKNOWLEDGMENTS

This paper is an outgrowth of my dissertation research at the University of Texas. I thank Drs. J. C. Maxwell, S. Mosher, M. Cloos, W. D. Carlson, R. G. Coleman, F. W. McDowell and Gary Gray for assistance and encouragement during my research, and for helpful discussions, comments and criticisms of earlier versions of this paper. The manuscript was improved through reviews by Brian Patrick, Mary Donato and an anonymous reviewer, and by discussions with E. H. Brown and B. W. Evans. Juan de la Fuente of the Klamath National Forest provided base maps and logistical support during my stays in the Klamaths, and Lee Garrett and Dave Helper served as field assistants. Field and laboratory studies were supported by grants from the University of Texas Geology Foundation, California Division of Mines and Geology and the Geological Society of America.

REFERENCES CITED

Barnes, C. G., Jachens, R. C., and Donato, M. M., 1982, Evidence for basal detachment of the western Paleozoic and Triassic belt, northern Klamath Mountains, California [abstr.]: Geological Society of America Abstracts with Programs, v. 14(4), p. 147.

Barrows, A. G., Jr., 1969, Geology of the Hamburg-McGuffy Creek area, Siskiyou County, California, and the petrology of the Tom Martin Ultramafic Complex [Ph.D. thesis]: Los Angeles, California, University of California at Los Angeles, 301 p.

Bell, A. M., 1981, Vergence: an evaluation: Journal of Structural Geology, v. 3, p. 197–202.

Brown, E. H., 1971, Phase relations of biotite and stilpnomelane in the greenschist facies: Contributions to Mineralogy and Petrology, v. 31, p. 275–299.

—— 1974, Comparison of the mineralogy and phase relations of blueschists from the North Cascades, Washington and greenschists from Otago, New Zealand: Geological Society of America Bulletin, v. 85, p. 333–344.

—— 1977, The crossite content of Ca-amphibole as a guide to pressure of metamorphism: Journal of Petrology, v. 18, p. 53–72.

—— 1983, Field guide to the Shuksan Metamorphic Suite: Guidebook for Geological Society of America Penrose Conference on Blueschists and related Eclogites, Bellingham, Western Washington University, 28 p.

Brown, E. H., and O'Neil, J. R., 1982, Oxygen isotope geothermometry and stability of lawsonite and pumpellyite in the Shuksan Suite, North Cascades, Washington: Contributions to Mineralogy and Petrology, v. 80, p. 240–244.

Brown, E. H., Wilson, D. L., Armstrong, R. L., and Harakal, J. E., 1982, Petrologic, structural, and age relations of serpentinite, amphibolite, and blueschist in the Shuksan Suite of the Iron Mountain-Gee Point area, North Cascades, Washington: Geological Society of America Bulletin, v. 93, p. 1087–1098.

Cloos, M., 1985, Thermal evolution of convergent plate margins: thermal modeling and reevaluation of isotopic Ar-ages for blueschists in the Franciscan Complex: Tectonics, v. 4, p. 421–433.

Coleman, R. G., 1971, The Colebrook Schist of southwestern Oregon and its relation to the tectonic evolution of the region: U.S. Geological Survey Bulletin 1339, 61 p.

Coleman, R. G., 1972, Blueschist metamorphism and plate tectonics: 24th International Geological Congress, Section 2, p. 19–26.

Coleman, R. G., and Lanphere, M. A., 1971, Distribution and age of high-grade blueschists, associated eclogites, and amphibolites from Oregon and California: Geological Society of America Bulletin, v. 82, p. 2397–2412.

Coleman, R. G., and Helper, M. A., 1983, The significance of the Condrey Mountain Dome in the evolution of the Klamath Mountains, California and Oregon [abstr.]: Geological Society of America Abstracts with Programs, v. 15, no. 5, p. 294.

Coleman, R. G., Helper, M. A., and Donato, M. M., 1983, Geologic Map of the Condrey Mountain Roadless Area, Siskiyou County, California: U.S. Geological Survey, Miscellaneous Field Study Map MF-1540-A, Scale 1:50,000.

Davis, G. A., Monger, J.W.H., and Burchfiel, B. C., 1978, Mesozoic construction of the Cordilleran "collage," central British Columbia to central California, *in* Howell, D. G., and McDougall, K. A., eds., Mesozoic Paleogeography of the Western United States: Society of Economic Paleontologists and Mineralogists, Pacific Section Paleogeography Symposium 2, p. 1–32.

Davis, G. H., 1983, Shear-zone model for the origin of metamorphic core complexes: Geology, v. 11, p. 342–347.

Dewey, J. F., and Bird, J. M., 1970, Mountain belts and the new global tectonics: Journal of Geophysical Research, v. 75, p. 2625–2647.

Donato, M. M., Coleman, R. G., and Kays, M. A., 1980, Geology of the Condrey Mountain Schist, California and Oregon: Oregon Geology, v. 42, p. 125–129.

Ernst, W. G., 1975, Systematics of large-scale tectonics and age progression in Alpine and circum-Pacific blueschist belts: Tectonophysics, v. 17, p. 255–272.

—— 1979, Coexisting sodic and calcic amphiboles from high-pressure metamorphic belts and the stability of barroisitic amphibole: Mineralogical Magazine,

v. 43, p. 269–278.

—— 1984, California blueschists, subduction, and the significance of tectonostratigraphic terranes: Geology, v. 12, p. 436–440.

Etheridge, M. A., and Hobbs, B. E., 1974, Chemical and deformational controls on recrystallization of mica: Contributions to Mineralogy and Petrology, v. 43, p. 111–124.

Gray, G. G., 1985, Structural, geochronologic, and depositional history of the western Klamath Mountains, California and Oregon: implications for the early to middle Mesozoic tectonic evolution of the western North America Cordillera [Ph.D. thesis]: Austin, Texas, University of Texas at Austin, 168 p.

Hamilton, W., 1969, Mesozoic California and the underflow of Pacific mantle: Geological Society of America Bulletin, v. 80, p. 2409–2430.

Harper, G. D., 1980, The Josephine ophiolite—Remains of a Late Jurassic marginal basin in northwestern California: Geology, v. 8, p. 333–337.

Harper, G. D., and Wright, J. E., 1984, Middle to Late Jurassic tectonic evolution of the Klamath Mountains, California-Oregon: Tectonics, v. 3, p. 759–772.

Haugerud, R. A., Morrison, M. L., and Brown, E. H., 1981, Structural and metamorphic history of the Shuksan blueschist terrane in the Mount Watson and Gee Point areas, North Cascades, Washington: Geological Society of America Bulletin, v. 92, p. 374–383.

Heinrich, W., and Althaus, E., 1980, Die obere stabilitagrenze von lawsonit plus albit bsw. jadeit: Fortschritte der Mineralogie, v. 58, p. 49–50.

Helper, M. A., 1985, Structural, metamorphic, and geochronologic constraints on the origin of the Condrey Mountain Schist, north central Klamath Mountains, northern California [Ph.D. thesis]: Austin, Texas, The University of Texas at Austin, 209 p.

—— 1983a, Deformation-metamorphism relationships in a regional blueschist-greenschist facies terrane, Condrey Mt. Schist, north-central Klamath Mts., N. California [abstr.]: Geological Society of America Abstracts with Programs, v. 15(5), p. 427.

—— 1983b, Subduction related deformation and metamorphism in the regional blueschist-greenschist terrain of the Condrey Mountain Window, Klamath Mts., northern California [abstr.]: Geological Society of America Abstracts with Programs, v. 15(6), p. 594.

Henley, R. W., 1975, Metamorphism of the Moke Creek Lode, Otago, New Zealand: New Zealand Journal of Geology and Geophysics, v. 18, p. 229–237.

Hoffman, C., 1972, Natural and synthetic ferroglaucophane: Contributions to Mineralogy and Petrology, v. 34, p. 135–159.

Holland, T.J.B., 1979, Experimental determination of the reaction paragonite = jadeite + kyanite + H_2O, and an internally consistent thermodynamic data set for part of the system Na_2O-Al_2O_3-SiO_2-H_2O, with application to eclogites and blueschists: Contributions to Mineralogy and Petrology, v. 68, p. 293–301.

Hotz, P. E., 1967, Geologic map of the Condrey Mountain quadrangle, and parts of the Seiad Valley and Hornbrook quadrangles, California: U.S. Geological Survey Geologic Quadrangle Map GQ-618, scale 1:62,500.

—— 1971, Geology of lode gold districts in the Klamath Mountains, California and Oregon: U.S. Geological Survey Bulletin 1290, 91 p.

—— 1973, Blueschist metamorphism in the Yreka-Fort Jones area, Klamath Mountains, California: U.S. Geological Survey Journal of Research, v. 1, p. 53–61.

—— 1979, Regional metamorphism in the Condrey Mountain quadrangle, north-central Klamath Mountains, California: U.S. Geological Survey Professional Paper 1086, 25 p.

Hsui, A. T., and Toksöz, M. N., 1979, The evolution of thermal structures beneath a subduction zone: Tectonophysics, v. 60, p. 43–60.

Irwin, W. P., 1960, Geologic reconnaissance of the northern Coast Ranges and Klamath Mountains, California, with a summary of the mineral resources: California Division of Mines and Geology Bulletin 179, 80 p.

Irwin, W. P., 1972, Terranes of the western Paleozoic and Triassic belt in the southern Klamath Mountains, California: U.S. Geological Survey Professional Paper 800-C, C103–C111.

—— 1981, Tectonic accretion of the Klamath Mountains, *in* Ernst, W. G., ed., The geotectonic development of California: Englewood Cliffs, New Jersey, Prentice-Hall, p. 29–49.

Johannes, W., and Puhan, D., 1971, The calcite-aragonite transition reinvestigated: Contributions to Mineralogy and Petrology, v. 31, p. 28–38.

Kays, M. A., and Ferns, M. L., 1980, Geologic field trip through the north-central Klamath Mountains: Oregon Geology, v. 42, p. 23–35.

Klein, C. W., 1977, Thrust plates of the north-central Klamath mountains near Happy Camp, California: California Division of Mines and Geology Special Report 129, p. 23–26.

Lanphere, M. A., Irwin, W. P., and Hotz, P. E., 1968, Isotopic age of the Nevadan orogeny and older plutonic and metamorphic events in the Klamath Mountains, California: Geological Society of America Bulletin, v. 79, p. 1027–1052.

Mattinson, J. M., and Echeverria, L. M., 1980, Ortigalita Peak gabbro, Franciscan Complex: U-Pb dates of intrusion and high-pressure, low-temperature metamorphism: Geology, v. 8, p. 589–593.

McDowell, F. W., Lehman, D. H., Gucwa, P. R., Fritz, D., and Maxwell, J. C., 1984, Glaucophane and ophiolites of the northern California Coast Ranges; isotopic ages and their tectonic implications: Geological Society of America Bulletin, v. 95, p. 1373–1382.

Medaris, L. G., Jr., and Welsh, J. L, 1980, Prograde metamorphism of serpentinite in the western Paleozoic and Triassic belt, Klamath Mountains Province [abstr.]: Geological Society of America Abstracts with Programs, v. 12, p. 120.

Misch, P., 1966, Tectonic evolution of the northern Cascades of Washington, *in* Gunning, H. C., ed., A symposium of the tectonic history and mineral deposits of the western Cordillera in British Columbia and neighboring parts of the United States: Canadian Institute of Mining and Metallurgy Special Volume 8, p. 101–148.

—— 1977, Bedrock geology of the North Cascades, *in* Brown, E. H. and Ellis, R. H., eds., Geologic excursions in the Pacific Northwest: Bellingham, Western Washington University, p. 1–62.

Miyashiro, A., 1972, Metamorphism and related magmatism in plate tectonics: American Journal of Science, v. 272, p. 629–656.

Mortimer, N., 1983, Deformation, metamorphism and terrane amalgamation, NE Klamath Mountains, CA, *in* Howell, D. G., Cox, A., Jones, B. L., and Nur, A., eds., Proceedings of the Circum Pacific Terrane Conference, Stanford University.

Mortimer, N., and Coleman, R. G., 1984, A Neogene structural dome in the Klamath Mountains, California and Oregon, *in* Nilsen, T. H., ed., Hornbrook Formation Fieldtrip Guidebook: Society of Economic Paleontologists and Mineralogists, p. 36–44.

Moore, D., 1984, Metamorphic history of a high-grade blueschist exotic block from the Franciscan Complex, California: Journal of Petrology, v. 25, p. 126–150.

Nitsch, K. -H., 1980, Reaktion von bariumfeldspat (celsian) mit H_2O zu cymrite unter metamorphen bedingungen: Fortschritte der Mineralogie, v. 58, p. 98–99.

Ramsay, J. G., 1967, Folding and Fracturing of Rocks: New York, McGraw-Hill, 568 p.

Robinson, P., Spear, F. S., Schumacher, J. C., Laird, J., Klein, C., Evans, B. W., and Doolan, B. L., 1982, Phase relations of metamorphic amphiboles: Natural occurrence and theory: *in* Veblen, D. R. and Ribbe, P. R., eds., Amphiboles: Petrology and experimental phase relations: Mineralogical Society of America Reviews in Mineralogy, v. 9B, p. 1–227.

Saleeby, J. W., Harper, G. D., Snoke, A. W., and Sharp, W. D., 1982, Time relations and structural-stratigraphic patterns in ophiolite accretion, west central Klamath Mountains, California: Journal of Geophysical Research, v. 87, p. 3831–3848.

Saleeby, J. W., Blake, M. C., and Coleman, R. G., 1984, U/Pb zircon ages of thrust plates of the west central Klamath Mountains and Coast Ranges, Northern California and Southern Oregon [abstr.]: EOS (American Geophysical Union Transactions), v. 65, p. 1147.

Scott, S. D., 1974, Sphalerite geobarometry of regionally metamorphosed terrains [abstr.]: Geological Society of America Abstracts with Programs, v. 6, p. 946–947.

Steiger, R. H., and Jager, E., 1977, Conventions on the use of decay constants in geo- and cosmochronology: Earth and Planetary Science Letters, v. 36, p. 359–362.

Suppe, J., and Armstrong, R. L., 1972, Potassium-Argon dating of Franciscan metamorphic rocks: American Journal of Science, v. 272, p. 217–233.

Suppe, J., and Foland, K. A., 1978, The Goat Mountain schists and Pacific Ridge complex: a redeformed but still intact Late Mesozsoic Franciscan schuppen complex, *in* Howell, D. and McDougall, K., eds., Mesozoic Paleogeography of the Western United States: Society of Economic Paleontologists and Mineralogists Pacific Coast Paleogeography Symposium 2, p. 431–451.

Vance, J. A., 1957, The geology of the Sauk River area in the northern Cascades of Washington [Ph.D. thesis]: Seattle, Washington, University of Washington, 313 p.

Wang, C. Y., and Shi, Y., 1984, On the thermal structure of subduction complexes: A preliminary study: Journal of Geophysical Research, v. 89, p. 7709–7718.

Williams, P. F., and Campagnoni, R., 1983, Deformation and metamorphism in the Bard area of the Sesia Lanzo Zone, Western Alps, during subduction and uplift: Journal of Metamorphic Geology, v. 1, p. 117–140.

Wood, B. L., 1971, Structure and relationships of late Mesozoic schists of northwest California and southwest Oregon: New Zealand Journal of Geology and Geophysics, v. 14, p. 219–239.

Wood, R. M., 1979, The iron-rich blueschist facies minerals: I. Deerite: Mineralogical Magazine, v. 43, p. 251–259.

—— 1982, The Laytonville Quarry (Mendocino County California) exotic block: iron-rich blueschist-facies subduction-zone metamorphism: Mineralogical Magazine, v. 45, p. 87–99.

Wright, J. E., 1982, Permo-Triassic accretionary subduction complex, southwestern Klamath Mountains, northern California: Journal of Geophysical Research, v. 87B, p. 3805–3818.

MANUSCRIPT ACCEPTED BY THE SOCIETY JULY 29, 1985

Geological Society of America
Memoir 164
1986

Geology of the Shuksan Suite, North Cascades, Washington, U.S.A.

Edwin H. Brown
Department of Geology
Western Washington University
Bellingham, Washington 98225

ABSTRACT

The Shuksan Metamorphic Suite (Misch 1966) and correlative Easton Schist (Smith 1903) occur on the western flank of the North Cascades of Washington as large fault-bounded fragments (10s of kms long) imbricated with other rock units in a north-south belt extending more than 180 km. The Shuksan Suite is dominated by greenschist, blueschist, quartzose carbonaceous phyllite, and quartzofeldspathic semi-schist. A Jurassic, oceanic, near-arc tectonic site of deposition is hypothesized. Metamorphism began with production of Late Jurassic high pressure amphibolites in a contact aureole localized near peridotite, and was followed by regional blueschist facies metamorphism that lasted into the Early Cretaceous. Phase assemblages of the regional metamorphism include (in addition to quartz, chlorite, phengite, and sphene) lawsonite + albite in pelitic and psammitic schists, actinolite + albite + epidote + pumpellyite in greenschists, and crossite + albite + epidote + iron oxide in blueschists. The protolith rock type markedly controlled the development of the metamorphic index minerals. Temperatures of regional metamorphism ranged from approximately 330 to 400°C; pressures, not easily estimated, were perhaps 7 to 9 kilobars. Deformation during the blueschist metamorphism was directed approximately normal to the continental margin, as evidenced by the regional pattern of shear lineations. Uplift and imbrication of the Shuksan Suite with other units occurred by fault motion parallel to the continental margin. The metamorphic and uplift events are correlated with periods of high angle and low angle convergence respectively between the Farallon and North American plates.

INTRODUCTION

A regionally extensive assemblage of Late Jurassic to Early Cretaceous blueschists, greenschists, and quartzose phyllites occurring along the western flank of the North Cascades is the subject of this report (Figure 1). This rock unit was named the Shuksan Metamorphic Suite by Misch (1966). It is part of a more extensive blueschist belt that includes the Easton Schist of Smith (1903), which occurs in the central Washington Cascades. In total, blueschist rocks are distributed as fault-bounded fragments along a zone some 180 km in length.

In this review paper, the phase petrology, structure, and geologic history of the Shuksan Suite are summarized, with the goals of providing insight to general questions of com-positional controls of blueschist facies index minerals and the structural and tectonic evolution of blueschist belts.

REGIONAL SETTING

The Shuksan Suite occurs as the dominant structural element in an imbricate complex which also includes the following other rock units, shown on Figure 1: 1) ultramafic tectonite, most notably the Twin Sisters dunite; 2) Precambrian felsic gneiss with intrusive lower Paleozoic gabbros and diorites, known as the Yellow Aster Complex; 3) Permian barroisitic amphibolite and associated subordinate blueschist of the Vedder Complex; 4) upper Paleozoic, arc-derived sedimentary and volcanic rocks of

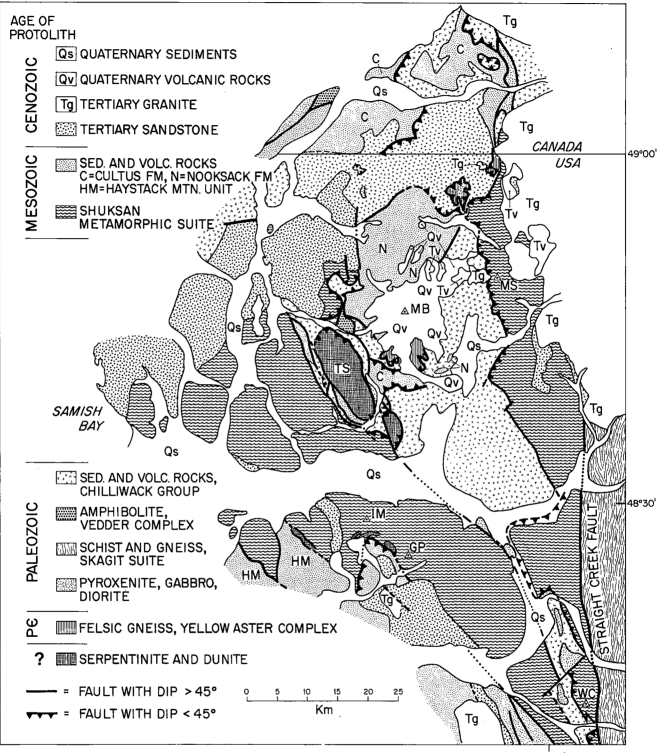

Figure 1. Regional geologic map of the Shuksan Metamorphic Suite, based on work by Vance (1957), Misch (1966, 1977), Monger (1966), Bechtel Report (1979), Vance and others (1980), Rady (1981), Frasse (1981), R. Lawrence (personal communication, 1981), Johnson (1982), Brown and others (1982), Blackwell (1983), Sevigny (1983), Jones (1984), Jewett (1984), and unpublished mapping by E. H. Brown, D. Silverberg, C. Ziegler, and P. Leiggi, all of Western Washington University. IM = Iron Mountain, MB = Mt. Baker, TS = Twin Sisters, MS = Mt. Shuksan, GP = Gee Point, WC = White Chuck Mountain.

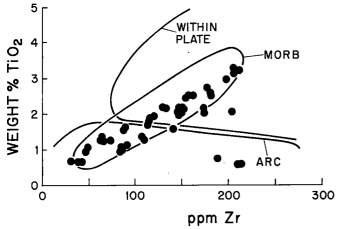

Figure 2. TiO_2 vs. Zr in basic rocks of the Shuksan Suite and related Easton Schist. Data from Ashleman (1979), Street-Martin (1981), and Dungan and others (1983). Field boundaries from Pearce (1982).

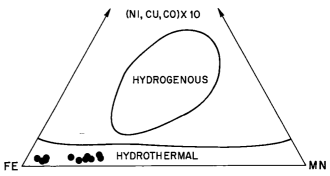

Figure 3. (Ni+Cu+Co) ×10 vs. Fe vs. Mn diagram of Bonatti and others (1976) differentiating between hydrogeneous (e.g. Mn nodules) and hydrothermal metalliferous sediments. Data from Street-Martin (1981).

the Chilliwack Group; and 5) Triassic to lower Cretaceous, arc-derived sedimentary and volcanic rocks of the Cultus Formation, Nooksack Formation (and associated Wells Creek Volcanics) and Haystack Unit. Misch (1966) interpreted the structure to be dominated by the Shuksan thrust and associated root zone, along which a broad plate of the Shuksan Suite has overridden all other pre-Tertiary rock units. More detailed mapping (unpublished information by Brown and M.S. students at Western Washington University) requires some revision of this concept in showing that high angle faults are more abundant than was previously thought, that these high angle faults are of strike-slip not dip-slip motion (i.e. they are not root zone faults), and that at least in one area (Gee Point-Iron Mountain, Figure 1) the Shuksan Suite is not the structurally highest unit. Thus, the structure of the imbricate zone appears to be more complex than previously envisaged. More study is needed before the geometry and kinematics of these faults will be well understood.

The imbricated complex described above is bounded on the east by the Straight Creek Fault, a right-lateral strike-slip fault with approximately one hundred kilometers of offset (Vance 1957, 1985). East of the Straight Creek Fault is the Skagit Metamorphic Suite of Misch (1966), which has Barrovian style facies and an upper Cretaceous to early Tertiary metamorphic age (Mattinson 1972).

LITHOLOGIES AND PROTOLITHS

Present lithologies of the Shuksan Suite are greenschist, blueschist, quartzose carbonaceous phyllite, quartzofeldspathic phyllonite and mica schist, Fe-Mn quartzose metasediment, rare magnesian schists interstratified with the carbonaceous phyllites, and rare metaplutonic rocks. Along the eastern outcrop belt of the Shuksan Suite, basic schist consisting mostly of greenschist with lesser blueschist is the predominant lithology; black carbonaceous, quartzose phyllite is also present. Misch (1966) named

these mappable units the Shuksan Greenschist and Darrington Phyllite, respectively. Contacts between these major units are faulted in many places, but appear locally to be primary. In the Gee Point area, Morrison (1977) noted that the Fe-Mn metasediments are typically developed in the contact area between apparently conformable Shuksan Greenschist and Darrington Phyllite, suggesting an ocean crust stratigraphy from bottom up of basite-metalliferous, sediment-quartzose pelitic sediment. Supporting this interpretation, chemical analyses of the metabasites by Ashleman (1979), Street-Martin (1981), and Dungan and others (1983) show them to be comparable to oceanic tholeiitic basalt (Figure 2), and the chemical composition of the Fe-Mn rocks matches that of oceanic hydrothermal deposits on active ridges (Figure 3). It follows from this stratigraphic interpretation that the entire eastern belt of the Shuksan Suite, from Mt. Shuksan to White Chuck Mountain (Figure 1), may be inverted, as the metabasites presently overly the metasediments. Alternatively, the greenschist might be thrust over the phyllite. However, in some places (e.g. north side of Mt. Shuksan) the contact appears to be unfaulted, and thus the inversion hypothesis is favored. The sequence appears to be right-way up in the Iron Mountain to Gee Point area (cf. Haugerud and others 1981).

The blueschist and greenschist parts of the Shuksan Greenschist unit coexist in apparent metamorphic equilibrium. The only differences in mineralogy are that the greenschist contains actinolite, and, in places, pumpellyite, whereas the blueschist lacks these minerals and contains crossite and commonly also magnetite and/or hematite. The mineralogic differences can be accounted for in terms of rock composition (Misch 1959; Brown 1974; Dungan and others 1983) and do not require a difference in P and T of formation. The blueschist is more oxidized and has a somewhat higher ratio of Na_2O/CaO. Blueschist and greenschist lithologies occur in layers up to a hundred meters or more in thickness, large enough to be mappable (on a local scale of a few km^2) above tree-line where exposures are excellent; these rocks are also interlayered on the outcrop scale. Whether the cause of compositional variation is related to primary basalt compositions or is the result of submarine alteration is not entirely clear, although the latter factor can be inferred for observed flattened relict pillows with green cores and blue rims.

The magnesian schists occur as beds up to a few meters thick, interstratified with and gradational into the quartzose phyllites. They are interpreted to be some type of clastic serpentinite deposit, and appear to imply the existence of oceanic mantle fragments exposed in the depositional basin.

The western parts of the Shuksan Suite, west of the Twin Sisters (Figure 1), have a preponderance of semi-schist, derived from quartzofeldspathic siltstone and sandstone. Locally intrusive or faulted into this rock are plutonic bodies ranging in composition from trondhjemite to diorite to gabbro to pyroxenite. These plutonic bodies bear the same metamorphic mineral assemblages and metamorphic fabric elements as the host schists, and thus were emplaced prior to the blueschist metamorphism. Greenschists and blueschists are rare in this region, and metalliferous metasediments are absent. It is possible that these rocks are not related to the Shuksan Suite to the east. However, the carbonaceous phyllite is common to both areas, the inferred metamorphic conditions are virtually identical, and the metamorphic ages are similar. Therefore, the western rocks seem best interpreted as a variant of the Shuksan protolith dominated by some type of terrigenous, arc-derived(?) material. Thus, the overall Shuksan assemblage of lithologies might best be interpreted as representing near-arc oceanic crust, as perhaps in a marginal basin (also suggested by Vance and others 1980).

GEOCHRONOLOGY

Metamorphic Age

Two episodes of metamorphism can be recognized in the Shuksan Suite. Rocks recording the earlier event comprise a relatively high grade (amphibolite facies) aureole associated with serpentinized peridotite in the vicinity of Gee Point and Iron Mountain (Brown and others 1982). K/Ar ages of hornblende and muscovite range from 144 to 164 Ma, with the muscovites giving the older ages (Armstrong 1980).

Rocks of regional extent in the Shuksan Suite that are dated by K/Ar on both whole-rock and mineral separates give ages ranging from 105 to 130 Ma; the average for eight localities being 120 Ma (Misch 1963, 1964; Bechtel Report 1979; Armstrong dates reported in Haugerud 1980 and in Brown and others 1982). Rb/Sr dates for the regional schists in the Gee Point area define a 128 ± 6 Ma isochron which is concordant with the K/Ar age for the same specimens (Brown and others 1982). A less accurate 105 ± 10 Ma isochron is defined by two specimens from the Mt. Watson area (Haugerud 1980).

Fission track ages of apatites from the Shuksan are 60 Ma and younger as determined by Sleeper (personal communication 1984). These data can be interpreted in terms of either a relatively late uplift of Shuksan metamorphic rocks through the 100°C annealing temperature of apatite fission tracks or an early uplift

followed by reheating. The samples studied are not in detectable proximity to Tertiary intrusive rocks.

Protolith Age

Low initial strontium ratios, 0.704-0.705, together with the Early Cretaceous metamorphic age, were interpreted by Armstrong (1980) to indicate a Jurassic age for the Shuksan protolith materials in the vicinity of Iron Mountain and Gee Point (Figure 1). A more direct determination of protolith age was recently obtained by Nicholas Walker (1984, University of California, Santa Barbara) on a single zircon separate from Shuksan diorite occurring west of the Twin Sisters. The age is 163.5 ± 2 Ma based on Pb^{206}/U^{238}.

The protolith of the regional schists of the Shuksan Suite appears to be coeval with 1) the high grade metamorphic aureole associated with ultramafic rock at Gee Point; 2) rocks of arc affinity in the region of Fidalgo Island (immediately west of the Figure 1 map area); 3) the Cultus Formation of southern British Columbia; and 4) the arc-related Wells Creek Volcanics underlying the Nooksack Group (Figure 1). In the future, more petrologic, geochronologic, and structural data may allow reconstruction of the tectonic setting of Shuksan protolith rocks relative to these other units of the region.

METAMORPHIC HISTORY

The oldest known metamorphic elements of the Shuksan suite are the 144-164 Ma amphibolites and related rocks associated with serpentinites in the Gee Point-Iron Mountain area. These rocks include amphibolite and interbedded quartzose mica schist, barroisite schist, blueschist, rare eclogite, talc - tremolite - chlorite metasomatic rock, and serpentinite. Field relations and mineralogy suggest juxtaposition at high confining pressure of hot ultramafic rock (peridotite) with oceanic sedimentary and volcanic rocks, producing amphibolites and rare eclogites in a thermal aureole. Subduction of oceanic crust into hot upper mantle material is envisaged as a possible tectonic setting for this metamorphism (Brown and others 1982).

The regionally developed lower grade rocks, which constitute the bulk of the Shuksan Suite, bear evidence of a polymetamorphic history. Epidote glomeroblasts and large pods (up to 20 cm long) are broken and cracked and have crossite grains growing in the cracks, and thus have been interpreted (by Haugerud and others 1981) as predating the main blueschist metamorphism. These epidotes overprint and replace an igneous, felty plagioclase lath texture that is preserved by patterns of dusty hematite inclusions. The epidotes appear to have crystallized in a static environment, under hydrothermal conditions promoting mobility of epidote and plagioclase-bearing elements. A sea floor environment is hypothesized. Similarly, Dungan and others (1983) have inferred sea floor alteration from major and trace element chemical analyses from the Shuksan Greenschist unit. The blueschist event that followed this early metamorphism

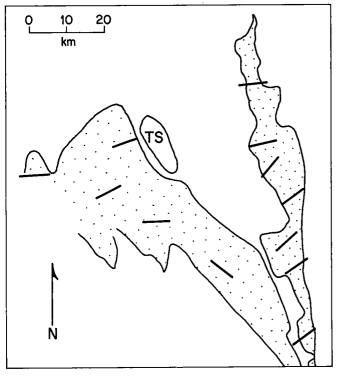

Figure 4. Early metamorphic lineations in the Shuksan Suite. Data from Haugerud (1980), Leiggi (unpublished), Silverberg (unpublished) and Brown (unpublished). Stippled area represents the Shuksan Suite. TS = Twin Sisters.

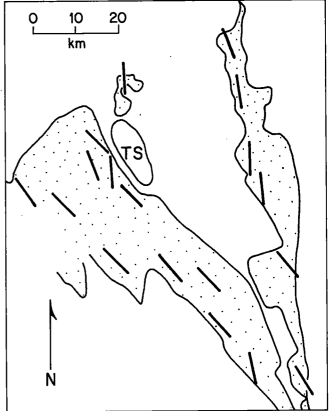

Figure 5. Late metamorphic fold axes and related lineations in the Shuksan Suite. Data from Morrison (1977), Haugerud (1980), Leiggi (unpublished), Silverberg (unpublished), and Brown (unpublished). Stippled area represents the Shuksan Suite. TS = Twin Sisters.

caused nearly complete recrystallization. The only vestiges of primary mineralogy and texture in basic rocks are rare augite phenocrysts, albitized plagioclase phenocrysts, chlorite-epidote amygdules, and rare flattened pillows. Relict but deformed bedding is observable in some of the metasedimentary rocks.

The main stage of metamorphism produced a strong deformational fabric, along which the blueschist index minerals crossite, jadeite-bearing aegirine-augite, and lawsonite, as well as chlorite, muscovite, stilpnomelane, pumpellyite, and other minerals developed. The mineral parageneses (described more fully below) indicate blueschist conditions, but at relatively high T/P for this facies. This metamorphism is the 120–130 Ma Early Cretaceous event dated by K/Ar and Rb/Sr. Where the regional rocks border the older amphibolite-serpentinite part of the Shuksan Suite, crossite and actinolite grains have barroisite and hornblende cores, suggesting that the regional metamorphism overprinted the earlier high-grade contact metamorphism. Brown and others (1982) speculated that the contact rocks represent the initiation of a subduction orogenic event and that the metamorphism that produced the regionally developed schists was a continuation of this process.

The predominant deformational features associated with the regional blueschist metamorphic event include: a pronounced foliation (S_1); an early lineation (L_1) defined by sheared relict clasts and phenocrysts, and by alignment of metamorphic amphiboles;

late metamorphic folds (F_2) that deform S_1 but have been accompanied by recrystallization (Misch 1969: 61), and range in amplitude from millimeters to hundreds of meters (Haugerud and others 1981); and crenulation lineations and minor quartz rodding (L_2) associated with F_2. Preliminary observations of the trend of L_1 suggest that it averages approximately N70E and is moderately consistent on a regional scale (Figure 4). The trend of L_2, also regionally consistent, averages N20W (Figure 5). It is statistically normal to L_1. Discounting possible post-metamorphic rotations of regional nature, the L_1 lineation is interpreted to represent the direction of plate convergence during blueschist metamorphism (following the arguments of Escher and Watterson 1974 and Shackleton and Ries 1984). The F_2 folds and associated L_2 lineations are interpreted to have formed under essentially the same tectonic-metamorphic regime as S_1 and L_1 and to represent late metamorphic flexural-slip folding which utilizes the S_1 surface. Investigation of the metamorphic deformational features is in progress.

MINERALOGY AND PHASE RELATIONS

Mineral assemblages in the main regional metamorphic

TABLE 1. REPRESENTATIVE MINERAL ASSEMBLAGES IN THE SHUKSAN SUITE
(Additional minerals not listed are apatite, pyrite, chalcopyrite, and barite)

	Greenschists			Blueschists				Fe—Mn Fm.				Phyllite and Mica Schist		Magnesian Schists					
Quartz	x	x	x	x	x	x		x	x	x	x	x	x	x	x	x	x		x
Albite	x	x	x	x	x	x	x					x		x		x			
Epidote	x	x	x	x	x	x		x		x		x							
Lawsonite												x							
Pumpellyite	x																		
Chlorite	x	x	x	x	x							x	x	x	x			x	x
Muscovite	x			x	x							x	x						
Paragonite													x						
Stilpnomelane		x	x	x				x	x										
Actinolite	x	x	x																x
Crossite				x	x	x	x	x	x	x									
Aegirine-augite					x					x	x								
Spessartine				x	x				x		x	x							
Cymrite					x														
Deerite								x											
Magnetite		x						x	x	x	x								
Hematite			x						x		x								
Sphene	x	x	x	x		x						x							
Calcite/Arag.		x	x	x				x					x			x			
Magnesite														x	x	x			
Dolomite															x	x			
Talc														x		x	x	x	x
Antigorite																		x	
Graphite													x						
Type Specimen	55-17	67E	40E	27Q	56D	76-13	Sh-10	39f	15	27-0	37-5	84	RH-E1	545A	545B	553	565A	563C	563D

phase of the Shuksan Suite define a metamorphic facies intermediate between the high P/T Franciscan type blueschists and the
greenschist facies. Representative assemblages in the various lithologies are listed in Table 1. Mineral compositions are plotted on
Figures 6 through 13 and are discussed below.

Epidotes range in Fe/(Fe+Al) from .12 to .35; no compositional gap is observed (Figure 6). A close relationship exists between the composition of epidote and species of coexisting
mineral, as shown on Figures 6 and 7. Zoning is common and is

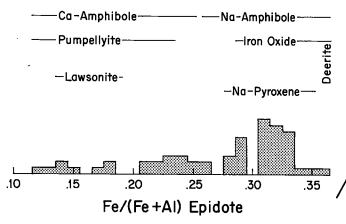

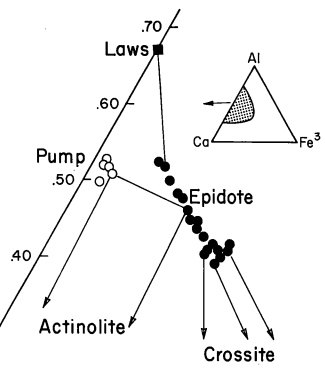

Figure 6. Composition of epidote in the Shuksan Suite and correlation
with coexisting mineral species.

Figure 7. Phase relations of epidote and pumpellyite on a Ca-Al-Fe^{3+}
diagram, projected from quartz, albite, chlorite, and H$_2$O.

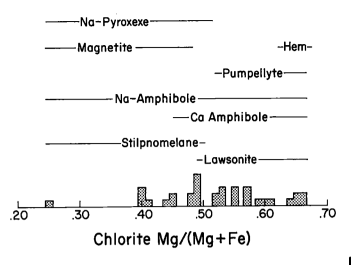

Figure 8. Mg/(Mg+Fe) ratio of chlorite in the Shuksan Suite and correlation with coexisting mineral species.

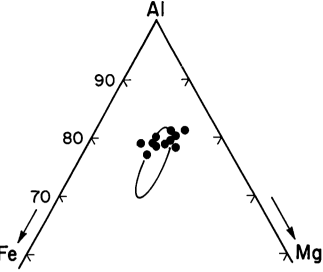

Figure 9. Al-Fe-Mg plot of compositions of pumpellyites in the Shuksan Suite. The composition range of pumpellyites in the Franciscan South Fork Mountain Schist is shown by the line enclosure for comparison (from Brown and Ghent, 1983).

almost invariably toward an iron-poorer rim. Typically, the zones are sharply bounded; a Becke line can be observed between zones in some grains. This pattern probably represents an overgrowth of the main stage blueschist facies epidote on earlier, more iron-rich, hydrothermal epidote.

Muscovites are phengitic. The average tetrahedral site occupancy for 10 samples is 6.86 Si per 8 cations.

Chlorites have widely varying Mg/(Mg+Fe), a compositional feature that correlates with mineral association (Figure 8). It is noteworthy that Fe-rich chlorites are stable instead of almandine garnet.

Pumpellyite is relatively aluminous (Figure 9). It crystallized near its upper temperature limit of stability, being replaced in parts of the Shuksan Suite by its high temperature breakdown products of iron-poor epidote + actinolite. Phase relations are shown on Figure 7.

Amphiboles range continuously in composition from actinolite to crossite (as described by Misch 1959; Brown 1974; and in Dungan and others 1983); the intermediate amphiboles in this series being rare. The sodic amphiboles occur in relatively oxidized rocks as evidenced by the high Fe^{3+}/Al content of coexisting epidote (Figure 6) and the common association of magnetite or hematite (Table 1). Dungan and others (1983) have also shown that the crossite-bearing rocks have higher Na_2O/CaO than actinolite-bearing rocks. Comparing the compositions of actinolite and crossite, one sees that the octahedral trivalent ions, Fe^{3+} and Al, substitute almost exclusively for Mg rather than Fe^{2+} (Figure 10), a relationship that agrees with the interpretation that the M2 position is occupied by Mg, Fe^{3+} or Al, but not Fe^{2+} (Ernst 1968). The sodic amphiboles range widely in composition on the Miyashiro diagram from glaucophane to magnesioriebeckite, but show a pronounced correlation of composition with mineral association (Figure 11).

Metamorphic pyroxenes are dominantly aegirine or aegirine augite (Figure 12). Jadeite content averages about 20 mole

percent. The pyroxenes are restricted to rocks with iron-rich epidote and are most common in the Fe-Mn formation.

Lawsonite is near its high temperature stability limit, being restricted to relatively aluminous metasediments (Table 1). Phase relations are shown on Figure 7; note that lawsonite does not coexist with amphiboles.

Garnets contain 30 or more mole percent of the spessartine component and less than 2 percent pyrope, even on the rims (Figure 13).

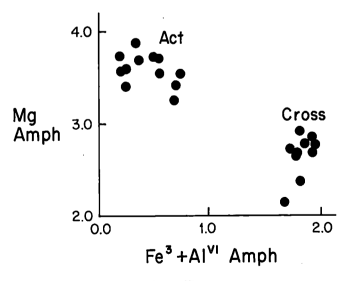

Figure 10. Plot of Mg vs. $(Fe^{3+} + Al^{VI})$ in amphiboles of the Shuksan Suite showing Mg $\rightleftharpoons (Fe^{3+} + Al^{Vi})$ substitution in the composition range from Ca to Na amphiboles. Fe^{2+} is approximately constant.

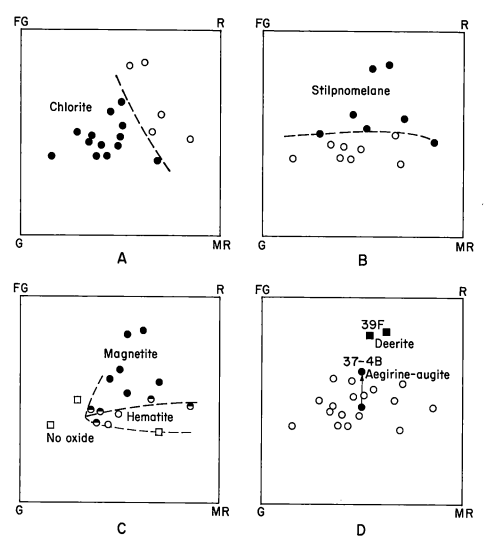

Figure 11. The composition of sodic amphibole as related to mineral association. A. Sodic amphibole with chlorite, solid circle; without chlorite, open circles. B. Sodic amphibole with stilpnomelane, solid circles; without stilpnomelane, open circles. C. Sodic amphibole with magnetite, solid circles; with magnetite plus hematite, half-solid circles; with hematite, open circles; without iron oxide, open squares. D. Sodic amphibole with deerite, solid squares; with aegirine-augite, solid circles; without deerite or aegirine-augite, open circles. Arrow points in the direction of zoning from core to rim (from Brown, 1974).

Cymrite ($BaAl_2Si_2O_6H_2O$) was identified (by microprobe analysis, refractive index measurements, and x-ray diffraction) in a blueschist (Table 1) with rather abundant spessartine garnets. The cymrite occurs as white tabular porphyroblasts easily seen in hand specimen. Cymrite is a high pressure hydrated equivalent of celsian, found in other blueschist belts and recently studied experimentally by Nitsch (1980).

Deerite occurs in the iron formation associated with the most Fe^{3+}-rich epidote and Fe^{2+}-rich crossite observed in the Shuksan Suite (Figures 6 and 11).

Carbonate minerals include calcite, aragonite, magnesite, and dolomite (Table 1). Of the $CaCO_3$ polymorphs, calcite is greatly predominant; however, schists west and south of the Twin Sisters contain aragonite as a phase apparently crystallized during

the main metamorphic event (this study; Evans and Misch 1976). The preservation of aragonite in schists inferred to have crystallized at T≥350°C (see later discussion) is not in agreement with experimental studies of the aragonite → calcite inversion, which suggest a maximum metamorphic temperature of 200-250°C for aragonite preservation (Carlson and Rosenfeld 1981).

Phase relations among common minerals of the Shuksan Suite are portrayed in Figure 7. Bulk rock composition exerts a very strong control on the distribution of the metamorphic index minerals. Lawsonite occurs only in pelites; pumpellyite occurs only in Fe^{3+}-poor psammites and basites; and Na-amphibole occurs only in Fe^{3+}-rich basites. As an indication of the significance of this effect, the blueschist facies index minerals lawsonite and Na-amphibole are absent in the Shuksan Suite west of the Twin

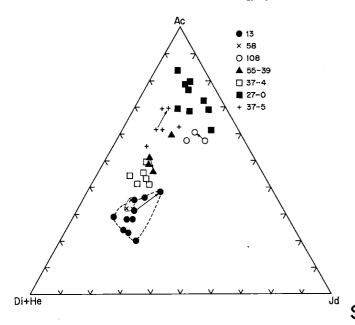

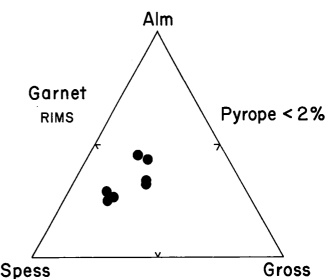

Figure 12. Composition of metamorphic pyroxenes from the Shuksan Suite. Pyroxene formulas were calculated from microprobe analyses on the basis of 6 oxygens and Fe^{3+} = Na - Al. Mineral assemblages in each specimen are as follows: 13, quartz + albite + epidote + crossite; 58, quartz + albite + epidote + chlorite; 108, quartz + crossite + epidote + magnetite; 55-39, quartz + crossite + epidote + chlorite; 37-4, quartz + magnetite + hematite; 27-0, quartz + epidote; 37-5, quartz + magnetite + hematite. Arrows point from core to rim on zoned grains (from Brown and O'Neil 1982).

Figure 13. Composition of garnets from the Shuksan Suite. Each spot represents an average of 3–6 rim analyses in one rock.

Sisters; however, the presence of aragonite and phase assemblages of pumpellyite + epidote + actinolite testify to P-T conditions of the blueschist facies. In this area the bulk rock composition is inappropriate for development of lawsonite or Na-amphibole.

Pressure-Temperature Conditions of Metamorphism

The pressure of metamorphism of the Shuksan Suite in the area from White Chuck Mountain to Gee Point (Figure 1) has been estimated to be approximately 7 kilobars based on the jadeite content of metamorphic pyroxene which coexists with quartz + albite and temperature of 330–400°C, as discussed below (Brown and O'Neil 1982). However, a recent study by Holland (1983) of activity-composition relations in jadeite-bearing pyroxenes indicates that $a_{jd}^{px} > X_{jd}^{px}$ for low jadeite concentrations. Thus the value of 7 kilobars, which was calculated based on $a_{jd}^{px} = x_{jd}^{px}$, is undoubtedly too low; Shuksan pyroxenes (with X_{jd} = .20) could possibly have crystallized at pressures up to 9 kb. The presence of aragonite in Shuksan rocks west of the Twin Sisters suggests pressure greater than 8 kilobars for this region.

Oxygen isotope fractionations between coexisting quartz and magnetite indicate temperatures ranging from 330° to 400°C, increasing from the White Chuck Mountain area northwesterly to the Skagit Valley (Figure 14; Brown and O'Neil 1982). This

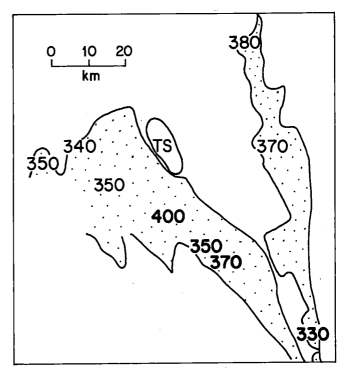

Figure 14. Temperatures inferred for the Shuksan Suite. Larger numbers are oxygen isotope temperatures; smaller numbers are temperatures based on epidote compositions in the reaction assemblage epidote + pumpellyite + chlorite + actinolite + quartz, as discussed in Brown and O'Neil (1982).

gradient is also marked by the disappearance of pumpellyite through the continuous reaction:

pumpellyite + chlorite + quartz = actinolite + clinozoisite + H_2O.

A temperature calibration of this equilibrium has been estimated based on thermodynamic calculations (Nakajima and others 1977) and determination of oxygen isotope temperatures for the reaction assemblage in the Shuksan Suite (Brown and O'Neil 1982). Utilizing this geothermometer, temperatures for the Shuksan Suite over a broad area are inferred from epidote composition in rocks containing the reaction assemblage (Figure 14). Rather little temperature variation is observed.

EMPLACEMENT HISTORY

The time of emplacement of Shuksan rocks into their present imbricated structural setting is bracketed by the facts that the faulting post-dates the Lower Cretaceous metamorphism (radiometric ages range from 105-164 m.y.) and predates intrusion of the Upper Cretaceous Spuzzum Batholith in southern British Columbia, dated as 75-100 m.y. (Richards and McTaggart 1976; Gabites 1983). The geometric relations of the imbricating faults are complex, as indicated in the introduction to this paper. Fault attitudes range from high to low angle. High angle faults are observed to cut low angle faults, to be cut by low angle faults, and to bend abruptly into low angle faults. Fault zones are characterized by the development of mylonite, typically 1 to 2 m thick, and showing minor new crystallization of quartz, chlorite, muscovite, stilpnomelane, and calcite. No high pressure phases have been recognized as newly crystallized in the fault zones. Sheared porphyroclasts in the mylonites define a lineation, which from preliminary measurement is, on the average, subhorizontal and north trending in both low and high angle fault segments (Figure 15). The lineations suggest that the high angle faults are of strike-slip nature. Augen asymmetry indicates a right-lateral sense of motion (following criteria outlined in Simpson and Schmid 1983). The movement sense on low-angle faults is dominantly, but not exclusively, lower plate north. Thus, based on mylonite fabric analysis, the Shuksan Suite appears to have been emplaced along a complex system of related thrust and strike-slip faults, all of which moved in a north-south direction parallel to the continental margin and length of the orogen. The sense of motion is such that the more westerly (outboard) rocks have been translated relatively to the north. This kinematic interpretation differs from that of Misch (1966) in which the Shuksan Suite is inferred to have been thrust to the west out of a root zone in the core of the Cascades.

TECTONIC EVOLUTION OF THE SHUKSAN SUITE

A tentative tectonic history of the Shuksan Suite, based on the foregoing observations and arguments, and subject to revision pending results of research in progress, is as follows: Protolith materials were deposited in a marginal basin along the North American coast in Middle to Late Jurassic time. Subduction

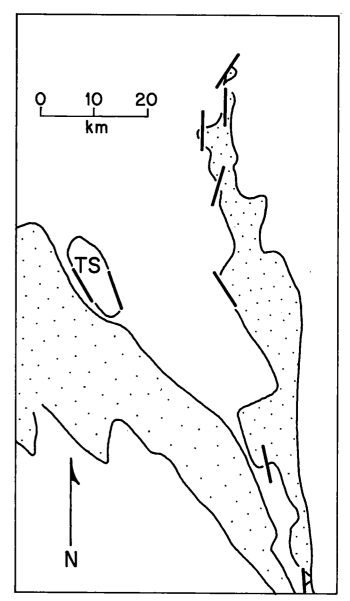

Figure 15. Lineations in mylonites of fault zones related to emplacement of the Shuksan Suite. Data from Jewett (1984), Silverberg (unpublished), and Brown (unpublished).

and resultant high pressure metamorphism began in the Late Jurassic and lasted through Early Cretaceous in a zone of high angle plate convergence. In the Upper Cretaceous, the blueschist assemblage was uplifted and imbricated with other rock units along faults with motions directed northward, parallel to the continental margin and representing transform plate interaction. The timing and directions of these inferred motions correlate with a change in direction of convergence of the Farallon and North American plates from high to low angle, deduced by Engebretson and others (1985). By this model, the blueschists are uplifted and emplaced in response to a change in plate motion.

ACKNOWLEDGMENTS

This work has been supported by grants from the National Science Foundation (EAR7917934) and Western Washington University. The microprobe studies upon which the phase compositions are based were carried out at the University of Washington and at Cambridge University, England. Over the past twelve years that this study has been underway, a great number of students at Western Washington University have made important contributions to the project, as field and laboratory assistants and as M.S. researchers. I am very much indebted to them. My Shuksan research has particularly benefited from discussions with Bernard Evans and Joe Vance. Dick Armstrong's collaborative work has been vital to unravelling chronologic problems, as has also been the recent help of Nick Walker. Reviews of this paper by Don Hyndman, Brian Patrick, Jim Talbot and Joe Vance have led to substantial improvements. The excellent pioneering studies of Peter Misch have formed an important basis for the present work.

REFERENCES CITED

Armstrong, R. L., 1980, Geochronology of the Shuksan Metamorphic Suite, North Cascades, Washington: Geological Society of America Abstracts with Programs, v. 12, p. 94.

Ashleman, J. C., 1979, The geology of the western part of the Kachess Lake quadrangle, Washington [M.S. thesis]: Seattle, University of Washington, 88 p.

Bechtel Report, 1979, Report of geologic investigations in 1978-79: Skagit Nuclear Power Project, for Puget Sound Power & Light Company, Seattle, Washington, v. 1 and 2.

Blackwell, D. L., 1983, Geology of the Park Butte-Loomis Mountain area, Washington (eastern margin of the Twin Sisters dunite) [M.S. thesis]: Bellingham, Western Washington University, 253 p.

Bonatti, E., Zerbi, M., Kay, R., and Rydell, H., 1976, Metalliferous deposits from the Appenine ophiolites: Mesozoic equivalents of modern deposits from oceanic spreading centers: Geological Society of America Bulletin, v. 87, p. 83-94.

Brown, E. H., 1974, Comparison of the mineralogy and phase relations of blueschists from the North Cascades, Washington, and greenschists from Otago, New Zealand: Geological Society of America Bulletin, v. 85, p. 334-344.

Brown, E. H., and Ghent, E. D., 1983, Mineralogy and phase relations in the blueschist facies of the Black Butte and Ball Rocks areas, northern California Coast Ranges: American Mineralogist, v. 68, p. 365-372.

Brown, E. H., and O'Neil, J. R., 1982, Oxygen isotope geothermometry and stability of lawsonite and pumpellyite in the Shuksan Suite, North Cascades, Washington: Contributions to Mineralogy and Petrology, v. 80, p. 240-244.

Brown, E. H., Wilson, D. L., Armstrong, R. L., Harakal, J. E., 1982, Petrologic, structural, and age relations of serpentinite, amphibolite, and blueschist in the Shuksan Suite of the Iron Mountain-Gee Point area, North Cascades, Washington: Geological Society of America Bulletin, v. 93, p. 1087-1098.

Carlson, W. D., and Rosenfeld, J. L., 1981, Optical determination of topotactic aragonite-calcite growth kinetics: Metamorphic implications: Journal of Geology, v. 89, p. 615-638.

Dungan, M. A., Vance, J. A., and Blanchard, D. P., 1983, Geochemistry of the Shuksan greenschists and blueschists, North Cascades, Washington: Variably fractionated and altered metabasalts of oceanic affinity: Contributions to Mineralogy and Petrology, v. 82, p. 131-146.

Engebretson, D. C., Gordon, R. G., and Cox, A., 1985, Relative motions between oceanic and continental plates in the Pacific Basin: Geological Society of America Special Paper 206.

Ernst, W. G., 1968, Amphiboles: New York, Springer-Verlag, 125 p.

Escher, A., and Watterson, J., 1974, Stretching fabrics, folds and crustal shortening: Tectonophysics, v. 22, p. 223-231.

Evans, B. W., and Misch, P., 1976, A quartz-aragonite-talc schist from the lower Skagit Valley, Washington, American Mineralogist, 61, p. 1005-1008.

Frasse, F. I., 1981, Geology and structure of the western and southern margins of the Twin Sisters Mountains, North Cascades, Washington [M.S. thesis]: Bellingham, Western Washington University, 87 p.

Gabites, J. E., 1983, Geology and geochronology east of Harrison Lake, British Columbia: Geological Association of Canada, Abstracts with Programs, v. 8, p. A25.

Haugerud, R. A., 1980, The Shuksan Metamorphic Suite and Shuksan Thrust, Mt. Watson area, North Cascades, Washington [M.S. thesis]: Bellingham, Western Washington University, 125 p.

Haugerud, R. A., Morrison, M. L., and Brown, E. H., 1981, Structural and metamorphic history of the Shuksan blueschist terrane in the Mount Watson and Gee Point areas, North Cascades, Washington: Geological Society of America Bulletin, v. 92, p. 374-383.

Holland, T.J.B., 1983, The experimental determination of activities in disordered and short-range ordered jadeitic pyroxenes: Contributions to Mineralogy and Petrology, v. 82, p. 214-220.

Jewett, P. D., 1984, The structure and petrology of the Slesse Peak area, Chilliwack Mountains, British Columbia, Canada [M.S. thesis]: Bellingham, Western Washington University, 164 p.

Johnson, S. Y., 1982, Stratigraphy, sedimentology and tectonic setting of the Eocene Chuckanut Formation, northwest Washington [Ph.D. dissertation]: Seattle, University of Washingotn, 191 p.

Jones, J. T., 1984, The geology and structure of the Canyon Creek-Church Mountain area, North Cascades, Washington [M.S. thesis]: Bellingham, Western Washington University, 125 p.

Mattinson, J. M., 1972, Ages of zircons from the Northern Cascade Mountains, Washington: Geological Society of America Bulletin, v. 83, p. 3769-3784.

Misch, P., 1959, Sodic amphiboles and metamorphic facies in Mountain Shuksan belt, northern Cascades, Washington [abs.]: Geological Society of America Bulletin, v. 70, p. 1736-1737.

—— 1963, New samples for age determinations from the Northern Cascades, *in* Kulp, J. L., senior investigator, and others, Investigations in isotopic geochemistry: Columbia University, Lamont Geological Observatory (U.S. Atomic Energy Commission NYO-7243), Rept. 8, p. 26-40, App. K., p. 1-4.

—— 1964, Age determinations on crystalline rocks of Northern Cascade Mountains, Washington, *in* Kulp, J. L., senior investigator, and others, Investigations in isotopic geochemistry: Columbia University, Lamont Geological Observatory (U.S. Atomic Energy Commission NYO-7243), Rept. 9, App. D, p. 1-15.

—— 1966, Tectonic evolution of the northern Cascades of Washington, *in* Gunning, H. C., ed., A symposium on the tectonic history and mineral deposits of the western Cordillera in British Columbia and neighboring parts of the United States: Canadian Institute of Mining and Metallurgy Special Volume 8, p. 101-148.

—— 1969, Paracrystalline microboudinage of zoned grains and other criteria for synkinematic growth of metamorphic minerals: American Journal of Science, v. 267, p. 43-63.

—— 1977, Bedrock geology of the North Cascades, *in* Brown, E. H., and Ellis, R. C., eds., Geological excursions in the Pacific Northwest: Guidebook, Bellingham, Western Washington University, p. 1-62.

Monger, J.W.H., 1966, Stratigraphy and structure of the type area of the Chilliwack Group, southwest British Columbia [Ph.D. dissertation]: Vancouver, University of British Columbia, 158 p.

Morrison, M. L., 1977, Structure and stratigraphy of the Shuksan Metamorphic Suite in the Gee Point-Finney Peak area, North Cascades [M.S. thesis]: Bellingham, Western Washington University, 69 p.

Nakajima, T., Banno, S., and Suzuki, T., 1977, Reactions leading to the disappearance of pumpellyite in low-grade metamorphic rocks of the Sanbagawa metamorphic belt in central Shikoku, Japan: Journal of Petrology, v. 18, p. 263–284.

Nitsch, K. H., 1980, Reaktion von Bariumfeldspat (Celsian) mit H_2O zu Cymrit unter metamorphen Bedingungen: Fortschritt Mineral, v. 58, p. 98–99.

Pearce, J. A., 1982, Trace element characteristics and lavas from destructive plate boundaries: *in* R. S. Thorpe, ed., Andesites: New York, John Wiley and Sons, p. 525–548.

Rady, P. M., 1981, Structure and petrology of the Groat Mountain area, North Cascades, Washington [M.S. thesis]: Bellingham, Western Washington University, 133 p.

Richards, T. A., and McTaggart, K. C., 1976, Granitic rocks of the southern Coast Plutonic Complex, and northern Cascades of British Columbia: Geological Society of America Bulletin, v. 87, p. 935–953.

Sevigny, J. H., 1983, Structure and petrology of the Tomyhoi Peak area, North Cascades, Washington [M.S. thesis]: Bellingham, Western Washington University, Bellingham, 203 p.

Shackleton, R. M., and Ries, A. C., 1984, The relation between regionally consistent stretching lineations and plate motions: Journal of Structural Geology, v. 6, p. 111–117.

Simpson, C., and Schmid, S. M., 1983, An evaluation of criteria to deduce the sense of movement in sheared rocks: Geological Society of America Bulletin, v. 95, p. 1281–1288.

Smith, G. O., 1903, The geology and physiography of central Washington, *in* Smith, G. O., and Willis, B., Contributions to the Geology of Washington: U.S. Geological Survey Professional Paper 19, 101 p.

Street-Martin, L. V., 1981, The chemical composition of the Shuksan metamorphic suite in the Gee Point-Finney Creek area, North Cascades, Washington [M.S. thesis]: Bellingham, Western Washington University, 76 p.

Vance, J. A., 1957, The geology of the Sauk River area in the northern Cascades of Washington [Ph.D. dissertation]: Seattle, University of Washington, 313 p.

——1985, Early Tertiary faulting in the North Cascades: Geological Society of America Abstracts with Programs, v. 17, no. 6, p. 415.

Vance, J. A., Dungan, M. A., Blanchard, D. P., and Rhodes, J. M., 1980, Tectonic setting and trace element geochemistry of Mesozoic ophiolitic rocks in western Washington: American Journal of Science, v. 280-A, p. 359–388.

MANUSCRIPT ACCEPTED BY THE SOCIETY JULY 29, 1985

Geological Society of America
Memoir 164
1986

Phase petrology of eclogitic rocks
in the Fairbanks district, Alaska

Edwin H. Brown
Department of Geology
Western Washington University
Bellingham, Washington 98225

Robert B. Forbes
Department of Geological Sciences
University of Washington
Seattle, Washington 98195

ABSTRACT

The eclogitic terrane near Fairbanks, Alaska, consists of interlayered basic, calc-magnesian, quartzose, and pelitic schists, providing an opportunity to evaluate mineral parageneses in a diverse suite of high pressure metamorphic rocks. The terrane is interpreted to have equilibrated at 600 ± 25°C and 15 ± 2 kbars based on: 1) the jadeite content of omphacite coexisting with quartz + albite and 2) the pelitic assemblage garnet + chloritoid + staurolite + kyanite + quartz. The Ellis and Green (1979) calibration of K_D-Fe/Mg of garnet/pyroxene also gives a 600°C temperature for basic schists where $X_{Ca}^{Ga} = .30$. However, the application of this geothermometer to calc-magnesian schists, where X_{Ca}^{Ga} ranges up to .47, gives temperatures that are much too high > (700°C). The K_D values show some scatter, but do not systematically vary with the jadeite content of pyroxene in the range from 8 to 45%, or with the grossularite content of garnet from 25 to 47%. Temperatures derived from biotite-garnet K_Ds with the calibration of Ferry and Spear (1978), are also approximately 600°C. At such temperatures, glaucophane in the Fairbanks eclogites exceeds the maximum stability limit defined by the experimental studies of Maresch (1977).

Comparison of the phase relations and inferred P-T of formation of the Fairbanks rocks with those from other high pressure terranes leads to the derivation of pressure-dependent equilibria that express the transition from amphibolites of the Sanbagawa belt to the substantially higher pressure kyanite eclogites of the Tauern Window. The Fairbanks eclogitic terrane is intermediate in this progression.

INTRODUCTION

The eclogites and related rocks discussed in this paper are exposed in a relatively small area (ca. 5 × 13 km) about 20 km north of Fairbanks, Alaska (Figure 1). As previously described by Swainbank and Forbes (1975), the eclogite-bearing terrane is surrounded by rocks of different lithology and metamorphic grade (lower P and T), and the terrane is probably a tectonic window or klippe. The terrane appears to be truncated on the northwest by the Chatanika Lineament, a possible splay of the Tintina Fault that extends hundreds of kilometers to the southeast through Alaska and the Canadian Yukon (Figure 1). Similar eclogitic rocks have been discovered recently northeast of the Chatanika, in a structural setting similar to that discussed in this paper (Laird and others in press), and several other blocks of exotic eclogitic rocks have been found along the cratonal margin in the Yukon near the Tintina Fault (Tempelman-Kluit, 1970; Erdmer and Helmstaedt 1983). This suite of eclogitic fragments is

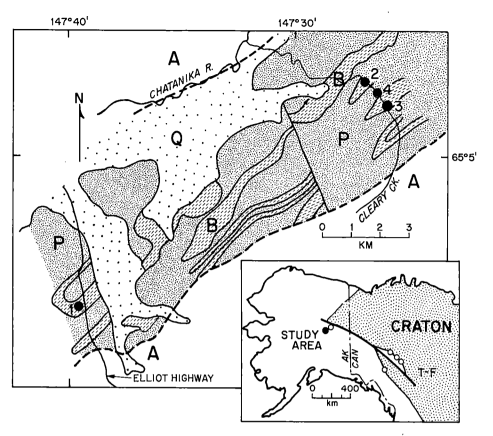

Figure 1. Locality map for specimens of this study. B = basic and calc-schists of the eclogitic block; P = pelitic and quartzose schists of the eclogitic block; A = regional amphibolite facies schists; Q = surficial deposits. Map simplified from Figure 2 of Swainbank and Forbes (1975). Inset map modified from Erdmer and Helmstaedt (1983) and Jones and others (1982). Open circles on inset map indicate other eclogite localities, TF = Tintina Fault.

interpreted by Tempelman-Kluit (1979) and Erdmer and Helmstaedt (1983) to record an event of tectonic accretion along the western margin of the North American craton.

The Fairbanks eclogite terrane contains a remarkable range of interlayered metasediments and metabasites providing an opportunity for the study of high pressure mineral parageneses in a compositionally diverse suite of rocks. The purpose of the present study is to establish the P-T conditions of metamorphism, to evaluate relevant metamorphic geothermometers, and to compare and contrast the phase associations with other eclogites and

TABLE 1. ABBREVIATIONS USED

Qtz	= quartz	Ab	= albite	Ep	= epidote
Ch	= Chlorite	Mu	= muscovite	Pg	= paragonite
Bi	= biotite	Px	= pyroxene	Ac	= acmite
Di	= diopsite	He	= hedenbergite	Jd	= jadeite
Om	= omphacite	CaA	= Ca-amphibole	Hb	= hornblende
Wi	= winchite	Tr	= tremolite	Act	= actinolite
Ba	= barroisite	Gl	= glaucophane	Alm	= almandine
Py	= pyrope	Gr	= grossularite	Ga	= garnet
Ct	= chloritoid	St	= staurolite	Ky	= kyanite
Cc	= calcite	Do	= dolomite	Sp	= sphene
Ru	= rutile	Mag	= magnetite	Hem	= hematite
Ilm	= ilmenite	Lith	= lithology	Loc	= locality

amphibolites in order to determine reaction relations and phase stability fields among eclogitic assemblages.

Swainbank and Forbes (1975) previously attempted to establish P-T parameters for the crystallization of the eclogite-bearing terrane based on wet chemically analyzed minerals and whole rocks. They estimated that the eclogites had crystallized at temperatures between 540°–620°C, at pressures exceeding 5 kb. At that time, the diagnostic pelitic and glaucophane-bearing metabasite assemblages reported in this paper had not been found and microprobe data were not available.

In this work, we have used specimens from the Swainbank and Forbes collection, and those recollected from four localities (Figure 1), adding an additional 150 specimens to our data base. The new sample suite contains several important mineral assemblages that were not previously recognized.

Exposures and the availability of fresh rock from natural outcrops are very poor in this region. Many of our samples come from a borrow pit along the Elliot Highway (locality 1, Figure 1). This excavation reveals the interlayered nature of various lithologies, including eclogite, amphibolite, impure marble, pelitic schist, calc-magnesian schist, and quartz-mica schists. Some of the freshest and most interesting samples come from dredge tailing piles,

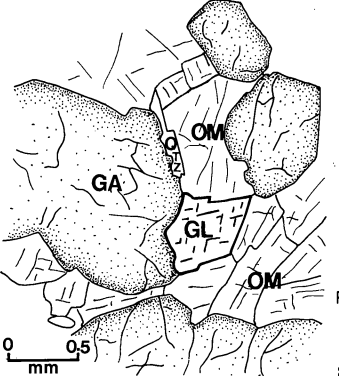

Figure 2. Textural relations of garnet, omphacite, glaucophane and quartz, suggesting an equilibrium assemblage. Specimen 2D. Abbreviations are explained in Table 1.

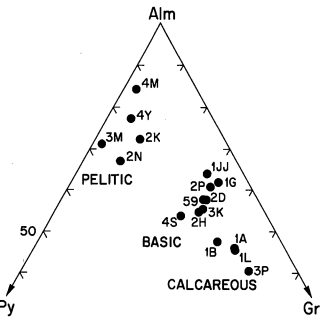

Figure 3. Rim compositions of garnets and host rock type.

just off the Steese Highway (localities 2, 3, 4, Figure 1), along Cleary Creek. In this creek, angular rock on the top of each pile was excavated by bucket line from bedrock under the alluvial deposits.

We have found no mineralogic evidence to indicate a significant variation of P and T across the sample area, although the estimated error for P-T determinations is about ±2–3 kbars and ±25°C.

Abbreviations used in this report are given in Table 1.

ANALYTICAL PROCEDURES

All mineral analyses were performed with an automated ARL EMX electron probe microanalyzer in the Department of Geological Sciences at the University of Washington. Operating voltage was 15 kv. Sample current and spot-size were different for different minerals in order to optimize counting and resolution and to avoid loss of volatile elements. All elements were measured in one run. Analyses were calibrated with mineral standards similar to the unknowns, and by on-line application of the Bence-Albee correction.

MINERALOGY

Many petrographic details, including modes, are given in

Swainbank and Forbes (1975) for representative rocks from the terrane. Textures indicate to us that the rocks are bimetamorphic. Initial eclogitic metamorphic assemblages are variably overprinted by epidote-amphibolite facies assemblages characterized by hornblende, albite, epidote, biotite, and chlorite. However, many rocks show little or no evidence of the amphibolite facies overprint, and texturally appear to represent equilibrium assemblages (Figure 2). Such assemblages were selected for microprobe study. Grain size is typically in the .2–.5 mm range although some coarse-grained schists contain garnets several millimeters in diameter. Mineral assemblages for rocks with analyzed specimens are listed in Table 2. All are considered to represent equilibrium associations. Representative mineral compositions are listed in Table 3.

Garnet rims range continuously in composition from almandine-rich to those with equal proportions of almandine and grossularite (Figure 3), depending on the host-rock type. Mn content is low on the rims ($X_{Mn}^{Ga} < .05$). Zoning is relatively weak, but of the normal type, with Mg increasing and Mn decreasing from core to rim (Figure 4). Poikiloblastic texture is common and in some rocks a snowball structure indicates syntectonic garnet growth.

Pyroxene ranges in composition from diopside 92-jadeite 8 to diopside 45-jadeite 45-acmite 10 (Figure 5). An enstatite-ferrosilite component was not detected. The more diopsidic pyroxenes are in relatively calcareous rocks and tend to be associated with grossular-rich garnets. The important problem of Fe^{2+} - Fe^{3+} determination was dealt with by assuming that Fe^{3+} = Na-Al, and Fe^{2+} = Fe_{total} - Fe^{3+}. Careful attention was given to the analyses for Na and Al to optimize the accuracy of this calculation. However, the Fe_{total} in omphacites is generally low

Brown and Forbes

TABLE 2. ASSEMBLAGES IN ROCKS CONTAINING ANALYZED MINERALS
(Abbreviations given in Table 1)

Spec.	1A	1B	1G	1L	1JJ	2D	2H	2I	2K	2N	2O	2P	3F	3H	3K	3M	3P	4M	4S	4Y	59	104	157	169
Qtz						X		X	X	X	X		X	X		X		X		X				X
Ab						X	X				X	X								X				
Ep			X			X	X					X					X							X
Chl							X		X	X	X							X		X			X	
Mu		X		X				X	X	X	X		X	X		X		X		X	X	X	X	X
Pq						X	X	X				X				X		X		X				
Bi														X					X			X	X	
Di	X	X															X							
Om			X	X	X	X						X		X					X	X				X
Tr-Act	X	X					X								X							X	X	
Ba																								
Gl				X										X					X		X			
Ga	X	X	X	X	X	X	X	X	X	X		X	X	X	X	X	X	X	X	X	X	X	X	X
Ct									X							X		X						
St																X								
Ky																X								
Cc	X	X	X	X	X	X							X								X	X		X
Do	X	?		X									X	X			X							
Sp	X		X	X	X								X	X		X	X		X				X	X
Ru						X	X	X	X			X			X			X	X	X	X			
Mag						X				X		X				X								
Hem									X	X						X								
Ilm							X		X			X						X		X				
Loc	1	1	1	1	1	2	2	2	2	2	2	2	3	3	3	3	3	4	4	4	1	1	1	1

TABLE 3. REPRESENTATIVE MINERAL COMPOSITIONS

Specimen	Diopside 1A	Omphacite 2D	Glaucophane 2D	Barroisite 4S	Tremolite-Actinolite 1A	Muscovite 3M	Paragonite 3M
SiO_2	54.3	56.9	59.3	51.3	55.0	49.6	46.9
Al_2O_3	1.94	10.3	10.6	8.54	2.75	29.3	39.0
TiO_2	0.01	0.06	0.00	0.25	0.01	0.22	0.08
Fe_2O_3*	0.00	3.22	1.89	3.65	3.68		
FeO*	5.98	2.95	6.08	4.94	4.14	3.64	1.11
MnO	0.04	0.02	0.07	0.02	0.05	n.d.	n.d.
MgO	13.5	7.64	12.2	15.7	18.5	2.64	0.07
CaO	21.6	12.0	1.32	8.98	11.1	n.d.	n.d.
Na_2O	1.08	7.52	6.93	3.23	0.97	0.76	7.44
K_2O	n.d.	n.d.	n.d.	n.d.	n.d.	9.95	1.03
Total	98.5	100.6	98.8	96.6	96.2	96.2	95.6
Si	2.02	2.01	7.95	7.26	7.76	6.60	6.00
Al	0.085	0.429	1.74	1.43	0.457	4.59	5.89
Ti	0.00	0.00	0.00	0.027	0.001	0.022	0.007
Fe^{3+}	0.00	0.086	0.191	0.389	0.391	--	--
Fe^{2+}	0.186	0.088	0.681	0.585	0.489	0.405	0.118
Mn	0.001	0.001	0.008	0.003	0.007	--	--
Mg	0.748	0.403	2.44	3.31	3.89	0.523	0.014
Ca	0.860	0.455	0.190	1.36	1.68	--	--
Na	0.078	0.515	1.80	0.886	0.265	0.196	1.85
K	--	--	--	--	--	1.69	0.169

*Calculated from analyzed Fe_{total} using formula constraints in pyroxene and amphibole, see text. All Fe given as FeO in mica.
n.d. = not determined

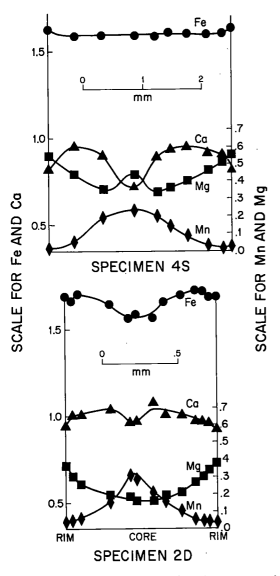

Figure 4. Garnet zoning profiles. Compositions shown are atomic proportions based on a formula with 12 oxygens.

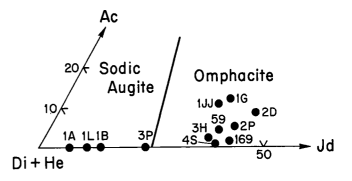

Figure 5. Compositions of pyroxenes. Fields of sodic augite and omphacite from Essene and Fyfe (1967).

(Fe_t ≤.15/formula unit) compared to Na and Al (≥.4 each) so the uncertainty of the calculation due to analytical error is significant. The problem is not as serious for diopside-rich compositions.

Amphiboles range from tremolite (Fe/(Fe + Mg) = .05 - .20), associated with diopside; to barroisite, associated with omphacite; to glaucophane, associated with omphacite plus albite (Figures 6 and 7). The compositional range from tremolite to barroisite appears to be continuous (Figure 7); however, a compositional gap between barroisite and glaucophane is suggested by the absence of compositions in this range, and the apparent stable coexistence of these two amphiboles in one specimen (Figure 8). Fe^{3+} and Fe^{2+} in the amphiboles were estimated with the constraints that the total cationic charge = 46, and that the sum of Mg + Fe^{2+} + Fe^{3+} + Ti + Mn + Al + Si = 13.

Muscovite in pelitic schists is phengite, containing approxi-

mately 30–35% of the celadonite component ($KR^{2+}R^{3+}Si_4O_{10}$ $(OH)_2$). The paragonite component in muscovite that coexists with paragonite is rather uniform at .09 to .11 mole fraction Na in seven specimens from across the area. Muscovites in some basic and calcareous rocks are enriched in Cr_2O_3, and are best termed fuchsite. Paragonite occurs only in pelitic schist in coexistence with muscovite. It has .06 to .09 mole fraction of the muscovite end member, and no celadonite substitution. Chlorites are all relatively magnesian (Mg/(Mg + Fe) = .55–.70). Epidote in basic schist contains .15–.20 mole fraction Fe; zoisite (as distinct from clinozoisite) occurs in calc-schists; and pelitic schists generally lack an epidote mineral. A carbonate phase, calcite or dolomite, or both, is present in almost all basic and calc-schists. Probe analyses of one calcite-dolomite pair indicate much less mutual solid-solution than would be expected at this metamorphic grade, suggesting a late, low-temperature recrystallization. Chloritoid occurs only in pelitic schists and is not abundant; staurolite, also only in pelitic schists, is limited to two of our samples; and kyanite has been found in only one specimen. Fe/(Fe + Mg) ratios of coexisting pelitic schist minerals are: garnet = staurolite > chloritoid. Plagioclase is a relatively rare mineral, occurring in a few basic and pelitic schists; the anorthite component is less than 5%.

Iron oxide and titanium minerals include magnetite, hematite, ilmenite, rutile, and sphene. Assemblages among the Fe-Ti oxides are listed in Table 2. Hematite coexisting with ilmenite in specimen 2K has 30 ± 2 mole% ilmenite solid solution. This composition and the Fe-Ti$_{oxide}$ assemblages observed here are very similar to those of the high-grade zones of the Sanbagawa belt, as reported by Itaya and Otsuki (1978).

ESTIMATION OF PRESSURE AND TEMPERATURE

Jadeite in Pyroxene

The maximum jadeite content of pyroxene is approximately 45% in specimen 2D, where the coexisting minerals include quartz and albite. Based on the experimental work of Newton and Smith (1967) and Holland (1980, 1983) for the equilibrium

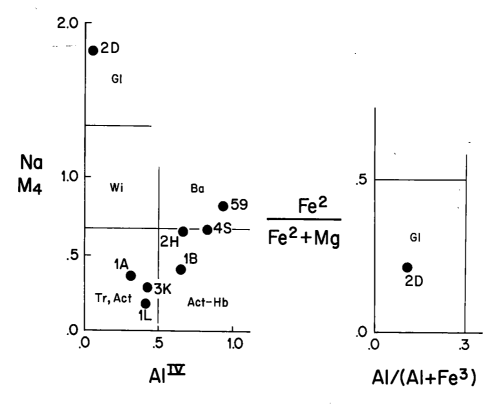

Figure 6. Compositions of amphiboles. Nomenclature from Leake (1978). Abbreviations are explained in Table 1.

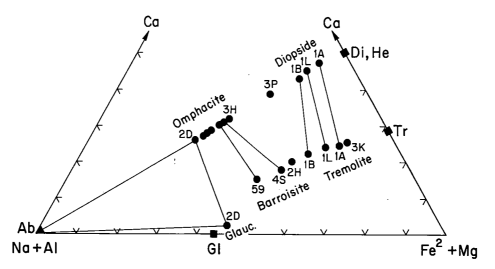

Figure 7. Compositions of coexisting amphiboles and pyroxenes. Note that acmite substitution in pyroxenes and tschermakite in amphiboles are not represented by the diagram.

jadeite + quartz = albite, for pure and impure jadeites, P-T stability conditions of the Fairbanks eclogitic rocks can be estimated to lie along line 1 on Figure 9, passing through 13.7 kbars at 600°C.

Pelitic Schist Assemblages

High pressure experimental studies of the stability relations among chloritoid, staurolite, kyanite, and garnet in the Fe end-member system have been conducted by Rao and Johannes (1979). They present results of two reversed equilibria: ct + ky = st + qtz + H_2O and st + qtz = alm + ky + H_2O. Intersection of these lines defines an invariant point (Figure 9). Pelitic schists of the present study contain these phases, and one specimen (3M) contains all the phases of the invariant point (ky, ct, st, alm, qtz).

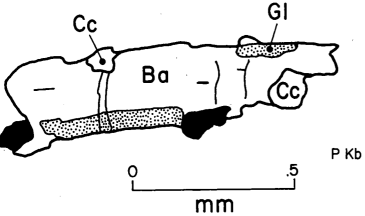

Figure 8. Composite amphibole grain consisting of barroisite and glaucophane. The boundary between amphiboles is sharp, being marked by a Becke line. Specimen 59.

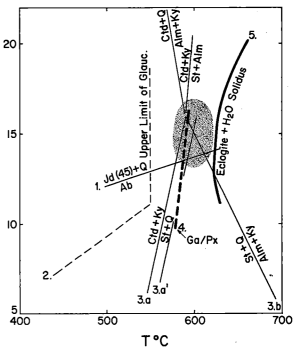

Figure 9. Estimated P-T conditions of formation of the Fairbanks eclogites, stippled area, and relevant experimental equilibria. Source of experimental data: (1) Holland (1983); (2) Maresch (1977); (3) Rao and Johannes (1979); (3á) calculated from Rao and Johannes for Mg/Fe ct >st; (4) Ellis and Green (1979) for $K_D = 17.6$ and $X_{Ca}^{Ga} = .30$; (5) Hill and Boetcher (1970). See text for discussion.

Addition of magnesium to the system shifts the equilibria in relation to the extent that Fe and Mg fractionate among the coexisting phases. For the present rocks, Fe and Mg are not fractionated between staurolite and garnet, thus the Rao and Johannes experimental line (3b, Figure 9) should be applicable. However, Mg is relatively concentrated in chloritoid over staurolite, so the equilibrium line relating these phases lies on the high temperature side of the experimental line (Figure 3a, Figure 9). Line 3á on Figure 9 shows this correction, as calculated from the entropy estimates and procedures of Holdaway (1978). The invariant point for mineral compositions of the Fairbanks eclogites is thus inferred to be at 15 kbars and 600°C, and we hypothesize that the presence of the mineral assemblage of the invariant point in our specimens indicates P-T conditions in the vicinity of these values.

It must be noted that controversy exists as to the accuracy of the experimental calibration of staurolite stability by Rao and Johannes (1979). For staurolite-bearing schists in the Shuswap terrane of British Columbia, Pigage and Greenwood (1982) found that the temperatures inferred for staurolite, based on the Rao and Johannes experiments, are higher than indicated by other minerals. Conversely, Yardley (1981) finds general agreement between temperatures based on staurolite equilibria and other geothermometers.

Pyroxene-Garnet K_D

We looked carefully at the pyroxene-garnet Fe/Mg K_D, to compare temperature estimates indicated by different calibrations of this geothermometer with the temperature indicated by the pelitic schist assemblage, and to examine possible covariance of K_D with compositions of pyroxene and garnet. Data used represent 5 to 10 analyses of rim compositions of adjacent pyroxene and garnet, which texturally appear to be in equilibrium (i.e. neither mineral looks resorbed or developed as a late replacement

mineral). The data are presented in Figures 10 to 12. Compositional ranges for each point shown on these diagrams represent the observed 1 sigma compositional variation of multiple analyses and are best interpreted as representing the departure from perfect equilibrium. The plots show that K_D is relatively constant over considerable compositional variation of the minerals, including Fe/Mg (Figure 10), Al (i.e. jadeite component) in pyroxene (Figure 11), and mole fraction of Ca in garnet (Figure 12). An inverse coupling of Al px and X_{Ca}^{Ga} exists (Figure 13), and this could obscure the affect of correlation of K_D with these compositional variations. However, there are wide departures from the general inverse correlation of Al px and X_{Ca}^{Ga}, and comparing Figures 12 and 13, we see that some samples with similar Al px and K_D have widely different X_{Ca}^{Ga} (e.g. samples 59 and 169).

Pyroxene-garnet composition relations in eclogitic rocks of the Sesia zone, Italy, were recently studied by Koons (in press) who found a covariance of $X_{Fe}^{Px}2$ (=Fe/Fe^2 + Mg) and X_{Ca}^{Px} (=Ca/Ca + Na), which in turn results in variation of K_D Fe/Mg Ga/Px with X_{Ca}^{Px} and X_{Fe}^{Px}. The pyroxenes exhibiting this behavior range from omphacite to jadeite. The Fairbanks pyroxenes, ranging from omphacite to diopside, do not exhibit these patterns.

The average K_D Fe/Mg for the Fairbanks rocks is 17.5. Applying the Raheim and Green (1974) calibration of K_D versus T, a temperature of 520°C at 15 kbars is indicated. This value is

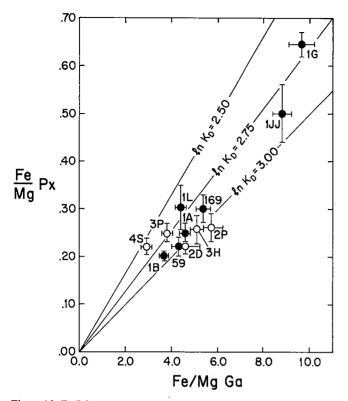

Figure 10. Fe/Mg pyroxene versus garnet. Error bars represent 1 sigma variation in composition for 5 to 10 analyses per sample. Solid symbols represent samples from locality 1.

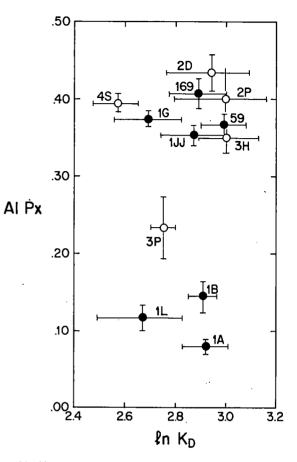

Figure 11. Al pyroxene versus ln K_D garnet/pyroxene Fe/Mg. Solid symbols represent samples from locality 1. Note that K_D does not vary systematically in the range from diopside to omphacite.

too low judged against the evidence from the pelitic schist assemblage. Looking at a plot of X_{Ga}^{Ca} versus ln K_D (Figure 12) with isotherms at 15 kbars calculated from the Ellis and Green (1979) equation, we see a very strong correlation between inferred temperature and X_{Ca}^{Ga}, with a range of 540–730°C for samples collected from the borrow pit locality. The 600°C isotherm defined by the pelitic schist passes approximately midway through the data points of Figure 12. The average K_D gives a 600°C temperature for the Ellis and Green equation where X_{Ca}^{Ga} = .31. If we apply the equation to all specimens assuming constant X_{Ca}^{Ga} = .31, then the calculated temperature range is reduced to 575–656°C. A reasonable conclusion is that temperatures calculated by the Ellis and Green (1979) equation are too high for grossular-rich garnets. For a common type of basic rock with X_{Ca}^{Ga} ≃ .30 the equation gives good results, but in view of the above difficulty, the form of the equation must be questioned.

Another problem presented by this geothermometer is the uncertainty of calculated K_Ds. Analytical error, mineral heterogeneity and error in estimation of Fe^{2+} and Fe^{3+} contribute to a rather wide range of K_D values for any given suite of rocks, as also noted in other studies (e.g. Holland 1979).

Biotite-Garnet K_D

The distribution coefficient between biotite and garnet in four samples and calculated temperatures based on three different

calibrations are given in Table 4. Results from application of the Ferry and Spear (1978) equation are in good agreement with the other geothermometers described above: T = 605 ± 40°C. The Goldman and Albee (1977) and Pigage and Greenwood (1982) equations give temperatures that are substantially too high. The Ferry and Spear equation has no correction factor for X_{Ca}^{Ga}, whereas the other equations do, and it would seem from this analysis that these are incorrect. The Pigage and Greenwood equation gives acceptable results for samples in their study in which the garnets contain only .04–.10 X_{Ca}^{Ga}. The pressure correc-

TABLE 4. TEMPERATURES CALCULATED FROM GARNET-BIOTITE K_D

Spec.	104	157	3K	4S
ln K_D	-1.77	-1.61	-1.90	-1.77
X_{Ca}^{Ga}	0.33	0.30	0.29	0.28
T°C at 15 kb				
F&S	600	660	560	600
G&A	724	763	667	721
P&G	816	870	785	779

Note: F&S = Ferry and Spear (1978); G&A = Goldman and Albee (1977); P&G = Pigage and Greenwood (1982).

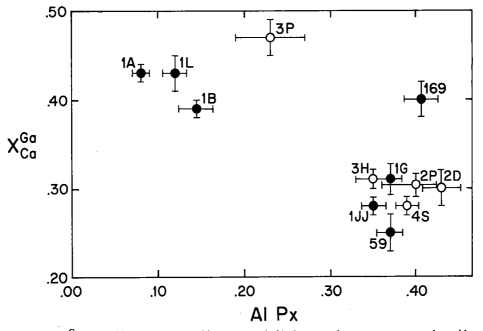

Figure 13. X^{Ga}_{Ca} versus Al_{px}. Note a general inverse correlation between these two parameters, but wide departures.

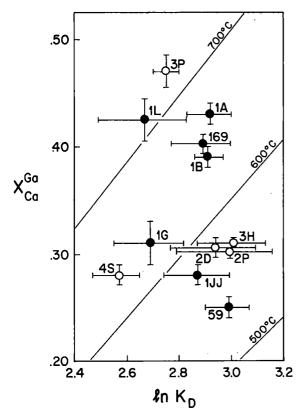

Figure 12. X^{Ga}_{Ca} versus $\ln K_D$ with isotherms calculated from Ellis and Green (1979) for 15 kb. Solid symbols represent samples from locality 1. Note the lack of systematic correlation of $\ln K_D$ with X^{Ga}_{Ca}, and the wide range of temperatures that would be inferred from the Ellis and Green calibration.

tions in the Ferry and Spear and Pigage and Greenwood equations are somewhat different, and could lead to temperature discrepancies. The Goldman and Albee equation has no pressure correction.

In a recent comparison of different calibrations of the biotite-garnet geothermometer as applied to pelitic rocks in the Northwest Territories, Canada, St-Onge (1984) has also concluded that the Ferry and Spear equation gives temperatures compatible with that of other indicators, whereas those derived from the Pigage and Greenwood equation are too high.

Muscovite-Paragonite

Geothermometry based on compositions of coexisting muscovite and paragonite is problematic owing to 1) uncertainty regarding the accuracy of the experimentally determined solvus (Eugster and others 1972) due to its derivation from synthesis runs; 2) the uncertainty of extrapolation of the experimentally determined 2 kilobar solvus to the 15 kilobar pressure inferred for the Fairbanks rocks; and 3) the known strong, inverse correlation of celadonite and paragonite substitution in muscovite (for example, Enami 1983).

Neglecting the celadonite affect, and applying the Eugster and others (1972) solvus with their proposed pressure correction (suggested to be valid for pressures only up to 10 kbars), a temperature of 590°C is determined for locality 3 and 550° for locality 2 of this study. In view of the difficulties of applying the experimental study, these values are surprising close to temperatures inferred by other means. However, application of this geothermometer to the data of Laird and others (in press) for muscovites in eclogites northeast of the Fairbanks locality, and that of Enami (1983) for Sanbagawa muscovites suggests temperatures significantly higher than those indicated by other geothermometers, a relationship also noted by Eugster and others (1972) for schists in Vermont.

Glaucophane Stability

Temperatures estimated for the Fairbanks eclogites are some 50°C higher than those for maximum glaucophane stability indicated by the widely cited experiments of Maresch (1977), as shown on Figure 9. This discrepancy could be due to incorrect interpretation of the crystallization temperatures of glaucophane-bearing rocks in the Fairbanks district. The localities of the glaucophane-bearing samples (1 and 2, Figure 1) are not the same as those of the staurolite samples (3 on Figure 1), for which a 600°C temperature is inferred. However, a similar temperature for all these localities is suggested by similar Fe/Mg K_D values for garnet-pyroxene (Fig. 12) and garnet-biotite (Table 4). Convergent temperatures are also indicated by similar mole fraction of Na in muscovite that coexists with paragonite: average X_{Na}^{Mu} = 0.09 in 3 samples at locality 2, and X_{Na}^{Mu} = 0.11 in 2 samples at locality 3. On the muscovite-paragonite solvus of Eugster and others (1972) this difference in X_{Na} reflects a ΔT of 40°C between the two localities. However, the muscovites from locality 2 have a slightly higher celadonite component, which according to the trend of inverse correlation between X_{Na} and celadonite shown by Enami (1983) suggests little or no temperature difference between localities 2 and 3.

Another possibility is that the glaucophane is late. Describing an eclogitic block along the Tintina lineament in the Yukon, which appears to be similar to the Fairbanks occurrence, Erdmer and Helmstaedt (1983) have described textures indicating that glaucophane formed after the main phase of eclogite crystallization. We do not find such textural relations in the Fairbanks rocks. For example, specimen 2D (Figure 2) shows apparent equilibrium relations among glaucophane, garnet, omphacite, and quartz (and also albite and epidote not in the field of the figure). The garnet-pyroxene K_D of grains in the field of Figure 2 indicate a temperature of 580°C at 15 kbars (with the Ellis and Green 1979, calibration).

Glaucophane has also been inferred to have formed above the stability limit observed by Maresch (1977) in eclogites of the Tauern Window that crystallized at a temperature estimated from several mineral equilibria to be 620 ± 30°C (Holland 1979).

The composition of the natural glaucophane used by Maresch (1977) is somewhat poorer in end-member glaucophane component than those of the Fairbanks or Tauern Window areas. Possibly small variations in composition strongly affect the amphibole stability.

CONCLUSIONS

Considering the mineral equilibria discussed above, a rea-

sonable estimate of the temperature and pressure conditions of metamorphism is T = 600 ± 25°C and P = 15 ± 2 kbar. Similar T and P were estimated by Laird and others (in press) for related eclogites northeast of the Fairbanks locality. A discrepancy between these inferred conditions and that of the experimentally determined stability of glaucophane is not resolved.

FLUID COMPOSITION

A recent calculation by Franz and Spear (1983) of phase relations of calcsilicate minerals at high pressures allows estimation of $X_{CO_2}^{fluid}$ in the Fairbanks eclogites. Their figure 4 shows P-T-X_{CO_2} relations, which for T = 600° and P = 15 kilobars, indicates that the observed Fairbanks assemblages in basic and calcareous rocks of di + do + tr + cc, di + tr + cc + qtz, and tr + cc + do + qtz, formed at X_{CO_2} = .05, .07 and .15 respectively. Thus, we infer that the metamorphic fluids were water-rich.

COMPARISON OF THE PHASE RELATIONS WITH THOSE OF OTHER AREAS

For ease of analysis, the multicomponent system of the Fairbanks eclogitic rocks is reduced in dimension to that of a ternary diagram with corners representing Al + Fe^{3+}, Mg, and Na. The rationale for this simplification is as follows. Projection is made from quartz, epidote, garnet and H_2O. Ti and Mn are considered to be insignificant. K is omitted, as muscovite is the only K-bearing phase in most eclogite terranes; thus, reaction relations involving biotite and muscovite for those terranes which contain both K-phases cannot be shown. Partitioning of Fe^{3+} and Al is ignored, but recognized as having the potential to shift hypothesized reaction lines. Oxygen is assumed to be present in stoichiometric amounts. Carbonate is considered to be an extra phase, the occurrence of which represents internally buffered $X_{CO_2}^{fluid}$. The resulting system portrayed by the diagram consists of the components: Si, Al + Fe^{3+}, Fe^{2+}, Mg, Ca, Na, and H. The inspiration for this ternary plot came from a quaternary diagram in Laird and Albee (1981: Fig. 16); their diagram is simply condensed here by projection from garnet. This plot is also very similar to a projection used by Schliestedt (1980). Variation in phase compositions through observed ranges affects the location of plotted points on the diagram to a certain degree, but does not change the essential tie line configurations.

Phase compatibilities in the Fairbanks rocks together with those of rocks from other high pressure terranes, and reactions relating different parageneses are shown on Figure 14. Pressure and temperature estimates of each terrane are taken from reports in the literature, as referenced in the caption for Figure 14. For the Tauern Window, we have extrapolated from the observed assemblage of kyanite + garnet + quartz + zoisite + omphacite + dolomite, to an assemblage stable at equal P and T but lower X_{CO_2}: kyanite + garnet + quartz + zoisite + omphacite + glaucophane, following reaction relations formulated by Holland (1979). Pressure estimates for the Sanbagawa and New Caledo-

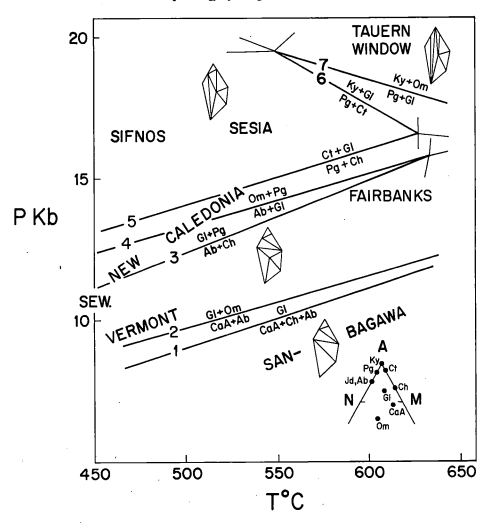

Figure 14. P-T grid relating parageneses of Fairbanks eclogites to those of other high pressure terranes. Reactions 1-7 are shown in full balanced form in Table 5. The ternary phase diagram has at its apices: A = Al + Fe^{3+}, N = Na, M = Mg. Phases on the diagram are projected from quartz, garnet, epidote, and H_2O. Sources of P-T data and parageneses for other areas: Sanbagawa, Japan–Banno and others (1978) and Enami (1983); Vermont-Laird and Albee (1981); New Caledonia-Brothers, and Yokoyama (1982); Sifnos-Greece, Schliestedt (1980); Tauern Window, Austria-Holland (1979); Sesia Zone, Italy-Desmons and O'Neil (1978), Koons (in press); SEW = Seward Peninsula, Alaska-Thurston (1985), Forbes and others (1984).

nia terranes are by no means certain. Eclogites of local occurrence in the Sanbagawa are interpreted by some workers (Brothers and Yokoyama 1982) as indicative of higher pressures than indicated on Figure 14; however, Takasu (1984) suggests that these eclogites are tectonically emplaced into the regional amphibolites. Mineralogic criteria in the amphibolites (e.g. Enami 1983) indicate the pressures of crystallization shown on Figure 14. The relatively high pressure inferred for New Caledonia rocks is based on uncertain interpretation of the paragenesis of early jadeite in partially retrograded eclogites (Yokoyama, personal communication, 1983). Assemblages reported for blueschists of the Seward Peninsula, Alaska (Thurston 1985) include ch + ab, ct + ab, pg + gl, and om + pg. If these assemblages all formed at approximately the same P and T, as is implied by Thurston, they suggest a

convergence of reaction lines 3, 4, and 5. The absence of jadeite and common occurrence of albite in this region indicate relatively low pressures.

The reactions are shown in complete form in Table 5. Slopes of the reaction lines are constrained by the inferred relative P and T of the various terranes and their respective assemblages, and by rough estimation of ΔS and ΔV following the procedures of Fyfe and others (1958) and Albee (1965). However, these two methods do not yield results that are in agreement everywhere. That is, the thermodynamic estimate of the slope for reaction 5 indicates a flat to slightly negative slope, whereas the occurrences of the assemblages and their estimated P and T require a positive slope, as shown.

Because the chemical system on which the grid is based is a

Brown and Forbes

TABLE 5. MINERAL COMPOSITIONS AND REACTION COEFFICIENTS
IN REACTIONS 1-7 OF FIGURE 14

		Qtz	Ep	Ga	H_2O	CaA	Ch	Ab	Gl	Om	Ct	Pg	Ky
Mineral Compositions	Si	1.00	3.00	3.00	00	7.12	6.32	3.00	7.95	2.00	2.00	6.01	1.00
	$Al+Fe^{3+}$	00	3.00	2.00	00	2.90	4.78	1.00	1.94	0.52	3.80	5.95	2.00
	Fe^{2+}	00	00	1.90	00	1.07	2.93	00	0.66	0.09	1.40	0.06	00
	Mg	00	00	0.30	00	2.68	6.99	00	2.45	0.40	0.90	0.02	00
	Ca	00	2.00	0.80	00	1.68	00	00	0.17	0.45	00	00	00
	Na	00	00	00	00	0.82	00	1.00	1.82	0.51	00	1.86	00
	H	00	1.00	00	2.00	2.00	8.00	00	2.00	00	4.00	4.00	00
Reaction Coefficients	1	1.18	-0.47	-0.18	-0.20	0.75	0.07	1.21	-1.00				
	2	1.05	-0.55	-0.21	0.22	1.05		1.34	-1.00	-0.74			
	3	1.95	0.39	-0.47	0.75		0.87	6.24	-2.41		-1.00		
	4	-3.01	1.29	-0.03	0.25			3.03	1.00	-6.05	-0.95		
	5	1.11	-0.29	0.93	-0.92		0.58		-1.02		-2.04	1.00	
	6	6.64	0.97	-2.22	-8.32				-1.00		3.44	0.98	-7.72
	7	-0.05	0.22	-0.01	-0.50				0.16	-1.0		0.11	-0.55

simplification of nature, the reaction lines shown are not strictly univariant. The most troublesome approximation is the lumping of Fe^{3+} and Al as one component. Partitioning of Fe^{3+} and Al between reactants and products could cause a P-T shift of the equilibria, particularly those involving Na-amphibole (e.g. Brown, 1974). However, Na-amphiboles in these high pressure rocks have low Fe^{3+}; they approach end-member glaucophane, and thus the partitioning effects are probably not significant.

The reactions of Figure 14 are for the most part not newly derived in this paper: (1) and (3) have long been hypothesized as glaucophane producing reactions (e.g. Ernst 1963); (2) was inferred to relate Sanbagawa amphibolites to Franciscan eclogites (Brown and Bradshaw 1979); (4) is recognized as an omphacite producing reaction in New Caledonia (Yokoyama, personal communication 1984); (7) has been derived to relate Tauern eclogites to lower grade assemblages (Holland 1979); and reactions 1, 2, and 3 are used by Brothers and Yokoyama (1982) to relate Sanbagawa and New Caledonia assemblages.

It must be emphasized that although the validity of most of the reactions of this grid is moderately well established through

several studies, arrangement of the reactions in Figure 14 is tentative and deserves further evaluation. The grid of Figure 14 provides a scheme for relating the relatively low pressure Sanbagawa assemblages dominated by albite, chlorite, and Ca-amphibole to the high pressure rocks of Sifnos, the Sesia zone, and the Tauern Window where paragonite, glaucophane, and omphacite prevail. In this scheme, the Fairbanks rocks are seen to comprise a relatively high temperature, low pressure eclogitic suite.

ACKNOWLEDGMENTS

In preparation of this report we have benefited considerably from discussion with S. Banno, B. W. Evans, and R. Swainbank and from reviews of the manuscript by B. W. Evans, S. Maruyama, J. Laird, and K. Yokoyama. E. Mathez assisted with the microprobe analyses, carried out at the University of Washington. Financial support was provided in part by National Science Foundation grants to Brown (EAR7919734) and Forbes (EAR8218471).

REFERENCES CITED

Albee, A. L., 1965, A petrogenetic grid for the Fe-Mg silicates of pelitic schists: American Journal of Science, v. 263, p. 512–536.

Banno, S., Higashino, T., Otsuki, M., Itaya, T., and Nakajima, T., 1978, Thermal Structure of the Sanbagawa metamorphic belt in central Shikoku: Journal of Physics of the Earth, v. 26, Suppl., 5345–5366.

Brothers, R. N., and Yokoyama, K., 1982, Comparison of the high pressure schist belts of New Caledonia and Sanbagawa, Japan: Contributions to Mineralogy and Petrology, no. 79, p. 219–229.

Brown, E. H., 1974, Comparison of the mineralogy and phase relations of blueschists from the North Cascades, Washington, and Greenschists from Otago, New Zealand: Geological Society of America Bulletin, v. 85, p. 333–344.

Brown, E. H., and Bradshaw, J. Y., 1979, Phase relations of pyroxene and amphibole in greenstone, blueschist and eclogite of the Franciscan Complex: Contributions to Mineralogy and Petrology, v. 71, p. 87–83.

Desmons, J., and O'Neil, J. R., 1978, Oxygen and hydrogen isotope compositions of eclogites and associated rocks from the Eastern Sesia Zone (Western Alps, Italy): Contributions to Mineralogy and Petrology, no. 67, p. 79–85.

Ellis, D. J., and Green, D. H., 1979, An experimental study of the effect of Ca upon garnet-clinopyroxene Fe-Mg exchange equilibria: Contributions to Mineralogy and Petrology, no. 71, p. 13–22.

Enami, M., 1983, Petrology of pelitic schists in the oligoclase-biotite zone of the Sanbagawa metamorphic terrain, Japan: Journal of Metamorphic Geology, v. 1, p. 141–161.

Erdmer, P., and Helmstaedt, H., 1983, Eclogite from central Yukon: A record of subduction at the western margin of ancient North America: Canadian Journal of Earth Sciences, v. 20, p. 1389–1408.

Ernst, W. G., 1963, Petrogenesis of glaucophane schists: Journal of Petrology, v. 4, p. 1–30.

Essene, E. J., and Fyfe, W. S., 1967, Omphacite in California metamorphic rocks: Contributions to Mineralogy and Petrology, no. 15, p. 1–23.

Eugster, H. P., Albee, A. L., Bence, A. E., and Thompson, J. B., Jr., 1972, The two-phase region and excess mixing properties of pargonite-muscovite crystalline solutions: Journal of Petrology, v. 13, p. 147–179.

Ferry, J. M., and Spear, F. S., 1978, Experimental calibration for the partitioning of Fe and Mg between biotite and garnet: Contributions to Mineralogy and Petrology, no. 66, p. 113–117.

Forbes, R. B., Evans, B. W., and Thurston, S. P., 1984, Regional progressive high-pressure metamorphism, Seward Peninsula, Alaska: Journal of Metamorphic Geology, v. 2, p. 43–54.

Franz, G., and Spear, F. S., 1983, High pressure metamorphism of siliceous dolomites from the central Tauern window, Austria: American Journal of Science, no. 283-A, p. 396–413.

Fyfe, W. S., Turner, F. J., and Verhoogen, J., 1958, Metamorphic reactions and metamorphic facies: Geological Society of America Memoir, no. 73, 259 p.

Goldman, D. S., and Albee, A. L., 1977, Correlation of Mg/Fe partitioning between garnet and biotite with O^{18}/O^{16} partitioning between quartz and magnetite: American Journal of Science, no. 277, p. 750–751.

Hill, R.E.T., and Boettcher, A. L., 1970, Water in the earth's mantle: Melting curves of basalt-water and basalt-water-carbon dioxide: Science, no. 167, p. 980–982.

Holdaway, M. J., 1978, Significance of chloritoid-bearing and staurolite-bearing rocks in the Picuris Range, New Mexico: Geological Society of America Bulletin, no. 89, p. 1404–1414.

Holland, T.J.B., 1979, High water activities in the generation of high pressure kyanite eclogites of the Tauern Window, Austria: Journal of Geology, v. 87, p. 1–27.

——1980, The reaction albite = jadeite + quartz determined experimentally in the range 600–1200°C: American Mineralogist, no. 65, 129–134.

——1983, The experimental determination of activities in disordered and short-range ordered jadeitic pyroxenes: Contributions to Mineralogy and Petrology, no. 82, p. 214–220.

Itaya, T., and Otsuki, M., 1978, Stability and paragenesis of Fe-Ti oxide minerals and sphene in the basic schists of the Sanbagawa metamorphic belt in central

Shikoku, Japan: Journal of the Japanese Association of Mineralogists, Petrologists and Economic Geologists, no. 73, p. 359–379.

Jones, D. L., Cox, A., Coney, P., and Beck, M., 1982, The growth of western North America: Scientific American, no. 247, p. 70–84.

Koons, P. O., Implications to garnet-clinopyroxene geothermometry of non-ideal solid solution in jadeitic pyroxenes, in press: Contributions to Mineralogy and Petrology.

Laird, J., and Albee, A. L., 1981, High-pressure metamorphism in mafic schist from northern Vermont: American Journal of Science, no. 281, p. 97–126.

Laird, J., Foster, H. L., and Weber, F. R., Amphibole eclogite in the Circle quadrangle, Yukon-Tanana Upland, Alaska, in press, in Conrad, W. L., and Elliot, R. L., eds., United States Geological Survey in Alaska: Accomplishments during 1981: U.S. Geological Survey Circular 868.

Leake, B. E., 1978, Nomenclature of amphiboles: Canadian Mineralogist, v. 16, p. 501–520.

Maresch, W. V., 1977, Experimental studies on glaucophane: An analysis of present knowledge: Tectonophysics, v. 43, p. 109–125.

Newton, R. C., and Smith, J. V., 1967, Investigations concerning the breakdown of albite at depth in the earth: Journal of Geology, v. 75, p. 268–286.

Pigage, L. C., and Greenwood, H. J., 1982, Internally consistent estimates of pressure and temperature: The staurolite problem: American Journal of Science, no. 282, p. 943–969.

Raheim, A., and Green, D. H., 1974, Experimental determination of the temperature and pressure of the Fe-Mg partition coefficient for coexisting garnet and clinopyroxene: Contributions to Mineralogy and Petrology, no. 48, p. 179–203.

Rao, B. B., and Johannes, W., 1979, Further data on the stability of staurolite + quartz and related assemblages: Neuen Jahrbuchs für Mineralogic Monatshefte, H-10, p. 437–447.

Schliestedt, M., 1980, Phasengleischgewichte in Hochdruckgesteinen von Sifnos, Griechenland. [Ph.D. thesis]: Technischen Universitat Carolo-Wilhlmina zo Braunschweig, West Germany, 143 p.

St-Onge, M. R., 1984, Geothermometry and geobarometry in pelitic rocks of north-central Wopmay Orogen (early Proterozoic), Northwest Territories, Canada: Geological Society of America Bulletin, v. 95, p. 196–208.

Swainbank, R. C., and Forbes, R. B., 1975, Petrology of eclogitic rocks from the Fairbanks district, Alaska, in R. B. Forbes, ed., Geological Society of America Special Paper 151, p. 77–123.

Takasu, A., 1984, Prograde and retrograde eclogites in the Sanbagawa metamorphic belt, Besshi District, Japan: Journal of Petrology, v. 25, p. 619–643.

Tempelman-Kluit, D., 1970, An occurrence of eclogite near Tintina Trench, Yukon: Geological Survey of Canada, Paper 70-1B, p. 19–22.

——1979, Transported cataclasite, ophiolite and granodiorite in Yukon: Evidence of arc-continent collision: Geological Survey of Canada, Paper 79-14.

Thurston, S. P., 1985, Structure, petrology and metamorphic history of the Nome Group blueschist terrane, Salmon Lake area, Seward Peninsula, Alaska: Geological Society of America Bulletin, v. 96, p. 600–617.

Yardley, B.W.D., 1981, A note on the composition and stability of Fe-staurolite: Neues Jahrbook Mineralogie Monatsheft, p. 127–132.

MANUSCRIPT ACCEPTED BY THE SOCIETY JULY 29, 1985

Geological Society of America
Memoir 164
1986

Field relations and metamorphism
of the Raspberry Schist, Kodiak Islands, Alaska

Sarah M. Roeske
Earth Sciences
University of California at Santa Cruz
Santa Cruz, California 95064

ABSTRACT

The glaucophane-bearing Raspberry Schist occurs as discontinuous, fault-bounded slivers along the northwest side of the Kodiak Islands in southern Alaska. The lithologies, predominantly metabasites and quartzites, experienced minor pre-metamorphic stratal disruption, extensive syntectonic deformation, and pervasive post-metamorphic disruption by faults. A major change in metamorphic assemblages and degree of recrystallization occurs across a fault zone, dividing the Raspberry Schist into units Js1 to the southeast and Js2 to the northwest. Js1 is characterized by relict igneous textures and phases and contains assemblages compatible with prehnite-pumpellyite, pumpellyite-actinolite, lawsonite-albite-chlorite, and lower greenschist transitional to blueschist facies. Rocks composing Js2 have been completely, dynamically recrystallized and contain assemblages described as transitional blueschist facies or high-temperature blueschist facies.

Rb-Sr and K-Ar dates from the Raspberry Schist indicate that it was metamorphosed in Early Jurassic time. Lower Jurassic calc-alkaline mafic volcanics and associated plutons lie immediately northwest of the schist, juxtaposed with the schist by a wide brittle shear zone. Another calc-alkaline suite composed of Lower to Middle Jurassic plutonic and volcanic rocks lies 100 km to the northwest on the Alaska Peninsula. The metamorphism of the Raspberry Schist may be related to one of these calc-alkaline suites, but more evidence of the type and timing of fault movements in the region is needed to prove this conclusion.

INTRODUCTION

The Raspberry Schist, previously informally named the Kodiak Island Schist terrane (Carden and Forbes, 1976), occurs as discontinuous, fault-bounded slivers along the northwest side of the Kodiak Islands, an archipelago including Kodiak, Afognak, Shuyak, and numerous small islands (Figures 1 and 2). Although blue amphibole-bearing rocks were reported on the southwest tip of the Kenai Peninsula more than sixty years ago (Figure 1) (Martin and others, 1915), none were reported on the Kodiak Islands until 1976 (Carden and Forbes, 1976). The similarity in age, metamorphism, and tectonic position of these two blueschist-bearing belts led to their correlation, extending the length of the blueschist belt to over 200 hundred km (Carden and Forbes, 1976; Carden and others, 1977). Since 1976, the age and tectonic

significance of these blueschist-bearing rocks has been debated, but the lack of a detailed study has provided few constraints. This paper summarizes field relations, lithology, and petrologic observations of the Raspberry Schist, presents preliminary microprobe analsyses and age data, and discusses regional correlations. Forthcoming papers by the author will present more detailed structural and petrologic data.

Regional Geology

The Raspberry Schist lies along the northwest side of one of the major tectonic elements of southern Alaska, the Border Ranges Fault. According to MacKevett and Plafker (1974) this

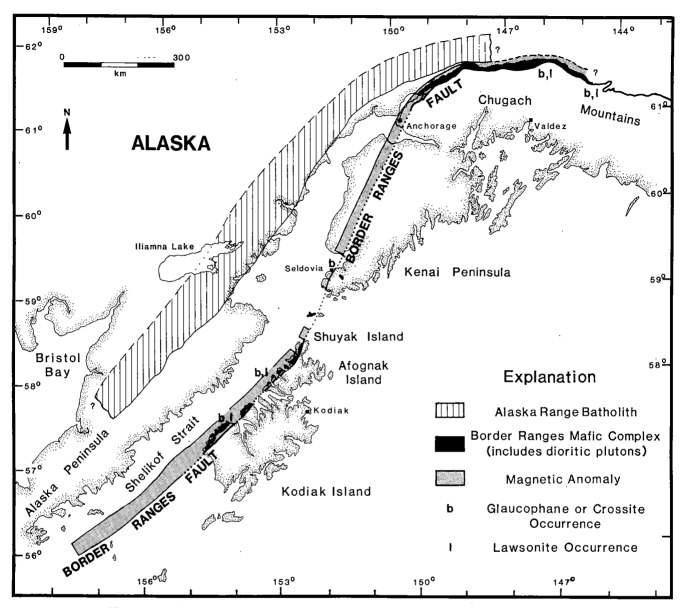

Figure 1. Location map of geologic and geographic features in southern Alaska, adapted from Fisher (1981), and Burns (1985).

fault, or fault zone, represents the site of a plate boundary during the Cretaceous. Southeast of the Border Ranges Fault lies an accretionary complex ranging from Cretaceous to Quaternary in age (Cowan and Boss, 1978; Moore and Connelly, 1979). Northwest of the fault on the Alaska Peninsula, rocks record three pulses of plutonism and associated calc-alkaline volcanism from Early Jurassic to middle Tertiary. Concordant K-Ar dates on the earliest phase of plutons range from 176–154 Ma (Reed and Lanphere, 1973). These rocks are referred to as the Alaska Range batholith and the region is interpreted as being the site of an Early to Middle Jurassic island arc (Reed and Lanphere, 1973; Reed and others, 1983) (Figure 1). Reed and others (1983) interpreted major element trends from a transect across the plutonic

belt as indicating that the arc formed over a south-dipping subduction zone.

On the Kodiak Islands the Border Ranges Fault separates Upper Triassic and Lower Jurassic crystalline rocks and volcaniclastics from a Cretaceous melange complex, the Uyak Complex (Figure 2) (Connelly, 1978; Connelly and Moore, 1979). Volcanic and volcaniclastic rocks, adjacent and probably related lithologies, comprise the two members of the Shuyak Formation. Late Triassic pelecypods in the volcaniclastic sediments constrain the age of the adjacent pillow basalts to Upper Triassic (Connelly, 1978). Locally, the pillow basalts are strongly deformed but generally not metamorphosed above prehnite-pumpellyite to lower greenschist facies (Hill, 1979; Roeske, 1984). The Afognak

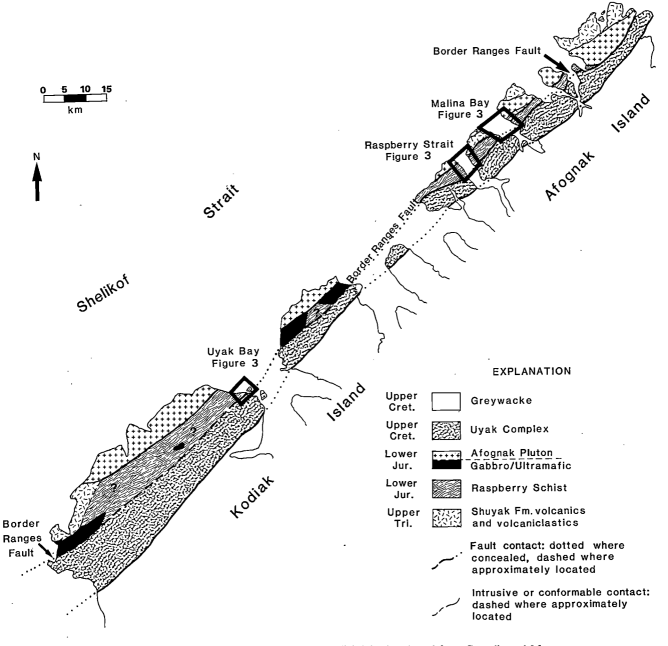

Figure 2. Geology of the northwest side of the Kodiak Islands, adapted from Connelly and Moore (1979).

pluton, a calc-alkaline dioritic to tonalitic body (Hill, 1979), intrudes the volcanic member of the Shuyak Formation, upgrading the volcanics to upper amphibolite facies immediately adjacent to the contact. The contact aureole extends up to 500 m into the volcanic rock but is not continuous, being broken by many late faults. K-Ar dates on the Afognak pluton and correlative diorites span ages from 184 to 192 Ma (Carden and others, 1977; Cowan and Boss, 1978; see Table 1).

K-Ar dates from the Raspberry and Seldovia schists, with the exception of dates obtained from crossite, are statistically identical to K-Ar dates from the Afognak pluton (Table 1). Based

on this coincidence of K-Ar dates, previous studies suggested that the pluton intruded the blueschists and reset the K-Ar age. This interpretation led some authors to conclude that the blueschist-bearing rocks formed during a metamorphic event prior to the intrusion of the pluton (Plafker and others, 1977; Pavlis, 1983; Carlson and Pavlis, 1984). The fact that the dates from the Seldovia blueschists are identical to those on the Kodiak Islands, yet the Seldovia Schist lies 15 km from any exposed plutonic rock, led other authors to interpret the K-Ar dates of the schist as reflecting the time of cooling to Ar retention of the blueschist facies assemblage, hence uplift or removal from a subduction

TABLE 1. K-Ar AGES

Rock Type	Mineral Dated	Age ± σ (m.y.)
Raspberry-Seldovia Schist*		
Seldovia		
Quartz-mica schist	Muscovite	190.4±5.7
Amphibole-mica schist	Muscovite	192.2±5.8
Greenschist	Actinolite	191.0±11.0
Greenschist	White mica	188.0±10.0
Blueschist	White mica	189.0±5.7
Amphibole-mica schist	Amphibole	184.4±5.5
Amphibole-mica schist	Amphibole	184.2±5.5
Greenschist	Chlorite	181.0±8.3
Blueschist	Crossite	162.9±4.9
Blueschist	Crossite	154.0±4.8
Port Graham		
Quartz-mica schist	Muscovite	192.2±5.8
Kodiak Islands		
Quartz-muscovite schist	Muscovite	192.1±5.8
White-mica-crossite schist	White mica	187.6±5.6
White-mica-crossite schist	Crossite	170.6±5.1
Blueschist	Crossite	161.4±19.4
Afognak Pluton*		
Dioritic migmatite	Hornblende	184.9±5.5
Dioritic migmatite	Hornblende	183.7±5.5
Hornblende diorite	Hornblende	188.4±5.7
Hornblende diorite		192.7±5.8
Chugach Mountains		
Valdez Quadrangle schist		
Blueschist	Crossite	175±5
Muscovite schist	Muscovite	154±5
Blueschist	Crossite	152±5
Lawsonite blueschist	Crossite	166±5
Blueschist**	Crossite	113±5
Blueschist**	Crossite	138±4
Actinolite-muscovite schist**	Actinolite	123±6
Valdez quadrangle mafic pluton		
Quartz gabbro	Clinopyroxene	185±19

*From Carden et al., 1977.
**Strongly sheared.
 From Winkler et al., 1981.

zone (Carden and others, 1977). Hill and Morris (1977) and Hill (1979) interpreted the blueschists as intruded by and coeval with the Afognak pluton. This led to the interesting interpretation that the Afognak pluton was an example of near-trench plutonism, possibly an early precursor to the main plutonic event of the Alaska Peninsula.

LITHOLOGIES

Mapping at a scale of 1:25,000 was conducted over two field seasons. The fjords along the northwest side of the Kodiak Islands cut approximately perpendicular to the trends of the rock units, providing excellent cross-sectional views. The exposure further inland is generally very poor and access is difficult, limiting the researchers to shoreline work.

The widest cross-section of the Raspberry Schist is exposed in Raspberry Strait. The schist is characterized here by an abrupt change in the degree of recrystallization across the trend of the unit. A 30 m wide fault zone separates partially recrystallized rocks, Js1 (Figures 3 and 4), from completely recrystallized schists, Js2, to the northwest. Unit Js1 is similar to unit Js2 in

both lithologic compositions and their relative proportions. The lithologies of the Raspberry Schist occur approximately in the following proportions: 50 percent metabasites, 30 percent quartzites, 10 percent pelites, and minor amounts of graphitic schist, ultramafics, and serpentinites. Because the lithologies of Js1 are less recrystallized than those of Js2 the protoliths are more readily interpretable. Consequently the following descriptions are from Js1 unless noted otherwise.

Metabasites in the Raspberry Schist frequently show well-preserved volcanic textures, indicating that the rocks were, in order of relative abundance, pillows, hyaloclastites, pillow breccias, massive flows, and felsic dikes. The hyaloclastites (aquagene tuffs) are generally fine-grained although locally interbedded with coarser volcanic breccias. Argillaceous layers interbedded with the hyaloclastites contain rounded fragments of igneous clinopyroxene and plagioclase. Preliminary microprobe analyses of the clinopyroxenes identifies them as titanaugites. The clinopyroxene compositions fall in the alkalic field on discriminatory plots, as opposed to the fields defined by clinopyroxenes from tholeiitic and calc-alkaline basalts (Letterier and others, 1982). Two small outcrops of metamorphosed ultramafics were found, both surrounded by sheared serpentinite. The relict mineral assemblages of the ultramafics are clinopyroxene and orthopyroxene, indicating the ultramafics were websterites. The adjacent metabasites do not show any blackwall assemblages or metasomatism, suggesting that their juxtaposition is postmetamorphic (Brown and others, 1982).

Calcareous rocks are seen only as irregular, discontinuous layers interbedded with metavolcanic and metavolcaniclastic lithologies. This absence of a limestone protolith distinguishes the Raspberry Schist on the Kodiak Islands from correlative blue and greenschists, the Seldovia Schist, on the Kenai Peninsula (Figure 1). The latter contains a block of pure limestone over 200 m long interlayered with metabasites.

Quartzites in the Raspberry Schist are typically very pure, with 2–10 cm thick quartz layers separated by 1–2 mm laminae of sheet silicates. The rhythmic layering and non-felsic nature of these sediments indicate that they are metacherts. The metacherts are interbedded with argillaceous, metavolcanic, and graphitic sediments on the scale of tens of centimeters. Two sites within Js1 contain coarse-grained sediments ranging from sandstone to pebble conglomerate in size. In Raspberry Strait a pebble conglomerate consists solely of quartz clasts with minor laminae of sheet silicates separating the clasts. This conglomerate is interbedded with metacherts and in fault contact with metabasites. In Uyak Bay gravel-sized metaconglomerate and interbedded coarse metagreywacke contain clasts of quartz and plagioclase. Sedimentary structures in the sandstone-size metasediment include graded bedding and ripples. These coarse sediments are in gradational contact with metacherts below and hyaloclastites above.

Environment of Deposition

The absence of relict potassium feldspar, plutonic clasts, and

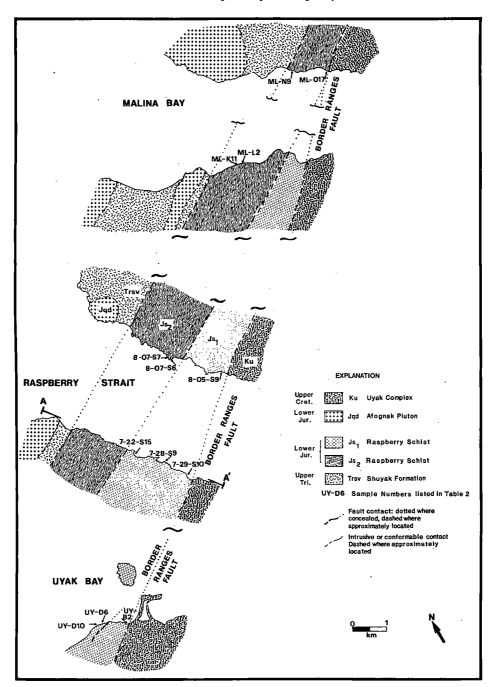

Figure 3. Metamorphic assemblages and geology of Uyak Bay, Raspberry Strait, and Malina Bay. The width of Malina Bay has been shortened along a line of N43W. See Figure 2 for location of the three sites. Note that the distance between Uyak Bay and Raspberry Strait has been considerably shortened.

metamorphic clasts in both the fine and coarse grained metasediments and the abundance of pelagic and hemipelagic sediments indicates a depositional environment removed from a continental source area. Clasts and mineral fragments in the coarser sediments and volcanic breccias could all be derived from protoliths within the schist. The abundant hyaloclastites, volcanic and volcaniclastic breccias, and minor coarse sediments with detrital clinopyroxene from an alkalic source suggest proximity to a top-

ographic high that included alkalic volcanics and cherts. If one assumes that the Raspberry Schist was metamorphosed in a convergent margin environment then it is quite possible that the narrow belt of rocks exposed today formed in a wide range of depositional environments and were tectonically juxtaposed.

Stratal Disruption

Each of the lithologies described above, with the exception

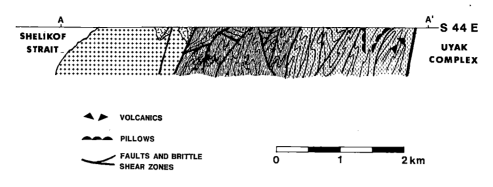

Figure 4. Schematic cross-section of Raspberry Strait. See Figure 3 for location of the cross-section line and key.

of the ultramafics, are seen in depositional contact with the other lithologies somewhere in the Raspberry Schist. The only evidence for tectonic mixing prior to or during metamorphism is the presence of rare lensoidal blocks of metabasite and metachert extended parallel to the foliation in an argillite matrix (Figure 5a). These blocks do not differ in metamorphic facies from the surrounding matrix. Any cataclastic fabric associated with their formation would have been obliterated in the extensive recrystallization that accompanied the relatively high P/T metamorphism. Pillow breccias and volcanic breccias have the appearance of a tectonic melange but the clasts are rounded and uniform in composition, indicating they formed as autoclastites (Lajoie, 1980). The contacts between the interlayered volcaniclastics, argillites, and cherts are irregular and wispy and have a swirled appearance. Metacherts interbedded with argillites often show minor boudinage (Figiure 5a) but lithologic layering is typically well-preserved and continuous on mesoscopic scale (Figure 5b).

In summary, the primary fabric is best described as predominantly coherent with minor zones of disrupted or obscure stratigraphy. Except for the lack of abundant argillite, the combination of mesoscopic fabrics observed in the schist and the variety of lithologies are similar to that described by Cowan (1985) as a Type II melange.

STRUCTURAL FABRICS

The first clearly tectonic deformation is seen in the formation of tight to isoclinal folds in both Js1 and Js2. The first generation of folds (F_1) ductilely fold lithologic layering but appear not to have folded the foliation. The second ductile fold generation, (F_2), transposed most of the earlier fabric into parallelism with F_2 axial planes, except in F_1 hinge regions. This second ductile folding produced most of the macroscopic folds, seen as folds 20 m or more in wavelength in the sea cliff exposures. These folds and their mesoscopic counterparts (Figure 5b) have axial planes that are typically near vertical to slightly overturned to the southeast or northwest (Figure 4). Axes of F_1 folds in both Js1 and Js2 plunge shallowly to moderately in a northeast-southwest trending girdle.

There is no direct evidence for two early ductile events in Js1 rocks. No refolded isoclinal folds are seen. The first fold event isoclinally folds bedding (Figure 5b) and the foliation appears to have formed axial planar to these folds.

Both early ductile events are overprinted by a locally strong crenulation or kinking event (Figure 5b). The kinks, F_3, often occur as conjugate pairs with wavelengths on the order of tens of cms and amplitudes of cms. The axes and axial planes of F_3 folds are oriented similarly to F_1 folds but show considerably more scatter. Little or no recrystallization is associated with this deformation.

Post-metamorphic Faults

Although the dominant structural grain (northeast) is fairly uniform throughout the Raspberry Schist the individual structures and lithologic units are typically discontinuous, broken up by a system of pervasive anastomosing faults (Figure 3). Most of the faults occur as discrete, steep to vertical planes with minor gouge development (Figure 5c). Relative displacements on faults with megascopic displacement, estimated from offset dikes, veins, and beds, and from slickenside striations on fault planes show predominantly dip-slip to oblique-slip movement.

In addition to the discrete fault planes, other faults are characterized by anastomosing brittle shear zones, ranging in width from meters to 100s of meters. These brittle shear zones form the contacts between Js1 and Js2 and between the Raspberry Schist and adjacent units both to the southeast and northwest. The intense cataclasis in the fault zones locally destroys the previous fabric, leaving the coherent schist as rounded or lenticular blocks floating in a scaly or brittly sheared matrix (Figure 5d). This fabric can be described as a melange and fits very well the description of a Type IV melange as outlined by Cowan (1985). The fabric formed in a brittle fault zone after the syntectonic deformation occurred and should not be confused with similar melange fabrics that may form during initial tectonic mixing in an accretionary complex (Cowan, 1985).

Determining the relationship between the kinks, discrete fault planes, and brittle shear zones would require more detailed

a.

b.

c.

d.

Figure 5. Photographs of structural features in the Raspberry Schist. a) The metachert block in the lower left is enclosed by a highly deformed argillite matrix that contains discontinuous beds and boudins of metachert; fault plane is marked by an arrow. Hammer for scale. The outcrop occurs within the brittle shear zone associated with the Border Ranges Fault in Uyak Bay. b) Metacherts in Js1. Axial planes of F_2 and F_3 folds designated with arrows. Notebook is 20 cm long. c) Discrete fault plane cutting metachert and argillite in Uyak Bay. d) Brittle shear zone in Uyak Bay. Cataclastic fabric developed in metabasite has isolated blocks and lenses of relatively undeformed metabasite in brittly sheared matrix. Hammer head in lower left for scale.

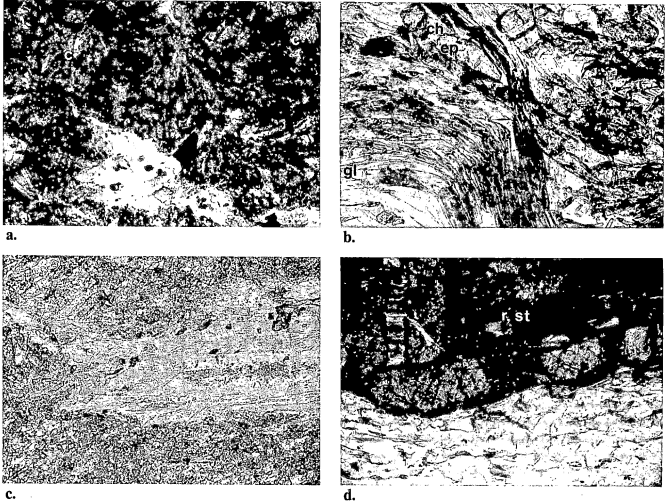

Figure 6. Photomicrographs of the Raspberry Schist. Field of view is 0.8 mm for all pictures. a) Js1 metabasite, sample #7-22-S15 (Table 2, Figure 3), with radiating laths of igneous plagioclase and relict clinopyroxene (above and to the right of letter c). Amygdule contains chlorite, epidote, and pumpellyite. b) Js2 metabasite, sample #M1-01 (Figure 7), containing glaucophane (gl), epidote (ep), and chlor (ch) in field of view. Crenulation of fabric is probably F_3 generation. c) Zoned amphibole with actinolite core and crossite rim. Groundmass is chlorite, epidote, albite, and amphibole. d) Ferruginous metachert, sample #Uy-D10 (Table 2, Figure 3), from epidote zone in Uyak Bay. Magnesio-riebeckite (r) and stilpnomelane (st) intergrown in extension fractures of epidote. The phases in the lower part of the thin section are quartz, magnesio-riebeckite, stilpnomelane, and phengite.

field work. Discrete fault planes are seen within, and are locally cut by, some brittle shear zones, indicating that at least some of the discrete faults pre-date the brittle shear zones. The fault plane orientations show considerable scatter but fall roughly in a northeast-southwest trending plane similar to the axial planes of the kinks. An argument against the relationship between the faults and kinks is provided by field observations of the Late Triassic and Early Jurassic volcanic and plutonic rocks northwest of the Raspberry Schist. The Shuyak volcanics and interbedded volcaniclastics do not show any evidence for F_1, F_2, or F_3 folds but do contain faults and brittle shear zones identical to those in the Raspberry Schist.

A constraint on the type of movement that occurred on these brittle shear zones is provided by stereonet analyses of slickenside lineations on blocks within the Border Ranges Fault zone. Plots of lineations measured in Uyak Bay indicate that the movement was predominantly dip-slip (Roeske, unpublished).

MICROSCOPIC PETROGRAPHY AND MICROPROBE ANALYSES

Introduction

Identification of the phases has been made by petrographic microscope and X-ray diffraction analyses on over 150 samples from Js2 and approximately 75 samples from Js1. Preliminary microprobe analyses have focused on amphiboles (Table 3 and Figure 7) and garnets (Figure 8) in Js2.

All analyses were done on the five-spectrometer JEOL 733 Superprobe at Stanford University. This machine is completely automated and uses the Bence-Albee correction program on Tracor Northern software for processing the data to weight percent oxides. Operating voltage was 15 kv and sample current was 14.9 to 15.3 namp. The spot size was 10 microns. Standards used were provided by Stanford University and include natural phases similar in composition to those studied. The major and minor oxides have a standard deviation of ±1 percent and 3 percent respectively of the individual totals. Duplicate analyses of the standards deviate by less than 1 percent of the total for each element.

The estimates of ferric iron proportions were done by the method of Papike and others (1974), with maximum ferrous, maximum ferric, and midpoint concentrations calculated. Brown (1977a) compared these three calculated compositions to the measured FeO content of medium to high pressure actinolitic and crossitic amphiboles. Brown's study showed that the maximum ferric iron calculation best fits the measured FeO content. This paper will present the glaucophane and actinolite compositions using the maximum Fe^{3+} calculated. Brown also noted, however, that with increasing Al^{IV} content the maximum Fe^{3+} calculation is no longer the best. For the one sample with $Al^{IV} > 0.7$ (8-07-S6) the data are presented as the midpoint calculation. The terminology used in classifying the amphiboles comes from the report of the I.M.A. Subcommittee on Amphiboles (Leake, 1978).

Units Js1 and Js2, as defined above by field observations, are petrographically distinct. Relict igneous textures and phases are common in Js1 (Figure 6a) whereas rocks from Js2 are completely, dynamically recrystallized (Figure 6b). Metabasites and pelites have been preferentially sampled, so that although metacherts compose approximately 30 percent of the Raspberry Schist, they represent approximately 20 percent of the samples collected.

Js1 Unit

Assemblages. The majority of metabasites contain the common mineral assemblage quartz + albite + chlorite ± hematite. Phases seen in addition to this assemblage, followed by the number of thin sections in which they are observed, are listed below.

    lawsonite-phengite-epidote (2)
    lawsonite-sodic amphibole-phengite (2)
    pumpellyite (6)
    pumpellyite-actinolite-epidote (1)
    actinolite-sodic amphibole-phengite (2)
    actinolite-sodic amphibole-phengite-epidote (2)
    epidote (12)

These metabasite assemblages can be broken into two categories: those containing abundant epidote versus those with minor or no epidote (Table 2). Assemblages containing abundant epidote, seen only in the northwest side of Js1 exposed in Uyak Bay

(Table 2 and Figure 3), do not contain pumpellyite or lawsonite. Js1 assemblages designated as minor or no epidote contain <5 percent epidote in thin section. In addition to these metamorphic phase assemblages, many of the metabasites contain relict clinopyroxene and plagioclase feldspar with albite and Carlsbad twins.

The metabasite compositions grade into quartzo-feldspathic sediments with increasing sedimentary component in the volcaniclastics. The quartzo-feldspathic rocks contain the same minimum assemblage as the metabasites, with the addition of the phases listed below.

    actinolite-phengite (6)
    actinolite-stilpnomelane (5)
    stilpnomelane (1)

The actinolite-stilpnomelane assemblages are restricted to the epidote zone of Uyak Bay where they occur in metasandstones and metaconglomerates.

Metacherts in Js1 are typically >85 percent quartz and contain the assemblage quartz + chlorite + phengite ± calcite ± hematite. Ferruginous metacherts, interbedded with metavolcanics in the epidote zone, contain magnesio-riebeckite, epidote, and stilpnomelane (Uy-D10, Table 2; Figure 6d).

Textures and Compositions of Individual Phases. Lawsonite, identified optically and with X-ray diffraction analyses, is a major component in the assemblages that contain it. It occurs as fine-grained needles replacing plagioclase and as small, idioblastic crystals intergrown with phengite and amphibole. The assemblage lawsonite-epidote may not be an equilibrium assemblage because the epidote is a very minor component (<1 percent) and not clearly in contact with lawsonite.

Sodic amphiboles in Js1 are a minor component in four metabasite samples and a major component in the ferruginous metacherts in Uyak Bay. The blue amphiboles in the metabasites have all been identified as crossite, based on optical properties and XRD scans. The crossite, intergrown with actinolite, occurs as very fine needle-shaped crystals in veins and on the rims of relict clinopyroxene. Textures in the coarser grained samples are compatible with the interpretation that the sodic amphibole formed as an overgrowth on the actinolite (Figure 6c). A miscibility gap is to be expected between these two amphiboles but proving that chemical equilibrium was attained is difficult in the Raspberry Schist. The phases are always intergrown or mantling each other and do not show any clear features of equilibrium as described by other authors for these amphibole pairs (Klein, 1969; Ernst, 1979; Ghose, 1981). The magnesio-riebeckite in the ferruginous metachert (Uy-D10, Tables 2 and 3) occurs as fine-grained crystals intergrown with stilpnomelane in extension fractures in the epidote (Figure 6d) and as feathery clusters around cores of winchite.

Pumpellyite in Js1 is restricted to metabasites poor in or lacking epidote. Both iron-rich and iron-poor pumpellyite have been identified optically, occurring as fine-grained mats in amygdules and replacing igneous plagioclase. X-ray diffraction and petrographic analyses have not revealed any paragonite, prehnite, or metamorphic pyroxene.

TABLE 2. MINERAL ASSEMBLAGES IN THE RASPBERRY SCHIST

Jsl

	non or minor epidote zone					epidote zone	
Qtz			X	X	X	X	X
Ab	X	X	X		X	X	
Chlor	X	X	X	X	X	X	
Pheng			X	X	X	X	X
Epi		X		X		X	X
Laws				X			
Pump	X	X					
Actin		X	X			X	X
Na-amph			X				X
Stilp							X
CC	X	X		X	X	X	
Sphene	X	X	X	X	X		X
Hem				X	X		X
Sample No.	Uy-B2	7-22-S15	7-28-S9	8-05-S9	7-29-S10	Uy-D6	Uy-D10
Lith	mbas	mbas	mbas	qfs	qfs	mbas	qtzt

Js2

	non or minor epidote zone					epidote zone
Qtz	X		X	X	X	X
Ab	X	X	X	X	X	X
Chlor	X	X	X		X	
Pheng	X	X		X	X	X
Epi	X	X	X			
Na-amph	X		X	X		X
Ca-amph		X	X			
Stilp				X		
Garnet					X	
Apatite	X			X	X	
CC	X	X	X	X		
Sphene	X	X				X
Rutile	X	X		X	X	
Hem						X
Sample No.	M1-N9	8-07-S6	8-07-S7	M1-L2	M1-K11	M1-017
Lith	mbas	mbas	mbas	pelite	qtzt	qtzt

Abbreviations used: Qtz = quartz, Ab = albite, Chlor = chlorite, Pheng = phengite, Epi = epidote, Laws = lawsonite, Pump = pumpellyite, Actin = actinolite, Na-Amph = sodic amphibole, Stilp = stilpnomelane, CC = calcite, Hem = hematite, Ca-amph = calcic amphibole.
Lith = lithology, mbas = metabasite, qtzt = quartzite, qfs = quartzofeldspathic.

Js2 Unit

Assemblages. The metabasites of Js2 can be described by Laird's (1980) "common" assemblage, consisting of amphibole + chlorite + epidote + plagioclase + quartz + Ti-phase ± Fe^{3+} oxide ± carbonate ± K-mica. Quartzites in Js2 contain the assemblage quartz + phengite + chlorite ± garnet ± calcite ± hematite ± apatite ± rutile (Table 2). In Malina Bay sodic amphibole is a minor component in metaquartzites and a major component in metapelites.

Textures and Compositions of Phases. The assemblages listed in Table 2 generally show equilibrium textures with several important exceptions: 1) zoning in amphiboles, 2) sphene rimming rutile, and 3) garnet altering to chlorite and opaques. Some of these disequilibrium textures are probably related to the late brittle deformation event and associated very low grade metamorphism (see below). Others, especially the zoning in some of the amphiboles, clearly formed during the main metamorphic event.

Most of the metabasites in Js2 are thinly-layered schists with alternating quartz-calcite and amphibole-chlorite-epidote-albite layers (Figure 6b). The amphiboles occur as fine- to medium-grained tabular crystals, typically intergrown with chlorite in the matrix or forming inclusions in albite poikiloblasts. Crossite and glaucophane, identified optically and by microprobe analyses, represent over 80 percent of the amphiboles in the metabasites and 100 percent of the amphiboles in the pelites and quartzites. Amphiboles in the pelites range from Fe-rich crossite to glaucophane (Figure 7). The minor amphibole in quartzites show little or no pleochroism and have analyses that classify them as extremely Fe^{2+}-poor glaucophane (Table 3 and Figure 7).

Actinolite is rare in Js2. It is always found with sodic amphibole, intergrown on a very fine scale with textures similar to those described in Js1. Blue-green amphiboles, seen in approximately 10 percent of the metabasites in Js2, have been identified by microprobe analyses as calcic crossite (M1-N9, Table 3) and barroisite bordering on actinolitic hornblende (8-07-S6, Table 3). The barroisite occurs as rims on hornblende cores in a coarse-

TABLE 3. MICROPROBE ANALYSES OF AMPHIBOLES

	ML-017	8-07-S7 rim	8-07-S7 core	UY-D10 rim mag.	ML-01	ML-M9	8-07-S6[†] core	8-07-S6[†] rim
	glaucophane	crossite	actinolite	riebeckite	crossite	crossite	hornblende	barroisite
SiO_2	56.28	55.40	54.16	54.84	56.38	55.07	46.30	50.82
Al_2O_3	12.16	5.89	2.85	1.80	7.18	8.26	12.13	6.75
$FeO*$	10.34	5.84	7.10	12.11	8.02	7.96	8.55	8.54
F_2O_3*	1.72	6.94	2.48	11.69	6.71	6.43	3.89	3.95
MgO	8.41	13.27	17.11	8.91	10.80	6.43	13.40	14.75
MnO	0.25	0.14	0.29	0.43	0.09	10.73	0.20	0.26
TiO_2	0.17	0.06	0.03	0.27	0.03	0.29	0.27	0.17
Cr_2O_3	0.11	0.02	0.05	0.00	0.00	0.20	0.06	0.07
CaO	0.62	4.33	11.30	2.74	1.63	2.62	9.23	8.26
Na_2O	7.04	5.03	1.16	5.58	6.37	6.14	3.45	3.45
K_2O	0.11	0.05	0.08	0.16	0.04	0.09	0.33	0.14
Anhydrous Sum	97.21	96.97	96.61	98.53	97.25	97.91	97.81	97.16
Si_{iv}	7.83	7.79	7.33	7.95	7.91	7.72	6.67	7.31
Al_{vi}	0.17	0.21	0.27	0.05	0.09	0.28	1.33	0.69
Al	1.82	0.77	0.20	0.26	1.10	1.08	0.73	0.45
Fe^{2+}	1.20	0.69	0.85	1.47	0.94	0.93	1.03	1.03
Fe^{3+}	0.18	0.73	0.26	1.27	0.71	0.68	0.42	0.43
Mg	1.74	2.78	3.64	1.93	2.26	2.24	2.88	3.16
Mn	0.03	0.02	0.03	0.05	0.01	0.04	0.02	0.03
Ti	0.02	0.01	0.00	0.03	0.00	0.02	0.03	0.02
Cr	0.01	0.00	0.01	0.00	0.00	0.01	0.01	0.01
$XOct$	0.01	0.00	0.00	0.01	0.02	0.00	0.12	0.12
Ca	0.09	0.65	1.73	0.43	0.25	0.39	1.42	1.27
$NaM4$	1.90	1.35	0.27	1.56	1.73	1.61	0.46	0.61
NaA	0.00	0.02	0.05	0.00	0.00	0.06	0.51	0.35
K	0.02	0.01	0.01	0.03	0.01	0.02	0.06	0.03

*Ferric/ferrous proportions calculated by method of Papike and others (1974). Maximum ferric calculation used unless otherwise noted.
[†]Calculation of ferric/ferrous proportion is midway between maximum ferrous and maximum ferric.

grained epidote-amphibolite. The contact between the overgrowths is sharp, as determined by detailed traverses of six grains.

Epidotes are commonly zoned, with inclusion-rich cores and inclusion-free rims. Optical properties and reconnaissance microprobe analyses show a range in Ps content from 33 to 25. Clinozoisite has been identified in only one sample (optically).

Garnets are found only in quartzites in Malina Bay and northeast Raspberry Strait where they compose less than 5 percent of the rock. Zoning in the garnets is minor, with a slight increase in almandine and decrease in spessartine from core to rim (Figure 8). Grossular and andradite components stay constant; the pyrope component makes up less than 5 percent of the garnet.

METAMORPHIC FACIES

The presence of abundant faults throughout the Raspberry Schist makes the determination of metamorphic facies difficult. Uncertainties on the amount of movement on these faults and the narrow range of rock compositions between faults prevents one from determining if rocks with different assemblages reflect differences in bulk rock composition or in P-T conditions during metamorphism. This section presents the constraints on the range of metamorphic facies within the Raspberry Schist but does not

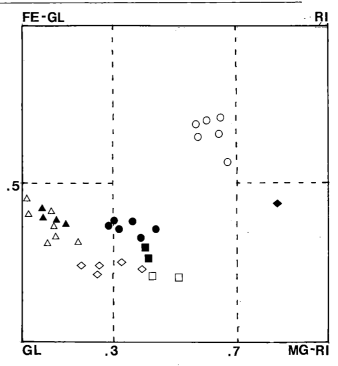

Figure 7. Sodic amphibole compositions for the Raspberry Schist. FE-GL = Iron end member glaucophane; MG-RI = Magnesium end member riebeckite. Uy-D10 ◆ ; Ml-017 △; Ml-09b ▲; Ml-N9 ● ; Ml-L2 ○ ; Ml-01 ■; 8-10-S3 ◇ ; 8-07-S7 □.

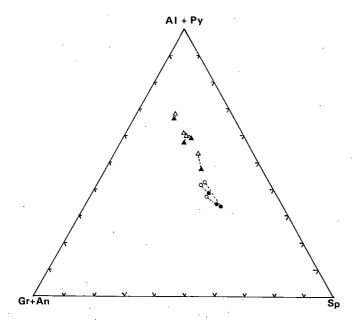

Figure 8. Garnet compositions of metacherts from the Js2 unit in Malina Bay. Al = almandine, Py = pyrope, Gr = grossular, An = andradite, Sp = spessartine. M1-K11 (core) ▲, (rim) Δ; M105 (core) ● , (rim) ○ See text for comment on pyrope, grossular, and andradite components.

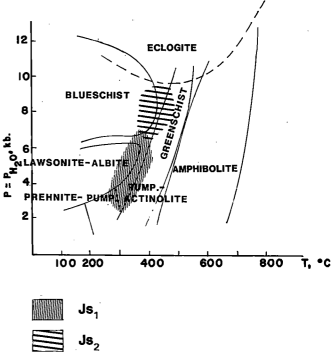

Figure 9. The range of metamorphic facies in which assemblages in the Raspberry Schist could occur. Outline of the facies is from Turner (1981).

attempt a narrow definition of the P-T conditions for the various assemblages. Facies terminology and definitions are taken from Turner (1981).

The local presence of lawsonite provides the firmest constraint on metamorphic facies in Js1. Any assemblage with lawsonite is by definition unique to the lawsonite-albite-chlorite or blueschist facies of Turner (1981). Most of the other assemblages seen in Js1 could occur in lawsonite-albite-chlorite facies (Brown, 1977b) but are also common in prehnite-pumpellyite, pumpellyite-actinolite, and lower greenschist facies. Nakajima and others (1977) and Coombs and others (1976) define the limits on prehnite-pumpellyite facies and pumpellyite-actinolite facies based on the compositions of pumpellyite and epidote. Lacking microprobe analyses of these phases, one must look at the trends of their abundances for evidence of a change in metamorphic facies. Assemblages in Js1 containing abundant epidote and lacking pumpellyite always occur northwest of those with pumpellyite and little or no epidote. The epidote and non-epidote assemblages as described above are not found interbedded or in close proximity to each other. The contact between the two may be gradational, in which case they could be interpreted as reflecting a change in metamorphic facies from pumpellyite-actinolite to lower greenschist. The absence of biotite, replaced by muscovite-stilpnomelane-actinolite or some combination of these phases (Brown, 1975), constrains these potential greenschist facies assemblages to lower greenschist facies. The scattered occurrence of sodic amphiboles is similar to the incipient development of sodic amphiboles as described in New Caledonia, Sanbagawa, and the South Fork Mountain Schist section of the Franciscan

Complex. These assemblages are characterized as bordering on blueschist facies (Turner, 1981). Based on the above constraints, a summary of the range of possible facies in Js1 is presented in Figure 9.

The problems in determining the metamorphic facies of rocks in Js2 are similar to those of Js1: a lack of phase assemblages characteristic of one facies and an abundance of faults. Constraints on the facies are provided indirectly by the absence of the phases aragonite, lawsonite, metamorphic pyroxene, and garnet in metabasites. The Raspberry Schist had been previously described as interlayered blueschist-greenschist facies (Carden and others, 1977). However, the scarcity of actinolite and the abundance of glaucophane or crossite in rocks of all compositions better fits descriptions of blueschist facies or rocks transitional to blueschist facies (Turner, 1981). The absence of lawsonite and pumpellyite may be related to bulk composition, but given the abundance of epidote, I prefer to interpret the assemblages of Js2 as forming in the high-temperature blueschist facies, outside of the P-T space for the formation of lawsonite. An argument against this interpretation is that other high-temperature blueschist facies assemblages lacking lawsonite and pumpellyite typically contain garnet and(or) omphacite (Sanbagawa, New Caledonia, Brothers and Yokoyama, 1982; Northwest Washington State, Brown and others, 1981).

The presence of barroisitic amphibole places some constraint on upper P-T limits of Js2. Ernst (1979) estimated the stability field of barroisite as between, and gradational with, glaucophane and actinolite, forming at temperatures above 300–350°C. The field Ernst described for barroisite falls near or above the high temperature limit of lawsonite (Nitsch, 1972), indicating that at least a portion of the Raspberry Schist formed at P-T conditions outside of the blueschist facies. Given the uncertainties on the upper temperature limit of glaucophane stability (Maresch, 1977) and its stability with respect to barroisite (Klein, 1969; Ernst, 1979), one has to leave open the possibility that many of the rocks containing glaucophane or crossite formed in the same P-T space as those with barroisite.

RELATION OF STRUCTURE TO METAMORPHISM

The different metamorphic assemblages of Js1 and Js2 may have formed under different pressure and temperature conditions, but the rocks throughout the Raspberry Schist show similar sequential structural and metamorphic events. Because the fabric in Js1 is poorly developed, the descriptions of microstructures below are restricted to Js2 assemblages.

The earliest metamorphic and deformation events are recorded in albite poikiloblasts. Inclusion trails of epidote, amphibole, and sphene locally show a fabric (S_i) at a high angle to that in the matrix (S_e). This internal fabric, S_i, records traces of folds that are poorly preserved in the matrix. These microstructures indicate that the development of the foliation in the matrix continued during and/or after the formation of the albite poikiloblasts (Vernon, 1978). The presence of crossite or glaucophane both inside and outside of the albite poikiloblasts suggests that the metamorphic conditions did not change dramatically between the formation of S_i, its folding, the growth of the poikiloblasts, or the formation of S_e.

The effect of the brittle shear zones and faults vary widely; some rocks show little or no evidence for cataclastic deformation whereas others have had the primary metamorphic fabric completely obliterated. The degree of deformation depends on the proximity to brittle shear zones associated with the major faults. Rocks within the fault zones are locally so altered and deformed as to be no longer recognizable as schistose metamorphic rocks. Rocks adjacent to the fault zones show both brittle and plastic deformation textures (Spry, 1983). Amphiboles and feldspars are clearly broken on discrete planes and quartzites contain both mortar and ribbony fabric. In a recently proposed terminology for fault-related rocks (Wise and others, 1984) the rocks in and adjacent to the wide fault zones would fall in the category of cataclasite or protomylonite.

The strongest deformation and recrystallization associated with the brittle faults occurs along the fault zone that separates the Raspberry Schist from the pluton and volcanics to the northwest (Figures 3 and 4). The main body of the pluton shows little deformation or alteration of the igneous phases. Within 300 m of the fault the plutonic rocks are extensively cataclastically deformed and altered. (Davies and Moore, 1984). The mineral assemblage epidote - albite - calcite - quartz - chlorite - white mica - prehnite completely replaces the most altered igneous rocks. These new phases do not have a preferred orientation. They occur as veins or form static overgrowths on the igneous minerals.

AGE AND FAULT DISPLACEMENT OF THE RASPBERRY SCHIST

Rb-Sr whole rock and phengite dates recently acquired from the Raspberry Schist give an age of approximately 196 Ma (R. Armstrong, personal communication, 1985), slightly older than the K-Ar dates from phengite separates (Table 1). The Rb-Sr dates form a well-defined isochron and show no evidence for more than one major thermal event. This age is also slightly older than the K-Ar dates for the Afognak Pluton (Table 1) and a recently reported U-Pb date of 184 Ma from a sample collected southwest of Uyak Bay (Hudson, 1985 and Hudson, personal communication, 1985). I collected two size categories of zircon separates from quartz diorites in Malina Bay that yielded concordant U-Pb dates of approximately 210 Ma (J. Mattinson, personal communication, 1984). Assuming that both U-Pb dates are reliable, the Afognak pluton may represent a series of intrusions spanning as much as 30 m.y.

The Shuyak metavolcanics and Afognak Pluton are separated from the Raspberry Schist by a brittle shear zone. The fact that the metamorphic assemblages and dates of the Raspberry Schist show no evidence for the contact metamorphism seen in the Shuyak metavolcanics indicates that substantial displacement must have occurred along this fault zone. A minimum age of movement on this fault is given by the age of dikes that intrude both the Raspberry Schist and Afognak Pluton. These dikes occur within 100 m of the fault between the schist and pluton but show no alteration or deformation. The dikes have K-Ar dates of 58 and 53 Ma (Davies and Moore, 1984; J. Decker, personal communication, 1984). A maximum age of movement is the age of metamorphism of the two units, between Latest Triassic and late Early Jurassic. Substantial displacement also appears to have occurred on at least one of the brittle shear zones within the Raspberry Schist, the one that separates Js1 from Js2. The discussion of assemblages (above) showed that Js1 and Js2 contain different metamorphic facies, with the higher temperature and/or pressure assemblages lying to the northwest (Js2).

The Border Ranges Fault, seen as a wide, poorly exposed brittle shear zone, parallels the main fault within the Raspberry Schist and appears to cut it off (Figure 3). Where the Raspberry Schist pinches out the Border Ranges Fault becomes the fault cutting the Shuyak volcanics and the Afognak Pluton (Figure 3). The similar style of deformation and the anastomosing trends of the Border Ranges Fault and the brittle shear zones suggests that the fault displacements within the Raspberry Schist may be related to one stage of movement of the Border Ranges Fault. If this interpretation is correct then it would be more accurate to describe the Raspberry Schist as a series of fault slivers caught up within the Border Ranges Fault zone, juxtaposed with rocks of similar age but completely different thermal regime.

CORRELATIVE UNITS ALONG THE BORDER RANGES FAULT

Border Ranges Mafic Complex

The Border Ranges Fault extends for at least 1500 km east and northeast of the Kodiak Islands (Figure 1). The most strikingly continuous feature associated with the Border Ranges Fault is a strong positive magnetic anomaly that continues almost uninterrupted for over 1000 km (Figure 1). The source for the anomaly has been ascribed to the Early Jurassic tonalitic plutons (Afognak and related rocks) (Fisher, 1981) and/or associated mafic and ultramafic rocks (Burns, 1982). The mafic-ultramafic rocks, named by Burns (1985) as the Border Ranges mafic complex, include ultramafic cumulates, gabbronorite cumulates, and massive gabbronorites. These occur as elongate pods on the scale of kms by tens of kms near the Border Ranges Fault, most commonly north of it. In the northern Chugach Mountains, Burns found the ultramafic and mafic rocks closely associated temporally and spatially with voluminous dioritic to tonalitic plutons. K-Ar dates from the associated plutons range from 189 to 194 Ma (Pavlis, 1982, 1983). On the Kodiak Islands gabbronorite and ultramafic cumulates lie within the Cretaceous melange immediately southeast of the Border Ranges Fault (Beyer, 1980) (Figure 2). Based on their similar age, chemistry, and tectonic position Burns (1985) correlates the Chugach Mountains mafic complex with the dioritic Afognak pluton and ultramafic cumulates on Kodiak Island.

Continuing east along the Border Ranges Fault, the magnetic anomaly remains strong to the central Valdez quadrangle, continuing to be associated with scattered bodies of ultramafic and gabbroic rock (Burns, 1982). Compositions of gabbroic rocks from the central Valdez quadrangle are similar to those from the northern Chugach Mountains. Preliminary K-Ar dates on the gabbros range from 154 to 188 Ma (Winkler and others, 1981). There is no evidence for continuation of the voluminous Early to Middle Jurassic tonalites and diorites.

The relation of the Early Jurassic Border Ranges mafic complex to the Alaska Range batholith is unclear. Burns (1985) suggests that the complex is not an ophiolite formed from mid-ocean ridge basalt, as suggested by most previous studies, but the plutonic base of an intraoceanic island arc. Most authors have not recognized the genetic relationship of the ultramafic rocks and voluminous diorites and have lumped the diorites with the Alaska Range batholith. Hudson (1979) separated the two plutonic belts, based on their differing ages and geographic distribution. Additional K-Ar and new U-Pb dates (Hudson, 1985) however, show that the youngest ages of the Border Ranges mafic complex overlap with the oldest ages of the Alaska batholith. Hudson revised his interpretation to accepting either his original view or the interpretation originally proposed by Hill (1979), that the Lower Jurassic plutonic rocks on the Kodiak Islands are a more mafic, near-trench precursor to the main batholith that formed in the Middle Jurassic on the Alaska Peninsula (Hudson, 1985). A new U-Pb date of approximately 210 Ma for the Afognak pluton (Mattinson, personal communication, 1984) suggests that sections of the Border Ranges mafic complex may be as much as 40 million years older than the Alaska Range batholith. If these two plutonic belts are related then the volcanism and associated plutonism covers a span of 60 m.y.

Correlative Blueschists

Various schists that crop out immediately northwest of the Border Ranges Fault further to the northeast of the Kodiak Islands have been proposed as correlative to the Raspberry and Seldovia Schists. Two sites in the Valdez quadrangle on the north side of the Chugach mountains (Figure 1) contain crossite and rare lawsonite; one of these sites yields ages similar to the Raspberry-Seldovia Schists (Table 1) (Winkler and others, 1981).

The easternmost site, known as the Liberty Creek terrane, lies directly in a zone of anastomosing faults associated with the Border Ranges Fault (Figure 1). The Liberty Creek terrane is more homogeneous in lithology and grade of metamorphism than the Raspberry Schist. The former lacks the abundant metacherts of the Raspberry Schist and is dominantly greenschist facies, with relatively scarce crossite and lawsonite (Winkler and others, 1981). Since no age data from the Liberty Creek terrane are available, it has been variously interpreted as related to the upper plate schists (Metz, 1976; Wallace, 1984) or to the Cretaceous melange to the south (Winkler and others, 1981). 50 km west of this site is an elongate, 4 by 40 km lens of blueschist, referred to here as the Valdez Blueschist, encompassed entirely by Cretaceous sediments of very low grade. The Valdez Blueschist contains lithologies and metamorphic assemblages typical of the Raspberry Schist with the addition that lawsonite is more common in the Valdez quadrangle (Winkler and others, 1981; Winkler, personal communication, 1983). The K-Ar ages show a wide range but can be broken into two categories based on whether the rocks are sheared or not. The ages of the unsheared rocks range from 175 to 152 Ma (Winkler and others, 1981), placing them within the analytical uncertainty of the Raspberry and Seldovia schists. Ages of 138 to 113 on strongly sheared rocks probably reflect a resetting due to metamorphism associated with movement on the Border Ranges Fault (Winkler and others, 1981).

Even if the Raspberry Schist does correlate with the Valdez Blueschist, the present length of over 1000 km may not reflect the original extent of the blueschist belt as numerous Tertiary vertical faults with unknown amounts of displacement extend throughout this region. The Border Ranges Fault in the Valdez quadrangle is interpreted as a major thrust fault (Plafker and others, 1985) and structural evidence from the Kodiak Islands shows that the movement was predominantly dip-slip (see Structural Fabrics). A detailed structural study of the Haley Creek terrane, adjacent to the Liberty Creek terrane, suggests that the schists were overprinted by a mylonitization associated with right lateral strike-slip movement (Crouse and Pavlis, 1985).

Other schistose rocks in the Western Chugach mountains have been proposed as possible correlations to the Raspberry and Seldovia Schist, primarily because of their similar tectonic position (Carden and Decker, 1977; Pavlis, 1983). Pavlis (1983) suggests that the amphibolites and metasediments intruded by the 188 to 194 Ma tonalitic plutons are correlative to the Raspberry and Seldovia blueschists. The former, however, show no indication of a previous high pressure metamorphism and instead clearly contain low pressure assemblages formed during static metamorphism (Pavlis, 1983).

REGIONAL GEOLOGIC CONCLUSIONS

The problem of the age and field relationship of the Raspberry Schist and Afognak Pluton, as described in the Introduction, has been solved. The blueschists and diorites formed approximately coevally but in distinct settings. The two units did not share a common history until after the Early Jurassic when they were juxtaposed by faults of unknown displacement.

The metamorphic assemblages and structural style of the Raspberry Schist and its tectonic position on the edge of an ancient plate boundary make it a strong candidate for recording a paleosubduction zone. One or two separate belts of Lower Jurassic calc-alkaline volcanics and plutons occur nearby and could be part of the arc associated with the subduction zone. The Alaska Range batholith is significantly younger than the blueschists but volcanics associated with the chain of plutons are coeval with the Early Jurassic metamorphic age of the schist. If the Raspberry Schist is associated with this calc-alkaline suite, then the original tectonic configuration of these two units need not have changed since Early Jurassic as these two belts are approximately 100 km apart (Figure 1). This interpretation would have to ignore the geochemical evidence suggesting that the Alaska Range batholith formed over a south-dipping subduction zone (Reed and others, 1983).

If the Afognak pluton and Shuyak volcanics are the arc associated with the blueschists, then major closure must have occurred across the arc-trench gap between the Early Jurassic and Paleocene. Some or all of this closure is represented today by a wide brittle shear zone.

More detailed tectonic reconstructions will depend on detailed stratigraphic studies of the Late Triassic and Early Jurassic units on the Alaska Peninsula and Kodiak Islands and on better constraints on the type and timing of movement on the series of faults associated with the Border Ranges Fault.

ACKNOWLEDGMENTS

I am grateful to the Geology Department and the Center for Materials Research at Stanford University for permission to use their microprobe facilities. Graduate students in Geology at Stanford were very generous in training me on the equipment and the computer. This manuscript was improved by the comments of J. Casey Moore, C. Carlson, M. Helper, and B. Evans and by discussions with J. G. Liou and E. H. Brown. The study was funded by a Sigma Xi Grants-in-Aid-of-Research, National Science Foundation Grant #EAR83-05883, Mobil, Arco, Union, the United States Geological Survey, and the Geological Society of America.

REFERENCES CITED

Beyer, B. J., 1980, Petrology and geochemistry of ophiolite fragments in a tectonic melange, Kodiak Islands, Alaska [Ph.D. thesis]: Santa Cruz, University of California, 227 p.

Brothers, R. N., and Yokoyama, K., 1982, Comparison of the high-pressure schist belts of New Caledonia and Sanbagawa, Japan: Contributions to Mineralogy and Petrology, v. 79, p. 219–229.

Brown, E. H., 1975, A petrogenetic grid for reactions producing biotite and other Al-Fe-Mg silicates in the greenschist facies: Journal of Petrology, v. 16, p. 258–271.

—— 1977a, The crossite content of Ca-amphibole as a guide to pressure of metamorphism: Journal of Petrology, v. 18, p. 53–72.

—— 1977b, Phase equilibria among pumpellyite, lawsonite, epidote and associated minerals in low grade metamorphic rocks: Contributions to Mineralogy and Petrology, v. 64, p. 123–136.

Brown, E. H., Bernardi, M. L., Christenson, B. W., Cruver, J. R., Haugerud, R. A., Rady, P. M., and Sondergaard, J. N., 1981, Metamorphic facies and tectonics in part of the Cascade Range and Puget Lowland of northwestern Washington: Geological Society of America Bulletin, v. 92, p. 170–178.

Brown, E. H., Wilson, D. L., Armstrong, R. L., and Harakal, J. E., 1982, Petrologic, structural and age relations of serpentinite, amphibolite, and blueschist in the Shuksan Suite of the Iron Mountain-Gee Point area, North Cascades, Washington: Geological Society of America Bulletin, v. 93, p. 1087–1098.

Burns, L. E., 1982, Gravity and aeromagnetic modelling of a large gabbroic body near the Border Ranges Fault, southern Alaska: United States Geological Survey Open-File Report 80-470, 72 p., 3 sheets, scale 1:250,000.

Burns, L. E., 1985, The Border Ranges ultramafic and mafic complex, south-central Alaska: cumulate fractionates of island-arc volcanics: Canadian Journal of Earth Sciences, v. 22, p. 1020–1038.

Carden, J. R., and Decker, J. E., 1977, Tectonic significance of the Knik River schist terrane, south-central Alaska, in Short notes on Alaskan geology: Alaska Division of Geological and Geophysical Surveys, Geologic Report no. 55, p. 7–9.

Carden, J. R., and Forbes, R. B., 1976, Discovery of blueschists on Kodiak Island, in Short notes on Alaskan geology: Alaska Division of Geological and Geophysical Surveys, Geologic Report no. 51, p. 19–22.

Carlson, C., and Pavlis, T. L., 1984, Significance of metamorphic complexes associated with the Peninsular terrane, southern Alaska, in Proceedings, Circum-Pacific terrane conference, Stanford, September 1983: Stanford, California, Stanford University Publications in Geologic Science, v. 18, p. 47–49.

Connelly, W., 1978, Uyak Complex, Kodiak Islands, Alaska: A Cretaceous subduction complex: Geological Society of America Bulletin, v. 89, p. 755–769.

Connelly, W., and Moore, J. C., 1979, Geologic map of the northwest side of the Kodiak and adjacent islands, Alaska: United States Geological Survey Miscellaneous Field Studies Map MF-1057, 2 sheets, scale 1:250,000.

Coombs, D. S., Nakamura, Y., and Vuagnat, M., 1976, Pumpellyite-actinolite facies schists of the Taveyanne Formation near Loeche, Valais, Switzerland: Journal of Petrology, v. 17, p. 440–471.

Cowan, D. S., 1985, Structural styles in Mesozoic and Cenozoic melanges in the western Cordillera of North America: Geological Society of America Bulletin, v. 96, p. 451–462.

Cowan, D. S., and Boss, R. F., 1978, Tectonic framework of the southwestern Kenai Peninsula, Alaska: Geological Society of America Bulletin, v. 89, p. 155–158.

Crouse, G. W., and Pavlis, T. L., 1985, Deformation of the Haley Creek terrane: Mesozoic transcurrent movement along the southern Alaskan margin: Geological Society of America Abstracts with Programs, v. 17, p. 350.

Davies, D. L., and Moore, J. C., 1984, 60 m.y. intrusive rocks from the Kodiak Islands link the Peninsular, Chugach, and Prince William terranes: Geological Society of America Abstracts with Programs, v. 16, p. 277.

Ernst, W. G., 1979, Coexisting sodic and calcic amphiboles from high-pressure metamorphic belts and the stability of barroisitic amphibole: Mineralogical Magazine, v. 43, p. 269–278.

Fisher, M. A., 1981, Location of the Border Ranges Fault southwest of Kodiak Island, Alaska: Geological Society of America Bulletin, v. 92, p. 19–30.

Forbes, R. B., and Lanphere, M. A., 1973, Tectonic significance of mineral ages of blueschists near Seldovia, Alaska: Journal of Geophysical Research, v. 78, p. 1383–1386.

Ghose, S., 1981, Subsolidus reactions and microstructures in amphiboles, *in* Veblein, D. R., ed., Amphiboles and other hydrous pyriboles - Mineralogy: Mineralogical Society of America Reviews in Mineralogy, v. 9A, p. 325–372.

Hill, M. D., 1979, Volcanic and plutonic rocks of the Kodiak-Shumagin shelf, Alaska: Subduction deposits and near trench magmatism [Ph.D. thesis]: Santa Cruz, University of California, 274 p.

Hill, M. D., and Morris, J. D., 1977, Near-trench plutonism in southwestern Alaska: Geological Society of America Abstracts with Programs, v. 9, p. 436–437.

Hudson, T. L., 1979, Mesozoic plutonic belts of southern Alaska: Geology, v. 7, p. 230–234.

——1985, Jurassic plutonism along the Gulf of Alaska: Geological Society of America Abstracts with Programs, v. 17, p. 362.

Klein, C., 1969, Two-amphibole assemblages in the system actinolite-hornblende-glaucophane: American Mineralogist, v. 54, p. 212–237.

Laird, J., 1980, Phase equilibria in mafic schist from Vermont: Journal of Petrology, v. 21, p. 1–37.

LaJoie, J., 1980, Volcaniclastic rocks, *in* Walker, R. G., ed., Facies Models: Geosciences Canada, Reprint Series 1, p. 191–200.

Leake, B. E., 1978, Nomenclature of amphiboles: Canadian Mineralogist, v. 16, p. 501–520.

Letterier, J., Maury, R. C., Thonon, P., Girard, D., and Marchal, M., 1982, Clinopyroxene compositions as a method of identification of the magmatic affinities of paleo-oceanic series: Earth and Planetary Science Letters, v. 59, p. 139–154.

MacKevett, E. M., Jr., and Plafker, G., 1974, The Border Ranges Fault in south-central Alaska: United States Geological Survey Journal of Research, v. 2, p. 323–329.

Maresch, W. V., 1977, Experimental studies on glaucophane: an analysis of present knowledge: Tectonophysics, v. 43, p. 109–125.

Martin, G. C., and Johnson, B. L., and Grant, U. S., 1915, Geology and Mineral Resources of Kenai Peninsula, Alaska: United States Geological Survey Bulletin 587, 243 p.

Metz, P. A., 1976, Occurrences of sodic amphibole-bearing rocks in the Valdez C-2 quadrangle, *in* Short notes on Alaskan geology: Alaska Division of Geological and Geophysical Surveys Geologic Report, no. 51, p. 27–28.

Moore, J. C., and Connelly, W., 1979, Tectonic history of the continental margin of southwestern Alaska: Late Triassic to earliest Tertiary, *in* Proceedings of the Sixth Alaska Geological Symposium, Anchorage: Anchorage, Alaska Geological Society, p. H-1-H-29.

Nakajima, T., Banno, S., and Suzuki, T., 1977, Reactions leading to the disappearance of pumpellyite in low-grade metamorphic rocks of the Sanbagawa metamorphic belt in central Japan: Journal of Petrology, v. 18, p. 263–284.

Nitsch, K., 1972, Das P-T-X CO_2 stabilitatsfeld von Lawsonit: Contributions to Mineralogy and Petrology, v. 34, p. 135–149.

Papike, J. J., Cameron, K. L., and Baldwin, K., 1974, Amphiboles and pyroxenes: Characterization of other than quadrilateral components and estimates of ferric iron from microprobe data: Geological Society of America Abstracts with Programs, v. 6, p. 1053–1054.

Pavlis, T. L., 1982, Origin and age of the Border Ranges Fault of southern Alaska and its bearing on the Late Mesozoic tectonic evolution of Alaska: Tectonics, v. 1, p. 343–368.

——1983, Pre-Cretaceous crystalline rocks of the western Chugach Mountains, Alaska: Nature of the basement of the Jurassic Peninsular terrane: Geological Society of America Bulletin, v. 94, p. 1329–1344.

Plafker, G., Jones, D. L., and Pessagno, E. A., Jr., 1977, A Cretaceous accretionary flysch and melange terrain along the Gulf of Alaska margin, *in* Blean, K. M., ed., The United States Geological Survey in Alaska: Accomplishments during 1976: United States Geological Survey Circular 751-B, p. B41-B43.

Plafker, G., Nokleberg, W. J., Fuis, G. S., Mooney, W. D., Page, R. A., Ambos, E. L., and Campbell, D. L., 1985, 1984 results of the Trans-Alaska Crustal Transect in the Chugach Mountains and Copper River basin, Alaska: Geological Society of America Abstracts with Programs, v. 17, p. 400.

Reed, B. L., and Lanphere, M. A., 1973, Alaska-Aleutian Range batholith: Geochronology, chemistry, and relation to circum-Pacific plutonism: Geological Society of America Bulletin, v. 84, p. 2583–2610.

Reed, B. L., Miesch, A. F., and Lanphere, M. A., 1983, Plutonic rocks of Jurassic age in the Alaska-Aleutian Range batholith: Chemical variations and polarity: Geological Society of America Bulletin, v. 94, p. 1232–1240.

Roeske, S. M., 1984, Metamorphic petrology of the glaucophane-bearing Raspberry Schist, Kodiak Islands, Southern Alaska [abs.]: EOS (American Geophysical Union Transactions), v. 65, p. 1146.

Spry, A., 1983, Metamorphic Textures: Oxford, Pergamon Press, 352 p.

Turner, F. J., 1981, Metamorphic Petrology, second ed.: New York, McGraw-Hill, 524 p.

Vernon, R. H., 1978, Porphyroblast-matrix microstructural relationships in deformed metamorphic rocks: Geologische Rundschau, v. 67, p. 288–305.

Wallace, W. K., 1984, Deformation and metamorphism in a convergent margin setting, northern Chugach Mountains, Alaska: Geological Society of America Abstracts with Programs, v. 5, p. 339.

Winkler, G. R., Silberman, M. L., Grantz, A., Miller, R. J., and MacKevett, E. M., Jr., 1981, Geologic map and summary geochronology of the Valdez quadrangle, southern Alaska: United States Geological Survey Open File Report 80-892-A.

Wise, D. U., Dunn, D. E., Engelder, J. T., Geiser, P. A., Hatcher, R. D., Kish, S. A., Odom, A. L., and Schamel, S., 1984, Fault-related rocks: Suggestions for terminology: Geology, v. 12, p. 391–394.

MANUSCRIPT ACCEPTED BY THE SOCIETY JULY 29, 1985

Geological Society of America
Memoir 164
1986

Rb-Sr and K-Ar study of metamorphic rocks of the Seward Peninsula and Southern Brooks Range, Alaska

Richard L. Armstrong
Joseph E. Harakal
Department of Geological Sciences
University of British Columbia
Vancouver, British Columbia V6T 2B4
Canada

Robert B. Forbes
Bernard W. Evans
Department of Geological Sciences AJ-20
University of Washington
Seattle, Washington 98195

Stephen Pollock Thurston
Chevron U.S.A., Inc.
P.O. Box 8200
Concord, California 94524

ABSTRACT

Blueschists of the Nome Group in the Seward Peninsula formed in Jurassic time (prior to ~160 Ma ago) in rocks of early Paleozoic to latest Precambrian age (approximately 360 to 720 Ma old). The Sr whole-rock isotopic signature on a plot of $^{87}Sr/^{86}Sr$ vs $^{87}Rb/^{86}Sr$ ratio—a fan shaped array of orthogneiss points lying between 720 and 360 Ma isochrons and paragneiss points showing a similar scatter *and* spread toward lower ages—is much like that of the Yukon Crystalline Complex and Cariboo-Omineca Belt in Canada; partial lithologic and historical similarity support the hypothesis of a common origin and tectonic setting marginal to Paleozoic North America. The areas were over-ridden during latest Triassic to Jurassic time by oceanic and exotic allochthons, and portions are studded with middle to Late Cretaceous plutons. At the same time all have experienced widespread resetting of K-Ar dates and regional uplift.

The southern Brooks Range shares many characteristics with the Seward Peninsula—late Precambrian to mostly(?) Paleozoic protoliths, including extensive pelitic and metavolcanic schists, Jurassic (prior to ~120 Ma) blueschist development, and comparable tectonic setting. A late Precambrian metamorphic mineral isochron date for muscovite schist (686 ± 116 Ma) in the Baird Mountains Quadrangle confirms previous K-Ar dating of the same rock by Turner and others (1979). This may be a tectonic fragment of an older blueschist terrane enclosed in a younger blueschist complex, but this area needs further study.

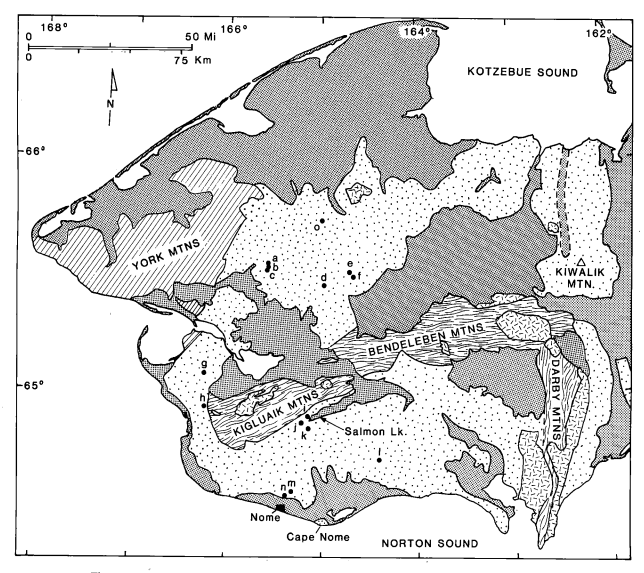

Figure 1. Generalized geologic map of the Seward Peninsula showing sample locations: (a) 80-113-1,
(b) 80-117-1, (c) 80-118-1, (d) 80-142-3, (e) SP81-333F and 80-19-4, (f) 80-15-5, (g) 80-151-1, (h)
80-155-1 (i) SL81-94-1, (j) SL81-5-2, (k) SL81-41-4, (l) 81-119, (m) 80-91-2, (n) 80-93-1 (o) 67-ASn-
595 (collected by C. L. Sainsbury). Light stipple: Nome Group and other low-grade metamorphic rocks;
dark stipple: Cenozoic sedimentary and volcanic rocks; diagonal striping: Paleozoic and Precambrian(?)
sedimentary rocks; wavy lines: gneiss complexes; granite symbol-hatchures: Cretaceous intrusive rocks.

INTRODUCTION

The blueschists of northwestern Alaska, less well known
because of their relative inaccessibility and poor exposure, occur
in two areas—the Seward Peninsula where they were first described in detail by Sainsbury and others (1970) and in the
southern Brooks Range Schist Belt where they were first reported
by Forbes and others (1971). Previous isotopic dating studies
have suggested that the major metamorphic event in the region
was of Precambrian age, but a Cretaceous thermal overprint was
observed that reset most mica Rb-Sr and K-Ar dates (Bunker and
others 1979; Turner and others 1979). To augment petrologic
studies of the high-pressure metamorphic rocks of the Seward

Peninsula by the University of Washington group (Forbes and
others 1984; Thurston 1985) and the U.S. Geological Survey
(Till 1983), we began a program of K-Ar and Rb-Sr dating of
white micas because prior experience (Suppe and Armstrong,
1972; Hunziker 1974) had shown that this was most likely to
yield the least disturbed age of metamorphism, free of ambiguities
about excess argon that had been observed to occur in low-K
minerals of the Brooks Range (Turner and others 1979). Samples
collected during University of Washington field studies were selected for dating after initial petrologic examination. Mineral separation and analytical work were done at the University of British
Columbia. As evidence for a Precambrian to early Paleozoic
protolith and Mesozoic metamorphism was revealed for the

Seward Peninsula, we decided to extend the Rb-Sr work to Brooks Range rocks, taking advantage of material previously collected and prepared for the K-Ar dating study of Turner and others (1979).

SEWARD PENINSULA

Schists of the Nome Group

Much of the Seward Peninsula is composed of schists of greenschist, blueschist, and albite-epidote-amphibolite facies grade. Pending further clarification of their stratigraphic relations (see Till 1982 and 1984), we include all these low-grade metamorphic rocks under the designation Nome Group (Moffit 1913), including possible allochthonous marbles and pelitic rocks that others have correlated with "slates of the York Region" (Sainsbury 1969, 1974). The depositional age of the Nome Group is poorly known; previous workers have suggested Precambrian (Sainsbury 1974; Sainsbury and others 1970), pre-Ordovician (Sainsbury 1969), Paleozoic (Moffit 1913) and pre-Devonian (Till 1983; Till and others 1983). Structurally and stratigraphically below the Nome Group are the paragneisses of the Kigluaik Group, exposed in the core of a broad arch that underlies the Kigluaik Mountains. Similar rocks underlie parts of the Bendeleben and Darby mountains (Figure 1). Suggested ages for the regional high-pressure metamorphism have ranged from Precambrian (Sainsbury and others 1970) to pre-Late Cretaceous (Hudson 1979). Basaltic (FeTi-rich and MgAl-rich) and acid magmas were injected into the Nome Group as (mostly) minor intrusions at some time prior to the high pressure regional metamorphism.

We have concentrated our efforts on rubidium-strontium analyses of whole-rocks and minerals separated from pelitic schist and metadiabase of the Nome Group since there are published Rb-Sr and K-Ar isotopic data available for the higher-grade Kigluaik Group rocks. A few K-Ar dates have also been done on selected white mica and amphibole. All samples were collected from locations (Figure 1) believed as free as possible from effects of the Late Cretaceous plutonic activity or uplift and cooling that has reset K-Ar systems in the Kigluaik and Bendeleben complexes to 81-87 Ma (Turner and others 1980; Wescott and Turner 1981; Sturnick 1984).

Rb-Sr and K-Ar Mineral Dates

Mineral dates based on Rb-Sr isochrons drawn between whole-rock and mineral compositions have been determined for phengite and paragonite in seven pelitic schists and one orthogneiss, and for paragonite in the glaucophanitic metadiabase (Tables 1 and 2; Figures 2 and 3). The results average 126 Ma, although it should be noted that the two oldest values are associated with large uncertainties (170 ± 80, 160 ± 26 Ma). Figure 2 shows what appears to be a correlation between date and whole-rock $^{87}Sr/^{86}Sr$ ratio. Samples with high $^{87}Sr/^{86}Sr$ ratio and old

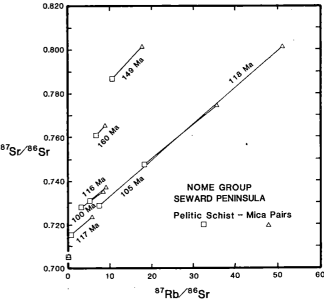

Figure 2. Rb-Sr Isochron diagram for Seward Peninsula whole-rock–mica pairs, Nome Group metasediments. The whole-rock–paragonite pair for metabasite 80-118-1 plots as one point on the abscissa.

whole rock date give somewhat older Rb-Sr mineral dates. Perhaps Cretaceous resetting has also mobilized Rb and Sr on the whole rock scale, thus producing the observed systematic pattern. K-Ar dates have been determined on one actinolite, one glaucophane, three phengites, and one paragonite (Table 3). An additional glaucophane was dated by D. L. Turner (unpublished; Table 3). The results range from 122 to 194 Ma, and average 159 Ma. Sturnick (1984) has recently reported K-Ar dates of 152 and 156 Ma for muscovite from Nome Group rocks in the eastern Kigluaik Mountains, in agreement with our results.

Four mineral separates (3 phengites and 1 paragonite) were analyzed by both techniques. They compare as follows, the Rb-Sr date of each pair is given first: 105 ± 4, 156 ± 5; 160 ± 26, 141 ± 5; 149 ± 6, 162 ± 6; 170 ± 80, 122 ± 4. The results of the two techniques overlap except in one case, where the K-Ar date is distinctly older. The average K-Ar date is older than the average Rb-Sr date (Figure 4) and, most significantly, the amphibole K-Ar dates are older than the mica dates. These observations suggest the presence of inherited Ar. In the Franciscan Complex of California, where inherited Ar is not a problem because protolith and metamorphic age are similar, the amphibole dates are invariably younger than mica dates (Coleman and Lanphere 1971; Suppe and Armstrong 1972). Except for the orthogneisses (81-94-1 and 81-5-2) and marble (80-155-1), our specimen locations (Figure 1) are distant from areas of known or likely Cretaceous thermal overprinting. Post-kinematic biotite occurs in medium-grade pelitic schists (with staurolite and garnet) close to the Kigluaik, Bendeleben, and Darby gneiss complexes, but these rocks are separated by faults from the Nome Group rocks we

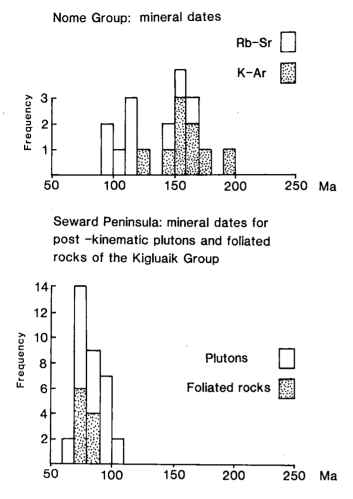

Figure 3. Histograms of mineral ages for Nome Group metamorphic rocks, post-kinematic plutons, and Kigluaik Group foliated rocks. Data from this study, Miller (1972), Sainsbury (1969, 1974), Miller and Bunker (1976), Miller and others (1976), Hudson (1979), Turner and others (1980), Wescott and Turner (1981), Hudson and Arth (1983), and Sturnik (1984).

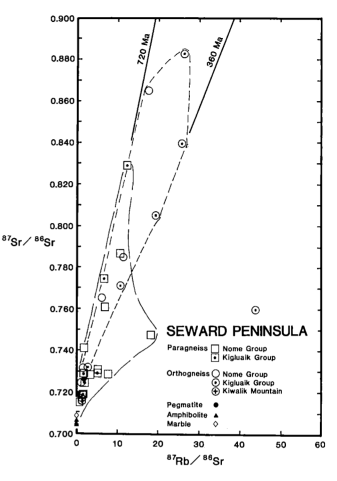

Figure 4. Rb-Sr Isochron diagram for Seward Peninsula whole rocks. Data from Bunker and others (1979) and this study.

sampled. Seven of the dated samples are glaucophane-bearing (Appendix I), and an additional two possess pseudomorphs after glaucophane. All samples, however, show at least some petrographic evidence for partial "overprinting" by greenschist-facies assemblages (chlorite + albite after glaucophane, chlorite + paragonite after chloritoid). This late alteration involves no new structural elements, and is thought to represent partial re-equilibration during decompression of the terrane (Thurston 1985). The age spectrum we have obtained seems to bear no relation to the intensity of this late greenschist alteration.

Recent studies (e.g. Altherr and others 1979; Frey and others 1983) have established a correlation between potassic white mica polytypes and the baric type of metamorphism: 3T (uniaxial) phengites characterize blueschists whereas 2M (biaxial) micas characterize medium and low-pressure terranes. Of the samples analyzed in this study, only those in orthogneiss from the Seward

Peninsula contain uniaxial phengite. Ironically, granitic orthogneiss sample SL 81-5-2 gave the youngest mineral date. Frey and others (1983) showed that, although they offer clues to former high-pressure metamorphism, 3T-potash micas are susceptible to radiogenic isotope re-equilibration in the absence of accompanying inversion to the 2M polytype.

It is obvious that we have not yet pinpointed the time of the blueschist-facies metamorphism in the Seward Peninsula. However, none of the sixteen mineral dates is older than Jurassic. We provisionally conclude that the high-pressure metamorphic cycle (Forbes and others 1984; Thurston 1985) reached a "maximum" some time before about 160 Ma (this is a reasonable upper bound of the older white mica K-Ar and Rb-Sr dates) and that this was followed by decompression and partial re-equilibration during the period 160 to about 100 Ma. The Cretaceous thermal overprint associated with Barrovian-style metamorphism in the Kigluaik and Bendeleben Mountains took place later, but before 100 to 80 Ma ago (Turner and others 1980: Wescott and Turner 1981). Plutonic activity in the Darby Mountains, which locally converted low-grade Nome Group rocks to hornfels (A. B. Till

TABLE 1. Rb-Sr DATA FOR ROCKS AND MINERALS OF THE NOME GROUP,
SEWARD PENINSULA

Sample Number	Rock Type	Material Analyzed*	Sr ppm	Rb ppm	$^{87}Rb/^{86}Sr$	$^{87}Sr/^{86}Sr$
Low-Grade Pelitic Schist						
SP81-333F	Gl-Ctd-Schist	WR	36.4	97.3	7.74	0.7283
		WM	17.2	300	51.1	0.8012
80-15-5	Gl-Ctd-Schist	WR	14.1	48.2	0.987	0.7155
		WM	88.4	182	5.96	0.7238
80-93-1	Q-Chl-Ab-Schist	WR	68.7	42.5	1.796	0.7407
80-142-3	Gl-Schist	WR	110	60.6	1.601	0.7187
80-19-4	Gl-Schist	WR	16.1	100	18.01	0.7478
		WM	31.0	380	35.7	0.7741
80-113-1	Gl-Schist	WR	85.9	149	5.03	0.7307
		PA	89.9	28.3	9.13	0.7375
81-119	Q-Chl-Ab-Schist	WR	130	136	3.023	0.7280
		WM	110	321	8.47	0.7357
80-91-2	"Gl"-Ctd-Schist	WR	45.3	166	10.69	0.7867
		WM	66.8	411	17.73	0.8016
80-41-4	"Gl"-Schist	WR	70.7	168	6.90	0.7609
		WM	108	331	8.92	0.7655
Metabasite and Marble						
80-118-1	Ga-Gl-Metabasite	WR	533	1.6	0.008	0.7053
		AM	155	0.3	0.005	0.7052
		PA	740	37.9	0.148	0.7056
80-117-1	Ga-Gl-Metabasite	WR	1041	6.0	0.017	0.7062
		AM	591	4.0	0.019	0.7059
80-151-1	Ab-Act-Metabasite	WR	271	6.5	0.070	0.7047
		AM	212	4.2	0.058	0.7046
80-155-1	Marble	WR	1081	29.0	0.077	0.7086**
Orthogneiss						
SL81-5-2	Gneiss	WR	37.1	220	17.46	0.8643
		WM	11.6	884	231	1.1559
SL81-94-1	Gneiss	WR	56.0	221	11.48	0.7848

*WR - Whole Rock, WM = White Mica, Am = Amphibole, PA = Pargonite.
**Initial ratio 550 Ma ago is 0.7080, which agrees with values of $^{87}Sr/^{86}Sr$ in early Paleozoic seawater. Burke and others (1982) report ratios of 0.7078 to 0.7091 for Cambrian-Ordovician time.
In the names of rocks: Ab - Albite, Act = actinolite, Chl = Chlorite, Ctd = chloritoid, Ga = garnet, Gl - glaucophane (if pseudomorphed, quote marks are added), Q = quartz.
For sample locations see Appendix 1.

personal communication), is dated at 110 to 90 Ma. Much of the uplift of the Nome Group terrane had occurred by 110 Ma ago.

Attempted U-Pb Dating

An attempt to date sphene from a typical glaucophanitic metabasite (80-118-1) by the U-Pb method (Mattinson 1981; this volume) was unsuccessful owing to high levels of non-radiogenic Pb and low U (J. M. Mattinson personal communication). It seems likely that this problem may be related to the presence of epidote, which contains an allanite component, and that lawsonitic blueschists may, in general, be more favorable candidates for U-Pb dating of sphene.

Rb-Sr Whole-Rock Analyses of Schists

Whole-rock Rb-Sr isotopic analyses of Seward Peninsula rocks are plotted in Figure 4. For the Nome Group we present new data for nine schists, three amphibolites, and one micaceous marble. The Kigluaik Group is represented by seven analyses of paragneiss and one of a conformable pegmatite from Bunker and others (1979). Nome Group orthogneiss is represented by two of our analyses and three of Bunker and others (1979), who also provide data for seven Kigluaik Group orthogneisses and two orthogneiss samples from Kiwalik Mountain.

Our own experience and that of others (e.g. Cordani and others 1978) have shown that pelitic schists may remain *nearly* closed systems during metamorphism so that pelite whole-rock isochrons approximate the age of deposition, but this observation must be further qualified. Isotopic features inherited from sediment provenance and Rb or Sr mobility during diagenesis and metamorphism can introduce considerable scatter and bias. Small specimens and samples of high Rb/Sr ratio are particularly susceptible to radiogenic Sr loss or gain. Low Rb/Sr ratio samples tend to gain radiogenic Sr but their $^{87}Sr/^{86}Sr$ ratios change only gradually because of their higher Sr content. This is shown

Armstrong and Others

Sample	Whole-Rock-Mineral Pair Date		K-Ar Date	Whole-Rock Date* assuming initial Sr =0.705	=0.708
Low-grade Pelitic Schist					
SP81-333F	118	± 7 @0.7176		212	184
80-15-5	117	± 4 @0.7139		745	533
80-93-1		---		1386	1270
80-142-3		---		600	469
80-19-4	105	± 4 @0.7210	156 ± 5	167	155
80-113-1	116	± 4 @0.7224		359	317
81-119	99.5	± 4 @0.7237		533	464
80-91-2	149	± 6 @0.7641	162 ± 6	536	517
80-41-4	160	± 26 @0.7452	141 ± 5	568	538
Metabasite					
80-118-1	170	± 80 @0.7053	122 ± 4	---	---
Orthogneiss					
SL81-5-2	96	± 8 @0.8405		636	628
SL81-94-1		---		488	470

*The uncertainity in this case cannot be evaluated solely from analytical data. The sensitivity of the date to the choice of initial ratio is indicated by comparison of the dates calculated for a typical volcanic-arc derived pelite initial ratio (0.705) and early Paleozoic seawater Sr isotopic composition (0.708).

by a large body of data for Belt-Purcell rocks compiled by Armstrong and others (1985). Apparent Rb-Sr dates of shale to schist samples scatter mostly toward the young side of a reference isochron for the beginning of deposition (based on U-Pb zircon dates). In that case even amphibolite facies metamorphism only reduces the model Rb-Sr dates by about 30% and most low-grade metamorphic samples lie in a fan within 20% of the reference age. Orthogneiss tends to be much more resistant to Rb and Sr redistribution. In general, metamorphism of any rock will rotate isochrons to produce lower dates and elevated initial ratios.

Nome Group schists have model dates (based on an assumed initial ratio of 0.705) ranging from 167 to 1386 Ma (Table 2). If we assume an initial ratio of 0.708 (as suggested by the marble sample) the calculated ages of five schists fall in the range 464 to 538 Ma, that is, early Paleozoic. Only three samples give much younger ages and were evidently drastically reset by the Mesozoic metamorphism. One low Rb/Sr ratio sample may be older than 1000 Ma or have gained radiogenic Sr during metamorphism, which seems more likely. The whole-rock data are consistent with an early Paleozoic to latest Precambrian age for the Nome Group. Samples collected by A. B. Till and J. A. Dumoulin (Till and others 1983; Dumoulin and Harris 1984; A. B. Till personal communication) have yielded the remains of conodonts and corals in Nome Group rocks, to which Ordovician, Silurian, and Devonian ages have been assigned, but schists underlying fossiliferous rocks could be somewhat older.

Rb-Sr Whole-Rock Analyses of Orthogneiss

Granitic orthogneiss bodies in sillimanite-grade schists of the

Kigluaik Mountains have been studied by Bunker and others (1979) and Till (1980, 1983). It is not known whether these are distinct from the orthogneiss bodies occurring in the Nome Group (to be discussed later). The apparent geochemical differences (e.g. higher K in the Kigluaik orthogneiss) may become blurred with additional work (cf. Till 1983). Bunker and others (1979) interpreted a 326 Ma isochron for five of six Kigluaik orthogneiss samples as the result of partial resetting of Precambrian ages by Cretaceous metamorphism. Till (1982) suggested that this Carboniferous age might represent an intrusive event.

Small lenticular bodies of tonalitic, granodioritic, and granitic orthogneiss, some with porphyroclastic and blastomylonitic texture and others with crystalloblastic texture are found among Nome Group metasediments. Owing to deformation and poor exposure, evidence as to their intrusive relationships is ambiguous. Present mineralogy (quartz, albite, microcline, 3T-phengite, chlorite, epidote, minor biotite, ±garnet) is compatible with blueschist and greenschist facies conditions.

Five Nome Group orthogneiss samples—three from Bunker and others (1979) and two analysed in this study from the Salmon Lake area (Thurston 1985)—are plotted in Figure 4. Also plotted are two points (from Bunker and others 1979) for a possibly correlative, much larger orthogneiss body at Kiwalik Mountain (Figure 1). A U-Pb zircon date of 381 Ma was obtained for the Kiwalik Mountain orthogneiss by J. Aleinikoff (Till 1982). The calculated initial $^{87}Sr/^{86}Sr$ ratio for the Kiwalik Mountain orthogneiss is relatively high (0.7081 to 0.7088) implying incorporation of older crust at the time of genesis.

The Nome Group orthogneisses scatter about an isochron with initial ratio of 0.720 and roughly 500 Ma age. They appear

TABLE 3. K-Ar MINERAL DATES, NOME GROUP, SEWARD PENINSULA

Sample Number	Rock	Mineral	%K(n)*	Radiogenic ^{40}Ar mol/gm x 10^{10}	$\frac{^{40}Ar\ rad.}{^{40}Ar\ total}$	Age (Ma ± σ)
80-19-4	Schist	Pheng.	7.87(2)	22.322	0.862	156 ± 5
80-91-2	Schist	Pheng.	6.93(2)	20.329	0.942	162 ± 6
80-41-4	Schist	Pheng.	6.35(2)	16.194	0.935	141 ± 5
80-117-1	Metabasite	Glauc.	0.113(2)	0.336	0.565	164 ± 6
80-118-1	Metabasite	Parag.	0.677(2)	1.487	0.573	122 ± 4
80-151-1	Metabasite	Actin.	0.146(2)	0.519	0.613	194 ± 7
67-ASn-595+	Metabasite	Glauc.	0.074(4)	0.237	0.608	175 ± 7

*(n) = number of replicate analyses.
+D.L. Turner, unpublished data. Sample described by Sainsbury and others (1970).
For location of samples see Appendix 1.

to be older than the Kigluaik orthogneisses. That is possible, but the data do not rule out the possibility of a single orthogneiss event with variable and high initial ratios in different bodies. One might suggest that the Kigluaik orthogneisses have suffered greater radiogenic Sr loss during prograde metamorphism associated with emplacement of Cretaceous plutons, and are thus of the same approximately early Paleozoic age as the Nome group orthogneisses, but the Kigluaik schist whole-rock Sr data show the opposite effect. Nome Group schists appear to be *more* reset than Kigluaik Group schists, so the younger dates for Kigluaik orthogneiss may hint at a true relative age relationship.

It seems reasonable to view one sample analysed by Bunker and others (1979) as a foliated Cretaceous intrusive (with model date of 88 Ma assuming a 0.705 initial ratio). The orthogneisses indicate the same general age bracket for Nome Group rocks as the schist analyses and fossil evidence, but, without multiple samples and zircon dates for each body, we can give no other firm conclusions.

AGE OF NOME GROUP METAMORPHISM

Bunker and others (1979) interpreted the Rb-Sr whole rock dates for Kigluaik Group schist and orthogneiss to indicate a Precambrian time of metamorphism and gneiss emplacement and argued that the isotopic data for a pegmatite-paragneiss pair supported that view. Their interpretation is, in our judgment, without justification. Metamorphism causes some Rb and Sr redistribution but does not normally erase crustal-residence model dates. A relatively non-radiogenic pegmatite in radiogenic paragneiss such as the one they describe could be as young as Cretaceous, with an initial $^{87}Sr/^{86}Sr$ ratio within the range we have observed for Cretaceous rocks in the Yukon. Only if the pegmatite was proven to originate in isotopic equilibrium with its country rock would the Precambrian age interpretation be persuasive. From the overprint of staurolite, kyanite, and sillimanite on probable Nome Group rocks on the south flank of the Kigluaik Mountains (Thurston 1985), the logical conclusion is that a significant part of the high-grade metamorphism of the Kigluaik Group is post-blueschist and thus of approximately Early Cretaceous age. This

metamorphism evidently did not greatly reset the Kigluaik paragneiss whole-rock Rb-Sr systems. A late Proterozoic age for the protolith of Kigluaik Group paragneiss would be consistent with its structural position beneath Nome Group rocks.

We must emphasize that there is *no* compelling evidence of Precambrian metamorphism in the Seward Peninsula, although we would have been glad to have found it and confirmed the earlier (and widely quoted) interpretation. In fact we originally set out to penetrate the Cretaceous metamorphic veil but found no stratigraphic or isotopic evidence of high-pressure or high-grade metamorphism prior to the Mesozoic. This is not to say that metamorphism and deformation did not occur earlier. Gardner and Hudson (1984) report folds with axial plane schistosity in older Nome Group rocks that are not observed in other rocks thought to be of Paleozoic age and conclude that they are Precambrian. On the other hand, rocks as young as Devonian show the regional blueschist overprint in the nearby Brooks Range (Hitzman and others 1982; and note the section of this paper on the Squirrel River area). Moreover, K-Ar dates for hornblende in ophiolites of the western Brooks Range and Yukon-Koyukuk regions are mostly reset in the range 138 to 164 Ma (Patton and others 1977) and garnet amphibolite at the sole of the ultramafic sheets gives hornblende K-Ar dates of 155 to 172 Ma (Patton 1984). This is consistent with Middle to Late Jurassic emplacement and rapid cooling of detached hot, and thus relatively thin, lithosphere (as described by Armstrong and Dick 1974) and suggests an integral role for that tectonic event in the formation of the northern Alaskan blueschists. The observations cited and our mineral dates are most easily reconciled with the principal episode of subduction-related high-pressure metamorphism taking place some time during the Jurassic, with the high-grade overprint observed in anticline cores occurring during the Early to Middle Cretaceous.

Cretaceous Plutons

The culmination of regional blueschist metamorphism in the Seward Peninsula in the Jurassic was followed by an approximately 50 Ma period of regional uplift accompanied by mag-

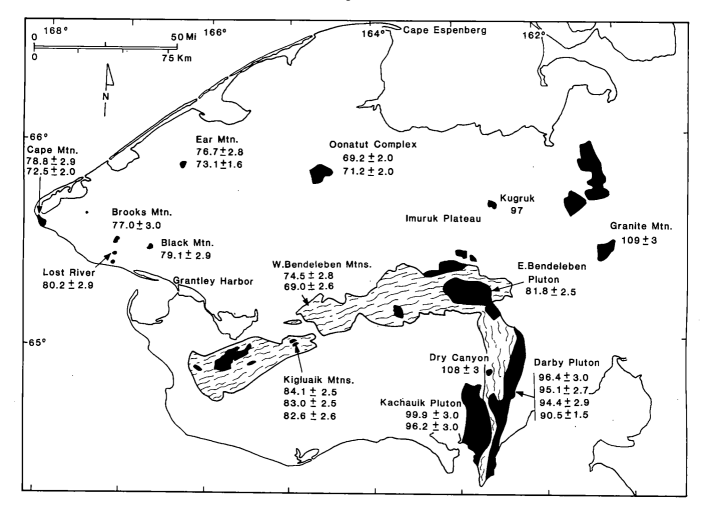

Figure 5. Map of the Seward Peninsula showing post-kinematic intrusive bodies and their ages. Sources are listed in the caption of Figure 3.

matic quiescence. Cretaceous plutonic activity (Figure 3 and 5) was initiated around 110 Ma ago by the emplacement of granitic bodies in the Darby Mountains and at Granite Mountain. This was followed by voluminous emplacement of granodioritic and granitic magmas in the Darby, Bendeleben, and Kigluaik Mountains. Granitic plutonism continued to about 70 Ma ago when the youngest of the small "tin granite" plutons was intruded into Nome Group schists and relatively unmetamorphosed Paleozoic rocks of the York Mountains.

Although the young "tin granites" and older plutons of the Darby Mountains were emplaced post-kinematically with respect to their enclosing rocks, granitic plutons in the Bendeleben and Kigluaik complexes are both post-kinematic and synkinematic. However, both structural types yield virtually concordant K-Ar uplift and cooling dates (80–87 Ma) in both complexes. Thermal metamorphism associated with the plutons developed Barrovian-type metamorphic zonation in adjacent country rocks, some of which belong to the Nome Group.

Cretaceous Reset K-Ar Dates

There are no published isotopic dates available for the metamorphic rocks of the Darby Mountains complex. However, the work of Bunker and others (1979), Wescott and Turner (1981), and Sturnick (1984) has provided dates related to uplift and cooling of the Kigluaik and Bendeleben complexes.

Within these gneiss complexes, there is ambiguity as to whether granitic "orthogneisses" represent intrusions metamorphosed during the Cretaceous or earlier times, or whether they are the result of deformation of Cretaceous plutonic rocks. Nevertheless, there is a remarkably tight grouping of K-Ar (cooling) dates among obviously metamorphic rocks (pelitic schists and gneisses, amphibolite), migmatitic gneisses, "orthogneiss," "gneissic granites," and post-kinematic rocks. K-Ar dates (eight biotites and two hornblendes) for the metamorphic and gneissic rocks on the north margin of the Kigluaik complex and the west end of the Bendeleben complex range from 81.4 to 87 Ma (Turner and others 1980; Wescott and Turner 1981; Sturnick 1984). As mentioned earlier, the post-kinematic intrusions also give dates in the range 82 to 84 Ma.

Only one Rb-Sr mineral date, 77 Ma for biotite, has been reported for paragneiss in the core of the Kigluaik complex (Bunker and others 1979). The blocking temperature for Rb-Sr in biotite is approximately the same as for K-Ar.

These concordant late Cretaceous dates from both intrusive and metamorphic rocks suggest that rapid uplift of the Kigluaik and Bendeleben gneiss complexes took place distinctly after emplacement and thermal equilibration of the late post-kinematic intrusions. Detrital staurolite and kyanite, presumably of Kigluaik–Bendeleben Mountains provenance, (as well as detrital glaucophane from the Nome Group) have been recognized in heavy mineral concentrates separated from sediments from the recently drilled COST exploratory well in Norton Sound, all first appearing in strata of Paleogene age (Turner 1983). An approximate cooling rate for the Kigluaik-Bendeleben complexes of 10° C/Ma is indicated and this would mean 0.5 km/Ma for a low thermal gradient (20°C/km) or 0.25 km/Ma for a relatively high (40°C/km) gradient. The actual rates in Cretaceous time could have been higher at first before declining to lower values.

Basaltic Volcanism

Cretaceous basaltic magmatism in the Seward Peninsula is indicated by a whole-rock K-Ar date of 74.9 Ma (Turner and others 1983) for a post-metamorphic basaltic dike in the western Bendeleben Mountains. The Cenozoic (29 Ma and younger) Imuruk Basalt Field is extensive (Turner and others 1980). The youngest eruptives on the Imuruk plateau (Figure 5) are of Holocene age (Hopkins 1963). Alkali basalt flows and vents north of Grantley Harbor (Figure 5) were dated by Turner and others (1980) at 2.5 to 2.7 Ma. Maar vents were active on the Espenberg Peninsula (Figure 5) during the late Pleistocene and Holocene. None of this basaltic magmatism is thought to have been a major factor in the thermal history of Seward Peninsula metamorphic rocks.

Isotopic Signature and Tectonic Correlations

Whole-rock Rb-Sr analyses of Seward Peninsula metamorphic rocks and orthogneisses, as plotted on Figure 4, lie in a fan with apex at a typical volcanic-arc initial $^{87}Sr/^{86}Sr$ ratio of 0.704 and bounding isochrons of about 720 and 360 Ma. A few schist samples and one gneiss sample lie toward the higher Rb/Sr, less radiogenic Sr side. Since the collection might include Paleozoic pelitic rocks that have lost radiogenic Sr during Mesozoic metamorphism and possibly a Cretaceous foliated granite, this scatter is not unexpected. What is notable is the very radiogenic character of Sr in most orthogneiss samples and the distinctive range of model whole-rock Rb-Sr dates.

In our experience such a regional isotopic signature has been

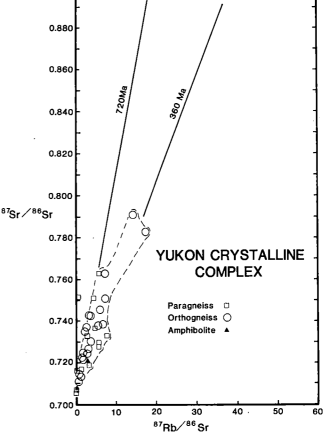

Figure 6. Rb-Sr Isochron diagram for amphibolite, schist, and orthogneiss of the Yukon Crystalline Complex, Canada. Data from Htoon (1979), Kuran and others (1982), Mortensen (1983), and unpublished analyses of rocks collected by C. I. Godwin and D. J. Tempelman-Kluit.

observed in one other nearby area of the northern Cordillera— the Yukon-Tanana Terrane (see Figure 6) and further to the south in the Cariboo structural culmination of the Omineca Belt in British Columbia (Blenkinsop 1972). This signature has been a cause of much puzzlement and is only now being understood with the aid of modern field studies and zircon U-Pb dating. The relatively radiogenic Sr indicates either Precambrian ages or unusually high initial ratios. In polymetamorphic terranes that ambiguity is not easily resolved.

In the Yukon Crystalline Complex of Alaska and British Columbia and in the Cariboo region a large volume of Windermere age—late Proterozoic, <770 Ma according to Armstrong and others, (1981) or <728 to 741 Ma according to Evenchick and others, (1984)—to early Paleozoic clastic sediments eroded from 2 Ga or older sources (Aleinikoff and others 1981; Dusel-Bacon and others 1983; Mortensen 1983; Evenchick and others 1984) and containing a few carbonate layers are country-rock for numerous lenticular orthogneiss bodies. Most of these orthogneisses turn out to be Devonian to Mississippian by U-Pb dating of zircon and some Rb-Sr dates (Tempelman-Kluit and Wanless

1980; Dusel-Bacon, and others 1983; Mortensen 1983) but gneissic Mesozoic plutons are also present. The Devonian to Mississippian granitic rocks are peculiar in having very radiogenic initial Sr—their whole-rock dates, calculated assuming a 0.704 initial ratio, are, in many cases late Proterozoic (unpublished UBC data). This can be rationalized by proposing that the gneiss is melted late Proterozoic sediment, or has assimilated an overwhelming amount of late Proterozoic sediment, thus acquiring its isotopic signature. It is still possible that some orthogneiss bodies are late Proterozoic but the simplest hypothesis, presently, is that almost all are mid Paleozoic with variably radiogenic initial Sr. The radiogenic Sr is itself a continental crust signature. The clastic plus orthogneiss section of the Yukon and Cariboo is overlain stratigraphically and/or tectonically by Devonian to Triassic volcanic, clastic, and carbonate rocks. Parts of this package suffered greatly during Late Triassic to Early Jurassic regional deformation and metamorphism (Tempelman-Kluit 1976, 1979; Cushing and others 1984) and was later overprinted by Jurassic to Mid Cretaceous granitic plutons, heat, and uplift, so that most K-Ar and Rb-Sr mineral dates are reset to about 100 Ma (Tempelman-Kluit and Wanless 1975; Aleinikoff and others 1981). The Seward Peninsula differs from the other areas in the greater abundance of metabasic rocks intrusive into late Proterozoic–early Paleozoic stratified rocks, the more pelitic character of its sediments, near absence of Carboniferous to Triassic supracrustal rocks, lack of Jurassic plutons, and higher pressure metamorphism later during the Jurassic.

The isotopic similarities of the clastic sedimentary rocks and orthogneiss bodies they contain are striking and lead us to suggest the close affinity of the Seward Peninsula with those other pre-Mesozoic North American continental-margin terranes. The Seward Peninsula contains more mafic igneous rocks intruding the finer clastic, more distal facies of the late Proterozoic to Paleozoic stratigraphic package outboard of North America and it was subducted more deeply during Jurassic time. In northern Alaska, as in parts of the Alps (Frey and others 1974; Desmons 1977; Chopin 1984), continental rather than oceanic crust has been subducted, and subsequently exhumed.

A rapid (~30 to 40 Ma) switch from blueschist metamorphism to doming, Barrovian metamorphism and arc plutonism has been described in the Aegean by Lister and others (1984) who point out the parallels there on Ios and Naxos with Cordilleran metamorphic core complexes. Low-angle normal faults that are characteristic of those areas may be a partial explanation for the irregular distribution of rock units in the Seward Peninsula. Such structures facilitate the uplift and rapid cooling of metamorphic terranes. However, obvious low-angle movement zones have not yet been discovered in the course of our mapping.

BROOKS RANGE

Blueschists occur at numerous localities in the southern Brooks Range, mostly within the Brooks Range Schist Belt, but also further west in the Hub Mountains, Kallarichuk Hills, and Squirrel River areas that are all in the larger metamorphic belt of the southern Brooks Range. Turner and others (1979) reported 76 K-Ar dates for minerals from the southern Brooks Range that were largely in the range of 86 to 185 Ma. Older dates for K-poor minerals (actinolite, tremolite, glaucophane, and paragonite) were in part, and probably correctly, attributed to excess radiogenic Ar. This is likely to occur wherever old rocks (in this case Paleozoic to Precambrian (?) sedimentary and volcanic rocks) undergo a much younger metamorphism (in this case at or before 120 to 180 Ma ago, essentially Jurassic) (Wanless and others 1970; Wilson 1972; Bocquet and others 1974). However, it was also felt by Turner and others (1979) that some of these older dates were relict or hybrid. In two areas in the Baird Mountains Quadrangle (Hub Mountain and Kallarichuk Hills) (Figure 7) consistent late Precambrian K-Ar dates were obtained. Since some of these dates were for relatively K-rich minerals, Turner and others (1979) were led to the conclusion that a Late Precambrian blueschist facies metamorphism had affected the schist belt, at least locally, and perhaps extensively.

Stratigraphic Evidence

The Brooks Range Schist Belt protoliths were argillaceous, quartzose, and calcareous-dolomitic sediments with intercalated mafic and felsic flows and sills. The rocks are variably metamorphosed to blueschist, greenschist, and lower amphibolite facies and intruded by granitic plutons with amphibolite facies contact aureoles. The plutons give Cretaceous K-Ar dates, but where dated by Rb-Sr and U-Pb they have been shown to be Paleozoic (e.g. Silberman and others 1979a; Dillon and others 1980). Paleozoic fossils and Devonian and late Proterozoic isotopic dates for felsic metaigneous rocks have been reported (Dillon and others 1980). Because early to mid-Paleozoic rocks are affected by blueschist metamorphism, Hitzman and others (1982) and Hitzman (1984) have argued that the blueschist event is post-Devonian. Observation of blueschist minerals in the Squirrel River area, which lies west of the Schist Belt and between it and the Seward Peninsula, also supports the geologic argument for a post-Devonian regional blueschist event. Sodic amphibole was recognized by Forbes in 1969 in samples of metabasite collected in the Squirrel River area by I. L. Tailleur. With his help the locality was revisited in 1974, and additional samples collected for petrography and analysis.

At the Squirrel River locality lensoid metabasite boudins about 0.3 m across occur in a 25 m long train parallel to well preserved bedding in Silurian-Devonian carbonate rocks (with bryozoan and trilobite remains). The mineral assemblage of the metabasite is glaucophane-chlorite-carbonate-quartz and the fabric incipiently schistose. A nearby basalt dike, evidently much younger, shows no metamorphic overprint. The metabasite is a highly sodic tholeiitic basalt, presumably a sill before deformation.

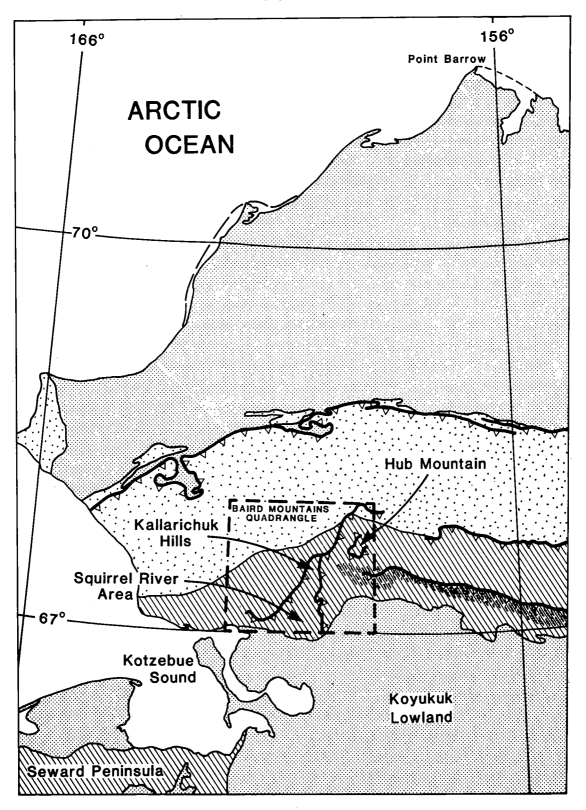

Figure 7. Map of northwestern Alaska (after Beikman 1980) showing the geologic setting of the areas discussed in the text. Diagonal lines show the extent of low-grade metamorphic rocks. The light stipple distinguishes unmetamorphosed Paleozoic to Jurassic rocks, the dense stipple Cretaceous and younger rocks.

TABLE 4. Rb-Sr DATA FOR ROCKS AND MINERALS OF THE BROOKS RANGE SCHIST BELT

Sample Number	Rock Type	Latitude Longitude	Material Analyzed*	Sr ppm	Rb ppm	^{87}Rb/^{86}Sr	^{87}Sr/^{86}Sr
Baird Mountains Quadrangle							
74 AF 145-1	Metasedimentary muscovite schist	67°39' 160°9.2'	WR	623	128	5.96	0.7598
			MS	114	347	8.87	0.803
74 AF 145-5	Amphibole schist	67°39' 160°9.2'	WR	169	16.8	0.287	0.7093
			H	102.45**	18.3**	0.517	0.7091
74 AF 144-3	Amphibole schist	67°10.3' 160°6.8'	WR	493	0.8	0.005	0.7054
			A	108	0.8	0.022	0.7050
74 Pe 2B	Biotite-muscovite gneiss	67°2.3' 160°10.2'	MS	22.7	338	43.4	0.7758
Amber River Quadrangle							
72 Pe 92	Biotite muscovite augen gneiss	67°13.6' 156°7'	WR	64.4	254	11.52	0.7969
			MS	12.4	826	198	1.059
73 RR 18 F	Biotite quartz muscovite schist	67°13.3' 156°42.2'	BI	6.1	463	228	1.046
			MS	2.6	283	339	1.298
A-1 CNM	Garnet glaucophane schist	67°13.4' 156°44.4'	P	816	4.8	0.017	0.7076
72E 100R	Quartz muscovite schist	67°20.2' 157°1.4'	MS	72.4	622	25.0	0.7616
72E 222	Coarse flaser granitic gneiss	67°19.1' 156°7.8'	WR	20.9	303	42.9	0.8918
			BI	6.1	2110	1175	2.451
			MS	6.5	1226	580	1.616
Survey Pass Quadrangle							
73 ATr 20.2	Quartz muscovite schist	67°8.3' 153°56'	WR	29.6	268	26.5	0.7819
			MS	48.6	520	31.2	0.7857
73 B 33	Muscovite feldspar schist	67°5.2' 155°1.2'	WR	39.8	140	10.20	0.7487
			MS	2.83	490	544	1.598
73 Pe 32	Calcareous muscovite schist	67°10.5' 153°58.8'	WR	21.4	151	20.5	0.7459
73 ATr 15.2	Biotite muscovite feldspar schist	67°6.2' 155°19'	WR	19.4	202	30.4	0.7929
			BI	6.0	1299	691	1.748
			MS	24.5	891	107	0.9068

*WR = Whole Rock, MS = Muscovite, BI = Biotite, H = Hornblende, A = Actinolite, P = Paragonite.
** Analyses by isotope dilution. All other concentrations are by X-ray fluorescence.

If the metabasite is not a tectonic sliver, and there is no evidence for that, then the metamorphism, which has also recrystallized and strained the enclosing carbonate, must be post-Devonian. No aragonite was found in the carbonate and the blue amphibole was too fine grained to prepare a mineral separate suitable for K-Ar dating, but the other evidence cited for post-Devonian blueschist metamorphism here is in accord with results from both the Seward Peninsula and Brooks Range schist belts, and links both into a nearly continuous belt of post-Devonian, and presumably Jurassic, blueschist occurrences across northern Alaska.

New Isotopic Data

We have dated by Rb-Sr many of the mineral separates analysed by Turner and others (1979). The results are presented in Tables 4 and 5. Where whole-rock material was obtainable we analysed both the whole-rock and mineral fractions. In some cases only small chips of rock could be retrieved and so may not be entirely representative of the whole rocks from which the minerals were extracted. That may explain some discordance between K-Ar and Rb-Sr dates. The high Rb/Sr ratios of most mineral separates make the dates rather insensitive to the initial ratio used so we have recorded both the rock-mineral pair (or rock-mineral isochron) date and the model date calculated assuming an assumed initial ^{87}Sr/^{86}Sr ratio of 0.705. This is about the lowest initial value to be expected in sedimentary rocks (Faure and Powell 1972) and will give maximum possible dates. This approach has proven useful in the Alps where micas from old basement give discordant whole-rock-mineral isochron dates (Satir 1974). We will first discuss our confirmation of Precambrian ages in the Hub Mountains and then the isotopic evidence

TABLE 5. Rb-Sr DATES (Ma) WITH ESTIMATED ONE SIGMA ERRORS FOR THE ROCKS AND MINERALS OF THE BROOKS RANGE SCHIST BELT

Samples	Isochron	Whole-Rock-Mineral Pair	Model Date Assuming Initial Sr = 0.705		K-Ar Date*
			Whole Rock	Mica	
Precambrian Metamorphism					
74 AF 145-1: MS ⎫		1026 ± 120 @ 0.672	645 ± 19	770 ± 23	653 ± 39
⎬ 686 ± 116 @ 0.7055 ± 0.0011 MSWD = 110					
74 AF 145-5: WR ⎭			---	---	
Mesozoic Metamorphism					
74 Pe 2B: MS	---	---		115 ± 4	107 ± 3
72 Pe 92: MS		99 ± 10 @ 0.812	560 ± 17	126 ± 8	116 ± 3
73 RR 18f: BI	---			105 ± 9	132 ± 4
MS	---			123 ± 24	130 ± 4
72E 100R: MS	---			159 ± 6	129 ± 4
72E 222: BI ⎫			306 ± 9	105 ± 9	96 ± 3
⎬ 96 ± 1.4 @ 0.833 ± 0.002 MSWD = 0.5					
MS ⎭				111 ± 7	101 ± 3
73 ATr 20.2: MS		60 ± 10 @ 0.762	204 ± 7	182 ± 6	116 ± 3
73 B 33: MS		112 ± 22 @ 0.732	301 ± 8	115 ± 20	113 ± 3
73 Pe 32: WR		---	141 ± 7	---	140 ± 4
73 ATr 15.2: BI ⎫			203 ± 11	106 ± 9	105 ± 3
⎬ 103 ± 2 @ 0.749 ± 0.001 MSWD = 0.6					
MS ⎭				133 ± 6	118 ± 4
3 Biotite	105 ± 5 @ 0.705 ± 0.028	MSWD = 0.3			
6 of 8 Muscovite	120 ± 7 @ 0.724 ± 0.005	MSWD = 25			

*Turner and others, 1979.

for a Jurassic age of blueschist metamorphism in other areas, starting in the west and moving eastward, progressively away from the Seward Peninsula.

Hub Mountain

Hub Mountain is an antiformal structure located in the northeastern part of the Baird Mountains Quadrangle, where the general east-west trends of the Brooks Range Schist Belt are sharply deflected to north-south (Figure 7). Hub Mountain lies northwest of the Brooks Range Schist Belt *sensu stricto*. The Hub Mountain half-window, or window as shown by Beikman (1980), exposes a core of metaplutonic, metavolcanic, and metasedimentary rocks. Surrounding the core, above what appears to be a thrust contact, is a crescent-shaped body of Ordovician (?) rocks. These are reported by Mayfield and others (1982) to include impure carbonate, shale, micaceous quartzite, and metabasite. Tailleur (personal communication to Forbes) has observed glaucophane-bearing metabasite enclosed in marble in the southwestern part of the Ordovician(?) rocks. At one locality north of the Tutuksuk River, well preserved Ordovician graptolites occur in a "sliver" of slate (Tailleur and Carter 1975). Well-formed muscovite in marble and glaucophane in metabasite occur in both core and structurally overlying peripheral metasedimentary rocks.

Study of core zone rocks is incomplete. Rock samples from only a few localities have been examined in thin section. Mayfield and others (1982) describe a late Precambrian pluton in the north

part of the core as hornblende meta-diorite and note that quartz monzonite, quartz diorite, and diorite also intrude the Hub Mountain area. Country rock to the plutons in highly contorted marble, pods of metabasite and layered schist, and quartzite.

The locality dated as Precambrian by Turner and others (1979) (74 AF 145) was first traversed by Hill Reiser of the U.S. Geological Survey and revisited by Forbes in 1974 after blueschist was recognized in the samples collected by Reiser. At that place layered schists crop out of high tundra in the core of the Hub Mountain antiform. Seven of the fourteen samples collected contained crossite or glaucophane. The blueschist layers are intercalated with micaceous schist, greenschist, and garnet amphibolite on a scale of cm to m. Some of the blueschist layers are carbonate-rich and clearly metasedimentary (Table 6 reports several assemblages from this locality). The K-Ar and Rb-Sr dated muscovite and hornblende separates come from layers intercalated with blueschist in the same outcrop. Sodic amphibole was not analysed because of the concern about excess Ar in low-K minerals.

Although Mayfield and others (1982) did not observe glaucophane in "several dozen" thin sections from the Hub Mountain area and were uncertain about the relationship between the high-K minerals dated as Precambrian and the blueschists, we have no doubt about the intimate spatial association. The K-Ar dates of 646 ± 19 Ma for muscovite and 729 ± 22 Ma for hornblende (Turner and others 1979) were confirmed by Mayfield and others (1982) who reported dates on three other muscovite and hornblende samples from other parts of the Hub

TABLE 6. MINERAL ASASEMBLAGES AT STATION 74 AF 145,
HUB MOUNTAIN, BROOKS RANGE

Sample	
-1 (K-Ar and Rb-Sr sample)	Muscovite-chlorite-quartz-carbonate
-2,3	Crossite-chlorite-carbonate-quartz-muscovite
-6,9	Amphibole-chlorite-epidote-albite-quartz-sphene
-5.1	Glaucophane-chlorite-albite-quartz-carbonate
-5.2 (K-Ar and Rb-Sr sample)	Garnet-amphibole-epidote-albite-quartz-sphene
-14	Muscovite-quartz-carbonate-albite

Mountain core zone ranging from 548 to 595 Ma. Our Rb-Sr analyses confirm a Precambrian age for one rock (mica schist sample 74 AF 145-1). In that case muscovite and whole-rock give model dates (assuming a 0.705 initial $^{87}Sr/^{86}Sr$ ratio) of 770 and 645 Ma, respectively. An isochron calculated using a nearby whole-rock amphibole schist and hornblende pair together with the whole-rock mica schist gives a date of 686 ± 116 Ma (Figure 8). The mafic sample (74 AF 145-5) ties the initial ratio to a reasonable value. The muscovite–whole rock (in this case a 30 gm chip) pair alone give a date slightly over 1 Ga with an impossibly low $^{87}Sr/^{86}Sr$ intercept (<0.68), so little significance can be attached to that date. Our hope to obtain a Rb-Sr date for the hornblende was frustrated by its low Rb content and the similarity of whole-rock and mineral Rb-Sr ratio. Our work confirms the Precambrian metamorphic age of that one sample, and removes any suspicion that it contains excess Ar and is in reality much younger. A protolith age similar to the metamorphic age is indicated but neither age is precisely determined.

Kallarichuk Hills

Turner and others (1979) reported twelve K-Ar dates for minerals of blueschists and other rocks of the Kallarichuk Hills, southwest of Hub Mountain (Figure 7) and also outside the Schist Belt, *sensu stricto*. Three low-K amphibole dates, 766 ± 45 Ma for glaucophane, and 661 ± 39 Ma and 632 ± 31 Ma for actinolite, are in the same range as the Precambrian dates for Hub Mountain but the possibility of excess Ar cannot be ruled out. Kallarichuk Hills micas gave K-Ar dates from 86 ± 2 to 195 ± 57 Ma and one of these we dated by Rb-Sr gave 115 ± 4 Ma. These are essentially the same as the results of K-Ar and Rb-Sr dating elsewhere in the Brooks Range Schist Belt.

Mesozoic Dates in the Brooks Range Schist Belt

In general muscovite and biotite from the Schist Belt give Rb-Sr dates between 96 and 159 Ma. The single mica with a possible date outside these limits came from sample 73 A Tr 20.2 for which whole-rock and muscovite had similar Rb/Sr ratios and the muscovite was not highly enriched in Rb and could be

anything from an uncertain 60 ± 10 Ma (whole-rock–muscovite isochron) to 182 ± 6 Ma old (model date assuming a 0.705 initial ratio). The one paragonite analysed lacked a whole-rock mate and was of such low Rb/Sr ratio that no age could be calculated for comparison with its 239 Ma K-Ar date.

The K-Ar and Rb-Sr dates are concordant, with overlapping 2σ errors. The average muscovite Rb-Sr date—126 or 115 Ma depending whether model or rock-mineral pair dates are

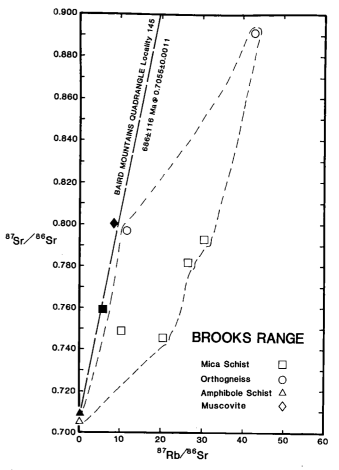

Figure 8. Rb-Sr Isochron diagram for Precambrian schist and amphibolite of the Baird Mountains Quadrangle (samples 145-1 and 145-5) and whole-rock samples from the southern Brooks Range Schist Belt.

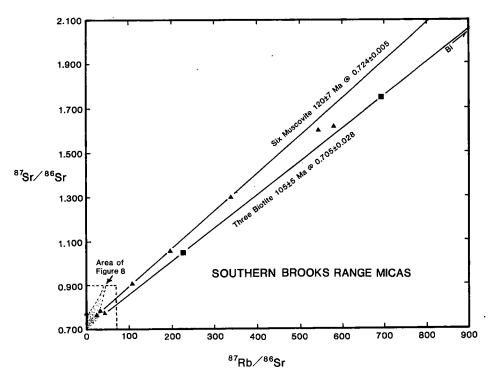

Figure 9. Rb-Sr Isochron diagram for micas from the southern Brooks Range Schist Belt.

favored—is identical to or slightly higher than the average K-Ar date, 116 Ma. Most (6 of 8) muscovites lie close to a 120 ± 7 Ma isochron with initial ratio 0.724 ± 0.005 (Figure 9). The closure temperature for Rb-Sr in muscovite is higher than that for K-Ar (see below); in cases where K-Ar dates have been lowered, the Rb-Sr often preserves evidence of earlier crystallization or slow cooling. Our muscovite Rb-Sr dates are consistent with a single, presumably Jurassic, metamorphic event that ended in Early Cretaceous time.

The average Rb-Sr biotite date, 105 or 101 Ma, is very close to the average K-Ar biotite date of 111 Ma. The three biotites alone lie on a 105 Ma isochron with initial $^{87}Sr/^{86}Sr$ ratio 0.705 $\pm$ 0.028 (Figure 9). The small difference between muscovite and biotite dates is consistent with their respective closure temperatures—biotite: 280–300°C for Ar, 300–320°C for Sr and muscovite: 350–380° C for Ar and 500°C for Sr (Hunziker 1974; Wagner and others 1977; Dodson 1979; Harrison and McDougall 1980). If excess argon is present it is more likely to show up in biotite than muscovite (Wanless and others 1970; Wilson 1972).

An approximation to a maximum cooling rate for the fall in temperature between muscovite and biotite isotopic diffusion blocking temperatures in the Schist Belt may be computed from the difference of 15 Ma between biotite and muscovite isochrons of Figure 9. The result is 13°C per Ma, and given a reasonable subduction-zone geothermal gradient of 10 to 20°C per km, an uplift rate of 1 km/Ma is obtained. A steeper thermal gradient would reduce this proportionately. The same calculation applied

to the difference in average biotite and muscovite K-Ar dates, 5 Ma, gives 15°C per Ma. With the same thermal gradient assumed, an identical uplift rate of 1 km/Ma is obtained. The results are in good agreement with typical Alpine rates of 0.2 to 1.4 km/Ma (Wagner and others 1977), and within a factor of two of the rate computed earlier for the Seward Peninsula over a much longer time span (~50 Ma).

Rb-Sr Whole-Rock Analyses

The whole-rock Sr results, plotted in Figure 8, show great scatter. Our model (assumed 0.705 initial ratio) whole-rock dates for orthogneiss are 306 and 560 Ma, consistent with an early to mid Paleozoic age. The remaining schists give younger dates spanning 301 to 141 Ma and are certainly partly reset by the regional metamorphic event(s). They give no hint of any pre-Paleozoic age and, although on average more radiogenic because of their higher Rb/Sr ratio, they appear to be younger than the Seward Peninsula schists. No more can be said without further studies involving multiple and larger samples of rock from the Schist Belt, and the application of other techniques such as whole rock Nd-Sm for crustal residence ages and U-Pb dating of zircons to give igneous emplacement ages and sedimentary provenance.

Tectonic Implications

There are Precambrian metamorphic rocks at Hub Mountain near the Brooks Range Schist Belt. The stratigraphic and isotopic evidence, from the Seward Peninsula to the Schist Belt, is for a Jurassic high-pressure metamorphic episode followed by Cretaceous uplift and cooling. The significance of the one Precambrian area is uncertain, even the extent of the Precambrian rocks is not yet well established. Late Precambrian K-Ar dates for micas have also been reported for the Ruby province by Silberman and others (1979b), so older rocks could be extensive in the metamorphic belts surrounding the Yukon-Koyukuk basin. The unique Baird Mountains Quadrangle locality deserves reexamination and detailed description. This may be a situation, as in the North Cascades (Armstrong and others 1983), where older blueschists (in that case Paleozoic) occur as exotic tectonic blocks in a Jurassic-Cretaceous blueschist terrane. It seems more logical to hold that hypothesis than to claim that all the coarse blueschists studied have been reset by heating during the Cretaceous plutonic episode and ignore the stratigraphic evidence that also supports a post-Devonian age. It remains to be conclusively established that the Precambrian event created blueschists. Mayfield and others (1982) could not confirm this but it is our impression at the one locality studied. Precambrian rocks may have merely been overprinted by a much later blueschist event or tectonically interleaved with younger blueschists. We leave the question open for future detailed work.

The Jurassic age for most blueschists of northern Alaska overlaps the ages for voluminous blueschists in California (Coleman and Lanphere 1971; Suppe and Armstrong 1972) and Washington (Brown and others 1982). Taken at face value the

K-Ar dates for Alaskan blueschist-assemblage minerals are in part a bit older than the Franciscan coarse exotic blocks or Shuksan blueschists (maximum 155 to 160 Ma, see also Mattinson, this volume); and more like those of pre-Nevadan Sierra Foothills blueschists (maximum 174-190 Ma) (Schweickert and others 1980).

It would be misleading to make exact correlations because of the remaining uncertainty concerning inherited Ar in Alaskan rocks and the potentially multiepisodic, variable duration, and time-transgressive nature of subduction and accretion episodes along western North America (Hamilton 1969; Coney et al. 1980) but the blueschist areas in northern Alaska deserve recognition as products of the Mesozoic subduction–accretion process that formed the Cordillera.

ACKNOWLEDGMENTS

We thank D. L. Turner of the University of Alaska for helping to provide the K-Ar mineral separates from his Brooks Range project, and for reviewing the manuscript; K. L. Scott for the Rb, Sr, and K analyses done at U.B.C.; J. Getsinger for editorial help; and A. B. Till of the U.S.G.S. Alaska Branch for multiple contributions to our work, including heroic efforts to keep us accurate, logical, and readable. Geochronometry research at U.B.C. has been supported by a Canadian Natural Sciences and Engineering Research Council operating grant to R. L. Armstrong. The work of the University of Washington group was supported by NSF grants EAR 80-08856 and EAR 82-18471.

APPENDIX I. MINERAL ASSEMBLAGES AND SAMPLE LOCATIONS FOR NOME GROUP,
SEWARD PENINSULA SAMPLES

		Latitude	Longitude
Pelitic Schists			
SP81-333F	ph(2M)-gl-ctoid-q-pa-ch-ap	65° 31.0'	164° 42.9'
80-15-5	ph(2M)-q-gl-ch-ctoid (→ch+pa)-ab-ep-(law?)*	65° 29.3'	164° 37.5'
80-93-1	ph(2M)-q-ch-ab-czo-tour-graph?	64° 33.9'	165° 22.4'
80-142-3	ph(2M)-q-gl-ga-ch-ep-sph-cc-ap	65° 27.8'	164° 58.1'
80-19-4	ph(2M)-gl-ctoid-ch-q-ap	65° 31.0'	164° 42.9'
80-113-1	pa + ph-q-gl-ch-ga-carb-sph	65° 33.2'	165° 34.4'
81-119	ph(2M)-q-ab-ep-ch-carb	64° 43.5'	164° 23.5'
80-91-2	ph(2M)-q-ctoid-(gl)*-ch-ab-czo-tour-bi	64° 34.7'	165° 18.9'
80-41-4	ph(2M)-ch-q-(gl)*-ab-ga-tour-czo-carb	64° 51.0'	165° 8.5'
Metabasites (metadiabases)			
80-118-1	gl-ga-pa-ep-sph-ab-ch-cc-q	65° 32.0'	165° 34.2'
80-117-1	gl-ga-pa-ep-ab-sph-ch	65° 32.2'	165° 33.9'
80-151-1	act-ch-czo-ab-sph-bi-rʰ-tour-cc	65° 5.0'	165° 12.9'
67-ASn-595+	gl-ga-sph-white mica	65° 42.3'	164° 59.8'
Marble			
80-155-1	cc-white mica-ab-q-sulfide	64° 57.1'	166° 11.9'
Orthogneiss			
SL81-5-2	ksp-ab-ph(3T)-q-chl-cc-bi	64° 52.9'	165° 13.1'
SL81-94-1	ksp-ab-ph(3T)-q-bi-ep-cc	64° 54.1'	165° 10.4'

*pseudomorphed minerals.
+information form Sainbury and others (1970).

APPENDIX II. ANALYTICAL PROCEDURES

```
     Rb and Sr concentrations were determined by replicate
analysis of pressed powder pellets using X-ray
fluorescence. U.S. Geological Survey rock standards were
used for calibration; mass absorption coefficients were
obtained from Mo K Compton scattering measurements. Rb/Sr
ratios have a precision of 2% (1σ) and concentrations a
precision of 5% (1σ). Sr isotopic composition was measured
on unspiked samples prepared using standard ion exchange
techniques. The mass spectrometer, Vacuum Generators
Isomass 54R, has data aquisition digitized and automated
using a Hewlett-Packard HP-85 computer. Experimental data
have been normalized to a 86Sr/88Sr ratio of 0.1194 and
adjusted so that the NBS standard SrCO3 (SRM 987) gives a
87Sr/86Sr ratio of 0.71020+2 and the Eimer and Amend
Sr a ratio of 0.70800+2. The precision of a single
87Sr/86Sr ratio usually is ≤ 0.00010 (1σ). Rb-Sr dates
are based on a Rb decay constant of 1.42 x 10⁻¹¹a⁻¹.
The regressions are calculated according to the technique
of York (1967).

     For samples with very low Sr or Rb concentrations or
extremely high Rb/Sr ratios conventional isotope dilution
techniques, using high purity 84Sr and 87Rb spikes, are
used to obtain concentrations with a precision of 1.5%. Sr
isotopic compositions measured on spiked samples and some
mineral separates may have a precision as low as 0.001.

     K is determined in duplicate by atomic absorption
using a Techtron AA4 spectrophotometer and Ar by isotope
dilution using an AEI MS-10 mass spectrometer and high
purity 38Ar spike. The constants used are:

     Kλe = 0.581 x 10⁻¹⁰a⁻¹,Kλβ=4.962 x 10⁻¹⁰a⁻¹,
     40K/K = 0.01167 atom percent.

     Decay constants are those recommended, by the IUGS
Subcommission on Geochronology (Steiger and Jäger 1977).
Errors reported are for one standard deviation or the
standard error of the mean unless otherwise noted.
```

REFERENCES CITED

Aleinikoff, J. N., Dusel-Bacon, C., Foster, H. L., and Futa, K., 1981, Proterozoic zircon from augen gneiss, Yukon-Tanana Upland, east-central Alaska: Geology, v. 9, p. 469–473.

Altherr, R., Schliestedt, M., Okrusch, M., Seidel, E., Kreuzer, H., Harre, W., Lenze, H., Wendt, I., and Wagner, G. A., 1979, Geochronology of high-pressure rocks on Sifnos (Cyclades, Greece): Contributions to Mineralogy and Petrology, v. 70, p. 245–256.

Armstrong, R. L., and Dick, H.J.B., 1974, A model for the development of thin overthrust sheets of crystalline rock: Geology, v. 2, p. 35–40.

Armstrong, R. L., Eisbacher, G. H., and Evans, P. D., 1981, Age and stratigraphic-tectonic significance of Proterozoic diabase sheets, Mackenzie Mountains, northwestern Canada: Canadian Journal of Earth Sciences, v. 19, p. 3126–323.

Armstrong, R. L., Harakal, J. E., Brown, E. H., Bernardi, M. L., and Rady, P. M., 1983, Late Paleozoic high pressure metamorphic rocks in northwestern Washington and southwestern British Columbia—The Vedder Complex: Geological Society of America Bulletin, v. 94, p. 451–458.

Armstrong, R. L., Parrish, R. R., van der Heyden, P., Reynolds, S. J., and Rehrig, W. A., 1985, Rb-Sr and U-Pb chronology of the Priest River metamorphic complex: Precambrian X basement and its Mesozoic-Cenozoic plutonic-metamorphic overprint, near Spokane, Washington, *in*, Geology of Washington: Washington Division of Geology and Earth Resources Publication, in press.

Beikman, H. M., compiler, 1980, Geologic Map of Alaska, 1:2,500,000 scale: U.S. Geological Survey.

Blenkinsop, John, 1972, Computer-assisted mass spectrometry and its application to rubidium-strontium geochronology [Ph.D. thesis]: University of British Columbia, Vancouver, 108 p.

Bocquet, J., Delaloye, M., Hunziker, J. C. and Krummenacher, D., 1974, K-Ar and Rb-Sr dating of blue amphiboles, micas, and associated minerals from the western Alps: Contributions to Mineralogy and Petrology, v. 47, p. 7–26.

Brown, E. H., Wilson, D. F., Armstrong, R. L., and Harakal, J. E., 1982, Petrologic structural and age relations of serpentinite, amphibolite, and blueschist in the Shuksan Suite, North Cascades, Washington: Geological Society of America Bulletin, v. 93, p. 1087–1098.

Bunker, C. M., Hedge, C. E., and Sainsbury, C. L., 1979, Radioelement concentrations and preliminary radiometric ages of rocks of the Kigluaik Mountains, Seward Peninsula, Alaska: U.S. Geological Survey Professional Paper 1129-C, p. 1–12.

Burke, W. H., Denison, R. E., Hetherington, E. A., Koepnick, R. B., Nelson, H. F., and Otto, J. B., 1982, Variation of seawater 87Sr/86Sr throughout Phanerozoic time: Geology, v. 10, p. 516–519.

Chopin, C., 1984, Coesite and pure pyrope in high-grade blueschists of the Western Alps: a first record and some consequences: Contributions to Mineralogy and Petrology, v. 86, p. 107–118.

Coleman, R. G., and Lanphere, M. A., 1971, Distribution and age of high-grade blueschists, associated eclogites, and amphibolites from Oregon and California: Geological Society of America Bulletin, v. 82, p. 2397–2412.

Coney, P. J., Jones, D. L., and Monger, J.W.H., 1980, Cordilleran suspect terranes: Nature, v. 288, p. 329–333.

Cordani, U. G., Kawashita, Koji, and Filho, A. T., 1978, Applicability of the rubidium-strontium method to shales and related rocks: American Association of Petroleum Geologists Studies in Geology, no. 6, p. 93–117.

Cushing, G. W., Foster, H. L., and Harrison, T. M., 1984, Mesozoic age of metamorphism and thrusting in the eastern part of east-central Alaska [abs.]: EOS—Transactions of the American Geophysical Union, v. 65, p. 290–291.

Desmons, Jacqueline, 1977, Mineralogical and petrological investigations of Alpine metamorphism in the internal French Western Alps: American Journal of Science, v. 277, p. 1045–1066.

Dillon, J. T., Pessel, G. H., Chen, J. H., and Veach, N. C., 1980, Middle Paleozoic magmatism and orogenesis in the Brooks Range, Alaska: Geology, v. 8, p. 338–343.

Dodson, M. H., 1979, Theory of cooling ages, in, E. Jager and J. C. Hunziker, editors, Lectures in Isotope Geology: Berlin, Springer Verlag, p. 194–202.

Dumoulin, J. A., and Harris, A., 1984, Carbonate rocks of central Seward Peninsula, Alaska [abs.]: Geological Society of America Abstracts with Programs, v. 16, p. 280.

Dusel-Bacon, C., Aleinikoff, J. N., Mortensen, J. K., and Foster, H. L., 1983, A belt of middle Paleozoic augen gneiss batholiths with evidence for an early Proterozoic component in the crust beneath the Yukon-Tanana terrane [abs.]: Geological Association of Canada Program with Abstracts, v. 8, p. 19.

Evenchick, C. A., Parrish, R. R., and Gabrielse, Hubert, 1984, Precambrian gneiss and late Proterozoic sedimentation in north-central British Columbia: Geology, v. 12, p. 233–237.

Faure, G. and Powell, J. L., 1972, Strontium isotope geology: New York, Springer, 188 p.

Forbes, R. B., Hamilton, T., Tailleur, I. L., Miller, T. P., and Patten, W. W., 1971, Tectonic implications of blueschist facies metamorphic terranes in Alaska: Nature Physical Sciences, v. 234, p. 106–108.

Forbes, R. B., Evans, B. W., and Thurston, S. P., 1984, Regional progressive high-pressure metamorphism, Seward Peninsula, Alaska: Journal of Metamorphic Geology, v. 2, p. 43–54.

Frey, M., Hunziker, J. C., Frank, W., Bocquet, J., Dal Piaz, G. V., Jäger, E., and Niggli, E., 1974, Alpine metamorphism of the Alps: A review: Schweizerische Mineralogische und Petrographische Mitteilungen, v. 54, p. 247–290.

Frey, M., Hunziker, J. C., Jäger, E., and Stern, W. B., 1983, Regional distribution of white K-mica polymorphs and their phengite content in the Central Alps: Contributions to Mineralogy and Petrology, v. 83, p. 185–197.

Gardner, M. C. and Hudson, T. L., 1984, Structural geology of Precambrian and Paleozoic metamorphic rocks, Seward Terrane, Alaska [abs.]: Geological Society of America Abstracts with Programs, v. 16, p. 285.

Hamilton, W., 1969, Mesozoic California and the underflow of Pacific mantle: Geological Society of America Bulletin, v. 80, p. 2409–2430.

Harrison, T. M. and McDougall, I., 1980, Investigations of an intrusive contact, northwest Nelson, New Zealand - I. Thermal, chronological and isotopic constraints: Geochimica et Cosmochimica Acta, v. 44, p. 1985-2003.

Hitzman, M. W., 1984, Geology of the Cosmos Hills and its relationship to the Ruby Creek copper-cobalt deposit [Ph.D. thesis]: Stanford University, Stanford, 266 p.

Hitzman, M. W., Smith, T. E., and Proffett, J. M., 1982, Bedrock geology of the Ambler District, southwestern Brooks Range, Alaska (map): Alaska Geological and Geophysical Survey, Geological Report 75.

Hopkins, D. M., 1963, Geology of the Imuruk Lake area, Seward Peninsula, Alaska: U.S. Geological Survey, Bulletin 1141-C, p. 1-101.

Htoon, Myat, 1979, Geology of the Clinton Creek asbestos deposit, Yukon Territory [M.S. thesis]: University of British Columbia, Vancouver, 194 p.

Hudson, T., 1979, Igneous and metamorphic rocks of the Serpentine Hot Springs area, Seward Peninsula, Alaska: U.S. Geological Survey Professional Paper 1079, 27 p.

Hudson, T., and Arth, J. G., 1983, Tin granites of Seward Peninsula, Alaska: Geological Society of America Bulletin, v. 94, p. 768-790.

Hunziker, J. C., 1974, Rb-Sr and K-Ar age determination and the Alpine tectonic history of the Western Alps: Memorie degli Insitute di Geologia e Mineralogia dell Universita di Padova, v. 31, 54 p.

Kuran, V. M., Godwin, C. I., and Armstrong, R. L., 1982, Geology and geochronometry of the Scheelite Dome tungsten-bearing skarn property, Yukon Territory: CIM Bulletin, v. 75, no. 838, p. 137-142.

Lister, G. S., Banga, Greetje, and Feenstra, Anne, 1984, Metamorphic core complexes of Cordilleran type in the Cyclades, Aegean Sea, Greece: Geology, v. 12, p. 221-225.

Mattinson, J. M., 1981, U-Pb systematics and geochronology of blueschists: preliminary results [abs.]: EOS—Transactions of the American Geophysical Union, v. 62, p. 1059.

Mayfield, C. F., Silberman, M. L., and Tailleur, I. L., 1982, Precambrian metamorphic rocks from the Hub Mountain terrane, Baird Mountains quadrangle, Alaska: U.S. Geological Survey Circular 844, p. 18-22.

Miller, T. P., 1972, Potassium-rich alkaline intrusive rocks of western Alaska: Geological Society of America Bulletin, v. 83, p. 2111-2128.

Miller, T. P., and Bunker, C. M., 1976, A reconnaissance study of the U and Th contents of plutonic rocks of the southeastern Seward Peninsula, Alaska: Journal of Research U.S. Geological Survey, v. 4, p. 367-377.

Miller, T. P., Elliott, R. L., Finch, W. I., and Brooks, R. A., 1976, Preliminary report on uranium-, thorium-, and rare-earth-bearing rocks near Golovin, Alaska: U.S. Geological Survey Open-File Report 76-710.

Moffit, F. H., 1913, Geology of the Nome and Grand Central quadrangles, Alaska: U.S. Geological Survey Bulletin 533, 140 p.

Mortensen, J. K., 1983, Age and evolution of the Yukon-Tanana Terrane, southeastern Yukon Territory [Ph.D. thesis]: University of California at Santa Barbara, 155 p.

Patton, W. W., Jr. 1984, Timing of arc collision and emplacement of oceanic crustal rocks on the margins of the Yukon-Koyukuk basin, western Alaska [abs.]: Geological Society of America Abstracts with Programs, v. 16, p. 368.

Patton, W. W., Jr., Tailleur, I. L., Brosge, W. P., and Lanphere, M. A., 1977, Preliminary report on the ophiolites of northern and western Alaska, *in* R. G. Coleman and W. P. Irwin, editors, North American Ophiolites: Oregon Department of Geology and Mineral Industries Bulletin 95, p. 51-57.

Sainsbury, C. L., 1969, Geology and ore deposits in the Central York Mountains, western Seward Peninsula, Alaska: U.S. Geological Survey Bulletin 1287, 101 p.

——1974, Geological map of the Bendeleben quadrangle, Seward Peninsula, Alaska: The Mapmakers, Anchorage, Alaska, 31 p.

Sainsbury, C. L., Coleman, R. G., and Kachadoorian, R., 1970, Blueschist and related greenschist facies rocks of Seward Peninsula, Alaska: U.S. Geological Survey Professional Paper 700-B, p. 33-42.

Satir, M., 1974, Rb-Sr-Altersbestimmungen an Glimmern der westlichen Hohen Tauern: Interpretation und geologische Bedeutung: Schweizerische Mineralogische und Petrographische Mitteilungen, v. 54, p. 213-228.

Schweickert, R. A., Armstrong, R. L., and Harakal, J. E., 1980, Lawsonite blueschist in the northern Sierra Nevada, California: Geology, v. 8, p. 27-31.

Silberman, M. L., Brookins, D. G., Nelson, S. W., and Grybeck, D., 1979a, Rubidium-strontium and potassium-argon dating of emplacement and metamorphism of the Arrigetch Peaks and Mount Igikpak plutons, Survey Pass quadrangle, Alaska: U.S. Geological Survey Circular 804 B, p. 18-19.

Silberman, M. L., Moll, E. J., Chapman, R. M., Patton, W. W., Jr., and Conner, C. L., 1979b, Precambrian age of metamorphic rocks from the Ruby province, Medfra and Ruby quadrangles—preliminary evidence from radiometric age data: U.S. Geological Survey Circular 804 B, p. 66-68.

Steiger, R. H., and Jäger, E., 1977, Subcommission on geochronology: Convention on the use of decay constants in geo- and cosmochronology: Earth and Planetary Science Letters, v. 36, p. 359-362.

Sturnik, M. A., 1984, Metamorphic petrology, geothermo-barometry and geochronology of the eastern Kigluaik Mountains, Seward Peninsula, Alaska [M.S. thesis]: University of Alaska, Fairbanks, 175 p.

Suppe, J., and Armstrong, R. L., 1972, Potassium-argon dating of Franciscan metamorphic rocks: American Journal of Science, v. 272, p. 217-233.

Tailleur, I. L., and Carter, R. D., 1975, New graptolite locality indicates Lower Ordovician rocks in southwestern Brooks Range: U.S. Geological Survey Professional Paper 975, 64 p.

Tempelman-Kluit, D. J., 1976, The Yukon Crystalline Terrane: Enigma in the Canadian Cordillera: Geological Society of America Bulletin, v. 87, p. 1343-1357.

——1979, Transported cataclasite, ophiolite, and granodiorite in Yukon: Evidence of arc-continent collision: Geological Survey of Canada Paper 79-14, 27 p.

Tempelman-Kluit, D. J. and Wanless, R. K., 1975, Potassium-argon age determinations of metamorphic and plutonic rocks in the Yukon Crystalline Terrane: Canadian Journal of Earth Sciences, v. 12, p. 1895-1909.

Tempelman-Kluit, D. J. and Wanless, R. K., 1980, Zircon ages for the Pelly Gneiss and Klotassin Granodiorite in western Yukon: Canadian Journal of Earth Sciences, v. 17, p. 297-306.

Thurston, S. P., 1985, Structure, petrology and metamorphic history of the Nome Group blueschist terrane, Salmon Lake area, Seward Peninsula, Alaska: Geological Society of America Bulletin, v. 96, p. 600-617.

Till, A. B., 1980, Crystalline rocks of the Kigluaik Mountains, Seward Peninsula, Alaska [M.S. thesis]: University of Washington, Seattle, 97 p.

——1983, Granulite, peridotite, and blueschist: Precambrian to Mesozoic history of Seward Peninsula, *in* Proceedings of the 1982 Symposium on Western Alaska, Geology and Resource Potential: Journal of the Alaska Geological Society, v. 3, p. 59-66.

——1984, Low-grade metamorphic rocks of Seward Peninsula, Alaska [abs.]: Geological Society of America Abstracts with Programs, v. 16, p. 337.

Till, A. B., Dumoulin, J. A., Aleinikoff, J., Harris, A., and Carroll, P. I., 1983, Paleozoic rocks of the Seward Peninsula: New insight [abs.]: Alaska Geological Society Symposium.

Turner, D. L., Forbes, R. B., and Dillon, J. T., 1979, K-Ar geochronology of the southwestern Brooks Range, Alaska: Canadian Journal of Earth Sciences, v. 16, p. 1789-1804.

Turner, D. L., Swanson, S. E., and Wescott, W., 1980, Continental rifting—a new tectonic model for the Central Seward Peninsula: Geophysical Institute, University of Alaska, 29 p.

Turner, R. F., ed., 1983, Geological and operational summary, Norton Sound COST No. 1 well, Norton Sound, Alaska: U.S. Geological Survey Open-File Report 83-124, 168 p.

Wagner, G. A., Reimer, G. M., and Jager, E., 1977, Cooling ages derived by apatite fission-track, mica Rb-Sr, and K-Ar dating: The uplift and cooling history of the Central Alps: Memorie degli Institute di Geologia e Mineralogia dell Universita di Padova, v. 30, 27 p.

Wanless, R. K., Stevens, R. D., and Loveridge, W. D., 1970, Anomalous parent-daughter isotopic relationships in rocks adjacent to the Grenville Front near Chibougamau, Quebec: Eclogae geologicae Helvetii, v. 63, p. 345-364.

Wescott, E., and Turner, D. L., editors, 1981, Geothermal reconnaissance survey of the Central Seward Peninsula, Alaska: Geophysical Institute University of Alaska Report UAG-R-284, 123 p.

Wilson, M. R., 1972, Excess radiogenic argon in metamorphic amphiboles and biotites from the Sulitjelma region, central Norwegian Caledonides: Earth and Planetary Science Letters, v. 14, p. 403–412.

York, D., 1967, The best isochron: Earth and Planetary Science Letters, v. 2, p. 479–482.

MANUSCRIPT ACCEPTED BY THE SOCIETY JULY 29, 1985

Geological Society of America
Memoir 164
1986

Caledonian high-pressure metamorphism in central western Spitsbergen

Yoshihide Ohta
Norsk Polarinstitutt
P.O. Box 158
1330 Oslo Lufthavn, Norway

Takao Hirajima
Department of Geology and Mineralogy
Faculty of Science, Kyoto University
Oiwakecho, Kitashirakawa, Sakyoku
Kyoto, 606 Japan

Yoshikuni Hiroi
Department of Geology, Faculty of Science
Chiba University, Yayoicho 1-33
Chibashi, 280 Japan

ABSTRACT

One of the oldest high-pressure metamorphic rocks, from the northernmost locality (Motalafjella in central western Spitsbergen, 73° 25′N and 13°E), is described, and an attempt is made to construct a model of Caledonian subduction from fragmentary evidence.

Glaucophane schist and eclogites in Spitsbergen have been known since 1957. These rocks have been studied by British and Norwegian geologists; recently Japanese petrologists have been involved in this study. Three important new discoveries have been made in Motalafjella and adjacent areas:

1) An unconformity has been found at the base of an Upper Ordovician-Silurian succession (the Bulltinden Formation). The basal conglomerate contains pebbles of stratigraphically-underlying, low-grade—high-pressure metamorphic rocks (Lower Unit of the Vestgötabreen Formation) and fossils in the matrix.

2) Lawsonite has been widely found in the low-grade metabasites of the Lower Unit, and jadeite-quartz-albite-bearing assemblages have been identified in the high-grade cherty rocks associated with eclogites. Thus, the Vestgötabreen Formation contains low- and high-grade rocks of the jadeite-glaucophane type metamorphic-facies series.

3) An upper [late] Riphean succession with shallow-marine-reef sedimentary rocks (the calc-argillo-volcanic formation) is intruded by metadiabase-gabbros, which have strong chemical similarities to oceanic basic rocks.

From 1) it can be inferred that the high-pressure metamorphism, i.e., the early Caledonian metamorphism, is definitely older than Upper Ordovician-Silurian beds, and discovery 2) provides evidence for a high-pressure metamorphic-facies series. The combination of oceanic basic rocks and the sedimentary rocks of shallow-marine-reef facies (3) is explained in terms of a change of tectonic domain, i.e., from that of a continental shelf during sedimentation to an active oceanic condition, e.g., an oceanic accretion zone and its related tectonic zones, by the time of basic intrusion. This change implies an active plate boundary between inferred continent and ocean plates and that this consuming boundary is responsible for the high-pressure metamorphism in an early Caledonian phase.

INTRODUCTION

Spitsbergen is located at the northwestern edge of the Eurasian plate and is considered to have been near northeast Greenland prior to the opening of the North Atlantic (Fig. 1-a). The rocks which suffered Caledonian orogeny are called the Hecla Hoek rocks, and range from Middle Riphean (possibly younger than 1,275 Ma; Edwards and Taylor, 1974) to Silurian in age. These rocks are variably metamorphosed, and are covered by unmetamorphosed Lower Devonian terrestrial sedimentary rocks in the northern and southern part of the island (Fig. 1-b).

After a long period of platform sedimentation from Carboniferous to Early Cretaceous time, early Tertiary diastrophism was superposed on the Caledonian structures, producing large-scale folds and thrusts in both the Hecla Hoek basement and the platform cover along the west coast of Spitsbergen (Lowell, 1972; Harland and Horsfield, 1974). Uplift and erosion followed, the Hecla Hoek rocks now being widely exposed in the Tertiary deformation zone (Fig. 1-c) (Flood et al., 1971).

This paper presents a short review of the studies on the high-pressure metamorphic rocks found in the central western Spitsbergen, and some new information from recent field and laboratory studies. In addition, a proposal is made to elucidate the Caledonian subduction zone.

PREVIOUS WORK

The blueschist of the Hecla Hoek succession in western Spitsbergen has been known among British geologists since 1957; D. G. Gee located it first at Motalafjella in 1962, and Horsfield found its northern extension on the ridge north of Vestgötabreen in 1968-69 (Harland et al., 1979). A description of the geology of the area was given by Horsfield (1972).

The Hecla Hoek rocks in the surrounding areas were studied by Hjelle et al. (1979) and Harland et al. (1979). Several formations are recognized: two formations of Upper Hecla Hoek - the Bulltinden Formation (500-1500 m thick Upper Ordovician-Silurian shallow marine sediments) and the Eocambrian tilloid formation (2-4 km); two formations of Middle Hecla Hoek - the calc-argillo-volcanic formation (1 km, the CAV formation hereafter) and the quartzite-shale formation (1 km, the Q-Sh formation herafter). The last two formations constitute the St. Jonsfjorden Group of Harland et al. (1979). These rocks form a 20-km-wide zone of NNW-SSE trend (Fig. 2-a) and have generally moderate W-dipping structures, including some folds. Both the east and west sides of this zone are bounded by Tertiary fault zones. The high-pressure metamorphic rock complex, the Vestgö-

tabreen Formation, and the Bulltinden Formation have been thrust to the NE. The former occurs above the latter as a core of overturned anticline, locally with thrust contact, about 10 km long and 300-400 m thick.

The high-pressure metamorphic rocks were initially described by Horsfield (1972) as the Vestgötabreen Suite, and they include glaucophane-bearing schists, epidote-actinolite greenstones and small masses of pyroxenite and serpentinite. Three whole-rock chemical analyses were presented in his paper; two are gabbroic and the third is sedimentary in composition. Five K-Ar ages are also given by the author; two muscovite and two whole-rock ages are in a range of 402-475 Ma, while one whole-rock age from clinopyroxene-bearing rock shows 621 Ma. The last one is thought by Horsfield to be unreliable since clinopyroxene may contain excess argon.

A pebble of a retrogressed schist found within the Bulltinden Formation contains pseudomorphs after garnet and is considered to be derived from the Vestgötabreen Suite (Horsfield, 1972). Hence, the high-pressure rocks are thought to be older than the Bulltinden Formation. This conforms with the Caledonian radiometric ages of the high-pressure metamorphic rocks. In view of the world-wide correlation between glaucophane schist belts and former compressive plate boundaries, Horsfield (1972) proposes the notion of a Caledonian subduction zone to form the high-pressure metamorphic rocks, and infers that at least one of the two colliding plates is oceanic.

Successive petrographic studies of the high-pressure metamorphic rocks were made by Manby (1978) and Ohta (1979), and some detailed chemical data were presented. Manby (1978) estimates the metamorphic temperature from the mineral assemblages in the green rocks of adjacent areas and the pressure from reconstruction of the thickness of sedimentary piles; a 15° C/km geothermal gradient was obtained. The gradient was then extrapolated to a depth of 20 km or more, corresponding to the estimated total thickness of the Hecla Hoek strata (Birkenmajer, 1975, Harland et al., 1979), and the conclusion is drawn that an eclogite could be formed at such a depth without introducing the idea of a subduction zone. The glaucophane is considered to be a retrogressive product formed during later thermal adjustment and uplift.

Ohta (1979) presents more whole-rock and mineral analyses of high-pressure metamorphic rocks. The main rock types of the epidote-actinolite greenstones are medium-K_2O, Na-alkaline and tholeiitic rocks (Irvine and Baragar, 1971), while the glaucophane-bearing rocks are thought to have been derived

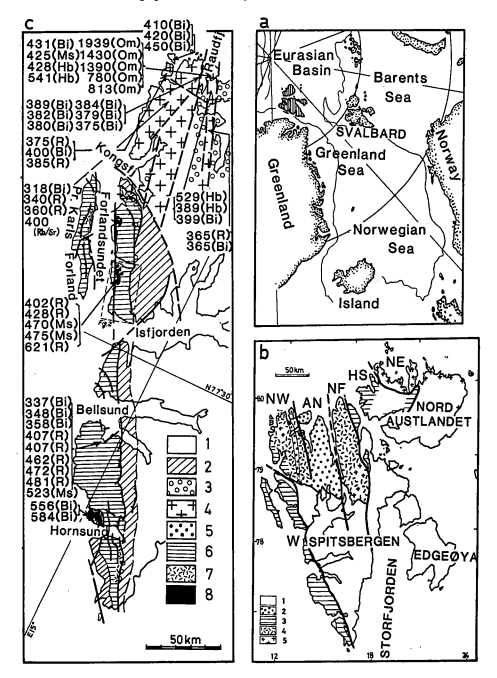

Figure 1. Simplified structural map of western Spitsbergen, with radiometric ages in Ma (mostly K-Ar ages). 1-a, Location of Svalbard at the present and pre-Tertiary position (lined). 1-b, Distribution of Caledonian rocks in Svalbard. W: the area discussed in the present paper, high-pressure and intermediate P-T facies series, NW: mainly gneisses-migmatites and granites, high-temperature-facies series, AN: Andree Land, Devonian molasse graben, unmetamorphosed, NF: Ny Friesland, metasedimentary rocks and high-grade metamorphic rocks, intermediate P-T facies series, HS: Hinlopenstretet area, open synclinorium of Upper and Middle Hecla Hoek successions with very low-grade metamorphism, NE: mainly migmatites and granites, high-temperature-facies series. 1: platform covers, 2: Devonian molasse, 3: low-grade metamorphic rocks, 4: gneisses and migmatites, 5: granites. 1-c, Tectono-metamorphic division of western Spitsbergen. 1 and 2: post-Devonian platform, weakly deformed (1) and strongly folded (2), 3: Devonian molasse, 4: Caledonian granites and migmatites, 5: Upper Ordovician-Silurian sediments, 6: shallow-marine sedimentary successions, metamorphosed under the low-grade conditions of intermediate P-T facies series, 7: same as 6, but highly metamorphosed, 8: high-pressure metamorphic complex and the structural units dominantly consist of possible oceanic basic rocks.

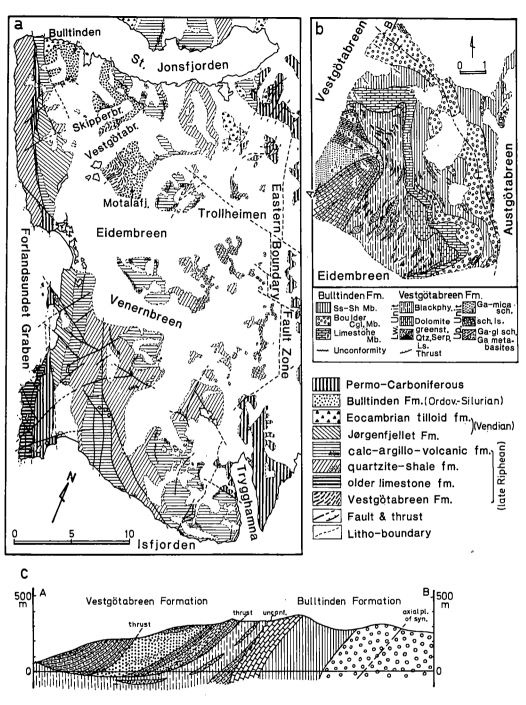

Figure 2. Geological maps of central western Spitsbergen. Map scales in kilometers. 2-a, Geological map of the area from St. Jonsfjorden to Isfjorden. 2-b, Detailed geological map of Motalafjella. E: eclogite, F: newly found fossil localities, A-B: position of geological cross section. 2-c, Geological cross section along the western part of Motalafjella. Same legend as Fig. 2-b.

from mixtures of basic rocks and argillaceous materials. The metamorphic condition was calculated from the Fe-Mg partitioning of the clinopyroxene-garnet pair to be about 10 kb and 540-570°C for the eclogitic rocks, and the coarse-grained glaucophane schist was considered to be a retrograde product. The prograde series of high-pressure metamorphism was not considered by

either Manby (1978) or Ohta (1979), since no low-grade, high-pressure indicator minerals had been found.

The question of a possible Caledonian subduction zone remains open, though a large variation of metamorphic facies series in different parts of the Svalbard Caledonides was noted by Ohta (1978) (Figs. 1-b and 3).

NEW EVIDENCE

A number of new discoveries were made by detailed mapping of Motalafjella and its adjacent areas in 1983. Systematic studies are being undertaken by different specialists, but we present here three main conclusions reached on the basis of results obtained at this stage of the studies.

Age of the High-Pressure Metamorphism

Two fossiliferous-limestone conglomerates have been reported by Scrutton et al. (1976) from the Bulltinden Formation at the northernmost and southeastern parts of Motalafjella (Fig. 2-b). The fossils, mainly corals, were identified as being of Silurian age.

Six more fossil localities were newly found in the middle valley and on the southern cliff of Motalafjella; three are in the distinct grey limestone member of the Bulltinden Formation occurring along the boundary between the Bulltinden and the Vestgötabreen Formations, while the rest are thin lenses interbedded in the boulder conglomerate member of the former formation (Fig. 2-b), and are themselves locally conglomeratic. Some hundreds of fossils, mainly gastropods and chain corals with some pentamerid brachiopods, crinoid fragments, cephalopods and several trilobites, are now being treated for identification by palaeontologists both in Oslo and Copenhagen. There is no doubt that they are of Silurian age, and possibly include some Middle-Upper Ordovician specimens, i.e., *Maclurites,* (written comm. Dr. J. S. Peel, 1984).

A limestone-conglomerate bed 3-5 m thick has been found at the base of a fossiliferous limestone member at two localities in the southern half of Motalafjella (Ohta et al., 1984). The bed is in contact with a massive dolomite of the Vestgötabreen Formation, and it contains sub-angular to round boulder pebbles of the dolomite and some metabasites. The matrix is a sandy gray limestone and incorporates gastropods similar to those described above. The dolomite near the contact shows a strongly weathered yellow color even on fresh-cut surfaces, and has many irregular cracks filled by the gray sandy limestone. The limestone member of the Bulltinden Formation is in contact with dolomite in the southern half of Motalafjella, unlike that of the northern half, where it occurs in conjunction with pelitic phyllites and low-grade metabasites. Thus, an angular unconformity of regional scale is observed between the Bulltinden and the Vestgötabreen Formation. The contact of these rocks was carefully examined on the ridges north of Vestgötabreen, north of Skipperbreen and on the southern end of Bulltinden (Fig. 2-a), and the unconformity has been confirmed at all these localities, although there are thrust faults in a few hundreds of metres structurally below the unconformity.

The limestone member of the Bulltinden Formation apparently underlies the dolomite, so the unconformity has been inverted.

A clear overturned syncline with a moderate W-dipping axial surface has been observed in the boulder conglomerate member of the Bulltinden Formation on the eastern cliff of Motalafjella (Figs. 2-b and -c). The inverted limestone member is the overturned upper limb of this syncline.

The discovery of the unconformity means that the Vestgötabreen Formation is definitely older than the Upper Ordovician-Silurian sediments of the Bulltinden Formation, i.e., the early Caledonian metamorphism is older than this formation, and there was a strong folding phase after the sedimentation of the formation. This evidence agrees with the radiometric ages (Horsfield, 1972) that the two muscovite K-Ar ages of the glaucophane-bearing schists (about 470 Ma), correspond to a later stage of the early Caledonian metamorphism, and that the two whole-rock ages of 402 and 428 may have partly been reset during the post-Bulltinden-Formation-folding period by formation of new cleavages. The latter age may correspond to the Haakonian deformation phase established in the Raudfjorden area, northwestern Spitsbergen (Fig. 1-c), where the folded rocks of possibly Silurian age are covered by Lower Devonian molasse sediments (Gee and Moody-Stuart, 1966; Gee, 1972; Harland, 1978). Harland (1973) proposed the name Forlandsund phase to refer to the latest Caledonian deformation, whatever its age, in the present area.

A Middle Cambrian stratigraphic hiatus was reported in the Hornsund area, southern Spitsbergen (Major and Winsnes, 1955) (Fig. 1-c) and termed the Hornsundian diastrophism (Birkenmajer, 1975). The K-Ar ages of biotites from the Lower Hecla Hoek mica schists of this area, 556 and 584 Ma (Gayer et al., 1966), may represent this early Caledonian deformation phase, and one whole-rock K-Ar age, 621 Ma (Horsfield, 1972), from an epidote-garnet-clinopyroxene-muscovite schist from Motalafjella, may be related to a similar phase. Two more stratigraphical gaps, one at the boundary between Eocambrian and Cambrian (the Jarlsbergian phase of Birkenmajer, 1975) and another in Middle Ordovician, have already been proposed in western Spitsbergen (Harland, 1978). The high-pressure metamorphism can certainly be correlated with one of these early Caledonian deformation phases.

Progressive Metamorphism in the High-Pressure Metamorphic Rocks

The Vestgötabreen Formation is divided into two units (Fig. 2-b). The Lower Unit (structurally) is comprised mainly of sericite-chlorite phyllites which include scattered lensoid masses of dolomite, quartzite, metabasite and serpentinite. The Upper Unit is predominantly composed of garnet-mica schists, schistose limestone and lensoid masses of garnet-glaucophanite and eclogite. These two units are separated by a thrust fault with a moderate W-dipping surface (Ohta et al., 1984) (Fig. 2-c).

Lower Unit. Idioblastic short prisms of lawsonite were found in the Lower Unit of the Vestgötabreen Formation (Hirajima et al., 1985). Some lawsonite grains are replaced by chlorite and epidote. In the garnet-bearing metabasites of the Lower Unit,

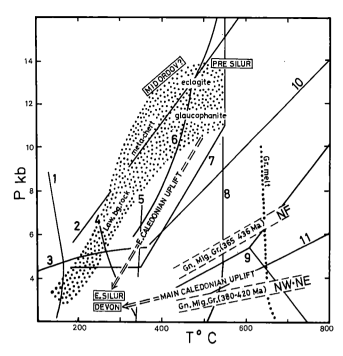

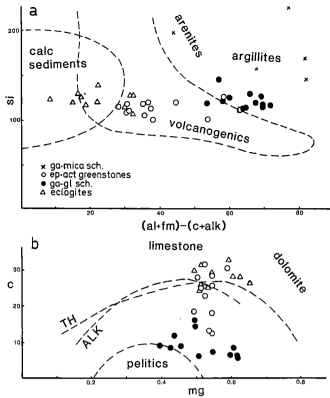

Figure 3. Inferred PT path of the high-pressure metamorphism of central western Spitsbergen. The path has been drawn from the data of the mineral assemblages of the Lower Unit (Hirajima et al., 1985), jadeite contents of metacherts (Kanat, 1984; Hirajima et al., 1985) and the Fe-Mg partitioning between garnet and clinopyroxene from glaucophane schists and eclogites (Ellis and Green, 1979, Kushiro, 1969, Holland, 1979-a, -b and 1980 and the data from Ohta, 1979 and in Fig. 6 of this paper) from the Upper Unit of the Vestgötabreen Formation. 1: Heul=-Law+Qt+Pl (Nitsch 1968), 2: lower stability limit of Jd100+Qt (Kushiro 1969), 3: lower stability limit of lawsonite to laumontite (Liou, 1971), 4: Preh+Cc+Qt=Pum+Act+V (Nitsch 1971F), 5: Pum+Cc+Qt=Ep+Act+V (Nitsch 1971), 6: Law+Alb=Zo+Parag+Qt (Holland 1979-a), 7: maximum stability limit of glaucophane (Maresch 1977) 8: Cht=St+Mus+Qt (Hoschek 1969), 9: alminosilicates triplepoint (Hoschek 1969), 10 and 11: lower limits of eclogite and garnet-granulite (Råheim and Green 1975). NF: Ny Friesland, NW: northwestern Spitsbergen, NE: Nordaustlandet; the locations are in Fig. 1-b.

Figure 4. Whole-rock composition of rocks from the high-pressure metamorphic complex of Motalafjella, showing possible contamination of pelitic sedimentary rocks into the basic high-pressure metamorphic rocks. Symbols are common in both 4-a and 4-b, and the coordinates are the Niggli-values. Data: Ohta, 1979 and 1984. 4-a, Niggli si- (al+fm)-(c+alk) diagram. Reference curves and fields after Simonen (1953) with some modifications by Ohta. 4-b, Niggli c-mg diagram. Reference curves and fields after Leake (1963). TH: differentiation trend of tholeiite (Karroo), ALK: that of Na-alkalic rocks (Ethiopia).

lawsonite occurs only as inclusions within garnet. Pumpellyite and glaucophane were also found in the lawsonite-bearing metabasites. They are fine-grained and irregular-shaped crystals. Glaucophane in the metabasites of the Lower Unit is everywhere replaced by chlorite-actinolite aggregates, in contrast to idiomorphic and tabular coarse-grained ones in the Upper Unit. Some chemical analyses of the constituent minerals are given in Hirajima et al. (1985).

These assemblages clearly reveal that the Lower Unit of the Vestgötabreen Formation has suffered low-grade high-pressure metamorphism. P-T conditions are estimated to be ~7 kb and 300°C (Fig. 3).

Upper Unit. Garnet-mica schists and schistose limestone are the main rocks in this unit and glaucophane schists, eclogites and metagabbros are included in the garnet-mica schists in the form of blocks and lenses of less than 1 km long.

The garnet-mica schists often contain large prisms of chlori-

toid and have strong diaphthoritic cleavages. These rocks are considered to be melange matrix of the Upper Unit, comparable to the sericite-chlorite phyllite in the Lower Unit.

Three distinct schistose limestone layers occur in this unit and wedge out towards the east (Figs. 2-b and -c), and a glaucophane schist-metagabbro body of more than 100 m thick occurs between the lower two limestones. Small lenses of eclogite and metachert are incorporated in the body.

A thick glaucophane schist occurs in the lower part of the unit in the middle-western slope of Motalafjella and extends to the middle of the ridge between Vestgötabreen and Skipperbreen. The rocks are strongly schistose and the cleavages are accentuated by white mica flakes. Some micaceous-cleavage surfaces show feather-amphibolite texture with long crystals of glaucophane. The glaucophane schists contain as much as 90% modal glaucophane, 8% large garnet and the rest is white mica, epidote-zoisite and opaque minerals. Thus, the rocks can be called garnet-glaucophanite.

The garnet-glaucophanites developed most distinctly near the borders between eclogitic metagabbros and garnet-mica

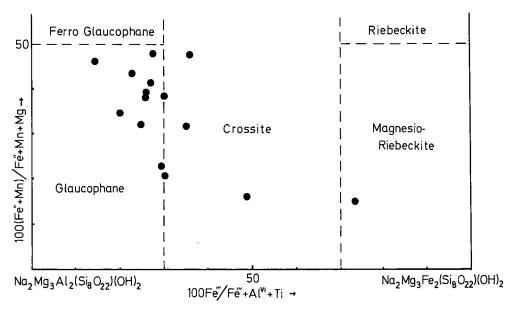

Figure 5. Chemical composition of the glaucophane from the high-pressure metamorphic rocks of Motalafjella. Data: Ohta (1979). Fe" ' was calculated as Al" ' + Fe" ' =2.00.

schists. This occurrence conforms with the bulk-chemical composition of the garnet-glaucophanites, which indicates strong contamination of pelitic materials in basic rocks (Fig. 4) (Ohta, 1979). The metagabbro blocks are enclosed in place by the garnet-glaucophane-mica schist, and this agrees with the occurrence of eclogite lenses in coarse-grained-garnet glaucophanite.

Large glaucophane crystals were observed along the cracks of the eclogite. These crystals were studied mineralogically by Prof. L. Ungaretti, Pavia University, Italy, at the request of the author (Ohta), with the results: a = 9.5456-9.5772Å, b = 7.7711-17.8132Å, c = 5.2991-5.3044Å and the unit cell volume = 874.41-879.38Å³. It is thus a typical crossite (Maresch, 1977). The chemical composition of other Na-amphiboles is given in Fig. 5.

The formation temperature for the garnet-glaucophanite by the Fe-Mg partitioning of garnet-clinopyroxene tends to be high, more than 600°C (after Ellis and Green 1979), and this is not consistent with the maximum stable temperature of glaucophane obtained by experimental study (Maresch, 1977).

Metagabbros occur in the form of lenses and layers with garnet-glaucophanite. An apparent ophitic-like texture is megascopically observed on the weathered surface of the metagabbros. However, the white prisms are actually epidote and zoisite-clinozoisite idioblasts, and the primary texture of the metagabbros has been totally destroyed by metamorphic recrystallization. Large dusty clinopyroxene grains are of the only relict minerals. Garnet forms large aggregates of irregularly elongated grains, which are enclosed in pools of albite aggregate. Albite, which has clearly recrystallized, fills the interstitial spaces between the garnet grains, and the cracks, and is locally included in the garnet. Chlorite is less abundant than nematoblastic actinolite. Epidote and zoisite-clinozoisite are granular, commonly as large idio-

blasts, and occur in very large amounts in the high-grade metagabbros. They show strong zonal structures in some rocks. New colorless clinopyroxene occurs as aggregates of small prismatic crystals, and can be distinguished from the relict grains by the absence of opaque pigments along the cleavages and grain boundaries. In some eclogite with more than 80% clinopyroxene, glaucophane, epidote, zoicite-clinozoisite and garnet are confined to occur in the triangular-shaped interstitial spaces of the clinopyroxene prisms. The garnet contains epidote and albite inclusions.

It is evident from the petrography of the metagabbros presented above that garnet grew at the expense of a primary plagioclase, and clinopyroxene developed from primary mafic constituents of the metagabbros to form the final product of eclogite. No strong cleavage developed in the metagabbros during this process. Typical eclogite consists of more than 80% modal omphacite, locally as much as 10% garnet, but totally absent in many places, and less than 5% of epidote-zoisite-clinozoisite and white mica. Minor amounts of rutile and other opaque grains are also present. The chemical compositions of garnet and clinopyroxene are shown in Figs. 6-a, -b and -c.

The formation condition calculated from the jadeite content of clinopyroxene and the Fe-Mg partitioning between co-existing garnet-clinopyroxene pairs is 14 kb and 575°C (Fig. 3). Fe" ' of clinopyroxene has been calculated after Neumann (1976), the temperature after Ellis and Green (1979) and the pressure after Kushiro (1969) and Holland (1979-a, -b and 1980). This result is preliminary in that the effect of relatively high Ca content in garnet has not been considered. Detailed mineral chemistry is now being studied in the laboratory of Prof. S. Banno, Kyoto University, Japan.

Thin metachert layers, having an individual thickness of less

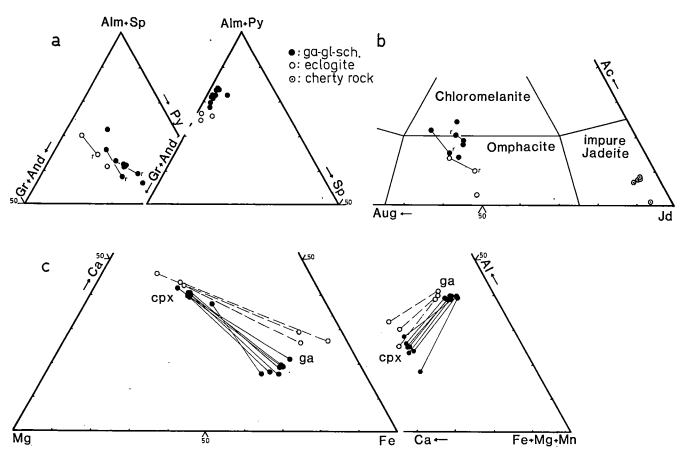

Figure 6. Compositions of garnet and clinopyroxene from the high-pressure metamorphic rocks. The formulas proposed by Ryburn et al. (1976) and Neumann (1976) are used for the calculation of the components. 6-a, Garnets. Data from Manby (1978), Ohta (1979) and some unpublished data of Ohta. r: rim of a grain, tie line connects the rim and core of the same grain. 6-b, Clinopyroxenes. Same legend as Fig. 6-a. Data; chloromelanite-omphacite: Manby (1978) and Ohta (1979) and some unpublished data from Ohta, impure jadeite: Kanat (1984) and Hirajima et al. (1985). 6-c, Garnet-pyroxene pairs. Same legend as Fig. 6-a, tie line shows the garnet and clinopyroxene from the same rock.

than 50 cm, are incorporated in the garnet-glaucophanite and metagabbros. The rocks are mainly red and white, but locally black. They are very fine-grained and dense rock, and reveal thin laminated structures. They are recrystallized into quartzite with garnet. Some pink metacherts have clinopyroxenes with jadeite contents 65-85% or more (Fig. 6-b). Jadeite-paragonite-albite-quartz and jadeite-almandine-phengite-hematite-quartz assemblages have been observed in the metacherts and the latter contain winchite, epidote, and albite. Kanat (1984) also reported clinopyroxenes with jadeite contents from 86.0 to 87.4% from silicic schists with a garnet-glaucophane-muscovite-chlorite-albite-quartz assemblage (Fig. 6-b), and calculated the metamorphic condition to have been 10-13 kb and 300-450°C (Fig. 3).

These metachert beds are commonly interlayered with garnet-glaucophane-mica schist and eclogitic metagabbros. All these three rocks, each 10-30 cm thick, exhibit gradational lithological transitions to each other.

A banded structure of eclogite and garnet-glaucophane-mica

schist, each band 1.0-1.5 m in thickness, has been observed between the two wedging schistose limestones (Fig. 2-b). This structure has been inherited from the original interlayering of basic rock and tuffaceous argillite beds.

Large idioblasts of chloritoid and garnet, both as much as 2 cm in size, together with glaucophane, occur in thin limited layers in the garnet-mica schists. These rocks are probably derived from mixtures of pelitic sediments and basic volcanic materials.

White mica flakes are abundant on the shiny cleavage surfaces of the schistose limestone. Chloritoid locally occurs in the layers rich in white mica. Thin layers of schistose limestone alternate with metachert, metagabbros and garnet-mica schists, and have gradational boundaries. This limestone has probably been derived from relatively homogeneous muddy limestone formed in a deep-sea environment.

Though detailed chemistries of the constituent minerals are still under study, the preliminary petrographic observations indicate that the metachert, garnet-glaucophanites, garnet-mica schists, schistose limestones, and eclogite developed from a deep-sea sedimentary-volcanic sequence. The micaceous schists of the Upper Unit and the pelitic phyllites of the Lower Unit are considered to be melange matrices. Strongly diaphthoritic cleavages were probably formed during later uplift, possibly related to the post-Bulltinden Formation folding, or even partly during the Tertiary deformation.

The metamorphic mineral assemblages in the rocks of the Vestgötabreen Formation are characterized as follows:

Lawsonite-glaucophane-actinolite-chlorite ......lower part of Lower Unit

Epidote-garnet-actinolite-chlorite, ...upper part of Lower Unit
epidote-garnet-glaucophane-chlorite

Omphacite-garnet-glaucophane-epidote-rutile, ...Upper Unit
glaucophane-phengite-paragonite and jadeite-quartz

These three progressive assemblages are very similar to the lawsonite-, epidote- and omphacite-zone of New Caledonia (Black, 1977; Brothers and Yokoyama, 1982).

Geochemistry of the Basic Rocks in the Adjacent Middle Hecla Hoek Successions

Basic igneous rocks occur both in the CAV and the Q-Sh formations of Middle Hecla Hoek to the east of Motalafjella (Fig. 2-a) (Harland et al., 1979; Hjelle et al., 1979). Those in the former formation are metadiabase-gabbros occurring as thin discontinuous sheets, while those in the latter are clearly effusive with a predominance of amygdaloidal lavas, agglomerates and tuffs. The metadiabase-gabbros have strong cleavages around the margins of the sheets, while the central parts are massive. Adjacent calcareous sedimentary rocks are folded conformably around the lense-shaped, dismembered sheets. These occurrences suggest that the emplacement of the metadiabase-gabbros is syn-late tectonic in relation to the main deformation.

The bulk chemistries of these metadiabase-gabbros and the effusive basic rocks reveal that they are clearly different from each other (Ohta, 1984). The metadiabase-gabbros are mainly tholeiitic, and have a moderate-high Fe concentration in the middle stages of differentiation, but the effusive rocks, the Trollheimen volcanics, are mainly Na-alkalic (hawaiite) with a small amount of calc alkaline rocks (Table 1). Alkali and alkali-earth trace elements of the metadiabase-gabbros show large variations owing to metamorphic recrystallization of feldspars and micas, but their immobile minor and trace elements, i.e., Ti, P and Zr, Cr, Ni, V, reveal that they have similar chemical characteristics of the ratios of these elements to ocean floor basalts or mid-oceanic ridge basalts as shown in Figs. 7-b-1, -b-2 and -b-3 (Pearce and Cann, 1973; Beccaluva et al., 1979). The Trollheimen volcanics are non-oceanic, for example, in terms of TiO_2-K_2O-P_2O_5 (Pearce and Cann, 1973; Ohta, 1984). Detailed data and discussions has been given in Ohta (1984).

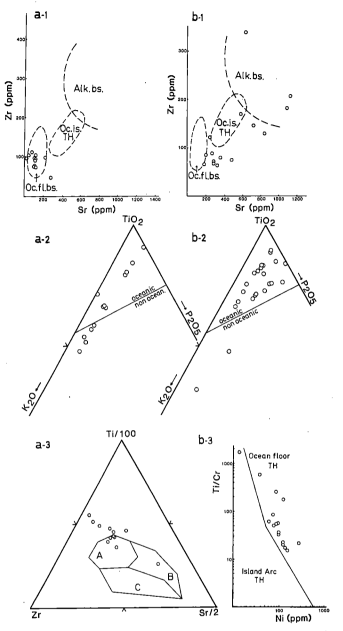

Figure 7. Trace element ratios of the basic rocks used to estimate their tectonic setting. 7-a, Metabasites of the Lower Unit of the high-pressure metamorphic complex of Motalafjella. Data: Ohta unpublished. a-1, Zr-Sr diagram (Pearce and Cann, 1973). a-2, TiO_2-K_2O-P_2O_5 diagram (Pearce and Cann, 1973). a-3, Ti/100-Zr-Sr/2 diagram (Pearce and Cann, 1973). A: ocean floor basalts, B: low-K tholeiites, C: calc-alkalic basalts. 7-b, Metadiabase-gabbros from the CAV formation (Ohta 1984). Data: Ohta, 1984. b-1, Zr-Sr diagram. b-2, TiO_2-K_2O-P_2O_5 diagram. b-3, Ti/Cr-Ni diagram (Beccaluva et al., 1979).

The Q-Sh formation, including the Trollheimen volcanics, has many sandy quartzite layers, as much as 50 m thick, in black sandy shales with relatively poor sorting. A few beds of oolitic muddy limestone occur in the lower part of the formation. Thus, the sedimentary environment is considered to be fore-reef ocean-

TABLE 1. CHEMICAL CLASSIFICATION OF THE BASIC ROCKS FROM THE HIGH-PRESSURE
METAMORPHIC COMPLEX AND ADJACENT MIDDLE HECLA HOEK FORMATIONS, SHOWING THE
SIMILARITY OF ROCK ASSOCIATIONS BETWEEN THE EPIDOTE-ACTINOLITE GREENSTONES
AND THE METADIABASE-GABBROS

			Rock Type		Rock Series			
		N	ultra basics	basalt	basic andesite	K-Alk.	Na-Alk.	Th	CA
Vestgötabreen Fm.	Eclogites	11	0	64	36	0	46	18	36
	Ga-gl sch.	12	0	92	8	25	8	59	8
	Ep-act greenstones	13	0	100	0	0	23	69	8
Mid Hecla Hoek Fms.	Metadb-gbs.	23	8	70	22	4	22	61	4*
	Trollheimen volcanics	20	45	50	5	20	75	0	5

N = number of analyzed rocks. Numbers, except for the N column, in percent.
*The ultrabasic rocks of the metadiabase-gabbros from the CAV formation are
excluded from the classification of rock series.
Chemical classification of rock series after Irvine and Baragar (1971), K-Alk =
potash alkalic rock series, Na-Alk = sodic alkalic, Th = tholeiitic, CA =
calc-alkalic.

side shelf, possibly near a shelf-edge fault-tectonic zone, and the volcanism might be localized along such faults.

The black calcareous, sandy shales of the CAV formation are similar to those of the Q-Sh formation, while many limestone beds, as much as 200 m thick, occur in the latter instead of quartzite. The limestones include large lenses of ancient reef as much as 100 m thick, as disclosed by the oolitic-pisolitic, algae-like and brecciated structures. These lenses behaved as solid units during later deformation. Thus, in conclusion, these sedimentary facies clearly represent a shallow shelf environment in a reef zone.

The metabasites of the Lower Unit of the high-pressure complex in Motalafjella have been chemically classified after the guide presented by Irvine and Barager (1971). They are dominantly tholeiitic (69% of the analysed rocks) and Na-alkalic (29%) in composition. This association of chemical groups is very similar to that of the metadiabase-gabbros of the CAV formation: tholeiite 61% and Na alkalic rocks 22% (Table 1). This similarity implies that the protoliths of high-pressure metabasites belong to the same igneous activity as the metadiabase-gabbros.

DISCUSSION: SYNTHESIS OF A PLATE TECTONIC MODEL

The new information described above seems to provide evidence for a Caledonian subduction model, summarized as follows:

The occurrence of oceanic/mid-oceanic-type basic rocks in the shallow marine CAV formation can be explained in terms of a change of tectonic domain. A terrestrial shelf condition on a continental type crust during the sedimentation periods of the Middle Hecla Hoek formations later changed its tectonic position

to a domain in which oceanic igneous activity ensued. In other words, an active oceanic tectonic domain, i.e., a part of Iapetus of Harland and Gayer (1972), replaced a previous continental shelf. This change of tectonic domain implies an active plate margin between the assumed oceanic and continental plates.

The timing of the oceanic igneous activity is not yet known, but it is at least older than Upper Ordovician. Some greenstone layers occur in the Eocambrian tilloid formation elsewhere in western Spitsbergen (Waddams, 1983), but their chemical characteristics have not been studied. The evidence for the timing of early Caledonian deformation phases is provided by stratigraphic hiatuses: the Jarlsbergian near the Eoambrian-Cambrian boundary, the middle Cambrian Hornsundian, and the Middle Ordovician (Major and Winsnes, 1955; Birkenmajer, 1975).

Harland (1961, 1978) defined the main Caledonian orogenic phase, the Ny Friesland phase, in post-Llanvirnian and possibly pre-Wenlockian time in central northern Spitsbergen (Fig. 1-b). Harland and Wright (1979) divided the Svalbard Caledonides into three longitudinal provinces and placed them far apart from each other, i.e., the Western province was placed near the northeast of Greenland, and the Eastern province was placed to the east of central eastern Greenland. These provinces were juxtaposed by supposed post-Devonian large-scale left-lateral strike-slip faults into the present configuration. Thus, the Western province was thought to be formed further north than the others, and its main orogenic phase was considered to be post-Silurian. In addition, the timing of the main deformation in the Western province was thought to be nearer to that of the Ellesmerian of Northern Greenland and the Canadian Arctic than to the Ny Frieslandian of the Eastern province of Svalbard. The latter is in conformity with the main orogenic phase of East

Greenland, Scandinavia and Scotland (Harland, 1978; Harland and Wright, 1979).

The discovery of the unconformity at the base of Upper Ordovician-Silurian sediments, the Bulltinden Formation, with the basal conglomerate including the pebbles of the high-pressure metamorphic complex, implies that the early Caledonian metamorphism in the Western province is at least older than Upper Ordovician age, and is similar to the Ny Frieslandian rather than to the Ellesmerian phase. Harland and Wright (1979) wrote that there is no record of Devonian in the Western province, but it does occur as thin fault slices in the north of St. Jonsfjorden (Fig. 1-c).

It is significant that the reported Caledonian high-pressure metamorphism from Anglesey, Wales (Gibbons, 1981), Knockormal, Ayrshire (Bloxam and Allen, 1958), and from northern Vermont in the Appalachian belt (Lanphere and Albee, 1974) are all of the Ordovician Taconic phase. The closure of the Iapetus Ocean, with subduction of some duration, is considered to have occurred in Cambrian-Silurian times (Bird and Dewey, 1970; Seyfert and Sirkin, 1979).

The progressive metamorphism of the high-pressure-facies series newly discovered in the Vestgötabreen Formation strongly suggests the existence of a subduction zone along the assumed active plate boundary. Massive oceanic basic rocks have been converted into lawsonite-bearing metabasites and eclogite, volcano-argillites into the garnet-bearing metabasites and garnet-glaucophanites, and the pelitic melange matrices into the black phyllites and garnet-mica schists. The rocks not involved in the subduction suffered metamorphism of an intermediate P-T-facies series, as revealed by the chloritoid-bearing greenschists of Prins Karls Forland (Atkinson, 1956; Manby, 1978) and the kyanite-staurolite schists of the Hornsund area (Smulikovski, 1965). These were briefly summarized by Ohta (1978). The high modal amounts of actinolite and epidote in the high-pressure metamorphic rocks of Motalafjella may show that an overprinting of a later intermediate P-T metamorphism occurred possibly in the post-Bulltinden Formation time.

The subducted rocks, and the rocks highly metamorphosed under the conditions of intermediate P-T facies series, were both cut into slices and were interwoven into a thrust complex during successive uplift. Juxtaposing of the different tectono-metamorphic blocks has therefore caused difficulties in correlating the stratigraphy of some adjacent structural units. This means that all successions observed fragmentally in western Spitsbergen do not necessarily belong to a single stratigraphic sequence, but they could be remnants of a number of separate ones, i.e., both oceanic and continental shelf sequences.

The overturned folds observed in the Bulltinden Formation were possibly formed in a last large folding phase of Caledonian (in post-Lower Silurian and pre-Devonian time). This phase corresponds to the Forlandsund phase of Harland (1973) and may be correlated with the Haakonian deformation phase defined in northwestern Spitsbergen, i.e., it is older than Devonian (Gee, 1972). The Devonian molasse graben developed after this phase.

The majority of the K-Ar ages of the gneiss-migmatites and intrusive granites of the Svalbard Caledonides are in the range of 380-420 Ma (Fig. 1-c) (Gayer et al., 1966), and this indicates that the orogenic uplift is Late Silurian-Early Devonian in age, associating it with the high-temperature facies metamorphism characterized by cordierite-garnet-sillimanite assemblages (Fig. 3). This phase is the same age as the Ellesmerian of Arctic Canada and North Greenland and the Scandian phase of Scandinavia.

ACKNOWLEDGMENTS

This is the third report from the co-operation project on the Caledonian high-pressure metamorphic rocks in Spitsbergen between Norsk Polarinstitutt and Kyoto University, Japan. We thank Prof. S. Banno, Kyoto University, and Mr. A. Hjelle, Norsk Polarinstitutt, for their discussions, Dr. J. S. Peel, Grønland Geological Survey, Denmark, for the fossil identification, Prof. L. Ungaretti, Pavia University, Italy, for the crystallographic analyses of glaucophane, and Mrs. Mai Britt Mørk, Geological Museum, Oslo, for the help to calculate the formation conditions of the rocks. We gratefully acknowledge financial support for the field work of the Japanese party in 1982 from the Polar Research Committee of Kyoikusha Co. Ltd., Tokyo.

REFERENCES CITED

Atkinson, D. J., 1956, The occurrence of chloritoid in the Hecla Hoek Formation of Prince Charles Foreland, Spitsbergen: Geological Magazine, v. 93, p. 63–71.

Beccaluva, L., Ohnerstetter, D. and Ohnerstetter, M., 1979, Geochemical discrimination between ocean-floor and Island-arc tholeiites. -an application to some ophiolites: Canadian Journal of Earth Science, v. 16, p. 1874–1882.

Bird, J. M. and Dewey, J. F., 1970, Lithosphere plate-continental margin tectonics and the evolution of the Appalachian orogen: Geological Society of America Bulletin, v. 81, p. 1031–1060.

Birkenmajer, K., 1975, Caledonides of Svalbard and plate tectonics: Bulletin of Geological Society of Denmark, v. 24, p. 1–19.

Black, P. M., 1977, Regional high-pressure metamorphism in New Caledonia: phase equilibria in the Ouegoa district: Tectonophysics, v. 43, p. 89–107.

Bloxam, T. W. and Allen, J. B., 1958, Glaucophane-schists, eclogite, and associated rocks from Knockormal in the Girvan-Ballantrae complex, south

Ayrshire: Transaction of Royal Society Edinburgh, v. 64, p. 1–27.

Brothers, R. N. and Yokoyama, K., 1982, Comparison of the high-pressure schist belts of New Caledonia and Sanbagawa, Japan: Contributions to Mineralogy and Petrology, v. 79, p. 219–229.

Edwards, M. and Taylor, P. N., 1974, A Rb/Sr age for granite-gneiss clasts from the late Precambrian Sveanor Formation, central Nordaustlandet: Norsk Polarinstitutt Årbok 1974, p. 255–258.

Ellis, D. J. and Green, D. H., 1979, An experimental study of the effect of Ca upon garnet-clinopyroxene Fe-Mg exchange equilibria: contributions to Mineralogy and Petrology, v. 71, p. 13–22.

Flood, B., Nagy, J. and Winsnes, T. S., 1971, Geological map of Svalbard, Sheet G1, Southern part: Norsk Polarinstitutt Skrifter, Nr. 154-A.

Gayer, R. A., Gee, D. G., Harland, W. B., Miller, J. A., Spall, H. R., Wallis, R. H. and Winsnes, T. S., 1966, Radiometric age determinations on rocks from Spitsbergen: Norsk Polarinstitutt Skrifter, Nr. 137, p. 1–39.

Gee, D. G., 1972, Late Caledonian (Haakonian) movements in northern Spitsbergen: Norsk Polarinstitutt Årbok 1970, p. 92–101.

Gee, D. G. and Moody-Stuart, M., 1966, The base of the Old Red Sandstone in central north Haakon VII Land, Spitsbergen: Norsk Polarinstitutt Årbok 1964, p. 57–68.

Gibbons, W., 1981, Glaucophanic amphibole in a Moinian shear zone on the mainland of North Wales: Journal Geological Society of London, v. 138, p. 139–143.

Harland, W. B., 1961, An outline structural history of Spitsbergen: in Raasch, G. O., ed., Geology of the Arctic, 1, p. 68–132, University of Toronto Press.

Harland, W. B., 1973, Tectonic evolution of the Barents Shelf and related plates: in Pitcher, M. G., ed., Arctic Geology, American Association of Petroleum Geologists, Memor. No. 19, p. 599–608.

Harland, W. B., 1978, The Caledonides of Svalbard: Geological Survey of Canada Paper, 78-13, p. 3–11.

Harland, W. B. and Gayer, R. A., 1972, The arctic Caledonides and earlier oceans: Geological Magazine, v. 109, (4), p. 289–314.

Harland, W. B. and Horsfield, W. T., 1974, West Spitsbergen orogeny: in Spencer, A. M., ed., Mesozoic-Cenozoic Orogenic belts, Data for orogenic studies, Geological Society of London Special Publication, No. 4, p. 747–755.

Harland, W. B., Horsfield, W. T., Manby, G. M. and Morris, A. P., 1979, An outline pre-Carboniferous stratigraphy of central western Spitsbergen: Norsk Polarinstitutt Skrifter Nr. 167, p. 119–144.

Harland, W. B. and Wright, N.J.R., 1979, Alternative hypothesis for the pre-Carboniferous evolution of Svalbard: Norsk Polarinstitutt Skrifter Nr. 167, p. 89–117.

Hirajima, T., Hiroi, Y., and Ohta, Y., 1985, Lawsonite and pumpellyite from Vestgötabreen Formation in Spitsbergen: Norsk Geologisk Tidskrift, v. 65, p. 267–274..

Hjelle, A., Ohta, Y. and Winsnes, T. S., 1979, Hecla Hoek rocks of Oscar II Land and Prins Karls Forland, Svalbard: Norsk Polarinstitutt Skrifter Nr. 167, p. 145–169.

Holland, T.J.B., 1979-a, Experimental determination of the reaction Paragonite+Jadeite+Kyanite+H_2O and internally consistent thermodynamic data for part of the system Na_2O-Al_2O_3-SiO_2-H_2O, with applications to eclogites and blueschists: Contributions to Mineralogy and Petrology, v. 68, p. 293–301.

Holland, T.J.B., 1979-b, Reversed hydrothermal determination of jadeite-diopside activities: EOS, v. 60, p. 405.

Holland, T.J.B., 1980, The reaction albite-jadeite+quartz determined experimentally in the range 600-1200°C: American Mineralogists, v. 65, p. 129–134.

Horsfield, W. T., 1972, Glaucophane schists of Caledonian age from Spitsbergen: Geological Magazine, v. 109, (1), p. 29–36.

Hoschek, G., 1969, The stability of staurolite and chloritoid and their significance in metamorphism of pelitic rocks: Contributions to Mineralogy and Petrology, v. 22, p. 208–232.

Irvine, T. N. and Baragar, W.R.A., 1971, A guide to the chemical classification of the common volcanic rocks: Canadian Journal of Earth Science, v. 8, p. 523–548.

Kanat, J., 1984, Jadeite from southern Oscar II Land, Svalbard. Mineralogical Magazine, v. 48, p. 301–303.

Kushiro, I., 1969, Clinopyroxene solid solutions formed by reactions between diopside and plagioclase at high pressures: Mineralogical Society of America Special Paper, 2, p. 179–191.

Lanphere, M. A. and Albee, A. L., 1974, $^{40}Ar/^{39}Ar$ age measurements in the

Worcester Mountains; evidence of Ordovician and Devonian metamorphic events in northern Vermont: American Journal of Science, v. 274, p. 545–555.

Leake, B. E., 1963, Origin of amphibolites from northwest Adirondacks, New York: Geological Society of America Bulletin, v. 74, p. 1193–1202.

Liou, J. G., 1971, P-T stabilities of laumontite, wairakite, lawsonite, and related minerals in the system $CaAl_2Si_2O_8$-SiO_2-H_2O: Journal of Petrology, v. 12, p. 379–411.

Lowell, L. D., 1972, Spitsbergen Tertiary orogenic belt and the Spitsbergen Fracture Zone: Geological Society America Bulletin, v. 83, p. 3091–3102.

Major, H. and Winsnes, T. S., 1955, Cambrian and Ordovician fossils from Sørkapp Land, Spitsbergen: Norsk Polarinstitutt Skrifter Nr. 106, p. 1–47.

Manby, G. M., 1978, Aspects of Caledonian metamorphism in central western Svalbard with particular reference to the glaucophane schists of Oscar II Land: Polarforschung, v. 48, (1/2), p. 92–102.

Maresch, W. V., 1977, Experimental studies of glaucophane: An analysis of present knowledge: Tectonophpysics, v. 43, p. 109–125.

Neumann, E.-R., 1976, Compositional relations among pyroxenes, amphiboles and other mafic phases in the Oslo Region plutonic rocks: Lithos, v. 9, p. 85–109.

Nitsch, K. H., 1968, Die Stabilitat von Lawsonit: Naturwissenschaft, v. 55, p. 388.

Nitsch, K. H., 1971, Stabilitatbeziehungen von Prehnit- und Pumpellyit -haltigen Paragenesen: Contributions to Mineralogy and Petrology, v. 30, p. 240–260.

Ohta, Y., 1978, Caledonian metamorphism in Svalbard, with some remarks on the basement: Polarforschung, v. 48, (1/2), p. 78–91.

Ohta, Y., 1979, Blue schists from Motalafjella, western Spitsbergen: Norsk Polarinstitutt Skrifter Nr. 167, p. 171–215.

Ohta, Y., 1984, Geochemistry of Precambrian basic igneous rocks between St. Jonsfjorden and Isfjorden, central western Spitsbergen: Polar Research, v. 3, n.s., p. 49–67.

Ohta, Y., Hiroi, Y. and Hirajima, T., 1984, Additional evidence of pre-Silurian high-pressure metamorphic rocks in Spitsbergen: Polar Research, v. 1, n.s., p. 215–218.

Pearce, J. A. and Cann, J. R., 1973, Tectonic setting of basic volcanic rocks determined using trace element analysis: Earth Planetary Science Letters, v. 19, p. 290–299.

Ryburn, R. J., Råheim, A., and Green, D. H., 1976, Determination of the P-T paths of natural eclogites during metamorphism - a record of subduction, A correction to a paper by Råheim and Green (1975): Lithos, v. 9, p. 161–164.

Råheim, A. and Green, D. H., 1975, P-T paths of natural eclogites during metamorphism. -a record of subduction: Lithos, v. 8, p. 317–328.

Scrutton, C. T., Horsfield, W. T. and Harland, W. B., 1976, Silurian fossils from western Spitsbergen: Geological Magazine, v. 113 (6), p. 519–523.

Seyfert, C. K. and Sirkin, L. A., 1979, Earth history and Plate tectonics: Harper and Row, New York.

Simonen, A., 1953, Stratigraphy and sedimentation of the Svecofennidic, early Archean supracrustal rocks in southwestern Finland: Bulletin de la Commission Geologique de Finlande, No. 160, p. 1–64.

Smulikovski, W., 1965, Petrology and some structural data of lowest metamorphic formations of the Hecla Hoek succession in Hornsund, Vestspitsbergen: Studia Geologica Polonica, v. 21, p. 97–107.

Waddams, P., 1983, The Late Precambrian succession in north-west Oscar II Land, Spitsbergen: Geological Magazine, v. 120, (3), p. 233–252.

MANUSCRIPT ACCEPTED BY THE SOCIETY JULY 29, 1985
NORSK POLARINSTITUTT CONTRIBUTION No. 239

Geological Society of America
Memoir 164
1986

A greenschist protolith for blueschist in Anglesey, U.K.

Wes Gibbons
Department of Geology
University College, Cardiff
P.O. Box 78
Cardiff CF1 1XL, Wales, U.K.

Mark Gyopari
Department of Environmental Sciences, Geology Division
Plymouth Polytechnic
Drake Circus, Plymouth
Devon PL4 8AA, England, U.K.

ABSTRACT

New exposures of Monian schists in SE Anglesey reveal a transition from weakly foliated, coarse, porphyroblastic actinolite greenschist to intensely foliated, fine-grained blueschist dominated by epidote and sodic amphibole. Textures are interpreted as recording an initial, near-static greenschist recrystallisation of a basic plutonic igneous rock, followed by deformation during continuing metamorphism at greenschist, then blueschist, grades. The sodic amphibole ranges in composition from glaucophane to ferro-glaucophane and crossite. A balanced reaction expressing the greenschist to blueschist transition is:

$$0.52 \text{ actinolite} + 1.67 \text{ albite} + 0.73 \text{ magnetite} + 0.15 \text{ chlorite} + 0.09 \text{ H}_2\text{O} =$$
$$1.0 \text{ crossite} + 0.4 \text{ epidote} + 0.44 \text{ hematite} + 0.32 \text{ quartz.}$$

During this reaction actinolite initially becomes increasingly barroisitic by glaucophane, Al-tschermakitic and edenitic coupled substitutions. Probe analyses indicate complete miscibility between actinolite and barroisite which then converts to sodic amphibole across a compositional gap spanning $\text{Ca}/(\text{Ca} + \text{Na})^{M4}$ values of 0.3-0.44. Epidotes in both greenschist and blueschist show Fe-depleted rims. Whereas blueschist epidotes show very high Fe content (79.5-92 mole% Al_2FeEp), those in the greenschist are lower and span an apparent compositional gap of 60-72 mole% Al_2FeEp. Low Na^{M4} values in unaltered actinolitic cores are similar to those in actinolites formed during sea-floor metamorphism. This suggests initially low-P metamorphic conditions for the greenschist, followed by deformation under rapidly increasing P which deflects the PT-time path away from the amphibolitic field into the blueschist facies. A preferred explanation for this unusual PT evolution is to invoke the subduction of young, hot oceanic crust previously recrystallised under low P during ocean-floor metamorphism.

INTRODUCTION

Blueschist facies rocks occur in SE Anglesey within a narrow strip of amphibolitic and micaceous schists some 20 km long and up to 5 km wide. The amphibolitic schists are dominated by sodic amphibole and epidote and occur as large lenticular masses within quartz-rich, phengite mica schists (Figure 1). These blueschists are exceptionally old, being certainly pre-Arenig and possibly late Precambrian (Gibbons, 1983a), although no radiometric data have yet been published on them (Dallmeyer and Gibbons, in prep.). Unfortunately, exposure across this small blueschist terrane is generally poor and outcrop towards the coast

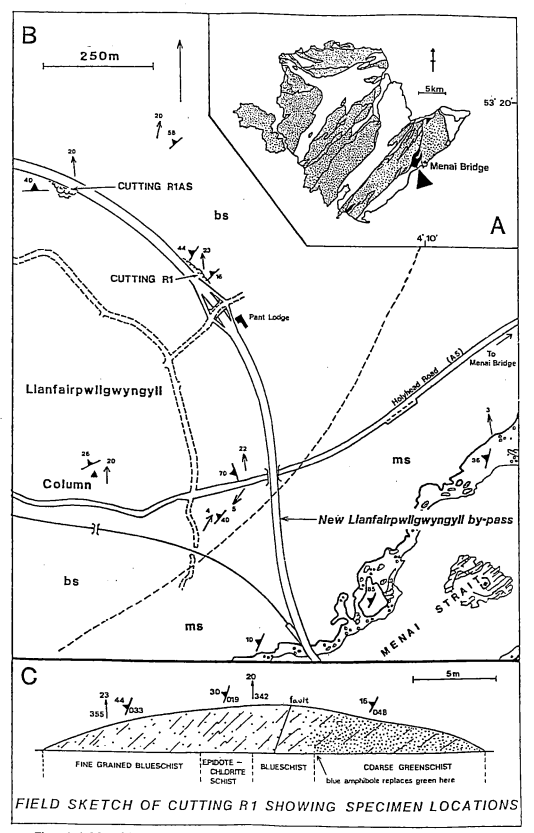

Figure 1. A: Map of Anglesey showing outcrop of Monian rocks (stippled) and basic blueschists (black).
B: Location map for road cutting exposure of blueschists, bs = blueschist; ms = mica schist. C: Sketch of
road cutting showing greenschist-to-blueschist transition.

is covered by Carboniferous sediments. Recently, however, a new road, bypassing the village of Llanfairpwllgwyngyllgogerych-wyrndrobwllllantyisiliogogogoch, was cut through the main blue-schist belt. Excellent, fresh exposures of blueschists were revealed at several places along the bypass. One of these new sections is of particular interest in temporarily exposing an apparent transition from typical Anglesey blueschist into a less foliated, coarser-grained greenschist (cutting R1 on Figure 1), thus potentially providing important information on the nature of the protolith to the basic blueschists. This paper reports the mineral chemistry of both blueschist and greenschist, and examines the transition between them.

PREVIOUS WORK

The existence of blueschists in Anglesey has been known since 1888 when the Rev. J. F. Blake reported the presence of a blue amphibole schist from the now famous exposures beneath the Marquis of Anglesey's Column just east of Llanfair p.g. (Figure 1). Using optical methods, Blake classified the blue amphibole as glaucophane and suggested a dioritic origin for the basic schists. The first whole-rock chemical analysis for this lithology (Washington 1901; see also Greenly 1919 pg. 117) shows, however, a basaltic rather than dioritic composition. This composition was later confirmed by Thorpe (1972) who interpreted the schists as metamorphosed ocean-floor tholeiites.

Greenly, in his classic Anglesey memoir (1919), again identified the sodic amphibole as "true glaucophane." He noted, however, the existence of a green core to some of the sodic amphiboles and identified this to be hornblende (also see Adye 1906, p. 39). Regarding a likely protolith for the blueschists, Greenly claimed that one could, in certain key sections, trace a gradation from low grade pillow lavas to blueschists. By contrast, in other places he recorded an apparent transition between the blueschists and coarse "hornblende schists," some of which preserved plutonic igneous textures (Greenly op. cit. p. 120).

No further work was attempted on these rocks until 1955 when Holgate separated and analysed the blue amphibole using wet-chemical methods. Although he notes that the cores of some blue amphiboles were different from the rim, he believes that this core amphibole had apparently differed sufficiently in specific gravity to be completely removed during the separation process. His results produce a formula interpreted by him to be crossite. However, according to Leake's 1978 classification, Holgate's amphibole contains too much Na in the A site and plots in the magnesio-arfvedsonite field (Macpherson 1983). Work by Bieler (1982), published in abstract form only, identifies the zoned amphiboles as possessing barroisitic or actinolitic hornblende cores and crossite rims. Finally, although epidote is the typical calc-silicate within the Anglesey blueschists, the localised occurrence of lawsonite co-existing with crossite at several localities on the eastern side of the Anglesey blueschist belt was recently confirmed by Gibbons and Mann (1983).

BLUESCHIST PETROLOGY

The phase assemblage of a typical Anglesey blueschist is dominated by sodic amphibole and epidote, with minor quartz, sphene, chlorite, magnetite and hematite. The rocks are strongly deformed, dense, and dark, possessing an intense, gently dipping foliation and a prominent N-S-trending mineral lineation. Both epidote and sodic amphibole are aligned to produce this lineation. Epidote occurs as either small, elongate (length <0.5 mm) crystals immersed in amphibole and chlorite, or as large, polycrystalline clots (up to 2 mm) around which wrap the main fabric. Some epidote and euhedral sodic amphibole also occurs within quartz pockets which lie enclosed within the main mass of blueschist. Chlorite needles can often be seen growing into quartz veins cutting the main fabric, often in en echelon trains, indicating post-deformational high-fluid pressures during metamorphic retrogression, with P_f transiently exceeding P_1.

Both epidotes and blue amphiboles are commonly zoned, with the epidotes exhibiting a cloudy yellow core and a clear rim, and the amphiboles showing a green core and a blue rim. Probe analyses for the blue sodic amphibole indicate a range of composition around the ferro-glaucophane/crossite/glaucophane boundary (Figure 2).

Fe^{3+} content was estimated by recalculating total cations to 13, exclusive of K, Na and Ca. Analyses were accepted only if both charge balance and crystal chemical criteria were satisfactorily fulfilled (e.g., K in A site only; Na in A and M4 site only; Ca in M4 site only; Si in T site only). Whether or not a sample shows consistent glaucophane compositions (e.g., specimen G3) is attributed to slight variations in bulk-rock chemistry affecting the recalculated $Mg/(Mg + Fe^{2+})$ ratio. $Mg/(Mg + Fe^{2+})$ values of 0.45-0.6 and $Fe^{3+}/(Fe^{3+} + Al^{VI})$ values of 0.15-0.41 place the amphiboles within the 'chlorite-coexistence' region of Okay (1980), which is consistent with the presence of both sodic amphibole and chlorite within these blueschists. Typical analyses of these sodic amphiboles are given in Table 1.

The green mineral forming the core to many sodic amphiboles is a barroisitic sodic-calcic amphibole (not hornblende as reported by Greenly 1919). Typical chemical formulae for core and rim compositions are shown below:

Barroisite core: $(Na_{1.01}Ca_{0.99})$ $(Fe^{2+}_{1.64}Mg_{1.83}Fe^{3+}_{0.82}Al^{VI}_{0.66})$
$(Al^{IV}_{0.85}Si_{7.15})O_{22}(OH)_2$
Crossite rim: $(Na_{1.77}Ca_{0.23})$ $(Fe_{1.05}Mg_{2.05}Fe^{3+}_{0.43}Al^{VI}_{1.27})$
$(Al^{IV}_{0.10}Si_{7.90})O_{22}(OH)_2$

Figure 3 and Table 1 illustrate the composition of these sodic-calcic amphiboles, most of which fall within the barroisite field with an $Mg/(Mg + Fe^{2+})$ value of 0.5-0.7. There is a clear compositional gap between these green barroisites and the sodic amphiboles, and there is a sharp optical boundary between them (Figure 4). Since the barroisites consistently occur as the green core to a blue rim, they are interpreted as recording the incomplete replacement of a sodic-calcic amphibole by a sodic amphi-

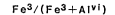

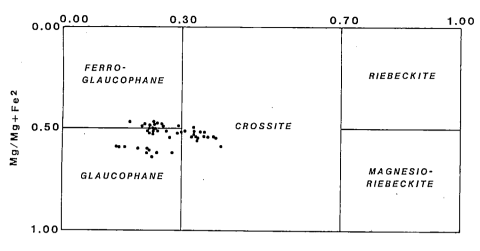

ALKALI AMPHIBOLES $Na_B \geq 1.34$ $(Na+K)_A < 0.50$

Figure 2. Composition of sodic amphiboles using the classification of Leake (1978).

TABLE 1. REPRESENTATIVE PROBE ANALYSES FOR SODIC AND CALCIC-SODIC AMPHIBOLES IN ANGLESEY BLUESCHIST*

	GL-CR 3B1.1 Core	FGL-GL 3B1.1 Rim	BA 3B1.2 Core	GL 2B1.2 Int	CR 3B1.2 Rim	BA 3B1.4 Int	FGL 3B1.4 Int	CR 3B1.4 Rim	BA 3B2.1 Core	FGL 3B2.1 Int	CR 3B2.1 Rim
SiO_2	56.29	56.24	51.68	56.29	56.25	49.32	54.91	55.97	50.28	54.36	55.80
Al_2O_3	6.79	8.95	7.76	8.84	7.09	9.05	8.70	7.27	9.22	9.19	7.56
TiO_2	–	–	–	–	–	0.22	–	–	0.25	0.36	0.26
FeO	17.62	17.03	19.02	16.65	17.89	19.97	16.86	17.58	19.60	17.82	17.54
MnO	0.19	–	0.26	0.16	0.21	0.35	0.18	0.18	0.25	0.16	0.19
MgO	8.34	7.65	8.77	7.69	8.09	7.93	7.64	8.25	8.50	7.35	8.17
CaO	2.43	1.64	5.99	1.59	1.97	5.94	2.00	1.93	6.06	2.59	2.13
Na_2O	6.07	6.59	4.94	6.49	6.46	4.30	6.58	6.25	5.44	5.91	6.06
K_2O	–	–	0.14	–	–	0.21	–	–	0.18	–	–
Total	97.91	98.10	98.56	97.71	97.96	97.29	96.87	97.43	99.78	98.00	97.71

FORMULAE BASED ON 23 OXYGENS
(normalized to 13 cations excluding K, Ca, and Na)

	GL-CR 3B1.1 Core	FGL-GL 3B1.1 Rim	BA 3B1.2 Core	GL 2B1.2 Int	CR 3B1.2 Rim	BA 3B1.4 Int	FGL 3B1.4 Int	CR 3B1.4 Rim	BA 3B2.1 Core	FGL 3B2.1 Int	CR 3B2.1 Rim
Si	7.99	7.91	7.46	7.93	7.96	7.21	7.86	7.94	7.21	7.73	7.89
Al^{iv}	0.01	0.09	0.54	0.07	0.04	0.79	0.14	0.06	0.79	0.27	0.11
Al^{vi}	1.12	1.39	0.78	1.40	1.14	0.77	1.32	1.15	0.79	1.27	1.15
Ti	–	–	–	–	–	0.02	–	–	0.03	0.04	0.03
Fe^{3+}	0.48	0.40	0.49	0.41	0.51	0.86	0.37	0.61	0.55	0.51	0.59
Mg	1.76	1.60	1.89	1.61	1.71	1.73	1.63	1.74	1.82	1.56	1.72
Fe^{2+}	1.61	1.60	1.81	1.55	1.60	1.58	1.65	1.47	1.80	1.61	1.48
Mn	0.02	–	0.03	0.02	0.02	0.04	0.02	0.02	0.01	0.02	0.02
Ca	0.37	0.25	0.93	0.24	0.03	0.93	0.30	0.29	0.93	0.40	0.32
Na	1.63	1.75	1.07	1.76	1.70	1.07	1.70	1.72	1.07	1.60	1.66
Na	0.04	0.05	0.31	0.01	0.07	0.14	0.13	–	0.44	0.03	–
K	–	–	0.03	–	–	0.04	–	–	0.33	–	–
$\frac{Mg}{Mg+Fe^{2+}}$	0.52	0.50	0.51	0.51	0.52	0.52	0.49	0.54	0.50	0.49	0.54
$\frac{Fe^{3+}}{Fe^{3+}+Al^{VI}}$	0.30	0.20	0.39	0.23	0.31	0.53	0.22	0.35	0.42	0.29	0.34

*All analyses reported in this paper were performed on microprobes at the Departments of Earth Sciences, Cambridge (energy dispersive) and Open University (wavelength dispersive). Accelerating potential was 20 kV with live counting times of up to 100s (80s at Cambridge). Peaks were processed and measured by iterative peak stripping and corrected using the method of Sweatman and Long (1969); GL = Glaucophane; CR = Crossite; FLG = Ferroglaucophane; BA = Barroisite.

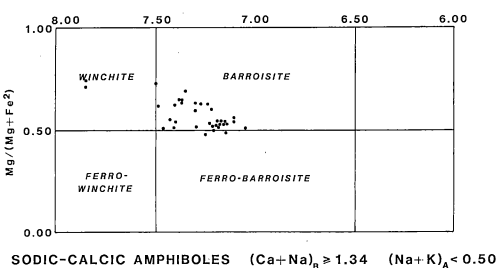

Figure 3. Composition of sodic-calcic amphiboles using the classification of Leake (1978).

bole i.e. the 2-amphibole assemblage was not generated under equilibrium conditions.

Mole% Al_2FeEp compositions of epidotes are given in Figure 5. All probed samples are Fe-rich and zoned crystals show a decrease in Fe from core to rim. The average mole % Al_2FeEp value for the blueschist epidote core analyses in Figure 5 is 92, whereas that for the rim is 79.5%. There is clearly no compositional gap at these high-Fe values under the blueschist metamorphism experienced by these Anglesey schists. The high-Fe content is further evidence (along with the presence of hematite in the phase assemblage) of a high oxidation state during metamor-

phism. In addition, the zoning exhibited by the Anglesey epidotes belongs to the strongly oxidised 'type 1' of Raith (1976), with a rather small total compositional variation (average core to rim difference of 12.6 mole% Al_2FeEp over 10 samples) and an apparent gradation in composition from core to rim.

GREENSCHIST PETROLOGY

A small exposure of greenschist was uncovered (and is still exposed) at the eastern end of a low bank on the north side of the new bypass (Figure 1). In hand specimen, the least deformed

Figure 4. Amphibole showing pale (blue) crossite rim and dark (green) barroisitic core. Note sharp optical boundary between the two. Scale bar = 100 μ.

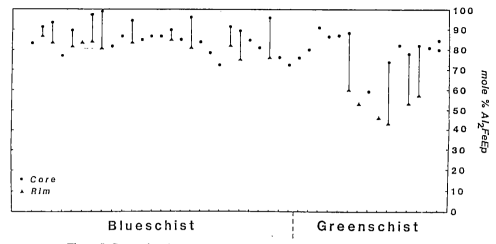

Figure 5. Comparison between epidote compositions in greenschist and blueschist.

samples possess a distinctive coarse texture with porphyroblasts of dark green amphibole (mostly 1-2 mm in size) set in a finer grained, foliated pale albitic matrix. The rock has the appearance of a deformed meta-gabbro or meta-diorite. Despite this meta-igneous appearance in hand specimen, in thin section the texture is entirely metamorphic. The porphyroblasts of pale green amphibole lie within a weakly foliated, finer grained assemblage of albite, epidote and finer amphibole, with minor sphene, quartz, magnetite and hematite. The magnetite occurs as cores to hematite rims. The amphibole is pleochroic in pale green to very pale green, with a bright 1st-order birefringence and a maximum c/Z extinction angle of 17°, all of which indicate an actinolitic composition. Small randomly orientated actinolites commonly grow within the large porphyroblasts, indicating partial recrystallisation in response to intracrystalline strain produced during deformation. Away from these least deformed greenschists, the S_1 fabric becomes increasingly dominant, producing an intensely foliated actinolite-epidote-chlorite-albite schist with the actinolite porphyroblasts now occurring as remnant augen around which the fabric is wrapped. The texture is therefore interpreted as recording an initial near-static recrystallisation (probably replacing a plutonic igneous texture) followed by deformation during continuing metamorphism at greenschist grade.

Probe data on these green amphiboles confirm an actinolitic composition (Table 2), with all analyses plotting within the actinolite field (Leake 1978) (Figure 6). Most porphyroblasts are both optically and chemically inhomogeneous, often showing variation in pleochroism from green to blue-green. This variation may be patchily developed, although often a consistent change from a green core to blue-green rim can be discerned. Probe analyses reveal the nature of this compositional change, with the amphibole rims being affected by both glaucophane and Al^{VI}-tschermak coupled substitutions. Typical core-rim compositions are given below:

Core: $(Na_{0.14}Ca_{1.86})$ $(Mn_{0.02}Fe^{2+}_{1.09}Mg_{3.44}Fe^{3+}_{0.29}Al^{VI}_{0.12})$ $(Al^{IV}_{0.36}Si_{7.64})$

Rim: $(Na_{0.23}Ca_{1.77})$ $(Mn_{0.03}Fe^{2+}_{1.27}Mg_{3.05}Fe^{3+}_{0.34}Al^{VI}_{0.34})$ $(Al^{IV}_{0.46}Si_{7.54})$

Epidote is less common in the greenschists than blueschists. Crystals are again often zoned with a relatively Fe-poor rim. A typical core-rim compositional pair is shown below:

Core: $Ca_2Al_{2.29}Fe^{3+}_{0.81}Mn_{0.01}Si_{3.14}(O, OH, F)_{13}$ (81 Mole% Al_2FeEp)

Rim: $Ca_{1.97}Al_{2.42}Fe^{3+}_{0.58}Mn_{0.01}Si_{3.01}(O, OH, F)_{13}$ (58 Mole% Al_2FeEp)

A consistent decrease in iron towards the rim of epidote crystals has been interpreted as reflecting increasing metamorphic grade (Strens 1965, Holdaway 1972, Raith 1976). In general, the Fe^{3+} content decreases with increasing temperature and pressure, and decreasing oxidation state. Immiscibility within the Al-Fe monoclinic epidote series has been suggested by many workers. Strens (1965) suggests a miscibility gap ranging from 39 to 72 mole% Al_2FeEp, and similar ranges are proposed by Holdaway (1972) (45-75 mole% Al_2FeEp) and Hietanen (1974) (35-75 mole% Al_2FeEp). Raith (1976) observed a gap within the range 72-53 Al_2FeEp for lower grade greenschists (c.400°C) reducing to 63-55 mole% Al_2FeEp for higher grade greenschists (c.500°C). Figure 5 gives a plot of the variations in epidote composition within the Anglesey blueschists and greenschists (and the transition between them). All blueschist Al_2FeEp values lie above the proposed miscibility gaps mentioned above. By contrast, the greenschist epidotes possess slightly lower Fe^{3+} values and span an apparent compositional gap of 60-72 mole% Al_2FeEp.

GREENSCHIST TO BLUESCHIST TRANSITION

The western margin of the greenschists is marked by a rapid transition into fine-grained, strongly foliated and lineated blueschist. The fabric in the greenschist passes insensibly into that of

TABLE 2. REPRESENTATIVE PROBE ANALYSES OF ACTINOLITES
FROM ANGLESEY GREENSCHIST

	1A1.1 Core	1A1.4 Core	1A2.4 Core	1A2.4 Rim	13.20 Core	13.20 Rim	13.21 Core	13.22 Rim
SiO_2	53.41	54.90	55.17	54.94	53.22	53.09	52.14	52.90
Al_2O_3	4.98	2.97	2.69	2.86	3.87	4.60	5.55	4.41
TiO_2	–	–	0.17	0.12	–	–	0.12	–
FeO	13.34	11.26	11.07	11.04	11.97	12.54	12.80	12.35
MnO	0.02	0.16	0.18	0.20	0.14	0.12	0.21	0.15
MgO	14.04	16.39	16.21	16.26	15.43	14.89	13.85	14.66
CaO	11.83	12.17	12.08	12.38	11.15	9.96	10.51	10.28
Na_2O	0.92	0.51	0.82	0.80	1.31	1.70	1.72	2.14
K_2O	–	–	–	–	–	0.09	–	–
Total	98.54	98.36	98.39	98.60	95.84	96.99	96.90	96.49

FORMULAE BASED ON 23 OXYGENS
(normalized to 13 cations excluding K, Ca, and Na)

	1A1.1 Core	1A1.4 Core	1A2.4 Core	1A2.4 Rim	13.20 Core	13.20 Rim	13.21 Core	13.22 Rim
Si	7.59	7.74	7.79	7.76	7.61	7.56	7.50	7.58
Al^{IV}	0.41	0.36	0.21	0.24	0.39	0.44	0.50	0.42
Al^{VI}	0.42	0.12	0.23	0.24	0.26	0.33	0.44	0.33
Ti	–	–	0.02	0.01	–	–	0.01	–
Fe^{3+}	0.14	0.24	0.05	0.01	0.35	0.58	0.41	0.30
Mg	2.97	3.44	3.41	3.42	3.29	3.16	2.79	3.16
Mn	0.02	0.02	0.02	0.02	0.02	0.01	0.03	0.02
Fe^{2+}	1.44	1.09	1.26	1.29	1.08	0.91	1.13	1.19
Ca	1.80	1.84	1.83	1.87	1.71	1.52	1.56	1.59
Na	0.20	0.14	0.17	0.13	0.29	0.48	0.44	0.41
Na	0.05	–	0.05	0.09	0.07	–	0.04	0.19
K	–	–	–	–	–	0.02	–	–

the blueschist and there is no evidence of an earlier greenschist foliation having been transposed into a new orientation during the imposition of blueschist conditions. Examination of specimens from this transition zone reveals the complex nature of actinolite breakdown. The pale green calcic amphibole is zoned firstly to deeper green then to a green-blue barroisite (Figure 7; Table 3) and finally to a blue sodic amphibole. Figure 8 shows an example of a transition zone amphibole preserving this 2-stage reaction, with pale green actinolite in the core converting outwards to deeper green sodic-calcic amphibole and pale blue sodic amphibole occurring as rim replacement. A similar amphibole zonation sequence has been recorded from Lower Ordovician metabasites in New Brunswick (Trzcienski et al., 1984). Optically, the change from actinolite to barroisite is gradational, whereas the boundary between barroisite and glaucophane is sharp. A plot of Ca/(Ca + Na^{M4}) against $Fe^{3+}/(Fe^{3+} + Al^{VI})$ shows compositional continuity between actinolites with Ca/(Ca + NaM4) values as high as 0.94, and barroisite with Ca/(Ca + NaM4) values down to 0.44 (Figure 9). By contrast, there is a clear compositional gap between the sodic and sodic-calcic amphiboles, reflecting a discontinuity between Ca/(Ca + NaM4) values of 0.3-0.44. A dominance of Al^{VI} over Fe^{3+} involved in the glaucophanic coupled substitution results in a consistent fractionation of Fe^{3+} across the compositional gap. The barroisitic core amphibole always contains a higher ferric iron/aluminum ratio than the sodic amphibole, a trend also noted in previous work on similar mineral pairs (Black 1973, Spear and others 1982). The ferric iron released by this fractionation in the Anglesey rocks is presumably utilized in the growth of new epidote.

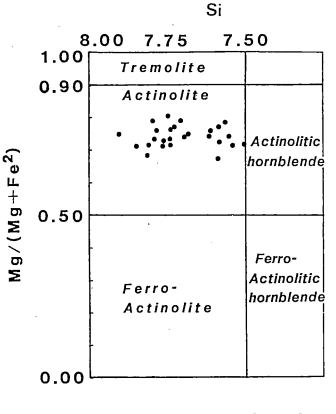

Figure 6. Composition of actinolites using the classification of Leake (1978).

Figure 7. Pale actinolite showing patchy replacement by dark barroisite. Scale bar = 100 μ.

The conversion of metabasite from actinolite greenschist to sodic-amphibole-bearing blueschist at the greenschist-blueschist transition is well documented (Laird 1982). Balanced reactions (for end-member phases) describing this change predict a loss of albite + calcic amphibole + chlorite + iron oxide in favour of sodic amphibole + epidote + quartz (Miyashiro & Banno 1958, Brown 1974, Laird 1980, 1982). The blueschist-forming reaction in Anglesey conforms well to this general model. Actinolite + albite + minor chlorite + magnetite are consumed and replaced by crossite/glaucophane + epidote + quartz + hematite. The reaction may be approximated as follows:

$$0.52 \ Na_{0.2}Ca_{2.0}Fe^{2+}_{0.2}Mg_{3.4}Fe^{3+}_{0.2}Al_{0.5}Si_{7.7}O_{22}(OH)_2 + 1.67 \ NaAlSi_3O_8 + 0.73 \ Fe_3O_4 + 0.15 \ (Mg_{2.9}Al_{1.3}Fe_{1.8})(Si_{2.7}Al_{1.3})O_{20}(OH)_8 + 0.09 \ H_2O = 1 \ Na_{1.77}Ca_{0.25}Fe^{2+}_{1.1}Mg_{2.2}Fe^{3+}_{0.4}Al_{1.4}Si_{7.9}O_{22}(OH)_2 + 0.4 \ Ca_2Fe^{3+}_{0.7}Al_{2.3}Si_3O_{12}(OH) + 0.32 \ SiO_2 + 0.44 \ Fe_2O_3.$$

0.52 actinolite + 1.67 albite + 0.73 magnetite + 0.15 chlorite + 0.09 H_2O = 1 crossite + 0.4 epidote + 0.32 quartz + 0.44 hematite.

METAMORPHIC CONDITIONS

Whole-rock XRF analyses of Anglesey blueschists and greenschist prove the bulk-rock chemistry to be essentially the same for both lithologies (Greenly 1919, Thorpe 1972, Gyopari 1985). The presence of hematite and the high Fe^{3+} values of epidotes in both lithologies indicate high FO_2 during both greenschist and blueschist metamorphism, and oxygen fugacity is discounted as a major influencing factor in the reaction from oxidised greenschist to oxidised blueschist. Similarly, the application of differential stress during blueschist metamorphism, as evinced by the intense LS fabric, probably acted as an important reaction catalyst but is not considered enough in itself to induce blueschist metamorphism. The reaction is rather interpreted as primarily a response to changing metamorphic PT conditions. The greenschist outcrop is interpreted as an isolated, disequilibrated remnant which, like the zoned amphibole in the surround-

TABLE 3. REPRESENTATIVE PROBE ANALYSES OF GREEN AMPHIBOLES FROM THE GREENSCHIST-TO-BLUESCHIST-TRANSITION ZONE

	ACT* WGf	ACT WGe	ACT WGb	ACT WGa	BA* WGk
SiO_2	54.65	54.07	52.07	54.20	51.81
Al_2O_3	2.07	2.51	4.38	3.95	6.07
TiO_2	0.04	0.06	0.09	0.05	0.01
FeO	11.14	11.46	13.13	12.96	13.30
MnO	0.23	0.26	0.26	0.28	0.22
MgO	16.48	15.74	14.18	14.68	13.08
CaO	11.05	10.30	9.41	9.08	8.22
Na_2O	1.21	1.73	2.43	2.64	3.31
K_2O	0.07	0.13	0.12	0.10	0.13
Total	96.94	96.26	96.07	97.92	96.15

FORMULAE BASED ON 23 OXYGENS
(normalized to 13 cations excluding K, Ca, and Na)

Si	7.79	7.78	7.55	7.68	7.50
Al^{iv}	0.21	0.22	0.45	0.32	0.50
Al^{vi}	0.14	0.20	0.30	0.34	0.54
Ti	–	0.01	0.01	0.01	0.01
Fe^{3+}	0.34	0.32	0.50	0.45	0.45
Mg	3.50	3.38	3.07	3.09	2.82
Fe^{2+}	0.99	1.06	1.09	1.05	1.16
Mn	0.03	0.03	0.03	0.03	0.03
Ca	1.69	1.59	1.64	1.38	1.27
Na	0.31	0.41	0.54	0.62	0.73
Na	0.02	0.07	0.14	0.10	0.20
K	0.01	0.02	0.02	0.02	0.02

*ACT = Actinolite; BA = Barroisite.

Figure 8. Amphibole basal section showing the remnant of a pale (green) actinolitic core surrounded by dark (green) barroisite with a rim of pale (blue) crossite. Scale bar = 100 μm.

ing blueschist, preserves a record of early, pre-blueschist metamorphic conditions. The imposition of synkinematic blue-schist metamorphism was not maintained for long enough entirely to eradicate evidence for the greenschist protolith.

An important pressure indicator in calcic amphiboles is the amount of Na present within the M4 site. This value increases with increasing pressure (Brown 1977). Although doubts have been raised as to the validity of this geobarometer in low and medium pressure terrains (Hynes 1982), there does appear to be a consistent increase in Na^{M4} occupancy in medium- to high-pressure terrains (e.g., note the Otago schist data of Brown 1977 in Hynes 1982, Figure 1d). A plot of Na^{M4} against Al^{IV} for the Anglesey rocks indicates increasing pressure from low-Na^{M4} actinolites to high-Na^{M4} barroisites (Figure 10). Superimposing Brown's 1977 tentative isobars upon this diagram gives a minimum pressure value of 240 MPa (2.4 kb) rising to >600 MPa (6 kb) for the sodic barroisites. Examples of actinolites generated under low-pressure-ocean floor metamorphic conditions (e.g., Bonatti et al. 1975, Girardeau and Mevel 1982, Ito and Anderson 1983) contrast in their low Na^{M4} content with most of the Anglesey examples. Significantly, however, the least altered porphyroblastic actinolite cores in Anglesey preserve Na^{M4} values similar to the ocean-floor examples (Figure 10). An implication of this is that the protolith for the blueschist may have undergone metamorphism under low-pressure greenschist conditions. A comparison of Al^{IV} values in the Anglesey actinolites with the tentative temperature estimates of Hietenan (1974) give minimum greenschist temperature values of around 400°C.

The maximum pressures recorded by these rocks were presumably responsible for the growth of the high-P blueschist assemblage. The position of the greenschist-blueschist transition in

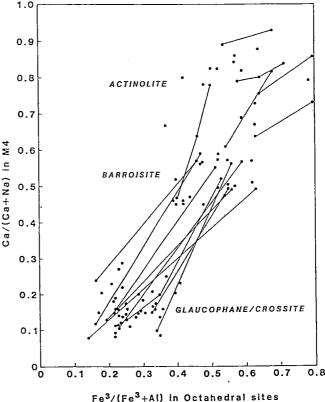

Figure 9. Ca: $(Ca + NaM4)$ v Fe^{3+}: $(Fe^3 + Al^{VI})$ plot of amphiboles from greenschist (actinolite and barroisite) and blueschist (barroisite to glaucophane/crossite). Na-rich core and Ca-rich rim analysis from the same crystal are connected by tie-lines.

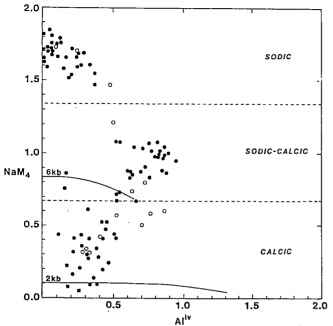

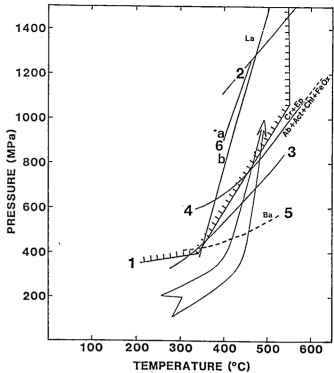

Figure 10. Amphiboles plotted with NaM4 v. tetrahedral Al. Closed circles = Anglesey amphiboles in greenschist (calcic and calcic-sodic) and blueschist (calcic-sodic and sodic). Open circles = Zoned amphiboles from the Tetagouche Group, New Brunswick (Trzcienski et al. 1984). Squares = low pressure actinolites produced by ocean floor metamorphism (data from Bonatti et al. 1974; Girardeau & Mevel 1982; Ito & Anderson 1983). Isobars from Brown (1977).

PT space is unfortunately not well defined, although one of the most recent estimates places it in the region of 400°C at 900 MPa (9 kb) (Brown and others 1984). The involvement of barroisitic amphibole in the Anglesey transition reaction suggests temperatures somewhat higher than this e.g., 430°-520°C at >6 kb (600 MPa) (Ernst 1979). A lower temperature limit is constrained by the presence of abundant epidote and the absence of lawsonite, indicating temperatures were c. 370-400°C at 600-800 MPa (Nitsch 1974; Holland 1979).

Figure 11 provides a rough petrogenetic grid which attempts to constrain the PT conditions recorded by the Anglesey blueschists. A possible prograde PT-time path envisaged for these rocks traces an initial increase in temperature at low pressure (geothermal gradient = 50-60°C/km) to produce the Na^{M4}-poor actinolitic greenschist. The replacement of this amphibole by barroisite and ultimately by glaucophane-crossite indicates rapidly increasing pressures (geothermal gradient = 4-7°C/km), deflecting the initial PT-time path away from the amphibolite facies and into the blueschist field. Alternatively this deflection may have taken place under decreasing temperature with the PT-t path showing a negative slope away from the barroisite stability field (c.f. Beiler 1982).

DISCUSSION

Barroisitic cores with sodic amphibole rims are common

Figure 11. Petrogenetic grid illustrating a possible prograde P-T-time path for the Anglesey schists. The path travels from actinolite greenschist through the barroisite stability field to a position where blue amphibole + epidote are stable. An alternative possibility is for the curve to show a counter-clockwise path with decreasing temperature during blueschist metamorphism. 1 = Upper T maximum stability limit of glaucophane from Maresch (1977); 2 = Jadeite + quartz = albite (Holland, 1980); 3 = phengite stability curve for Si^{+4} content of 3.4 (Velde 1967); 4 = greenschist-blueschist transition reaction given by Brown (1978); 5 = approximate lower limit of barroisite stability field (Ernst, 1979); 6 = upper stability of lawsonite from the reactions (6a) lawsonite + albite = zoisite + paragonite + quartz + H_2O (Holland, 1979), and (6b) lawsonite = zoisite + margarite + quartz + H_2O (Nitsch 1974).

elsewhere in the Anglesey blueschist belt, e.g., at the classic Marquis of Anglesey's Column locality. This suggests that many of the basic blueschists are derived from a greenschist protolith similar to that exposed in the new road cutting. This overprint of greenschist by blueschist is not a transition predicted by the subduction of cold material, where an initially rapid increase in pressure may produce blueschists which may then be overprinted by higher temperature conditions. One explanation for these Anglesey rocks would be the metamorphism of a basaltic protolith under greenschist temperatures at an oceanic spreading centre and then burial of the greenschist to a depth of at least 25 km. Such a model could invoke the termination of a transient oceanic spreading centre by its collision with a subduction system. Alternatively, the blueschists may have been generated by the propagation of nappes over newly generated oceanic crust, burying the latter to >25 km during plate collision without the direct involvement of a Benioff zone. Both models require a short time interval between the generation of the protolith and its metamorphism—a feature which accords well with the narrow age limits to Monian deformation and metamorphism (e.g., Gibbons 1983b, Beckinsale and others 1984).

The PT-time curve suggested for the Anglesey rocks is different from that derived using the procedure of Holland & Richardson (1979) from the similar amphiboles in the Tetagouche Group of New Brunswick (Trzcienski and others 1984). Remarkably, these latter transitional blueschists were, like the Anglesey examples, formed on the SE side of Iapetus, although the ages of the two units appear to be different. The Tetagouche blueschists have been interpreted as situated in the hangingwall of a SE-dipping Taconic subduction zone. Their unusual amphibole zonation, identical to that exhibited by the Anglesey amphiboles, has been tentatively interpreted as produced by two metamorphic events separated by a cooling episode. However, it is implied that the Tetagouche amphiboles define the main schistosity (i.e., both lower and higher P amphiboles grew parallel to the same fabric.) The interpretation of Trzcienski and others (op. cit.) would therefore require either the coincidence of two tectonic events each with closely similar stress conditions but very different PT-time paths or the complete transposition of any earlier S_1 fabric. Given the apparently "continuous path in P-T space" followed by the Tetagouche blueschists (Trzcienski and others op. cit.), we consider it simpler to interpret these rocks as having undergone initial metamorphism under a high geothermal gradient followed without a significant break by a rapid descent to a depth of 25 km into the crust during either subduction or nappe superimposition. A similar model is suggested to explain the Anglesey greenschist- to blueschist transition. The parallelism of both greenschist and blueschist metamorphic foliations, combined with a lack of evidence for any earlier discordant greenschist fabric, indicate continuous, synkinematic metamorphism affecting a medium- to coarse-grained (microgabbroic?) basic protolith.

The strong mineral-grain-shape lineation exhibited by the blueschists suggests a synmetamorphic transport direction towards either N or S. This is anomalous with respect to the NE-SW orientation of most structures within the British Caledonides. Indeed, all Monian exposures are quite different from other Caledonian units. This 'Monian Terrane' (Gibbons & Gayer 1985) is broadly interpreted as a series of fault-bounded slivers which provide a highly fragmentary record of tectonometamorphic events along an active plate margin during the late Precambrian to early Cambrian. The Anglesey blueschist belt, bounded by a major steep mylonite zone to the NW, and badly affected by faulting and thrusting to the east, forms one of these tectonic slices. The relationship between these remarkably well preserved blueschists and other Monian units however remains enigmatic. The coarse greenschist protolith identified in this study seems closer in character to the plutonic parent to Greenly's "hornblende schists" than the pillow-lava clasts typical of the Gwna melange. The poorly exposed contact between the Gwna and Penmynydd units in SE Anglesey accordingly remains a subject of continuing research.

ACKNOWLEDGMENTS

We thank Ned Brown, Walter Trzsienski, and Dave Bieler for reviewing the manuscript; Andy Buckley (Cambridge) and Andy Tindle (Open University) for their help with microprobes; Richard Bevins and Jana Horák (Nat. Mus. Wales) for much useful discussion, and Paula Westall (U.C.C.) for her efficiency and expertise on the word processor.

REFERENCES CITED

Adye, E. H., 1906, The twentieth century atlas of Microscopical Petrography: London, T. Murby, Publisher, p. 39.

Beckinsale, R. D., Evans, J. A., Thorpe, R. S., Gibbons, W., and Harmon, R. S., 1984, Rb-Sr whole-rock isochron ages, ^{18}O values and geochemical data for the Sarn Igneous Complex and the Parwyd gneisses of the Mona Complex of Llŷn, N. Wales: Journal of the Geological Society of London, v. 141, p. 701–710.

Bieler, D. B., 1982, P-T paths of Late Precambrian(?) glaucophane schists, Anglesey, N. Wales: Geological Society of America Abstracts with Programs, v. 14, p. 443.

Black, P. M., 1973, Mineralogy of New Caledonian Metamorphic rocks II. Amphiboles from the Ovégou district: Contributions to Mineralogy and Petrology, v. 39, p. 55–64.

Blake, J. F., 1888, The occurrence of a glaucophane-bearing rock in Anglesey: Geology Magazine, v. 5, p. 125–127.

Bonatti, E., Honnorez, J., Kirst, P., and Radicati, F., 1975, Metagabbros from the mid-Atlantic ridge at 06°N: Contact hydrothermal-dynamic metamorphism beneath the axial valley: Journal of Geology, v. 83, p. 61–78.

Brown, E. H., 1974, Comparison of the Mineralogy and Phase Relations of Blueschists from the North Cascades, Washington, and Greenschists from Otago, N.Z.: Geological Society of America Bulletin, v. 85, p. 333–344.

Brown, E. H., 1977, The Crossite content of Ca-Amphibole as a guide to pressure of metamorphism: Journal of Petrology, v. 18, p. 53–72.

Brown, E. H., 1978, A P-T grid for metamorphic index minerals in high pressure terrains; Geological Society of America Abstracts with Programs, v. 10, no. 7, p. 373.

Brown, E. H., Evans, B. W., Forbes, R. B., and Misch, P., 1984, Penrose conference report on blueschists and related eclogites: Geology, v. 12, p. 318–319.

Ernst, W. G., 1979, Coexisting sodic and clacic amphiboles from high-pressure metamorphic belts and the stability of barroisitic amphibole: Mines Magazine, v. 43, p. 269–278.

Gibbons, W., 1983a, Stratigraphy, subduction, and strike-slip faulting in the Mona Complex of North Wales - A review: Proceedings of the Geologists Association, v. 94, p. 147–163.

Gibbons, W., 1983b, The Monian "Penmynydd Zone of Metamorphism" in Llŷn, North Wales: Geological Journal, v. 18, p. 21–41.

Gibbons, W., and Gayer, R. A., 1985, British Caledonian Terranes, *in* Gayer, R. A., ed., The tectonic evolution of the Caledonide-Appalachian Orogen: Earth Evolution Sciences Monograph Series No. 1, p. 3–16.

Gibbons, W., and Mann, A., 1983, Pre-Mesozoic lawsonite in Anglesey, North Wales - the preservation of ancient blueschists: Geology, v. 11, p. 3–6.

Girardeau, J., and Melvel, C., 1982, Amphibolitized sheared gabbro from ophiolites as indicators of the evolution of the oceanic crust: Bay of Islands, Newfoundland: Earth and Planetary Science Letters, v. 61, p. 151–165.

Greenly, E., 1919, The Geology of Anglesey: London, Memoir of the Geological

Survey of the United Kingdom (2 vols).

Gyopari, M. C., 1985, A study of blueschist mineral chemistry and a new look at the Penmynydd-Gwna boundary in SE Anglesey [MSc thesis]: Cardiff, University College.

Hietanen, A., 1974, Amphibole pairs, epidote minerals, chlorite, and plagioclase in metamorphic rocks, Northern Sierra Nevada, California: American Mineralogist, v. 59, p. 22–40.

Holdaway, M. J., 1972, Thermal stability of Al-Fe-epidote as a function of FO_2 and Fe content: Contributions in Mineralogy and Petrology, v. 37, p. 307–340.

Holgate, N., 1951, On crossite from Anglesey: Mineralogical Magazine and Journal of the Mineralogical Society, v. 29, p. 792–987.

Holland, T.J.B., 1979, Experimental determination of the reaction Paragonite = Jadeite + Kyanite + H_2O, and internally consistent thermodynamic data for part of the system Na_2O-Al_2O_3-SiO_2-H_2O, with applications to eclogites and blueschists: Contributions to Mineralogy and Petrology, v. 68, no. 3, p. 293–301.

Holland, T.J.B., 1980, The reaction albite = jadeite + quartz determined experimentally in the range 600-1200°C: American Mineralogist, v. 65, p. 129–134.

Holland, T.J.B., and Richardson, S. W., 1979, Amphibole zonation in metabasites as a guide to the evolution of metamorphic conditions: Contributions to Mineralogy and Petrology, v. 70, p. 143–148.

Haynes, A., 1982, A comparison of amphiboles from Medium- and Low-pressure Metabasites: Contributions to Mineralogy and Petrology, v. 81, p. 119–125.

Ito, E., and Anderson, A. T., Jr., 1983, Submarine metamorphism of gabbros from the Mid-Cayman rise: Petrographic and mineralogic constraints on hydrothermal processes at slow-spreading ridges: Contributions to Mineralogy and Petrology, v. 82, p. 371–388.

Laird, J., 1980, Phase equilibria in mafic schist from Vermont: Journal of Petrology, v. 21, p. 1–37.

Laird, J., 1982, The transition from greenschist to glaucophane schist, *in* Reviews in Mineralogy, Vol. 9B, Amphiboles: Petrology and experimental phase relations: Mineralogical Society of America, p. 142–146.

Leake, B. E., 1978, Nomenclature of amphiboles: Mineralogical Magazine and Journal of the Mineralogical Society, v. 42, p. 533–563.

Macpherson, H. G., 1983, References for, and updating of, L. J. Spencer's 1st and 2nd supplementary list of British Minerals: Mineralogical Magazine and Journal of the Mineralogical Society, v. 47, p. 243–257.

Maresch, W. V., 1977, Experimental studies on glaucophane: An analysis of present knowledge: Tectonophysics, v. 43, p. 109–125.

Miyashiro, A., and Banno, S., 1958, Nature of glaucophanitic metamorphism: American Journal of Science, v. 256, p. 97–110.

Nitsch, K. H., 1974, Neve Erkenntnisse zur stabilitat von Lawsonit (abstract): Fortschritte der Mineralogie, v. 51, p. 34–35.

Okay, A. I., 1980, Mineralogy, petrology, and phase relations of glaucophane-lawsonite zone blueschists from the Tavsanh region, Northwest Turkey: Contributions to Mineralogy and Petrology, v. 72, p. 243–255.

Raith, M., 1976, The Al-Fe (III) epidote miscibility gap in a metamorphic profile through the Penninic Series of the Tauern Window, Austria: Contributions to Mineralogy and Petrology, v. 57, p. 99–117.

Spear, F. S., Robinson, P., and Schumacher, J. C., 1982, Calcic and sodic amphiboles, *in* Reviews in Mineralogy, Vol. 9B, Amphiboles: Petrology and experimental phase relations: Mineralogical Society of America, p. 82–87.

Strens, R.C.J., 1965, Stability and relations of the Al-Fe-epidotes: Mineralogical Magazine, v. 35, p. 464–475.

Sweatman, T. W., and Long, J.V.P., 1969, Quantitative electron-probe microanalysis of rock-forming minerals: Journal of Petrology, v. 10, p. 332–379.

Thorpe, R. S., 1972, Ocean floor basalt affinity of Precambrian glaucophane schist from Anglesey: Nature (Physical Science), v. 240, p. 164–166.

Trzcienski, W. E., Jr., Carmichael, D. M., and Helmstaedt, H., 1984, Zoned sodic amphibole: Petrologic indicator of changing pressure and temperature during tectonism in the Bathurst area, New Brunswick, Canada: Contributions to Mineralogy and Petrology, v. 85, p. 311–320.

Washington, H. S., 1901, A chemical study of glaucophane schists: American Journal of Science, v. 11, p. 42.

Velde, B., 1967, Si+4 content of natural phengites: Contributions to Mineralogy and Petrology, v. 14, p. 250–258.

MANUSCRIPT ACCEPTED BY THE SOCIETY JULY 29, 1985

Geological Society of America
Memoir 164
1986

Rb-Sr and U-Pb dating of the blueschists
of the Ile de Groix

Jean-Jacques Peucat
Laboratoire de Géochimie-Géochronologie
CAESS, CNRS, Institut de Géologie
Avenue du Général Leclerc
35042 Rennes Cedex, France

ABSTRACT

The blueschist metamorphism on the Ile de Groix has been dated between 420-400 Ma using three different isotopic systems: micaschists Rb-Sr total rock isochron, Rb-Sr phengite, and U-Pb zircon ages.

The behavior of the Rb-Sr, U-Pb and K-Ar systems can be summarized as follows:

Micaschists - 420 Ma (Rb-Sr total rock isochron)	>zircons - 400 Ma > (U-Pb, lower intercept)
Phengites - 396 to 349 Ma (Rb-Sr)	≥Phengites 358 - 340 Ma (K-Ar)
>Glaucophane 320 Ma (Ar^{39-40}, K-Ar)	>Barroisite and paragonite - 295 and 273 Ma (K-Ar)

The age of blueschist metamorphism is best preserved in the micaschists Rb-Sr total rock system (420 Ma). To explain this apparent large-scale isotopic homogenization of pelitic rocks without melting, a combined model is proposed: 1) a geochemical homogenization of the Rb-Sr ratio occurred during sedimentation by mechanical processes particularly in fine-grained rocks; 2) an isotopic homogenization occurred during the metamorphic event which induced an isotopic equilibration following the definition of Roddick and Compston (1977).

The highest Rb-Sr phengite ages and the U-Pb lower intercept age ($\approx$400 Ma) also provide a coherent indication of the blueschist metamorphic age. A later event, which only partially reset the Rb-Sr and K-Ar phengite systems, may be recorded in the K-Ar system of glaucophane (320 Ma). A lower retention of radiogenic argon in glaucophane than in the cogenetic phengite is observed. Often found in the W. American blueschists and in the Alps, this seems to be characteristic of glaucophane, which is thus very sensitive to loss of Ar during tectonic events of low grade.

The blueschist metamorphism of the Ile de Groix belongs to an early stage of the Variscan orogeny, well known in this region where high-pressure—high-temperature metamorphism (eclogites) and high-temperature—low-pressure metamorphism (migmatites) occur in various geodynamic situations at the same time (440 to 380 Ma). The mineral ages obtained from the minerals of Ile de Groix show that the blueschists had been uplifted before Hercynian times s.s. where all minerals record 300 Ma cooling ages.

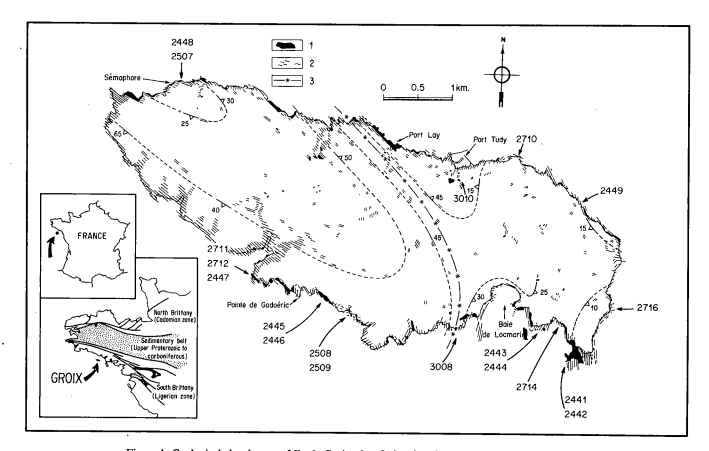

Figure 1. Geological sketch map of Ile de Groix after Quinquis (1980). 1 - various basic rocks, 2 - foliation in micaschists, 3 - western limit of garnet occurrence in metabasics (after Carpenter 1976). The Ile de Groix blueschists crop out south of the Armorican massif (insets) which is made up of three main blocks separated by shear zones. The northern zone is composed of 600 - 540 Ma (Cadomian orogeny) plutonic and metamorphic rocks containing 2000 Ma old plutonic relicts. The central zone is composed of upper Proterozoic to Carboniferous low-grade sedimentary rocks. The southern zone is a Paleozoic metamorphic belt made up of Ordovician orthogneisses, eclogites and migmatites 440 to 380 Ma old (Ligerian orogeny) involved in the Carboniferous orogeny by amphibolite grade metamorphism and leucogranite (S-type) emplacement. Major basic and (or) ultrabasic complexes of the Armorican Massif are in black on the map.

INTRODUCTION

Dating of amphibole and mica in blueschists can often resolve geochronological problems because the blocking temperatures of radiogenic Ar and Sr in these minerals are relatively concordant with the temperatures achieved in the high pressure-low temperature (H.P.-L.T.) assemblages. Consequently, it is possible to date a cooling event which shortly follows the high-pressure metamorphism. A large range of mineral ages is nevertheless an indication of a complex cooling history and therefore other radiogenic systems can be of great help to obtain more complete information. We attempt to compare Rb-Sr and K-Ar mineral ages with U-Pb zircon results and Rb-Sr whole rock data

collected from various rocks involved in the blueschist metamorphism.

GEOLOGICAL CONTEXT

The formations of the Ile de Groix belong to the Hercynian belt of Western Europe. They are essentially composed of an alternation of micaschists and metabasic layers sometimes tens of meters thick. Three major types of basic rock, those most frequently found on the island, are studied here: greenschists (prasinites), eclogites with glaucophane, and banded glaucophanites with millimeter-scale layers of glaucophane and epidote (Fig. 1).

The mineral assemblages have been described by numerous

TABLE 1. MAJOR ELEMENTS FROM THREE MAIN KINDS OF BASIC ROCKS*

	Eclogites 21 samples			Glaucophanite with epidote 47 samples			Greenschists 23 samples		
	Mean values (%)	Standard deviation (%)	Rejected values (%)	Mean values (%)	Standard deviation (%)	Rejected values (%)	Mean values (%)	Standard deviation (%)	Rejected values (%)
SiO_2	48.83	1.93	54.2	49.41	2.17	58.7-43.9	46.90	2.27	54.6-56.1
Al_2O_3	14.05	0.93	18.3	15.87	1.18	18.8-19.2-12.6	15.79	1.2	19.2-12.4
Fe_2O_3	7.42	1.97	14.5	8.64	2.27	14.2	6.51	2.7	
MgO	5.09	0.86	2.3-7.4	4.87	1.77	12.4-9.5	7.84	2.3	13.1
CaO	8.31	1.63		7.93	2.65	0.99	8.20	1.7	3.2
Na_2O	4.10	0.59	6.0	3.48	0.93	1.4-5.9	3.31	1.09	6.3
K_2O	0.19	0.15	0.85	0.50	0.37	2.2-1.8	0.34	0.23	1.4-1.2
TiO_2	2.06	0.83		1.89	0.46	5.1-0.5	1.81	0.42	0.65-0.5
P_2O_5	0.06	0.08	0.3-0.2	0.06	0.1	0.9	0.09	0.11	0.46
L.O.I.	1.29	0.40	3.09	2.5	0.76	4.7-4.8-5-5.1	3.79	1.24	8-9.5

*Rejected values are outside of mean values ± 2σ.

authors (Cogné, 1960; Félix, 1969, 1972; Makanjuola and Howie, 1972; Triboulet, 1974; Carpenter, 1976); they are typical of the H.P.-L.T. metamorphism of the blueschist facies. The P-T conditions of this metamorphism have been estimated by Triboulet (1974) and Carpenter (1976).

According to Triboulet, the blueschist metamorphism (M_1) can be divided into three zones of different P-T conditions: the eclogite facies with glaucophane: 8.5 kb; 530°C; the glaucophane, epidote and garnet facies: 8 kb; 500°C; the greenschist facies with blue-green amphiboles: 7.5 kb; 470°C.

A retrograde event (R) partially affects the previous assemblages in a facies with blue-green amphibole, epidote, albite and chlorite at 6.5 kb, 470°C.

According to Carpenter, M_1 can be divided into two zones based on the presence or absence of garnet. In the garnet zone (eastern part of the island) the pressure reached 8 to 9 kb at a temperature of 400-450°C. In the zone without garnet, pressure was around 6.5 to 8 kb at T = 350-400°C. During the retrograde event (R), lower pressures were accompanied by increase in temperature up to 500°C.

The geochemical nature of the basic rocks has been characterized using trace, major, and rare earth elements (Jahn and others, 1977; Carpenter and others, 1978; Bernard-Griffiths and others, in prep.). The REE patterns allow us to show an association of tholeiitic and alkali basalts. Table 1 presents the average values obtained for the major elements in these rocks. The values which are eliminated during statistical treatment show the great variation observed and the complexity of the geochemical processes.

Deformation in the Groix series as well as the relationship between deformation and crystallization have been studied by Jeannette (1965), Cogné and others (1966), Boudier and Nicolas (1976), Carpenter (1976) and Quinquis (1980). According to Quinquis, the first deformation D_1, characterized by a foliation deformed by sheath folds, is associated with the crystallization of M_1 minerals. During the retromorphic event, some M_1 minerals

(phengite - epidote) continue to crystallize, and chlorite and albite appear. Two phases of folding (D_2 + D_3) occur after the retrograde event and it is during these deformations that glaucophane is folded and broken.

The age of the H.P. - L.T. metamorphism cannot be determined by geological field methods alone. Because the H.P. rocks are insular and also allochthonous (Lefort and Segoufin, 1978), the blueschists of Groix are geologically isolated from recognized continental formations whose history goes back from 2,000 Ma to 300 Ma (Vidal, 1980). The polymetamorphic nature of the Armorican Massif and the preservation of glaucophane parageneses lead us to believe that this H.P. - L.T. metamorphism episode is not part of the most ancient history of the massif. Radiometric dating of these formations was undertaken using Rb-Sr and U-Pb methods and K-Ar in collaboration with J. C. Hunziker. Preliminary Rb-Sr results have been previously published (Peucat and Cogné, 1977).

Rb-Sr AND K-Ar PHENGITE AGES

The Rb-Sr analyses were performed on samples prepared from 100 mg of powdered whole rock. Strontium was separated using ion-exchange resins, and rubidium and strontium concentrations were determined (after XRF estimation) using isotope dilution methods. Samples were loaded onto oxidized tantalum filaments and analyzed on a TSN 206 S mass spectrometer (30 cm radius; 60° sector; 10 KV accelerating voltage). Data processing was handled on-line by a Hewlett-Packard 9810 computer. The NBS 987 standard yielded a value of 0.71016 ±3 during the period of the analyses. The probable error (2 σm) on a run is typically of the order 0.03% for the $^{87}Sr/^{86}Sr$ ratio and 2% for the $^{87}Rb/^{86}Sr$ ratio. Ages were calculated according to the method of York (1966), using the decay $\lambda^{87}Rb = 1.42 \times 10^{-11}$ yr^{-1} for isochron calculation. The 2 σ isochron-age error is multiplied by $\sqrt{MSWD}$ when MSWD > 1.

J.-J. Peucat

TABLE 2. Rb-Sr ANALYTICAL DATA FROM TOTAL ROCKS (T.R.) AND MINERALS, AND PREVIOUS K-Ar AGES

Samples		Rb (ppm)	Sr (ppm)	$^{87}Rb/^{86}Sr$	$^{87}Sr/^{86}Sr$	Rb-Sr ages in M.A.	K-Ar ages in M.A.
Micaschists							
R 55*	T.R.	84.9	30.3	8.13	0.7529		
R 56*	T.R.	196	65.1	8.74	0.7569		
R 96*	T.R.	128	60.6	6.11	0.7376		
2710	T.R.	52.6	50.0	3.05	0.72273		
Micaschists with glaucophane							
2711	T.R.	93.4	182	1.49	0.71483		
2712	T.R.	108	104	2.98	0.72309		
	Phengite	218	24.9	25.4	0.8369	358±10	
2714	T.R.	194	71.8	7.82	0.75340		
	Phengite	412	77.3	15.4	0.79619	396±18	
R 54*	T.R.	73.5	263	0.81	0.7094		
	Phengite	250	24.9	29.1	0.8551	362±10	
PM 6							
	Phengite	364	71.6	14.7	0.7767	342±11	327±10[§]
	Glaucophane	..	..	..	..	..	329∓33[§]
Greenschists							
2442	T.R.	2.44	136	0.052	0.70525		
	Paragonite	7.15	27.0	0.76	0.71333	n.d.	173±14[†]
2444	T.R.	141	52.7	7.74	0.75183		
	Phengite	270	36.9	21.2	0.82273	374±15	349±14[†]
2446	T.R.	15.8	108	0.42	0.70790		
2448	T.R.	1.39	195	0.02	0.70658		
2506	T.R.	9.86	231	0.123	0.70667		
2507	T.R.	11.2	183	0.178	0.70701		
Glaucophane-epidote schists							
2441	T.R.	22.0	258	0.25	0.71027		
	Phengite	359	8.31	125	1.3139	342±7	348±14[†]
2445	T.R.	43.6	704	0.18	0.70709		
	Phengite	187	14.0	38.8	0.90074	352±9	341±14[†]
2447	T.R.	31.5	379	0.24	0.70898		
2449	T.R.	1.47	184	0.023	0.70783		
LH 11	Glaucophane	..	..	..	..	..	318±34[§]
LH 13	Phengite	418	20.0	60.4	1.0254	367±6	336∓11[§]
	Epidote	..	..	..	0.71029		
H.M.	Glaucophane	..	..	..	..	..	320±10**
2509	T.R.	30.0	362	0.239	0.70701		
3008	T.R.	55.5	142	1.13	0.71136		
3010	T.R.	107	326	0.955	0.71357		
Eclogites							
2443	T.R.	4.2	252	0.048	0.70683		
	Phengite	217	12.2	51.3	0.9631	352±8	358±15[†]
PC10							
	Phengite	210	39.1	15.5	0.7808	344±10	339±11[§]
	Epidote	..	..	..	0.7055	..	
	Glaucophane	..	..	..	..	..	317±11[§]
	Barroisite	..	..	..	..	..	295∓14[§]

*analyses from Vidal, 1980, [†]unpublished data from Hunziker, [§]analyses from Carpenter 1976 and Carpenter and Civetta 1977, **analyses from Maluski 1977a.

Ten phengite samples from basic rocks and aluminous schists were analyzed by the Rb-Sr method. The ages (Table 2) were calculated from the phengite total rock pair in seven samples, from the phengite-epidote pair in two samples, and with an initial ratio of 0.705 for the last one. The results range from 342 ± 7 to 396 ± 18 Ma.

The K-Ar ages of the same samples obtained by Hunziker (Table 2) range from 341 ± 14 to 358 ± 15 Ma. These ages are equivalent or slightly younger than the Rb-Sr ages although the analytical errors are covered. The K-Ar ages obtained by Carpenter and Civetta (1976) are lower than Hunziker's (327 ±20 to 339 ±20 recalculated with errors of 2σ).

These ages cannot reflect a simple cooling phenomenon because the time span is too long (50 Ma) for a cooling rate compatible with the preservation of the H.P. - L.T. assemblages. This large range of mica ages suggests rather that there is a partial resetting. That is to say, the ages found here are intermediate, occurring between two geological events. This means that the H.P. - L.T. metamorphism (M_1) is as old as or older than 396 ± 18 Ma and that some event which occurred after 340 Ma gave rise to the partial resetting. This resetting must have taken place under conditions of low temperature because it neither totally opened the K-Ar system of the phengite (<350°C after Purdy and Jaeger, 1976) nor destabilized the glaucophane; it could not have been the retrograde event (R) reaching 470-500°C. Consequently R must have rapidly followed the H.P. - L.T. event.

GLAUCOPHANE AGES

Unlike the phengites, glaucophane only crystallized during metamorphism M_1 and is retrograded only slightly or not at all. Maluski (1977a) obtained a ^{39}Ar - ^{40}Ar age confirming the K-Ar ages obtained by Carpenter and Civetta (1976). They are all more recent than those obtained for the phengites. We shall retain the age obtained by Maluski (1977a) - 320 ± 10 Ma - as being the most accurate. It should be noted, however, that younger ages were found by Carpenter and Civetta (1976) for a barroisite (295 ± 14 Ma) and by Hunziker (Table 2) for a paragonite (273 ± 14 Ma).

The totality of the ages found in micas and amphiboles is distributed in the following way:

Phengite Rb-Sr≥Phengite K-Ar>Glaucophane K-Ar>Barroisite-paragonite K-Ar
(396-340 Ma)   (338-340 Ma)   (320 Ma)   (295-273 Ma)

The result is at first glance surprising, as the retentivity of amphiboles for Ar is often considered greater than that found in micas (Hanson and Gast, 1967). We may here observe an unusual inverse relationship of the amphibole-mica ages.

Several examples in the literature, however, show that the behavior of Ar in glaucophane is variable and may be different from that of other amphiboles.

Cases in which the ages of glaucophane are more ancient than those obtained on phengite are known in Corsica with 90 and 40 Ma, respectively (Maluski, 1977b), in New Caledonia, with 68 and 36 Ma, respectively (Black and others, 1977), and also in certain parts of the Alps (Grand Paradis) with 80 and 40 Ma respectively (Chopin, 1979). The differences in ages observed are too great to be explained by a simple cooling model, and the various authors explain their findings in terms of a high-pressure polymetamorphism (New Caledonia) or else the possibility of a thermal - or even tectonic - event which resulted in the re-opening of the phengites (Grand Paradis, Corsica) without involving the glaucophane.

The opposite case, in which the phengite ages are higher than the glaucophane age is rather widespread, especially in the schists of the Franciscan Terrain and in the American northwest up to Alaska (Coleman and Lanphere, 1971; Suppe and Armstrong, 1972; Forbes and Lanphere, 1973; Carden and others, 1971; Schweichert and others, 1980). Coleman and Lanphere (1971) observed that Ar is best retained in the phengites, and that glaucophane loses argon during tectonic transport or during a late metamorphic phase. This latter situation is also the case in certain areas of the western Alps (Hunziker, 1974), where three age ranges are found for glaucophane (80, 40 and 15 Ma). The phengite ages (50-40 Ma) are in accordance with the second of the glaucophane age groups. Hunziker (1974) does not believe that it is here a question of three periods of high-pressure meta-

morphism but rather that the glaucophane recrystallized "under metastable conditions in response to a subsequent tectonic and greenschist metamorphic overprint after the formation under high-pressure conditions." This tendency of glaucophanes to lose argon during tectonic events has been clearly shown by Maluski (1977b) in Corsica using the ^{39}Ar - ^{40}Ar method. Different plateaus, corresponding to different tectonic phases are evident, thus showing the sensitivity of these minerals "to tectonic events on sites with low or medium retentivity" (Maluski, 1977b).

In the case of the Ile de Groix (after the M_1 episode), the glaucophane suffered two phases of folding (D_2 and D_3) or breaking. This leads us to think that there was indeed a loss of Ar during D_2 or D_3. This seems to have been a total loss, since a single "plateau" was recorded by ^{39}Ar - ^{40}Ar. The age of 320 ± 10 Ma has a precise geological significance, that of the age of the tectonic event D_2 and/or D_3. This tectonic event was probably not associated with a high rise in temperature, since it only partially affected the phengite ages and since the glaucophane remained stable. The K-Ar ages of glaucophane therefore record an event of a principally dynamic nature.

Rb-Sr WHOLE ROCK DATES FROM MICASCHISTS AND BASIC ROCKS

The Micaschists

Seven of the eight sample analysed define an isochron with the micaceous greenschist 2444 (Table 2, Fig. 2a). The age calculated is 422 ± 16 Ma at $2\sigma \times \sqrt{MSWD}$ with initial ratio of 0.7050 ±8 (MSWD = 7).

The interpretation of a micaschist isochron is difficult, and several hypotheses may be considered. This age may have no geological significance and the alignment observed corresponds to a mixing line, but no correlation between isotopic ratios versus major or trace elements agrees with this hypothesis. This age may have a geological significance and in this case there are several possibilities: it could be an inherited age, the age of the sedimentation, the age of diagenesis, or a metamorphic age. We can presume that the sedimentary deposits and associated volcanism on Groix are pre-Ordovician in age if the analogy between Groix and Le Pouldu series is valid. These latter series outcrop 20 km away from Groix on the continent, and are composed of greenschists and micaschists cut by a 474 ± 5 Ma-old granite (Vidal, 1972; Calvez, 1976). Alternatively, using the $^{87}Sr/^{86}Sr$ initial-versus-time relationship, the oldest age of sedimentation in Groix may be estimated at between 560 - 430 Ma (Fig. 2b). If the preceding geological propositions are correct, the sedimentation occurs between 560 and 474 Ma. Therefore, the age at 420 Ma obtained by micaschist Rb-Sr isochron can only be the age of diagenesis or of a metamorphic event.

It is, however, difficult to imagine the mechanism by which the homogenization of Sr isotopes on a scale of kilometers may take place in sedimentary formations where there is no melting.

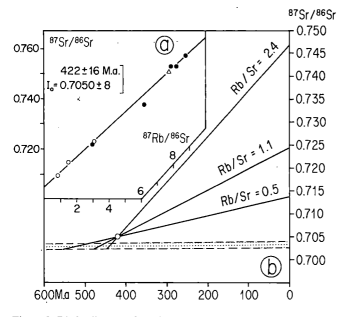

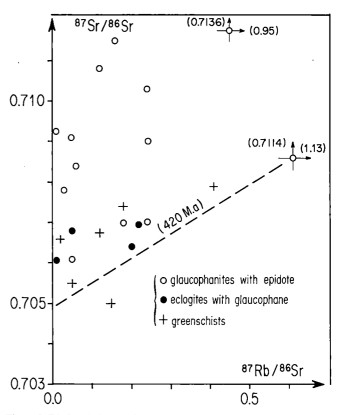

Figure 2. Rb-Sr diagrams for micaschists. a. Whole rock isochron diagram; as black circles, metapelitic schists; as open circles, micaschists with glaucophane; as triangle the 2444 micaceous greenschist. b. Isotopic evolution lines of micaschists drawn through the initial ratio (Io) of the isochron at 422 Ma for Rb/Sr = 2.4 (metapelitic schists) and 0.5 for micaschists with glaucophane: 1.1 is the mean value. They cut the mantle growth curve around 500 Ma, which may correspond with the maximum age for sedimentation. The same estimate has been given by Carpenter and Civetta (1976) for differentiation of basic rocks in spit of $^{87}Sr/^{86}Sr$ whole rock modifications.

Figure 3. Rb-Sr whole-rock isochron diagram of basic rocks. Additional values are from Carpenter 1976.

Nevertheless, we know that the isotope homogenization seems to be favored by such phenomena as the appearance of schistosity (Bath, 1974; Gebauer and Grünenfelder, 1974), a fine grained texture (Cordani and others, 1978) and the importance of a fluid phase (Watson, 1981).

From these observations we may imagine the following scenario: (1) the sedimentation is accompanied 1) by a geochemical homogenization (mechanical mixing), with the result that on a metric scale, the variations of Rb-Sr ratios are equivalent to variations seen on the kilometer scale in the sedimentary formation; and 2) by a diminution of isotopic heterogeneities. (2) When temperature rises during a metamorphic event, volume diffusion occurs on a scale of meters. It is favored by an extensive fluid phase, the appearance of schistosity, and the lack of isotopic heterogeneities occurring during sedimentation. (3) This homogenization occurs stepwise on the meter scale throughout the whole formation. As the variations in the Rb-Sr ratio are equal on the meter scale to those seen in the whole formation, this gives the impression of a large-scale isotopic homogenization. This model is an adaptation of the strontium isotopic equilibration model proposed by Roddick and Compston (1977) for plutonic rocks.

In this model, the age obtained corresponds to a metamorphic event. But in the case of Groix, we cannot exclude the possibility that these phenomena occur during events prior to M_1,

for example during a low-intensity metamorphism masked by the high-pressure one. In any case, this isochron age means that the blueschist metamorphism (M_1) cannot be older than 420 Ma. This gives:

$$422 \pm 16 \text{ Ma} \geqslant M_1 \geqslant 396 \pm 18 \text{ Ma}$$
$$\text{(isochron age)} \qquad \text{(phengite age)}$$

The low initial $^{87}Sr/^{86}Sr$ ratio of the isochron (0.705) shows the little participation of much older detrital component in these sediments which are probably partially derived from alteration of the associated basic rocks.

The Basic Rocks

The results are quoted in Table 2 and Fig. 3. The $^{87}Sr/^{86}Sr$ initial ratios recalculated at 420 Ma range between 0.705 and 0.709. The data from Carpenter (1976) show a range extending between 0.704 and 0.711.

The rocks with the lowest initial ratios, less than 0.707, are the greenschist and the eclogite, which are amongst the most massive and homogeneous rocks. Those rocks with the highest initial ratios are essentially heterogeneous lithologies formed of alternating millimeter- to centimeter-thick bands of glaucophane

TABLE 3. U-Pb ANALYTICAL DATA ON NONMAGNETIC SIZE FRACTIONS (in μ) OF ZIRCONS
FROM AN ALBITIC MICASCHIST (Sample 2716)

Sample 2716	Concentration (ppm)		Isotopic ratios measured			Apparent ages (in M.A.)		
	U	Pb rad.	$^{206}Pb/^{204}Pb$	$^{207}Pb/^{206}Pb$	$^{208}Pb/^{206}Pb$	$^{206}Pb/^{238}U$	$^{207}Pb/^{235}U$	$^{207}Pb/^{206}Pb$
1. >62	424	43.9	666	0.09516	0.18122	538	645	1039
2. 62-45	433	36.6	2,076	0.07036	0.13909	474	522	723
3. 45-37	493	38.4	770	0.07863	0.15832	423	452	601
4. <37	471	35.2	1,790	0.06781	0.12485	426	453	595

and epidote representing metamorphic foliation. The anisotropy of these latter rocks is linked to tectonometamorphic phenomena occurring during M_1 and it may be at this time that the original $^{87}Sr/^{86}Sr$ modification of the basalts took place. Minerals very rich in Sr, such as epidote (500 to 1,000 ppm), or even glaucophane (60 to 260 ppm) (Peucat, 1983), could have entrapped the Sr during metamorphism. This Sr could have come from micaschists before the completion of the isotopic homogenization during the circulation of the fluid phases.

In this way, "crustal" contamination would not occur during the emplacement of the basic magma but during metamorphism. This process would have two opposite results: (1) homogenization of the Sr isotopes within the sedimentary formations, (2) perturbation of the Sr of the basic rocks by the more radiogenic Sr from the sediments. This contamination would be greatest in the zones whose levels were subjected to the greatest modification, also corresponding to the zones of the most extensive fluid circulation (i.e. banded glaucophane-epidote schists in shear zones).

U-Pb RESULTS ON DETRITAL ZIRCONS

U-Pb analyses were performed on samples prepared from 1-10 mg of zircon separate, following the method of Krogh (1973). The total blank level is between 0.5 and 1.0 ng. Common lead is assumed to have the following isotopic ratios: $^{206}Pb/^{204}Pb = 18.0$; $^{207}Pb/^{204}Pb = 15.5$; $^{208}Pb/^{204}Pb = 37.0$

Analyses were performed on a Cameca mass spectrometer (30 cm radius, 60° sector, 10 kv accelerating voltage). Signals were recorded using an electron multiplier. Data processing was performed on-line by a Hewlett-Packard 9825 Computer. The error on the $^{207}Pb/^{206}Pb$ is 0.2% and on U-Pb ratios 2%, more analytical details can be found in Peucat and others (1981). Uranium decay constants used were:

$\lambda^{238}U = 1.55125 \times 10^{-10}$ an^{-1}, $\lambda^{235}U = 9.8485 \times 10^{-10}$ an^{-1} and $^{238}U/^{235}U = 137.88$.

Zircons were extracted from a sample of albitic-micaschist on a surface outcrop on the east coast of the island. In general, this mineral is scarce in the metasediments of Groix.

In the sample studied, the zircons are of a yellow color, often

elongated, having clear but rounded crystalline forms with an irregular surface probably due to corrosive factors, as has been described in albitised granites (Caruba, 1979).

These zircons are zoned and occasionally present rounded cores rich in inclusions which are themselves zoned. These two "generations" possibly originated prior to the metamorphism of the blueschists and doubtless correspond to one or several ancient magmatic events. Alternatively, the thin overgrowths may correspond to recrystallizations during metamorphism. The analytic results are reported in Table 3 and Figure 4. Four different size fractions with different magnetic susceptibilities were analysed. In the concordia diagram they describe a discordia whose upper intercept gives a primary age of 1808 ± 150 Ma and a secondary age of 399 ± 12 Ma with a MSWD of 5. The relation between the discordance, the grain-size and the U content are normal, although the two smallest fractions are in agreement within the limit of analytical error. The points are extremely discordant and the result obtained may not be explained by a model of continuous diffusion of radiogenic lead.

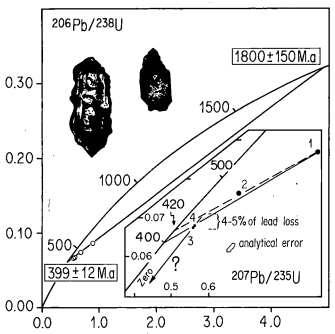

Figure 4. U-Pb Concordia diagram for zircons from an albitic micaschist. Zircon sizes (1 to 4) are quoted in Table 3.

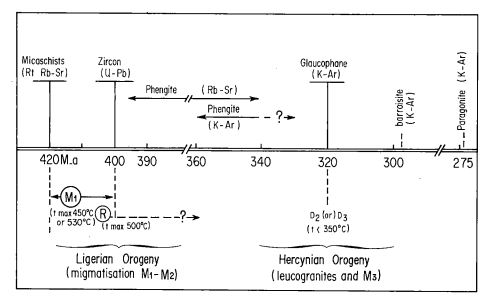

Figure 5. Summary of the geochronological results obtained from various chronometers in the blueschist of Ile de Groix. The Ligerian and Hercynian orogenies are defined on the adjacent mainland.

The 1800 Ma Age

The zircons analyzed are from detrital minerals and the age obtained corresponds with the average ages of the inherited zircons. Similar results have been obtained on zircons contained in Ordovician sandstones in Brittany (Vidal and others, 1980). They show the presence of an ancient basement (minimum age: 1800 Ma) from which the zircons were derived during one or more sedimentary cycles.

The 400 Ma Age

The high degree (more than 90%) of discordance of the zircons analysed shows that the zircons have lost a large part of their old Pb probably for the most part during an episodic loss at 400 Ma.

The four fractions are not, however, perfectly aligned in the concordia diagram (MSWD = 5). This may be due to the fact that the probable initial dispersion of the U-Pb system of the zircon in the diagram is not totally attenuated by the lead loss at a point of convergence of 400 Ma. In addition to the major phenomenon of episodic lead loss at 400 Ma, continuous loss from overgrowths is possible as shown in migmatites (Peucat, 1983). For the purposes of the calculation, the addition of 4-5% of Pb to the finest size fractions permit us to estimate an age of 420 Ma.

We may thus conclude that the age of 399 ± 12 Ma recorded for detrital zircons is the minimum age of an episode of the loss of old Pb. This result, correlated with that of the phengites which fixes the M_1 metamorphism at 396 ± 18 Ma or higher, leads us to think that 399 ± 12 Ma is a minimum age for the blueschist event.

CONCLUSIONS

The geochronological results of Rb-Sr, K-Ar and U-Pb obtained for minerals and total rocks in the complex of the blueschists of the island of Groix may be summarized as represented on Figure 5.

The interpretation of these ages is as follows: (1) the micaschist whole-rock isochron age of 422 ± 16 Ma is a maximum age limit for the high-pressure/low-temperature metamorphism M_1, (2) the age of 399 ± 12 Ma obtained for the zircons is a minimum age for this metamorphism, (3) the ages measured by Rb-Sr and K-Ar for the phengites are intermediate and represent the affect a tectonic disrupting event subsequent to 340 Ma. This disrupting event, which does not totally open the K-Ar system of phengites, occurred in a low temperature context (<350°C) but this cannot be the retromorphic event (R) which must follow rapidly after M_1, (4) the Hercynian event was recorded by the K-Ar system of glaucophane. This 320 Ma age refers to D_2 - D_3 tectonic episodes and is correlated with the disrupting event which partially opened the Rb-Sr and K-Ar systems of phengites in a low temperature context. It appears that the glaucophane lost argon more easily than the coexisting phengite and that these blue amphibole ages have here a tectonic significance (D_2-D_3), (5) the 295-273 Ma K-Ar ages obtained from barroisite and paragonite are interpreted as being final cooling ages of minerals with a lower Ar retention. The blueschists of the Ile de Groix belong to a Siluro-Devonian event prior to the Hercynian period which occurred during the Carboniferous.

This period is also characterized in Brittany by a H.P.-H.T. metamorphism with eclogites and basic granulites dated at 440 to 380 Ma (Peucat and others, 1982) and by a low to medium pressure-high temperature metamorphism (migmatisation and anatectic granite) dated between 420 and 380 Ma (Vidal, 1973; Peucat, 1983) and related (Peucat and others, 1978; Audren and Peucat, 1980) to the evolution of a paired metamorphic belt.

After 340 Ma, the Hercynian metamorphism sensu stricto reached the amphibolite facies and the melting of sediments with S-type granite emplacement related to a collision. During this time, the blueschists of Groix were probably partially uplifted (Quinquis and others, 1978; Matte and Mattauer, 1978; Quinquis and Choukroune, 1981) because they do not generally record the Hercynian thermal event which culminated near 500-550°C at 300 Ma (Peucat, 1983) within the rocks which outcrop today in the adjacent mainland.

ACKNOWLEDGMENTS

The author wishes to thank E. H. Brown, B. W. Evans, E. Jäger and J. M. Mattinson, who have reviewed this paper; J. C. Hunziker, who has performed K-Ar analyses on phengites; and M.S.N. Carpenter, J. Cogné and Ph. Vidal for helpful discussions. I also thank C. Robart for translation of the original manuscript, J. Cornichet and J. Macé for assistance in laboratory, and M. H. Fichet for typing.

REFERENCES CITED

Audren, C., and Peucat, J. J., 1980, Structure de la double ceinture dévonienne de Bretagne méridionale, *in* Roach, E. A., ed., Structure and Tectonic evolution of the Armorican Massif: Journal of the Geological Society of London, 137, p. 215.

Bath, A. H., 1974, Rb-Sr data on variably metamorphosed Paleozoic argillites: International Meeting for Geochronology Cosmochronologie and isotope geology: Paris, Abstract, 1 p.

Bernard Griffiths, J., Carpenter, M.S.M., Peucat, J. J., and Jahn, B. M., Geochemical and isotopic characteristics of blueschist facies metabasic rocks from the Ile de Groix, Armorican massif (France), in prep.

Black, M. C., Brothers, R. N., and Lanphere, M. A., 1977, Radiometric ages of blueschists in New Caledonia: International symposium on geodynamics in South-West Pacific, ed. Technip, Paris, p. 279-282.

Boudier, F., and Nicolas, A., 1976, Interprétation nouvelle des relations entre tectonique et métamorphisme dans l'île de Groix (Bretagne): Bulletin de la Société Géologique de France, ser. 7, v. 18, p. 135-144.

Calvez, J. Y., 1976, Comportement des systémes U-Pb et Rb-Sr dans les orthogneiss d'Icart et de Moëlan (Massif Armoricain), thése de 3éme cycle: Rennes, 74 p.

Carden, J. R., Connelly, W., Forbes, R. B., and Turner, D. L., 1971, Blueschist at the Kodiak islands, Alaska: An extension at the Seldovia schist terrane: Geology, v. 5, p. 529-533.

Carpenter, M.S.N., 1976, Petrogenetic study of the glaucophane schists and associated rocks from the Ile de Groix, Brittany, France [Ph.D. thesis]: Oxford University, 271 p.

Carpenter, M.S.N., and Civetta, L., 1976, Hercynian high pressure/low temperature metamorphism in the Ile de Groix blueschists: Nature, v. 262, p. 276-277.

Carpenter, M.S.N., Pivette, B., and Peucat, J. J., 1978, Apport de l'analyse discriminante sur l'étude géochimique de trois séries métabasiques du sud du Massif Armoricain: 6e Réunion annuelle des Sciences de la Terre, Orsay, Livre en dépôt Société Géologique de France, p. 90.

Caruba, R., 1979, Etude expérimentale de la cristallochimie, de la morphologie, de la stabilité et de la genése du zircon et des zirconosilicates en vue d'applications pétrogénétiques: Thése d'état, Nice, 143 p.

Chopin, C., 1979, De la Vanoise au massif du Grand Paradis. Une approche pétrographique et radiochronologique de la signification géodynamique du métamorphisme de haute pression: Thése de 3éme cycle, Paris 6, 145 p.

Cogné, J., 1960, Schistes cristallins et granites en Bretagne méridionale: Le domaine de l'anticlinal de Cornouaille: Mémoire du Service de la Carte géologique France, 382 p.

Cogné, J., Jeannette, D., and Ruhland, M., 1966, L'île de Groix. Etude structurale d'une série métamorphique á glaucophane en Bretagne méridionale: Bulletin du Service de la Carte géologique d'Alsace Lorraine, v. 19, p. 2397-2412.

Coleman, R. G., and Lanphere, M. A., 1971, Distribution and age of high-grade blueschists, associated eclogites, and amphibolites from Oregon and California: Geological Society of America Bulletin, v. 82, p. 2397-2412.

Cordani, U. G., Kawashita, K., and Filho, A. T., 1978, Applicability of the Rb-Sr method to shales and related rocks, *in* Cohee, G. U., Glaessner, M. F., and Hedberg, H. D., eds., Studies in Geology, No. 6, Contributions to the geologic time scale: American Association of Petroleum Geologists, Edward Brothers, Inc., p. 93-117.

Felix, C., 1969, Etude pétrographique des roches basiques de l'île de Groix (Bretagne méridionale): Annales de la Société Géologique de Belgique, v. 92, p. 359-370.

Felix, C., 1972, Etude structuro-minéralogique des pseudomorphes de présumée lawsonite des glaucophanoschistes de l'île de Groix (Bretagne, France): Considérations sur la possibilité d'une paragenése á glaucophane et lawsonite: Annales de la Société Géologique de Belgique, v. 95, p. 345-391.

Forbes, R. B., and Lanphere, M. A., 1973, Tectonic significance of mineral ages of blueschists near Seldovia, Alaska: Journal of Geophysical Research, v. 78, n. 8, p. 1383-1386.

Gebauer, D., and Grünenfelder, M., 1974, Rb-Sr whole rock dating of late genetic to anchimetamorphic, Paleozoic sediments in Southern France (Montagne Noire): Contributions to Mineralogy and Petrology, v. 47, p. 113-130.

Hanson, G. M., and Gast, P. W., 1967, Kinetic studies in contact metamorphic zones: Geochimica et Cosmochimica Acta, v. 31, p. 1119.

Hunziker, J. C., 1974, Rb-Sr and K-Ar age determination and the Alpine tectonic history of the Western Alps: Memorie degli Instituti di Geologia e Mineralogia dell' Universita di Padova, 31, 55 p.

Hunziker, J. C., 1979, Potassium Argon dating, *in* Jäger, E., and Hunziker, J. C., eds., Lectures in Isotopic Geology: Springer-Verlag Publ., p. 52-75.

Jahn, B. M., Peucat, J. J., and Carpenter, M.S.N., 1977, Géochimie des terres rares de glaucophanites et roches associées de l'île de Groix (France) et de Taiwan: 5e Réunion annuelle des Sciences de la Terre, Rennes, livre en dépôt Société Géologique de France, p. 280.

Jeannette, D., 1965, Etude tectonique de l'île de Groix (Morbihan): Thése de 3éme cycle, Strasbourg, 104 p.

Krogh, T. E., 1973, A low contamination method for decomposition of zircon and the extraction of U and Pb for isotopic age determinations: Geochimica et Cosmochimica Acta, v. 37, 3 p., 485-494.

Lefort, J. P. and Segoufin, 1978, Etude géologique de quelques structures magnétiques reconnues dans le socle périarmoricain submergé: Implications géodynamiques concernant la fracturation proto-atlantique et l'orogénése hercynienne: Bulletin de la Société Géologique de France, ser. 7, v. 20, 2, p. 185-192.

Makanjuola, A. A., and Howie, R. A., 1972, The mineralogy of the glaucophane schistes and associated rocks from île de Groix, Brittany, France: Contributions to Mineralogy and Petrology, v. 35, p. 83-118.

Maluski, H., 1977a, Intérêt de la méthode $^{40}Ar/^{39}Ar$ pour la datation des glaucophanes. Exemple des glaucophanes de l'île de Groix (France): Comptes Rendus Académie des Sciences Paris, v. 283, ser. D, p. 223-226.

Maluski, H., 1977b, Applications de la méthode $^{40}Ar/^{39}Ar$ aux minéraux des roches cristallines perturbées par des événements thermiques et tectoniques en Corse: Thése d'Etat U.S.T.L., Montpellier, 172 p.

Matte, P., and Mattauer, M., 1978, Tectonique des plaques et chaîne hercynienne: Les "schistes bleus" de la côte sud armoricaine sont-ils les témoins d'une obduction?: 6e Réunion annuelle des Sciences de la Terre, Orsay, livre en dépôt á la Société Géologique de France, p. 270.

Peucat, J. J. and Cogné, J., 1977, Geochronology of some blueschists from île de Groix, France: Nature, London, v. 268, p. 131–132.

Peucat, J. J., Le Metour, J., and Audren, Cl., 1978, Arguments géochronologiques en faveur de l'existence d'une double ceinture métamorphique d'age siluro-dévonien en Bretagne méridionale: Bulletin de la Société Géologique de France, Paris, sér. 7, v. 20, p. 163–167.

Peucat, J. J., Hirbec, Y., Auvray, B., Cogné, J., and Cornichet, J., 1981, Late Proterozoic zircon ages from a basic-ultrabasic complex: A possible Cadomian orogenic complex in the Hercynian belt of western Europe: Geology, v. 9, p. 169–173.

Peucat, J. J., Vidal, Ph., Godard, G., and Postaire, B., 1982, Precambrian U-Pb zircon ages of eclogites and garnet pyroxenites from South Brittany (France): An old oceanic crust in the West European Hercynian belt: Earth and Planetary Science Letters, v. 60, p. 70–78.

Peucat, J. J., 1983, Géochronologie des roches métamorphiques (Rb-Sr et U-Pb). Exemples choisis au Groënland, en Laponie, dans le Massif armoricain et en Grande Kabylie: Mémoire de la Société Géologique et Minéralogique de Bretagne, v. 28, 108 p.

Purdy, J. W., and Jäger, E., 1976, K-Ar ages on rock-forming minerals from the Central Alps: Memorie dell'Universita di Padova, v. XXX, 31 p.

Quinquis, H., 1980, Schistes bleus et déformation progressive. L'exemple de l'île de Groix (Massif Armoricain): Thése de 3éme cycle, Rennes, 145 p.

Quinquis, H., Audren, Cl., Brun, J. P., and Cobbold, P., 1978, Intense progressive shear in Ile de Groix blueschists and compatibility with subduction or obduction: Nature, v. 273, p. 43–45.

Quinquis, H., and Choukroune, P., 1981, Les schistes bleus de l'île de Groix dans la chaîne hercynienne: Implications cinématiques: Bulletin de la Société Géologique de France, ser. 7, v. 23, p. 409–418.

Roddick, J. C., and Compston, W., 1977, Strontium isotopic equilibration: A solution to a paradox: Earth and Planetary Science Letters, v. 34, p. 238–246.

Schweickert, R. A., Armstrong, R. L., and Harakal, J. E., 1980, Lawsonite blueschist in the Northern Sierra Nevada, California: Geology, v. 8, p. 27–31.

Suppe, J., and Armstrong, R. L., 1972, Potassium-argon dating of Franciscan metamorphic rocks: American Journal of Science, v. 272, p. 217–233.

Triboulet, C., 1974, Les glaucophanites et roches associées de l'île de Groix (Morbihan, France): Etude minéralogique et pétrogénétique: Contributions to Mineralogy and Petrology, v. 45, p. 65–90.

Vidal, Ph., 1972, L'axe granitique de Moëlan-Lanvaux (sud du Massif Armoricain): Mise en évidence par la méthode Rb-Sr de trois épisodes de plutonisme pendant le Paléozoique inférieur: Bulletin de la Société Géologique et Minéralogique de Bretagne, sér. 7, p. 75–89.

Vidal, Ph., 1973, Premiéres données géochronologiques sur les granites hercyniens du sud du Massif Armoricain: Bulletin de la Société Géologique de France, Paris, ser. 7, v. 15, p. 239–245.

Vidal, Ph., 1980, L'évolution polyorogénique du Massif Armoricain: Apport de la géochronologie et de la géochimie isotopique du strontium: Mémoire de la Société Géologique et Minéralogique de Bretagne, v. 21, 162 p.

Vidal, Ph., Peucat, J. J., and Lasnier, B., 1980, Dating of granulites involved in the Hercynian foid-belt of Europe: An example taken from the granulite facies orthogneisses at La Picherais, Southern Armorican Massif, France: Contributions to Mineralogy and Petrology, v. 72, p. 283.

Watson, E. B., 1981, Diffusion in magmas at depth in the earth: The effects of pressure and dissolved H_2O: Earth and Planetary Science Letters, v. 52, p. 291–301.

York, D., 1966, Least square fitting of a straight line: Canadian Journal of Physics, v. 44, p. 1079–1086.

MANUSCRIPT ACCEPTED BY THE SOCIETY JULY 29, 1985

Geological Society of America
Memoir 164
1986

Blue amphiboles in metamorphosed Mesozoic mafic rocks from the Central Alps

R. Oberhänsli
Labor für Mikroröntgenspektroskopie der phil. nat.
Fakultät der Universität Bern
Baltzerstrasse 1
CH-3012 Bern, Switzerland

ABSTRACT

Alkali amphiboles are widespread in the Swiss Alps. They are products of an 'Eoalpine' Cretaceous pressure-accentuated metamorphic event which has been obliterated in the Central Alps by the younger 'Lepontine' Tertiary metamorphism of Barrovian type. The regions west and east of the Lepontine metamorphic dome show different tectonometamorphic evolution. In the west, the Penninic basement, the ophiolites and Mesozoic sedimentary rocks exhibit a common, strong high-pressure metamorphism and were uplifted together. In the east, the basement was deeply subducted before the final continental collision and shows a strong eclogite facies overprint. The ophiolites and the Penninic Mesozoic sedimentary rocks contain few eclogitic relicts. Within the eastern part of Switzerland (Grisons), alkali amphibole compositions in metabasic rocks change from glaucophane in the west to magnesioriebeckite and riebeckite in the east. Depending on bulk composition (e.g. Fe_2O_3/FeO; H_2O), the metamorphic stability field of blueschists overlaps with that of the greenschist facies (e.g. cro* + epi + ank + chl + qtz is cofacial with alb + chl + act + epi + mus + qtz in the eastern Grisons) as well as with that of the eclogite facies (e.g. glc + gar + epi + par + chl + chd is cofacial with omp + gar + zoi/epi + rut in the western Grisons).

INTRODUCTION

In the Swiss Alps amphiboles of the glaucophane-riebeckite series are reported from several different geological settings.

In the *Penninic realm,* alkali amphiboles are reported both west and east of the 'Lepontine' Tertiary metamorphic area (Fig. 1). In the western part of the Swiss Alps, blueschists containing variable mineral assemblages occur in the pre-Triassic cover (assemblages IVa, Table 2) as well as the basement (assemblages IVb, Table 2) of the Bernhard Nappe (Grubenmann, 1906; Woyno, 1912; Schürmann, 1953; Schär, 1959; Bearth, 1963; Bocquet, 1974). In the eastern part, they occur within crystalline Adula Nappe (assemblages I, Table 2) (Plas, 1959; Heinrich, 1982). Additionally, blueschists occur on both sides of the Lepontine in the Penninic Mesozoic 'Bündnerschiefer' (calc-schists) sequences from the Lower Engadine tectonic window in

the east to the Zermatt–Saas Fee ophiolite sequence in the west (assemblages II–III, Table 2). These Mesozoic sediments and ophiolites are remnants of the Penninic realm of the Tethyan Ocean. They have undergone an 'Eoalpine' Cretaceous (100–65 Ma) (Jäger, 1973) high pressure and a 'Lepontine' Tertiary (38–35 Ma) (Hunziker, 1969) regional metamorphism. The Tertiary overprint has, to a great extent, erased early high pressure–low temperature parageneses in the Central Alps. Only east and west of the Lepontine area have blueschist occurrences partly or completely escaped the Tertiary metamorphic event. The metamorphic map of the Alps (Niggli, 1973) shows the spatial relations of the metamorphic events as well as their age relations.

In the *Helvetic realm,* the Lower Cretaceous 'Helvetic Kieselkalk' (siliceous limestone) of the Axen Nappe (Schindler, 1959, 1969; Frey et al., 1973) and the para-autochthonous cover of the Aarmassiv (Heim, 1910; Niggli et al., 1956) contain alkali

---

*See Table 1.

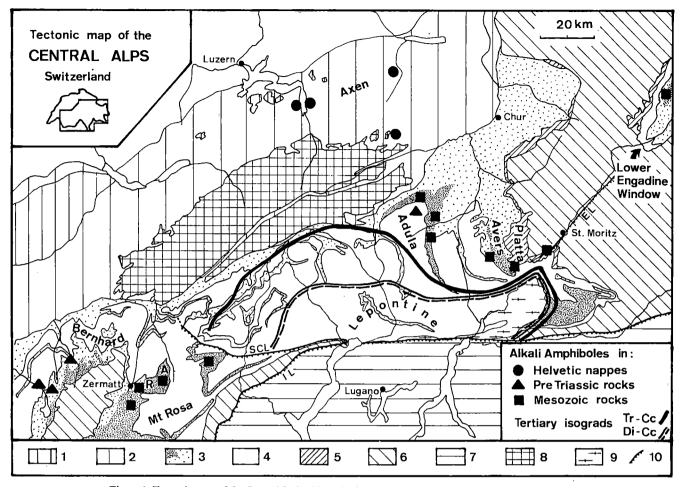

Figure 1. Tectonic map of the Central Swiss Alps with isograds of the Tertiary regional metamorphism
and major alkali amphibole deposits. 1) Prealpine nappes, 2) Helvetic nappes, 3) Penninic sediments
(Permian to Tertiary), particularly Mesozoic 'Bündnerschiefer'; dark: ophiolite bearing 'Bündnerschiefer,' 4) Penninic nappes (crystalline basement), 5) Penninic ophiolite nappes, 6) Eastern Alpine
nappes, 7) Southern Alps, 8) Central massives (Prepermian rocks), 9) Tertiary granites, granodiorites
and tonalites, 10) faults: SCL: Simplon–Centovalli line, IL: Insubric line, EL: Engadine line.

amphiboles. Magnesioriebeckite or riebeckite, coexisting with
glauconite, calcite, quartz and K-feldspar are reported from anchimetamorphic sediments as well as on slickensides.

Glaucophane was reported as detritus from the sediments of
the Upper Marine (Lower Chattian to Burdigalian) *Molasse* of
Savoy, France, (Mange and Oberhänsli, 1982) and Switzerland
(Maurer, 1982; Maurer et al., 1982).

The geographic distribution of alkali amphiboles in the
Swiss Alps has been compiled by Plas (1959), Niggli and Niggli
(1965), Niggli (1970) and Dietrich et al. (1974). In this paper, I
shall summarize the complex occurrences east and west of the
Lepontine in the context of their geologic and tectonic units. The
compositions of the alkali amphiboles and their parageneses are
linked to the Alpine tectonometamorphic history. Two areas

TABLE 1. LIST OF ABBREVIATIONS

act = actinolite	glc = glaucophane	phe = phengite
alb = albite	gra = graphite	prw = preiswerkite
amp = amphibole	hbl = hornblende	qtz = quartz
ank = ankerite	hem = hematite	ric = richterite
bio = biotite	ilm = ilmenite	rie = riebeckite
cro = crossite	kya = kyanite	rut = rutile
chl = chlorite	mgr = magnesioriebeckite	sph = sphene
chd = chloritoid	mag = magnesite	stp = stilpnomelane
cal = calcite	mgt = magnetite	spi = spinel
czo = clinozoisite	mus = muscovite	tlc = talc
epi = epidote	omp = omphacite	zoi = zoisite
gar = garnet	par = paragonite	

TABLE 2. MINERAL ASSEMBLAGES OF ECLOGITES, BLUESCHISTS, AND GREENSCHISTS
FROM DIFFERENT LOCALITIES IN THE SWISS ALPS

I   Adula Nappe, basement (Vals)   Plas, 1959; Heinrich, 1982
```
      omp + gar + qtz + czo + phe + hbl + plg
      omp + gar + qtz + epi + phe + glc + alb
      hbl + glc + chl + epi + qtz + phe + tit + gar ± bio
      gar + glc + phe + qtz
      omp + gar + hbl + zoi + qtz + dol + cal
      glc + epi + cal + chl + hbl + rut
```

IIa   Adula cover, NE Nappe front (Vals)   Oberhansli, 1977
```
      cro + alb + phe + gar + epi/czo + qtz + chl
      cro + epi/czo + chl + ank + mgt
      alb + hbl + chl + epi/czo ± qtz
      alb + act + epi/czo + chl + phe + qtz
```

IIb   Adula cover E of the Nappe (Neu Wahli)   Gansser, 1937; Oberhansli, 1977
```
      hbl + epi + alb + gar + chl + czo + tit
      glc + act/tre + gar + zoi + epi + cal
      glc + hbl + mus + czo + gar + alb + chl + epi
      gar + glc/cro + Fe - cal + mgt + omp
      epi + cal + alb + chl + act
      gar + cro + epi + alb + cal + mgt + tre
      cal + gar + omp + epi + qtz + mgt + act
      cal + gar + aeg(?) + omph + epi + qtz + mgt
```

IIc   Western Avers (Madrisa)   Oberhansli, 1977
```
      glc + cro + qtz + chl + ank + eip/czo + tit
      glc + cro + phe + chl + ank + epi/czo + rut
      alb + act + epi/czo + cal + chl + phe
```

IId   Eastern Avers (Jufer Horen)   Dietrich und Oberhansli, 1976
                                                       Oberhansli, 1977
```
      mgr + act + chl + epi/czo + sph
      mgr + act + ank + phe + rut + stp
      alb + act + chl + epi/czo + qtz + hem
```

IIe   Platta Nappe, radiolaritic metasedimentary rocks   Phillipp, 1982
```
      qtz + mgr + stp + gra
      qtz + alb + ric + stp + gra
      alb + mgr + stp
      qtz + alb + rie + stp
      alb + qtz + mgr + gra + stp + aeg aug
```

IIf   Platta Nappe, greenschists   Dietrich, 1969; Oberhansli, 1977
```
      alb + chl + epi + act + mus + qtz
      chl + act + mgr
```

IIg   Lower Engadine Window   Heugel, 1975; Oberhansli, 1977
```
      pum + act + chl + epi + alb
      pum + chl + epi + cal + qtz
      pum + chl + eip + alb
      chl + epi + alb + mgr
      chl + epi + stp + mus + qtz
      cal + chl + alb + qtz + mus
      cal + chl + epi + qtz + mgr + stp
```

IIIa   Zermatt-Saas Fee Zone (Allalin gabbro)   Bearth, 1967; Meyer, 1983
```
      jad + omp + gar + zoi + tlc + chl + mag
      omp + gar + kya + mg chd + zoi + tlc + chl + mag + rut
      omp + gar + zoi + czo/epi + glc + tlc + chl + par + phe + mag + rut
      zoi + czo/epi + hbl + chl + par + mar + prw + bio + alb + spi + ilm
      zoi + czo/epi + act/tre + chl + par + alb + sph + hem ± rut
      czo/epi + chl + alb + sph ± cal
```

IIIb   Zermatt-Saas Fee Zone (Rimpfischhorn)   Bearth, 1967; Oberhansli, 1980
```
      omp + gar + glc + epi + par ± qtz ± cal ± rut
      omp + gar + glc + epi + par ± phe ± qtz ± cal ± rut
      omp + gar + glc + hbl + epi + par ± qtz ± rut
      omp + gar + glc + epi + par + chd ± rut
      omp + gar + glc + chd + tlc
      gar + glc + epi + par ± alb ± qtz ± rut ± sph
      gar + glc + epi + par + chl + chd ± rut
      gar + glc + epi + chl + chd + tlc ± rut
      hbl + act + epi + phe + alb ± bio ± qtz ± cal ± sph
      hbl + act + epi + chl + alb ± sph
      alb + act + hbl + epi + phe + chl ± bio
```

IVa   Bernhard Nappe   Grubenmann, 1906; Woyno, 1912
      pre Triassic cover (Metallier unit)   Schurmann, 1953; Schar, 1959
                                               Bocquet, 1974
```
      qtz + alb + epi + phe + chl + cal
      qtz + alb + epi + chl + cal
      qtz + par + chl + chd
      qtz + alb + chl + glc
      alb + chl + glc
      alb + chl + epi + glc
```

IVb   Bernhard Nappe, basement (Siviez unit)   Bearth, 1963
```
      alb + chl + act + epi ± omp ± glc
      alb + chl + act + epi + gar ± omp ± glc
      chl + act + epi
      alb + chl + cal
```

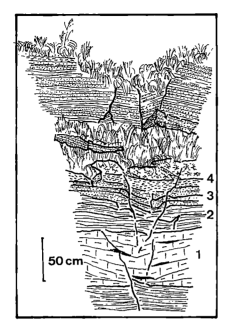

Figure 2. Thinly intercalated layers of greenschists and blueschists in the Averser 'Bündnerschiefer.' 1) 'Bündnerschiefer' (Mesozoic calcschists) with quartz lenses, 2) greenschists (prasinites), 3) blueschists, 4) coarsely grained layers of blueschists with glaucophane rosettes on schistosity planes.

where alkali amphiboles occur in Mesozoic rocks will be discussed in detail: 1) the Grisons and 2) the Zermatt–Saas Fee area. These areas are situated east and west with respect to the culmination of Tertiary regional metamorphism where the Eoalpine high-pressure event has not been completely obliterated (Fig. 1).

THE GRISONS

In the Grisons, the eastern part of the Swiss Alps, blueschists occur in the pre-Triassic crystalline masses of the Adula Nappe (Plas, 1959; Heinrich, 1982). They can also be found in the mafic rocks of the southwestern 'Bündnerschiefer' (calcschists) series (Oberhänsli, 1977, 1978). The 'Bündnerschiefer' series are considered to represent the sedimentary cover of the lower Pennine Nappes, e.g., the Adula Nappe. The ophiolite material bearing 'Bündnerschiefer' extend throughout the southwest Grisons from the Adula eastwards to the Lower Engadine tectonic window (Fig. 1). In the uppermost Penninic Platta Nappe, calcic amphiboles occur throughout the nappe in Mesozoic sediments, mafic volcanic rocks (e.g. pillow breccias), and gabbros, whereas alkali amphiboles are restricted to the southern part of the nappe. There, magnesioriebeckite forms blue layers in greenschists (Dietrich, 1969). In the basal parts of the Platta Nappe, alkali amphiboles occur at the contact between radiolaritic metasediments and serpentinites (Philipp, 1982).

The 'Bündnerschiefer' include late Mesozoic calcareous marbles, quartz-rich calcschists and graphitic phyllites with intercalations of volcanosedimentary tuffaceous layers, bodies of mas-

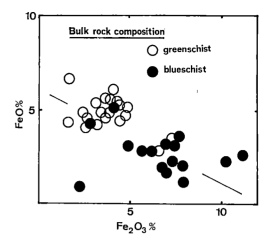

Figure 3. Distribution of FeO versus Fe_2O_3 from greenschists and blueschists of the southwest Grisons. Blueschists show higher total Fe contents as well as higher Fe_2O_3 values.

sive pillow lavas, pillow breccias, and gabbros or serpentinites. The interstratified metatuffaceous greenschist layers, which may be followed laterally up to hundreds of meters, often contain thin layers (5 to 50 cm) of blueschists (Fig. 2). These Mesozoic metasediments show no unequivocal trace of high pressure-low temperature indicators, e.g. relict jadeite or omphacite together with quartz, or lawsonite. Usually glaucophane-rich alkali amphiboles occur; rarely do Fe-rich chloritoids and aegirine-augitic pyroxenes occur. The intercalation of thin blueschist and greenschist layers (Fig. 2) underwent the same metamorphic grade or overprint.

In the frontal (northern) part of the *Adula* Nappe, Mesozoic mafic rocks occur along with calc-mica schists within the crystalline basement (assemblages I, Table 2) as well as the overlying 'Bündnerschiefer' series (assemblages IIa, Table 2). There, alkali amphiboles occur together with garnet + quartz + omphacite in glaucophane eclogites and garnet glaucophanites (Plas, 1959; Heinrich, 1982). They are crossitic in composition and rimmed by blue-green tschermakitic hornblende and later actinolite. West of the Adula Nappe, only one locality with alkali amphibole is found within the 'Bündnerschiefer' (Gansser, 1937). Inhomogeneous basaltic pillow breccias form lenses of blueschists (assemblage IIb, Table 2) resting in greenschist-layers. Crossite is surrounded first by glaucophane, then sometimes by barroisitic hornblende. Actinolite or tremolite comprise the outer rim.

Chemical investigations in the *Avers* area (Oberhänsli, 1977) show that the metatuffaceous layers have high alumina basalt composition and that the blue and green coloured layers differ in their alkali and total iron content. Blueschists show higher values for Na_2O, total Fe, and especially Fe_2O_3 (Fig. 3) than the greenschist layers. Similar observations have been made by Brown (1974) on blueschists from the Cascades. This has been interpreted as effects of submarine weathering or oceanic alteration of the surfaces of tuffaceous layers, creating premetamorphic differences in bulk chemistry which led to the production of different metamorphic assemblages between the green and blue layers. Alkali amphiboles in the mafic layers of the 'Bündnerschiefer' (assemblage IIc, Table 2) grow across schistosity planes, often as big crystals (up to 2 cm). They show inclusion-rich cores with epidote, clinozoisite, sphene and oxides, indicating slight rotation during synkinematic growth. Rims of identical composition, but free of inclusions indicate the postkinematic stability of alkali amphiboles. In the western part of the Avers, glaucophane and crossite are stable (assemblage IId, Table 2). Occasionally glaucophane rimmed by crossite is found. In the eastern part, towards the Platta Nappe, zoning of actinolite around magnesioriebeckite and inverse zoning of magnesioriebeckite around actinolite occurs.

In the southern *Platta* Nappe, magnesioriebeckite, riebeckite and richterite was formed in a matrix of quartz and albite (assemblages IIe, Table 2) in acid metasediments (Cornelius, 1935; Philipp, 1982). After the formation of the alkali amphiboles, aegirine-augite overgrowing magnesioriebeckite and later stilpnomelane was formed. Magnesioriebeckite was formed syn- to post-tectonically. There is complete miscibility between magnesioriebeckite and actinolite as manifested by continuous zonations. Actinolitic cores are rimmed by magnesioriebeckite. In some places riebeckite is rimmed by actinolite. K-Ar ages of blue amphibole yield 90 to 60 ma (Deutsch, 1983) and put a minimum age on the last penetrative deformation in the Engadine.

In the mafic metavolcanic rocks of the southern Platta Nappe, magnesioriebeckite + actinolite + chlorite form blue layers in greenschists (assemblages IIf, Table 2). Magnesioriebeckite occurs as euhedral unoriented crystals and is not zoned. From this area, buffered equilibrium parageneses with amp + chl + alb + epi + qtz have been described (Dietrich, 1969).

In the *Lower Engadine* window, riebeckitic alkali amphiboles are found in metahyaloclastites belonging to a tholeiitic pillow sequence, as well as a metabasic sill and its adjacent carbonaceous metasediments. These rocks have undergone pumpellyite-to-greenschist-facies metamorphism (assemblage IIg, Table 2). They show pumpellyite and actinolite. Stilpnomelane grows in bundles across schistosity planes (Heugel, 1975). Riebeckite occurs in the assemblage chl + epi + alb + rie and shows no zonation.

Chemical compositions of alkali amphiboles fit into a regional distribution pattern (Fig. 4): in the western part of the southwest Grisons (Adula, W - Avers), glaucophane and crossite (assemblages IIa, IIb, IIc, Table 2) occur, whereas in the eastern parts (E - Avers, Platta Nappe, Lower Engadine Window), magnesioriebeckite to riebeckite (assemblages IId, IIf, IIg, Table 2) is found. This change in amphibole composition takes place within one rock type, the metabasic layers. The amount of iron-rich minerals (e.g. mgt, hem, epi, stp) varies along the profile.

In the west, where Tertiary overprinting is significant, alkali amphiboles are rimmed by calcic amphiboles. Well away from the area of intense Tertiary overprint, in the southeastern part of Avers and in the Platta Nappe, both calcic amphibole rimmed by

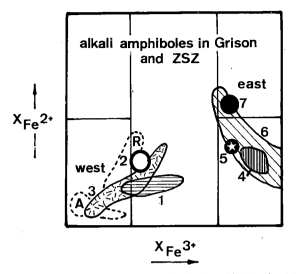

Figure 4. Chemical variation diagram for alkaliamphiboles after Miyashiro (1957). The chemical composition of alkali amphiboles from metabasic layers in the 'Bündnerschiefer' (Mesozoic calcschists) of the southwest Grisons change regionally from glaucophane in the west (Adula cover) to riebeckite in the east (Lower Engadine Window). 1) Cover of the Adula nappe (NE - nappe front) (assembl. IIa, Tab. 2), 2) Cover of the Adula nappe (E of the nappe) (assembl. IIb, Tab. 2), 3) W – Avers (assembl. IIc, Tab. 2), 4) E – Avers (assembl. IId, Tab. 2), 5) Platta nappe: ophiolites (assembl. IIf, Tab. 2), 6) Platta nappe: radiolarian sedimentary rocks (Philipp, 1982), 7) Lower Engadine Window (assembl. IIg, Tab. 2). In the Zermatt - Saas Fee Zone west of the Lepontine Area glaucophane and crossite occur in the eclogitic mafic rocks, A) Allalin gabbro complex (Meyer, 1983), R) Rimpfischhorn pillow lava complex.

alkali amphibole and the inverse zonation pattern can be found. Further east no zonation is observed.

These observations can be compared with the general Alpine metamorphic history: a 100-60 ma (Jäger, 1973) Eoalpine high-pressure event which was followed by a 38-35 ma (Hunziker, 1969) Tertiary regional-metamorphic event of Barrovian type. Miyashiro (1957) suggests that glaucophane destabilizes with decreasing pressure or increasing temperature, while magnesioriebeckite becomes the stable phase. Due to the lack of critical mineral assemblages (e.g. gar + cpx + qtz), neither temperature nor pressure can be estimated for the Eoalpine metamorphism in the blueschists of the Mesozoic 'Bündnerschiefer' in the southwest Grisons. The occurrence of glaucophane and relict aegirine-augitic pyroxene only in the western parts leads to the assumption of slightly higher pressures in the 'Bündnerschiefer' of the western than of the eastern Grisons. Jäger (1973) suggests flat temperature gradients with a slight increase towards east during the Eoalpine phase.

The Tertiary Lepontine regional metamorphic overprint is much stronger in the western part (Adula) than in the eastern (Engadine), as indicated by the concentric pattern of the Lepontine metamorphic isograds mapped by Niggli (1960) and Wenk (1962). Due to the updoming of the Central Alps, deeper portions of the nappe pile are exposed in the western than in the eastern Grisons (Fig. 1). Pressure and temperature estimates for Tertiary metamorphism in the central Adula Nappe are reported as 525 °C/4-6 kb by Thompson (1976), Klein (1976) and Baumgartner and Löw (1983) and are in agreement with estimates derived from the mineral assemblages found in Mesozoic rocks (Teutsch, 1982). In the eastern part, pressures are much lower and have been estimated from lithostatic overburden to have reached 2-3 kb in the Platta Nappe (Cornelius, 1935; Dietrich, 1969). The temperature influence of the Tertiary metamorphism fades out towards the east.

In the Avers (central SW–Grisons), Eoalpine (100–60 ma) and Tertiary (38–35 ma) temperatures and/or pressures must have been quite similar to each other in order to allow further growth of glaucophane after deformation and to maintain glaucophane as a stable phase during Tertiary metamorphism. This metamorphism produced stilpomelane in adjacent rocks.

Petrographic observations and petrologic studies (Oberhänsli, 1977) lead to the conclusion that in the blueschists, which occur within the 'Bündnerschiefer' of the southwestern Grisons, glaucophane was formed due to favourable chemical-bulk compositions, Fe_2O_3/FeO ratio, and low T/P (°C/kb) ratios (T/P < 80; typical T/P ratios for eclogites range from 30 to 60 and for Barrovian type metamorphism from 100 to 120). It seems that in the Mesozoic 'Bündnerschiefer', high pressures never have been attained since no high-pressure relicts can be found in the Mesozoic rocks (Oberhänsli, 1977; Teutsch, 1982). However, in the underlying crystalline nappes, especially in the Adula Nappe, zones with bodies of glaucophane - eclogites and kyanite - hornblende - eclogites document pressures of 10 to 20 kb (Heinrich, 1982).

The rarity of high-pressure relicts in the Mesozoic cover in contrast to the abundance of high-pressure relicts in the pre-Mesozoic rocks of the crystalline nappes as well as the Mesozoic rocks incorporated into the basement could be explained by the following mechanism:

In Cretaceous time, the ophiolite-bearing Mesozoic 'Bündnerschiefer' of the southwestern Grisons as well as the uppermost Penninic Nappes (e.g. Platta Nappe) experienced different metamorphic conditions than the crystalline basement nappes. Thus, at present, the 'Bündnerschiefer' sit allochthonous on the crystalline basement. While the crystalline nappes and, in the western Grisons, a thin para-autochthonous 'Bündnerschiefer' cover have been subducted or buried at depth where high-pressure assemblages were produced, the overlying Mesozoic sediments never reached the required depth to produce eclogitic parageneses in their mafic layers. They followed a metamorphic pathway with low T/P ratios without having been affected by high pressures. Wherever the chemical compositions of the rocks were suitable, glaucophane was produced.

THE ZERMATT–SAAS FEE ZONE

In the western part of the Central Alps, blueschists occur mainly in the Mesozoic ophiolite-bearing complexes of the Zer-

matt area (assemblages IIIa, IIIb, Table 2) and in the pre-Triassic cover of the Bernhard Nappe (assemblages IVa, Table 2). Within the Mesozoic rocks of the Zermatt area, they are located in a lower, ophiolitic complex, the Zermatt–Saas Fee Zone (ZSZ). This unit is overlain by the Combin Zone, consisting of a sequence of intercalated calcschists ('Bündnerschiefer') and greenschists but no blueschists. The ZSZ underwent a high-pressure metamorphic Eoalpine event producing eclogites and was later affected by the Tertiary Barrovian type metamorphism of the Lepontine phase. The ZSZ is composed of rocks typical for a major ophiolite complex: serpentinites, metagabbros, metapillow lavas, metatuffaceous layers and deep sea sediments. A complete ophiolite sequence is not preserved, however. Different tectonic slices with ophiolitic material are separated by serpentinites. Blueschists occur in the metapillow lava and metapillow breccia sequences and glaucophane is also widespread in massive metagabbroic rocks. Metatuffaceous layers very often are completely retromorphosed to greenschist assemblages (Bearth, 1959, 1965, 1966, 1967, 1973, 1974; Ernst and Dal Piaz, 1978; Ganguin, 1983). In contrast to the southwest Grisons, many eclogite relicts are found in the ZSZ.

Recent investigation of the Allalin metagabbro (Meyer, 1983), a major gabbroic body of the ZSZ, shows that glaucophane is a postkinematic high-pressure mineral in the eclogitic metagabbros. This author suggests that the gabbro, after its intrusion, was directly brought to greater depth (subducted?), producing dry garnet-clinopyroxene coronas around olivine in a granulitic stage (18-20 kb; 650-725 °C). Then a first deformative high-pressure stage (>20 kb; 500-600 °C), with eclogitic assemblages, was followed by a second, post-deformative, high-pressure phase (14-19 kb; 480-550 °C), producing glaucophane (assemblages IIIa, Table 2). Later, emplacement and Tertiary metamorphism overprinted the eclogitic metagabbro in greenschist facies (? 4 kb; 500 °C) to a saussuritized flasergabbro. Only one glaucophane generation is found in the Allalin metagabbro and represents products of hydration during the second high-pressure phase. These glaucophanes are generally richer in Mg than glaucophanes from the metabasites of the ZSZ (Fig. 4) (Meyer, 1983).

In the eclogitic metapillow lavas of the Rimpfischhorn (assemblages IIIb, Table 2), a major mafic complex of the ZSZ, two generations of glaucophane can be clearly observed (Bearth, 1973; Chadwick, 1974; Oberhänsli, 1982; Ganguin, 1983). The first generation (glc I) grows along and in primary volcanic structures (pillow rims), and a second generation (glc II) is aligned into the main schistosity. Glaucophane occurs within the eclogitic pillow cores (glc II) as well as the glaucophanitic pillow rims (glc I + II).

In this complex, metabasic rocks were transformed to eclogites and later retromorphosed to garnet amphibolites. Locally, well-preserved metapillow lavas show eclogitic pillow cores and glaucophanitic pillow rims (Fig. 5). Possible hyaloclastites and tuff series are represented by garnet amphibolites with pseudomorphs after lawsonite, or by prasinites (chlorite-albite-epidote

Figure 5. Metapillow lava from the Rimpfischhorn near Zermatt, showing deformed pillows with eclogitic cores and glaucophanitic rims, resting in greenschists. 1) Eclogitic pillow core, 2) Epidote-rich light border, 3) Dark border or matrix with glaucophane, omphacite, white mica, ankerite and garnet, 4) Matrix consisting of glaucophane, white mica and ankerite, 5) Prasinitic layer (albite-apidote-actinolite-chlorite schist).

-actinolite schists). Eclogites with omp + gar + zoi + phe + glc and glaucophanites consisting of glc + gar + phe + hbl + chl + Mg-rich chd + tlc occur together as relicts of the high-pressure metamorphism within albite-epidote-garnet amphibolites or prasinites. No mappable limits can be drawn between eclogites, glaucophanites, garnet amphibolites or prasinites. All these rocks belong to a single tectonic unit and clearly have suffered the same P-T history but amphibolites and prasinites exhibit different degrees of retromorphism.

The chemical composition of the metabasic rocks from Zermatt compares well with actual plate margin basalts (Bearth and Stern, 1979). For eclogites and glaucophanites, Bearth and Stern (1971) assumed isochemical starting material and suggested a complex metasomatic transformation from eclogite to glaucophanite with gain of Mg and H_2O and loss of Ca. But, metasomatism follows the primary volcanosedimentary structures and therefore must have taken place during devitrification of the glassy pillow matrix along with oceanic metamorphism.

The P-T history of the metabasic rocks of the ZSZ started with the production of pillow lavas. Submarine alteration hydrated and oxidized the glassy pillow rims as well as the hyaloclastitic or tuffaceous layers. Thus, three mafic bulk compositions, differing largely in their H_2O content, were produced prior to the Alpine metamorphic events. During subduction, the more or less dry pillow cores were transformed to eclogites, whereas the hydrated pillow rims were transformed to blueschists. In the tuffaceous layers lawsonite was produced.

Pressure and temperature estimates on eclogite relicts yielded a peak of metamorphism at about 14 kb (minimum pressure) and around 600 °C (Oberhänsli, 1980). Wherever zoned minerals occur, prograde conditions from core to rim are found.

In a second, deformative high-pressure stage, blue amphiboles were also formed in pillow cores. Later again, Tertiary greenschist overprint followed penetrative schistosity and transformed the complete rock sequence, leaving eclogite relicts in the pillow cores and blueschists in the rims. The hyaloclastic matrix around the pillow lavas reacted mainly to form garnet amphibolites with pseudomorphs after lawsonite or prasinites, depending on the intensity of the overprint.

According to Hunziker (1974), K/Ar isotopes of glaucophanes of the ZSZ indicate a maximum age of about 80 ma (77.9 ± 9.3). Most of the isotopically analyzed glaucophanes have been isotopically reset by later overprints and therefore the ages scatter between 40 to 50 ma and between 15 to 30 ma.

In conclusion, even at very high pressures the production of blueschists seems to be controlled by geochemical parameters such as the original H_2O content and oxidation state of the protolith (Bearth, 1959). Whether these compositional differences are of premetamorphic origin as in the pillow lavas of the Rimpfischhorn or are produced by metamorphic events similar to the hydration of the Allalin gabbro (Meyer, 1983) seems of lesser importance.

SUMMARY AND CONCLUSIONS

As shown above, alkali amphiboles are widespread in the Swiss Alps and occur in many geologic settings and tectonic units. Riebeckite and magnesioriebeckite occur as neoformations in the Mesozoic anchimetamorphic sediments of the Helvetic Nappes or in metabasic rocks in the eastern Grisons. Glaucophane and crossite occur generally in metabasic rocks and can be correlated with a pressure-accentuated (subduction type) metamorphism. This 'Eoalpine' Cretaceous metamorphic event has been obliterated in the Central Alps by the younger 'Lepontine' Tertiary metamorphism of Barrovian type.

The regions east and west of the Lepontine metamorphic dome show different tectonometamorphic evolutions. The blueschist assemblages have been visualized schematically in a NCMA tetrahedron projected from qtz and H_2O (Fig. 6). Only 'type' parageneses taken from Table 2 are plotted for each region discussed above. Intersections in some tetrahedrons result more likely from the fact that the assumptions for projections in the model system NCMA have been violated (e.g. the chemical potential of H_2O may be internally controlled) rather than from changing P-T conditions.

In the east: in the SW - Grisons, only the pre-Alpine continental basement was deeply subducted during the continental collision and shows a strong high-pressure overprint. The ophiolites and the Penninic Mesozoic sediments of the 'Bündnerschiefer' contain no eclogitic relicts. They exhibit weak pressure-accentuated greenschist-facies overprint, which formed glc and cro, but generally no eclogite-facies assemblages such as

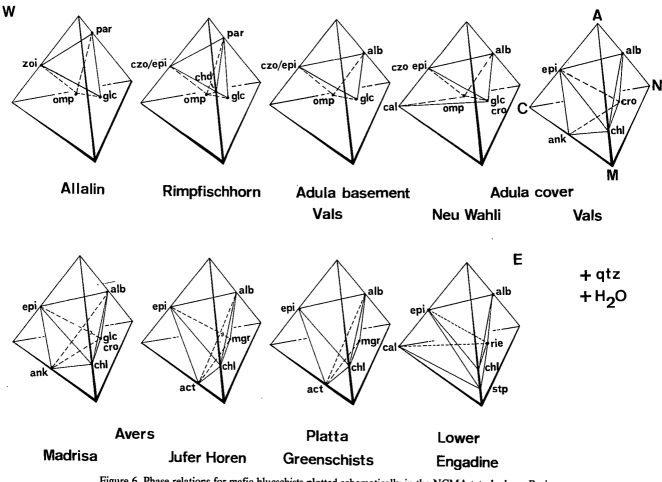

Figure 6. Phase relations for mafic blueschists plotted schematically in the NCMA tetrahedron. Projection from quartz and H$_2$O.

$$A = Al_2O_3 + Fe_2O_3 + Cr_2O_3; C = CaO$$
$$N = Na_2O + K_2O; M = MgO + FeO + MnO$$

jadeite, omphacite or lawsonite. Probably they have only been subducted to a shallow depth.

Within the southwest Grisons, alkali amphibole compositions in metabasic rocks change from glaucophane in the west (Adula) to magnesioriebeckite in the east (Lower Engadine Window). This change of composition is observed in rocks of similar bulk composition and therefore indicates regional differences in recorded P-T conditions.

In the west, the Penninic basement of the Bernhard and the Monte Rosa Nappes, the ophiolites and the Mesozoic sediments of the Zermatt–Saas Fee Zone exhibit a common, strong high-pressure–low-temperature metamorphism. They were subducted and uplifted together.

In high pressure–low-temperature metamorphic terrains, the occurrence of glaucophane or crossite is controlled by the

bulk rock chemistry. Glc + gar + epi + par + chl + chd is favoured relative to omp + gar + zoi/epi + rut by a high Fe$_2$O$_3$/FeO and high H$_2$O-content in the protolith at metamorphic conditions in the range 8–12 kb/400–500 °C for the Adula Nappe and 14–19 kb/400–650 °C for the ZSZ. In pressure-accentuated greenschist facies, cro + epi + ank + chl + qtz is stabilized relative to alb + chl + act + epi + mus + qtz by similar differences in premetamorphic composition.

ACKNOWLEDGMENTS

I greatly appreciate the help and critical review of my wife H. Oberhänsli, C. A. Heinrich, L. Kanat, V. Dietrich, J. Meyer, E. H. Perkins and A. B. Thompson. This work has been partly supported by the schweizerischer Nationalfonds project 2.497-0.75 and project 2.484-0.79.

REFERENCES CITED

Baumgartner, L., and Löw, S., 1983, Deformation und Metamorphose der Adula-Decke südwestlich San Bernardino: Schweiz. mineral. petrogr. Mitt. 63, 215–232.

Bearth, P., 1959, Über Eklogite, Glaukophanschiefer und metamorphe Pillow-laven: Schweiz. mineral. petrogr. Mitt. 39, 267–286.

Bearth, P., 1963, Contribution a la subdivision tectonique et stratigraphique du cristallin de la nappe du Gran - St - Bernard dans le Valais (Suisse): Livre Mém. Fallot 2, 407-418 (Soc. géol. France).

Bearth, P., 1965, Zur Entstehung alpinotyper Eklogite: Schweiz. mineral. petrogr. Mitt. 45, 179-188.

Bearth, P., 1966, Zur mineralfaziellen Stellung der Glaukophangesteine der Westalpen: Schweiz. mineral. petrogr. Mitt. 46, 13-23.

Bearth, P., 1967, Die Ophiolite der Zone von Zermatt-Saas Fee: Beitr. geol. Karte der Schweiz NF 132.

Bearth, P., 1973, Gesteins- und Mineralparagenesen aus den Ophioliten von Zermatt: Schweiz. mineral. petrogr. Mitt. 53, 299-334.

Bearth, P., 1974, Zur Gliederung und Metamorphose der Ophiolite der Westalpen: Schweiz. mineral. petrogr. Mitt. 54, 385-397.

Bearth, P., and Stern, W., 1971, Zum Chemismus der Eklogite und Glaukophanite von Zermatt: Schweiz. mineral. petrogr. Mitt. 51, 349-359.

Bearth, P., and Stern, W., 1979, Zur Geochemie von Metapillows der Region Zermatt - Saas: Schweiz. mineral. petrogr. Mitt. 59, 349-373.

Bocquet, J., 1974, Blue amphiboles of the western Alps. Chemistry and physical conditions: Schweiz. mineral. petrogr. Mitt. 54, 425-448.

Brown, E. H., 1974, Comparison of the mineralogy and phase relations of blueschists from the North Cascades, Washington, and greenschists from Otago, New Zealand: Bull. Geol. Soc. Amer. 85, 333-344.

Chadwick, B., 1974, Glaucophane fabric in the cover of the Monte Rosa nappe, Zermatt-Saas Fee, southwestern Switzerland: Bull. Geol. Soc. Amer. 85, 907-910.

Cornelius, H. P., 1935, Geologie der Err Julier Gruppe. Teil I, Das Baumaterial: Beitr. Geol. Karte der Schweiz NF 70.

Deutsch, A., 1983, Datierungen an Alkaliamphibolen und Stilpnomelan aus der südlichen Platta-Decke (Graubünden): Eclogae geol. Helv. 76, 295-308.

Dietrich, V., 1969, Die Ophiolithe des Oberhalbsteins (Graubünden) und das Ophiolithmaterial der ostschweizerischen Molasseablagerungen, ein petrographischer Vergleich: Europ. Hochschulschr. 17/1, Lang & Co., Bern.

Dietrich, V., Vuagnat, M., and Bertrand, J., 1974, Alpine metamorphism of mafic rocks: Schweiz. mineral. petrogr. Mitt. 54, 291-332.

Dietrich, V., and Oberhänsli, R., 1976, Der Gabbro der Jufer-Horen (Avers, GR): Schweiz. mineral. petrogr. Mitt. 56, 481-500.

Ernst, W. G., and Dal Piaz, G. V., 1978, Mineral parageneses of eclogitic rocks and related mafic schists of the Piemonte ophiolite nappe, Breuil-St. Jacques area, Italian Western Alps: Am. Miner. 63, 621-640.

Frey, M., Hunziker, J. C., Roggweiler, P. and Schindler, C., 1973, Progressive niedriggradige Metamorphose glaukonitführender Horizonte in den helvetischen Alpen der Ostschweiz: Contr. Min. Petrol. 39, 185-218.

Ganguin, J., 1983, Etude géologique et pétrographique de la Täschalp: Unpubl. diploma thesis Univ. Bern.

Gansser, A., 1937, Der Nordrand der Tambodecke: Schweiz. mineral. petrogr. Mitt. 17, 291-523.

Grubenmann, U., 1906, Über einige schweizerische Glaukophangesteine: Festschr. H. Rosenbusch Ref.: Eclogae geol. Helv. 9 (1907), 612-614.

Heim, A., 1910, Über die Stratigraphie der autochtonen Kreide und des Eozäns am Kistenpass: Beitr. geol. Karte der Schweiz NF 24, 21-45.

Heinrich, C. A., 1982, Kyanite-eclogite to amphibolite facies evolution of hydrous mafic and pelitic rocks, Adula nappe, Central Alps: Contr. Min. Petrol. 81, 30-38.

Heugel, W., 1975, Die Geologie des Piz Mundin: Unpubl. diploma thesis Univ. Bern.

Hunziker, J. C., 1969, Rb-Sr-Altersbestimmungen aus den Walliser Alpen. Hellglimmer und Gesamtgesteinsalterswerte: Eclogae geol. Helv. 62, 527-542.

Hunziker, J. C., 1974, Rb-Sr and K-Ar age determination and the Alpine tectonic history of the Western Alps: Mem. Ist. Geol. Min. Univ. Padova vol. XXXI, 1-55.

Jäger, E., 1973, Die alpine Orogenese im Lichte der radiometrischen Altersbestimmungen: Eclogae geol. Helv. 66, 11-21.

Klein, H. H., 1976, Metamorphose von Peliten zwischen Rheinwaldhorn und Pizzo Paglia (Adula-und Simano Decke): Schweiz. mineral. petrogr. Mitt. 56, 457-479.

Mange, M., and Oberhänsli, R., 1982, Detrital lawsonite and blue sodic amphibole in the Molasse of Savoy, France and their significance in assessing Alpine evolution; Schweiz. mineral. petrogr. Mitt. 62, 415-436.

Maurer, H., 1982, Oberflächentexturen an Schwermineralien aus der unteren Süsswassermolasse (Chattien) der Westschweiz: Eclogae geol. Helv. 75, 23-31.

Maurer, H., Gerber, E., Nabholz, W., and Funk, H. P., 1982, Sedimentpetrographie und Lithostratigraphie der Molasse im Einzugsgebiet der Langete (Aarwangen - Napf, Oberaargau): Eclogae geol. Helv. 75, 381-413.

Meyer, J., 1983, Mineralogie und Petrologie des Allalingabbros: Unpubl. Ph. D. thesis Univ. Basel.

Miyashiro, A., 1957, The chemistry, optics and genesis of the alkali amphiboles: J. Fac. Sci. Univ. Tokyo II/11, 57-83.

Niggli, E., 1960, Mineralzonen der alpinen Metamorphose in den Schweizer Alpen: Int. Geol. Congr. XXI Sess., Norden, Copenhagen part B, 132-138.

Niggli, E., 1970, Alpine Metamorphose und alpine Gebirgsbildung: Fortschr. Miner. 47, 16-26.

Niggli, E., 1973, Metamorphic map of the Alps. 1:1'000'000: UNESCO, Paris.

Niggli, E., Brückner, W., and Jäger, E., 1956, Über Vorkommen von Stilpnomelan und Alkali-Amphibol als Neubildung der alpidischen Metamorphose in nordhelvetischen Sedimenten am Ostrande des Aarmassives: Eclogae geol. Helv. 49, 469-480.

Niggli, E. and Niggli, C., 1965, Karten der Verteilung einiger Mineralien der alpidischen Metamorphose in den Schweizer Alpen (Stilpnomelan, Alkali-Amphibol, Chloritoid, Staurolith, Disthen, Sillimanit): Ecolgae geol. Helv. 58, 335-368.

Oberhänsli, R., 1977, Natriumamphibol-führende metamorphe, basische Gesteine aus den Bündnerschiefern Graubündens: Unpubl. Ph. D. thesis ETH-Z Nr. 5982, Zürich.

Oberhänsli, R., 1978, Chemische Untersuchungen an Glaukophan-führenden basischen Gesteinen aus den Bündnerschiefern Graubündens: Schweiz. mineral. petrogr. Mitt. 58, 139-156.

Oberhänsli, R., 1980, P-T Bestimmungen anhand von Mineralanalysen in Eklogiten und Glaukophaniten der Ophiolite von Zermatt: Schweiz. mineral. petrogr. Mitt. 60, 215-235.

Oberhänsli, R., 1982, The P-T history of some pillow lavas from Zermatt: Ofioliti 2/3, 431-436.

Phillipp, R., 1982, Die Alkaliamphibole der Platta-Decke zwischen Silsersee und Lunghinpass (Graubünden): Schweiz. mineral. petrogr. Mitt. 62, 437-455.

Plas, L. van, 1959, Petrology of the northern Adula region, Switzerland: Leidse Geol. Meded. 24, 415-602.

Schindler, C., 1959, Zur Geologie des Glärnisch: Beitr. geol. Karte der Schweiz NF 108.

Schindler, C., 1969, Neue Aufnahmen in der Axen-Decke beidseits des Urner-Sees: Eclogae geol. Helv. 62, 155-171.

Schär, J. P., 1959, Géologie de la partie septentrionale de l'évantail de Bagnes: Arch. Sci. Genève 12, 473-620.

Schürmann, H.M.E., 1953, Beiträge zur Glaukophanfrage II: N. Jb. Min. Abh. 87, 303-394.

Teutsch, R., 1982, Alpine Metamorphose der Misoxer-Zone: Unpubl. Ph.D. thesis Univ. Bern.

Thompson, P. H., 1976, Isograd patterns and pressure-temperature distributions during regional metamorphism; Contr. Min. Petrol. 57, 277-295.

Wenk, E., 1962, Plagioklas als Indexmineral in den Zentralalpen. Die Paragenese Calzit-Plagioklas: Schweiz. mineral. petrogr. Mitt. 42, 139-152.

Woyno, T. J., 1912, Petrographische Untersuchungen der Casannaschiefer des mittleren Bagnestales (Wallis): N. Jb. Min. Beil. Bd. 33, 136-209.

MANUSCRIPT ACCEPTED BY THE SOCIETY JULY 29, 1985

Geological Society of America
Memoir 164
1986

Early Alpine eclogite metamorphism in the
Penninic Monte Rosa-Gran Paradiso basement nappes
of the northwestern Alps

Giorgio V. Dal Piaz
Istituto di Geologia
via Giotto 1
Padova, Italy

Bruno Lombardo
Centro di Studio sui problemi dell'orogeno
delle Alpi Occidentali
via Accademia delle Scienze 5
Torino, Italy

ABSTRACT

Eclogites occur in the Monte Rosa and Gran Paradiso Penninic nappes of the Western Alps as small bodies within metasedimentary rocks and, in Monte Rosa, rarely, as boudinaged dikes within late-Variscan granitoids. Minor but characteristic constituents of these eclogites are glaucophane and paragonite. Bulk composition of the protoliths ranges from basaltic to Fe-Ti basaltic.

High-pressure assemblages comprising chloritoid, phengite, garnet ± kyanite, talc, Mg-chlorite and glaucophane also occur in micaschists of both nappes. The Early Alpine high-pressure assemblages postdate the intrusion of late-Variscan granitoids and are overprinted by the greenschist, mid-Tertiary Lepontine metamorphism.

Garnet-clinopyroxene geothermometry and phase compatibilities suggest high-pressure metamorphic temperatures ranging from 440°C to 530°C in the Monte Rosa nappe, with pressure in excess of 8-10 Kb but lower than approximately 14 Kb. These conditions are in the same range as established for the overlying Piedmont ophiolite nappe and for parts of the Austroalpine basement nappe, implying a similar tectonometamorphic evolution during Early Alpine time. Garnet-omphacite pairs from the amphibole eclogites of the Gran Paradiso nappe yield lower equilibration temperatures (T ≃400°C), possibly recording a later stage in the P-T-t path of this unit.

INTRODUCTION

A specific feature of the Western Alps is the occurrence of a high-pressure metamorphism of Cretaceous age (Dal Piaz and others, 1972; Hunziker, 1974) not only in metasedimentary sequences and in rock derived from former ocean crust but also in large, coherent tracts of continental basement. In the Austroalpine (Paleo-African) Sesia-Lanzo unit (Dal Piaz and others, 1972; Compagnoni and others, 1977), the continental basement was subjected to eclogite-facies metamorphism under such high pressures that jadeite-bearing, plagioclase-free parageneses are ubiquitous in former late-Variscan granitic rocks and pre-granitic gneisses of high grade. In the Penninic (Paleoeuropean) Monte Rosa-Gran Paradiso nappes, eclogites are associated with meta-granites and polymetamorphic gneisses in which the conditions for plagioclase breakdown were generally not attained.

It is the purpose of this contribution to summarize new data on the eclogites of the Monte Rosa-Gran Paradiso nappes, review the evidence of high-pressure transformations in the country rocks and, finally, discuss the distribution and tectonic implications of this type of metamorphism in the Northwestern Alps.

GEOLOGICAL SETTING AND LITHOSTRATIGRAPHY OF THE MONTE ROSA-GRAN PARADISO BASEMENT NAPPES

Metamorphic rocks with HP assemblages are exposed in the Western Alps along an arcuate belt (Fig. 1) bordering the Piedmont plain, from the Sestri-Voltaggio Line, which divides the eclogitized meta-ophiolites of the Gruppo di Voltri from the anchimetamorphic units of the Northern Apennines, to the Ossola-Tessin gneiss region, where they are progressively obliterated by the younger Lepontine metamorphism of mid-Tertiary age (Frey and others, 1974; Ernst, 1973).

Unlike the HP metamorphism in other blueschist belts such as the California Coast Ranges or New Caledonia, the HP metamorphism in the Western Alps affected large, coherent bodies of the pre-Alpine continental crust, which are now exposed as a pile of basement nappes, each surrounded by an envelope of thrust sheets consisting of dismembered ophiolite complexes and Mesozoic metasediments. The tectonic history leading to such an intricate arrangement of tectonic units with widely different paleogeographical provenances and histories began in Early Mesozoic times with the rifting of the Variscan continent, the development of a Paleoeuropean (Helvetic and Penninic) margin and of a Paleoafrican (Austroalpine) margin of thinned and dismembered continental crust and, finally, with the Jurassic opening of the Western Tethys ocean basin. During the Cretaceous, most of the oceanic crust (now represented by the Piedmont ophiolite nappe system) and some parts of both continental margins were subjected to high-pressure, low- to medium-temperature conditions, presumably in a subduction zone dipping under the Paleoafrican margin (Ernst, 1971; Dal Piaz and others, 1972). The continent-continent collision ensuing from the consumption of the intervening ocean basin eventually arrested the subduction process and allowed the return of subducted material to shallower crustal levels. Both subduction and tectonic exhumation were accompanied by strong imbrication of slices of thinned continental crust with sheets derived from the Western Tethys oceanic crust and mantle.

Continuing ductile deformation in the overthrust nappe pile, first under HP conditions and later under greenschist- to amphibolite-facies conditions during the Lepontine metamorphism, absorbed much of the subsequent shortening and caused the folding of the nappe boundaries so clearly displayed in the classical structure profiles of Argand (1911).

The intrusion of quartz diorite and monzonite bodies and strong uplift ended the plutonic history of the Western Alps in Late Oligocene time. The uplift is still continuing and much of the topographical relief we see today is probably not older than the Pliocene.

Lithostratigraphy of the Monte Rosa Nappe

The Monte Rosa basement complex consists of high-grade metapelites with rare intercalations of basic rocks, intruded by porphyritic granites, and of the younger Furgg Formation, a strongly deformed and heterogeneous sequence which crops out at different structural levels in the nappe and comprises albite gneisses and micaschists with bodies of metabasic rocks and marbles (Bearth, 1952, 1953, 1954, 1957, 1958; Dal Piaz, 1966, 1971; Laduron, 1976; Wetzel, 1972). These lithologies are partly to completely overprinted by polyphase Alpine mineral assemblages and related deformations. The metapelites crop out mostly in the southern and western parts of the nappe (Fig. 2), whilst the augen gneisses deriving from the porphyritic granites are preponderant in the eastern part.

Typical relict assemblages in the metapelites are biotite-sillimanite-garnet-quartz-K-feldspar-plagioclase and garnet-biotite-muscovite-sillimanite-plagioclase (Bearth, 1952, 1957; Dal Piaz, 1971; Laduron, 1976). Cores of brown hornblende which occur exceptionally in amphibolites of the Vanzone area, Valle Anzasca (Laduron, 1976, p. 34–36) are the only mineral relics of high grade in the metabasic rocks.

The metapelites are veined by streaks and pods of pegmatite with fibrolite and nodules of altered cordierite. Planar fabrics and folds developed during the high-grade metamorphism recorded by these assemblages and are transected by discordant bodies and dikes of granite, as shown by the superb rock exposures on the western ridge of the Dufour Spitze, on the southern face of the Castore, and on the eastern face of Monte Rosa (Bearth, 1952; Dal Piaz, 1966).

The granitic rocks comprise coarse-grained types and younger aplites. The first were emplaced at 310 ± 20 Ma, as shown by a total-rock Rb-Sr isochron on six samples, whilst the intrusion of the latter occurred at 260 Ma, contemporaneously with a late-Variscan thermal event (Hunziker, 1970, 1974; Frey and others, 1976). The high-grade metamorphism of the metapelites predates the emplacement of both granite types and is thus of Variscan age or older (Bearth, 1952).

Both the metapelites and the Variscan granites were subjected to a HP metamorphism of Cretaceous age and then to a pervasive medium-grade event in the mid-Tertiary. Evidence for the first event will be discussed in this paper and is given by (1) the occurrence of eclogites in the metapelites, in the Variscan granites and in the Furgg Formation, (2) the overprinting of the high-grade assemblages by HP mineral transformations and the conversion of granite rocks along shear zones into chloritoid-phengite $\pm$ kyanite schists (Dal Piaz, 1971), (3) Rb-Sr isotope re-homogenization at 125 ± 20 Ma in albitized granite gneisses sited in the frontal (northwestern) part of the nappe (Hunziker, 1970; Frey and others, 1976).

The mid-Tertiary Lepontine event culminated at 38 Ma (Hunziker, 1969) and reached the highest P-T conditions (amphibolite facies) in the northeastern (and deepest) parts of the nappe (Bearth, 1958; Laduron, 1976), whilst in the rest of the nappe, mineral assemblages belong to the greenschist facies (Bearth, 1952; Dal Piaz, 1964, 1966).

Lithostratigraphy of the Gran Paradiso Nappe

Like Monte Rosa, the structurally equivalent Gran Paradiso

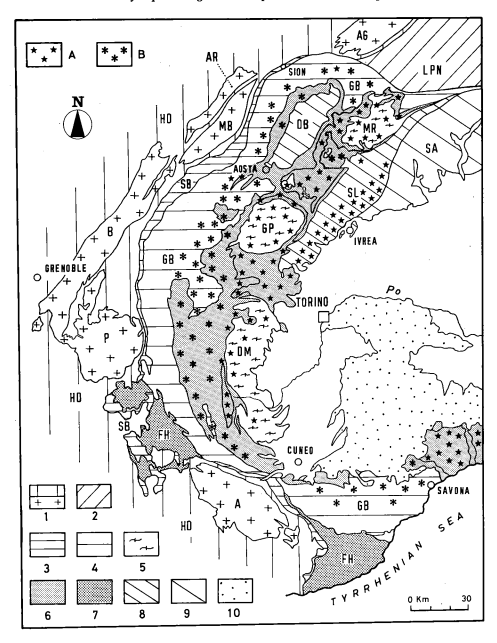

Figure 1. Tectonic map of the Western Alps showing the location of the Monte Rosa-Gran Paradiso nappes and the generalized distribution of the HP-LT assemblages. Units derived from the Paleoeuropean margin: 1) Helvetic-Dauphinois zone, including decollement nappes and sedimentary covers (HD) and the uplifted basement massifs of Argentera (A), Pelvoux (P), Belledonne (B), Mont Blanc (MB), Aiguilles Rouges (AR) and Aar-Gotthard (AG); 2) Lower Penninic nappes of the Ossola-Tessin region (LPN); 3) Subbriançonnais units and Sion-Courmayeur zone (SB); 4) Gran San Bernardo nappe (GB); 5) Upper Penninic Monte Rosa (MR), Gran Paradiso (GP) and Dora Maira (DM) nappes. Units derived from the Piedmont-Ligurian oceanic basin and related marginal edges: 6) Composite Piedmont ophiolite nappe, including the Gruppo di Voltri (GV); 7) Helmintoides flysch units (FH). Units derived from the Paleoafrican margin: 8) Austroalpine Sesia-Lanzo (SL) and Dent Blanche (DB) nappes; 9) Southern Alps (SA), including the Ivrea zone. 10) Late-orogenic sedimentary sequences of Monferrato and Langhe, south of Torino. High pressure assemblages: A) eclogite and B) low- to high-grade blueschists.

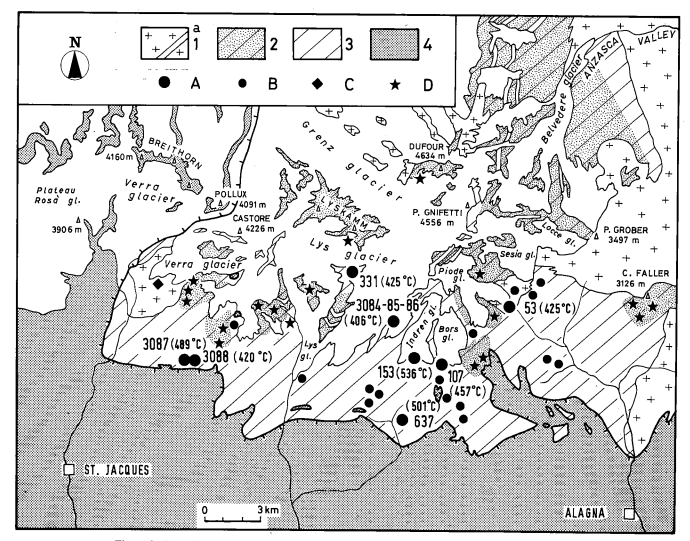

Figure 2. Generalized map of the southwestern part of the Monte Rosa nappe and distribution of Early-Alpine assemblages. Monte Rosa nappe: 1) Augen gneisses, metagranites and (a) large dikes from late-Variscan intrusives; 2) Pre-granitic high-grade metapelites weakly overprinted by the Alpine metamorphic events; 3) High-grade metapelites partly to completely reworked by the Early-Alpine and Lepontine events, including the undifferentiated Furgg Formation; 4) Composite Piedmont ophiolite nappe. Early-Alpine assemblages: A) location of the analyzed eclogite metabasites (in brackets: temperatures estimated for garnet-omphacite pairs from the calibration of Ellis and Green, 1979, assuming a nominal pressure of 10 Kb; B) other occurrences of eclogite metabasites; C) phengite-chloritoid-Mg-chlorite-kyanite assemblages in mylonitic granites; D) kyanite pseudomorph after sillimanite and kyanite-phengite-garnet assemblages in metapelites.

nappe consists of a pre-granitic complex of gneisses and meta-basic rocks, of Alpine augen gneisses derived from Variscan porphyritic granites, and of rare younger metasedimentary sequences (Fig. 3). The relics of high-grade assemblages in the gneisses, however, are less abundant than in Monte Rosa, occurring only at a few localities in the northern part of the nappe (Compagnoni and others, 1974). Typical assemblages are: biotite-plagioclase (oligoclase or andesine)-K-feldspar-fibrolite and biotite-muscovite-plagioclase-garnet-sillimanite-andalusite (?) (Callegari and others, 1969; Compagnoni and Prato, 1969; Compagnoni and others, 1974). Relict amphibolites containing a brown horn-

blende of composition $(K_{0.1})$ $(Na_{0.2}Ca_{1.7})$ $(Mg_{3.3}Fe''_{1.2}Mn_{0.1}$ $Ti_{0.1}Al_{0.4}^{VI})$ $(Al_{0.8}^{IV} Si_{7.2})$ occur locally in the central part of the nappe (Lombardo, unpubl. data). Contact metamorphism of the metapelites adjacent to the Variscan granites is recorded by rare relics of hornfelses containing red-brown biotite, plagioclase, K-feldspar-sillimanite, andalusite (?), corundum and green spinel (Callegari and others, 1969; Compagnoni and Prato, 1969; Lombardo, 1970).

The younger sequences are the Bonneval Gneiss and the Money Complex. The Bonneval Gneiss is a series of K-feldspar-albite-phengite gneisses possibly deriving from late-Paleozoic acid

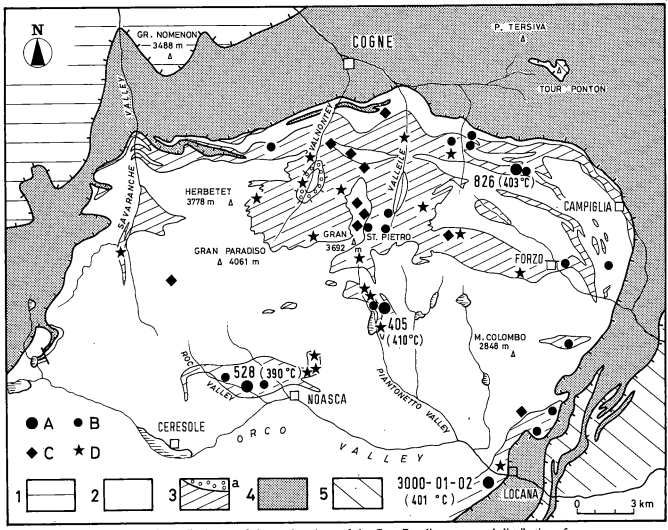

Figure 3. Generalized map of the northern part of the Gran Paradiso nappe and distribution of Early-Alpine assemblages. 1) Gran San Bernardo nappe; Gran Paradiso nappe: 2) Augen gneisses and metagranites from late-Variscan intrusives; 3) Pre-granitic high-grade metapelites weakly to completely overprinted by the Early-Alpine and Lepontine tectonometamorphic events; 3a) late-Paleozoic meta-clastic deposits of Money complex; 4) Composite Piedmont ophiolite nappe; 5) Austroalpine Sesia-Lanzo unit and Tour Ponton Klippe. Early-Alpine metamorphism: A) location of the analysed eclogite metabasites (in brackets: temperatures estimated for garnet-omphacite pairs from the calibration of Ellis and Green, 1979, assuming a nominal pressure of 10 Kb); B) other occurrences of eclogite metabasites; C) phengite-chloritoid -Mg-chlorite ± kyanite ± glaucophane assemblages in mylonitic granites; D) kyanite pseudomorphs after sillimanite and phengite-chloritoid ± kyanite ± garnet assemblages in metapelites.

volcanics and volcaniclastic sediments (Bertrand, 1968) or from late-Variscan leucogranites, which crop out only in the southwestern part of the nappe. The metaclastic Money complex (Compagnoni and others, 1974) comprises albite micaschists and gneisses with horizons of quartz-rich metaconglomerates and is probably of Late Carboniferous age. Neither sequence shows evidence of high-grade metamorphism.

The tectonic and metamorphic history of the Gran Paradiso basement during the Alpine orogeny is rather similar to that of Monte Rosa, with a first event producing eclogites and HP transformations in the high-grade metapelites and the Variscan gran-

ites, followed by extensive overprinting of the earlier assemblages under greenschist-facies conditions in late Eocene time (Chopin and Maluski, 1978, 1980; Krummenacher and Evernden, 1960). The age of the HP metamorphism is Early-Alpine, as HP phengites apparently unaffected by the Eocene event give Ar^{39}-Ar^{40} ages around 60-75 Ma (Chopin and Maluski, 1978, 1980). By analogy with the HP phengites of the Sesia-Lanzo unit (Hunziker, 1974; Oberhänsli and others, 1982, 1985) these ages are probably to be considered as cooling ages, dating the closure of the phengite system to Ar diffusion and only setting an upper limit to the age of the HP metamorphism.

TABLE 1. XRF BULK ROCK ANALYSES OF ECLOGITES
FROM THE MONTE ROSA-GRAN PARADISO BASEMENT NAPPES

Sample No.	MRO 3084	MRO 3085	MRO 3087	MRO 3088	MRO 3000	MRO 3001	MRO 3002	GP 405
SiO_2	46.20	45.16	49.30	48.01	47.65	46.14	49.18	48.54
TiO_2	4.40	4.19	1.88	2.06	2.53	1.51	1.25	2.36
Al_2O_3	12.92	13.40	17.78	14.73	13.97	13.75	14.72	14.02
Fe_2O_3	2.83	6.28	3.06	4.04	4.41	1.84	2.16	1.72
FeO	14.27	11.64	7.14	6.99	6.88	9.02	6.67	13.99
MnO	0.28	0.32	0.19	0.23	0.21	0.18	0.18	0.34
MgO	5.27	4.82	6.14	7.78	6.00	8.88	6.72	5.65
CaO	8.21	9.21	10.60	10.35	12.00	9.42	12.31	9.14
Na_2O	3.17	3.37	4.48	4.03	4.52	4.65	5.07	3.47
K_2O	0.22	0.19	0.05	0.10	0.28	0.08	0.03	0.06
P_2O_5	0.66	0.76	0.06	0.30	0.28	0.05	0.07	0.40
L.O.I.	1.34	0.45	1.26	1.20	1.38	4.55	1.50	0.80
Total	99.77	99.79	101.94	99.82	100.11	100.07	99.86	100.49

Sample location (Figures 2 and 3):
Monte Rosa: 3084, 3085 - Capanna Gnifetti, Gressoney valley; 3087, 3088 - Pian di
Verra, Ayas valley.
Gran Paradiso: 3000, 3001, 3002 - Montigli (Locana); 405 - Bocchetta di Valsoera
(Piantonetto valley).

ECLOGITES OF THE MONTE ROSA - GRAN PARADISO NAPPES

Eclogites and omphacite rocks were discovered in the Monte Rosa nappe by Franchi (1903) during the first detailed mapping of the Italian Western Alps and are mentioned as "eclogites, chloromelanitites and jadeitites" in the legends of Monte Rosa, Varallo and Domodossola sheets of the 1:100.000 scale Geological Map of Italy (Franchi and others, 1912a, 1927; Novarese and Stella, 1913). The eclogites found by Franchi occur as boudins and small bodies in the garnet micaschists of the southern slope of Monte Rosa (Fig. 2) and are extensively altered to garnet-bearing albite amphibolites. Other eclogite bodies, probably boudinaged dikes, were found by Dal Piaz and Gatto (1963) in the granitic orthogneisses of the upper Sesia valley (Fig. 2). Layers and boudins up to a few meters long of garnet-epidote amphibolite with pods of relict eclogite are common in the Furgg Formation (Dal Piaz, 1966; Wetzel, 1972), but altered eclogites have been reported also from the pre-granitic paragneisses of the eastern ridge of Monte Rosa (Bearth, 1952, p. 79) and of the Portjengrat zone, the northernmost tectonic element of the Monte Rosa nappe (Wetzel, 1972; Klein, 1978).

Eclogites were found in the Gran Paradiso region much later than in Monte Rosa (Prato, 1971; Compagnoni and Lombardo, 1974; Pennacchioni, 1982-83) and the metabasic rocks of this tectonic unit are described only as amphibolites by Michel (1953) and in the legend of sheet Gran Paradiso (Franchi and others, 1912b). The Gran Paradiso eclogites occur as boudins, boudinaged bands and sub-concordant bodies in the paragneisses of the pre-granitic complex. Similar to the Monte Rosa occurrences, completely unaltered eclogite is rare, and the metabasic rocks of this complex are more characteristically garnet-bearing albite amphibolites in which the omphacite is partly to completely re-

placed by symplectites of albite and actinolite, or more commonly, they are albite-epidote-biotite amphibolites (Prato, 1971; Compagnoni and Lombardo, 1974). The largest metabasic body preserving eclogite assemblages is 2.5 km long and 100 m thick and crops out in the Roc Valley (Prato, 1971).

The Monte Rosa-Gran Paradiso eclogites are currently being re-investigated (Dal Piaz and Lombardo, in progress) and some petrological results of this study will be summarized here. The location of the eclogite bodies examined by these authors is shown in Figs. 2 and 3.

Bulk compositions of eight Monte Rosa-Gran Paradiso eclogites are mostly basaltic (Table 1 and Fig. 4), but some eclogite bodies are enriched in Fe, Ti and P and may derive from ferrobasaltic protoliths. These iron-rich varieties occur both in the Monte Rosa nappe (Capanna Gnifetti: samples MRO 3084-3085) and in the Gran Paradiso nappe (Piantonetto valley: GP 405). Garnet-clinopyroxene rocks with abundant epidote ± quartz, carbonate and scheelite, which are possibly Ca-rich volcaniclastic layers of pre-Variscan age, occur in the Noasca-Roc Valley area of the Gran Paradiso nappe (Zago, 1981).

The typical mineral assemblage in both tectonic units is garnet-omphacite-amphibole-rutile-white mica ± quartz ± zoisite ± Cu-Fe sulfides. Banded garnet-omphacite-zoisite eclogites occur locally in the Gran Paradiso nappe (M. Ballèvre, pers. comm.), whilst epidote and calcite are major constituents in some parts of the eclogite body of the Roc Valley. Hornblende, albite, chlorite, yellow-brown biotite and sphene are abundant in the more retrogressed varieties. Metabasic rocks consisting of colourless amphibole with cores of brown hornblende, garnet, rutile ± epidote and chlorite, occur both in Monte Rosa (Laduron, 1976, p. 36) and in Gran Paradiso (Lombardo, 1970). The close similarity of these rocks to pre-Alpine amphibolites re-equilibrated under eclogite conditions (Ungaretti and others, 1983) suggests

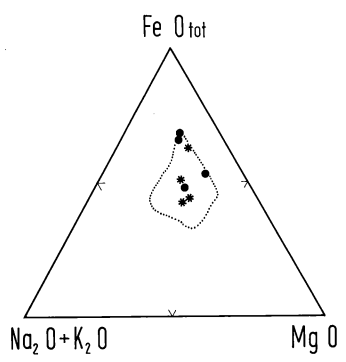

Figure 4. (Na$_2$O + K$_2$O) - FeO$_{tot}$ - MgO diagram for the eclogites from Monte Rosa (dots) and Gran Paradiso (stars). The composition field of eclogites from the Austroalpine Sesia-Lanzo unit is shown for comparison.

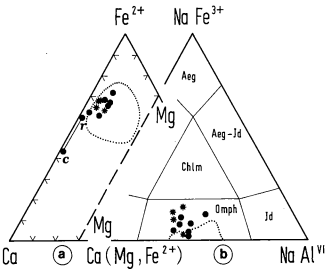

Figure 5. Composition plots of garnets (a) and omphacites (b) from the eclogites of Monte Rosa (dots) and Gran Paradiso (stars); c and r are the compositions of cores and rims of garnet from a fine-grained orthogneiss of the Gran Paradiso nappe (Tartarotti, 1984). The dotted lines outline the composition fields of garnets (a) and omphacites (b) from eclogites of the Austroalpine Sesia-Lanzo unit (Ungaretti and others, 1983). The dashed line (a) outlines the composition boundary preferred by Coleman and others (1965) for garnets from type-C eclogites (Smulikowski, 1964; Coleman and others, 1965). Composition boundaries in the sodic pyroxene triangle (b) are after Essene and Fyfe (1967): omph = omphacite, jd = jadeite, aeg = aegirine, chlm = chloromelanite.

that this mineral assemblage may be actually coeval with the more abundant garnet-omphacite assemblages.

The electron microprobe study of representative samples (Tables 2 and 3) has shown that the garnets are almandines (Alm 60-70 mole percent) with high grossular (20-37%), low pyrope (3-10%) and very low spessartine contents (1-3%) (Fig. 5). The omphacites have jadeite contents ranging from 34 mole percent (GP528, Roc Valley, Gran Paradiso) to 49 mole percent (MRO 637, Alpe Indren, and MR03088, Capanna Gnifetti, Monte Rosa) and acmite contents up to 14 mole% (GF 528, Gran Paradiso) but mostly between 3 and 10 mole % (Fig. 5). Fe-Mg partition coefficients in coexisting garnet-omphacite pairs are distinctly high, ranging from 55 to 85 in the Gran Paradiso eclogites and from 22 to 60 in those from the Monte Rosa nappe (Tables 2 and 3).

The amphibole is a glaucophane of the pale-colored aluminous variety known as "gastaldite" in the older literature (Fig. 6) in most eclogite bodies, but a colorless variety of sodic-calcic composition in eclogites GP280 (same locality as GP 826), MR03088 (Capanna Gnifetti) and MR0331 (Lyskamm Nose, Monte Rosa). According to the I.M.A. classification (Leake, 1978), these latter amphiboles are barroisite in the eclogites from Monte Rosa and winchite in the Gran Paradiso eclogite, both types being characteristic of rather high-pressure conditions (Fig. 6). Similar, but less sodic and aluminous, compositions occur as colourless patches and rims in the brown amphiboles of

the relict amphibolites from Gran Paradiso (Lombardo, unpub. data).

The white mica is paragonite in all analyzed eclogites, with approximately 0.6 wt % K$_2$O, corresponding to a muscovite-paragonite solution of composition Pa$_{95}$Mu$_5$.

Textural evidence suggests that in some eclogite bodies glaucophane and paragonite may have been in equilibrium with garnet and omphacite, but in most Monte Rosa-Gran Paradiso eclogites, glaucophane appears to have grown later than omphacite and garnet, yet before the breakdown of omphacite into albite-amphibole symplectite. A later stage in the amphibole evolution is represented by iron-rich blue-green barroisite, mantling glaucophane and coexisting with albite; this Na-Ca amphibole evidently formed from the breakdown of garnet and omphacite under lower pressure conditions.

HP TRANSFORMATIONS IN THE METASEDIMENTS AND IN THE ORTHOGNEISSES

As in other nappes of the Western Alps, strain heterogeneity during the Alpine metamorphism preserved in the Monte Rosa-Gran Paradiso nappe portions of the pre-Alpine basement in which the deformation at the microscopic scale was very weak or nonexistent and new minerals grew only as reaction rims around, or as pseudomorphs, on the pre-Alpine minerals. These relict portions occur within volumetrically much larger tectonite rocks,

TABLE 2. COMPOSITIONAL PARAMETERS OF GARNET-OMPHACITE PAIRS FROM THE MONTE ROSA
ECLOGITES, Fe^{2+}/Mg PARTITION COEFFICIENTS AND TEMPERATURES ESTIMATED FROM THE
CALIBRATION OF ELLIS AND GREEN (1979) FOR P = 10 Kb

Sample No. MRO	53	107	153	331	637	3086	3087	3088	T16
Garnet									
$X_{Fe^{2+}}$	0.62	0.65	0.61	0.65	0.67	0.69	0.60	0.66	0.61
X_{Mg}	0.03	0.10	0.09	0.10	0.07	0.09	0.07	0.13	0.07
X_{Mn}	0.01	0.02	0.01	0.01	0.01	0.03	0.01	0.01	–
X_{Ca}	0.340	0.234	0.290	0.240	0.250	0.192	0.324	0.200	0.320
Omphacite									
X_{Al}	0.47	0.39	0.46	0.39	0.49	0.40	0.44	0.49	0.43
$X_{Fe^{3+}}$	0.02	0.11	0.03	0.05	0.13	0.08	0.06	0.05	0.08
$X_{Fe^{2+}}$	0.13	0.09	0.12	0.08	0.11	0.09	0.10	0.06	0.14
X_{Mg}	0.38	0.41	0.39	0.48	0.27	0.43	0.40	0.40	0.35
K_D^{Fe-Mg}	60.4	29.6	22.0	39.0	23.5	36.6	34.3	33.8	21.8
$T^{o}C$	425	457	536	425	501	406	489	420	556

Sample location in Figure 2; T16: eclogite from Furggtal, Switzerland (Wetzel, 1972).

in which the original textures have been partly to completely obliterated by pervasive deformation, and relict minerals are not present or occur only as porphyroclasts within new foliations defined by the Alpine minerals. Such local preservation of earlier textural relations and mineral assemblages also occurred in the different stages of the Alpine metamorphic evolution and offers the possibility of deciphering the characteristics of the earlier HP metamorphism through the mask of the Lepontine recrystallizations.

HP Transformations in the Pre-granitic Gneiss Complex

These transformations have been described by Bearth (1952) and Dal Piaz (1971) in the Monte Rosa nappe and by Compagnoni and Prato (1969) and Compagnoni and Lombardo (1974) in the Gran Paradiso nappe. Static transformations (Fig. 7) comprise (1) the replacement of prismatic sillimanite by pseudomorphs of fine-grained kyanite, (2) the replacement of an unknown Al-silicate by brown pseudomorphs of cryptocrystalline kyanite + garnet ± chloritoid, (3) the growth of garnet rims around pre-Alpine garnet and biotite, especially at quartz-biotite interfaces, and (4) the replacement of biotite by white mica + rutile + opaques. Cordierite is replaced in the pegmatite streaks and pods by pseudomorphs of fine-grained mica (talc?) + garnet + muscovite. Jadeite pseudomorphs after plagioclase, such as those described in similar high-grade metapelites of the Austroalpine Sesia-Lanzo unit (Dal Piaz and others, 1972, 1973), are not present and the original plagioclase may be still preserved.

Mineral assemblages in dynamically-recrystallized paragneisses (Bearth, 1952; Dal Piaz, 1964, 1966, 1971; Prato, 1971) comprise phengite, chloritoid, garnet, kyanite and, rarely, also glaucophane as characteristic minerals of the HP stage, whilst

new biotite and porphyroblastic albite are typical of the Lepontine overprinting.

HP Transformations in Pelitic Hornfelses of the Gran Paradiso Nappe

The static transformations induced in these rocks by the HP metamorphism are very similar to those already described for the high-grade metapelites: (1) prismatic sillimanite is replaced by fine-grained kyanite, (2) pseudomorphs of cryptocrystalline kyanite + garnet replace an Al-silicate, which was possibly andalusite according to Compagnoni and Prato (1969) but was more probably cordierite, (3) biotite is partly replaced by coronite garnet and white mica, and (4) garnet rims grow around the opaques and corundum (Compagnoni and Prato, 1969; Callegari and others, 1969; Lombardo, 1970). As in the paragneisses, the original plagioclase (An_{23-37}) is locally preserved.

HP Minerals in the Bonneval Gneiss Complex

Aegirine and riebeckite occur at a few localities in quartz-rich metasedimentary layers of the Bonneval Gneiss (Saliot, 1978; Vearncombe, 1983) and jadeite of composition $Na_{0.98}Ca_{0.02}Al_{0.96}Fe'''_{0.02}Fe''_{0.01}Mg_{0.01}Si_2O_6$ was reported by Saliot (1979) from a leucocratic rock also containing quartz, K-feldspar, two generations of phengite, green biotite, albite and aegirine. The green biotite is not in equilibrium with the first generation of phengite (Si = 3.60) but only with the second, less silicic (Si = 3.24) generation.

HP Transformations in Metagranites and Granitic Orthogneisses

No evidence was found so far in undeformed Monte Rosa-

TABLE 3. COMPOSITIONAL PARAMETERS OF GARNET-OMPHACITE PAIRS FROM THE GRAN PARADISO ECLOGITES, Fe^{2+}/Mg PARTITION COEFFICIENTS AND TEMPERATURES ESTIMATED FROM THE CALIBRATION OF ELLIS AND GREEN (1979) FOR P = 10 Kb.

Sample No.	GP405	GP528	GP826	MRO3002
Garnet				
$X_{Fe^{2+}}$	0.64	0.64	0.58	0.62
X_{Mg}	0.05	0.08	0.03	0.10
X_{Mn}	0.01	0.01	0.02	0.01
X_{Ca}	0.302	0.275	0.371	0.270
Omphacite				
X_{Al}	0.42	0.34	0.42	0.45
$X_{Fe^{3+}}$	0.08	0.14	0.04	0.06
$X_{Fe^{2+}}$	0.09	0.06	0.10	0.05
X_{Mg}	0.41	0.46	0.44	0.44
K_D^{Fe-Mg}	58.3	61.3	85.0	54.6
T°C	410	390	403	401

Sample location in Figure 3; analysis of sample GP826 from Tartarotti (1984).

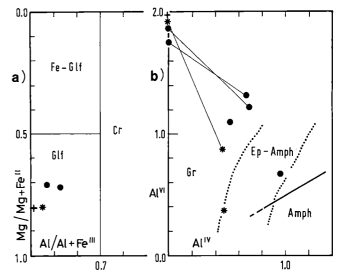

Figure 6. Composition of sodic and sodic-calcic amphiboles from the Monte Rosa-Gran Paradiso nappes in Miyashiro (1957) diagram (a) and the Al^{VI}/Al^{IV} diagram (b). Amphiboles from the Monte Rosa eclogites: dots; from the Gran Paradiso eclogites: stars; from a "silvery micaschist" (GP1083) of the Bardoney valley (Gran Paradiso): cross. Data from Dal Piaz, and Lombardo (in prep.) and Ungaretti and others (1981). Tie-lines in b) identify blue-green amphiboles rimming glaucophane. The solid line in b) separates low Al^{VI} (low pressure) amphibole compositions from high Al^{VI} (high pressure) amphibole compositions (Raase, 1974). The composition field of amphiboles from the greenschist facies (Gr), the epidote-amphibolite facies (Ep-Amph) and the amphibolite facies (Amph) are also shown (Zakrutkin, 1968). Note that only the left part of Miyashiro's diagram is shown in a); Glf = glaucophane; Fe-Glf = ferro-glaucophane; Cr = crossite.

Gran Paradiso metagranites of HP transformations like those described by Compagnoni and Maffeo (1973) and by Dal Piaz and others (1972, 1973) in the similar Mucrone metagranite of the Austroalpine Sesia-Lanzo unit, where the original plagioclase is replaced by pseudomorphs of jadeite + quartz + zoisite. Both in Monte Rosa and Gran Paradiso, the original plagioclase of the tectonically least affected metagranites apparently survived meta-stably the HP metamorphism or was simply replaced by saussurite or albite and sericite (Bearth, 1952; Bertrand, 1968; Callegari and others, 1969; Dal Piaz, 1971; Prato, 1971). Mineral assemblages of the orthogneisses in the Monte Rosa nappe quartz, microcline, albite + epidote (or oligoclase in the deepest parts), biotite, phengite (Si = 3.40), sphene and garnet of composition close to $Alm_{50}Gro_{50}$ (Bearth, 1952, 1958; Frey and others, 1976; Laduron, 1976). Rb-Sr mineral isochrons point to a mid-Tertiary age for these assemblages (Frey and others, 1976).

Mineral assemblages in the Gran Paradiso orthogneisses are the same as in Monte Rosa and likewise include a grossular-rich garnet ($Alm_{43}Sp_3Gro_{49}And_5$ to $Alm_{60}Sp_2Py_1Gro_{31}And_6$ at the rim) in garnet coexisting with albite from a fine-grained orthogneiss analyzed by Tartarotti (1984). As in Monte Rosa the isotopic age of these assemblages is mid-Tertiary (Chopin and Maluski, 1978, 1980; Krummenacher and Evernden, 1960), but some minerals, phengite in particular, are possibly isotopically re-equilibrated relics of the earlier HP event (Chopin and Maluski, 1980).

The "Silvery Micaschists"

The best evidence of HP transformations in granitic rocks is

given by these peculiar rocks which occur both in the Monte Rosa nappe (Bearth, 1952; Reinhardt, 1966; Dal Piaz, 1971) and in Gran Paradiso (Bertrand, 1968; Prato, 1971; Compagnoni and Lombardo, 1974), generally as lens-shaped bodies (boudinaged blastomylonites) within granitic orthogneisses, but also as concordant bands within metamorphosed volcaniclastic sequences of acid composition from the Bonneval Gneiss Complex (Bertrand, 1968). Fine-grained chloritoid-kyanite micaschists similar to the silvery micaschists, but commonly containing garnet, occur as thin layers within the micaschists derived from dynamically-recrystallized Monte Rosa and Gran Paradiso paragneisses (Bearth, 1952; Dal Piaz, 1971; Compagnoni and Lombardo, 1974).

Three outcrops of silvery micaschists (Mt. Séti, Lécharenne and Pointe des Eaux Rousses), all in the Gran Paradiso nappe, have been recently studied in detail by Chopin (1981) and the following description is mainly based on his work. The silvery micaschists of Mt. Séti and Pointe des Eaux Rousses contain the characteristic assemblage quartz-chlorite-chloritoid-talc-phengite, whilst garnet and glaucophane are major additional phases at Lécharenne. Kyanite-bearing varieties occur both in Monte Rosa (Bearth, 1952; Dal Piaz, 1971) and in Gran Paradiso (Compagnoni and Lombardo, 1974). Margarite was found by Chopin (1981) only as inclusions in chloritoid and may not have been in

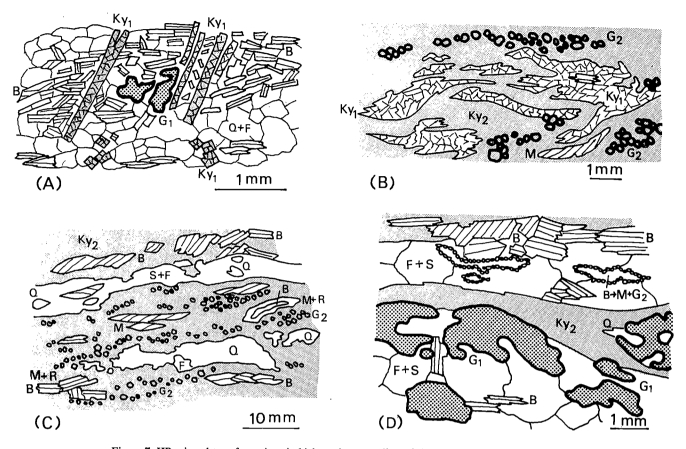

Figure 7. HP mineral transformations in high-grade metapelites of the Monte Rosa nappe (Dal Piaz, 1971). Relict minerals of pre-Alpine age are biotite (B), quartz (Q), feldspar (F) and an older generation of garnet (G_1). Early-Alpine HP minerals are kyanite, as polycrystalline paramorphs on prismatic sillimanite (Ky1) and micro-cryptocrystalline pseudomorphs on an earlier Al-Silicate (Ky2), white mica (M) and rutile (R) which partly replace biotite, sericite (S) and a new generation of garnet (G2), rimming biotite or growing in the kyanite pseudomorphs. The high grade metapelites are from Punta Perazzi (A), Pizzo Bianco (B), NE of Capanna Q. Sella (C) and Corno Faller (D).

equilibrium with the rest of the mineral assemblage. Bulk compositions of these rocks are characterized by very low CaO (0.1-0.2 wt %), low K_2O and Na_2O (0.2-2.5 wt %), high FeO (7-12 wt %) and Al_2O_3 (16-24%) and exceptionally high MgO (10-23 wt%). Mg/(Mg + Fe) ratios are high, between 0.6 and 0.8. The mineral chemistry is obviously controlled by this strongly magnesian bulk composition, with up to 45 mole % of the magnesian end member in chloritoid. Talc contains up to 15 mole % minnesotaite, and the glaucophane is a magnesian variety, with very little Fe^{3+} (Fig. 6), but nearly 10 atom % of Li in the M4 site (Ungaretti and others, 1981).

Phase relations in the silvery micaschists are depicted in Fig. 8 after Chopin (1981) and are characterized by the stability of talc + phengite ± Mg-chlorite, the coexistence of talc with chloritoid ± garnet, and the instability of chlorite + quartz. The formation of these peculiar assemblages has been ascribed to prograde high-pressure reactions involving the disappearance of Fe-Mg chlorite in rocks of pelitic composition (Chopin, 1981). An alternative view is that they result from metasomatic reactions within

HP shear zones affecting the metagranitic rocks (Bearth, 1952; Reinhardt, 1966; Dal Piaz, 1971). The proposed reactions are of the type:

—biotite + K-feldspar + H_2O = chloritoid + Mg-chlorite + kyanite + quartz + K in solution

—biotite + K-feldspar + plagioclase = white mica + kyanite + quartz

—biotite + plagioclase + Mg = white mica + chloritoid + Mg-chlorite + Na in solution

and imply that the metasomatic alteration of the granitic rocks was synchronous with the HP metamorphism. A complementary hypothesis is that the silvery micaschists are low-temperature, chlorite-rich mylonites of pre-Alpine age similar to granite mylonites described in some Pyrenean massifs (Carreras and others, 1980), which have been recrystallized under HP conditions.

The evidence, both geological and petrographical, of the progressive microstructural transition from granitic augen gneiss to silvery micaschists (e.g. in the upper Ayas Valley, Dal Piaz, 1971) does not favour in most instances the hypothesis that these rocks were derived from sediments, but is better explained by derivation from granites undergoing HP heterogeneous deformation or from HP recrystallization of previously sheared granites.

P-T CONDITIONS DURING THE HP METAMORPHISM

A detailed petrological analysis of mineral compatibilities in the Monte Rosa-Gran Paradiso nappes, like those proposed by Koons (1982) for the Sesia-Lanzo unit or Kienast (1984) for the metamorphic ophiolites of the Western Alps, is beyond the scope of this review and only a few mineral equilibria which are relevant to the HP metamorphism will be dealt with here.

The following mineral equilibria may be used to constrain the P-T conditions of the Early Alpine metamorphism in the Monte Rosa-Gran Paradiso nappes (Fig. 9): 1) The reaction albite = jadeite + quartz, after Holland (1980). 2) The stability of omphacite, $Jd_{50}Di_{50}$, coexisting with quartz estimated from 1) using the method of Holland (1980), and assuming that $a_{Jd} = X_{Jd}$ in natural P2/n omphacites as suggested by Holland (1983). 3) The reaction lawsonite + albite = zoisite + paragonite + quartz + H_2O as calculated by Holland (1979). 4) The reaction chloritoid + quartz = staurolite + almandine + H_2O after Rao and Johannes (1979). 5) The maximum stability field of a natural glaucophane (Maresch, 1977). 6) The Al-silicate phase diagram after Holdaway (1971).

Reaction 3), which in P-T space is close to the lines for the reactions lawsonite = zoisite + margarite + quartz (Nitsch, 1974) and lawsonite = zoisite + kyanite + quartz + H_2O (Newton and Kennedy, 1963), provides a lower temperature limit for the assemblages in metabasic rocks, whilst reaction 4) provides an upper temperature limit for the assemblages of metapelitic rocks. Reaction 2) can be taken approximately as the lower pressure limit for the stability of eclogite assemblages, whilst reaction 1) sets the upper pressure limit for the HP metamorphism of most granitic rocks, where jadeite is absent. The jadeite occurring in the Bonneval Gneiss of Gran Paradiso is enclosed in K-feldspar and may have grown from disordered feldspar at lower pressures than those determined experimentally for reaction 1) (Saliot, 1979). An alternative explanation for this puzzling occurrence of jadeite is possibly tectonic overpressure in rocks subjected to P-T conditions close to the albite breakdown reaction.

Assuming that $P_{H_2O} = P_{total}$, the mineral equilibria listed above define a field in P-T space ranging from approximately 420°C and 9 Kb to 550 °C and 15 Kb (Fig. 9). In order to refine this rather loose estimate and investigate possible regional differences in the Monte Rosa-Gran Paradiso nappes, we have estimated equilibration temperatures in some eclogitic rocks using the fractionation of Fe and Mg in coexisting garnet-omphacite pairs.

In the garnet-clinopyroxene geothermometer developed by Ellis and Green (1979) from data obtained in experiments at high T and P on basaltic compositions, the compositional dependence of the partition coefficient is accounted for by considering the Ca

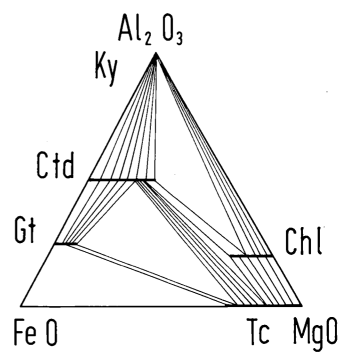

Figure 8. Phase relations in the system $FeO-MgO-Al_2O_3-SiO_2-H_2O$ as deduced from mineral assemblages in the Gran Paradiso nappe (Chopin, 1981). The assemblages containing talc have been described by Chopin (1981), whilst the kyanite-bearing assemblages were inferred by the same author from the report of Compagnoni and Lombardo (1974). Tc = talc, Chl = chlorite, Ky = kyanite, Ctd = chloritoid, Gt = garnet. Quartz is an additional phase in all the assemblages and a_{H_2O} is fixed.

content of garnet (X_{Ca}^{gar}), whilst the garnet-clinopyroxene geothermometer developed by Saxena (1979) from compositional data on coexisting phases in the system $CaO-MgO-FeO-MnO-Al_2O_3-Na_2O-SiO_2$ also attempts to account for the influence on the partition coefficient of the clinopyroxene composition, which is approximated by an asymmetric solution model involving the four components Ca, Mg, Fe and Al. Both thermometers neglect the possible influence of the acmite component in the clinopyroxene and have been calibrated on jadeite-poor pyroxenes equilibrated under conditions of high T and P which involve long extrapolation to the low-temperature conditions estimated for the Early Alpine metamorphism. Moreover, in amphibole eclogites such as those investigated here, where the amphibole is in some instances surely post-eclogitic, the Fe-Mg exchange may have outlasted the climax of the HP metamorphism, possibly recording a later stage of equilibration, in particular the compositional re-adjustments resulting from amphibole-producing reactions.

In spite of these shortcomings, both thermometers generate temperature estimates consistent with phase equilibria even in glaucophane eclogites, as shown in test calculations on garnet-omphacite pairs from ophiolite eclogites of the Western Alps (Kienast, 1984; Benciolini and others, 1984).

For its greater simplicity and in order to compare the equilibration temperatures obtained on the Monte Rosa-Gran Paradiso

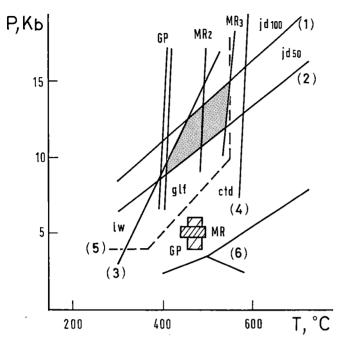

Figure 9. Petrogenetic grid for the Early-Alpine metamorphism in the Monte Rosa-Gran Paradiso nappes. Mineral equilibria 1) to 6) are discussed in text. Also shown is the variation of equilibration temperature with pressure as calculated with the relation proposed by Ellis and Green (1979) in garnet-omphacite pairs from the eclogites of the Gran Paradiso nappe (GP) and of the Monte Rosa nappe (samples MR03087 = MR2 and MR0153 = MR3). Dotted field: P-T conditions of the HP-LT metamorphism in the Monte Rosa nappe. Bars: P-T conditions assumed for the Lepontine overprinting in the Monte Rosa (MR) and Gran Paradiso (GP) nappes by Frey and others (1976) and Chopin (1979), respectively.

eclogites with previous estimates on eclogites of neighboring tectonic units, in this paper we have utilized the relation of Ellis and Green (1979). Note, however, that the results obtained with this relation in glaucophane eclogites tend to fall at the lower end of the total range obtained utilizing also the gt-cpx thermometers of Saxena (1979) and Ganguly (1979), and are thus to be regarded generally as minimum estimates only (Benciolini and others, 1984).

The compositional parameters of coexisting garnet-omphacite pairs from the Monte Rosa and Gran Paradiso eclogites and the temperatures estimated with the calibration of Ellis and Green, assuming a nominal pressure of 10 Kb, are given in Tables 2 and 3, whilst the variation of equilibration temperatures with pressure is shown in Fig. 9. The data may be separated into three groups: the first comprises the eclogites from the Gran Paradiso nappe, with equilibration temperatures clustering around 400°C, the second comprises most of the Monte Rosa eclogites, with T ranging from approximately 410° to 490°C, and the third comprises three Monte Rosa eclogites with T clustering around 530°C. In view of the assumptions underlying these results and the uncertainties discussed above, these values should be regarded as order-of-magnitude figures, whose main interest lies in suggesting a trend of progressively higher equilibration temperatures

during the HP event in the Penninic continental crust from the Gran Paradiso nappe to some parts of the Monte Rosa nappe. It is encouraging to note that the P-T conditions deduced from garnet-clinopyroxene thermometry and simple phase relations are in good agreement with the P-T estimates obtained by Chopin (1981) with more sophisticated petrological reasoning from mineral composition data combined with experimental data on the system $MgO-Al_2O_3-SiO_2-H_2O$ for the phengite-talc-chloritoid rocks of the Gran Paradiso nappe (T = 400-450°C; P = 7-10 Kb). From a regional viewpoint, the data discussed above might indicate that the Gran Paradiso nappe was metamorphosed under lower T and P conditions than the Monte Rosa nappe, some parts of which have reached T conditions close to the highest values estimated for basement nappes in the Western Alps, such as the Austroalpine Sesia-Lanzo unit (but without reaching such high-pressure conditions as those recorded in this unit). However, the observations 1) that the Gran Paradiso omphacites are not particularly biased towards either the acmite or the jadeite component, which might conceivably affect Fe-Mg partition coefficients, and 2) that the low equilibration temperatures recorded in the Gran Paradiso eclogites are all from gt-omph pairs coexisting with glaucophane or winchite, suggest that a variation in time of nominal equilibration temperatures is equally likely, the low temperatures of the Gran Paradiso eclogites recording the compositional changes induced in gt-omph pairs by amphibole-producing reactions.

An alternative possibility is that metamorphic temperatures during the climax of the HP event may not have been the same throughout a single tectonic unit, depending more on local factors such as shear-heating than on the overall thermal budget. In this case the high equilibration temperatures of some Monte Rosa eclogites are possibly not representative of the conditions prevailing in this unit during the HP event.

From the evidence currently available, we consider the hypothesis of post-eclogitic re-equilibrations as the most likely explanation for the low equilibration temperatures in the Gran Paradiso eclogites. In this case temperatures attained in the Gran Paradiso nappe at the climax of the HP metamorphism remain unspecified. The closely similar tectonometamorphic evolution observed in the Gran Paradiso and Monte Rosa nappes suggests that these temperatures may not have been significantly different in the former unit from those recorded in the Monte Rosa eclogites.

REGIONAL DISTRIBUTION AND TECTONIC IMPLICATIONS OF THE HP METAMORPHISM

The occurrence of a HP-LT metamorphism of Early Alpine age in the Northwestern Alps was first described by Dal Piaz and others (1972) and Ernst (1971), who also proposed plate-tectonics models to explain its peculiar distribution in the tectonic units of this region. Subsequent contributions discussed the HP-LT metamorphism at the scale of the whole Western Alps (Ernst, 1973; Dal Piaz, 1974; Desmons, 1977; Caby and others, 1978) or focused on specific facets of this topic, exploring its petrological (Kienast, 1984), structural (Gosso and others, 1979), geochronological (Hunziker, 1974; Bocquet and others, 1974) and paleotectonic (Homewood and others, 1980) implications.

Accruing petrological data on single tectonic units of the Northwestern Alps has meanwhile made it possible to discuss the characters and distribution in this region of the HP metamorphism in a more quantitative, yet still rather crude, fashion. Two block diagrams appear particularly informative in this respect: one from the Ossola Valley through Monte Rosa to Aosta valley (Fig. 10A), the second from the Aosta Valley through Gran Paradiso to the Orco Valley (Fig. 10B). The tectonic units with HP assemblages exposed along these cross-sections and the P-T conditions assumed for the culmination of the HP metamorphism are the following:

1) The tectonically composite, Austroalpine Sesia-Lanzo unit, with T = 550°C and P > 15 Kb in the lower tectonic element (Ungaretti and others, 1983; Lardeaux and others, 1982; Desmons and O'Neil, 1978; Desmons and Ghent, 1977; Compagnoni, 1977), T = 500°C and P ≈13-15 Kb in the marginal parts of the overlying Second Diorite-Kinzigite unit (Lardeaux, 1981) and again T and P similar to those estimated for the lower element in the Plaida-Tillio thrust sheet, the uppermost tectonic element of the Sesia-Lanzo unit (Lardeaux and others, 1982a).

2) The Austroalpine Mt. Emilius Klippe, where high-grade rocks have been subjected to around 470°C and P exceeding 11 Kb (Dal Piaz and others, 1983) and the northern units ofthe Dent Blanche system, for which P-T conditions apparently did not exceed values appropriate for greenschist-facies compatibilities, but where the occurrence of Alpine aegirine and glaucophane (Pillonet Klippe; Dal Piaz, 1976) and of rare metamorphic pyroxene in some magmatic rocks (Arolla unit; Dal Piaz and Govi, 1968; Ayrton and others, 1982) and the abundance of phengite may suggest also HP conditions.

3) The Piedmont ophiolite nappe system, with T ranging from 450° to 540°C in metamorphic ophiolite sheets (Ernst and Dal Piaz, 1978; Oberhänsli, 1980; Martin, 1983; Baldelli and others, 1982, 1985), and T and P presumably not exceeding values appropriate for greenschist-facies compatibilities in the dominantly metasedimentary and tectonically composite Combin unit, where, however, thrust sheets with blueschist assemblages also exist (Baldelli and others, 1983).

4) The Penninic Monte Rosa-Gran Paradiso nappes discussed in previous sections of this paper, with maximum T around 530°C and P between 10 and 14 Kb.

5) The Penninic Gran San Bernardo nappe, with high-grade blueschist assemblages (glaucophane, garnet, chloritoid ± lawsonite; Bearth, 1962, 1963; Frey and others, 1974) which, however, appear on stratigraphical grounds to be younger than late Cretaceous-early Eocene (?) (Ellenberger, 1952; Bearth, 1963; Marthaler, 1984).

Some points which deserve further comment are the following:

(a) Metamorphic zonation of Early Alpine age is possibly preserved within single tectonic units such as the Sesia-Lanzo nappe, where jadeite-bearing assemblages in metagranitic rocks are replaced towards the NE by albite + omphacite compatibilities (Lardeaux and others, 1982b), or the ophiolite nappe in which eclogites from the Breuil-Zermatt-Saas region (that is the frontal part of the nappe) apparently recrystallized at 70°C higher T than the eclogites from more southern parts of the nappe, where gt-cpx geothermometry yields equilibration temperatures around 450-460°C (Baldelli and others, 1982 and 1985). It/is unlikely, however, that any eclogitic zoneography could survive undisrupted the large-scale transposition caused by the two major phases of ductile deformation which can be recognized on structural ground as later than the HP culmination (Gosso and others, 1979; Lardeaux and others, 1982a; Gosso and others, 1982).

(b) Eclogite units may be coupled with units showing similar eclogite assemblages and the same tectonometamorphic evolution (e.g. Monte Rosa-Gran Paradiso and the overlying Zermatt-Saas ophiolite nappe) or they may be juxtaposed with tectonic units with blueschist assemblages of possibly younger age (e.g. the Zermatt-Saas ophiolite nappe and the Gran San Bernardo nappe).

(c) Although age determinations may in the Northwestern Alps still be fraught by problems of interpretation (Desmons and others, 1982; Chopin and Maluski, 1982), there is, in our opinion, already a considerable body of geochronological data suggesting that eclogite conditions may have persisted in some tectonic units of the Northwestern Alps throughout the whole Cretaceous, starting from approximately 130 Ma for the eclogite recrystallizations in the Sesia-Lanzo unit (Oberhänsli and others, 1982, 1985) and in Monte Rosa (Hunziker, 1970, 1974) to end at 60 Ma with the closure of the HP-phengite systems; the Sesia-Lanzo unit (Hunziker, 1974) and in Gran Paradiso (Chopin and Maluski, 1980). Such a long time span assigned to the eclogite metamorphism may be puzzling on stratigraphical grounds, as sediments of Cretaceous age are common in the Northwestern Alps, but it does not conflict with the stratigraphical evidence presently known for the Sesia-Lanzo unit or the Monte Rosa-Gran Paradiso nappes.

(d) A post-Paleocene age for the high-grade blueschist assemblages of the Gran San Bernardo nappe implies that this nappe has been subjected to shallower and cooler HP conditions during the late stages of exhumation and HP evolution of the eclogitic units.

From a more general viewpoint, the observations discussed above suggest that:

1) The eclogite metamorphism may have taken place in the Western Alps over a much longer interval than is assumed in models involving the subduction of coherent lithospheric slabs. Hence, simple isotherm configurations derived from these models may not adequately represent the P-T-t path to which any particular tectonic unit was subjected during the Early Alpine tectonometamorphic event. Subduction of relatively cold thrust sheets derived from disrupted portions of both oceanic and continental crust and tectonic exhumation along oblique thrust planes may be

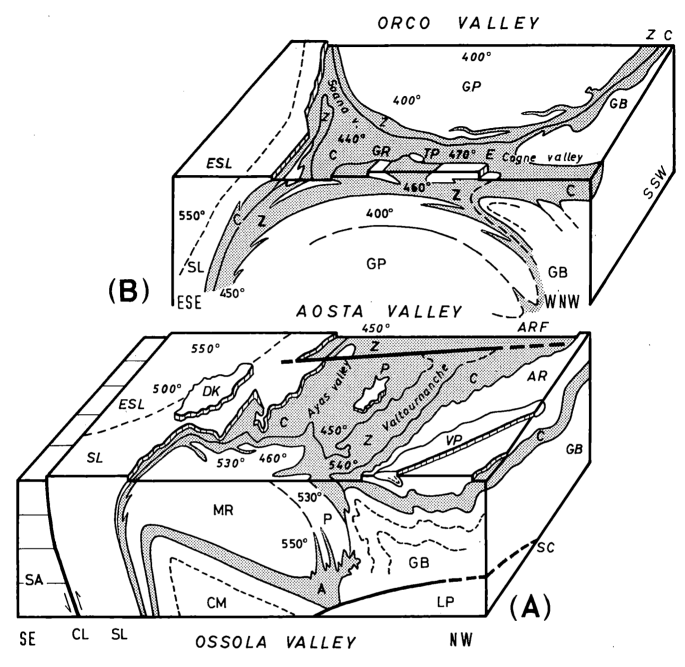

Figure 10. Regional distribution of garnet-omphacite equilibration temperature in the Early-Alpine eclogite units of the internal Northwestern Alps as calculated with the relation of Ellis and Green (1979) assuming a nominal pressure of 15 Kb in the Sesia-Lanzo unit and 10 Kb in all other units. Paleoafrican tectonic units: Southern Alps (SA), Austroalpine II diorite-kinzigite (DK) and Valpelline (VP) units, Austroalpine greenschist (SL) to eclogitic Sesia-Lanzo (ESL), Pillonet (P), Arolla (AR), Glacier-Rafray (GR), Tour Ponton (TP) and Mt. Emilius (E) units. Composite Piedmont ophiolite nappe: meta-ophiolite sheets of the eclogitic Zermatt-Saas (Z) and Antrona (A) units; dominantly metasedimentary sheets of the greenschist Combin composite unit (C). Paleoeuropean tectonic units: eclogitic Monte Rosa (MR) and Gran Paradiso (GP) nappes; Gran San Bernardo nappe (GB) including Alpine glaucophane-garnet-chloritoid ± lawsonite assemblages; Lower Penninic nappes (LP). Tectonic lines: Canavese (CL), Simplon-Centovalli (SC) and Aosta-Ranzola (ARF).

a tectonic mechanism adequate to explain how eclogite conditions may persist for a few tens of million years.

2) The last stages of the HP evolution in eclogite units may have been accompanied in both continental and oceanic units at shallower structural levels by blueschist metamorphism, the change from eclogite- to blueschist conditions possibly marking the transition from tectonic processes dominated by subduction to processes characteristic of continental collision.

ACKNOWLEDGMENTS

Financial support for this study was provided by the "Centro di studio sull'Orogeno delle Alpi occidentali," CNR-Torino, and by the Ministero della Pubblica Istruzione. The mineral compositional data were collected in the CNR electron microprobe laboratory of the Mineralogy and Petrology Institute, University of Modena. We thank W. G. Ernst, B. W. Evans, J. C. Hunziker and R. Oberhänsli for helpful reviews.

REFERENCES CITED

Argand, E., 1911, Les nappes de recouvrement des Alpes Pennines et leurs prolongements structuraux: Mat. Carte Géol. Suisse, v. 31, 26 p.

Ayrton, S., Bugnon, Ch., Haarpaintner, T., Weidmann, M., and Frank, E., 1982, Géologie du front de la nappe de la Dent-Blanche dans la région des Monts-Dolins, Valais: Eclogae Geol. Helv., v. 75, p. 269–286.

Baldelli, C., Dal Piaz, G. V., and Lombardo, B., 1982, Ophiolite eclogites from Verres, Western Alps, Italy: Terra Cognita, v. 2, p. 305, Abstr., and Chemical Geol., v. 50 (1985), p. 87–98.

——, ——, and Polino, R., 1983, Le quarziti a manganese di Varenche-St. Barthélemy una sequenza di copertura oceanica della falda piemontese: Ofioliti, v. 8, p. 207–221.

Bearth, P., 1952, Geologie und Petrographie des Monte Rosa: Beitr. Geol. Karte Schweiz, v. 96, 94 p.

——1953, Geologischer Atlas der Schweiz 1:25,000, Blatt Zermatt, mit Erläuterungen: Schweiz. Geol. Kommission, Kümmerly & Frey, Bern.

——1954, Geologischer Atlas der Schweiz 1:25.000, Blatt Saas: Schweiz. Geol. Kommission, Kümmerly & Frey, Bern.

——1957, Die Umbiegung von Vanzone (Valle Anzasca). Eclogae geol. Helv., v. 50, p. 161–170.

——1958, Ueber einen Wechsel der Mineralfazies in den Wurzelzone des Penninikums: Schweiz. min. petr. Mitt., v. 38, p. 363–373.

——1962, Versuch einer Gliederung alpinmetamorpher Serien der Westalpen: Schweiz. min. petr. Mitt., v. 42, p. 127–137.

——1963, Contribution à la subdivision tectonique et stratigraphique du cristallin de la nappe du Grand-St-Bernard dans le Valais (Suisse): Geol. Soc. France, Mém. h. sér., v. 2, p. 407–418.

Benciolini, L., Tartarotti, P., Dal Piaz, G. V., De Vecchi, P., and Polino, R., 1984, La geotraversa Gran Paradiso-Zona Sesia in alta Val Soana: Geol. Soc. Ital., 72th Congress, Abstr., p. 30–31.

Bertrand, J. M., 1968, Etude structurale du versant occidental du Massif du Grand Paradis (Alpes Graies): Géol. Alpine, v. 44, p. 57–87.

Bocquet, J., Delaloye, M., Hunziker, J. C., and Krummenacher, D., 1974, K-Ar and Rb-Sr dating of blue amphiboles, micas and associated minerals from the Western Alps: Contrib. Mineral. Petrol., v. 47, p. 7–26.

Caby, R., Kienast, J.-R., and Saliot, P., 1978, Structure, métamorphisme et modèle d'évolution tectonique des Alpes occidentales: Revue Géogr. Phys. Géol. Dyn., v. 20, p. 307–322.

Callegari, E., Compagnoni, R., and Dal Piaz, G. V., 1969, Relitti di strutture intrusive erciniche e scisti a sillimanite nel Massiccio del Gran Paradiso: Geol. Soc. Ital. Boll., v. 88, p. 59–69.

Carreras, J., Julivert, M. and Santanach, P., 1980, Hercynian mylonite belts in eastern Pyrenees: an example of shear zones associated with late folding: Journ. Struct. Geol., v. 2, p. 5–9.

Chopin, C., 1979, De la Vanoise au Massif du Grand Paradis (Thesis): Univ. P. et M. Curie, Paris, 145 p.

——1981, Talc-Phengite: a widespread assemblage in high-grade pelitic blueschists of the Western Alps: Journ. Petrology, v. 22, p. 628–650.

——and Maluski, M., 1978, Resultats preliminaires obtenus par la méthode $^{39}Ar/^{40}Ar$ sur des mineraux alpins du massif du Grand Paradis et de son enveloppe: Geol. Soc. France Bull., v. 20, p. 745–749.

——, ——1980, $^{39}Ar/^{40}Ar$ dating of high pressure metamorphic micas from the Gran Paradiso area (Western Alps): evidence against the blocking temperature concept: Contrib. Mineral. Petrol., v. 74, p. 109–122.

——, ——1982, Unconvincing evidence against the blocking temperature concept: A reply: Contrib. Mineral. Petrol., v. 80, p. 391–394.

Coleman, R. G., Lee, D. E., Beatty, L. B., and Brannock, W. W., 1965, Eclogites and eclogites: their differences and similarities: Geol. Soc. Amer. Bull., v. 76, p. 483–508.

Compagnoni, R., 1977, The Sesia-Lanzo Zone: high pressure-low temperature metamorphism in the Austroalpine continental margin: Min. Petr. Soc. Ital. Rend., v. 33, p. 335–374.

——and Prato, 1969, Paramorfosi di cianite su sillimanite in scisti pregranitici del Massiccio del Gran Paradiso: Geol. Soc. Ital. Boll., v. 88, p. 537–549.

——and Maffeo, B., 1973, Jadeite-bearing metagranites l.s. and related rocks in the Mount Mucrone Area (Sesia-Lanzo Zone, Western Italian Alps): Schweiz. min. petr. Mitt., v. 53, p. 355–378.

——and Lombardo, B., 1974, The Alpine age of the Gran Paradiso eclogites: Min. Petr. Soc. Ital. Rend., v. 30, p. 223–237.

——Elter, G., and Lombardo, B., 1974, Eterogeneità stratigrafica del complesso degli "Gneiss minuti" nel Massiccio cristallino del Gran Paradiso: Geol. Soc. Ital. Mem., v. 13/1, p. 227–239.

——Dal Piaz, G. V., Hunziker, J. C., Gosso, G., Lombardo, B., and Williams, P., 1977, The Sesia-Lanzo Zone, a slice of continental crust with Alpine high pressure-low temperature assemblages in the Western Italian Alps: Min. Petr. Soc. Ital. Rend., v. 33, p. 281–334.

Dal Piaz, G. V., 1964, Il cristallino antico del versante meridionale del Monte Rosa. Paraderivati a prevalente metamorfismo alpino: Min. Soc. Ital. Rend., v. 20, p. 101–136.

——1966, Gneiss ghiandoni, marmi ed anfiboliti antiche del ricoprimento Monte Rosa nell'alta Val d'Ayas: Geol. Soc. Ital. Boll., v. 85, p. 103–132.

——1971, Nuovi ritrovamenti di cianite alpina nel cristallino antico del Monte Rosa: Min. Petr. Soc. Ital. Rend., v. 27, p. 437–477.

——1974, Le métamorphisme de haute pression et basse température dans l'évolution structurale du bassin ophiolitique alpino-apenninique: Geol. Soc. Ital. Boll., v. 93, p. 437–468, and Schweiz. min. petr. Mitt., v. 54, p. 399–424.

——1976, Il lembo di ricoprimento del Pillonet, falda della Dent Blanche nelle Alpi occidentali: Ist Geol. Min. Univ. Padova Mem., v. 31, 60 p.

——and Gatto, G., 1963, Considerazioni geologico-petrografische sul versante meridionale del Monte Rosa: Accademia Naz. Lincei Roma, Phys. Sci. Rend., v. 34, p. 190–194.

——and Govi, M.,1968, Lo stilpnomelano in Valle d'Aosta: nuovi ritrovamenti: Geol. Soc. Ital. Boll., v. 87, p. 91–108.

——Hunziker, J. C., and Martinotti, G., 1972, La Zona Sesia-Lanzo e l'evoluzione tettonico-metamorfica delle Alpi nordoccidentali interne: Geol. Soc. Ital. Mem., v. 11, p. 433–460.

——, ——, ——1973, Excursion to the Sesia Zone of the Schweiz. Mineralogische und Petrographische Gesellschaft, September 30th to October 3rd, 1973: Schweiz. min. petr. Mitt., v. 53, p. 477–490.

——Lombardo, B., and Gosso, G., 1983, Metamorphic evolution of the Mt. Emilius Klippe, Dent Blanche nappe, Western Alps: Amer. J. Sci., v. 283-A,

p. 438-458.

Desmons, J., 1977, Mineralogical and petrological investigations of Alpine metamorphism in the internal French Western Alps: Amer. J. Sci., v. 277, p. 1045-1066.

——and Ghent, E. D., 1977, Chemistry, zonation and distribution coefficients of elements in eclogitic minerals from the eastern Sesia Unit, Italian Western Alps: Schweiz. min. petr. Mitt., v. 57, p. 397-411.

——and O'Neil, J. R., 1978, Oxygen and hydrogen isotope compositions of eclogites and associated rocks from the eastern Sesia Zone (Western Alps, Italy): Contrib. Mineral. Petrol., v. 67, p. 79-85.

——Hunziker, J. C., and Delaloye, M., 1982, Unconvincing evidence against the blocking temperature concept. Comments on: "^{39}Ar/^{40}Ar dating for high-pressure metamorphic micas . . ." by C. Chopin and H. Maluski: Contrib. Mineral. Petrol., v. 80, p. 386-390.

Ellenberger, F., 1952, Sur l'âge du métamorphisme dans la Vanoise: Geol. Soc. France C. R. Somm., p. 318-321.

Ellis, D. J., and Green, D. H., 1979, An experimental study of the effect of Ca upon garnet-clinopyroxene Fe-Mg exchange equilibria: Contrib. Mineral. Petrol., v. 71, p. 13-22.

Ernst, W. G., 1971, Metamorphic zonations on presumably subducted lithospheric plates from Japan, California and Alps: Contrib. Mineral. Petrol., v. 34, p. 45-59.

——1973, Interpretative synthesis of metamorphism in the Alps: Geol. Soc. Amer. Bull., v. 84, p. 2053-2088.

——and Dal Piaz, G. V., 1978, Mineral parageneses of eclogitic rocks and related mafic schists of the Piemonte Ophiolite nappe, Breuil-St. Jacques area, Italian Western Alps: Amer. Mineral., v. 63, p. 621-640.

Essene, E. J., and Fyfe, W. S., 1967, Omphacite in Californian metamorphic rocks: Contrib. Mineral. Petrol., v. 15, p. 1-23.

Franchi, S., 1903, Sul rinvenimento di nuovi giacimenti di roccie giadeitiche nelle Alpi occidentali e nell'Appenino ligure: Geol. Soc. Ital. Boll., v. 22, p. 130-134.

——Mattirolo, E., Novarese, V. and Stella, A., 1921a, Carta Geologica d'Italia, Foglio 29 M.te Rosa: Servizio Geol., scale 1:100,000.

——,——,——,——1912b, Carta Geologica d'Italia, Foglio 41 Gran Paradiso: Servizio Geol., scale 1:100.000.

——,——,——,——1927, Carta Geologica d'Italia, Foglio 30 Varallo: Servizio Geol., scale 1:100.000.

Frey, M., Hunziker, J. C., Frank, W., Bocquet, J., Dal Piaz, G. V., Jaeger, E. and Niggli, E., 1974, Alpine metamorphism of the Alps. A review: Schweiz. min. petr. Mitt., v. 54, p. 247-290.

——,——O'Neil, J. R. and Schwander, H. W., 1976, Equilibrium-disequilibrium relations in the Monte Rosa granite, Western Alps: petrological, Rb-Sr and stable isotope data: Contrib. Mineral. Petrol., v. 55, p. 147-179.

Ganguly, J., 1979, Garnet and clinopyroxene solid solutions, and geothermometry based on Fe-Mg distribution coefficients: Geochim. et Cosmochim. Acta, v. 43, p. 1021-1029.

Gosso, G., Dal Piaz, G. V., Piovano, V. and Polino, R., 1979, High pressure emplacement of early-alpine nappes, postnappe deformations and structural levels (Internal Northwestern Alps): Ist. Geol. Min. Univ. Padova Mem., v. 32, 15 p.

Gosso, G., Kienast, J. R., Lardeaux, J. M. and Lombardo, B., 1982, Replissement intense et transposition (en climat métamorphique de haute pression) des contacts tectoniques majeurs dans l'édifice supérieur des nappes alpines (zone Sésia-Lanzo): Acad. Sci. Paris C. R., v. 294, p. 343-348.

Holdaway, M. J., 1971, Stability of andalusite and the aluminum silicate diagram: Amer. J. Sci., v. 271, p. 97-131.

Holland, T.J.B., 1979, Experimental determination of the reaction paragonite = jadeite + kyanite + H_2O, and internally consistent thermodynamic data for part of the system Na_2O-Al_2O_3-SiO_2-H_2O, with applications to eclogites and blueschists: Contrib. Mineral. Petrol., v. 68, p. 293-301.

——1980, The reaction albite = jadeite + quartz determined experimentally in the range 600°-1200°C: Amer. Mineral., v. 65, p. 129-134.

——1983, The experimental determination of activities in disordered and short-

range ordered jadeitic pyroxenes: Contrib. Mineral. Petrol., v. 82, p. 214-220.

Homewood, P., Gosso, G., Escher, A. and Milnes, A., 1980, Cretaceous and Tertiary evolution along the Besançon-Biella traverse (Western Alps): Eclgoae geol. Helv., v. 73, p. 635-649.

Hunziker, J. C., 1969, Rb-Sr Altersbestimmung aus den Walliser Alpen: Eclogae geol. Helv., v. 62, p. 527-542.

——1970, Polymetamorphism in the Monte Rosa, Western Alps: Eclogae geol. Helv., v. 63, p. 151-161.

——1974, Rb-Sr and K-Ar age determinations and the Alpine tectonic history of the Western Alps: Ist. Geol. Min. Univ. Padova Mem., v. 31, 55 p.

Kienast, J.-R., 1984, Le métamorphisme de haute pression et basse température (eclogites et schistes bleus): données nouvelles sur la pétrologie des roches de la croûte océanique subductée et des sédiments associés [Ph.D. Thesis]: Université P. et M. Curie, Paris, 484 p.

Klein, I., 1978, Post-nappe folding southeast of the Mischabelrückfalte (Pennine Alps) and some aspects of the associated metamorphism: Leidse Geol. Med., v. 51, p. 233-312.

Koons, P. O., 1982, An investigation of experimental and natural high-pressure assemblages from the Sesia-Zone, Western Alps, Italy [Thesis]: E.T.H. Zürich, n. 7169, 260 p.

Krummenacher, D. and Evernden, J. F., 1960, Déterminations d'âge isotopique faites sur quelques roches des Alpes par la méthode Potassium-Argon: Schweiz. min. petr. Mitt., v. 40, p. 267-277.

Laduron, D., 1976, L'antiforme de Vanzone. Etude pétrologique et structurale dans la valle Anzasca (Province de Novara, Italie): Ist. Géol. Univ. Louvain Mém., v. 28, 121 p.

Lardeaux, J. M., 1981, Evolution tectono-métamorphique de la zone nord du massif de Sesia-Lanzo (Alpes occidentales). Un exemple d'eclogitisation de la croûte continentale [Thesis]: Univ. P. et M. Curie, Paris, 226 p.

——Gosso, G., Kienast, J.-R., and Lombardo, B., 1982a, Relations entre le métamorphisme et la déformation dans la Zone de Sesia-Lanzo (Alpes Occidentales) et le problème de l'éclogitisation de la croûte continentale: Geol. Soc. France Bull., v. 24, p. 793-800.

——Lombardo, B., Gosso, G., and Kienast, J.-M., 1982b, Découverte de paragenèses à ferro-omphacite dans les orthogneiss de la Zone Sesia-Lanzo septentrionale (Alpes Italiennes): Acad. Sci. Paris C. R., v. 296, p. 453-456.

Leake, B. E., 1978, Nomenclature of amphiboles: Min. Crist. Soc. France Bull., v. 101, p. 453-467.

Lombardo, B., 1970, Studio geologico-petrografico del Complesso degli Gneiss minuti nella Valle di Piantonetto (Massiccio del Gran Paradiso) [Unpubl. Thesis]: Ist. Petrografia Univ. Torino, 150 p.

Maresch, W. V., 1977, Experimental studies on glaucophane: an analysis of present knowledge: Tectonophysics, v. 43, p. 109-125.

Marthaler, M., 1984, Géologie des unités penniques entre le val d'Anniviers et le val de Tourtemagne (Valais, Suisse): Eclogae geol. Helv., v. 77, p. 395-448.

Martin, S., 1983, La mine de Praborna (Val d'Aoste, Italie): une série manganesifère metamorphisée dans le facies éclogite [Thesis]: Univ. P. et M. Curie, Paris, 254 p.

Michel, R., 1953, Les Schistes Cristallins des Massifs du Grand Paradis et de Sesia-Lanzo (Alpes Franco-Italiennes): Sci. de la Terre, Nancy, 287 p.

Miyashiro, A., 1957, The chemistry, optics and genesis of the alkali-amphiboles: Journ. Fac. Sci. Univ. Tokyo, v. 11, p. 57-83.

Newton, R. C., and Kennedy, G. C., 1963, Some equilibrium reactions in the join $CaAl_2$-SiO_2-H_2O: J. Geophys. Res., v. 68, p. 2967-2983.

Nitsch, K. H., 1974, Neue Erkenntnisse zur Stabilität von Lawsonit: Fortschr. Miner., v. 51, p. 34-35.

Novarese, V., and Stella, A., 1913, Carta Geologicica d'Italia, Foglio 15 Domodossola: Servizio Geol., scale 1:100.000.

Oberhänsli, R., 1980, P-T Bestimmung anhand von Mineralanalysen in Eklogiten und Glaucophaniten der Ophiolite von Zermatt: Schweiz. min. petr. Mitt., v. 60, p. 215-235.

——Hunziker, J. C., Martinotti, G., and Stern, W. B., 1982, Mucronites: an example of Eo-alpine eclogitisation of Permian granitoids, Italy: Terra Cognita, v. 2, p. 325, Abstr., and Chemical Geol., v. 50 (1985).

Pennacchioni, G., 1982-83, Studio geologico della dorsale tra le basse Valnontey e Valleile, Cogne, Valle d'Aosta [Unpubl. Thesis]: Ist. Geol. Univ. Padova, 212 p.

Prato, R., 1971, Il settore centro-occidentale del Massiccio del Gran Paradiso: Acc. Sci. Torino Atti, v. 105, p. 453–467.

Raase, P., 1974, Al and Ti contents of hornblende, indicators of pressure and temperature of regional metamorphism: Contrib. Mineral. Petrol., v. 45, p. 231–236.

Rao, B. B., and Johannes, W., 1979, Further data on the stability of staurolite + quartz and related assemblages: Neues Jb. Mineral. Mh., 1979, p. 437–447.

Reinhardt, B., 1966, Geologie und Petrographie der Monte Rosa-Zone, der Sesia-Zone und des Canavese im Gebiet zwischen Valle d'Ossola und Valle Loana (Prov. Novara, Italien): Schweiz. min. petr. Mitt., v. 46, p. 553–678.

Saliot, P., 1978, Le métamorphisme dans les Alpes françaises [Ph.D. Thesis]: Univ. Paris-Sud, Orsay, 183 p.

——1979, La jadéite dans les Alpes françaises: Bull. Mineral., v. 102, p. 391–401.

Saxena, S. K., 1979, Garnet-clinopyroxene geothermometer: Contrib. Mineral. Petrol., v. 70, p. 229–235.

Smulikowski, K., 1964, Le problème des éclogites: Geol. Sudetica, v. 1, p. 13–52.

Tartarotti, P., 1984, Studio geologico e petrografico della falda pennidica del Gran Paradiso nell'alta Val Soana, Piemonte [Unpubl. Thesis]: Ist. Geol. Univ. Padova, 140 p.

Ungaretti, L., Mazzi, F., Rossi, G. and Dal Negro, A., 1981, Crystal-chemical characterization of blue amphiboles: Proc. XI Gen. Meeting I.M.A. Novosibirgsk, Rock-forming Minerals, p. 82–110.

——Lombardo, B., Domeneghetti, C. and Rossi, G., 1983, Crystal-chemical evolution of amphiboles from eclogitised rocks of the Sesia-Lanzo Zone, Italian Western Alps: Bull. Minéral., v. 106, p. 645–672.

Vearncombe, J. R., 1983, High pressure-low temperature metamorphism in the Gran Paradiso basement, Western Alps: J. Metam. Geol., v. 1, p. 103–115.

Wetzel, R., 1972, Zur Petrographie und Mineralogie der Furgg-Zone (Monte Rosa-Decke): Schweiz. min. petr., v. 52, p. 161–236.

Zago, R., 1981, Studio geologico e giacimentologico di un settore del Gran Paradiso a nord di Noasca, Valle dell'Orco [Unpubl. Thesis]: Ist. Geol. Univ. Padova, 98 p.

Zakrutkin, V. V., 1968, The evolution of amphiboles during metamorphism: Zap. Yses. Mineral. Obsch., v. 96, p. 13–23.

MANUSCRIPT ACCEPTED BY THE SOCIETY JULY 29, 1985

Geological Society of America
Memoir 164
1986

Blueschist-facies metamorphism of manganiferous cherts: A review of the alpine occurrences

Annibale Mottana
Dipartimento di Scienze della Terra
Cattedra di Mineralogia
Città Universitaria
I-00185 Roma, Italy

ABSTRACT

Manganiferous cherts occurring in the subducted portion of the alpine belt (eclogite to blueschist facies metamorphism) present four types of structural settings: a) in stratigraphic coherency with the paleooceanic crust (ophiolites); b) in volcano-sedimentary sequences derived from disrupted ophiolite + chert + sediment associations; c) in reworked volcano-sedimentary sequences deposited on shelf or near the continent; d) in insufficiently studied terranes. In the literature, more than 100 occurrences are mentioned, but standard petrographic data are available for only about 40. Modern geochemical, mineralogical, or petrological data are limited to the 6 following occurrences: Haute Maurienne in France; Alagna, St. Marcel and the upper Valtournanche in Italy; Zermatt in Switzerland; and Andros in Greece. Modification of the bulk composition occurred during metamorphism, by homogenization of the chert sequence (radiolarites and their shale partings), as well as by addition of elements through a circulation of the fluid phase as far down as the underlying ophiolites.

The mineralogy includes about 100 identified species; of these, 28 groups are reviewed, with special attention to their crystal chemistry and the compositional differences derived from bulk chemical control, oxidation conditions, and changing P, T, $f(O_2)$ of the polyphasal alpine metamorphism. It is noteworthy that glaucophane never occurs in cherts, although it is present in the nearby mafites.

The analysis of phase compatibility points to three different assemblages: a) "oxidized," where silicate minerals contain Mn^{3+} in addition to Fe^{3+}: manganic varieties of sodic clinopyroxenes and micas, piemontite, ardennite, braunite; b) "neutral," where silicates contain Mn^{2+} next to Fe^{3+}, but the oxides can contain Mn^{3+}: spessartine, rhodonite, pyroxmangite, pyrophanite, andradite, riebeckite, acmite but also hausmannite; c) "reduced," where Mn is entirely divalent, and Fe^{2+} is present as well as Fe^{3+}. Reducing assemblages are typical of carbonate-bearing partings, where kutnahorite and rhodochrosite coexist with tephroite, rhodonite, Mn-humites; and of rare "aluminous" partings, containing jacobsite and galaxite. Numerous veins, containing minerals of any type, testify to the continuous fracture flow percolating through the metamorphosed chert sequence. Published estimates on the equilibration conditions of blueschist-facies alpine cherts vary widely: P = 5 to 14 kbar; T = 300 to 525°C; $f(O_2) = 10^{-5}$ to 10^{-25} bar. Such a scatter is due to the yet insufficient experimental knowledge on the Mn-Si-Al-O-OH system, including the nearly complete lack of determinations of the influence of other elements (e.g. Fe, Ca, As) on the reaction topology, and the still limited study of the natural phase relations in certain critical outcrops.

INTRODUCTION

Cherts[1] are essential constituents of layer 1 of the oceanic crust, both in present and in past sequences. Although they may be found close to the oceanic spreading centers, trapped between pillow lavas, usually cherts constitute a thin, discontinuous horizon sandwiched between pillow lavas and overlying pelagic sediments. A typical chert sequence is composed at the base of silica-rich horizons ("radiolarites") with thin shale partings; these increase in thickness and frequency toward the top so as to grade into dominantly clayey strata, initially of biogenic origin, but grading laterally and upwards to volcaniclastic, or detrital, due to addition of materials from the continental crust. The constant association of gabbros (layer 3), pillow basalts (layer 2) and cherts (layer 1) became established as a typical indicator of an abyssal oceanic environment thanks to Steinmann (1927: Steinmann's Trinity): cherts acquired the role of marker beds in all the ophiolite suites, even when metamorphosed under low-temperature—high-pressure (LT-HP) conditions in former zones of plate collision.

A peculiar feature of cherts associated with ophiolites is their high Mn content coupled with Fe depletion; that of the closely associated metabasalts contains Fe-Cu sulphide deposits. Such a joint feature was recognized throughout the Alps and Apennines by Debenedetti (1965) but it found an explanation only after Bonatti et al. (1976) applied to Ligurian ophiolites a model originally proposed by Elder (1965) to explain in plate-tectonic terms the metallogenesis occurring at oceanic-spreading centers.

According to this model (recently reconsidered and clearly summarized by Bonatti (1980)), ocean water is heated up to 350°C by circulating as deep as several km in the accreting basaltic mid-ocean ridge. There it leaches elements such as Si, Ba, Mn and Fe. These are subsequently discharged during the ascent, either at shallow depth within the basalts by segregation favoured by decreasing temperature and increasing sulphur activity (pods and stockworks of Cu-Fe sulphides in the ophiolites), or directly into the ocean water. Here the hot solutions precipitate together with silica-rich biogenic materials that are the normal sediments at this water depth. The resulting cherts would then include not only elements derived from subsea hydrothermal solutions (Si, Mn, Ba, Li) but also other elements that biogenic silica scavenges from sea water during slow deposition (Cu, Ni, Zn, U, Co). The composition of chert deposits would change with increasing distance from the metallogenic source because of decreasing hydrothermal supply and increasing "dilution" with normal oceanic sediments of terrigenous, biogenous and authigenous origin (cp. Nisbet and Price, 1974). The ratio of certain elements to others is

a function of distance of deposition from the spreading axis (Sugisaki et al., 1982).

In the deep seaway separating the paleo-European from the paleo-African plate, a belt of pelagic sediments containing discontinuous cherts with the above-mentioned characteristics was deposited during late Jurassic (Bosellini and Winterer, 1975), and locally Tithonian time (Folk and McBride, 1978). They were later involved in the subduction process. Most of them maintained their close association with ophiolites, so that the sequences may be regarded as marker horizons for the ancient oceanic crust (Pennidic and Ligurian troughs).

On the other hand, other cherts appear to be unrelated to ophiolites: they occur within the continental crust as a part of the sedimentary sequence deposited on the continental shelf and are over- and underlain by carbonatic rocks. Mafic and ultramafic rocks may be present in the sequence, but with no precise stratigraphic relations nor vertical connections with the metal-bearing cherts. The difference is best explained in the Pennine Alps where the ophiolitic-related sequence is known as the Zermatt-Saas Unit and the continental sequence as the Combin Unit (Dal Piaz et al., 1979; Bearth and Schwander, 1981).

The aims of the present investigations are therefore the following:

a) to review the existing data on sedimentary and metamorphic cherts related to ophiolites in the Alpine belt, in order to establish their peculiar geochemical characteristics, the extent of the chemical transformation undergone during the subduction metamorphism, and to identify the elements suitable to act as geochemical tracers for other outcrops where no evidence of oceanic vs. continental derivation survived the blueschist-facies metamorphism;

b) to review the mineralogy and phase petrology so as to help clarify the application of experimental studies in the synthetic Mn-Si-Ca-C-O-H system currently underway (Abs-Wurmbach and Peters, 1981; Abs-Wurmbach, 1980; Maresch and Mottana, 1976; Keskinen and Liou, 1979; Peters et al., 1973; 1974). This will eventually widen the use of rocks easily recognizable in the field from the mere condition of structural markers to that of indicators of equilibration within the high-pressure-facies sequence.

SOURCE OF DATA

A complete list of the Mn mineralizations in the alpine belt is not yet available. Castello (1981) compiled the inventory for Val d'Aosta, but for other localities in Piemonte, and in the French and Swiss front parts of the Pennidic domain, the most complete survey still dates back to Huttenlocher (1934). For other regions of the alpine belt the information on HP- metamorphic Mn mineralizations is even less systematic. Only in the structurally analogous (although affected by metamorphism at a later time) Median Aegean Crystalline Belt of the Hellenides are Mn-occurrences well documented (Reinecke, 1983).

Tables 1 to 4 list as completely as possible the occurrences,

---

[1]The term "chert" is used here in the broad sense of "sedimentary rock composed mainly of authigenic silica, probably of biogenic derivation." It includes, therefore, both chert s.s., according to Smith's (1960) definition, and other rocks such as phtanite, diatomite, porcelanite, hornstone, flint, and radiolarite. In the "chert sequence" other rocks, such as the ubiquitous shale partings ± metal-bearing) ochres, umbers, metalliferous nodules and even volcanogenic and/or carbonatic debris are included.

TABLE 1. ALPINE OCCURRENCES OF BLUESCHIST-FACIES "OCEANIC" CHERTS IN A COHERENT
STRATIGRAPHIC RELATIONSHIP WITH OPHIOLITES

No.	Locality, country	Detected minerals	P/T conditions	Reference
1	Täsch, CH*	Brn, Sp, Pm, Qz, Phg, (Cam, Phl, Tur)	n.g.	Bearth & Schwander, 1981
2	Saas Fee, CH*	Brn, Sp, Pm, Qz, Phg	n.g.	Abs-Wurmbach et al., 1983
3	NE Plan Maison B, I*	Brn, Hm, Qz, Sp, Pm, Phg, Ard	10/450	Dal Piaz et al., 1979
4	Plan Tendre, I*	Brn, Sp, Pm, Lcp, Ves, Phg, (Cc, Tur)	10/450	
5	Cignana Lake, I*	Brn, Sp, Pm, Qz, Phg, (Chl, Cam, Phl)	10/450	Abs-Wurmbach et al., 1983
6	St. Marcel, I*	Brn, Qz, Pm, Sp, Rhd, Pxm, Phg, ...	10-15/450-480	Mottana & Griffin, 1979 Martin-Vernizzi, 1982
7	Fénis, I*	Brn, Qz, Sp, Pm, Phg	n.g.	Castello, 1981
8	Liconi, I*	Qz, Sp, Brn, Phg, Pm	n.g.	Castello, 1981
9	Olbicella, I*	Qz, Gnt, Phg, Chl, Ep, Brn, Hm, Cam	10-11/430	Chiesa et al., 1976, 1977
10	Rio Bruxé, I*	Qz, Gnt, Phg, Chl, Ep, Brn, Hm, Ctd	10-11/430	Chiesa et al., 1976, 1977
11	Monte Spassoja, I*	Qz, Gnt, Pm, Nam, Phg, Chl, Brn, Ru	8/400	Chiesa et al., 1976, 1977
12	Carpenara, I*	Qz, Gnt, Pm, Nam, Phg, Chl, Brn, Hm	8/400	Chiesa et al., 1976, 1977

*CH = Switzerland, I = Italy

TABLE 2. ALPINE OCCURRENCES OF BLUESCHIST-FACIES "OCEANIC" CHERTS IN REWORKED VOLCANO-SEDIMENTARY
SEQUENCES, NOT COHERENT BUT EVIDENTLY RELATED TO OPHIOLITES

No.	Locality, country	Observed minerals	P/T conditions	Reference
1	Bonneval, F*	Rdc, Pxm, Tep, Nam, Sp, Rhd, Qz, Son,...	6-8/450-470	Chopin, 1978
2	Bessans, F*	..Alg, Frd, Ktn, Pm, Ard, Brn, Hm	6-8/450-480	Chopin, 1978
3	Tinos, GR*	Qz, Pm	n.g.	Reinecke, 1983
4	Eubea, GR*	Brn, Pm, Qz	n.g.	Reinecke, 1983
5	Vitali, Andros, GR*	Qz, Pm, Thu, Cc, Ab, Ms, Cym, Cs	>9.5/400⟶	Reinecke, 1983
6	Petalon, Andros, GR*	Qz, Sp, Cpx, Ard, Pm, Hm, Nam, etc.	⟶5-6/350-450	Reinecke, 1983
7	Apikia, Andros, GR*	Qz, Pm, Sp, Ard, Hm, Cpx, etc.	"	Reinecke, 1983
8	Rdisa-Vori, Andros, GR*	Brn, Rhd, Rdc, Qz, Ab, etc	"	Reinecke, 1983

*F = France; GR = Greece.

TABLE 3. ALPINE OCCURRENCES OF BLUESCHIST-FACIES "OCEANIC" CHERTS IN VOLCANO-SEDIMENTARY SEQUENCES
PRESUMABLY OF SHELF OR CONTINENTAL ORIGIN

No.	Locality, country	Observed minerals	P/T conditions	Reference
1	Bec Forciù, I*	Qz, Pm, Sp, Ep, Cam, Brn, Hm	3.5/450	Dal Piaz et al., 1979
2	NE Plan Maison A., I*	Qz, Gnt, Phg, Pm, Brn, Hm	3.5/450	Dal Piaz et al., 1979
3	Motta di Pleté, I*	Qz, Sp, Pm, Phg, Brn, Hm, Nam, Cpx, Cc	3.5/450	Dal Piaz et al., 1979
4	Belvedere di Alagna, I*	Qz, Phg, Sp, Pm, Brn, Hm, Chl	3.5/450	Dal Piaz et al., 1979
5	Varanche, I*	Brn, Sp, Pm, Ard, Phg, Rdc	n.g.	Millosevich, 1906; Pelloux, 1946
6	Cogne, Alpe Pila, I*	Qz, Sp, Acm, Nam, Phg, Ep, Carb	n.g.	Castello, 1981
7	Salbertrand, I*	Brn, Qz, Sp, Hm, Rhd, Carb±Chl	n.g.	Lincio, 1922
8	Combe Bremond, F*	Cc, Brt, Pyrosmalite, Brn, Bixbyite, Hus, Rhd	n.g./300-400	Bourbon & Fonteilles, 1972
9	Val Ceresolo, I*	Qz, Cdt, Ep, Phg, Ab, Tur	10/400	Chiesa et al., 1976, 1977
10	Scortico, I*	Rdc, Pxm, Qz, Tep, Sp, Hm, Pyrite	7-8/330-365	Di Sabatino, 1967
11	M. Corchia, I*	Cc, Qz, Pm	n.g.	Battaglia et al., 1978
12	Moramanno, I*	Qz, Sp, Rhd, Phg, Hus, Brn	6-9/250-340	Busato, 1984

*I = Italy.

separated according to their field relation with ophiolites. In fact four major types are known:

1. Cherts in stratigraphic coherency with ophiolites according to the Steinmann's Trinity, and therefore of undisputed "oceanic" derivation.

2. Cherts in ophiolite-bearing disrupted volcano-sedimentary sequences, the relationship of which with the Steinmann's Trinity, and therefore with the "oceanic" environment of deposition, is reasonably well established.

3. Cherts contained in continental or shelf sequences, which may or may not include mafites; the "oceanic" derivation of these cherts may sometimes be suspected, but they may also have deposited very near the continent.

4. Scattered occurrences of cherts the origin of which can-

TABLE 4. ALPINE OCCURRENCES OF BLUESCHIST-FACIES CHERTS OF UNSTATED OR DUBIOUS AFFINITY

No.	Locality, country	Observed minerals	P/T conditions	Reference
1	Feglievec, Alagna, I*	Rdc, Tep, Pxm, Rhd, Ktn, Qz, Sp, Brn	14/550 5/525	Peters et al., 1978 Abs Wurmbach et al., 1983
2	Molino di Roje, I*	Qz, Phg, Sp, Pm, Nam, Brn	14/550	Dal Piaz et al., 1979
3	Sparone, I*	Qz, Sp, Phg, Ep, Nam, Stm, Cc	15/300-400	Minnigh, 1979
4	P. Forcola, Corio, I*	Brn, Sp, Ard, Pm, Rhd, Phg	n.g.	Gennaro, 1925
5	Voragno, Ceres, I*	Brn, Qz, Pm, Sp, Ard, Phg	n.g.	Zambonini, 1922 Gennaro, 1925
6	Chiaves, I*	Brn, Rhd, Qz	n.g.	Roccati, 1906
7	Villar Dora, I*	Rhd, Qz	n.g.	Roccati, 1906
8	Viù, I*	Rhd, Brn, Qz	n.g.	Roccati, 1906
9	Acceglio, I*	Sursassite, Qz	n.g.	Lombardo (pers. comm.)

*I = Italy.

not be even guessed, mostly because of the pervasive effects of metamorphism.

Cherts listed in Tables 1 to 4 belong only to sequences bearing evidence of having undergone blueschist-facies metamorphism, although they may also show evidences of re-equilibration under lower-pressure and/or higher-temperature conditions. For numerous cherts of the alpine belt metamorphosed under conditions other than the blueschist facies, or where the HP imprint was completely obliterated by the later LP (Lepontine) overprint, the reader is referred to the recent works of Peters et al. (1980), Cortesogno et al. (1979) and Schreyer et al. (1975). A number of these, and all the localities mentioned in Tables 1 to 4 are plotted in the tectono-metamorphic sketch-map of Fig. 1.

GEOCHEMISTRY

The chemical composition of deep-sea opal of biogenic precipitation, as given by Donnelly and Merrill (1977) (Table 5, column 5), consists of 88% SiO_2, 11% H_2O and ~1% of other oxides trapped in the hydrated silicate framework. These other oxides are responsible for the stabilization of the disordered cristobalite-like structure characteristic of opal-A (Jones and Segnit, 1971). Cherts recovered from the oceans have compositions diverging from that of pure opaline matter for a variety of reasons:

a) biogenic silica was diluted at the spreading axis by volcanogenic debris derived from basalts, hyaloclastites, and palagonites; the proportion depends on the distance from the spreading axis, the eruptive capacity of the volcanoes, and the turbulence of the sea bottom;

b) metallic elements of "hydrothermal" origin coprecipitated with biogenic silica in decreasing amounts with increasing distance from hydrothermal vents;

c) "hydrogenous" elements coprecipitated with biogenic silica, in quantities depending on the speed and time of the sedimentation process as well as on the depth of the scavenged water column;

d) various elements of continental, or even extraterrestrial derivation, "diluted" the biogenic materials by the combined effect of submarine water flows and distance from the continental source;

e) diagenetic processes concurrent with the transformation of opal-A to opal-CT (Jones and Segnit, 1971), or even to chalcedony, leached away exchangeable cations together with an increasing amount of water; such a "maturation" is favored by time and temperature (Ernst and Calvert, 1969) and it is contrasted by the impurities initially present in the inconsolidated sediment (Robertson, 1977).

The composition of bottom-sampled cherts is influenced mostly by the first four factors. On the other hand, the composition of cherts sampled on land, and therefore in an advanced stage of lithification and diagenesis, depends on their maturation. Even in cherts still preserving opal as a major phase the amount of H_2O is down to 5-1%. Conversely, silica and the other metals increase, partly vicariously, partly as the result of the chemical interaction with shale partings driven by the migration of water.

A clear distinction between cherts s.s. (radiolarites) and their shale partings, believed to have formed as ambient muds accumulated slowly in the periods between the deposition of the radiolarian-rich turbidity currents (cf. Nisbet and Price, 1974; Barrett, 1982) is possible and easy in sedimentary and diagenetic sequences up to the zeolite facies (e.g. eastern Liguria, Barrett, 1982; Folk and McBride, 1978; Thurston, 1972). The distinction fades away progressively at higher grades, except in very special cases.

In fact there is a continuous chemical change in the overall composition of metacherts with metamorphic grade. I interpret this as due mostly to the homogenization of the chert sequence by elemental migration in the fluid phase during the dehydration reactions accompanying progressive metamorphism. Table 5 and Fig. 2 show that such an homogenization occurs mainly with leaching out of SiO_2 and entry of MgO, CaO, and MnO into the system with increasing metamorphic grade; other components, although showing irregular variations (that may depend in part on statistical reasons) also tend to increase.

Trace elements are imperfectly known in HP-metamorphic cherts (Table 5) and their statistics are biased because most data do in fact refer to ore prospecting areas. With the

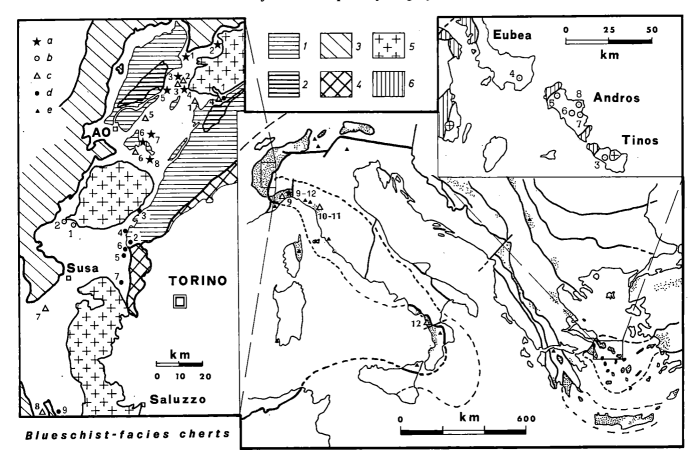

Figure 1. Structural sketch-map of the central portion of the Alpine Belt (Alps, Apennines and Helle-nides) showing areas affected by blueschist-facies metamorphism (stippled) and the localities of chert outcrops. The inset to the left is the portion of the western Alps where most cherts are located; that on the right is an enlargement of the northern Cycladic Islands in Greece. Symbols: 1 = subducted continental crust (Sesia Zone and Dent Blanche Nappe); 2 = Second Diorito-kinzigitic Zone and Valpelline series; 3 = Brianconnais units s.l.; 4 = Ivrea-Verbano Zone and ultramafic massifs; 5 = Granitoid massifs (partially subducted); 6 = Attica autochthonous units; a (stars), oceanic cherts as in Table 1; b (open circles), oceanic cherts as in Table 2; c (open triangles), reworked cherts as in Table 3; d (dots), cherts of dubious derivation (Table 4); e (triangles), cherts not metamorphosed under blueschist facies. Numbers as in Tables 1 to 4.

exception of Cu, trace elements are very low in amount. In particular, base metals (Co, Ni, Zn) are present in such low amounts as to suggest a clear distinction between the mode of formation of Mn-enrichment in cherts and that in the manganese nodules occurring in the oceanic deeps (Bonatti, 1980; see also Brown, this volume).

Blueschist-facies cherts are enriched in easily mobile trace elements such as Sr and Ba with respect to cherts metamorphosed under the zeolite facies, or merely diagenetic; moreover they contain a little more Zr, Y and Pb. This suggests a continuous process of homogenization of the chert sequence, with trace elements originally in the metalliferous clay partings migrating into radiolarites. Moreover, there is a contribution from the underlying metabasites undergoing prograde metamorphic reactions involving release of water-rich fluids previously incorporated in the basalts during the oceanic stage.

Strangely enough, when compared to the vigorous actualistic trend of modern geology, very little has been done to apply to subducted cherts the same geochemical parameters that proved fruitful in permitting fine distinctions on origin mixing and maturation of sedimentary cherts. The available data are mostly scattered, particularly for the sequences undoubtedly connected with ophiolites because they maintain a stratigraphic coherency even after the polyphasal alpine metamorphism. Such an obvious geological situation might in fact be the cause of the lack of interest. This is also possibly the reason why systematic work on HP cherts has been carried out on reworked and/or redeposited sequences.

The Andros sequence (Table 2) has compositional variations that were interpreted by Reinecke (1983) as derived from a sedimentary mixture of biogenic opal (or detrital quartz: Si corner of Fig. 3) with a pelagic clay, rich in Al as well as Mn derived

TABLE 5. AVERAGE CHEMICAL COMPOSITION (ON A VOLATILE-FREE BASIS)
OF BLUESCHIST-FACIES CHERTS AND OTHER CHERTS OF VARIOUS GRADES

		BS		GS	Z	D	S
	n	$\bar{x}$	σ	$\bar{x}$(9)	$\bar{x}$(129)	$\bar{x}$(12)	

Major elements (wt.%)

SiO_2	38	72.60	9.54	73.20	85.81	89.96	>99
Al_2O_3	38	7.61	4.05	13.88	5.45	5.75	0.19
Fe_2O_3*	38	9.31	5.78	4.23	5.41	1.97	0.10
(FeO)	27	(2.73)	(3.15)	(1.81)	(0.77)	-	-
MnO	38	4.36	5.70	2.49	0.14	0.00	0.00
MgO	38	1.50	1.38	1.43	1.23	0.57	0.03
CaO	38	1.97	2.17	0.89	0.38	0.25	0.00
Na_2O	38	1.23	2.11	1.28	(0.07)	-	0.39
K_2O	38	1.02	0.95	1.70	1.31	1.18	0.09
TiO_2	38	0.30	0.18	0.79	0.21	0.23	0.03
P_2O_5	38	0.09	0.05	0.12	-	-	-

Trace elements (ppm)

As	7	375	420	-	-	422	-
Ba	5	960	1560	-	263	110	9
Co	11	126	107	-	50	25	2
Cr	11	68	38	-	52	23	8
Cu	24	1667	1712	-	118	24	29
La	5	44	35	-	-	-	1
Ni	24	133	64	-	48	34	28
Pb	11	145	85	-	7	11	-
Rb	5	15	6	-	-	61	5
Sr	5	118	49	-	28	36	-
Y	11	29	11	-	5	10	-
Zn	14	253	125	-	53	54	-
Zr	11	180	72	-	33	33	-

Notes: BS = blueschist facies; GS = greenschist facies; Z = zeolite
facies; D = diagenetic cherts; S = deep-sea opaline silica.
*total iron oxide expressed as Fe_2O_3.

from the halmyrolytic alteration of basaltic tuffs. The Pm+Sp quartzites (abbreviations of minerals are explained in the Appendix) were derived from these by blueschist facies metamorphism. Reinecke (1983), moreover, interpreted as a mixture of opal and authigenic smectites (nontronite and Fe-rich montmorillonite) the source rock giving way during the HP-LT metamorphism to the Npx- or Crs-bearing quartzites. He justified the high Na_2O content of these rocks as due to absorption in the smectites during sea-water deposition.

The MnO/TiO_2 ratios calculated from Reinecke's data (1983) (Fig. 4) imply a formation of the Andros chert sequence in the deep ocean floor at a distance from land >1000 km and at a water depth of 6000 m (Sugisaki et al., 1982).

MINERALOGY

Over 100 mineral species have been identified in HP metamorphic cherts. Several of them are rare or unique species; others are varieties having Mn as a significant component. Many are secondary, as they occur in hydrothermal veins or are the product of retrogressive transformation from the blueschist facies assemblage under decreasing P (and increasing H_2O activity). Yet the number of primary phases is rather high, when compared with the limited number of phases occurring in common rocks like pelitic schists or quartzo-feldspathic gneisses.

Such a rich mineralogy, and the related complex mineral equilibria, make cherts very promising potential markers for P, T reconstructions, to supplement P-T grids derived for metabasic or metapelitic rocks. However, this is not so at the present moment, for lack of adequate calibration.

In order to review such a large number of mineral species in a concise, yet effective way, I have tried to follow a rigorous scheme. The minerals are listed according to crystal-chemical criteria, and each description includes appearance, association, chemical variation, valence states of Mn and Fe, and potential value for the establishment of metamorphic conditions.

The latter point is the most difficult; the present mineralogy of the alpine cherts is in fact the result of several phases of metamorphism, so that it includes HP-LT as well as LT or LP phases. Not all minerals give inherent clues to decide whether or not they are interesting to the blueschist facies. Moreover, the bulk rock oxidation conditions, at constant P and T, constrain the growth or breakdown of some minerals, so that it is often difficult to state whether an observed phase is indeed a HP phase, or simply is an indicator of $f(O_2)$.

Broadly speaking, three types of assemblages can be distinguished in cherts on the basis of the bulk oxidation ratios: 1)

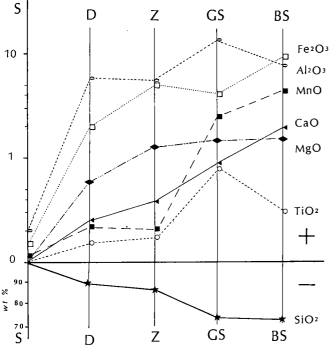

Figure 2. Variation of major chemical elements in cherts with increasing maturation/metamorphic grade. Symbols: S = present deep-sea opaline silica; D = diagenetic chert sequence, Tuscany; Z = zeolite facies chert sequence, eastern Liguria; GS = greenschist-facies cherts, Combin Zone of the Alps, and other alpine localities; BS = blueschist facies cherts, Zermatt-Saas Zone of the Alps and Andros. Averages compiled from various sources and scattered individual data (see Table 5).

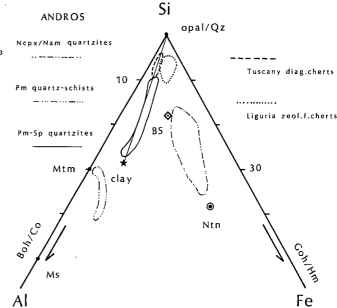

Figure 3. Compositional variation of Andros blueschist facies cherts in terms of Si-Al-Fe^{tot} (atomic percent). The compositional variation of diagenetic and zeolite facies cherts is also shown (compiled from Reinecke, 1983; Barrett, 1981; etc.). Symbols: BS = average blueschist-facies chert (see Table 5); Clay = average deep-sea clay (Reinecke, 1983); for other symbols see abbreviation list.

oxidized, i.e. containing Fe^{3+} and Mn^{3+}; 2) reduced, with Fe^{2+} and Mn^{2+}; 3) neutral, with Mn^{2+} and Fe^{3+}. More on this will be discussed in the section on phase compatibility.

Quartz

It is the commonest mineral occurring in cherts; however, because it is usually disregarded in the descriptions no chemical or physical data are available.

Most primary quartz is fine-grained, granoblastic to xenoblastic, and shows wavy extinction. Large undeformed crystals occur in veins and in pressure shadows around braunite and garnet. Chalcedony (fibrous quartz) is restricted to late veins. No opal survived metamorphism, or formed under the final, post-deformational conditions.

Garnet

Although not the commonest component, garnet is a conspicuous field indicator of metamorphosed manganiferous cherts. Usually it occurs in fine-grained, euhedral crystals associated with quartz, forming irregular patches of an orange to yellow colour. Less commonly, when associated with pyroxenes or hydrous minerals, it occurs as large euhedral crystals, visibly zoned in

different shades of yellow. Under the microscope a weak birefrangence can often be seen.

The MnO content of garnet varies from a maximum of 41.0 wt% (Table 6) down to about 15% with a large predominance of values around 35%. Charge balance calculations are equivocal, but optical absorption spectra and theoretical considerations (Frentrup and Langer, 1981) unequivocally suggest that Mn is almost totally divalent, whereas Fe may be either trivalent or divalent. Thus high-Mn garnets can be computed essentially as binary solid solutions of Sp with And (±Gro), while low Mn-garnets appear to be solid solutions of Sp with Alm (± Andr ± Gro) (Fig. 5). Textural evidence and zoning profiles show that Alm-Sp garnets formed later than Sp-And garnets. This feature is well evidenced at Andros, where there is no continuity between different compositional fields. The St. Marcel garnets are richer in Fe than in other chert occurrences, usually as the Alm component (Table 6, n. 2). Fe is maximum in garnets from a peculiar Gnt + Mt-quartzite associated with cherts at Andros (Table 6, n. 6). These garnets can be calculated essentially as solid solutions of And (Fe as Fe^{3+}) and calderite (Mn as Mn^{2+}) having as much as 41 mol% of the latter unusual component.

The zonation profiles of garnets depend essentially on f(O$_2$) variations during metamorphism: where f(O$_2$) increases, the rims are usually And-richer than the cores; where f(O$_2$) is constantly low, rims are either Sp-rich over Gro-rich cores, or are Alm-rich over Sp-rich cores. This is usually the case in Piemonte.

There is no evidence that the Sp content increases with

A. Mottana

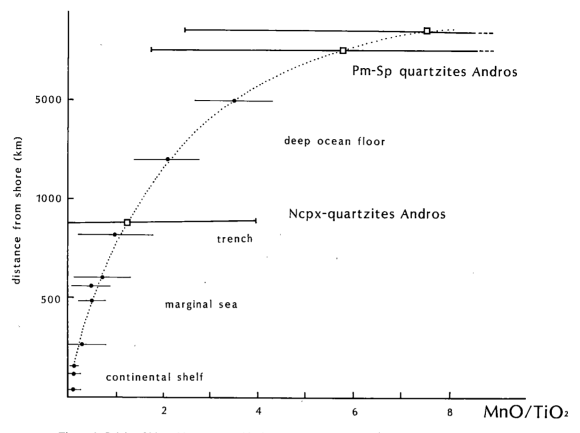

Figure 4. Origin of blueschist metamorphic Andros cherts on the basis of the MnO/TiO$_2$ metal ratio calibrated by Sugisaki et al. (1982) on sedimentary cherts of known deposition.

increasing oxidation ratio of the host rock, as suggested by Hsu (1968). In fact, the purest spessartines are often found in the "reduced" assemblages (Table 6, n. 3). In these HP-LT garnets TiO$_2$ and MgO are minor: the former never exceeds 0.40 wt% except when in association with carbonates, where it may approach 1 wt% (Griffin, unp. res.); the latter may reach as much as 8.55 wt% in some unusual Alm-poor rims (Table 6, n. 4) but it is usually limited to <2 wt%, and may even fall below the detection limit (<0.2 wt%, Reinecke, 1983). Chopin (1978) points out that MgO, although minor, is a sensitive indicator of f(O$_2$) conditions; it falls between 0.05 and 0.20 wt% in "reduced assemblages," while it varies between 1 and 2 wt% in "oxidized assemblages."

A peculiar Uv-rich andraditic garnet (Table 6, n. 7) occurs in a special horizon at St. Marcel, in association with other Cr-rich minerals (see below).

Tephroite

It is present only in certain carbonate-rich lenses included in the chert sequence (Peters et al., 1978; Chopin, 1978), in reduced assemblages dominated either by rhodochrosite or by humite-group minerals (Chopin, 1978). Its composition is commonly rather pure (Table 7, n. 1-3), with amounts of MgO, FeO, and CaO barely summing up to 2 wt% (Peters et al., 1978; Reinecke,

1983). Only a few tephroites of the Haute-Maurienne reach about 10 mol% of Fo + Fa end-members (Table 7, n. 3).

Titanite

The typical titanite occurring in the blueschist assemblage is the pink variety greenovite (Mottana and Griffin, 1979; Table 7, n. 4-5). This is not only present in equilibrium with omphacite (St. Marcel) but it also occurs as large euhedral crystals in late piemontite veins and in patches where omphacite has broken down to albite + diopside (St. Marcel). Some greenovites replace rutile. Greenovite owns its pink colour to the high Mn: Fe ratio; both cations are divalent and replace Ca in the 7-fold coordination. Green titanite (Table 7, n. 6), when present, is always retrograde or fills late veins.

Mn-Humites

These minerals are rare in the metachert sequence, having been described only twice (Chopin, 1978; Dal Piaz et al., 1979). However, if present, they are important markers, since they are limited to thin layers of peculiar composition, characterized by very high MnO content, undersaturation in SiO$_2$, and very low f(O$_2$) ("reduced assemblages," Chopin, 1978). Alleghanyite is the

TABLE 6. REPRESENTATIVE ANALYSES OF GARNET

	1	2	3	4	5	6	7	8
SiO_2	36.17	37.71	35.9	38.4	36.8	34.4	35.57	36.73
TiO_2	0.42	0.25	0.18	-	0.1	-	-	0.20
Al_2O_3	18.84	20.19	20.4	21.8	18.7	0.4	0.62	19.67
Fe_2O_3	2.0	-	0.42	-	3.2	29.7	22.22	1.80
FeO	-	16.75	-	0.56	11.1	0.4	2.51	2.46
MnO	41.03	16.39	40.0	25.3	24.3	17.1	3.94	34.82
MgO	0.34	1.94	0.20	8.55	0.2	0.2	-	0.44
CaO	1.75	8.57	2.2	3.81	6.0	18.0	27.13	5.06
Cr_2O_3	-	-	-	-	-	-	7.81	-
Total	100.55	101.80	99.3	98.4	100.4	100.0	99.80	101.18

1. St. Marcel (Griffin, unp.; SM67): unzoned, in assemblage with Rhd, Pxm, Hm, etc.
2. St. Marcel (Martin-Vernizzi, 1982; 2/11: rim, in equilibrium with Cpx (see Table 10, n. 7).
3. Haute-Maurienne (Chopin, 1978; 84-9): in equilibrium with Son in a Carb-poor reduced assemblage.
4. Lago Cignana (Dal Piaz et al., 1979; 4138-13r): rim, enriched in Py-component.
5. Andros (Reinecke, 1983; 79/292): in equilibrium with Npx (see Table 10 n. 8).
6. Andros (Reinecke, 1983; 80/234): calderite-rich Gnt in Mt-bearing quartzite.
7. St. Marcel (Colomba, 1910): Uv-rich Gnt in equilibrium with Kos-rich Cpx (see Table 9 n. 6).
8. Average of 174 garnets in metamorphic cherts of all grades.

most common species found. In the Haute-Maurienne it is usually associated with sonolite (Table 7, n. 7-8) in the absence of the intermediate member manganhumite. The intergrowths between the two have been explained by Chopin (1978) as an equilibrium relation due to the preferred incorporation of TiO_2 in sonolite; however, they may also derive from the breakdown of manganhumite, as suggested by Winter et al. (1983).

The alleghanyite described by Dal Piaz et al. (1979), unusual in that it is rich in CaO + MgO (~5 wt%) and Al_2O_3 (up to 1.83 wt%), has been shown by X-ray powder diffraction to be leucophoenicite (Winter et al., 1983), a dimorph of Mn-humite.

Friedelite

This rare mineral has been detected only once (Chopin, 1978), as a late-formed phase in a carbonatic humite-bearing reduced assemblage. It is strictly associated with rhodochrosite and/or sonolite. Compositionally, it is extremely close to the theoretical formula. It forms, most probably, because it is the only phase able to incorporate the Cl present in the rock, while the coexisting Mn-humites incorporate F. Pyrosmalite, a mineral of the same group as friedelite, has been detected in an oxydized carbonatic assemblage (Bourbon and Fonteilles, 1972).

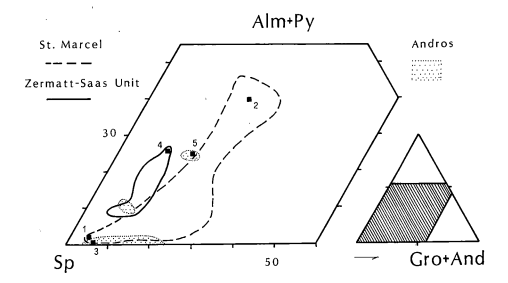

Figure 5. Compositional variation of garnets in the triangle Mn^{2+} (Sp)–Mg^{2+} + Fe^{2+} (Alm + Py)–Ca^{2+} (Gro + And) substitutions (components). Numbered dots refer to samples in Table 6.

TABLE 7. REPRESENTATIVE ANALYSES OF TEPHROITE, TITANITE, AND Mn-HUMITES

	1	2	3	4	5	6	7	8
SiO_2	28.42	29.7	29.9	30.9	30.8	30.5	24.4	26.6
TiO_2	1.19	–	–	38.2	38.5	38.8	0.38	3.2
Al_2O_3	0.00	–	0.12	1.3	0.72	1.1	0.12	0.12
Fe_2O_3*	–	–	–	0.40	0.11	0.52	–	–
FeO*	0.29	0.99	4.6	–	–	–	0.09	1.47
MnO	68.89	68.9	64.2	0.34	0.99	0.42	68.8	62.7
MgO	1.21	0.70	1.70	–	–	–	1.13	1.71
CaO	0.00	0.16	0.12	28.4	26.9	27.2	0.64	0.16
Na_2O	–	–	–	–	0.09	0.04	–	–
F	–	–	–	–	–	0.2	–	–
Total	100.00	100.45	100.6	99.44	98.11	98.78	95.6	96.0

1. Tephroite, St. Marcel (Griffin, unp.; SM19): in a reduced assemblage with Sp, Rhd, Pxm, Rdc.
2. Tephroite, Alagna (Peters et al., 1978; Mn 3K III): in a reduced assemblage with Rhd and Ktn.
3. Tephroite, Haute-Maurienne (Chopin, 1978, 84-2): "the most substituted one," in reduced carb. assemblage.
4. "Greenovite," St. Marcel (Mottana and Griffin, 1979; SM96): pale pink inter-grown with Omp and Ab.
5. "Greenovite," St. Marcel (Mottana and Griffin, 1979; SM95): redbrown, in a vein, intergrown with Pm.
6. Titanite, St. Marcel (Mottana and Griffin, 1979; SM94): green, in an Omp-Ab-Carb-schist.
7. Alleghanyite, Haute-Maurienne (Chopin, 1979, 163): in association with Glx, Ilm, and Rbk.
8. Sonolite, Haute-Maurienne (Chopin, 1978; 84-9): with Sp (see above) in a reduced carb-poor assemblage.
*Total iron oxide expressed either as Fe_2O_3 or FeO.

Epidotes

Colour and pleochroism make piemontite one of the most conspicuous minerals of metacherts, although it is not the most common one. Its presence is in fact restricted to "oxidized assemblages."

The chemistry of epidotes in metacherts (Table 8 and Fig. 6)

appears to depend on the availability of certain cations rather than on the P,T conditions of metamorphism, as suggested by Miyashiro and Seki (1958), and it is critically controlled by the partial pressure of oxygen. Piemontites are richest in Mn^{3+} when coexisting with other Mn-bearing phases. Extreme Mn^{3+}-bearing piemontites are to be found in equilibrium with braunite (Chopin, 1978) or ardennite (Reinecke, 1983); those coexisting

TABLE 8. REPRESENTATIVE ANALYSES OF EPIDOTE GROUP MINERALS

	1	2	3	4	5	6	7	8
SiO_2	35.83	34.78	37.2	36.40	37.27	38.44	39.4	39.2
TiO_2	–	–	–	0.01	–	0.19	–	–
Al_2O_3	17.28	16.42	21.4	18.92	22.19	28.71	31.8	32.0
Fe_2O_3	0.59	1.59	0.6	5.56	14.39	6.86	0.1	1.0
Mn_2O_3	20.43	19.21	15.3	13.24	0.86	0.12	2.0	1.1
MnO	–	0.15	0.9	2.58	0.77	–	–	–
MgO	0.11	0.00	–	0.19	0.00	0.10	–	0.2
CaO	22.18	17.03	22.6	20.47	21.36	24.04	24.6	24.6
SrO	0.78	8.53	–	–	1.89	0.30	–	–
Total	97.20	97.71	98.0	97.37	98.31	98.76	97.9	97.9

1. Piemontite, St. Marcel (Mottana and Griffin; in prep., SM88), (sum includes BaO = 0.00).
2. Piemontite, St. Marcel (Mottana and Griffin; in prep., SM7), (sum includes BaO = 0.10).
3. Piemontite, Andros (Reinecke, 1983; 79/526); coex. with Thl n. 7.
4. Piemontite, St. Marcel (Smith et al.; 1982); Na_2O = 0.02.
5. Pistacite, St. Marcel (Mottana and Griffin; in prep., SM62), (sum includes BaO = 0.00).
6. Clinozoisite, St. Marcel (Mottana and Griffin; in prep., SM304), vein.
7. Thulite, Andros (Reinecke, 1983; 79/529), coex. with Pm n. 3.
8. Thulite, Andros (Reinecke, 1983; 79/500).

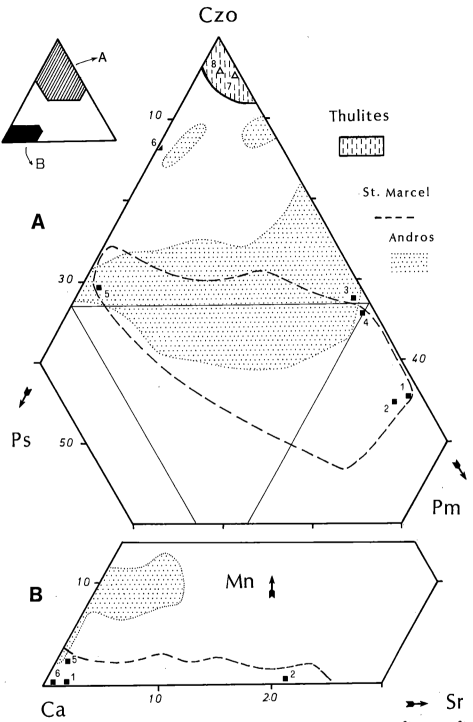

Figure 6. Compositional variation of epidotes. (A) trivalent substitutions in the triangle Al^{3+} (Czo)–Fe^{3+} (Ps)–Mn^{3+} (Pm). (B) divalent substitutions in the triangle Mn^{2+}–Ca^{2+}–Sr^{2+}. Numbered dots refer to samples in Table 8.

with spessartine are significantly depleted in Mn^{3+} and enriched in Al and/or Fe^{3+}, and may even contain a little Mn^{2+} substituting for Ca in the A site.

Another peculiarity of piemontites in blueschist-facies meta-cherts, as against those of other environments, is the possibility of significant Sr (up to 23 at.% of the total A site at St. Marcel, Table 8, n. 2, and Fig. 6B) beside Mn^{2+} (up to 15 at.%; Reinecke, 1983) substituting for Ca. In general, early-formed piemontites are richer in Sr than late ones, while being at the same time not so rich in Mn^{3+}. Vein piemontites of latest formation are darkest in

A. Mottana

	1	2	3	4	5	6	7	8
SiO_2	31.0	30.5	37.12	35.89	36.9	52.91	52.89	54.60
TiO_2	0.1	–	–	0.12	0.41	0.07	1.04	0.02
As_2O_5	7.4	8.4	–	–	–	–	–	–
Al_2O_3	23.2	22.4	31.89	29.89	15.6	0.57	4.77	2.55
Fe_2O_3	3.6	3.6	–	–	–	11.715	8.44	13.71
Cr_2O_3	–	–	–	–	–	11.95	0.00	0.00
FeO	–	–	3.27	2.83	4.70	0.559	0.00	–
MnO*	22.8	21.2	0.61	3.25	2.45	0.58	0.96	1.78
MgO	3.6	4.2	10.26	9.61	2.69	4.41	9.91	8.70
CaO	3.5	5.0	0.45	0.66	34.6	6.63	16.11	11.51
Na_2O	–	–	3.40	3.02	0.03	9.82	5.91	6.54
Total	100.35	100.53	87.00	85.27	97.4	"99.213"	100.03	99.56

1. Ardennite, Andros (Reinecke, 1983; 80/158): in quartzite, coex. with Gnt (sum
   includes 0.05 V_2O_5; 5.2 H_2O).
2. Ardennite, Andros (Reinecke, 1983; 80/146): in a vein, coex. with Qz and Hm (sum
   includes 0.03 V_2O_5; 5.2 H_2O).
3. Tourmaline, Plan Maison (Dal Piaz et al., 1979; 392): in quartzite, coex. with
   Brn and Pm.
4. Tourmaline, Saas Fe (Bearth and Schwander, 1981; PVB 1731): in quartzite, with
   Gnt and Pm.
5. Vesuvianite, Plan Tendre (Dal Piaz et al., 1979; DBL 761).
6. Clinopyroxene, chromian, St. Marcel (Martin-Vernizzi, 1982; 362/342).
7. Clinopyroxene, titanian, St. Marcel (Martin-Vernizzi, 1982; 200/25-3).
8. Clinopyroxene, "schefferite," St. Marcel (Bondi et al., 1978; To-A): in vein,
   (sum includes K_2O = 0.02).
*Total manganese oxide expressed as MnO.

colour and reach as much as 43 at.% of Mn^{3+} in the octahedral site (Table 8, n. 1), while often being completely devoid of Sr and Mn^{2+}. The multiple generation of piemontite, as a member of the HP assemblage as well as a mobilized-vein constituent and a product of the breakdown of certain pyroxenes, is testified also by the zonation profiles that show either dark-brown Mn^{3+}-rich cores rimmed by light-pink Mn^{3+}-poor rims and even outer rims of green epidote (pistacite: Table 8, n. 5), or cores of allanite and pistacite continuously rimmed by light and/or dark piemontite, rarely with a fringe of pistacite (Mottana and Griffin, 1982).

At Andros (Reinecke, 1983), piemontite may be rimmed also by another, Mn-poor member of the epidote group, the orthorhombic thulite (Table 8, n. 7-8), which occurs as discrete grains or as euhedral rims on allanite grains. Another epidote phase occurring as the outer rim of certain piemontites, or as discrete grains filling veins, is clinozoisite (Table 8, n. 6).

Allanite, the REE-bearing member of the monoclinic epidote group, is present in many cherts, where it acts as nucleus to the growth of piemontite or pistacite (Chopin, 1978) and even thulite (Reinecke, 1983). The Mn content is variable but usually significant. In the case of Andros (Reinecke, 1983), Mn even exceeds the amounts of Al + Fe in the Y position and can be partly reduced to Mn^{2+} substituting for Ca and REE in the A position, as well as in the Y position, by the coupled substitution $^{VIII}Ca^{2+} + {}^{VI}Al^{3+} \rightleftharpoons {}^{VIII}REE^{3+} + {}^{VI}Mn^{2+}$ (Reinecke, 1983).

Ardennite

This rare mineral occurs both as rock-forming in "oxidized assemblages," in equilibrium with piemontite, braunite and spes-

sartine, and in veins and segregations together with quartz and hematite. The complex structure allows the entry of unusual metal ions like Cu (up to 1.1 wt% as CuO, Reinecke, 1983) and As (up to 9.8 wt% as As_2O_5, Reinecke, 1983). V may be present (1 wt% V_2O_5, Chopin, 1978). Two analyses are given in Table 9.

Reinecke (1983) could show a wide substitution of As (+V) for Si (+Al), probably in the form of the coupled substitution (As, V)$^{5+}$ + $Mn^{2+} \rightleftharpoons Si^{4+} + Mn^{3+}$, and a partial substitution of O^{2-} with OH^-, F^-, Cl^-. Furthermore, Mn^{2+} may also be present, substituting for Ca in 7-fold coordination and for Mg and Cu in 6-fold coordination.

The formation of ardennite appears to be controlled not only by the availability of those unusual elements, but also by the intensive conditions of metamorphism; the mineral requires a particularly high $f(O_2)$, since it contains pentavalent As and V in addition to Mn^{3+}.

Tourmaline

Manganiferous tourmaline is present in almost all blueschist-facies cherts, where it acts as the sink for boron derived either from submarine hydrothermal activity or from marine illitic clays. The rare analytical data available (Table 9, n. 3-4) indicate that Mn can be present both in the divalent and in the trivalent state, substituting either for Mg or for Al in the octahedral sites.

Vesuvianite

Regional metamorphic occurrences of this mineral are

TABLE 10. REPRESENTATIVE ANALYSES OF CLINOPYROXENE GROUP MINERALS

	1	2	3	4	5	6	7	8
SiO_2	54.76	57.23	54.52	55.51	58.38	54.6	53.54	51.9
TiO_2	0.0	0.0	0.0	0.16	0.03	–	0.0	<0.1
Al_2O_3	11.86	18.75	5.30	7.74	11.51	0.35	3.24	3.4
Fe_2O_3	17.51	5.70	11.61	2.32	–	5.43	17.82	26.1
FeO	0.0	0.0	0.0	0.0	0.89	–	0.74	3.9
MnO*	0.26	0.66	0.79	3.38	0.85	0.24	1.57	0.5
MgO	0.93	1.93	8.18	9.57	9.16	15.4	6.43	0.3
CaO	1.27	2.58	11.69	13.76	12.79	22.1	10.18	2.2
Na_2O	13.68	13.44	7.59	6.99	6.78	1.88	8.02	11.7
Total	100.27	100.29	99.68	99.43	100.39	99.7	101.54	100.0

1. Aegirinaugite, St. Marcel (Griffin and Mottana, 1982; SM 89 core), pale yellow.
2. Impure jadeite, St. Marcel (Griffin and Mottana, 1982; SM 89 rim), violet.
3. Chloromelanite, St. Marcel (Griffin and Mottana, 1982; SM 36), light voilet.
4. Chloromelanite, St. Marcel (Griffin and Mottana, 1982; SM 102 rim), dark violet.
5. Omphacite, St. Marcel (Brown et al., 1978; # 2), deep violet.
6. Impure diopside, St. Marcel (Dal Piaz et al., 1979; ST. M1).
7. Aegirinaugite, St. Marcel (Martin-Vernizzi, 1982; 69/11) in equilibrium with Gnt (Table 6 n. 2).
8. Impure aegirine, Andros (Reinecke, 1983; 79/292), green, coexisting with Gnt (Table 6 n. 5).
*Total manganese oxide expressed as MnO.

rather rare. In metacherts there is only one demonstrated find (Table 9, n. 5), in an oxidized assemblage.

Pyroxenes

Clinopyroxenes, dominantly falling in the triangle Di-Jd-Acm, are present at a few localities (St. Marcel, Andros, Alagna), where blending with the ophiolite debris and/or the shale partings as well as migration of fluids during the metamorphism allows the chert to acquire the suitable Na_2O-rich compositions. Usually they occur at the top and bottom of the chert sequences. The first clinopyroxenes form with compositions controlled by the initial bulk rock composition; then metasomatic exchanges alter the bulk composition and favor the crystallisation of other cpx, either by breakdown or by new growth controlled by the preexisting phase and/or phase assemblage. A detailed account of the stratigraphic control on the pyroxene chemistry is given by Martin-Vernizzi (1982) for St. Marcel, while Griffin and Mottana (1982) and Reinecke (1983) emphasize the aspects of metasomatic exchange and phase-assemblage control.

Despite these and other studies, the minerogenetic sequence, the solvus relationships and even the systematics of these clino-pyroxenes remain fairly confused. The major problem, still incompletely solved, is the oxidation state of Mn. Charge balance calculations indicate that in over 85% of the purple-blue pyroxenes ("violan") all Fe is present as Fe^{3+} (acmite component) and the major part of Mn as Mn^{3+} (manganacmite). On the contrary, in other colored varieties (green and brown "schefferite") the charge balance is reached by making all Mn divalent and sometimes also a portion of Fe. This implies a substantial variation in $f(O_2)$, i.e. that different buffering conditions were acting during the metamorphic evolution.

The St. Marcel occurrence is particularly complex in this respect, but also exceptionally rewarding: beside proving the existence of the manganacmite component up to substantial amounts (13 mol.%: Table 10, n. 4), the presence of terrestrial Kos-dominated clinopyroxene solid-solutions (up to 36 mol.% Kos: Table 9, n. 6) and the widening of the possible solubility of Ti at HP-LT conditions (up to 1.04 wt.% TiO_2: Table 9, n. 7) could be established here. In the St. Marcel and Andros chert sequences the first formed clinopyroxenes are either green aegirinaugites (Table 10, n. 7) or yellow-brown aegirinjadeites (Table 10, n. 1) or emerald-green aegirinkosmochlors (Table 9, n. 6), depending on the bulk composition of the rock, which is uniformly Fe^{3+}-rich but selectively either Al- or Cr-bearing. The Mn content of these pyroxenes is low (~0.50 wt.% MnO^t) and charge-balance is reached without oxidizing it entirely to Mn^{3+}.

The early-formed aegirinjadeites are mantled by dark purple rims of impure jadeite (Table 10, n. 2), or they are overgrown, or new discrete grains form by other pyroxenes of chloromelanitic composition (Table 10, n. 3), usually purple-blue in color because they are substantially enriched in Mn (2-3 wt.% MnO^t), in the form of manganacmite component.

Chloromelanites are zoned and grade continuously to omphacitic compositions (Table 10, n. 5) by mere decrease of the acmitic components; the change takes place by reduction of Mn first and Fe afterwards, as shown by charge-balance calculation. In such a case these elements are added to Aug in the form of Joh and Hd components.

The final, post-blueschist stage develops with the cpx breaking down into a more or less coarse intergrowth consisting of diopsidic pyroxene + albite ± acmitic pyroxene; depending on the presence of the latter, Mn is absorbed preferentially in this or it remains bound, and is therefore indirectly enriched, in the diop-

Acm(±MnAcm±Kos)

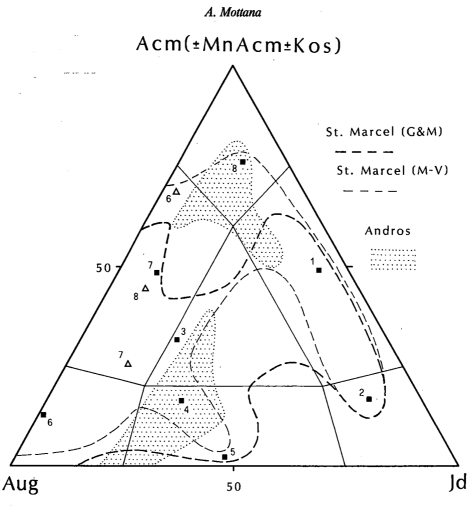

Figure 7. Compositional variation of clinopyroxenes in the triangle Jd-Aug-Acm (±MnAcm ± Kos). Fields for St. Marcel clinopyroxenes determined by Griffin and Mottana (1982: G&M) and Martin-Vernizzi (1982: M-V) are distinguished. Numbered triangles refer to samples listed in Table 9 and dots to samples listed in Table 10.

sidic product (Table 10, n. 6). In addition, cherts are crossed by final veins of brown aegirinaugite ("schefferite"), containing Mn prevalently in the divalent state (Table 10, n. 7).

A second problem is the width of the solvus in the Di-Jd-Acm system under the conditions of the blueschist facies.

Contrary to the indications of Brown et al. (1978), Griffin and Mottana (1982) argue for the non-existence of a solvus between diopside (Jd_3 to Jd_{12}) and omphacite (Jd_{35} to Jd_{50}) and maintain that the only solvus possible is between omphacite (J_{55}) and jadeite (Jd_{85}), closing up to ca. Acm_{20}. Their supporting evidence, supplemented with data from Andros (Reinecke, 1983), is given in Fig. 7.

Pyroxenoids

These minerals do not occur in cherts s.s. but in the carbonate-bearing inlayers, in reduced assemblages containing tephroite and/or a member of the Mn-humite group. Rhodonite and pyroxmangite are often present together, in bladed inter-

growths the equilibrium conditions of which are difficult to decide.

At Alagna (Peters et al., 1978) rhodonite and pyroxmangite are interpreted to be in equilibrium (Table 11, n. 4-5). In the Haute-Maurienne (Chopin, 1978) there are indications that rhodonite precedes pyroxmangite, since it forms the core of the crystals of the latter (Table 11, n. 6-7).

The equilibrium between the two polymorphs is divariant, since rhodonite concentrates Ca relative to pyroxmangite (8-15 mol.% $CaSiO_3$ vs. 3-9% respectively); the latter is enriched in Fe + Mg by a factor of 2 (Table 11 and Fig. 8). The absolute amounts of ions substituting for Mn vary from one locality to the other, probably reflecting the type of carbonate present, the bulk composition of the mother-rock (Maresch and Mottana, 1976; Peters et al., 1980), as well as the local availability of certain elements (Fig. 8). An exceptionally pure pyroxenoid was measured at St. Marcel by Griffin (Table 11, n. 1). Compositionally it falls in the field of pyroxmangite, but it is impossible to confirm this because the grain is entirely included in a grain of hausmannite. Conver-

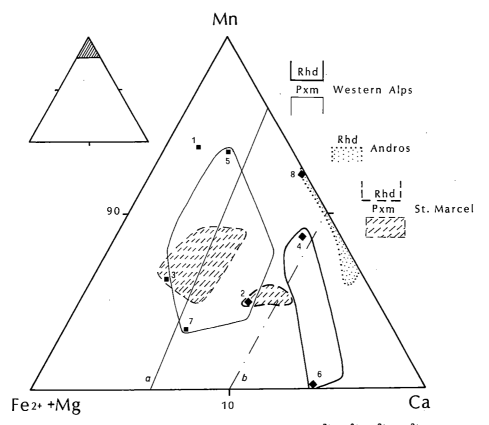

Figure 8. Compositional variation of pyroxenoids in the triangle Mn^{2+}–Ca^{2+}–Fe^{2+} +Mg^{2+}. Numbered dots refer to samples listed in Table 11. The field for rhodonites and pyroxmangites of St. Marcel and other localities of the western Alps are distinguished by diagonal ruling of the former. Lines *a* and *b* mark the compositional region where pyroxenoids can present either the pyroxmangite (left) or the rhodonite (right) structure according to different bulk compositions or P, T conditions (cp. Maresch and Mottana, 1976).

TABLE 11. REPRESENTATIVE ANALYSES OF PYROXENOIDS

	1	2	3	4	5	6	7	8
SiO_2	45.58	46.52	46.43	47.3	46.8	46.5	47.0	46.1
TiO_2	–	–	–	–	–	0.19	0.18	–
Al_2O_3	0.0	0.0	0.0	–	–	0.09	0.09	–
FeO	0.0	0.42	0.78	0.38	0.47	2.0	1.67	<0.2
MnO	52.28	47.86	48.96	48.0	51.1	43.4	46.0	49.3
MgO	1.47	1.88	2.79	0.40	0.68	1.36	2.4	<0.2
CaO	0.59	3.70	1.57	3.89	1.33	5.1	2.7	4.5
Total	99.92	100.42	100.61	99.97	100.38	98.6	100.0	99.9

1. Pyroxenoid included in hausmannite, St. Marcel (Griffin, unp.; SM 8).
2. Rhodonite, St. Marcel (Griffin, unp.; SM 67) coexisting with Pxm and Sp (see Table 6, n. 1).
3. Pyroxmangite, St. Marcel (Griffin, unp.; SM 67) coexisting with Rhd and Sp (see Table 6, n. 1).
4. Rhodonite, Alagna (Peters et al., 1978; Mn 3C VIII) coexisting with Pxm and Rdc.
5. Pyroxmangite, Alagna (Peters et al., 1978; Mn 3C VIII) coexisting with Rhd and Rdc.
6. Rhodonite, Haute-Maurienne (Chopin, 1978; Rh 2/84-3) coexisting with Pxm in a reduced carb. assemblage.
7. Pyroxmangite, Haute-Maurienne (Chopin, 1978; Px 2/84-3) coexisting with Rhd in a reduced carb. assemblage.
8. Rhodonite, Andros (Reinecke, 1983; 80/117) near tephroite in a Hus-bearing rock.

A. Mottana

TABLE 12. REPRESENTATIVE COMPOSITIONS OF AMPHIBOLE GROUP MINERALS

	1	2	3	4	5	6	7	8
SiO_2	54.1	51.6	57.8	59.0	56.68	56.74	55.71	56.74
TiO_2	0.11	0.13	-	-	-	-	-	0.06
Al_2O_3	0.09	0.65	3.1	7.1	0.62	0.43	0.59	0.59
FeO^*	12.7	23.9	11.15	7.55	6.36	3.96	7.52	0.00
MnO^*	11.5	8.1	1.6	0.6	0.43	4.28	3.19	0.60
MgO	17.0	11.1	14.9	14.8	20.61	21.44	18.58	22.89
CaO	0.98	0.83	1.3	0.4	5.05	10.53	10.23	6.94
Na_2O	0.20	1.1	6.8	7.6	5.29	0.54	0.50	4.10
K_2O	0.05	0.09	-	-	0.91	0.04	0.03	4.88
Total	96.7	97.5	96.65	97.05	97.95	97.96	96.35	98.83

1. Tirodite, Haute-Maurienne (Chopin, 1978; 84/8) in a carbonatic reduced assemblage with Pxm, Rhd, Tep, Gnt.
2. Dannemorite, Haute-Maurienne (Chopin, 1978; 265) in a Gnt-bearing quartzite.
3. Mg-riebeckite, Andros (Reinecke, 1983; 80/8 in a Gnt-bearing quartzite (total iron given as Fe_2O_3 = 12.4).
4. Crossite, Andros (Reinecke, 1983; 80/16), in a Piem-Gnt-quartzite (id. Fe_2O_3 = 8.4).
5. Winchite, St. Marcel (Martin-Vernizzi, 1982; 55/25-1), vein, in massive Braunite ore.
6. Actinolite, St. Marcel (Martin-Vernizzi, 1982; 47/26), in a Brn-free, Gnt + Phg-bearing quartzite.
7. Actinolite, Sparren (Bearth and Schwander, 1981; 1617) in a Gnt-quartzite.
8. Potassiumrichterite, St. Marcel (Mottana and Griffin, in prep.), pink, in a vug with Qz (sum includes SrO = 0.63 and F = 1.40).
*Total iron oxide or total manganese oxide expressed as equivalent.

sely, the most Ca-rich rhodonite (Table 11, n. 8) coexists with tephroite in a reduced carbonatic assemblage.

Amphiboles

A variety of amphiboles occurs in blueschist-facies metamorphic cherts, and glaucophane itself is frequently mentioned in petrographic descriptions. However, no glaucophane composition has ever been detected by microprobe; rather a manganiferous crossite, even in cherts directly overlying metabasites containing true glaucophane, as it is in the case of the St. Marcel occurrence.

Most amphiboles are retrograde. The few ones relating to blueschist-facies conditions show a wide spread of compositions.

In the reduced carbonate-bearing assemblages of the Haute-Maurienne (Chopin, 1978), the only amphibole present is a colourless member of the tirodite-dannemorite series, very rich in Mn^{2+} (Table 12, n. 1). In the oxidized assemblages of the same area there is a similar type of amphibole, less Mn and Mg-rich and displaced toward the ferroan end-member (dannemorite, Table 12, n. 2), and occasionally containing tetrahedral Al (Chopin, 1978). More commonly, however, the amphiboles are alkaline, with a composition close to riebeckite (Chopin, 1978). In most alpine occurrences, in fact, alkali-amphiboles are mostly Mg-rich crossites and riebeckites, sometimes zoned from the former to the latter. They are never particularly Mn-bearing (<1.5% wt. MnO); rather they contain iron that can be calculated entirely as Fe^{3+} (Table 12, n. 3-4). Where the bulk composition is enriched in Ca, or depleted in Na, first generation amphiboles are

members of the calcium-sodium series like winchite (Table 12, n. 5; see also Reinecke, 1983) or richterite (Martin-Vernizzi, 1982), usually rather poor in Mn (<0.5% wt. MnO). During the progress of metamorphism both alkali and calcium-alkaline amphiboles become rimmed by calcium amphiboles of the tremolite-actinolite series (Table 12, n. 6-7), seldom trending toward edenite (Martin-Vernizzi, 1982; see also Rondolino, 1936). A typical peculiarity of these rims is the increased content in Mn, reaching the maximum of 4.65 wt% MnO in a ferroactinolite of Zermatt (Bearth and Schwander, 1981), and the decrease of the oxidation ratio with increasing Ca/Ca + Na. All these amphiboles appear to be nearly silica-saturated. In alkali-amphiboles charge balance calculations indicate that most iron is Fe^{3+}, while Mn is Mn^{2+} and enters the Y position. In calcium amphiboles, again most Fe is Fe^{3+}, but in the case of actinolites, and particularly of the ferroactinolites of Zermatt, (Bearth and Schwander, 1981) Fe^{2+} becomes dominant and even exclusive. In such a case Mn is also calculated as divalent but it must be moved from Y to the M4 site to fill this position as diadoch for Ca.

Amphiboles occurring in the veins are fibrous, colourless tremolites (Rondolino, 1936). However, recently potassium richterites containing significant Sr and F have also been discovered (Table 12, n. 8).

Micas

Micas are absent in the "reduced assemblages" (Chopin, 1978), but they are fairly common in the "oxidized" ones, where they occur in multiple generations, both primary and within

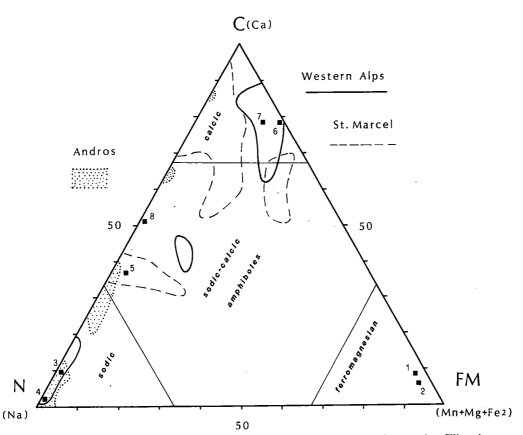

Figure 9. Compositional variation of amphiboles on the basis of the dominant cation filling the structural M4 site. Symbols: N = Na; C = Ca; Fm = Mn + Mg + Fe^{2+}. Numbered dots refer to samples listed in Table 12.

veins. Most mica is phengitic, typical of blueschists and eclogites. A trioctahedral mica is also present ("manganophyllite"). In most cases, this is the destabilization product of phengite, but possibly a primary type also exists (Brown et al., 1978).

Phengite is either colourless, or pink ("alurgite"), or brownish (Dal Piaz et al., 1979; Martin-Vernizzi, 1982), depending on the Mn content. Colourless phengite in the kink-bands of alurgites or in felts in the pyroxene typically contains less than 0.6-0.7% wt. MnO^{tot} (Table 13, n. 2); alurgite has MnO^t variable from very low (~0.2) to a maximum of 2.59 wt.% (Table 13, n. 3) and dominantly in the form of Mn^{3+}. Brownish phengite (grown as lamellae in the cleavage of alurgite, or as fringes at the contact between alurgite and feldspar) contains 1.5-2.0 wt.% MnO^{tot} (Table 13, n. 4). All these micas are strongly celadonite-substituted. In particular the field of brownish phengites and some pink phengites is close to the boundary with celadonite (Fig. 10B). Alurgite n. 1 in Table 13, having Cel = 72.5, is the most celadonitic mica fully characterized thus far from a metamorphic terrane. These micas show a large spread of the paragonitic substitution that is usually as low as zero but may reach as much as 20 mol.% (Table 13, n. 5). This feature is in marked contrast with the known chemical properties of phengites in HP-LT metamorphic terranes. It has been tentatively explained (Martin-Vernizzi, 1982) by the high Na$_2$O content of the metasomatic fluids.

Some phengites are also high in Ti (up to 2.03 wt.% TiO$_2$: Martin-Vernizzi, 1982).

At Andros, white mica is exceptional not only for being a muscovite but also because it contains a high amount of barium (up to ~7 wt.% BaO: Reinecke, 1983). This is certainly due to the anomalous bulk-rock composition, reflected by the presence of other phases like cymrite and celsian (see below).

Yellow phlogopite ("manganophyllite") marginally replacing phengite or growing as fans in the veins crossing the braunite ore is characterized by high TiO$_2$ (up to 1.33 wt.%), variable Na$_2$O (up to 0.89 wt.%), and a variable MnO content, constantly higher than the replaced phengite (2 to 7 % wt.; Table 13, n. 8). The FeO^t content is also very variable (from 0.3 up to 5.9 wt.%; Martin-Vernizzi, 1982). FeO^t + MnO^t correlates negatively with MgO, making it likely that they are, at least partly, in the divalent state. An unusual Si-rich, Al-poor oxyphlogopite in equilibrium with phengite was described by Brown et al. (1978) (Table 13, n. 6). Martin-Vernizzi (1982) questioned the finding, but the mineralogy of St. Marcel is far too variable and the description too complete, also from the crystallographic viewpoint, to suspect an analytical error.

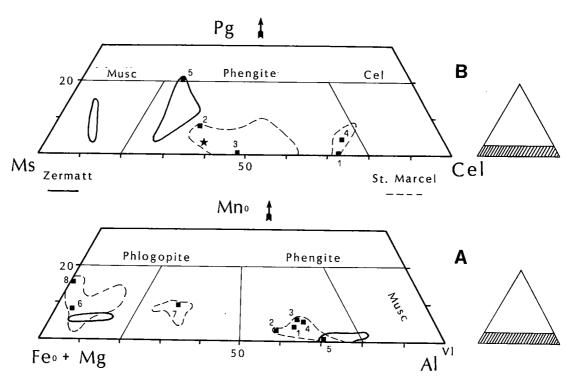

Figure 10. Compositional variation of phengitic and phlogopitic micas. (A) trioctahedral micas in the triangle Al^{VI}–Mn–Fe + Mg; (B) dioctahedral micas in the triangle Pg (=100 Na/Na + K)–Cel (=100 (Si-3))–Ms (=100K/Na + K). Numbered dots refer to samples listed in Table 13; the star is the Ba-bearing muscovite of Andros.

Chlorites

These minerals are frequently mentioned in petrographic descriptions, but very rarely chemical analyses are presented, and even the identification of the 14 Å vs 7 Å structure is often neglected. Chlorites are primary in the "reduced assemblages"; they are mostly secondary in the "oxidized" ones. They are widespread in the late veins, in the form of fans or rosettes.

In the "reduced assemblage" of Haute-Maurienne, Chopin (1978) detected an almost entirely Mn^{2+}-substituted 14 Å thuringite (Table 14, n. 1) in equilibrium with alleghanyite, and marginally replacing galaxite. In the veins crossing the rocks, and in the "oxidized assemblages" chlorites are poor in Mn and span the entire Fe-Mg series (Table 14, n. 2-3).

Talc

The equilibrium assemblage talc + phengite (±chlorite ± phlogopite), was examined in detail by Chopin (1981) in the "oxidized" schists of Haute-Maurienne (Table 14, n. 4) and by Reinecke (1983) at Andros. Such an unusual assemblage had been described by Abraham and Schreyer (1976) in the cherts of Brezovica, Yugoslavia and shown to be a typical indicator of HP metamorphism with little or no superimposed LP retrogradation.

Cymrite

Cymrite (Table 14, n. 8) replaces celsian in both the reduced

and the oxidized assemblages of Andros (Reinecke, 1982, 1983). Texturally it is late in comparison with HP Mn-bearing phases like braunite and piemontite; yet it is a HP-LT mineral, in comparison with the chemically identical assemblage celsian + water.

Feldspars

Albite is present quite often both as a groundmass component and as a breakdown product of HP phases (mainly pyroxenes). It is also a normal constituent of the veins. Compositionally, albite is always very pure, substitutions being limited to 1-2 mol.% An and 1 mol.% Or (Table 14, n. 5). A K-feldspar, indicated as "microcline" but actually not characterized by crystallographic methods, has also been reported; its composition is usually more substituted than albite (Ab_{3-8} An_{0-1}: Table 14, n. 6). Some K-feldspars are Ba-rich (Martin-Vernizzi, 1982; Griffin and Mottana, 1982) and are regarded as early products of HP metamorphism.

Nearly pure celsian (with 1-4 mol.% Or in solid solution: Table 14, n. 7) has been found only at Andros, in discrete grains always rimmed by cymrite.

Braunite

Braunite is the ubiquitous manganese opaque phase in blueschist-facies metacherts. It forms early during diagenesis or in the zeolite facies (Bonatti et al., 1976) by reaction of poorly

TABLE 13. REPRESENTATIVE ANALYSES OF MICA GROUP MINERALS

	1	2	3	4	5	6	7	8
SiO_2	56.10	50.70	51.64	55.99	48.74	44.76	39.44	40.13
TiO_2	0.24	1.14	0.53	0.18	0.21	0.38	0.50	0.73
Al_2O_3	18.98	22.19	21.65	19.40	27.20	10.47	15.16	12.37
Fe_2O_3*	0.63	2.65	2.36	0.33	3.71	0.31	1.72	0.94
MnO*	1.52	1.22	2.59	2.28	0.25	4.09	5.99	7.47
MgO	6.75	6.76	5.05	6.31	4.22	24.10	21.39	22.51
CaO	–	0.13	0.28	0.09	0.23	–	0.13	0.24
Na_2O	0.03	0.58	0.00	0.35	1.55	0.15	0.44	0.42
K_2O	11.20	10.29	11.24	10.88	9.25	9.99	10.11	11.28
F	0.05	–	–	–	–	0.55	–	–
Total	95.50	95.66	95.34	95.81	95.36	94.80	94.88	96.09

1. Manganoan phengite, pink "alurgite," St. Marcel (Brown et al., 1978; 5) polytype 3T (partitioned as Mn_2O_3 1.29 + MnO 0.36 by the authors).
2. Colorless phengite, St. Marcel (Martin-Vernizzi, 1982; 16/25-3).
3. Pink phengite, "alurgite," St. Marcel (Martin-Vernizzi, 1982; 42/346).
4. Brownish phengite, St. Marcel (Martin-Vernizzi, 1982; 27/25-3).
5. Pink mica, paragonite-rich, Plan Maison (Dal Piaz et al., 1979; $\lambda 39_1$), Brn-free, Pm+Gnt-quartzite.
6. Manganoan phlogopite (yellow mica), St. Marcel (Brown et al., 1978; 4) polytype 1M.
7. Yellow phlogopite, St. Marcel (Griffin, unp.; VM-7).
8. Light brown phlogopite, St. Marcel (Martin-Vernizzi, 1982; 7/25-3).
*Total iron oxide or total manganese oxide expressed as equivalent.

TABLE 14. REPRESENTATIVE ANALYSES OF CHLORITE, TALC, CYMRITE, AND FELDSPAR GROUP MINERALS

	1	2	3	4	5	6	7	8
SiO_2	19.7	29.55	26.1	61.04	67.35	63.03	33.0	31.0
TiO_2	0.12	–	–	0.01	–	–	–	–
Al_2O_3	27.0	19.12	20.9	0.22	18.92	18.52	26.9	26.0
Fe_2O_3	0.73	–	–	0.0	0.0	0.0	–	–
FeO	–	0.87	28.2	–	–	–	–	–
MnO	30.7	1.08	(0.07)	0.07	0.0	0.0	–	–
MgO	7.1	34.38	15.4	29.05	0.06	–	0.0	0.0
CaO	0.04	–	0.03	–	0.09	–	0.0	0.0
BaO	–	–	–	–	–	–	39.7	38.7
Na_2O	–	–	–	–	11.52	0.58	0.0	0.0
K_2O	–	–	–	–	0.15	16.07	0.48	0.22
Total	85.4	85.00	90.6	"90.99"	98.09	98.20	100.08	95.92

1. Mn-thuringite, Haute-Mauriene (Chopin, 1978; 263-2); reduced assemblage with Glx and Alg.
2. Chlorite, Plan Maison (Dal Piaz et al., 1979; $\lambda 39_2$); Gnt-free, Pm+Brn-quartzite.
3. Chlorite, Plan Maison (Dal Piaz et al., 1979; DBL 702-1).
4. Talc, Haute-Maurienne (Chopin, 1981; 6-255) coex. Phg, Chl, Phl (sum includes 0.2 CoO, 0.2 NiO, 0.2 ZnO).
5. Albite, St. Marcel (Martin-Vernizzi, 1982; 3/25-3).
6. Microcline, St. Marcel (Martin-Vernizzi, 1982; 23/25-3).
7. Celsian, Andros (Reinecke, 1982; 192/2) in assemblage with Qz, Ab, Pm, Brn, Cc.
8. Cymrite, Andros (Reinecke, 1982; 192/2) replacing n. 7 above.

crystalline oxide and hydroxide phases with opal. This results in a sudden decrease of $f(O_2)$, since the precursors contain Mn as Mn^{4+} while braunite contains Mn^{2+} and Mn^{3+}. Nevertheless braunite is one of the most typical mineralogical indicators of high $f(O_2)$ ("oxidized assemblages"). Only a few, very late veins and certain reduced carbonate-bearing layers do not contain braunite (Martin-Vernizzi, 1982).

Braunite usually occurs in the form of xenomorphic crystals, in textural equilibrium with quartz. Hausmannite is another phase that may or may not participate in this equilibrium (Martin-Vernizzi, 1982); in one case hollandite has also been detected (Brown et al., 1978). In the absence of contacting quartz, braunite shows apparent equilibrium relationships with other manganese phases, like rhodonite (or pyroxmangite) and tephroite (Abs-Wurmbach et al., 1983). Due to deformation and/or the influence of circulating fluids, braunite may recrystallize in the form of

TABLE 15. REPRESENTATIVE ANALYSES OF BRAUNITE, SPINEL, AND NEOTOCITE

	1	2	3	4	5	6	7	8
SiO_2	9.80	10.2	10.5	9.83	0.30	0.15	–	46.1
TiO_2	0.00	0.04	–	0.99	0.13	0.53	0.1	0.15
Al_2O_3	0.07	1.44	0.1	0.25	48.0	10.1	–	0.04
Fe_2O_3	0.62	15.2	7.6	2.86	8.9	53.6	102.1	3.44
MnO	81.39	60	71.2	71.98	39.9	33.5	1.2	35.5
MgO	1.19	2.05	0.4	0.06	1.79	0.10	–	2.4
CaO	0.44	0.07	0.6	4.62	–	0.28	–	0.64
CuO	–	–	2.6	–	–	–	–	–
Na_2O	–	–	–	–	–	–	–	0.01
K_2O	–	–	–	–	–	–	–	–
Total	93.51	n.g.	93.0	90.59	99.0	98.3	103.3	88.0

1. Braunite, St. Marcel (Abs-Wurmbach et al., 1983), in assemblage with Jd and Cc.
2. Braunite, Cignana Lake (Dal Piaz et al., 1979, 4138).
3. Braunite, Andros (Reinecke, 1983; 79/557) next to hollandite.
4. Néltnerite, St. Marcel (Martin-Vernizzi, 1982; 21/25-3) fine grained, recrystalized over braunite.
5. Galaxite, Haute-Maurienne (Chopin, 1978; 263) average analyses, associated with jacobsite.
6. Jacobsite, Haute-Maurienne (Chopin, 1978; 263) average analyses, associated with galaxite.
7. Magnetite, Andros (Reinecke, 1983; 80/233) coex. with Hm in a Cpx-quartzite.
8. Neotocite, Haute-Maurienne (Chopin, 1978; 84) carbonatic assemblage with sonolite.

fine grains arranged as beads around the relics of the original crystal. Primary braunite has a composition remarkably constant in Si (1.000 ± 0.003 atoms p.f.u.). Among the elements substituting for Mn^{3+} in octahedral coordination, Al is usually very low, as is Ti, while Fe fluctuates from almost zero (Table 15, n. 1) up to ~20 mol.% (Table 15, n. 2). Such a variability appears not to be related to metamorphic grade (as suggested by Dasgupta and Manickavasagam, 1981), but rather to the availability of iron in the rock and to the composition of the coexisting phases, being maximum when hematite is present. Fine-grained braunites of second generation are enriched in Fe by a factor of two (Dal Piaz et al., 1979). They may also be enriched in Ti (up to 3.82 wt.% TiO_2, Martin-Vernizzi, 1982).

The divalent substitution for Mn^{2+} is also dependent essentially on bulk composition and recrystallization. As for composition, typical is the case of certain Andros braunites that are so rich in Cu as to plot well off the normal field in Fig. 11B. Mg is usually very low, or even absent, in the HP braunites equilibrated at low T, in contrast with those found (Abraham and Schreyer, 1975) in contact-metamorphic cherts. Here again, there are conspicuous exceptions: the Lago di Cignana braunites contain up to 30 mol.% Mg substituting for Mn (Table 15, n. 2). Since there are no indications of increased temperature of metamorphism in the area, their unusual chemistry can be related only to bulk compositional factors. The substitution of Ca for Mn^{2+} is clearly related to the same increased activity of Ca in the late fluids, combined with the lowering of T, producing breakdown of the omphacites into diopside (see p. 24), and it is therefore maximal in braunites of second generation. One of these, from St. Marcel (Table 15, n. 5), even exceeds 50 mol.% Ca, and it is in fact a néltnerite.

Bixbyite

This important indicator of a very high $f(O_2)$ has been detected only once (Bourbon and Fonteilles, 1972) in an unusual assemblage where, next to oxidized minerals like braunite and hausmannite, typical reduced phases are also present.

Spinels

Three Mn spinels have been detected in blueschist-facies metacherts: galaxite, jacobsite, and hausmannite. The first two have been found only in the reduced assemblages of Haute-Maurienne; the third one is present in several oxidized assemblages, in constant connection with braunite. Jacobsite may occur alone in some Fe-rich, Al-poor layers, the metamorphosed equivalents of shale partings of peculiar composition. More commonly, however, jacobsite and galaxite occur together in the carbonate-bearing reduced horizons and show textural relationships suggesting equilibrium.

Hausmannite occurs as a discrete phase only in oxidized assemblages, in textural equilibrium with braunite. It may also occur as exsolutions in recrystallized braunite. Hausmannite alters rather readily into pyrolusite or manganite, and eventually to cryptomelane. The present rarity of hausmannite may therefore turn out to be a secondary feature.

The iron spinel, magnetite, has been positively detected only once, in association with hematite (Table 15, n. 7). Chromite has been described at St. Marcel, in the horizon containing chromian pyroxene and garnet (Colomba, 1910), but it has not been confirmed by modern methods.

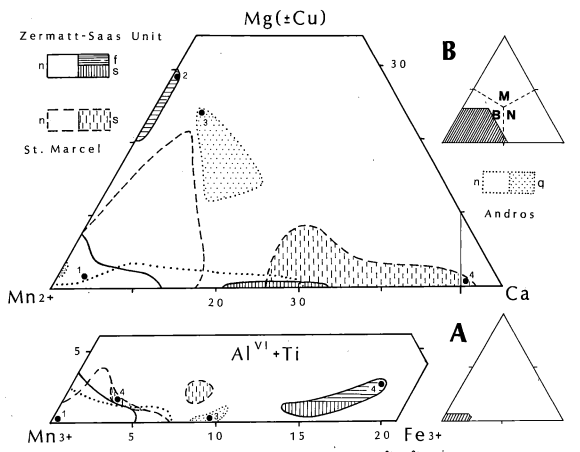

Figure 11. Compositional variation of braunites. (A) in the triangle $Mn^{3+}-Fe^{3+}-Al^{vi}$ (+Ti); (B) in the triangle $Ca-Mg$ ($\pm Cu$)$-Mn^{2+}$. The fields for normal primary (n), ferromagnesian primary (f), secondary (s), and copper-rich (q) braunites are distinguished. In inset (B) the compositional fields for braunite s.s., B, néltnerite, N, and the hypothetical Mg-braunite, M, are shown. Numbered dots refer to samples listed in Table 15.

Hollandite

It has been described at St. Marcel (Brown et al., 1978) and Andros (Reinecke, 1983). In the first occurrence it appears to be in textural equilibrium with typical high pressure phases; in the second it occurs within calcite and quartz veins, and may therefore be interpreted to be a late product. However, Reinecke (1982, 1983) considers hollandite a relic of the HP metamorphic stage.

Pyrophanite

It is the common Ti-bearing phase of the carbonatic-reduced assemblages (Table 16, n. 1), while it is very rare in the oxidized ones, where its place is taken by rutile.

Rutile

It is rather common in the oxidized assemblages, but it can also be present, occasionally, in the reduced ones in association with pyrophanite (Table 16, n. 3).

Ilmenite

It is a fairly rare constituent of oxidized assemblages, where it does not coexist with pyrophanite but where it incorporates as much as 34 mol.% of the $MnTiO_3$ component (Table 16, n. 2).

Hematite

Although rarely characterized precisely, this is a very important constituent because it is simultaneously an indicator of temperature as well as oxygen fugacity. In cherts this is particularly important: hematite forms at the very beginning of diagenesis and remains stable throughout all stages of metamorphism. Thus a careful analysis of its compositional profile may give clues on changes of $f(O_2)$ during the progress of metamorphism.

From the different solubility of TiO_2 and MnO in hematite, Reinecke (1983) was able to distinguish rocks recrystallized under comparatively low oxidation conditions from others recrystallized at comparatively high $f(O_2)$. In the first ones hematite equilibrated in the presence of a Mn^{2+}-bearing phase (usually spessartine), so that its composition is dominated by the solid

TABLE 16. REPRESENTATIVE ANALYSES OF PYROPHANITE, ILMENITE, RUTILE, AND HEMATITE

	1	2	3	4	5	6	7	8
SiO_2	0.14	0.20	0.07	–	–	0.0	0.46	–
TiO_2	53.1	49.3	98.95	2.3	0.2	1.73	0.18	0.71
Al_2O_3	0.08	0.01	0.0	–	–	0.0	0.29	0.49
Fe_2O_3	–	4.8	–	95.6	98.5	95.02	97.30	98
FeO	1.4	28.7	0.37	1.4	–	–	–	–
MnO	46.3	15.8	0.18	0.7	1.3	1.18	1.08	2.04
MgO	0.05	0.03	0.0	–	–	0.25	0.16	0.09
CaO	–	0.10	0.20	–	–	0.0	0.13	–
Na_2O	–	0.04	–	–	–	0.0	0.28	–
Total	100.1	98.9	99.77	100.0	100.0	98.18	99.96*	101.33

1. Pyrophanite, Haute-Maurienne (Chopin, 1978; 84) in a sonolite-bearing carbonatic assemblage.
2. Ilmenite, Haute-Maurienne (Chopin, 1978; 263) in a riebeckite-quartzite.
3. Rutile, St. Marcel (Martin-Vernizzi, 1982; 320) retrograded Cpx+Gnt-quartzite (coex. Pyr).
4. Hematite, Andros (Reinecke, 1983; 80/208) "reduced" assemblage: Sp-quartzite.
5. Hematite, Andros (Reinecke, 1983; 79/256) "oxidized" assemblage: Pm-quartzite.
6. Hematite, St. Marcel (Griffin; unp. SM 67) Brn-free, Sp+Rhd+Pxm+Pyr-quartzite.
7. Hematite, St. Marcel (Martin-Vernizzi, 1982; 2/25-2) Npx(Ac-Jd)+Pm+Phg-rock (sum includes K_2O = 0.08).
8. Hematite, Lake Cignana (Dal Piaz et al., 1979; 4138-6B).

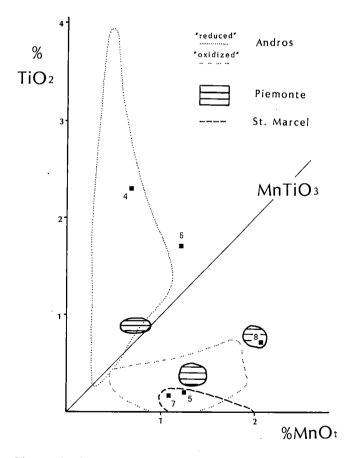

Figure 12. Compositional variation of hematites. Line labelled "$MnTiO_3$" divides the fields of reduced (above) from oxidized (below) assemblages as observed at Andros. Numbered dots refer to samples listed in Table 16.

solution toward pyrophanite Fe_2O_3-$MnTiO_3$; in the second ones hematite equilibrated in the presence of a Mn^{3+}-bearing silicatic phase (usually piemontite, but also alurgite, etc.), and the solid solution system Fe_2O_3-Mn_2O_3 dominates. From Fig. 11 it is clear that cherts of different alpine localities equilibrated at different $f(O_2)$. St. Marcel is a typically oxidized occurrence on the basis of the composition of hematites (Table 16, n. 7); however, it also contains rare braunite-free reduced assemblages with rhodonite and pyroxmangite, the hematite of which falls in the reduced field (Table 16, n. 6). In the rocks where hematite coexists with ilmenite, or contains exsolutions of it (Chopin, 1978), the mineral is oversaturated in Ti independently of the oxygen fugacity; in such a case, Ti-rich hematites can be found also in oxidized rocks (Reinecke, 1983).

Carbonates

Deep-sea cherts do not contain carbonates, because they are deposited below the "compensation level." Nevertheless, carbonate-bearing rocks are often found intermingled with cherts on top of the ophiolite sequence, either because of a quick addition during sedimentation (turbidity currents) or because of a gradual upheaval of the sea-bottom with increasing distance from the spreading axis (see Introduction). An added amount of carbonate can be brought in by the circulating fluids during the post-depositional and metamorphic stages.

Manganese carbonates fall into two distinct fields in the $CaCO_3$-$MnCO_3$ solid solution system (Fig. 13): near the rhodochrosite apex (Table 17, n. 6) and in the middle, around the composition of kutnahorite (Table 17, n. 4-5). They often coexist, separated by a compositional gap, the dependence of which on temperature and phase assemblage has been dealt with in detail by Peters et al. (1978, 1980).

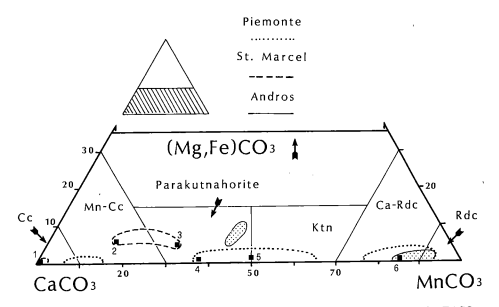

Figure 13. Compositional variation of carbonates in the triangle $CaCO_3$–$MnCO_3$–(Mg, Fe)CO_3. Numbered dots refer to samples listed in Table 17.

At St. Marcel and elsewhere in Piemonte a Ca-poor kutnahorite, similar to the parakutnahorite present in many deposits of eastern Europe (Table 17, n. 3), and a Mn-bearing calcite (Table 17, n. 2) are also present. Calcite not only is a rock-forming mineral in equilibrium with silicates, but frequently occurs as filler of the late veins. The rock-forming calcites are the most manganiferous (Table 17, n. 2). The vein calcite may be very pure (Table 17, n. 1). However, Sr-bearing calcite was also detected at St. Marcel (Brown et al., 1978), and interpreted as a primary phase derived from the inversion of aragonite. At Andros similar Sr-bearing calcites were interpreted by Reinecke (1983) as late depositions from fluids enriched in Sr as a result of the solution of preexisting silicate phases. Despite the Sr content and the observed biaxiality, the evidence that aragonite was present under blueschist-facies conditions as the stable $CaCO_3$ phase is considered to be "not conclusively proven" by Brown et al. themselves.

Neotocite

Chopin (1978) classified as neotocite on the basis of the microprobe analysis a semi-amorphous material occurring in the Haute-Maurienne sequence. The petrological significance of this ill-defined mineral is obscure. In agreement with its manner of formation in the type locality (Dasgupta and Manickavasagam, 1980), in the alpine cherts neotocite may be a retrograde deposit due to the alkaline metasomatic fluids permeating the sequence.

Other Minerals

The review of the blueschist-facies minerals in cherts would not be complete without mention being made of a large number of other minerals, more or less identified at different times. The following minerals are reported with sufficient reliability: gold (Millosevich, 1906); kämmererite (Colomba, 1910); romeite (Pelloux, 1913); manganberzeliite (Pelloux, 1946); stilpnomelane (Chiesa et al., 1976); kellyite, siegenite, pentlandite, millerite, chalcocite (Chopin, 1978); fluorapatite (Brown et al., 1978); baryte, todorokite (Reinecke, 1983). Moreover, other silicates occur in HP-metamorphic quartzites, sometimes regarded as cherts, but more likely to be of detrital derivation. The fact that some of them may show local enrichments of Mn supports their interpretation as the product of shallow-water stripping and resedimentation of an original ophiolite + chert sequence.

Concluding Remarks

Not all the minerals occurring in cherts produce information useful to petrology, nor is their information equally significant. The most interesting among them are those changing in composition not only as a function of bulk-rock composition, but as a result of the intensive parameters of metamorphism at nearly constant bulk composition. Unfortunately, the last condition is rarely met in alpine cherts, affected by metasomatism driven by migrating fluids.

Minerals providing maximum information are the following: a) pyroxenes: their composition is sensitive mainly to P and $f(O_2)$, in the limited-temperature span of the blueschist facies, but it may be strongly affected by metasomatic exchange, particularly when this induces the reduction of Fe and Mn; b) braunite: it is a sensitive indicator of addition of Ca and of decreasing $f(O_2)$, while remaining unaffected by pressure and temperature; c) phengitic mica: it changes in composition with P, but in an uncontrolled way, due to lack of appropriate calibrations in the Mn-bearing system; d) hematite: it reflects essentially the changes in $f(O_2)$.

Other minerals have their stabilities controlled by bulk

A. Mottana

TABLE 17. REPRESENTATIVE ANALYSES OF CARBONATES

	1	2	3	4	5	6
FeO	0.0	0.0	0.0	0.07	0.13	0.2
MnO	0.13	10.28	18.96	25.9	32.6	53.3
MgO	0.20	2.18	1.81	0.26	0.38	0.2
CaO	55.77	40.14	32.68	34.1	26.2	7.2
Total	56.10	52.60	54.35	60.33	59.31	60.8

1. Calcite, St. Marcel (Martin-Vernizzi, 1982; 2/25-1) in equilibrium with oxidized phases.
2. Mn-calcite, St. Marcel (Martin-Vernizzi, 1982; 5/26), Brn-free, Gnt+Carb-rock.
3. Kutnahorite, St. Marcel (Griffin, unp; SM8), in Gnt+Ab+Rhd+Nam+Hus+Brn-rock.
4. Kutnahorite, Alagna (Peters et al., 1978; Mn 3 2 II) coex. with Tep (NB: calculated back from % wt. carbonate).
5. Kutnahorite, Alagna (Peters et al., 1978; Mn 3 k III) coex. with Rhd and Tep (NB: as above).
6. Rhodochrosite, Andros (Reinecke, 1983; 80/117), in Rhd+Tep+Hus-rock.

composition, i.e. they are restricted to carbonatic rocks (tephroite, humites) or to aluminous rocks (jacobsite, galaxite). A few minerals appear to be irrelevant to the aim of petrological reconstruction within the blueschist facies, although they are important in other metamorphic facies: e.g. garnets. The real petrological significance of epidotes and ardennites is poorly known; they definitively reflect very high oxidizing conditions, but the few calibrations tried so far are hardly applicable to the P, T conditions of the blueschist facies.

PHASE COMPATIBILITY

A chert sequence is always made up of a number of thin layers, each different from the next in chemical bulk composition, mineral content and oxidation state, thus becoming also macroscopically different in colour and weathering properties. These differences are believed to be inherited from the primary sediment (see Kawachi et al., 1983). However, they become somewhat blurred during metamorphism due to an intensive elemental exchange driven by the fluids mobilized by prograde reactions involving dehydration. These fluids also deposit in the cracks of the contracting dehydrating layers, so that the chert sequence is characterized by swarms of veins. In addition, cherts involved in the alpine metamorphism also experience multiple shearing and fracturing. Consequently, the overall megascopic effect is an apparently chaotic association of multicoloured blocks. However, each block (and, on a larger dimension, a few best preserved outcrops) maintains a coherent stratigraphy and each thin layer its own phase assemblage. In fact, on a small size, metamorphism turns out to be so conservative that even original differences in oxidation state between adjacent layers are maintained. Not without reason Chopin (1978) could then conclude that oxygen did not behave as an "entirely mobile component"; actually, that be-

tween adjacent layers oxygen definitively behaves as an "entirely inert component."

Consequently, in his Haute-Maurienne chert sequence Chopin (1978) defines several types of assemblages characterized by different oxidation states and bulk compositions:

a) *oxidized assemblages,* where Mn is dominantly trivalent in the minerals braunite, piemontite, ardennite, and phengite. Spessartine is also present and incorporates all available Mn^{2+}, hematite is the characteristic iron-bearing phase, and rutile the Ti-phase; hausmannite is in equilibrium with braunite;

b) *reduced assemblages,* where Mn is only divalent. These assemblages can be divided in two subgroups:

1) carbonatic assemblages, where Mn^{2+} is present mostly in rhodochrosite and kutnahorite, but also in pyroxenoids, humites, and tephroite; pyrophanite is the typical Ti-bearing phase; hematite is present, sometimes associated with magnetite or ilmenite;

2) aluminous assemblages, where Mn^{2+} is mostly divalent but is combined in spinels and chlorites; humites are also present but carbonates and pyrophanite are rare; braunite may be present, and concentrates all the available Mn^{3+} together with the spinels (hausmannite in s.s.). Garnets, when present, are Sp-poor, And-rich solid solutions.

Chopin's system has been echoed by most later investigators; in particular, it received a chemical support on the basis of the *'oxidation ratio'* of the whole rock,

$$\text{ox.r.} = \frac{2Fe_2O_3 \times 100}{2Fe_2O_3 + FeO} + \frac{2Mn_2O_3 \times 100}{2Mn_2O_3 + MnO}.$$

At Andros, Reinecke (1983) could show that rocks with ox.r. exceeding 100 invariably contain oxidized assemblages (in Chopin's meaning), whereas those in which it is well below 100 invariably contain reduced assemblages. There are, however, a number of rocks with ox.r. around 100 (approximately 95-110), which may contain minerals typical of either type of assemblage.

These rocks may be called "neutral" from the point of view of their redox conditions. However, many of them do in fact contain disequilibrium assemblages, where early-formed oxidized minerals were not retrograded entirely during a later phase of metamorphism (usually at lower P, but also at higher T: the Lepontine phase of the Alpine cycle).

When comparing mineralogy and chemistry of alpine blueschist-facies cherts, it is obvious that the oxidized assemblages are typical of the silica-rich layers (radiolarites, or cherts s.s.), while the reduced ones are typical of the carbonatic and aluminous layers (the partings, either calcareous or clayey). Neutral assemblages are most common in the reworked layers, and as such they occur mainly in the reworked or resedimented sequences (Tables 2 and 3). They also occur in those layers where there are the best evidences of bulk chemical change induced by migration of metasomatic fluids.

Since cherts are initially deposited with both Mn and Fe in their highest oxidation states (Burns and Burns, 1977), their present oxidation conditions are not so much inherited from the sediments (as suggested by Kawachi et al., 1983), but rather they were acquired during the lithification stage (cf. Bonatti et al., 1976). During the earliest stage of lithification Mn^{4+} was reduced to Mn^{3+} in the silica-rich beds, and to Mn^{2+} in the calcareous and clayey partings, probably because these were rich in strong reducing components (CO_2 and S, respectively). The apparent richness in oxidized Mn minerals of blueschist-facies cherts is, according to this interpretation, not at all due to the HP-LT conditions of the blueschist facies; rather, it results from the maintenance of the initial diagenetic $f(O_2)$ in the individual layers due to the rather rapid development of the metamorphism in the subduction zone. Where similar cherts were involved in a regional metamorphic regime, the slow progress in T and P favored equilibration of $f(O_2)$ among adjacent layers and a widespread formation of "neutral" assemblages.

The mineral assemblages observed and/or interpreted in alpine chert sequences are listed in Tables 18 to 20. Attempts at constructing compatibility diagrams for the system Ca-Mn-Fe-Al-Si-C-O-H or related subsystems have been made by various authors (e.g. Abs-Wurmbach et al., 1983; Abs-Wurmbach and Peters, 1981; Dasgupta and Manickavasagam, 1981; Peters et al., 1974, 1978, 1980). They have been successful to a certain extent, especially when applied to specific occurrences. However, they are far from being exhaustive with respect to the complex evidence of all the alpine cherts, as given in Tables 18 to 20. Thus they were not reproduced here.

The veins crossing the chert outcrops are not only relevant because they contain Mn minerals, but because they give insight in the overall geochemical balance of the chert sequence. These veins formed under very different P, T conditions: during the prograde stages of the subduction metamorphism, at the metamorphic peak, and during the declining stages. They carry very different minerals, as a result not only of the different P-T conditions acting at the moment of the vein filling, but also of the change in composition of the circulating fluids.

A complete list of the assemblages observed in veins is given in Table 21.

To understand the petrogenetic evolution of cherts another observation is significant: the reaction relationships among different phases actually observed in thin section. A list of these is given in Table 22. This list is clearly insufficient in relation to the known phase stability and compatibility in synthetic system, but with the exception of Martin-Vernizzi (1982) and Reinecke (1983), few researchers have actually provided petrographic evidence for the observed reactions.

CONDITIONS OF METAMORPHISM

Direct determinations of the conditions at which cherts successively equilibrated during their metamorphism are scarce. Usually, conditions are determined on the underlying metabasites and applied to cherts on the assumption that metamorphism affected homogeneously the entire unit. A better insight is usually detected from the oxidation conditions, since they differ between cherts and surrounding rocks and in adjacent chert layers.

Determinations of T and P

Direct determinations are hindered by the yet incomplete knowledge of phase relations and elemental partitions in the relevant manganese-bearing systems. This is typified by the procedure followed by Brown et al. (1978) for St. Marcel. They suggest $P = 8 \pm 1$ kbar on the basis of the univariant assemblage Omph + Qz + Ab with known Jd component of the pyroxene, corrected for the Fe^{3+} content; therefore, they completely disregard the Mn^{2+} and Mn^{3+} contents, on the assumption that manganese components in the solid solution behave in the same way as iron components. Their equilibration temperature $T = 300 \pm 50°C$ was established using the Kfs-Ab pair (Fig. 14, box B); again in a Mn-free system. Practically, only their determination of $f(O_2) = 10^{-17}$ is based on thermodynamic calculations on manganese minerals, the reaction Qz + Braun = Rhd + O_2.

In discussing the results of Brown et al. (1978), Griffin and Mottana (1982) use similar procedures, although they produce different results: their proposed maximum P, T conditions for St. Marcel are 500°C, 14 kbar (Fig. 14, star).

Chopin (1978), Peters et al. (1978) and Dal Piaz et al. (1979) estimate or infer P, T conditions only by reference to nearby areas or even from general considerations on the multistage development of the alpine metamorphism. Therefore, Martin-Vernizzi (1982) should be given credit for the first attempt at determining a point in the P, T field from data intrinsic in the cherts themselves, i.e. from the equilibrium of two manganiferous minerals. Using the Ellis and Green (1979) geothermometer for the contacting rims of a manganous aegirinaugite (Table 10, n. 7) and a spessartine-almandine garnet (Table 6, n. 2) she obtains T = 467° C for P = 15 kbar, or T = 457° C for P = 10 kbar (Fig. 14, box M). My calculation of the same data, but taking into consideration charge balance requirements (see Grif-

TABLE 18. OXIDIZED ASSEMBLAGES OBSERVED IN THE ALPINE BLUESCHIST-FACIES CHERTS

Qz + Clm	Qz + Brn + Hus
Qz + Clm + Aln	Qz + Jd (Mn³⁺) + Pm
Qz + Clm + Pm	Qz + Phg (Mn³⁺) + Tc
Qz + Di + Pm + Brn	Qz + Jd (Mn³⁺) + Gnt + Ru
Qz + Omp (Mn³⁺) + Ap + Cc	Qz + Cam + Tit + Brn
Ab + Clm + Pm + Brn	Pm + Cpx (Acm) + Tit + Nam
Qz + Pm + Brn + Kfs + Cc	Qz + Cpx (Acm - Jd) + Pm + Phg (Mn³⁺) + Ru
Kfs + Pm + Brn + Omp (Mn³⁺) + Phg (Mn³⁺)	Ab + Jd (Mn³⁺) + Phg (Mn³⁺) + Brn + Prs
Qz + Pm + Brn + Hus + Cpx(Acm-Jd) + Phg (Mn³⁺)	Qz + Pm + Brn + Hus + Ard + Phg (Mn³⁺)
Qz + Pm + Brn + Hm + Tc + Gnt	Qz + Pm + Cpx + Gnt + Rbk + Dan
Qz + Ab + Kfs + Clm + Phg (Mn³⁺) + Tit	Qz + Ab + Pm + Brn + Hm + Clm + Phg (Mn³⁺)
Qz + Ab + Pm + Brn + Hm + Ard + Phg (Mn³⁺) + Gnt	Qz + Ab + Brn + Hus + Tit + Clm + Phg (Mn³⁺)
Qz + Ab + Brn + Hus + Cpx (Di - Kos) + Tit + Cc	Qz + Ab + Pm + Omp (Mn³⁺) + Phg (Mn³⁺) + Cc + Ap
Qz + Pm + Brn + Hm + Gnt + Aln + Tc + Ru	

TABLE 19. NEUTRAL ASSEMBLAGES OBSERVED IN
THE ALPINE BLUESCHIST-FACIES CHERTS

Qz + Rhd	Qz + Gnt + Hm
Qz + Cpx (Acm) + Hm	Qz + Gnt + Cpx (Acm) + Hm
Qz + Gnt + Hm + Chl	Qz + Gnt (Cald - And) + Mt + Ap
Qz + Ab + Hm + Cam	Qz + Hm + Nam + Ep
Qz + Cpx (Acm) + Nam + Ms	Qz + Cpx (Acm) + Nam + Ms + Mt
Qz + Gnt + Hm + Nam + Cpx (Acm)	Qz + Gnt + Chl + Ms + Hm
Qz + Gnt + Hm + Nam + Ms	Qz + Gnt + Hm + Nam + Ms + Ep
Qz + Gnt + Hm + Nam + Ms + Ep + Chl	Qz + Gnt + Rhd + Pxm + Hm + Pyr

TABLE 20. REDUCED ASSEMBLAGES OBSERVED IN THE ALPINE BLUESCHIST-FACIES CHERTS

a) Carbonatic assemblages

Ktn + Rhd	Ktn + Rhd + Rdc
Ktn + Pxm	Ktn + Rhd + Gnt
Ktn + Rhd + Tep	Ktn + Rhd + Brn
Rdc + Rhd + Pxm	Rdc + Rhd + Brn + Gnt
Rdc + Rhd + Brn + Gnt + Qz	Rdc + Cc + Rhd + Gnt
Rdc + Cc + Rhd + Gnt + Qz	Rdc + Cc + Rhd + Gnt + Oz
Rdc + Rhd + Tep + Gnt + Brt + Hus	Rdc + Rhd + Pxm + Gnt + Tep + Prs
Rdc + Rhd + Pxm + Gnt + Tep + Prs + Tir + Qz	Rdc + Rhd + Gnt + Tep + Son + Alg + Prs
Rdc + Rhd + Pxm + Gnt + Tep + Son + Alg + Prs	Rdc + Gnt + Tep + Son + Alg + Frd + Prs
Rdc + Gnt + Tep + Son + Alg + Frd + Rhd + Prs	Son + Prs + Jcb + Neo

b) Aluminous assemblages

Ktn + Jcb	Gnt + Cc
Son + Alg + Jcb + Glx	Son + Alg + Jcb + Glx + Chl

fin and Mottana, 1982), gives T = 376° C at P = 10 kbar (or T = 385° C at P = 15 kbar). The difference (80° C) may appear to be large, but lies within the accepted range of errors for most geo-thermometric methods based on elemental partition. Its real value is however questionable, since the Fe and Mg contents in the gnt + cpx pair considered are relatively minor. Martin-Vernizzi (1982) further tried calculating pressure by the method of Holland (1980), using three manganic aegirinjadeites. She obtains values varying from 8.5 to 10 kbar (for T = 450-480° C). Gasparik and Lindsley's (1980) geobarometer suggests pressures between 10 and 12 kbar (at T ~400°). For the retrograde metamorphic stages Martin-Vernizzi (1982) calculates T = 395-409° C and P = 8 kbar for the blueschist-facies post-eclogitic stage (Fig. 14, box M') and can only infer T ⩽450° C, P < 6 kbar for the final greenschist-facies stage (Fig. 14, box M").

For the reduced carbonatic assemblages of Alagna, Peters et al. (1980) estimate T = 525° C (Fig. 14, box A), by applying to coexisting manganese carbonates the solvus of Goldsmith and Graf (1957), while at Andros Reinecke (1983) could only estimate a T significantly below the critical temperature (~540° C) of the Knt-Rdc solvus of De Capitani and Peters (1981) (Fig. 14, box G). Using the equilibrium cymrite=celsian + H_2O experimentally determined by Nitsch (1980), he also obtains 400° C and 5 kbar; this point in the P-T field refers to the retrometamorphic greenschist facies stage (Fig. 14, box G').

Determination of $P(H_2O)$ and $X(CO_2)$

Most of the equilibration conditions given above are based on the assumption $P_{tot} = P(H_2O)$. In fact this is certainly not the case for the carbonatic reduced assemblages, and it is also unlikely for other assemblages containing phases like titanite, or

TABLE 21. ASSEMBLAGES OBSERVED IN THE VEINS CROSSING
ALPINE BLUESCHIST-FACIES CHERTS

a) Interpreted as formed under high pressure
   Qz + Ard   Qz + Ard + Hm
   Ab + Clm + Pm + Cc   Ab + Kfs + Clm + Pm + Tit
   Qz + Ab + Clm + Brn + Tit   Qz + Ab + Kfs + Clm + Brn + Tit + Pm

b) Interpreted as formed during the retrogressive stage
   (or of unstated origin)
   Cc   Qz + Ab + Kfs + Hm + Prs + Krp
   Cc + Todorokite   Qz + Cc + Pm + Brn + Hol
   Ab + Oz   Qz + Kfs + Pm + Cam
   Ab + Kfs   Qz + Rhd + Rdc + Cc
   Qz + Hol   Qz + Ab + Cam + Brn
   Ab + Cpx (Di-Acm)   Qz + Ab + Cpx(Di-Acm) + Tit
   Qz + Gnt + Cpx(Acm)   Qz + Ab + Cam + Cc
   Qz + Ab + Cpx(Di-Acm)   Qz + Cs + Kfs
   Ab + Chl + Cam   Qz + Ab + Kfs + Hm + Prs + Krp + Tit + Brn + Pm

TABLE 22. OBSERVED REACTIONS DOCUMENTED TO OCCUR IN ALPINE SCHISTS

a) eclogite to blueschist facies
   Brn$_I$→Brn$_{II}$
   Clm ——→Omp + Ab + Phg

b) blueschist to greenschist facies (or with addition of volatiles
   Rhd ——→Rdc + Qz   Cpx ——→Cam
   Clm ——→Di + Hm   Cpx ——→Phl
   Phg ——→Phl + Hm (?)   Cam ——→Chl
   Pm ——→Phl   Cpx ——→Phl + Hm
   Omp + Qz + H_2O + CO_2 ——→Cam + Ab + Cc + O_2
   Pm + Brn + CO_2 ————→Gnt + Cc + Qz + Hm + H_2O + O_2
   Pm + Brn + Ru + CO_2 ——→Gnt + Cc + Qz + Hm + Tit + H_2O + O_2

pyrophanite, that form in the presence of CO_2 (Hunt and Kerrick, 1977).

For the carbonatic assemblages of Alagna, Peters et al. (1978, 1980) construct a set of compatibility diagrams and isothermal-isobaric sections showing the change of phase boundaries with arbitrary changes of $X(CO_2)$, assuming constant T = 500° C and P_{tot} = 5 kbar, and $f(O_2)$ buffered by the assemblage Rdc + Tep + Hausm. They do not suggest precise values; however, they point out that the presence of crossing tie-lines indicates different $X(CO_2)$ in adjacent layers.

At St. Marcel, the Ti minerals point out a continuous decrease of X (CO_2) with time in the rocks, although calcite is present even in the last veins. In typical eclogite-facies assemblages, rutile is the stable Ti-phase. In HP veins and in the post-eclogite blueschist-facies assemblages greenovite becomes stable and replaces rutile, but it is still in equilibrium with piemontite (Mottana and Griffin, 1979). Finally, in the greenschist facies assemblages and latest veins, the typical titanite is green or colourless. Hunt and Kerrick's (1977) experimental data for the reaction calcite + quartz + rutile = sphene + CO_2 extrapolated to the P,T conditions of the blueschist-facies metamorphism, indicate that $X(CO_2)$ of the fluid phase was initially in excess of the buffering value (0.02) and dropped later to lower values. However, substituting Mn for Ca should affect the calculated $X(CO_2)$ by an unknown amount.

$P(H_2O)$ must be lower than P_{tot} because H_2O is not only diluted by CO_2 but also by other volatile components, such as B

(present in tourmaline), Cl (in friedelite), and F (in titanites, amphiboles and humites).

Determination of f(O₂)

Chopin (1978) was the first to suggest brackets for the oxidation state of the various blueschist-facies assemblages occurring in the alpine cherts. Little but nuances can be added to his method, but the preferred values are different due to the new experimental studies carried out in recent years. In particular, the correction for pressure, disregarded by Chopin, proves to be significant, at least for the reaction 2 Braun + 12 Qz $\rightleftharpoons$ 14 Rhd + O_2 (Abs-Wurmbach et al., 1983). The revised tentative T, $f(O_2)$ diagram is given in Fig. 15, modified and corrected to better fit the P, T range of blueschist facies from the analogous tentative diagrams of Abs-Wurmbach et al. (1983).

By definition (see Phase Compatibility, above) oxidized assemblages are those where Mn^{3+} is present in prominent minerals such as braunite, piemontite, and hausmannite. Recent experimental data by Abs-Wurmbach et al. (1983) in the pressure range 1 bar-15 kbar show that braunite is limited at high $f(O_2)$ by the reaction (1) 7 Pyrolusite + 1 Quartz = 1 Braunite + O_2. Pyrolusite is usually secondary in blueschist cherts; however, reaction (1) is certainly realized in the early stages of chert recrystallization. Under atmospheric pressure, reaction (1) occurs in the $f(O_2)$ range 10^{-3} –10^{-1}, and at higher values the pressure is higher. In the case of a solid solution of Fe^{3+} for Mn^{3+}, the

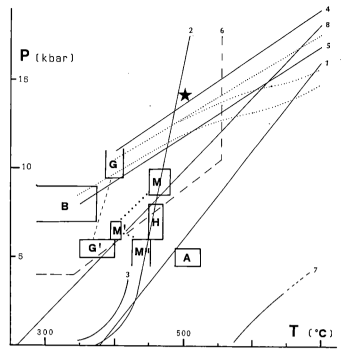

Figure 14. Constraints to P, T conditions (lines) and suggested P, T brackets for various alpine cherts equilibrated in the blueschist facies. Lines: 1. = Pxm-Rhd reaction in the pure $MnSiO_3$ composition (Maresch and Mottana, 1976); 2. = breakdown of natural carpholite (Mottana and Schreyer, 1977); 3. = Synthesis of Sp from Mn-Chl + Qz (Hsu, 1968); 4. = Breakdown of Ab into Jd + Qz (Holland, 1980); 5. = Cc to Arg polymorphic inversion (Carlson, 1980); 6. = Maximum possible stability field of glaucophane (Maresch, 1977); 7. = Maximum possible stability field for piemontite Pm33 (Keskinen and Liou, 1979); 8. = Johannsenite to bustamite transformation (Angel, 1984). Dotted lines: maximal and minimal stabilities for Omp with compositions from Jd50 to Jd80 and C or P structure respectively (Holland, 1980). Boxes: A = Alagna (Peters et al., 1978); B = St. Marcel (Brown et al., 1978); G and G′ = Andros (Reinecke, 1983); H = Haute Maurienne (Chopin, 1978); M, M′ and M″ = St. Marcel (Martin-Vernizzi, 1982). Star = maximum possible conditions for St. Marcel (Griffin and Mottana, 1982). For further informations see text.

reaction would be (2), bixbyite + quartz ⇌ ferric braunite + O_2; $f(O_2)$ would drop by an unspecified amount. This reaction is not likely to be verified in cherts, since Fe and Mn deposit separately for geochemical reasons (see Introduction). However, it may have played a role in a few occurrences, like Lake Cignana (Table 15, n. 2), where primary braunite is unusually rich in Fe.

A confirmation of the fact that $f(O_2)$ in most cherts must be lower than the values for reaction (1) lies in the widespread occurrence of the braunite + hausmannite assemblage. Experimentally, hausmannite forms from partridgeite at a $f(O_2)$ of 10^{-8} -10^{-6} (here and from now on the oxygen fugacity is calculated for the T interval 350-450° C and constant P = 10 kbar). The hausmannite + braunite assemblage has a stability field lying entirely in the braunite + quartz $f(O_2)$ field (Huebner, 1976). Since no partridgeite has ever been found in blueschist facies

cherts, it is easy to infer that $f(O_2)$ never exceeded 10^{-5} in any assemblage or metamorphic stage.

The quartz- and braunite-consuming reactions (3), 2 Brn + 12 Qz = 14 Rhd/Pxm + $3O_2$, and (4), 2 Brn = 4 Hus + 2 Rhd/Pxm + O_2, mark another step toward low $f(O_2)$, but still in oxidized assemblages. At low P the two reactions occur very close together in the $f(O_2)$ range 10^{-15} -10^{-13} but diverge at P > 10 kbar since reaction (4) shifts from below to above the conditions defined by the CuO/Cu_2O buffer (Abs-Wurmbach et al., 1983). When Fe is present in the bulk composition, it enters as Fe^{3+} in braunite and hausmannite, and it is rejected by rhodonite[2]. This allows their stabilization towards lower $f(O_2)$ (Dasgupta and Manickavasagam, 1981; Bhattacharyya et al., 1984), so that they may be sparsely found in neutral or reduced assemblages.

In particular, hausmannite was detected at Andros in a rhodochrosite + rhodonite + tephroite assemblage classified as "reduced" on petrographic and chemical criteria. This assemblage was shown to be a buffer in the Mn-Si-C-O system at P = 3 kbar, T = 600° C and $f(O_2)$ = 10^{-8} (Peters et al., 1973), but there are no experiments under blueschist-facies conditions.

Within the broad range of $f(O_2)$ for oxidized assemblages, a distinction should be made for the conditions of layers that, in addition to braunite, contain piemontite and/or ardennite, and layers devoid of these, carrying spessartine as the typical Mn-phase. The former are well-known to represent higher oxidation conditions than the latter (Smith and Albee, 1967; Kawachi et al., 1983). The experimental data of Keskinen and Liou (1979) and Anastasiou and Langer (1977) imply $f(O_2)$ higher than 10^{-20} -10^{-10} for piemontite with 33 mol% Pm. Thus, an even higher $f(O_2)$ must be assumed for the stabilization of Pm-rich piemontites such as those occurring at St. Marcel. Epidotes containing low Pm-component are stable at lower $f(O_2)$, in the presence of spessartine, but in any case above the conditions of the HM-MT- buffer (Keskinen and Liou, 1979).

Assemblages classified as "neutral" (Table 19) are defined by oxidation ratios implying that all Mn is present as Mn^{2+}, and Fe is present as Fe^{3+}. Their upper limit is represented by reaction (5), 6 Rhd/Pxm + 2 Hus = 6 Tep + O_2 which, in the P-T conditions of the blueschist facies, lies in the $f(O_2)$ range $10^{-17}-10^{-14}$. The lower limit follows the HM/MT buffer, that in the pure system occurs at $f(O_2)$ 10^{-25} − 10^{-21}. Magnetite is present very rarely in cherts, and possibly owes its presence to Mn^{2+} substituting for Fe^{2+}, enlarging its $f(O_2)$ stability towards slightly higher values.

Reduced assemblages (Table 20) are defined on the basis of the presence of Mn only in the divalent state, while Fe is mostly in the divalent state and in solid solution in carbonates and silicates. Only in the per-aluminous compositions (Chopin, 1978) may traces of Mn^{3+} and Fe^{3+} be present in solid solution within spinels.

---

[2]According to the experimental data of Maresch and Mottana (1976), reactions (3) and (4) should lie entirely in the stability field of pyroxmangite at the P,T conditions under discussion, but Ca stabilizes rhodonite and makes it a participating phase (Dal Piaz et al., 1979; Martin-Vernizzi, 1982; Griffin, unp.).

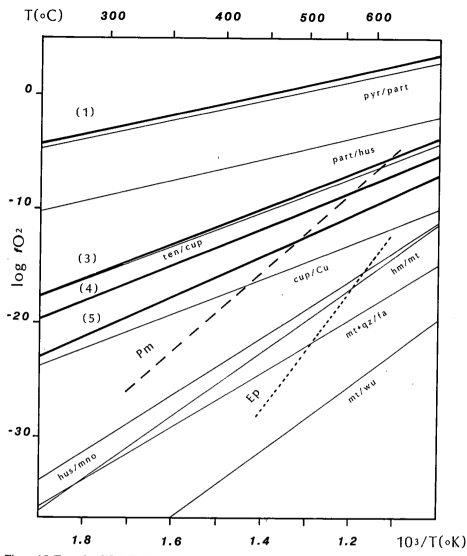

Figure 15. Tentative $f(O_2)$-T diagram showing the approximate position of reaction curves relevant to manganiferous cherts calculated at P = 10 kbar. Thick lines refer to reactions as numbered in text; thin lines and abbreviations to well-known buffers (from top to bottom: pyrolusite/partridgeite; partridgeite/hausmannite; tenorite/cuprite; cuprite/copper; hausmannite/manganosite; hematite/magnetite; magnetite + quartz/fayalite; magnetite/wustite). Broken lines are the maximum $f(O_2)$-T stabilities for Pm and Ep as given by Keskinen and Liou, 1979).

Carbonatic-reduced assemblages cannot be evaluated in the simple Mn-Si-O system because of the critical importance of the C-O-H gas equilibria. Moreover, where humites are present, the activities of SiO_2 as well as other volatiles must be taken into consideration (Winter et al., 1983). Experimental data available for the Mn-C-Si-O-H system derive mostly from decarbonation reactions at P substantially lower than those typical of blueschists (mostly 2000 bars: Peters et al., 1973, 1974). Their extrapolations to P = 10 kbar introduce a large degree of uncertainty. This is possibly the reason why the latest papers of Peters et al. (1980, 1978) give schematic phase relations, rather than numeric values. Moreover, in decarbonation reactions, $X(CO_2)$ and $X(H_2O)$ cannot be neglected: e.g. the stability field of rhodochrosite is strongly reduced at low $X(CO_2)$ and moves towards lower T and lower $f(O_2)$. Since the lowest possible $f(O_2)$ is defined by the absence of graphite, the lowest possible $f(O_2)$ limit for reduced carbonatic assemblages may be suggested as 10^{-30} -10^{-28}. On the other hand, from the presence of silicates such as tephroite and rhodonite, and even of the Mn^{3+}-bearing phase braunite in rhodochrosite-bearing rocks, an upper limit in the order of 10^{-11} -10^{-15} may be deduced for these kinds of rocks.

For reduced aluminous assemblages Chopin (1978) proposed a $f(O_2)$ in the range 10^{-11} -10^{-18}. In fact the presence of jacobsite brackets the oxidation conditions between those of its reduction to manganwustite and its oxidation to hematite + hausmannite, (6) $MnFe_2O_4 \rightleftharpoons (Mn, Fe) O + \frac{1}{2}O_2$, and (7)

$3MnFe_2O_4$ + ½ O_2 ⇌ $3Fe_2O_3$ + Mn_3O_4. Neither reaction is properly calibrated and in addition jacobsite may incorporate Mn^{3+} and Fe^{2+}, thus leading to divariant equilibria. However, it has been inferred from field studies (Essene and Peacor, 1983) that reaction (6) lies one order of magnitude higher than the reduction of pure magnetite to wustite, and reaction (7) about two orders of magnitude higher than the oxidation of pure magnetite to hematite. Thus the range of $f(O_2)$ for aluminous reduced assemblages lies in the broad range 10^{-25} – 10^{-22}.

CONCLUSIONS AND SUGGESTIONS

At the end of a review paper, it is customary to draw conclusions on the status of the art and to put forward suggestions for future work. In the case of the alpine blueschist-facies metamorphic cherts, this is particularly opportune. In fact, it has been only a decade since interest in them shifted from the mere source of unusual minerals to the scientifically much higher status of potentially significant marker horizons, useful for geological as well as petrological interpretations.

As with most of the recent geological investigations in the Alps, such a change was induced by the new meaning cherts have acquired in a geological reconstruction based on the plate-tectonic theory, as they are representatives of layer 1 of the oceanic crust of the Earth. Furthermore, the change was facilitated by the introduction of microprobe as a routine working tool. Only point-analysis (lately combined with single-crystal X-ray diffraction studies) can clarify the complex crystal-chemistry of chert minerals despite the yet unresolved limitation that Mn, besides Fe, is present in two different oxidation states. Also, each grain or crystal displays variations in chemistry that depend on (and therefore indicate) the variation of pressure, temperature, oxygen fugacity and chemical potential in the continuously re-equilibrating polyphasal alpine metamorphism.

Not without reason mineralogy is the branch of geological science that was prodded first and most intensively during the last years. By this, the crystal chemical knowledge of the major rock-forming minerals of cherts can be considered satisfactory for such mineral groups as the epidotes, the clinopyroxenes, and the garnets, and for a number of minor, but petrologically significant, phases. Other mineral groups still deserve to be investigated systematically, such as the micas and pyroxenoids. The available data on these minerals indicate that microprobe analysis is not sufficient for their complete characterization; single-crystal X-ray work is needed in view of the structural complexity (polytype, periodicity) that reflects on the petrological behavior.

A point that weakens the use of cherts as petrologically significant marker horizons is the small number of occurrences studied in detail so far, in comparison with the extension of the blueschist-facies terrains in the Alpine Belt. Luckily the three best-studied localities were accidentally chosen to provide an overall idea of the behavior of cherts within a subduction zone and to develop sensible petrological models. St. Marcel is a coherent "oceanic" chert + ophiolite sequence, containing oxidized assemblages, and affected by at least three phases of the alpine metamorphism. The Haute-Maurienne occurrence is an "oceanic" sequence reworked in a volcano-sedimentary environment so as to contain contaminated bulk compositions characterized by both oxidized and reduced assemblages. Moreover, its location next to the front part of the belt submitted to blueschist-facies metamorphism allows study of slightly different P-T conditions than at St. Marcel. Andros lies in a continental sequence and contains even more geochemically-complex bulk compositions, resulting in the formation of very unusual phases, besides certain manganiferous phases that had been previously known only at St. Marcel.

In the future, several other occurrences should be worked out in detail in areas where clear-cut results have been achieved in our understanding of the alpine metamorphism through the study of basites and pelites. The reconnaissance studies carried out so far allow pointing out Plan Maison, Lake Cignana, Ceres and possibly M. Spassoja and Carpenara (eastern border of the Voltri Group) as the localities (Table 1) suitable for maximal result. The latter two localities are particularly important in that they correlate directly with the well-known sequence of manganiferous cherts on top of the ophiolites of Val Graveglia, eastern Liguria, metamorphosed under zeolite facies conditions (Cortesogno et al., 1979).

The final outcome of systematic mineralogical studies should be a complete knowledge of the phase assemblage and compatibility in the various stages of re-equilibration of the Alpine Belt during uplift, as well as a possible zonation of metamorphism both within and outside the blueschist facies. In turn, such a detailed knowledge is essential to allow a correct application of the results of experimental investigation into the Mn-bearing systems that are being pursued more or less independently at centers like Bochum or Bern.

If mineralogy enjoys a fairly satisfactory status, which is the forerunner of developments for petrology, much less satisfactory is the condition of geochemical studies. Only few random determinations are available; a work such as that performed by Reinecke (1983) at Andros, or much more systematically by Barrett (1981) in the zeolite-facies cherts of eastern Liguria and Tuscany, is highly desirable. This is particularly so in those localities where, as at St. Marcel, the Steinmann's Trinity is still preserved despite the intense polyphase deformation accompanying the alpine metamorphism. Here is the need and opportunity to study not only the control of the initial bulk composition on the phase assemblage, but also the extent of modifications undergone by addition of elements mobilized during metamorphism, from levels as far down as the underlying ophiolites.

As has been successfully done over the last fifteen years with basic rocks, a study should be made of cherts to identify immobile elements useful as indicators of their original depositional environments, and to identify the elements mobilized under different conditions of metamorphism. The ascertained mobilization of copper and chromium from the underlying ophiolites is a good example of such an approach; however, better results will certainly arise from the study of REE as well as of certain isotopes. Geochemical investigations on this line are currently underway elsewhere in the world, and it is hoped they will start soon also in the Alpine Belt, which is still the best studied but also the most debatable geological puzzle.

ACKNOWLEDGMENTS

This review benefited by information and ideas developed throughout the years during joint investigations carried out with M. Bondi, G. Ferraris, W. L. Griffin, G. Kurat, G. Liborio, and G. Rossi. Bill Griffin deserves my additional thanks for allowing publication of analytical results that constitute a part of our future papers on St. Marcel. Visits to metacherts outside the Alps were also beneficial: in particular those in Shikoku with S. Banno, M. Iwasaki and T. Suzuki, in Washington with E. H. Brown, B. W. Evans and P. Misch, and in California with W. G. Ernst and R. G. Coleman. The Penrose Conference was essential to me in clarifying several points as well as in compelling me to organize data and thoughts.

Investigations on manganese minerals were made possible by operating grants of the Italian National Research Council (C.N.R., Rome), and a grant-in-aid of the C.N.R. made the author's attendance possible at the Penrose Conference.

LIST OF ABBREVIATIONS AND THE THEORETICAL FORMULAE OF MINERALS MENTIONED IN THE TEXT

Ab	albite, $Na[AlSi_3O_8]$
Acm	acmite, $NaFe[Si_2O_6]$
Alg	alleghanyite, $Mn_5[(OH, F)_2|(SiO_4)_2]$
Alm	almandine, $Fe_3Al_2[SiO_4]_3$
Aln	allanite, $Ca(Ce,Th) (Fe^{3+}, Mg, Fe^{2+}) Al_2 [O|OH|SiO_4|Si_2O_7]$
Alu	alurgite, $K(Al, Mn^{3+})_2 [(OH)_2|AlSi_3O_{10}]$
An	anorthite, $Ca[Al_2Si_2O_8]$
And	andradite, $Ca_3Fe_2[SiO_4]_3$
Ap	apatite, $Ca_5[(F, OH)|(PO_4)_3]$
Ard	ardennite, $Mn_4^{2+} (Mg, Al, Fe^{3+}, Mn^{3+})_2Al_4 [(OH)_6|(As, V) O_4|(SiO_4)_2|Si_3O_{10}]$
Arg	aragonite, $Ca [CO_3]$
Au	gold, Au
Aug	augite, $Ca(Mg, Al, Fe) [(Al, Si) SiO_6]$
Bez	berzeliite, $(Ca, Na)_3 (Mg, Mn)_2)_2 (AsO_4)_3$
Boh	bohemite, $AlOOH$
Brn	braunite, $Mn^{2+}Mn_6^{3+}SiO_{12}$
Brt	baryte, $Ba[SO_4]$
Bt	biotite, $K(Mg, Fe)_3 [(OH)_2|(Al, Fe^{3+}) Si_3O_{10}]$
Bxb	bixbyite, $(Mn, Fe)_2O_3$
Cald	calderite, $Mn_3Fe_2[SiO_4]_3$
Cam	Ca-amphibole, —
Carb	carbonate, —
Cc	calcite, $Ca[CO_3]$
Cel	celadonite, $K(Al, Fe)_2 [(OH)_2|(Al, Fe^{3+}) Si_3O_{10}]$
Chc	chalcocite, Cu_2S
Chl	chlorite, —
Chr	chromite, $(Fe, Mg) Cr_2O_4$
Clm	chloromelanite, $(Ca, Na) (Mg, Al, Fe^{3+}) [Si_2O_6]$
Cpx	clinopyroxene, —
Crs	crossite, $Na_2 (Fe^{2+}, Mg)_3 (Al, Fe^{3+})_2 [OH|Si_4O_{11}]_2$
Cs	celsian, $Ba [Al_2Si_2O_8]$
Ctd	chloritoid, $Fe^{2+}Al_2 [(OH)_2|O|SiO_4]$
Cym	cymrite, $Ba [Al_2Si_2O_8] \cdot H_2O$
Czo	clinozoisite, $Ca_2Al_3[O|OH|SiO_4|Si_2O_7]$
Dan	dannemorite, $Mn_2Fe_5^{2+} [OH|Si_4O_{11}]_2$
Di	diopside, $CaMg [Si_2O_6]$
Edn	edenite, $NaCa_2Mg_5 [Si_7AlO_{22}|(OH)_2]$
Ep	epidote, —
Fa	fayalite, $Fe_2 [SiO_4]$
Fo	forsterite, $Mg_2 [SiO_4]$
Frd	friedelite, $Mn_8 [(OH, Cl)_{10}|Si_6O_{15}]$
Glx	galaxite, $MnAl_2O_4$

Gnt	garnet, —
Goh	goethite, $FeOOH$
Gp	glaucophane, $Na_2Mg_3Al_2 [OH|Si_4O_{11}]_2$
Gro	grossular, $Ca_3Al_2 [SiO_4]_3$
Hol	hollandite, $BaMn_8O_{16}$
Hm	hematite, Fe_2O_3
Hus	hausmannite, Mn_3O_4
Ilm	ilmenite, $FeTiO_3$
Jcb	jacobsite, $MnFe_2O_4$
Jd	jadeite, $NaAl [Si_2O_6]$
Kam	kämmererite, $(Mg, Fe^{2+})_5 (Cr, Al) [(OH)_8AlSi_3O_{10}]$
Kel	kellyite, $Mn_6^{2+} [(OH)_8Al_2Si_2O_{10}]$
Kfs	K-feldspar, $K [AlSi_3O_8]$
Kos	kosmochlore, $NaCr [Si_2O_6]$
Krp	kryptomelane, $K_2Mn_8O_{16}$
Krt	potassium-richterite, $KCaNaMg_5 [OH|Si_4O_{11}]_2$
Ktn	kutnahorite, $CaMn [CO_3]_2$
Lcp	leucophoenicite, $Mn_7 [(OH, F)_2|(SiO_4)_3]$
Mh	manganhumite, $Mn_7 [(OH, F)_2|(SiO_4)_3]$
Mc	microcline, $K [AlSi_3O_8]$
Mil	millerite, NiS
MnAcm	manganacmite, $NaMn^{3+} [Si_2O_6]$
Mnph	manganophyllite, $KMn_3 [(OH)_2|AlSi_3O_{10}]$
Ms	muscovite, $KAl_2 [(OH)_2|AlSi_3O_{10}]$
Mt	magnetite, Fe_3O_4
Mtm	montmorillonite, $(Al_{1.67}Mg_{0.33}) [(OH)_2|Si_4O_{10}] \cdot Na_{0.33} (H_2O)_4$
Nam	Na-amphibole —
Nlt	néltnerite, $CaMn_6^{3+}SiO_{12}$
Npx	Na-clinopyroxene, —
Ntc	neotocite, $(Mn, Fe) SiO_3 \cdot H_2O$
Ntn	nontronite, $Fe_2^{3+}[(OH)_2|Al_{0.33}Si_{3.67}O_{10}] \cdot Na_{0.33} (H_2O)_4$
Omp	omphacite, $(Na, Ca) (Mg, Al) [Si_2O_6]$
Or	orthoclase, $K[AlSi_3O_8]$
Part	partridgeite, Mn_2O_3
Pg	paragonite, $NaAl_2 [(OH)_2|AlSi_3O_{10}]$
Phg	phengite, —
Phl	phlogopite, $KMg_3 [(OH)_2|AlSi_3O_{10}]$
Pl	plagioclase, —
Pm	piemontite, $Ca_2Mn_3^{3+}[O|OH|SiO_4|Si_2O_7]$
Pnt	pentlandite, $(Ni, Fe)_9S_8$
Prs	pyrolusite, MnO_2
Ps	pistacite, $Ca_2Fe_3^{3+} [O|OH|SiO_4|Si_2O_7]$
Py	pyrope, $Mg_3Al_2 [SiO_4]_3$

Pyr	pyrophanite, $MnTiO_3$
Pys	pyrosmilite, $(Fe^{2+}, Mn)_8 [(OH, Cl)_{10}|Si_6O_{15}]$
Pxm	pyroxmangite, $Mn_7 [Si_7O_{21}]$
Qz	quartz, SiO_2
Rbk	riebeckite, $Na_2Fe_3^{2+}Fe_2^{3+} [OH|Si_4O_{11}]_2$
Rct	richterite, $Na_2CaMg_5 [OH|Si_4O_{11}]_2$
Rdc	rhodochrosite, $Mn [CO_3]$
Rhd	rhodonite, $Mn_5 [Si_5O_{15}]$
Rom	romeite, $(Ca, Mn, Na)_2 (Sb, Ti)_2 (O, F, OH)O_6$
Ru	rutile, TiO_2
Sie	siegenite, $(Co, Ni)_3S_4$
Sp	spessartine, $Mn_3Al_2 [SiO_4]_3$
Son	sonolite, $Mn_9 [(OH, F)_2|(SiO_4)_4]$

Stp	stilpnomelane, $(Ca, Na, K, H) (Fe, Mg, Al, Mn)_8 [Al_{1.5} Si_{10.5}O_{36}] \cdot nH_2O$
Tc	talc, $Mg_3 [(OH)_2|Si_4O_{10}]$
Tep	tephroite, $Mn_2 [SiO_4]$
Thl	thulite, $(Ca, Mn)_2Al_3 [O|OH|SiO_4|Si_2O_7]$
Tir	tirodite, $Mn_2Mg_5 [OH|Si_4O_{11}]_2$
Tit	titanite, $(Ca, Mn) [O|TiO_4]$
Tod	todorokite, MnO_2
Tr	tremolite, $Ca_2Mg_5 [(OH)|Si_4O_{11}]_2$
Tur	tourmaline, $NaMn_3Al_6 [(OH)_4|(BO_3)_3|Si_6O_{18}]$
Uv	uvarovite, $Ca_3Cr_2 [SiO_4]_3$
Ves	vesuvianite, $Ca_{10} (Mg, Fe)_2Al_4 [(OH)_4|(SiO_4)_5|(Si_2O_7)_2]$
Win	winchite, $CaNaMg_4Al [Si_8O_{22}|(OH)_2]$

REFERENCES CITED

Abraham, K. and Schreyer, W., 1975, Minerals of the viridine hornfels from Darmstadt, Germany: Contrib. Mineral. Petrol., v. 49, p. 1-20.

Abraham, K. and Schreyer, W., 1976, A talc-phengite assemblage in piemontite schists from Brezovica, Serbia, Yugoslavia: Jour. Petrology, v. 17, p. 421-439.

Abs-Wurmbach, I., 1980, Miscibility and compatibility of braunite, $Mn^{2+} Mn_6^{3+}O_8/SiO_4$, in the system Mn-Si-O at 1 atm in air: Contr. Mineral. Petrol., v. 71, p. 393-399.

Abs-Wurmbach, I., and Peters, Tj., 1981, Experimentelle und petrographische Untersuchungen über die Al-Verteilung zwischen koexistierenden Phasen in Manganlagerstätten: Fortschr. Mineral., v. 59, p. 4-5.

Abs-Wurmbach, I., Peters, Tj., Langer, K. and Schreyer, W., 1983, Phase relations in the system Mn-Si-O: an experimental and petrologic study: Neues Jahrbuch Miner. Abh., v. 146(3), p. 258-279.

Anastasiou, P. and Langer, K., 1977, Synthesis and physical properties of piemontite $Ca_2Al_{(3-p)}Mn_p^{3+} (Si_2O_7/SiO_4/O/OH)$: Contrib. Mineral. Petrol., v. 60, p. 225-245.

Angel, R. J., 1984, The experimental determination of the johannsenite-bustamite equilibrium inversion boundary: Contr. Mineral. Petrol., v. 85, p. 272-278.

Barrett, T. J., 1981, Chemistry and mineralogy of Jurassic bedded chert overlying ophiolites in the North Apennines, Italy: Chem. Geol., v. 34, p. 289-317.

Barrett, T. J., 1982, Stratigraphy and sedimentology of Jurassic bedded cherts overlying ophiolites in the North Apennines, Italy: Sedimentology, v. 29, p. 353-373.

Battaglia, S., Nannoni, R. and Orlandi, P., 1978, La piemontite del Monte Corchia (Alpi Apuane): Atti Soc. Tosc. Sc. Nat. Mem., s. A., v. 84, p. 174-178.

Bearth, P. and Schwander, H., 1981, The post-Triassic sediments of the ophiolite zone Zermatt-Saas Fee and the associated manganese mineralisations: Eclogae geol. Helv., v. 74, p. 189-205.

Bhattacharyya, P. K., Dasgupta, S., Fukuoka, M. and Roy, S., 1984, Geochemistry of braunite and associated phases in metamorphosed non-calcareous manganese ores of India: Contrib. Mineral. Petrol., v. 87, p. 65-71.

Bonatti, E., 1980, Metal deposits in the oceanic lithosphere, in Emiliani, C., ed., The Sea, Wiley, chap. 17, p. 639-686.

Bonatti, E., Zerbi, M., Kay, R. and Rydell, H., 1976, Metalliferous deposits from the Apennine ophiolites: Mesozoic equivalents of modern deposits from ocean spreading centers: Geol. Soc. Amer. Bull., v. 87, p. 83-94..

Bondi, M., Mottana, A., Kurat, G. and Rossi, G., 1978, Cristallochimica del violano e della schefferite di St. Marcel (Valle d'Aosta): Rend. Soc. It. Min. Petr., v. 34, p. 15-25.

Bosellini, A. and Winterer, E. L., 1975, Pelagic limestone and radiolarite of the Tethyan Mesozoic: a genetic model: Geology, v. 3, p. 279-282.

Bourbon, M. and Fonteilles, M., 1972, Présence de pyrosmalite et de rhodonite dans un horizon manganésifère oxydé du Crétacé supérieur briançonnais: Bull. Soc. fr. Minéral. Cristallogr., v. 95, p. 623-624.

Brown, P., Essene, E. J. and Peacor, D. R., 1978, The mineralogy and petrology of manganese-rich rocks from St. Marcel, Piedmont, Italy: Contrib. Mineral., Petrol., v. 67, p. 227-232.

Burns, R. G. and Burns, V. M., 1977, Mineralogy of manganese nodules in Glasby, G. P., ed., Marine manganese deposits, Elsevier, chap. 7, p. 185-248.

Busato, S., 1984, Minerogenesi del giacimento manganesifero di Mormanno (Calabria settentrionale, Italia): Thesis, Univer. Rome, 420 p.

Castello, P., 1981, Inventario delle mineralizzazioni a magnetite, ferro-rame e manganese del complesso piemontese dei calcescisti con pietre verdi in Valle d'Aosta: Ofioliti, v. 6(1), p. 5-46.

Chiesa, S., Cortesogno, L. and Lucchetti, G., 1976, Gli scisti quarzitici del gruppo di Voltri: caratteri stratigrafici, petrografici e mineralogici: Ofioliti, v. 1, p. 199-216.

Chiesa, S., Cortesogno, L. and Forcella, F., 1977, Caratteri e distribuzione del metamorfismo alpino nel Gruppo di Voltri e nelle zone limitrofe della Liguria occidentale con particolare riferimento al metamorfismo di alta pressione: Rend. Soc. Ital. Miner. Petr., v. 33, p. 253-279.

Chopin, C., 1978, Les paragenèses réduites ou oxidées de concentrations manganésiphères des "schistes lustrés" de Haute-Maurienne (Alpes françaises): Bull. Minéral., v. 101, p. 514-531.

Chopin, C., 1981, Talc-phengite: a widespread assemblage in high-grade pelitic blueschists of the western Alps: Jour. Petrology, v. 22, p. 628-650.

Colomba, L., 1910, Sopra un granato ferri-cromifero di Praborna (St. Marcel): Rend. R. Accad. Lincei, s. 5, v. 19, p. 146-150.

Cortesogno, L., Lucchetti, G. and Penco, A. M., 1979, Le mineralizzazioni a manganese nei diaspri delle ofioliti liguri: mineralogia e genesi: Rend. Soc. It. Miner. Petr., v. 35, p. 151-197.

Dal Piaz, G. V., Di Battistini, G., Kienast, J. R. and Venturelli, G., 1979, Manganiferous quartzitic schists of the Piemonte ophiolite nappe in the Valsesia-Valtournanche area (Italian Western Alps): Mem. Scienze Geol., v. XXXII, p. 3-24.

Dasgupta, H. C. and Manickavasagam, R., 1981, Regional metamorphism of non-calcareous manganiferous sediments from India and the related petrogenetics for a part of the system Mn-Fe-Si-O: Jour. Petrology, v. 22(3), p. 363-396.

Debenedetti, A., 1965, Il complesso radiolariti-giacimenti di manganese-giacimenti piritoso-cupriferi-rocce a fuchsite, come rappresentante del Malm nella Formazione dei Calcescisti. Osservazioni nelle Alpi piemontesi e della Val d'Aosta: Boll. Soc. Geol. It., v. 84, p. 131-163.

De Capitani, C. and Peters, Tj., 1981, The solvus in the system $MnCO_3$-$CaCO_3$: Contrib. Mineral. Petrol., v. 76, p. 394-400.

Di Sabatino, B., 1967: Su una paragenesi del giacimento manganesifero di Scortico (Alpi Apuane): Periodico Mineral., v. 36, p. 965-992.

Donnelly, T. W. and Merrill, I., 1977, The scavenging of magnesium and other chemical species by biogenic opal in deep-sea sediments: Chem. Geol., v. 19, p. 167-186.

Elder, J. W., 1965, Physical processes in geothermal areas: Am. Geophys. Union Monograph, v. 8, p. 211-239.

Ellis, D. J. and Green, D. M., 1979, An experimental study of the effect of Ca upon garnet-clinopyroxene Fe-Mg exchange equilibria: Contrib. Mineral. Petrol., v. 71, p. 13-22.

Ernst, W. G. and Calvert, S. E., 1969, An experimental study of the recrystallization of porcellanite and its bearing on the origin of some bedded cherts: Amer. Jour. Sci., v. 267A, p. 114–133.

Essene, E. J. and Peacor, D. R., 1983, Crystal chemistry and petrology of coexisting galaxite and jacobsite and other spinel solutions and solvi: Amer. Miner., v. 68, p. 449–455.

Ewing, J., Windisch, C., and Ewing, M., 1970, Correlation of horizon A with JOIDES bore-hole results: J. Geophys. Res., v. 75, p. 5645–5653.

Folk, R. L. and McBride, E. F., 1978, Radiolarites and their relations to subjacent "oceanic crust" in Liguria, Italy: Jour. of Sedim. Petrology, v. 48(4), p. 1069–1102.

Frentrup, K. R. and Langer, K., 1981, Mn^{3+} in garnets: optical absorption spectrum of a synthetic Mn^{3+} bearing silicate garnet: N. Jb. Miner. Mh. Jg. 1981 (6), p. 245–256.

Gasparik, T. and Lindsley, D. H., 1980, Phase equilibria at high pressure of pyroxenes containing monovalent and trivalent ions *in* Prewitt, C. T., ed., Pyroxenes, Reviews Mineral., v. 7, p. 309–339.

Gennaro, V., 1925, Micascisti a piemontite nelle Valli di Lanzo (Alpi piemontesi): Rend. R. Acc. Lincei, s. 6, v. 2, p. 508–510.

Goldsmith, J. R. and Graf, D. L., 1957, The system $CaO-MnO-CO_2$, the solid-solution and decomposition relations: Geoch. Cosm. Acta, v. 11, p. 310–334.

Griffin, W. L. and Mottana, A., 1982, Crystal chemistry of clinopyroxenes from the St. Marcel manganese deposit, Val d'Aosta, Italy: Amer. Miner., v. 67, p. 568–586.

Holland, T.J.B., 1980, The reaction albite = jadeite + quartz determined experimentally in the range 600-1200 °C: Amer. Miner., v. 65, p. 129–134.

Hsu, L. C., 1968, Selected phase relationships in the system Al-Mn-Fe-Si-O-H: a model for garnet equilibria: Jour. Petrology, v. 9, p. 40–83.

Huebner, J. S., 1976, The manganese oxides - a bibliographic commentary *in* Rumble, D., ed., Oxide minerals, Reviews Mineral., v. 3, p. 1–17.

Hunt, J. A. and Kerrick, D. M., 1977, The stability of sphene: experimental redetermination and geologic implications: Geoch. Cosm. Acta, v. 41, p. 279–288.

Huttenlocher, H. F., 1934, Die Erzlagerstättenzonen der Westalpen: Schw. Min. Petr. Mitt., v. 14, p. 22–149.

Jones, J. B. and Segnit, E. R., 1971, The nature of opal. I. Nomenclature and constituent phases: J. Geol. Soc. Austr., v. 18, p. 56–68.

Kawachi, Y., Grapes, R. H., Coombs, D. S. and Dowse, M., 1983, Mineralogy and petrology of a piemontite-bearing schist, western Otago, New Zealand: J. metamorphic Geol., v. 1(4), p. 353–372.

Keskinen, M. and Liou, J. G., 1979, Synthesis and stability relation of Mn-Al piemontite, $Ca_2MnAl_2Si_3O_{12}(OH)$: Amer. Miner., v. 64, p. 317–328.

Lincio, G., 1922, Note mineralogiche sui giacimenti di manganese e di pirite di Salbertrand (Alta Valle della Dora Riparia): Atti Soc. Nat. Mat. Modena, s. 5, v. 7, p. 29–32.

Maresch, W. V. and Mottana, A., 1976, The pyroxmangite-rhodonite transformation for the $MnSiO_3$ composition: Contr. Miner. Petrol., v. 55(1), p. 69–79.

Martin-Vernizzi, S., 1982, La mine de Praborna (Val d'Aoste, Italie): un série manganesifère metamorphisée dans la facies eclogite: Thesis, Univ. Paris VI, pp. 215.

Millosevich, F., 1906, Sopra alcuni minerali di Val d'Aosta: Rend. R. Acc. Lincei, s. 5, v. 15, p. 317–321.

Minnigh, L. D., 1979, Petrological and structural investigations of the Sparone area in the Orco Valley (Southern Sesia-Lanzo border zone, western Italian Alps): Thesis, Univ. Leiden, p. 119.

Miyashiro, A. and Seki, Y., 1958, Enlargement of the composition field of epidote and piemontite with rising temperature: Amer. Jour. Sci., v. 256, p. 423–430.

Mottana, A. and Griffin, W. L., 1979, Pink titanite (greenovite) from St. Marcel, Valle d'Aosta, Italy: Rend. Soc. It. Miner. Petr., v. 35, p. 135–143.

Mottana, A. and Griffin, W. L., 1982, The crystal chemistry of piemontite from the type-locality (St. Marcel, Val d'Aosta, Italy): Rep. 13th IMA Meeting Varna, in press.

Nisbet, E. G. and Price, I., 1974, Siliceous turbidites: bedded cherts as redeposited ocean ridge-derived sediments *in* Hsü, K. J., and Jenkins, H. C., eds., Pelagic sediments: on land and under the Sea, Int. Ass. Sed. Spec. Publ. No. 1, p. 351–356.

Nitsch, K. H., 1980, Reaktion von Bariumfeldspat (Celsian) mit H_2O zu Cymrit unter metamorphen Bedingungen: Fortschr. Mineral., v. 58, Bh. 1, p. 98–100.

Pelloux, A., 1913, Nuove forme della romeina di St. Marcel in Valle d'Aosta: Ann. Mus. Civ. St. Nat. Genova, 3, v. 6, p. 22–24.

Pelloux, A., 1946, Contributo alla mineralogia della Valle d'Aosta: Rend. Soc. Min. It., v. 3, p. 188–206.

Peters, Tj., Schwander, H. and Trommsdorff, V., 1973, Assemblages among tephroite, pyroxmangite, rhodocrosite, quartz: experimental data and occurrences in the Rhetic Alps: Contrib. Mineral. Petrol., v. 42, p. 325–332.

Peters, Tj., Valarelli, J. V. and Candia, M. A., 1974, Petrogenetic grids from experimental data in the system Mn-Si-C-O-H: Revista Brasil. Geosciencias, v. 4, p. 15–26.

Peters, Tj., Trommsdorff, V. and Sommerauer, J., 1978, Manganese pyroxenoids and carbonates: critical phase relations in metamorphic assemblages from the Alps: Contr. Mineral. Petrol., v. 66, p. 383–388.

Peters, Tj., Trommsdorff, V. and Sommerauer, J., 1980, Progressive metamorphism of manganese carbonates and cherts in the Alps *in* Varentsov, I. M. and Grasselley, G. Y., eds., Geology and Geochemistry of manganese, v. 1, Budapest, Akademiai Kiadò.

Reinecke, T., 1982, Cymrite and celsian in manganese-rich metamorphic rocks from Andros island, Greece: Contr. Mineral. Petrol., v. 79, p. 333–336.

Reinecke, T., 1983, Mineralogie und Petrologie der Mangan und Eisenreichen Metasedimente von Andros/Kykladen/Griechenland: Thesis, Univ. Braunchweig, pp. 256.

Robertson, A.H.F., 1977, The origin and diagenesis of cherts from Cyprus: Sedimentology, v. 24, p. 11–30.

Roccati, A., 1906, Rodonite di Chiaves e di altre località delle Valli di Lanzo: Atti R. Accad. Sci. Torino, s. 2, v. 41, p. 487–496.

Rondolino, R., 1936, Sopra alcuni anfiboli manganesiferi di Praborna (San Marcello, Valle d'Aosta): Per. Miner., v. 7, p. 109–121.

Schreyer, W., Abraham, K. and Trochim, H. D., 1975, Piemontit-Schiefer von Brezovica, Südserbien: Acta Geologica (Zagreb) VIII/13, v. 41, p. 251–268.

Smith, W. E., 1960, The siliceous constituents of chert: Geol. Mijnbouw, v. 39, p. 1–8.

Smith, D. and Albee, A. L., 1967, Petrography of a piemontite-bearing schist, San Gorgonio Pass, California: Contr. Mineral. Petrol., v. 16, p. 189–203.

Smith, G., Hålenius, U. and Langer, K., 1982, Low temperature spectral studies of Mn^{3+}-bearing andalusite and epidote type minerals in the range 30000-5000 cm^{-1}: Phys. Chem. Minerals, v. 8, p. 136–142.

Steinmann, G., 1927, Die ophiolitischen Zonen in den mediterranen Kettenbergen: C.R. 14th Int. Geol. Congress, Madrid, v. 2, p. 637–668.

Sugisaki, R., Yamamoto, K. and Adachi, M., 1982, Triassic bedded cherts in central Japan are not pelagic: Nature, v. 298, p. 644–647.

Thurston, D. R., 1972, Studies on bedded cherts: Contr. Mineral. Petrol., v. 36, p. 329–334.

Winter, G. A., Essene, E. J. and Peacor, D. R., 1983, Mn-humites from Bald Knob, North Carolina: mineralogy and phase equilibria: Amer. Miner., v. 68, p. 957–959.

Zambonini, F., 1922, Ardennite di Ceres in Val d'Ala (Piemonte): Rend. R. Acc. Naz. Lincei, s. 6, v. 30, p. 147–157.

MANUSCRIPT ACCEPTED BY THE SOCIETY JULY 29, 1985

Geological Society of America
Memoir 164
1986

The blueschist facies schistes lustrés of Alpine Corsica:
A review

Wes Gibbons
Colin Waters
Department of Geology
University College, Cardiff
P.O. Box 78
Cardiff CF1 1XL, Wales, U.K.

John Warburton
BP Petroleum Development of Spain, S.A.
Edificio AGF
C/Albacete, 5 - 5ª, Sur Izquierda
28027 Madrid, Spain

ABSTRACT

Blueschists in Corsica represent a continuation of the internal Alpine Penninic Zone
schistes lustrés nappe, metamorphosed during the Eoalpine (late Cretaceous) collision
of a Tethyan subduction complex with the European continental plate. Although extremely complex, the Corsican *schistes lustrés* preserve some semblance of a coherent
stratigraphy and have been divided into several units, the most important of which are
the Castagniccia, Inzecca, and Santo Pietro Groups. The metasedimentary and ophiolitic
rocks associated with the Inzecca and Castagniccia Groups have been correlated with
similar Liguro-Piemontais units in the Western Alps and interpreted as representing
Tethyan oceanic basement and cover. The nature of the basement to the Santo Pietro
Group is, however, more controversial and could be oceanic, continental or both.

The Corsican blueschists have suffered syn-metamorphic polyphase folding and
thrusting, generally externally (WNW) directed, although SSW-directed nappe displacement has been described in western Cap Corse. The onset of high-pressure metamorphism is at least mid-Cretaceous in age, but a spread of younger radiometric dates
has led authors to suggest that blueschist metamorphism has continued through the late
Cretaceous and even into the Eocene, when the Adria microplate collided with Corsica.
The highest metamorphic grades are recorded by undated eclogite relics which are
overprinted by a pervasive blueschist metamorphism. The highest blueschist pressures
are recorded along the eastern margin of Alpine Corsica. A later phase of (late Eocene?)
SE-verging folding and backthrusting occurred under greenschist facies conditions. Late
(post-Miocene) upright folding about N-S axes has greatly influenced the present outcrop pattern.

Most recent work on these blueschists has concentrated on Cap Corse in NE
Corsica where several nappes have been identified. Two of these, the Ersa-Centuri and
Farinole Nappes, have been correlated with Austro-Alpine basement, with the Ersa-
Centuri nappe being emplaced from the west. Various tectonic models have attempted to
refine the 'Eoalpine obduction' model, but there remains considerable disagreement over
the timing of individual tectonometamorphic events. There is as yet no record of blue-

schist metamorphism having occurred prior to the involvement of the European continental basement, the upper part of which has also suffered blueschist metamorphism. An analogy with these Eoalpine events is provided by the present-day collision of Australia with the Banda Arc.

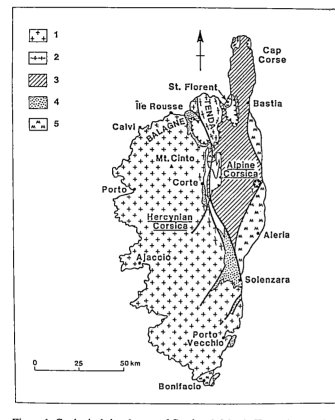

Figure 1. Geological sketch map of Corsica. 1. Mostly Hercynian granite basement; 2. sheared and metamorphosed granitic basement; 3. Eoalpine *schistes lustrés* nappe of Alpine Corsica (HP/LT metasedimentary rocks and meta-ophiolites) 4. autochthonous Tertiary sedimentary cover on 1. Thrust teeth pointing into colourless areas define nappes of Triassic-Eocene rocks unaffected by Eoalpine metamorphism. 5. Miocene sediments. T = Tox. Star = Jadeite locality at Sant' Andrea de Cotone (Autran 1964; Caron and others 1981).

INTRODUCTION

The island of Corsica, lying in the Western Mediterranean 160 km SE of Nice, is geologically divisible into a western area dominated by Hercynian granites and an eastern area characterized by metamorphic rocks produced during Alpine plate collision (Figure 1). The latter area, known as Alpine Corsica, represents a continuation of the *schistes lustrés* of the internal Western Alps (Figure 2). Like much of the *schistes lustrés* on the European mainland, the Corsican schists include metasediments and metaophiolites which preserve high P-low T blueschist mineral assemblages. The Corsican blueschist outcrop extends over some 1,800 sq km., forming rough, often inhospitable terrain which reaches a maximum height of 1766 m at San Petrone. The varied nature of both the metamorphic grade and the protoliths

within this blueschist unit has resulted in a spectacular diversity of phase assemblages. This paper provides a review of previous work relating to the stratigraphy, structure and metamorphism of the Corsican blueschists and compares the various tectonic models which have been advanced to explain the late Cretaceous to Eocene evolution of this part of the Alpine orogen.

"L'évolution des idées sur les Schistes lustrés est longue et confuse" (Durand-Delga, 1978, p. 19). The correlation between the Corsican schists and the Alpine *schistes lustrés* was made by Haug as early as 1896. Prior to 1950, over 260 publications made at least a passing reference to Corsican geology (Gauthier 1976). However, the first detailed modern petrological account of the blueschists was provided by Brouwer and Egeler in 1951. Important work on the tectonic contact between the *schistes lustrés* and the underlying metamorphosed Hercynian basement in NE Corsica was presented about this time by Stam (1952) and Delcey & Meunier (1966). Early uncertainty and disagreement regarding the age of the Corsican *schistes lustrés* reached a peak in 1963 with the publication of a paper by Lapadu-Hargues & Maisonneuve proposing a pre-Hercynian age for the schists, unlike all other workers who envisaged a Mesozoic (and possibly Tertiary) age. However, despite some continuing uncertainty and a paucity of evidence, the age range of the *schistes lustrés* in Corsica and the Western Alps is now generally accepted to be Carnian (Mid-Triassic) to Cenomanian.

The publication of detailed maps (1:80,000) of Corsica greatly increased knowledge of the schist complex (e.g. Routhier 1964). The 1970s saw numerous papers by Amaudric du Chaffaut (see reference list), including an important attempt to collate all known metamorphic data (Amaudric du Chaffaut and others 1976). Other detailed work produced about this time includes Peterlongo (1968), Saliot & Carron (1971), Ohnenstetter and others (1976), Sauvage-Rosenberg (1977), and Caron (1977), and culminated in the publication of a comprehensive Masson guide (Durand-Delga 1978). The thesis work of Caron (1977) produced the most detailed attempt at a stratigraphic subdivision of the Corsican *schistes lustrés* (Caron and Delcey, 1979; see also Péquignot and others 1984). With the development of plate tectonic theory, Corsica has become recognised as a classic area recording the obduction of an ophiolitic and metasedimentary nappe (Alpine Corsica) over the leading edge of "European" continental crust (Hercynian Corsica) during late Cretaceous (Eoalpine) plate collision (Mattauer & Proust 1976; Mattauer and others 1977, 1981; Zacher 1979; Faure & Malavieille 1980, 1981; Cohen and others 1981; Warburton 1983; Malavieille 1983; Harris 1985). Gibbons and Horák (1984), emphasising the apparent similarity of most Eoalpine radiometric dates in both metamorphosed European basement and the overlying *schistes lustrés*, argue that the blueschist metamorphism occurred during

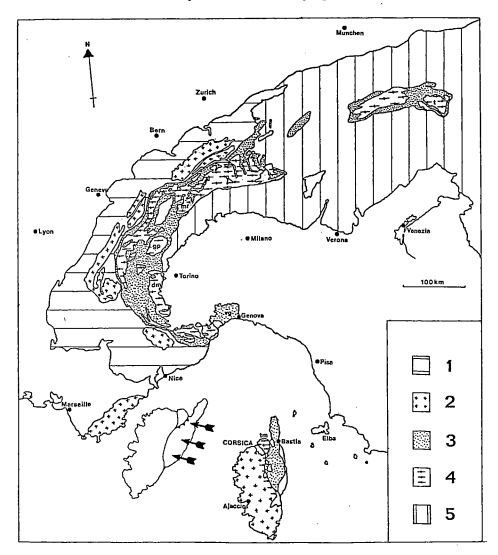

Figure 2. Geological sketch map of the Alpine system (after Frey and others 1974; Moullade 1978, Gibbons & Horák 1984). 1. Permo-Mesozoic-Tertiary cover to External Western Alps (Helvetics); 2. Pre-Alpine crystalline basement (includes Permo-Carboniferous cover in Corsica and S. Provence); 3. Penninic domain: Permo-Mesozoic to Tertiary cover, including the Mesozoic high-P metasedimentary rocks and ophiolites of the *schistes lustrés*; 4. Penninic domain: Pre-Triassic basement beneath Penninic schists (affected by high-P metamorphism); 5. Austroalpine domain and Southern Alps; tm = Tenda Massif, dm = Dora Maira, gp = Gran Paradiso, mr = Monte Rosa, t = Tauern window, vg = Voltri Group. The approximate position of Corsica prior to Oligocene rotations is also shown. Arrows mark transport direction of the main *schistes lustrés* nappe over the granitic crystalline basement during Eoalpine plate collision.

and in response to the actual collision rather than during earlier intraoceanic subduction.

STRATIGRAPHIC SETTING

The Alpine *schistes lustrés* preserve some semblence of a stratigraphic succession and may be viewed as an essentially coherent blueschist terrane. Several stratigraphic interpretations have been proposed for the *schistes lustrés* of Alpine Corsica (Mattauer & Proust 1976; Caron 1977; Durand-Delga 1977,

1978; Warburton 1983; Péquignot and others 1984). However, the lack of geochronological evidence, and the extreme complexity resulting from polyphase folding, thrusting, and metamorphism, have combined to ensure that so far all attempts at stratigraphic classification remain tentative. The *schistes lustrés* have been separated broadly into upper and lower ophiolite-bearing units. Most of the lower unit (the *schistes lustrés inférieur* of Durand-Delga 1977, 1978) corresponds to the série de la Castagniccia (Castagniccia Group) of Caron (1977). It crops out mainly along the eastern margin of Alpine Corsica in the cores of broad antiforms. The overlying unit (*schistes lustrés supérieure* of Durand-Delga 1977, 1978) include the série de l'Inzecca (Inzecca Group) of Caron (1977) and is widely exposed in SW Alpine Corsica along the contact with the Hercynian basement. Both units consist of metasediments (pelites, calc-schists, marbles, metaquartzites) and are associated with abundant meta-ophiolites (serpentinites, metagabbro, metabasites, radiolarian cherts). The ophiolitic rocks represent dismembered and variably metamorphosed Tethyan oceanic crust, (Beccaluva and others 1977, Venturelli and others 1981), dated as Oxfordian (Ohnenstetter and others 1975) and occurring as isolated masses and imbricate slices within a metasedimentary matrix.

Another important stratigraphic unit within the Corsican *schistes lustrés* is the Santa Pietro Group (série de Santo-Pietro-de-Tenda of Delcey 1974). This unit is dominated by calc-schists, marbles and metaquartzites capped by metabasite. In the type locality around Santo Pietro di Tenda (Figure 3), these rocks rest with tectonic contact upon mylonitised Hercynian granitic basement. The nature of the original basement to this group is controversial. Whereas some workers have interpreted the Santo Pietro sequence as resting on a Tethyan oceanic basement, others believe a European continental basement to be more likely. The 'oceanic basement' interpretation rests, firstly, upon an apparently conformable contact of Santo Pietro Group on pillow lavas at Tox (Figure 1) in southern Alpine Corsica, and secondly, on the local presence of manganiferous quartzites in the Santo Pietro Group (Caron and Delcey 1979). Manganiferous metasediments have been described elsewhere, for example in the *schistes lustrés* of the Zermatt-Saas Zone of the Swiss Alps (Bearth and Schwander 1981), where they are considered to indicate deep oceanic conditions. The 'continental basement' interpretation, preferred by Mattauer and others (1981) and Warburton (1983) emphasises the existence at Monte Asto (in NW Alpine Corsica) of an apparently basal conglomerate (with granitic basement clasts) grading up through carbonate conglomerates into typical Santo Pietro Group metasediments (Varenkamp 1957).

Péquignot and others (1984) have identified a new *schistes lustrés* unit - la série du Monte Piano Maggiore (the Maggiore Group) - which they interpret as transitional between the Inzecca and Santo Pietro Groups. These authors envisage the Santo Pietro Group as having been deposited at the edge of the European continental margin with the Maggiore and Inzecca Groups being formed progressively oceanward in deeper waters. Later piggyback thrusting towards the continental margin emplaced the deepest water Inzecca Group over the Maggiore Group which was thrust in turn over the Santo Pietro Group.

STRUCTURE

An intense, often sub-horizontal, layer-parallel LS foliation (S1), with mineral grain shape lineations (L1) occurring parallel to the axes of tight to isoclinal folds, are characteristic of the *schistes lustrés* and the immediately underlying Hercynian basement (Faure and Malavieille 1980; Mattauer, Faure and Mala-

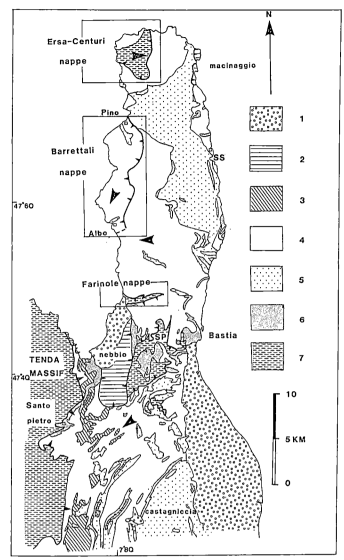

Figure 3. Geological sketch map of northern Alpine Corsica: 1. Miocene - Quaternary sediments; 2. allochthonous unmetamorphosed Mesozoic sedimentary and basic volcanic rocks. 3. Inzecca Group; 4. Ophiolites; 5. Castagniccia Group; 6. Santo Pietro Group; 7. sheared granitic basement, including allochthonous slices. Large arrows mark transport direction of nappe units; the Ersa-Centuri, Barrettali and Farinole nappes are enclosed within boxes; teeth on thrusts indicate overriding unit. SS = Santa Severa; SP = Serra di Pigno.

vielle 1981; Caron and others 1981; Malavieille 1983; Warburton 1983, Harris 1985). The S1 fabric is often mylonitic in character, particularly where the protolith is coarse grained, as in the granitic basement and the ophiolitic gabbros. A generally shallow plunging L1 lineation is assumed to be parallel to the nappe transport direction (Mattauer and Proust 1975, and all later authors) and is usually described as trending N070-080°E. Quartz c-axis plots and micro-rotational criteria have been used to interpret the S1 foliation as having developed in response to a westerly directed simple shear (Mattauer and others 1981). Im-

TABLE 1. THE RELATIONSHIP OF STRUCTURES TO TECTONOMETAMORPHIC EPISODES IN CORSICA;
A COMPARISON BETWEEN WARBURTON (1983), HARRIS (1984), MALAVIEILLE (1983),
AND MATTAUER AND OTHERS (1981).

	Warburton	Harris	Malavieille	Mattauer and others
HP/LT Eoalpine Metamorphism (Mid-Cretaceous to ?Eocene)	**D1** Early: Foliation (S1) Mineral Lineation (L1) Late: Thrusts	**D1** Early: Foliation (S1) axial planar to folds Lineation (L1) Late: SSW directed Barretalli nappe thrust over main schistes lustrés nappe	**D1** Foliation (S1) axial planar to sheath folds (F1) Lineation (L1) Includes W. directed emplacement of main schistes lustrés nappe and E. directed Ersa-Centuri nappe	**D1** Early: Folds (F1a) axial planar foliation (S1)
	D2 Early: Sheath folds (F2) with axially planar fabric (S2) and mineral lineation (L2) Late: Thrusts	**D2** Eastward back-thrusting to Ersa-Centuri nappe with associated (F2) folds verging toward E. (mid-late Eocene)		Late: Sheath folds (F1b) with axes parallel to lineation (L1). Folds (F1c) with axes sub-perpendicular to L1
Greenschist Facies Metamorphism (Late Eocene)	**D3** SE-directed back-thrusting predating and locally synchronous with SE-verging F3 folds	**D3** Rare upright folds (F3), SE-verging axial planar cleavage (S3)	**D2** Folds (F2), SE-verging with axial planar cleavage	**D2** SE-verging folds (F2) with axial planar cleavage (S2)
	Emplacement of high level nappes (Nebbio & Macinaggio)		Nebbio & Macinaggio gravity nappes	Gravity gliding tectonics. Late Eocene
	D4 N-S Open, upright Folds (F4); (Late Miocene-Quaternary)			

portant exceptions to this general rule have been recorded from three areas in Cap Corse (Figure 3). Two of these areas represent klippen of continental basement thrust over the *schistes lustrés*: the Ersa - Centuri Nappe and the Farinole Nappe (Figure 3). Caby and others (submitted) (1984) have suggested these klippen are remnants of Austro-Alpine basement, basing this deduction upon marked lithological dissimilarities between these rocks and the 'European' basement of Western Corsica and the Alps. An alternative interpretation is made by Faure and Malavieille (1981) who mapped the Farinole Nappe as a northerly extension of a slice of Hercynian (European) basement present in the *schistes lustrés* to the south (around Serra di Pigno) (Figure 3). The Farinole Nappe displays a NE-SW trending L1 lineation (Mattauer and others 1981) with a SW transport direction (Harris 1984). By contrast, the Ersa-Centuri nappe displays an E-W L1 lineation and has been interpreted by Malavieille (1983) as thrust eastward, rather than westward, over the *schistes lustrés*. The ophiolitic Barretalli Nappe ("zone B" of Harris 1984) is

interpreted as a SSW directed nappe within the Cap Corse ophiolite stack (Figure 3).

Various authors have attempted to elucidate the minor structures produced during the early (Eoalpine) deformation, which was synchronous with high P/low T metamorphism (Table 1). Sheath folds are common and have been classed as F1 structures by Malavieille (1983), Faure and Malavieille (1980), and Harris (1984). Mattauer and others (1981) and Warburton (1983) prefer to classify their sheath folds as second generation structures (F1b and F2, respectively), coaxially refolding F1 isoclines. Mattauer and others (1981) further subdivide their early deformation event to include westerly verging (F1c) folds with fold axes normal to L1. It is likely that the many complex early fold patterns displayed by the *schistes lustrés* were produced during a prolonged history of progressive shearing (Cobbold & Quinquis 1980; Mattauer and others 1981; Warburton 1983). The latter author notes that his D2 shearing shows a more restricted distribution along discrete planes than those produced during D1. He suggests, therefore, that D2 developed along localised thrust surfaces during progressive uplift. Finally, Harris (1984) identifies F2 folds in the *schistes lustrés* beneath the Ersa-Centuri Nappe and interprets them as both eastward verging and synchronous with blueschist metamorphism.

The structural history of the *schistes lustrés* has been greatly complicated by an important later deformation event which occurred after the high P/low T metamorphism. This event has been variously described as D2 (Cohen and others 1981; Mattauer and others 1981) or D3 (Caron and others 1981; Warburton 1983, Harris 1984) (Table 1). All authors agree, however, that the deformation produced SE-verging, gently plunging folds. An axial planar foliation is found associated with many of the folds, typically as a spaced pressure solution striping in marbles and metaquartzites, or as a crenulation cleavage in pelites and metabasites. These structures developed during (and just prior to) a late Alpine greenschist facies metamorphic overprint, often interpreted as late Eocene to early Oligocene (Maluski 1977; Carpena and others 1979; Cohen and others 1981; Maluski and Schaeffer 1982).

Warburton (1983) has recognised a major system of late thrusts with a SE directed backthrust geometry. The majority of these thrust surfaces strike parallel to the hinges of the SE verging folds and are often, but not always, folded by them. Warburton attributes both the folds and thrusts to an Eocene D3 event. The thrust system is interpreted as having effected considerable further imbrication of the *schistes lustrés*. Furthermore, he regards this event as responsible for the emplacement of several mylonitized granitic basement nappes, including the Ersa-Centuri Nappe, over the *schistes lustrés*. This is a phenomenon previously interpreted as being a result of high P-low T Eoalpine thrusting alone (Cohen and others 1981; Mattauer and others 1981). By contrast, a late Eocene syn-greenschist metamorphism thrusting event is not recognised by Harris (1984) in the *schistes lustrés* of northern Cap Corse.

During the Oligocene, the Corsardinian continental microplate split off from the European craton and rotated some 30° anticlockwise (Nairn and Westphal 1968; Alvarez 1972; Auzende and others 1973; Westphal and others 1973, 1976; Chabrier and Mascle 1975; Arthaud and Matte 1976; Vandenberg and Zijderveld 1982). A final, late, mild compressional phase affecting Alpine Corsica produced a series of open, upright, large scale, N-S trending folds. These structures, which affect rocks as young as Miocene (Burdigalian to Lower Langian) often have had a pronounced effect upon the present day outcrop pattern in NE Corsica. The narrow, mountainous isthmus of Cap Corse, for example, reflects the N-S trend of a broad antiform flanked by synclines within which Tertiary rocks are preserved.

METAMORPHISM

The highest metamorphic grades within the Alpine schists are preserved as eclogite relics recorded from E and NE Corsica. The former locality occurs along the eastern margin of the Cas-

tagniccia near Sant' Andrea di Cotone (Caron and others 1981), whereas the latter occurs in western Cap Corse (Dal Piaz & Zirpoli 1979; Guiraud 1982; Harris 1984). In the Sant' Andrea di Cotone occurrence, the relict eclogite assemblage of Na-poor, Fe-rich omphacite (Jd 18-30) and almandine-rich, pyrope-poor garnet is described in iron-rich metabasites. In Cap Corse, lenses of Fe-Ti gabbro occur along a thrust zone running from Marine d'Albo to Pino (Figure 3) and contain an assemblage of garnet, omphacite (Jd 8-30) and rutile (Harris 1984). Omphacite-garnet assemblages are also recorded from the base of the Farinole Nappe by Caby and others (1984). The age of this metamorphism is unknown, although the eclogites formed before the D1 high P/low T event recorded by the *schistes lustrés*.

The eclogite minerals generally show extensive overprinting by blueschist assemblages. Caron and others (1981) describe post-eclogite blueschist assemblages including jadeite (Jd_{90}), aegirine, spessartine-rich garnet (as rims on eclogitic garnets), lawsonite, sodic amphibole and deerite. The occurrence of nearly pure jadeite in equilibrium with quartz at Sant' Andrea di Cotone is particularly notable (Autran 1964, Essene 1969). In Cap Corse, Harris (1984) has recorded a blueschist assemblage of glaucophane/crossite + epidote ± phengite ± calcite overprinting the eclogite minerals.

The blueschist metamorphism was synchronous with the first major phase of deformation, presumably induced as the rocks were being transported upwards from the eclogite P-T field. Folding within blueschists is often associated with axial planar growth of new blueschist minerals such as sodic amphibole, lawsonite or epidote, and phengite (Caron and others 1981, Warburton 1983, Harris 1984). Mattauer and others (1981) describe pre- to syntectonic HP-LT Eoalpine parageneses, with some rocks exhibiting undeformed and weakly oriented sodic amphiboles. They conclude that polyphase crystallisation occurred continuously throughout the HP-LT Eoalpine deformation.

Several authors note an overall increase in Eoalpine metamorphic pressure from west to east across Alpine Corsica (Amaudric du Chaffaut and others 1976; Mattauer and others 1981). Metabasite blueschist assemblages including lawsonite, glaucophane and garnet are typical throughout eastern Alpine Corsica. This may be contrasted with crossite + epidote assemblages common in NW Alpine Corsica. Gibbons and Horák (1984) demonstrate how the sheared granitic basement of the Tenda massif in NW Corsica (Figure 3) also experienced blueschist metamorphism. Crossite, epidote, phengite and albite are typical products of this syntectonic Eoalpine metamorphism. At deeper levels in the basement, over 1000 m from the contact with the *schistes lustrés*, actinolitic amphibole is stable in place of crossite. This change is interpreted by Gibbons and Horák (1984) as the result of increasing temperature downwards into the granitic basement during Eoalpine metamorphism. Unlike the latter study, Amaudric du Chaffaut and others (1976) indicate that lawsonite (instead of epidote) is present in the sheared granitic basement, whereas Warburton (1983) describes lawsonite pseudomorphed by white mica and zoisite.

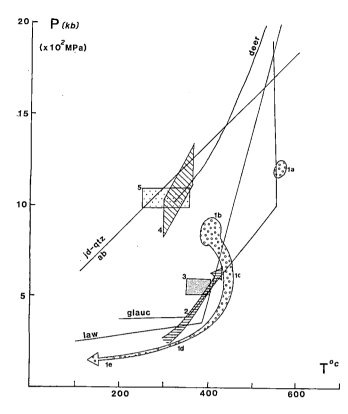

Figure 4. Petrogenetic grid outlining PT conditions for Alpine Corsica estimated by various authors: 1. Harris (1984); 1a. eclogite relicts along thrust contact between Barrettali and main *schistes lustrés* nappes; 1b. glaucophane-lawsonite facies metamorphism during first major deformation phase; 1c. greenschist facies overprint; 1d. resetting of zircons at 36-38 My; 1e. resetting of apatites at 30 My; 2. Blueschist metamorphism affecting the granitic basement beneath the *schistes lustrés* (Gibbons and Horák 1984); 3. Blueschist metamorphism Santa Severa (see Fig. 3), Eastern Cap Corse (Amaudric du Chaffaut and others 1976); 4. Blueschist metamorphism in basement slice within *schistes lustrés* Sant' Andrea di Cotone (Caron and others 1981. NB: the lower PT estimate is preferred by these authors); 5. Blueschist metamorphism at Sant' Andrea di Cotone (Autran 1964). stability curves for: deerite (Lattard and Schreyer 1981); jadeite + quartz (Newton and Kennedy 1968); lawsonite (Nitsch 1974); Upper T stability limit of glaucophane (Maresch 1977).

The age of Eoalpine blueschist metamorphism in Corsica is generally taken to be mid-Cretaceous (Albian - Cenomanian). This age is based mainly upon the radiometric data of Cohen and others (1981) who used $^{87}Rb/^{86}Sr$ whole-rock methods on the Tenda basement to produce an age of 105 ± 8 Ma. Maluski (1977) dates a glaucophane separate from the Tenda as 90 Ma (using $^{40}Ar/^{39}Ar$). Maluski also produces a more controversial age of 40 ± 2 Ma for blue amphiboles in the granitic basement near Popolasca in western Alpine Corsica. Other $^{40}Ar/^{39}Ar$ dates, on phengites, K-feldspars and biotites (Maluski 1977; Maluski and Schaeffer 1982), and fission track dates on apatites and zircons (Carpena et al. 1979) have also produced young ages, ranging from mid-Eocene to earliest Oligocene.

Overprinting of metabasite blueschist minerals by green-

schist assemblages is common in Alpine Corsica. At Sant' Andrea di Cotone late (D3) folding was locally accompanied by the growth in metabasites of albite, pumpellyite and epidote (Caron and others 1981). In Cap Corse, albite and chlorite porphyroclasts overgrow earlier foliations formed during HP/LT metamorphism to produce typical Alpine 'prasinites'. According to Warburton (1983) this static greenschist overprint is often concentrated near zones of SE-directed backthrusting interpreted as being of Eocene age.

Figure (4) compares the pressures and temperatures estimated for various assemblages in the Corsican *schistes lustrés* and underlying Hercynian basement. Estimates for the early eclogitic metamorphism (in Cap Corse) have been provided by Guiraud (1982) and Harris (1984) using garnet-clinopyroxene, garnet-phengite and plagioclase - sodic pyroxene exchange thermobarometry (Ellis & Green 1979; Krogh & Raheim 1978; Currie & Curtis 1976). Guiraud (1982) derives PT conditions of 625°C at 12 kb (1200 MPa) for the rocks preserved in a shear zone at the base of the Farinole Klippe. Harris (1984) places his eclogite nodules as having formed at 535°C at 12 kb (1200 MPa) minimum pressure.

PT estimates for the syntectonic blueschist metamorphism vary considerably across Alpine Corsica. At Sant' Andrea di Cotone, Caron and others (1981) used dolomite - calcite geothermometry to estimate temperatures of 300°C - 350°C within a poorly constrained pressure estimate of 8-13 kb (800 - 1300 MPa). The presence of jadeite + quartz, lawsonite, glaucophane, Mn-rich garnet and deerite at this locality indicates conditions well into the high P/low T blueschist field. The earlier work of Autran (1964) at this quarry estimated PT conditions of 250-350°C at 10-11 kb (1000-1100 MPa).

In Cap Corse a PT estimate from glaucophane - lawsonite - garnet - omphacite metagabbros on the east coast was given by Amaudric du Chaffaut and others (1976) as 350-400°C at 5-6 kb (500-600 MPa). A more recent estimate for blueschist metamorphism on Cap Corse is provided by Harris (1984) as 360-425°C at 6.7-8.6 kb (670-860 MPa). Further west, within the Hercynian basement of the Tenda massif, Gibbons and Horaĺk (1984) suggest PT conditions of 3903°-490°C at 6-9 kb (600-900 MPa). These estimates imply a tendency towards both increasing temperature and decreasing pressure of blueschist metamorphism moving westwards across northern Corsica.

BLUESCHIST TECTONICS IN ALPINE CORSICA

The Corsican *schistes lustrés* have been modelled in terms of an intraoceanic subduction system terminated by the arrival of European continental crust (e.g. Mattauer and Proust 1976, Mattauer and Tapponnier 1978). During the obduction of Tethyan oceanic crust, intense shearing under blueschist metamorphic conditions occurred in the uppermost part of the Corsican (Eoalpine) basement and lower part of the overriding oceanic crust. Mattauer and others (1981) interpret granitic basement slices found within *schistes lustrés* as having been derived from the

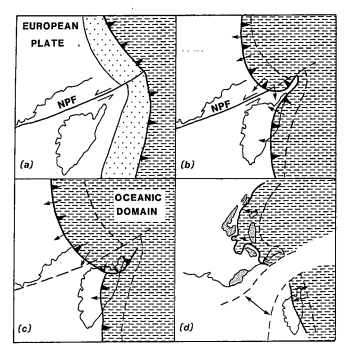

Figure 5. Model proposed by Malavieille (1983) showing the progressive closure of Western Tethys and obduction direction of the Tethyan ophiolite. The salient features of the diagram include the formation during the late Cretaceous of a 'continental promontory' controlled by the position of the North Pyrenean Fault (NPF) (fig. 5b) and opposing emplacement directions of *schistes lustrés* (fig. 5c). Fig. 5d portrays the Oligocene anticlockwise rotation of Corsica away from the European mainland producing the present day plate geometry. Stipple = European continental margin. Horizontal dashed lines = overriding *schistes lustrés* nappe.

subducting continent and transported externally (i.e. toward the west) for distances of over 20 Km. An alternative for at least some of these basement slices is that they may have been emplaced during Warburton's inferred SE directed backthrusting event.

The *schistes lustrés* nappe now occupies a position along the contact between the Corsardinian microplate (formerly part of the leading edge of the European plate) and the Adria microplate (Warburton, this volume). The timing of collision between the Adria continental microplate and Corsica is generally considered to have been mid-Tertiary (e.g. Kligfield 1980; Reutter and others 1980). Warburton (this volume), developing the Reutter and others scenario, models the leading edge of the Adria plate as oceanic crust (or an island arc). Obduction of this 'Adria' leading edge is held responsible for thrusting higher pressure rocks over lower pressure units. Thus, in this model, the Eoalpine blueschist metamorphism results from the obduction of a Tethyan subduction system, followed firstly by the obduction of the Adria plate over the *schistes lustrés* and secondly by SE directed "retrocharriage' backthrusting. The deformation is considered to be progressive, occurring over a time span of some 75 Ma, from late Cretaceous to late Eocene. Support for this model is provided by the wide range of blueschist ages recorded from Corsica and the

Alps (Dal Piaz and others 1972, Bocquet (Desmons) 1974, Hunziker 1974, Delaloye and Desmons 1976, Maluski 1977). Alternatively, evolution of the Corsican *schistes lustrés* may be modelled as having involved two distinctly separate events (e.g. Zacher 1979). In this hypothesis, a late Cretaceous Eoalpine blueschist metamorphism is followed by a long break in obduction collision. It has been suggested (e.g. Reutter and others 1980) that this period marked a time of subduction polarity change, with westward subduction occurring beneath Corsica. The mid-Tertiary deformation and greenschist metamorphism has been attributed to the Oligocene anticlockwise rotation of the Corse-Sardinian block which caused E and SE directed thrusting of Alpine Corsica towards the Adria plate (Zacher 1979).

The recent correlation of the Ersa-Centuri and Farinole Nappes as allochthonous fragments of Austro-Alpine basement (Caby and others, in press) has been explained in terms of a backthrusting event (table 1). Both Warburton and Harris envisage this backthrusting to have taken place after the obduction of the Adria microplate. One difference between the interpretations of these two scientists is that Warburton places great emphasis on D3 backthrusts affecting the *schistes lustrés*, whereas Harris describes only SE verging folds beneath the backthrusting Ersa-Centuri Nappe. Another point of difference is that in northern Cap Corse Harris has identified blueschist minerals associated with the Ersa-Centuri backthrusting, while in southern Cap Corse Warburton identifies his backthrusting event as partly synchronous with greenschist metamorphism.

Malavieille (1983) proposes a very different model to explain the anomalous eastward thrust direction of the Ersa-Centuri nappe. He suggests that, during the late Cretaceous, Corsica existed as a promontory separated from the European mainland by the transcurrent North Pyrenean Fault. Interaction of sinistral transcurrent movement with the Eoalpine obduction front resulted in the local superposition of E and SE directed nappes over the previously westward thrust *schistes lustrés* in NE Corsica (Figure 5). Unlike the hypothesis of Harris (1984), the eastward thrusting of the Ersa-Centuri Nappe is attributed to this deformation phase (i.e. late Eoalpine) rather than a mid-Tertiary event. Another possible influence of transpressive deformation on the emplacement of the *schistes lustrés* nappes is suggested to explain the anomalous SSW transport direction of the Barretalli Nappe (Harris 1984). A greater component of sinistral strike-slip movement along the Eoalpine obduction front could produce this observed change in transport direction.

As yet there is no radiometric record of HP/LT metamorphism having occurred in the Corsican *schistes lustrés* prior to the involvement of the European basement with the Tethyan subduction system (Gibbons and Horák 1984). Whether or not HP/LT remnants of a former intraoceanic subduction system are preserved in the *schistes lustrés*, it is clear that blueschist metamorphism was active during the collision of European continental crust with the former intraoceanic subduction system. An analogy with this Eoalpine collision is provided by the present day collision of Australian continental crust with the formerly intraoce-

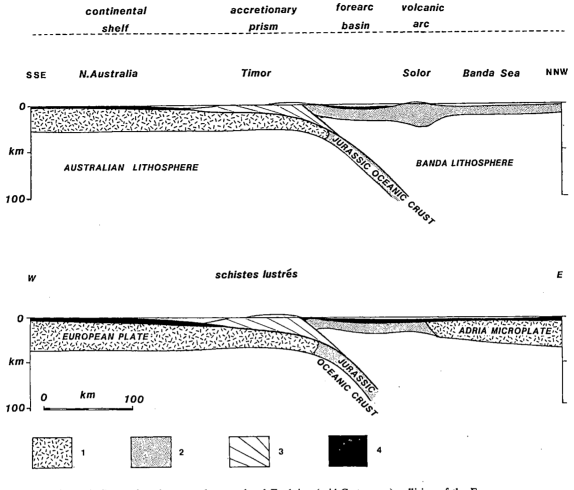

Figure 6. Comparison between the postulated Eoalpine (mid-Cretaceous) collision of the European plate (Corsica) and the Adria microplate, and the present day collision of Australia with the Banda Arc (after Hamilton, 1979). Two notable differences are a: the subducted oceanic crust was younger in the Eoalpine collision; and b: the Adria microplate included continental crust. 1. continental crust, 2. oceanic crust. 3. accretionary prism; 4. sediments in forearc basin and on continental shelf.

anic Banda Arc (Hamilton 1979). The leading edge of the Australian continent, subducted beneath the formerly intraoceanic accretionary prism to a depth of over 30 Kms, is presumably at blueschist facies PT conditions. Continued collision may result in the obduction of the volcanic arc (and the back-arc ocean crust of the Banda Sea) over the presently obducting accretionary prism (Figure 6).

In conclusion, it is clear that there is general acceptance of the Eoalpine obduction model postulated and refined by Mat-

tauer and others over the last ten years. Only recently, however, have more detailed analyses of Corsican blueschist nappe geometries and kinematics been attempted, and there is still considerable disagreement over the timing of individual tectonometamorphic events. Despite the fact that the Corsican *schistes lustrés* provide some of the best and most accessible exposures of Eoalpine blueschists in the internal Alpine belt, mapping these extremely complex rocks is difficult, radiometric dates remain few, and it is likely to be some time before a widely accepted detailed model emerges for the evolution of this part of the Alpine orogen.

REFERENCES CITED

Alvarez, W., 1972, Rotation of the Corsica Sardinia microplate: Nature, v. 235, p. 103–105.

Amaudric du Chaffaut, S., 1975, L'unite de Corté: un temoin de "Piémontais externe" en Corse: Bulletin de la Société géologique de France, v. 17, p. 739–745.

Amaudric du Chaffaut, S., Caron, J. M., Delcey, R., and Lemoine, M., 1972, Données nouvelle sur la stratigraphie des *schistes lustrés* de Corse: La série

de l'Inzecca. Comparaisons avec les Alpes Occidentales et l'Appenin ligure: Compte Rendu de l'Academie des sciences. Paris, v. 275.

Amaudric du Chaffaut, S., Kienast, J. R., and Saliot, P., 1976, Repartition de quelques minéraux du métamorphisme alpine en Corse: Bulletin de la Société géologique de France, v. 18, p. 1179–1180.

Amaudric du Chaffaut, S., and Lemoine, M., 1974, Découverte d'une série jurassico - crétacée Briançonnaises transgressive sur la marge interne de la Corse granitique: Compte Rendu de l'Académie des Sciences. Paris, v. 278, p. 1317–1320.

Amaudric du Chaffaut, S., and Saliot, P., 1979, La Région de Corte: secteur clé pour la compréhension du métamorphism alpin en Corse: Bulletin de la Société géologique de France, v. 21, p. 149–154.

Arthaud, F., and Matte, P., 1975, Les décrochements tardi-hercyniens du Sud-Ouest de l'Europe Géometrie et essai de reconstitution des conditions de la déformation. Tectonophysics, v. 25, p. 139–71.

Arthaud, F., and Matte, P., 1976, Arguments géologiques en faveur de l'absence de mouvements relatifs de la Corse par rapport à la Sardaigne depuis l'orogenese hercyniene: Compte Rendu de l'Académie des Sciences. Paris, v. 283, p. 1011–1014.

Autran, M. A., 1964, Description de l'association jadéite + quartz et des paragenèses minerales associées dans les *schistes lustrés* de Sant 'Andrea di Cotone: Bulletin de la Société Française de Minéralogie et de Cristallographie, v. 87, p. XLIII–XLIV.

Auzende, J. M., Bonnin, J., and Olivet, J. L., 1973, The origin of the Western Mediterranean basin: Quarterly Journal of the Geological Society of London, v. 129, p. 607–620.

Bearth, P., and Schwander, H., 1981, The post-Triassic sediments of the ophiolites zone Zermatt-Saas Fee and the associated manganese mineralisations: Eclogae geologicae Helvetiae, v. 74, p. 189–205.

Beccaluva, L., Ohnenstetter, D., Ohnenstetter, M., and Venturelli, G., 1977, The trace element geochemistry of Corsican ophiolites: Contributions to Mineralogy and Petrology, v. 64, p. 11–31.

Bocquet (Desmons), J., 1974, Ètudes minéralogiques et pétrologiques sur les métamorphismes d'âge alpin dans les Alpes Françaises: Thesis, University of Granoble. 489 p.

Brouwer, H. A., and Egeler, C. G., 1951, Sur le métamorphisme à glaucophane dans la nappe des *schistes lustrés* de la Corse: Proceedings Koninklijke Nederlandsche Akademie Van Weten schappen, serie B, v. 54, p. 130–139.

Brown, E. H., 1978, A P-T grid for metamorphic index minerals in high pressure terranes: Geological Society of America abstracts with programs, Annual Meeting, v. 10, No. 7, p. 373.

Caby, R., Kienast, J. R., Harris, L., and Guiraud, M., 1984, Les klippes de socle ante-alpin du Cap Corse: Arguments petrogenetique et structuraux pour une origine sud-alpine: (Submitted to Bulletin de la Société géologiques de France) *in* Harris, L. B., 1984, Déformations et déplacements dans la châine alpine: l'example des *schistes lustrés* du Cap Corse: Thesis, Rennes (in press).

Caron, J. M., 1977, Lithostratigraphie et tectonique des *Schistes lustrés* dans les Alpes Cottiennes Septentrionales et en Corse orientales: Thesis, Strasbourg, 315 p.

Caron, J. M., and Delcey, R., 1979, Lithostratigraphie des *schistes lustrés* Corses: diversité des series post-ophiolitiques: Compte Rendu de l'Academie des Sciences. Paris, v. 288, series D, p. 1525–1528.

Caron, J. M., Kienast, J. R., and Triboulet, C., 1981, High pressure - Low temperature metamorphism and polyphase Alpine deformation at Sant 'Andrea di Cotone (Eastern Corsica, France): Tectonophysics, v. 78, p. 419–451.

Carpena, J., Mailhe, D., Naeser, C. W., and Poupeau, G., 1979, Sur la datation par traces de fission d'une phase tectonique d'age éocène supérieure en Corse: Compte Rendu de l'Academie des Sciences, Paris, v. 289, p. 829–832.

Chabrier, G., and Mascle, G., 1975, Comparaison des evolutions géologiques de la Provence et de la Sardaigne (a partir d'examples de la région toulonnaise et de la Nurra Sarde): Revue de Geographie Physique et de Géologie Dynamique, v. 1;7, p. 121–135.

Cobbold, P. R., and Quinquis, H., 1980, Development of sheath folds in shear-regimes: Journal of Structural Geology, v. 2, p. 119–126.

Cohen, C. R., Schweickert, R. A., and Odom, A. L., 1981, Age of emplacement of the *Schistes Lustrés* nappe, Alpine Corsica: Tectonophysics, v. 73, p. 267–284.

Currie, K. L., and Curtis, L. W., 1976, An application of multicomponent solution theory to jadeitic pyroxenes: Journal of Geology, v. 84, p. 179–194.

Dal Piaz, G. V., Hunziker, J. C., and Martinotti, G., 1972, La zona Sesia-Lanzo e l'evoluzione tettonico - metamorfica delle Alpi nordoccidentali interne: Memorie della Società Geologica Italiana, Roma, v. 11, p. 433–460.

Dal Piaz, G. V., and Zirpoli, G., 1979, Occurrence of eclogite relics in the ophiolite nappe from Marine d'Albo, Northern Corsica: Neues Jahrbuch fur Mineralogie, ML. 3, p. 118–122.

Delaloye, M., and Desmons, J., 1976. K-Ar radiometric age determinations of white micas from the Piemont zone, French-Italian Western Alps: Contributions to Mineralogy and Petrology, v. 57, p. 297–303.

Delcey, R., 1974. Données sur deux nouvelles séries lithostratigraphiques de la zone des *schistes lustrés* de la Corse nord-orientale: Compte Rendu de l'Academie des Sciences, Paris, v. 279, p. 1693–1696.

Delcey, R., and Meunier, A., 1966, Le Massif du Tenda (Corse) et ses bordures: La series volcano-sedimentaire, les gneiss et les granites; leurs rapports avec les *schistes lustrés*: Bulletin Carte géologique de la France, No. 278, v. 61, p. 237–51.

Durand-Delga, M., 1977, La réunion extraordinaire 1976 de la Société Géologique de France en Corse: Bulletin de la Société d'Histoire Naturelle de Corse, v. 622, p. 39–50.

Durand-Delga, M., 1978, Corse. Guides géologique régionaux. Edited by Masson, Paris.

Ellis, D. J., and Green, D. H., 1979, An experimental study of the effect of Ca upon garnet-clinopyroxene Fe-Mg exchange equilibria: Contributions to Mineralogy and Petrology, v. 71, p. 13–22.

Essene, E. J., 1969, Relatively pure jadeite from a siliceous Corsican gneiss: Earth and Planetary Science Letters, v. 5, p. 270–272.

Faure, M., and Malavieille, J., 1980, Les plis en fourreau du substratum de la nappe des *schistes lustrés* de Corse. Signification cinématiques: Compte Rendu de l'Academie des Sciences, Paris, v. 290, p. 1349–1352.

Faure, M., and Malavieille, J., 1981, Ètude structurale d'une cisaillement ductile: le charriage ophiolitique corse, dans la région de Bastia: Bulletin de la Société géologique de France, v. 23, p. 335–343.

Frey, M., Hunziker, J. C., Frank, W., Bocquet (Desmons), J., Dal Piaz, G. V., Jager, E., and Niggli, E., 1974, Alpine metamorphism of the Alps. A review: Schweizerische Mineralogische und Petrographische Mitteilungen, v. 54, p. 247–290.

Gauthier, A., 1976, Essai de bibliographie géologique de la Corse: Bulletin de la Société d'Histoire Naturelle de Corse, v. 621, p. 79.

Gibbons, W., and Horák, J., 1984, Alpine metamorphism of Hercynian hornblende granodiorite beneath the blueschist facies *schistes lustrés* nappe of N.E. Corsica: Journal of Metamorphic Geology, v. 2, p. 95–113.

Guiraud, M., 1982, Gèothermobarométrie du faciès schiste vert à glaucophane modelisation et applications (Afghanistan, Pakistan, Corse, Bohème): Thesis, Montpellier.

Hamilton, W., 1979, Tectonics of the Indonesian region: U.S. Geological Survey Professional Paper No. 1078, 345 p.

Harris, L. B., 1984, Déformations et déplacements dans la châine alpine: l'example des *schistes lustrés* du Cap Corse: Thesis, Rennes.

Harris, L. B., 1985, Progressive and polyphase deformation of the *schistes lustrés* in Cap Corse, Alpine Corsica. Journal of Structural Geology, v. 7, p. 637–650.

Haug, E., 1896, Ètudes sur la tectonique des Alpes suisses: Bulletin de la Société géologiques de France, v. 24, p. 535–594.

Hunziker, J. C., 1974, Rb-Sr and K-Ar age determinations and the Alpine tectonic history of the western Alps: Memoire degli Istituti di Geologia e Mineralogia dell Universita di Padova, 31, 55 p.

Kligfield, R., 1980, Structural-Geochronological constraints on plate tectonic models for the Corsica-North Apennine collision, *in* Evolution and Tectonics of the Western Mediterranean and surrounding areas. Instituto Geographico Nacional, Special Publication No. 201, Madrid, 1980, p. 163–178.

Krogh, E. J., and Raheim, A., 1978, Temperature and pressure dependence of Fe-Mg partitioning between garnet and phengite, with particular reference to eclogites: Contributions to Mineralogy and Petrology, v. 66, p. 75–80.

Lapadu-Hargues, P., and Maisonneuve, J., 1963, L'Age des *Schistes lustrés* de la

Corse: Bulletin de la Société Géologique de France, v. 5, p. 1012–1028.

Lattard, D., and Schreyer, W., 1981, Experimental results bearing on the stability of the blueschist - facies minerals deerite, howieite and zussmanite, and their petrological significance: Bulletin de Minéralogie, v. 104, p. 431–440.

Malavieille, J., 1983, Ètude tectonique et microtectonique de la nappe de socle de Centuri (zone des *schistes lustrés* de Corse); consequences pour la géométrie de la châine alpine: Bulletin de la Société Géologique de France, v. 25, p. 195–204.

Maluski, H., 1977, Application de la méthode $^{40}Ar/^{39}Ar$ aux mineraux des roches cristallines perturbées par les èvènements therimiques et tectoniques en Corse: Montpellier. 113 p.

Maluski, H., and Schaeffer, O. A., 1982, ^{40}Ar - ^{39}Ar laser probe dating of terrestrial rocks: Earth and Planetary Science Letter, v. 59, p. 21–27.

Maresch, W. V., 1977, Experimental studies on glaucophane: an analysis of present knowledge: Tectonophysics, v. 43, p. 109–125.

Mattauer, M., Faure, M., and Malavieille, J., 1981, Transverse lineation and large-scale structures related to Alpine obduction in Corsica: Journal of Structural Geology, v. 3, p. 401–409.

Mattauer, M., and Proust, F., 1975, Données nouvelles sur l'évolution structurale de la Corse Alpine: Compte Rendu de l'Academie des Sciences, v. 281, p. 1681–1684.

Mattauer, M., and Proust, F., 1976, La Corse alpine: un modèle de genèse du métamorphisme hau;te pression par subduction de croûte continentale sous du matérial océanique: Compte Rendu de l'Academie des Sciences, Paris, v. 282, p. 1249–1252.

Mattauer, M., Proust, F., and Etchecopar, A., 1977, Linéations "a" et méchanisme de cisaillement simple liés au chevauchement de la nappe des *schistes lustrés* en Corse: Bulletin de la Société géologique de France, v. 14, p. 841–847.

Mattauer, M., and Tapponier, P., 1978, Tectoniques des plaques et tectonique intracontinentale dans les Alpes franco-italiennes: Compte Rendu de l'Academie des Sciences, Paris, v. 287, p. 899–902.

Moullade, M., 1978, The Ligurian Sea and the adjacent areas, *in* Nairn, A.E.M., and Kames, W. H., eds., The Ocean Basins, vol. 4B, The Western Mediterranean, p. 67–148.

Nairn, A.E.M., and Westphal, M., 1968, Possible implications of the palaeomagnetic study of late Palaeozoic igneous rocks of NW Corsica: Palaeogeography, Palaeoclimatology, Palaeoecology, v. 5, p. 179–204.

Newton, R. C., and Smith, J. V., 1967, Investigations concerning the breakdown of albite at depth in the earth: Journal of Geology, v. 75, p. 268–286.

Newton, R. C., and Kennedy, G. C., 1968, Jadeite, analcite, nepheline and albite at high temperatures and pressures: American Journal of Science, v. 266, p. 728–735.

Nicolas, A., 1967, Géologie des Alpes piémontaises entre Dora Maira et Grand Paradis: Travaux du Laboratoire de Géoligue de la Faculté des Sciences de l'Université de Grenoble, v. 43, p. 139–167.

Nitsch, K. H., 1974, Neue Erkenntnisse zur Stabilität von Lawsonit (Abs):

Ohnenstetter, D., Ohnenstetter, M., and Rocci, G., 1975, Tholeiitic cumulates in a high pressure metamorphic belt: Petrologie, v. 1, No. 4, p. 291–317.

Ohnenstetter, D., Ohnenstetter, M. and Rocci, G., 1976, Ètude des métamor-

phismes successifs des cumulats ophiolitiques de Corse: Bulletin de la Société géologiques de France, v. 18, p. 115–134.

Peterlongo, J. M., 1968, Les ophiolites et le métamorphissme à glaucophane dans le massif de l'Inzecca, et la région de Vezzani, Corse: Bulletin du Bureau de Recherches Géologiques et Miniéres. Paris, 2nd series, Section 4, no. 1, p. 17, 17–94.

Péquignot, G., Potdevin, J-L., Caron, J-M, and Ohnenstetter, M., 1984, Détritisme ophiolitique dans les *Schistes lustrés* corses et paléogeographie du domaine piémontais: Bulletin de la Société de France, v. 26, p. 913–920.

Reutter, K. J., Giese, P., and Closs, H., 1980, Lithospheric split in the descending plate: Observations from the Northern Apennines: Tectonophysics, v. 64, T1–T9.

Routhier, P., 1964, Présentation générale des deuxiémes éditions des Feuilles Bastia et Luri, au 80 000e: Bulletin du Service de la Carte géologique de la France: v. 61, p. 253–277.

Saliot, P. and Carron, J. P., 1971, L'évolution des roches plutonique de Corse méridionale dans les conditions d'un métamorphisme à prehnite et pumpellyite de Fiable pression: Compte Rendu de l'Academie des Sciences, Paris, v. 272, p. 2772–2280.

Sauvage-Rosenberg, M., 1977, Tectonique et microtectonique des *Schistes lustrés* et ophiolites de la vallée de la Golo (Corse Alpine): Thesis, Montpellier, 91 p.

Stam, J. C., 1952, Geologie de la région du Tenda septentrionale (Corse): Thesis, Amsterdam, 96 p.

Vandenberg, J., and Zijderveld, J.O.A., 1982, Palaeomagnetism in the Mediterranean area, *in* Berckhemer, H., and Hsu, K. J., eds., Alpine Mediterranean Geodynamics: American Geophysical Union Geodynamics series 7, p. 83–112.

Varenkamp, H., 1957, Géologie et pétrology du Tenda central (Corse). Thesis, Amsterdam.

Velde, B., 1967, Si^{4+} content of natural phengites: Contributions to Mineralogy and Petrology, v. 14, p. 250–258.

Venturelli, G., Thorpe, R. S., and Potts, P. J., 1981, Rare earth and trace element characteristics of ophiolitic metabasalts from the Alpine-Apennine belt: Earth and Planetary Science Letters, v. 53, p. 109–123.

Warburton, J., 1983, The tectonic setting and emplacement of Ophiolites. A comparative study of Corsica and the Western Alps. Thesis, University College Swansea (Unpublished), 433 p.

Westphal, M., Bardon, C., Bossert, A., and Hamzeh, R., 1973, A computer fit of Corsica and Sardinia against Southern France: Earth and Planetary Science Letters, v. 18, p. 137–140.

Westphal, M., Orsini, J., and Vellutini, P., 1976, Le microcontinent Corso-Sarde, sa position initiale: Données palaeomagnétiques et raccords géologiques: Tectonophysics, v. 30, p. 141–157.

Zacher, W., 1979, The geological evolution of NE Corsica, *in* Van der Linden, W.J.M., ed., Fixism, mobilism or relativism: Van Bemmelen's search for harmony: Geologie en Mijnbouw, v. 5;8, p. 135–138.

MANUSCRIPT ACCEPTED BY THE SOCIETY JULY 29, 1985

Printed in U.S.A.

Geological Society of America
Memoir 164
1986

The ophiolite-bearing Schistes lustrés nappe in Alpine Corsica: A model for the emplacement of ophiolites that have suffered HP/LT metamorphism

John Warburton
BP Petroleum Development of Spain, S.A.
Edificio AGF
C/Albacete, 5 - 5ª, Sur Izquierda
28027 Madrid, Spain

ABSTRACT

The Adria microplate has been thrust westward onto the Corsican microplate by a distance of at least 150 km. This crustal duplication followed the Late Cretaceous easterly directed subduction of the eastern Corsican continental margin after consumption of intervening Tethyan oceanic crust. The suture zone separating the two microplates is delineated by the ophiolite-bearing Schistes lustrés nappe, which suffered a Late Cretaceous HP-LT metamorphism. The Corsican blueschists crystallised during a major thrusting event at depths probably in excess of 35 km, and were overprinted by post-tectonic greenschist facies assemblages at 15–20 km during Late Eocene times. Eventual uplift and exposure of the Schistes lustrés and ophiolites were probably the result of isostatic compensation and erosion during and following their emplacement onto continental basement.

It is proposed that the major thrusting event occurred within a subducted oceanic portion of the Corsican microplate, in the footwall of a leading, overriding oceanic portion of the Adria microplate. Thrusts probably propagated in a dominantly piggy-back sequence within the subducted oceanic slab, first involving the most deeply buried rocks, then progressively the less deeply buried rocks. This is evident from the fact that the ophiolites and Schistes lustrés that preserve the highest metamorphic "grade" are the most easterly, and structurally highest. The model deduced for emplacement of the Corsican ophiolite nappe may be applicable to the Schistes lustrés nappe that forms a major tectonic unit in the internal zones of the western Alps, and to the metamorphic ophiolite-bearing nappes in western Liguria.

INTRODUCTION

The study presented here provides a detailed description of structures that developed during the course of a progressive deformation related ultimately to the emplacement of the Schistes lustrés nappe with ophiolites. Throughout this study the phrase "ophiolite emplacement" means the translation of an ophiolite-bearing nappe onto a continental margin; this is perhaps a more general term than "obduction" (Coleman 1971) because ophiolites emplaced at depth (maybe in zones of previous lithospheric subduction) are generally not thought of as obducted. Obduction implies a rather high level mode of ophiolite emplacement. The work also demonstrates the difficulty in using a conventional "D number" structural categorisation in high grade metamorphic rocks, where progressive deformation resulted in several generations of superimposed foliations, folds, and thrusts. Such a discussion is generally of interest to those working in such metamorphic terranes as the Moine and Dalradian high grade complexes of the Scottish Highlands, or in the internal zones of the western Alps.

GEOLOGICAL SETTING AND TECTONIC UNITS

The Schistes lustrés nappe is the main ophiolite-bearing tectonic unit in Corsica as well as in the internal zones of the western Alps. A detailed discussion of the Schistes lustrés nappe in the western Alps is beyond the scope of this paper, hence the reader is referred to several relevant pieces of work which themselves present further references (Ramsay 1963; Dal Piaz et al. 1975; Caron 1977, Compagnoni et al. 1977; Ernst and Dal Piaz 1978, Homewood et. al 1980; Milnes et al. 1981; Warburton 1983). A general review of Corsican geology is provided by Gibbons et al. in this volume.

Prior to the rotation of Corsica away from the southern European mainland during Oligocene times, the Schistes lustrés of Corsica and the western Alps formed a continuous tectonic unit (Westphal et. al. 1973; Arthaud and Matte 1976 1977; Westphal and others 1976; Orsini and others 1980; Warburton 1983). The main geological features of Corsica (Figure 1) had probably developed prior to this rotation.

The Schistes lustrés are monotonous calcareous micaceous metasediments that occupy much of Alpine Corsica and the internal zones of the western Alps. The Schistes lustrés have typically suffered polyphase deformation and metamorphism such that all sedimentary structures have been obliterated. Only bedding, preserved as compositional layering, remains. For many years the lithological and structural complexity of the Schistes lustrés precluded any detailed lithostratigraphy until Caron (1977) mapped out several lithotectonic "units" in the Schistes lustrés of Corsica and the internal zones of the western Alps.

In Corsica the components of a dismembered Tethyan ophiolite, dated at $161 < 3$ m.y. (Ohnenstetter et al. 1975), form sub-rounded pods and extensive imbricate slices that are set in the metasedimentary calcareous micaschist "matrix" of the Schistes lustrés (Figure 1). All members of the ophiolite suite exist in Corsica (ultramafics, gabbros, sheeted dykes, extrusive basalts, and red radiolarian pelagic cherts) but no ophiolite sequence is preserved. The sequence has been reconstructed by considering the degree of fractionation of the different ophiolitic rock types relative to that of a typical tholeiitic magma (Ohnenstetter et al. 1975; Rocci et al 1976). Ophiolitic breccias have been described in the Corsican Schistes lustrés at Rospigliani (Figure 1) by Ohnenstetter (1979). It is suggested that the breccias record activity along fossil transform zones. The breccias have been compared with similar rock associations in the Eastern Ligurian ophiolites of northern Italy (Barrett and Spooner 1977; Cortesogno et al. 1978; Cortesogno et al. 1981).

The Schistes lustrés nappe in Corsica consists of an easterly dipping and thickening wedge of metasediments and ophiolites essentially constituting a tectonic melange that was emplaced as a stack of easterly dipping thrust sheets. The nappe overlies the basement and original sedimentary cover of the Corsican continental margin (Hirn and Sapin 1976; Morelli et al. 1977). Locally, imbricate slices of Alpine metamorphosed basement, which were originally mainly crystalline and granodioritic, have

been "included" into the Schistes lustrés nappe (Figures 1 and 2). It is considered in this study that these basement slices are distributed along major (D_3) backthrusts. The most important basement thrust slices are the Oletta-Sierra di Pigno unit (and its southern continuation in the form of several smaller slices) and the Centuri basement slice of northern Cap Corse (Figure 3).

Several lithotectonic units have been recognised in the Schistes lustrés metasediments of Corsica (Caron 1977). Each one has an analog in the internal zones of the western Alps (Amaudric du Chaffaut et al. 1972; Delcey 1974; Tricart 1974; Caron 1977; Bourbon et al. 1979; Gracianski et al. 1981; Lahondere 1981; Warburton 1983). One of these units, the *Unit of Inzecca* (which forms much of the Schistes lustrés of the Upper and Middle structural units (Sauvage-Rosenburg 1977), Figures 3 and 4) has received considerable attention (Caron 1977; Rieuf 1980) and has been compared with very similar ophiolite related lithologies in the Bracco zone of the Ligurian Apennines (Decandia and Elter 1972; Caron 1977; Rocci et al. 1980).

The metasedimentary rocks of the lithotectonic units range in age from Mid-Triassic to Cenomanian (Warburton 1983). Some of these rocks were apparently deposited upon Jurassic (Tethyan) oceanic crust (now ophiolites), for example, the Inzecca unit, while others (for example the *Unit of Santo Pietro*, Figure 2) represent the sedimentary cover to a thinned continental margin. The ophiolite-bearing Schistes lustrés nappe in Corsica was emplaced onto the highly deformed Unit of Santo Pietro that consists predominantly of a marble/quartzite unit directly above metamorphic basement (Delcey 1974; Warburton 1983; Figure 2). Similarly, in the western Alps the Schistes lustrés nappe was thrusted onto a marble/quartzite unit of Triassic age (Warburton 1983).

The Schistes lustrés nappe in Corsica, as well as the subjacent crystalline basement and its sedimentary cover, which is dominantly quartzite and marble, suffered a pre- to syntectonic early Alpine HP/LT metamorphism that is related to an easterly or south-easterly dipping subduction zone active during "mid" to Late Cretaceous times (Maluski 1977, Cohen et. al. 1981). The intensity and "grade" of early Alpine metamorphism in Corsica and the western Alps increase from west to east (for Corsica see Amaudric du Chaffaut et al. 1976 and Scius 1981; for the western Alps see Velde 1965; Saliot 1973; Velde and Kienast 1973; Bocquet (Desmons) 1974; and Saliot and Velde 1982). This implies that the highest thrust sheets (preserved in the east) were buried deepest. An essentially post-tectonic, late Alpine greenschist facies metamorphism overprinted the early Alpine parageneses mainly during Late Eocene times in both Corsica and the western Alps (Maluski 1977; Chopin and Maluski 1980).

The most westerly tectonic units in Alpine Corsica form a steeply dipping complex stack of thrust imbricate slices between the basement of western Corsica and the Schistes lustrés nappe (Figures 3 and 4). This thrust imbricate stack is traditionally referred to as the *Ecailles de Corté*. Toward the east the Ecailles de Corté are exposed in a tectonic window beneath the Schistes lustrés nappe and are known as the *Sant Angelo–Pedani Unit*

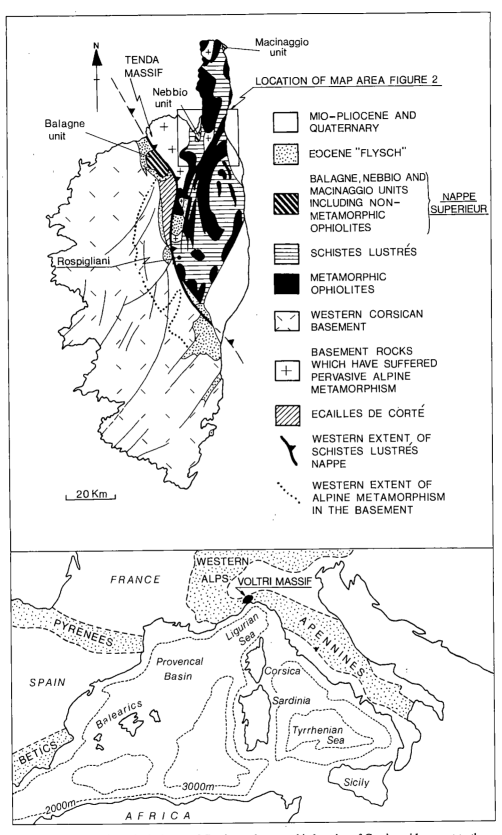

Figure 1. Schematic geological map of Corsica and geographic location of Corsica with respect to the major Alpine mountain belts and Western/Southern Europe. The ophiolitic Voltri Massif of western Liguria is outlined.

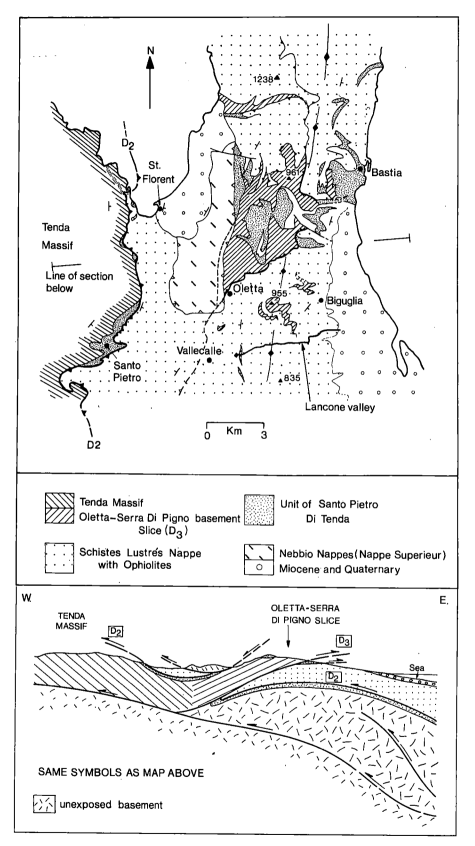

Figure 2. Simplified geological map of the area west of Bastia. Location shown as inset in Figure 1. The schematic cross section shows the relationship between the major tectonic units discussed in the text.

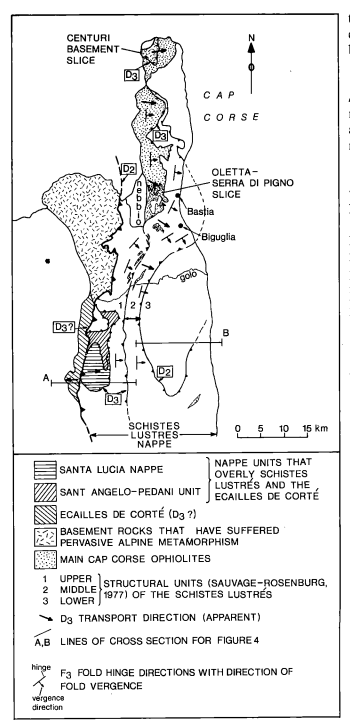

SANTA LUCIA NAPPE

SANT ANGELO-PEDANI UNIT

ECAILLES DE CORTÉ (D3 ?)

NAPPE UNITS THAT OVERLY SCHISTES LUSTRÉS AND THE ECAILLES DE CORTÉ

BASEMENT ROCKS THAT HAVE SUFFERED PERVASIVE ALPINE METAMORPHISM

MAIN CAP CORSE OPHIOLITES

1 UPPER
2 MIDDLE   STRUCTURAL UNITS (SAUVAGE-ROSENBURG,
3 LOWER   1977) OF THE SCHISTES LUSTRÉS

D3 TRANSPORT DIRECTION (APPARENT)

A,B LINES OF CROSS SECTION FOR FIGURE 4

hinge
    F3 FOLD HINGE DIRECTIONS WITH DIRECTION OF
    FOLD VERGENCE
vergence
direction

Figure 3. Map showing the apparent direction of D₃ thrust displacement (however, see pop-up model in Figure 13), the orientation of F₃ fold hinges, and the major D₂ and D₃ thrust surfaces. A and B are the cross section lines for Figure 4.

(Ollé 1981). A major backthrust emplaces a part of the Ecailles de Corté unit over the Schistes lustrés and Sant Angelo-Pedani Unit. The backthrust sheet is known as the *Santa Lucia Nappe* (Figures 3 and 4). Each of these three tectonic units involves basement, plus Mesozoic and Lower Tertiary sedimentary rocks

that appear to have formed the cover to the basement (Amaudric du Chaffaut 1975). In addition, each of the units has suffered both early and late Alpine metamorphism (Rieuf 1980; Ollé 1981).

The highest structural unit of Alpine Corsica is the *Nappe Superieur*, which is poorly preserved and restricted to only three main localities (Figure 1). The relationships among rock types attributed to this nappe are often not clear. The Nappe Superieur is non-metamorphic and forms the following tectonic elements:

1) *The Macinaggio Klippe*. This unit is preserved on the eastern tip of the Cap Corse peninsula where it overlies the Schistes lustrés with tectonic discordance (Durand-Delga and Vellutini 1977). Rock types include Upper Triassic and Lias dolomites and limestones and Upper Cretaceous flysch.

2) *The Nebbio Unit*. This "nappe" occupies the topographic depression between Oletta and St. Florent (Figure 2). Rock types include pre-Carboniferous schists, Upper Jurassic to Eocene limestones, and Upper Triassic to Lias dolomites and limestones. Radiolarites and pelagic limestones form the sedimentary cover to possibly ophiolitic pillow lavas. The structure of the Nebbio nappe is unclear because of poor exposure but like the Macinaggio unit it is surrounded by and overlies the Schistes lustrés.

3) *The Balagne Nappe*. This unit forms much of the low hilly country directly to the west of the Tenda Massif (Figure 1). The nappe is predominantly composed of ophiolites (ultramafics, gabbros, extrusive basalts) with associated pelagic sediments and "mid" to Upper Cretaceous flysch. The basalts of the Balagne nappe have been compared and equated geochemically with ophiolitic basalts of the northern Apennines (Beccaluva et al. 1979) and the pelagic sediments with the Bracco series ophiolite-related sediments of the Apennines and the Helminthoid Flysch nappes of the western Alps (Nardi 1968; Elter 1975; Caron 1977). The Balagne nappe was emplaced onto Eocene conglomerates and turbiditic (flysch) sediments. It occupies a lower structural level than the Nebbio and Macinaggio units that overlie the Schistes lustrés. It is suggested that the non-metamorphic ophiolites of the Balagne and Nebbio nappes may represent remnants of a formerly more continuous slice of oceanic lithosphere, possibly equivalent to the eastern Ligurian ophiolites (Barret 1982) that were part of the leading edge of the Adria microplate (see later discussion).

THE TECTONOMETAMORPHIC EVOLUTION OF ALPINE CORSICA

In Alpine Corsica, a sequence of distinct structural events has been recognised (Warburton 1983). Each event produced structures of different style, and all were superimposed during the Alpine metamorphism. It is proposed here that these structures reflect the temporal changes that occurred during emplacement of the Schistes lustrés nappe onto the Corsican continental margin.

It is possible to distinguish between two tectonometamorphic episodes in the metamorphic rocks of Alpine Corsica (Figure

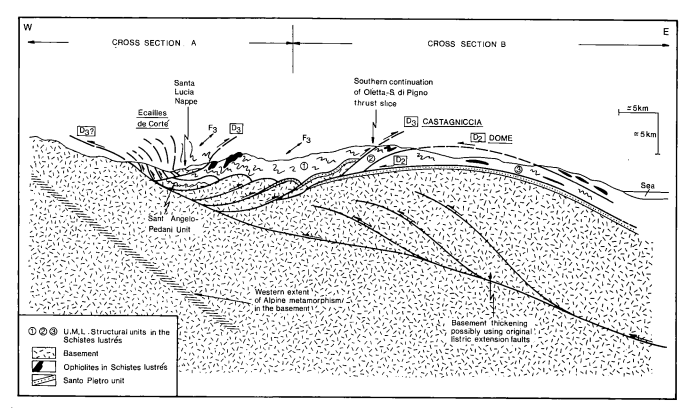

Figure 4. Sketch composite cross section across Alpine Corsica. The lines of section are shown on Figure 3. The main features of the section are the Castagniccia dome, which is interpreted to be due to basement thrusting (possibly rejuvenation of early extensional margin faults), the new interpretation of the Santa Lucia Nappe and Ecailles de Corté as possibly constituting a D_3 pop-up, and the southern continuation of the Oletta-Sierra di Pigno (D_3) structure. The topography of the basement to the west of the Castagniccia dome is after Rieuf (1980).

5). The first episode consisted of two deformation "events" (D_1 and D_2) that both occurred during the early Alpine HP/LT metamorphism. The second episode consisted of a single deformation "event" (D_3) that occurred prior to, and during late Alpine greenschist facies metamorphism, but after D_1, D_2, and the early Alpine metamorphism.

Although the two episodes are distinguished fundamentally on the basis of metamorphic conditions, D_1 and D_2 structures have similar orientations, which are different from the D_3 structures. Since several generations of structures developed during each of the tectonometamorphic episodes (Figure 5), each episode represents a continuous "phase" of deformation where early formed structures were overprinted by later ones during the same metamorphism. This means that, from a mechanistic point of view, the assignation of "D numbers" (implying separate deformation events) may carry little significance.

The majority of detailed field study was carried out by the author in the hill country between Bastia and St. Florent (Figure 3), and structures from that area are described in the following sections. However, much work was also undertaken in a large number of localities across the whole of Alpine Corsica. This study therefore serves to integrate current ideas concerning geo-

logical structure and history, and to build up a regional tectonic model for Alpine Corsica.

The First Tectonometamorphic Episode (D_1 and D_2)

D_1. A layer-parallel mylonitic planar/linear (LS tectonite) foliation (S_1), with mineral grain shape lineation (L_1), is ubiquitous in the metamorphic basement, Unit of Santo Pietro, and in the Schistes lustrés with ophiolites. The grain shape lineation (L_1) trends approximately N070°E across Alpine Corsica (Figure 6). Microstructural studies such as quartz C-axis plots (Mattauer et. al. 1977) and micro-rotational criteria parallel to XZ of the finite strain ellipsoid (Malavieille 1982) suggest that the foliation developed in response to southwesterly directed simple shear with the nappes being transported towards the southwest. This shearing event accompanied emplacement of the Schistes lustrés nappe and was contemporaneous with the early Alpine HP/LT metamorphism (Cohen et. al. 1981; Faure and Malavieille 1981; Malavieille 1982; Warburton 1983). Petrographic details concerning the S_1 foliation are fully discussed by Warburton (1983). To be geologically precise, the present N070°E orientation of mineral grains has to be rotated 30 degrees clockwise, to N100°E,

TECTONOMETAMORPHIC EPISODES AND

STRUCTURAL ELEMENTS

FIRST TECTONOMETAMORPHIC EPISODE.	HP/LT EARLY ALPINE METAMORPHISM (90 m.y.)	D_1	EARLY	FOLIATION (S_1) MINERAL GRAIN SHAPE LINEATION (L_1)
			LATE	THRUSTS
		D_2	EARLY	FOLDS (F_2) WITH AXIAL PLANAR SCHISTOSITY (S_2) AND MINERAL ELONGATION LINEATION (L_2)
			LATE	THRUSTS
SECOND TECTONOMETAMORPHIC EPISODE.	LATE ALPINE GREENSCHIST FACIES METAMORPHISM (40 m.y.)	D_3	EARLY	FOLDS (F_3) (WHICH PREDATE THRUSTS) AND HAVE AN AXIAL PLANAR (S_3) CLEAVAGE. BACK THRUSTS.
			LATE	FOLDS (F_3) WITH (S_3) AXIAL PLANAR CLEAVAGE.
		D_4		FOLDS (F_4).

Figure 5. Structural elements and tectonometamorphic episodes in the metamorphic rocks of Alpine Corsica. This scheme is compared with that of other workers by Gibbons et al., this volume.

to account for the post-nappe emplacement rotation of Corsica during Oligocene times. However, in this paper I will discuss the direction of emplacement and other directional criteria in terms of the present day orientation.

D_2. The earliest folds to deform the S_1 foliation are designated F_2. The F_2 folds occasionally develop an axial planar (S_2) fabric along which HP/LT minerals are aligned. F_2 folds occur in all lithologies but are less common in the metamorphic basement rocks. The majority of F_2 folds are tight to isoclinal (Figures 7 and 8) and have flat-lying axial planes parallel to the regional S_1 foliation and the base of the Schistes lustrés nappe. F_2 hinges show all declinations, but many parallel the L_1 lineation (Figure 9). F_2 sheath folds are common, with long axes parallel to L_1. The maximum demonstrable amplitude of F_2 isoclines is 12 m, but such folds are asymmetric and may be parasitic on larger structures. Sheath folds may also exist on a large scale. Although

commonly asymmetric, there is no unique direction of F_2 fold vergence.

Thrusting During the First Episode (D_1 and D_2)

In lithologically layered tectonites (such as the Corsican Unit of Santo Pietro) where no original sedimentary sequence is preserved, it is virtually impossible to demonstrate where a thrust has turned flat and travelled parallel to bedding (or S_1), or indeed whether all, or any of the lithological layers are thrust bounded. Frequently, imbricate slices involve basement and demonstrate thrusting at or near the basement—"cover" interface. Such imbricates often have syntectonic HP/LT minerals along their mylonitised margins, which demonstrate that some thrusting occurred during the first episode. In the area to the west of Bastia two generations of First Episode thrusts can be distinguished (Figure

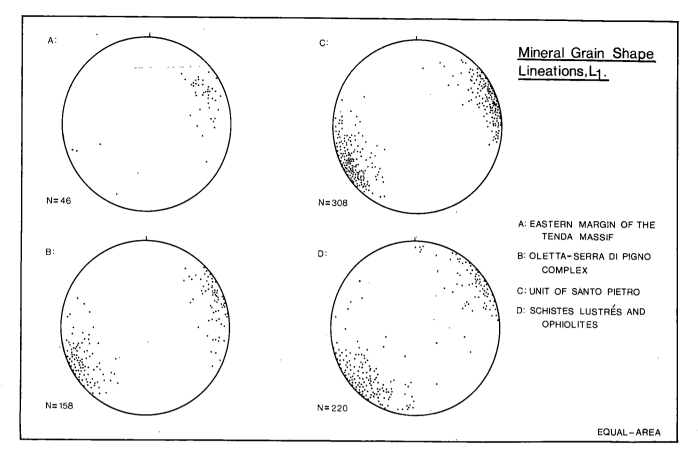

Figure 6. Orientation of mineral grain shape lineations (L_1) in the tectonic units that crop out to the east of the Tenda Massif (in the area shown in Figure 2).

5). D_1 thrusts are essentially layer parallel and are folded by F_2 folds. They have a high flat to ramp length ratio and are similar to elongate shear "pods" separated by anastomosing mylonitic zones. Occasionally D_1 thrusts post-date the S_1 mylonitic foliation but pre-date F_2 (Figure 10). D_2 thrust structures post-date F_2 folds and rather than having the geometry of elongate ductile shear pods (such as the D_1 thrusts) they display the more typical features of foreland thrust belt structures (Boyer and Elliott 1982). For example, in the structure shown at Monte Canarinco (Figure 11) it is possible to demonstrate lateral hangingwall cut-offs of the S_1 mylonitic foliation in a slice of metamorphic basement that rests on a footwall flat in marbles and quartzites of the Santo Pietro Unit. The culmination wall affects other more internal and higher thrust structures further to the northeast, demonstrating that this basement imbricate slice was emplaced below earlier formed thrusts that were carried piggy-back toward the southwest (parallel to transport direction indicated by the L_1 lineation and sense of mineral C-axis rotation). The northeast-southwest orientation of the lateral hangingwall cutoffs of the S_1 foliation in the basement imbricate slice as well as the bulging-up of higher (more north-easterly) thrust sheets confirms the south-westerly D_2 transport direction.

Early Alpine Metamorphic Conditions

The rocks of Alpine Corsica suffered peak metamorphic conditions in the blueschist facies about 90 m.y. (Maluski 1977), at a depth of approximately 35 km. This depth is calculated from the highest pressure estimates of 7–9 kb, at 250–350°C for Corsican blueschists and metamorphic rocks (cited in Ohnenstetter et al. 1976, and Caron et. al. 1981 respectively; see also Warburton 1983). The early Alpine metamorphism is related to "mid" to Late Cretaceous oceanic subduction (Mattauer and Proust 1976).

The Second Tectonometamorphic Episode (D₃)

F_3 Folds. F_3 folds deform all lithologies in Alpine Corsica, except for the Miocene (Burdigalian) molassic sediments and the nappe units of the Balagne and Nebbio (Figure 1). In the Bastia–St. Florent area, F_3 folds are essentially upright or steeply inclined and verge south easterly (Figure 12). Interlimb angles are generally close, but tight and gentle forms also occur (Figures 7 and 8). The maximum amplitude of F_3 folds is 300 m (in the Lancone Valley, Figure 2). In the area to the west of Bastia, F_3 hinges trend N030°E, but further south (near Biguglia) they

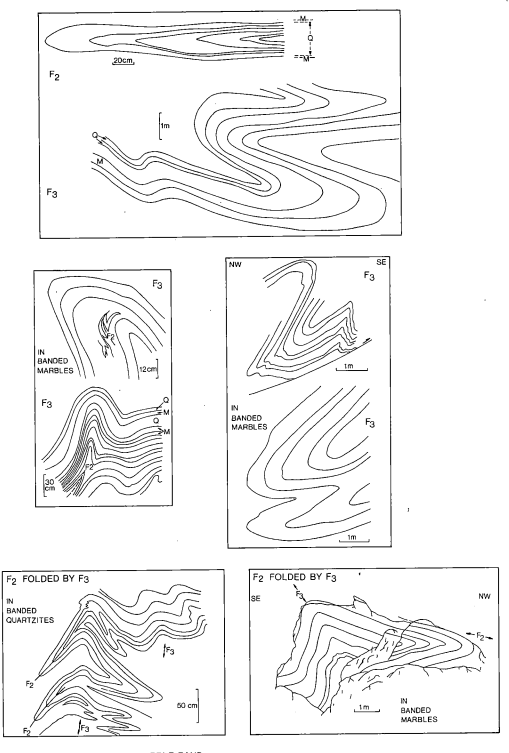

Q = QUARTZITE BAND , M = MARBLE BAND

ALL SECTIONS ARE PROFILE SECTIONS, INCLUDING COAXIAL REFOLDS OF F_2 BY F_3

Figure 7. Profile sections traced from photographs of F_2 and F_3 folds in marbles and quartzites of the Unit of Santo Pietro (area of Figure 2). Where F_2 and F_3 occur in the same outcrop only coaxial refolds are illustrated.

J. Warburton

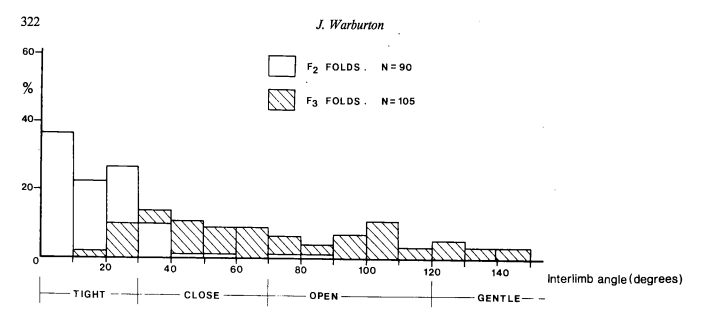

Figure 8. Histogram to show the variations in F_2 and F_3 interlimb angles in the Unit of Santo Pietro in the area of Figure 2.

swing to trend N080°E (Figure 3). Along the Cap Corse Peninsula, and to the south in central Corsica, F_3 hinges are oriented N-S, and folds verge eastwards.

The majority of F_3 folds have an axial planar (S_3) cleavage that is a pressure solution striping in marbles and quartzites, and a crenulation cleavage in calcareous micaschists and ophiolitic prasinites. Late Alpine metamorphic minerals (such as albite, epidote, chlorite, and actinolite) are usually undeformed and grow across S_3, but occasionally albite porphyroblasts are flattened parallel to S_3.

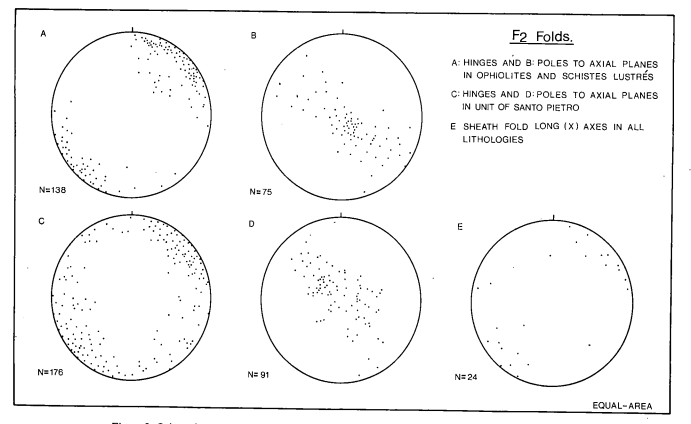

Figure 9. Orientation data for F_2 folds in the area to the east of the Tenda Massif (the area shown in Figure 2).

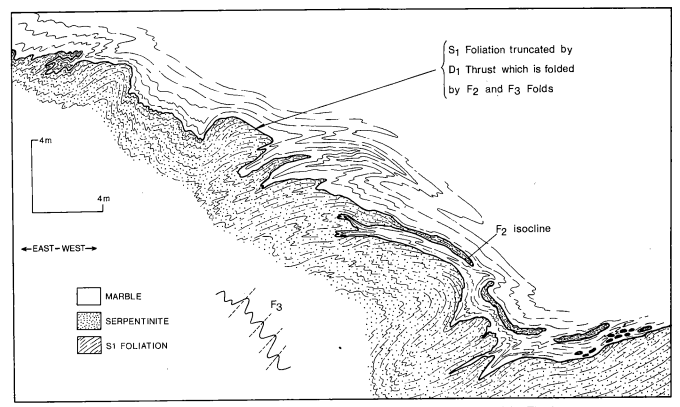

Figure 10. D$_1$ thrust that juxtaposes marbles of the Unit of Santo Pietro with serpentinite. The thrust truncates the S$_1$ foliation but is folded by F$_2$ and F$_3$ folds. (Locality approximately 5 km west of Bastia). Redrawn from a field sketch.

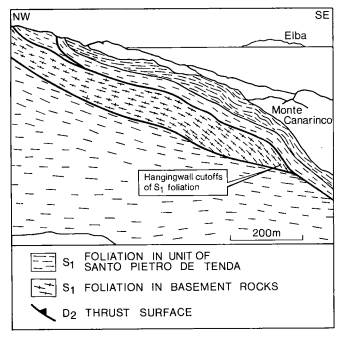

Figure 11. D$_2$ thrust structure at Monte Canarinco (4.5 km SW of Bastia). A lateral culmination wall is developed above an imbricate slice of metamorphic basement. Thrust transport direction was obliquely out of the page to the left (southwestward) as indicated by the orientation of hangingwall cutoffs of the S$_1$ foliation in the basement slice, and the folding of higher (more northeasterly thrusts) by the culmination wall.

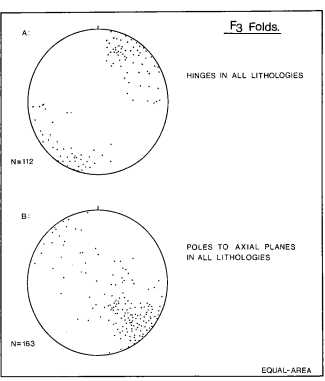

Figure 12. Orientation data for F$_3$ folds in the area of Figure 2, west of Bastia.

J. Warburton

Thrusting During the Second Episode (D₃)

The majority of large thrust structures in Alpine Corsica developed during the D_3 deformation. They generally have back-thrust geometry and overprint all structures formed during the first episode. Numerous authors have considered them as late structures, post-dating HP/LT metamorphism and the emplacement of the Schistes lustrés nappe (Mattauer and Proust 1976; Mattauer and Tapponier 1978; Faure and Malavieille 1981; Malavieille 1982). D_3 thrusts have a complex geometry, but bear greater resemblance to high level thrust belt structures than do the D_1 thrusts. However, the D_3 thrust geometry is similar to that of some D_2 thrusts, and often a distinction is only possible from examination of metamorphic mineralogy within mylonitic zones that bound thrust slices. For example, syntectonic greenschist facies minerals occur along the thrust surface at the base of the Oletta-Serra di Pigno basement slice in the area between Bastia and St. Florent (Figure 2). This indicates D_3 thrusting during late Alpine metamorphism. The major D_3 thrusts of Alpine Corsica are shown on Figure 3.

Regional D_3 thrust transport direction was from west to east across Alpine Corsica (Figure 3). In the area to the west of Bastia, it was towards the southeast. This area is therefore anomalous with respect to the regional D_3 transport direction.

The evidence for D_3 thrust transport direction in the area west of Bastia is as follows:

1) D_3 shear zones at the surfaces of imbricate slices (often involving rock 2–3 m on either side of the thrust surface) show that the hangingwalls moved southeasterly over the underlying footwalls.
2) D_3 frontal ramps are oriented northeast-southwest and dip towards the northwest. Northwest-southeast oriented slickensides occur occasionally on these ramp structures.
3) Numerous northwest-southeast oriented steep faults with wrench sense occur. These are interpreted as D_3 lateral ramps.
4) F_3 folds both predate and postdate D_3 thrusts (Figure 5). Since both sets verge southeasterly (or easterly along the Cap Corse peninsula and to the south of the Bastia area) it is likely that D_3 thrust transport was in the same direction.

The detailed geometry of D_3 thrusts is described elsewhere (Warburton 1983).

Late Alpine Metamorphic Conditions

Pressure estimates on greenschist facies assemblages from Alpine Corsica (cited in Ohnenstetter et al. 1976; Amaudric du Chaffaut et al. 1976) imply metamorphism at 10–20 km depth. Ages obtained from Late Alpine metamorphic minerals range from Late Eocene to Early Oligocene, from 46 ± 5 to 34 m.y. (Maluski 1977; Carpena et al. 1979; Maluski and Schaeffer 1982). The late Alpine metamorphic peak occurred at approximately 40 m.y. (Maluski 1977).

RELATIONSHIP BETWEEN THE FIRST AND SECOND TECTONOMETAMORPHIC EPISODES

So far, the first and second tectonometamorphic episodes have been treated separately. However, they represent temporal changes in structural style that are associated with a progressive "Alpine deformation" and progressively changing metamorphic conditions.

Relative to the southwesterly directed thrusting during the first episode, the D_3 thrusts on a regional scale are effectively "backthrusts," since, as discussed in the previous section, they appear to show a generally easterly displacement. In the area west of Bastia the story is more complex since the D_3 transport direction was approximately perpendicular to that during the first episode. This particular problem has been discussed in detail by Warburton (1983). These observations have prompted previous authors to consider the "backthrusting" as a separate deformation, later than the emplacement of the Schistes lustrés nappe (Mattauer and Proust 1976). However, recent advances in our understanding of thrust belt geometry (e.g., Dahlstrom 1969, Boyer and Elliott 1982, Butler 1982) have provided a means of integrating so-called backthrusts into progressive thrust models. The majority of attempts to integrate deformation events have been performed in the external zones of mountain belts, like the classic studies in the Canadian Rockies. Metamorphic belts such as those in eastern Corsica (or the internal zones of the western Alps) present a particular problem, because successive structural "events" are often bracketed in time by the growth of different generations of metamorphic minerals. As a result, evolutionary models are tightly constrained. Backthrusts and backfolds are commonly developed on all scales in belts less metamorphosed than Corsica. They can be interpreted as a consequence of foreland directed thrusting (Figure 13) and need not form during a separate deformation. There need be little or no physical translation of rock towards the hinterland during the formation of backthrusts or backfolds. The D_3 structures in Corsica can also be related to a backthrust-backfold mechanism that operated during the southwesterly directed emplacement of the Schistes lustrés nappe.

The changing tectonic style throughout the first episode, as well as the entire Alpine deformation, were from conditions conducive to mylonite development (S_1 foliation formation), to thrusting with a recognisable and familiar high level structural geometry. D_3 thrusting had probably almost ceased prior to, or during the late Alpine metamorphism. It was pointed out earlier that the late Alpine metamorphism probably occurred at depths between 10 and 20 km, within the greenschist facies. I will assume, therefore, that since D_3 thrusting terminated in the greenschist facies, it did so at a depth of 15 km. If, as mentioned earlier, the Corsican blueschists crystallised at depths approaching 35 km, then the combined D_1, D_2 and D_3 thrust displacements were potentially responsible for a vertical component of about a 20 km uplift of the most deeply buried rocks. However, since thrusting

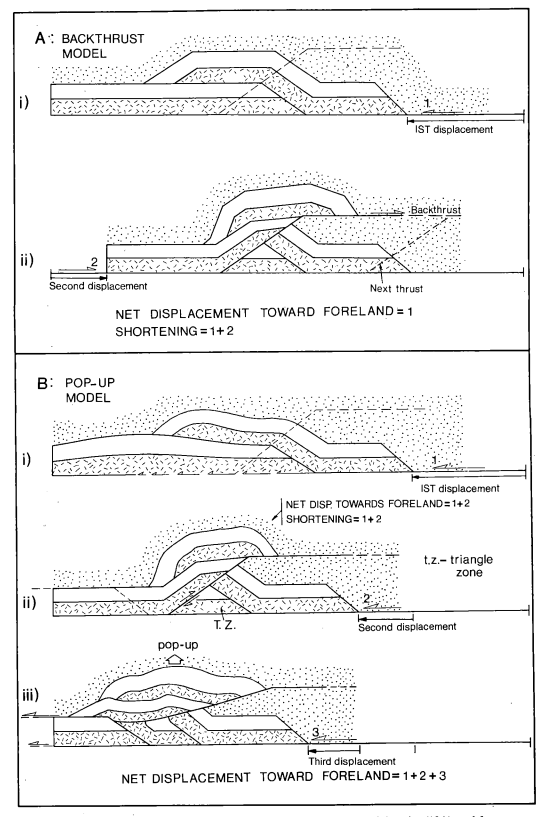

Figure 13. Comparison of the total displacements resulting from true backthrusting (13A), and from "underthrusting" (pop-up model) (13B). With the pop-up model, apparent backthrusts can be interpreted as the consequence of foreland directed transport alone. Stages A (ii) and B (ii) should be compared.

results in crustal thickening, isostatic recovery and erosion at the surface must be considered as important parameters in the ascent of blueschist terranes during, as well as after, thrusting (Roeder 1980; Warburton 1983).

The "Ecailles de Corté" are extensive imbricate slices that formed (possibly during D_3) at the basement/cover interface beneath the Schistes lustrés nappe at its western limit (Amaudric du Chaffaut 1971, 1975). These structures have a foreland thrust belt type geometry, different from the elongate ductile shear pod type of D_1 imbricates exposed at the basement/"cover" interface mainly in the east of Alpine Corsica. Durand-Delga (personal communication 1981) considers this a reflection of structural level; the structures exposed at the basement contact on the eastern margin of the Tenda Massif formed deeper and earlier than those in the west. The Ecailles de Corté form a hinterland dipping imbricate stack, and are therefore not backthrusts. As mentioned earlier, part of the Ecailles de Corté have been thrust back over the Schistes lustrés and Sant Angelo-Pedani units (to form the Santa Lucia Nappe) and the simplest interpretation is that they constitute the frontal imbricates of a D_3 pop-up (Figure 4). Further work is required to tie down the interrelations among these units.

The change in structural style across Alpine Corsica is consistent with the regional variation in Alpine metamorphic conditions. The "grade" and intensity of early Alpine metamorphism appear to decrease from east to west, and some blue amphiboles on the western margin of the Schistes lustrés nappe have yielded late Alpine ages (H. Maluski, cited in Amaudric du Chaffaut and Saliot 1979; M. Durand-Delga, personal communication 1981).

Regionally then, the style of structures changes, from ductile D_1 imbrication in the east (near Bastia), to higher level (possibly D_3) foreland directed thrusting in the west (at Corté). D_3 backthrusts affect earlier (D_1 and D_2) structures across the entire width of Alpine Corsica. Since the early Alpine metamorphism increases in "grade" and intensity toward the east, the structurally highest, easterly dipping pre-backthrusting thrust sheets have probably been uplifted from the greatest depths. These features are consistent with dominantly piggy-back style thrust propagation during uplift and emplacement of the Schistes lustrés nappe.

Summary

Figure 14 illustrates the proposed evolution of Alpine Corsica during the uplift and emplacement of the Schistes lustrés nappe. Development of the S_1 mylonitic foliation was followed by D_1, then D_2 thrusting, reflecting the early stages of nappe emplacement during the early Alpine HP/LT metamorphism. After considerable uplift (resulting from the combined effects of isostatic compensation and erosion at the surface following progressive crustal thickening) the D_2 structural style was replaced by D_3 backthrusting and possibly continued D_3 foreland directed thrusting of the Ecailles de Corté. This change in style is reflected in a geological traverse at the present day, from east to west across Alpine Corsica (Figure 4). Thrusting ceased in the green-

schist facies at a depth of approximately 15 km. The exposure of blueschists at the surface in Corsica was ultimately due to isostatic uplift and erosion.

This model provides a possible solution for the observed superimposition of structures with different styles formed under progressively changing metamorphic temperatures and pressures. Instead of using a traditional approach where the formation of superimposed structures is considered due to separate deformation events, the approach adopted here describes a progressive, constantly changing deformation regime related ultimately to ophiolite nappe emplacement and a single phase of mountain building.

Western Alps and Ligurian Apennines

The tectonometamorphic scheme for Corsica presented here may be applicable to the internal zones of the western Alps and also to the western Liguria area. The units of central-western Liguria (the Voltri Group/Sestri Voltaggio Zone) form the major boundary between the western Alps and northern Appennine domains (Cortesogno et al. 1979). The Voltri Group (Figure 1) forms the main ophiolite-bearing metamorphic unit in western Liguria and comprises Schistes lustrés and ophiolites that were thrust westwards onto the western Ligurian crystalline basement during "Alpine" metamorphism (Piccardo et al. 1979). Cortesogno et al. (1979) have compared tectonometamorphic events between the western Alpine internal zones and the western Ligurian units, and have related metamorphic mineralogy and chemistry to the progressive exhumation of the major western Ligurian nappe units from a subduction zone. As in the Internal zones of the western Alps and in Alpine Corsica, the ophiolitic metabasic rocks of the Voltri Group suffered an early HP/LT, followed by a later greenschist facies metamorphism (Piccardo et al. 1979). In western Liguria, eastward directed backthrusts (equivalent broadly to the D_3 structures in Corsica) developed during Oligocene times, and propagated eastward by Late Miocene/Pliocene times to emplace the Tuscany nappe units (Piccardo et al. 1979; Reutter et al. 1980; Figure 15). D_3 backthrust structures in Corsica developed prior to and during the late Alpine greenschist facies metamorphism, during Late Eocene to Oligocene times (Warburton 1983).

It seems that the tectonometamorphic histories of the Schistes lustrés nappe in Corsica and the western Alps, and of the metamorphic ophiolitic nappes in western Liguria, were similar and closely related in time during the collision of the European plate and Adria microplate.

TECTONIC SETTING OF OPHIOLITE EMPLACEMENT IN CORSICA

The Schistes lustrés nappe in Corsica occupies a position along the contact of the Corso-Sardinian microplate (formerly part of the European Plate) and the Adria microplate (Giese and Prodehl 1976; Morelli et al. 1977). Geophysical studies have

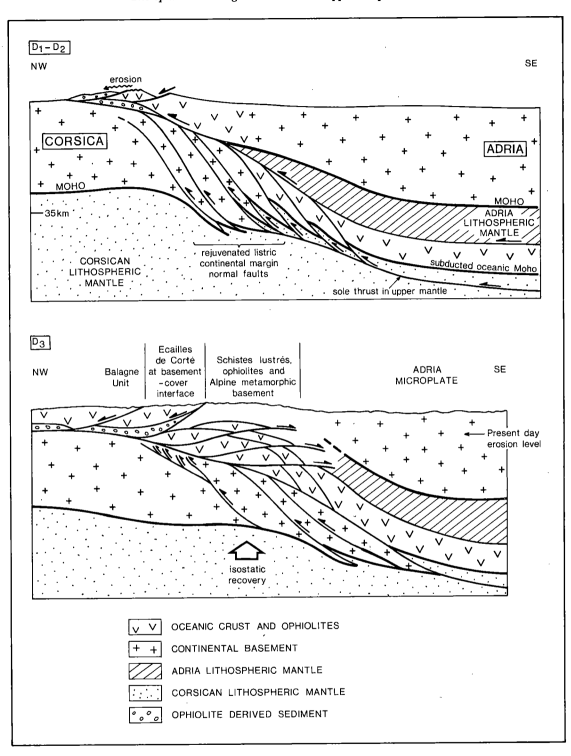

Figure 14. Evolution of Alpine Corsica in the context of geodynamics of the Western Mediterranean. D_1-D_2 represents the initial imbrication and thrusting of subducted oceanic lithosphere onto the Corsican continental margin. The major sole thrust was at some depth within the Corsican lithospheric mantle (see text). D_3 occurred after sticking of the major foreland propagating thrusts at a relatively high level and superimposed pop-up backthrusts onto the emplaced oceanic crustal imbricate slices. Isostatic recovery, uplift and erosion provided an important uplift mechanism for the emplaced ophiolites. The approximate present day level of erosion is indicated on the lower diagram that should be compared broadly with Figure 15. The main geological features of Alpine Corsica are shown.

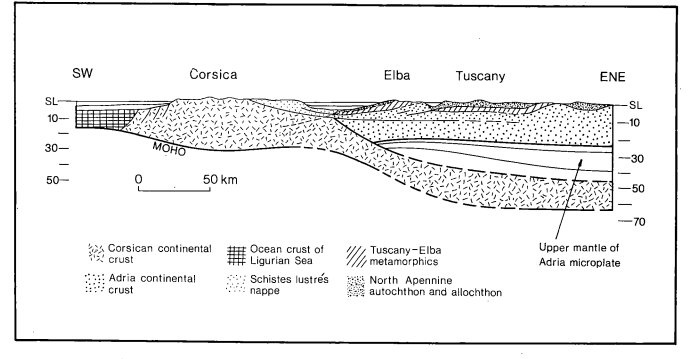

Figure 15. Deep structure of the Western Mediterranean from geophysical studies of Morelli et al., 1977.

shown that the western margin of the Adria microplate was thrust over the European and Corsican continental margins prior to the Oligocene rotation of Corsica and Sardinia to their present position (Figure 15). It is proposed here that thrusting occurred within the subducted oceanic crust of the Corsican microplate in the footwall of an overriding oceanic portion of the Adria microplate. Within the subducted slab, thrusts probably propagated in dominantly piggy-back sequence that first involved deeply buried (subducted) rocks, then progressively less deeply buried rocks. Imbricates were thus eventually thrust onto the subducted continental margin (Figure 14). The Balagne and Nebbio nappes (both of which contain non-metamorphic ophiolites and Upper Cretaceous spilite-bearing conglomerates) may represent remnants of emergent, obducted oceanic crust of the Adria microplate. The geometry of the Adria microplate oceanic crust may have resembled that exposed in Papua New Guinea at the present day, prior to extensive Tertiary and recent erosion responsible for the present day exposure level in Corsica. It is possible that extensive erosion of the Papuan ophiolite in the future may reveal an underlying metamorphic belt similar to that exposed in Corsica at the present day.

PREVIOUSLY PROPOSED MODELS—DISCUSSION

Several basic concepts appear to be accepted in the recent published literature to explain the emplacement of HP/LT ophiolites and their emergence at the surface. Roeder (1980) suggests that the ascent of eclogites to the surface requires thrusting parallel to the zone of previous subduction with a considerable amount

of erosion that has to be maintained at the thrust front. The initiation of thrusting is probably a response to subduction of relatively buoyant continental crust and blockage of the subduction system (Mattauer and Proust 1976; Welland and Mitchell 1977). Such a mechanism is consistent with the "upwedging" model proposed by Dewey (1976) and followed by Mattauer and Tapponier (1978), and by Mattauer et al. (1981) to explain emplacement of the internal zone nappes in the western Alps and Corsica respectively. Dewey (1978) envisaged that the continuation of the upwedging process resulted in the expulsion of large sheets of oceanic crust onto the continental margin and ultimately in the steepening and overturning of structures towards the hinterland. Cohen (1982) presented a similar model to explain the backthrust structures in Corsica.

Theoretical models (Nicolas and Le Pichon 1980; Boudier et al. 1982) and fault plane solutions of earthquakes in active trenches (Issacks and Molnar 1971; Chen and Forsyth 1978), as well as seismic sections (Prince and Kulm 1975; Hussong et al. 1976; Scholl et al. 1980; Speed and Larue 1982), show that active thrusting occurs within the upper parts of subduction zones. Such thrusts may be transported down a subduction zone to be rejuvenated later during the obduction process. The model proposed here, although essentially similar to the "upwedging" model of Dewey (1976), is more refined in that the constraints of presently accepted thrust tectonic models have been applied. The concept of a generally piggy-back style of thrust propagation within the metamorphic belt is evident from field relationships in Corsica (for example Figure 11) and on a large scale from the overall metamorphic pattern across the belt.

CONCLUSIONS

1) Ophiolites and associated metasediments (collectively the Schistes lustrés nappe) in Corsica preserve blueschist facies assemblages that probably formed in a subduction zone at depths of approximately 35 km. Their presence at the surface today can be explained by a combination of their thrust emplacement onto the Corsican continental margin, accompanied and followed by isostatic readjustment and erosion.

2) The S_1 foliation, F_2 sheath folds, and D_1 and D_2 thrusts formed during the early stages of nappe emplacement, coevally with the early Alpine HP/LT metamorphism. D_3 thrusting reflects the "late" stages of the Alpine deformation prior to (and possibly during) the late Alpine greenschist facies overprint.

3) The observed structural history in Alpine Corsica, beginning with the development of the S_1 mylonitic foliation, followed by D_1, D_2, then D_3 thrusting records the progressively changing processes and deformation mechanisms that operated during the emplacement of an ophiolite nappe that has suffered HP/LT metamorphism.

4) The emplacement of the Schistes lustrés nappe in Corsica involved thrusting within the subducted oceanic crust of the Corsican microplate in the footwall of the overriding Adria microplate.

ACKNOWLEDGMENTS

I would like to thank R. H. Graham and N. Fry for discussion in the field and for criticism of the Ph.D. thesis from which much of the material presented here was taken. I also thank E. H. Rutter, J. R. Vearncombe, B. A. Martin and numerous colleagues in BP Aberdeen and in the Universities of Swansea and Leeds for discussions on thrust tectonics. I wish to extend my gratitude to professor M. Durand-Delga of the Department of Géologie Mediterranéenne, Toulouse, for inviting me to visit the institute, and for frequent correspondence. The management of BP Petroleum Development Ltd., London, are thanked for their permission to publish the work, as is Keith Greenwood of the BP office in Madrid, who drafted the figures. The comments of two referees and Professor E. H. Brown were much appreciated.

REFERENCES CITED

Amaudric du Chaffaut, S., 1971, Etude geologique de la région de Solaro—Poggio-di-Nazza (Corse sud-orientale) [Thesis]: Paris, 128 pp.

——1975. L'unitéde Corté: un temoin de "Piémontais externe" en Corse: Bulletin dela Société Géologique de France, v. 7, no. 17, no. 5, p. 739–745.

Amaudric du Chaffaut, S., Caron, J. M., Delcey, R., and Lemoine, M., 1972, Données nouvelle sur la stratigraphie des Schistes lustrés de Corse: La série de l'Inzecca. Comparaisons avec les Alpes Occidentales et l'Apennin ligure: Compte Rendu de l'Académié des Sciences, Paris, v. 275, p. 2611–2614.

Amaudric du Chaffaut, S., Kienast, J. R., and Saliot, P., 1976, Repartition de quelques minéraux du métamorphisme alpine en Corse: Bulletin de la Société Geologique de France, v. 7, no. 18, p. 1179–1180.

Amaudric du Chaffaut, S., and Saliot, P., 1979, La Région de Corté: secteur clé pour la compréhension du métamorphism alpin en Corse: Bulletin dela Société Géologique de France, v. 7, no. 21, no. 2, p. 149–154.

Arthaud, F., and Matte, P., 1976, Arguments géologiques en faveur de l'absence de mouvements relatifs de la Corse par rapport à la Sardaigne depuis l'orogenèse hercynienne: Compte Rendu de l'Academie des Sciences, Paris, v. 283, p. 1011–1014.

——1977, Détermination de la position initiale de la Corse et de la Sardaigne à la fin de l'orogènese hercynienne grâce aux marquers géologique antémésozoiques: Bulletin de la Société Géologique de France, v. 7, no. 19, no. 4, p. 833–840.

Barret, T. J., 1982, Review of stratigraphic aspects of the ophiolitic rocks and pelagic sediments of the Vara Complex, North Apennines, Italy: Ophioliti, v. 7, p. 3–46.

Barrett, T. J., and Spooner, E.T.C., 1977, Ophiolitic breccias associated with allochthonous oceanic crustal rocks in the east Ligurian Apennines, Italy—A comparison with observations from rifted oceanic ridges. *Earth and Planetary Science Letters, v. 35,* p. 79–91.

Beccaluva, L., Ohnenstetter, D., and Ohnenstetter, M., 1979, Geochemical discrimination between ocean-floor and island-arc tholeiites; application to some ophiolites. Canadian Journal of Earth Sciences, v. 16, no. 9, p. 1874–1882.

Bocquet (Desmons), J., 1974, Etudes minéralogiques et pétrologiques sur les métamorphismes d'âge alpin dans les Alpes francaises [Thesis]: University of Grenoble, 489 pp.

Boudier, F., Nicolas, A., and Bouchez, J. L., 1982, Kinematics of oceanic thrusting and subduction from basal sections of ophiolites, Nature, no. 296, p. 825–828.

Bourbon, M., Caron, J. M., Lemoine, M., and Tricart, P., 1979, Stratigraphie des Schistes lustrés piemontais dans les Alpes cottiennes (Alpes occidentales franco-italiennes): nouvelle interpretation et conséquences géodynamiques. Compte Rendu Sommáire de Seances Societé Géologique de France, v. 4, p. 180–182.

Boyer, S. E., and Elliott, D., 1982, Thrust Systems, Bulletin of the American Association of Petroleum Geologists, v. 66, no. 9, p. 1196–1230.

Butler, R.W.H., 1982, The terminology of structures in thrust belts, Journal of Structural Geology, v. 4, p. 239–245.

Caron, J. M., 1977, Lithostratigraphie et tectonique des Schistes lustrés dans les Alpes Cottiennes septentrionales et en Corse orientale [Thesis]: Strasbourg, 315 p.

Caron, J. M., Kienast, J. R., and Triboulet, C., 1981, High-pressure–Low-temperature metamorphism and polyphase Alpine deformation at Sant'Anrea di Cotone (Eastern Corsica, France), Tectonophysics, v. 78, p. 419–451.

Carpena, J., Maillhe, D., Naeser, C. W., and Poupeau, G., 1979, Sur la datation par traces de fission d'une phase tectonique d'age éocène supérieure en Corse: Compte Rendu de l'Academie de Sciences de Paris, v. 289, p. 829–832.

Chen, T., and Forsyth, D. W., 1978, A detailed study of two earthquakes seaward of the Tonga trench: implications for mechanical behaviour of the oceanic lithosphere: Journal of Geophysical Research, 83, p. 4995.

Chopin, C., and Maluski, H., 1980, ^{39}Ar-^{40}Ar dating of high pressure metamorphic micas from the Gran Paradiso area, (Western Alps): evidence against the blocking temperature concept: Contributions to Mineralogy and Petrology, v. 74, p. 109–122.

Cohen, C. R., 1982, Model for a Passive to Active Continental Margin Transition: Implications for Hydrocarbon Exploration: Bulletin of the American Association of Petroleum Geologists, v. 66, no. 6, p. 708–718.

Coleman, R. G., 1971, Plate tectonic emplacement of upper mantle peridotites along continental edges: Journal of Geophysical Research, v. 76, p. 1212–1222.

Cohen, C. R., Schweickert, R. A., and Odom, A. L., 1981, Age of emplacement of the Schistes lustrés nappe, Alpine Corsica: Tectonophysics, v. 73, p. 267–283.

Compagnoni, R., Dal Piaz, G. V., Hunziker, J. C., Gosso, G., Lombardo, B., and Williams, P. F., 1977, The Sesia-Lanzo zone, a slice of continental crust with Alpine High-pressure–Low-temperature assemblages in the western Italian Alps: Rendicotti Societa di Mineralogia e Petrologia Italiana, v. 33 (1), p. 281–334.

Cortesogno, L., Calbiati, B., Principi, G., and Venturelli, G., 1978, Le brecce ofiolitiche della Liguria orientale: nuovi dati e discussione sui modelli palaeogeografici: Ophioliti, v. 3 (2/3), p. 99–160.

Cortesogno, L., Galbiati, B., and Principi, G., 1981, Descrizione dettagliata di alcuni caratteristici affioramenti di brecce serpentinitiche della Liguria orientale ed interpretazione in chiave geodinamica: Ophioliti, v. 6, no. 1, p. 47–76.

Cortesogno, L., Grandjacquet, C., and Haccard, D., 1979, Contribution a l'étude de la liason Alpes-Apennins; evolution Tectonometamorphique des principaux ensembles ophiolitiques de Ligurie (Apennins du Nord): Ophioliti, v. 4, (2), p. 157–172.

Dahlstrom, C.D.A., 1969, The upper detachment in concentric folding: Bulletin of Canadian Petroleum Geology, v. 17, no. 3, p. 326–346.

Dal Piaz, G. V., Raumer, J. von, Sassi, F. P., Zanettin, B., and Zanferrari, A., 1975, Geological Outline of the Italian Alps: in Squyres, C. H., ed., Geology of Italy: Earth Sciences Society of the Libyan Arab Republic.

Decandia, F. A., and Elter, P., 1972, La "zona" ofiolitifera del Bracco nel settore compreso fra Levanto ela Val Gravegna (Apennino Ligure). *66e Congresso della Societa Geologica Italiana,* Pisa, p. 37–64.

Delcey, R., 1974, Données sur deux nouvelles séries lithostratigraphiques de la zone des Schistes lustrés de la Corse nord-orientale: Compte Rendu de Academie de Sciences Paris, v. 2;79, p. 1693–1696.

Dewey, J. F., 1976, Ophiolite obduction: Tectonophysics, v. 31, p. 93–120.

Durand-Delga, M., and Vellutini, P., 1977, Problèmes posés par le Sédimentaire allochtone de Macinaggio (Corse) et par l'origine de ses détritus: Geologie Méditerraneene, v. 4, no. 4, p. 271–280.

Elter, P., 1975, Introduction à la géologie de l'Apennin septentrional: Bull. Soc. Géol. France, v. 7, no. 17, p. 956–962.

Ernst, W. G. and Dal Piaz, G. V., 1978, Mineral parageneses of eclogitic rocks and related mafic schists of the Piemonte ophiolite nappe, Breuil-St. Jacques area, Italian Western Alps: American Mineralogist, v. 63, p. 621–640.

Faure, M., and Malavieille, J., 1980, Les plis en fourreau du substratum de la nappe des Schistes lustrés de Corse. Signification cinématiques: Compte Rendu de l'Academie de Sciences, Paris, v. 290, p. 1349–1352.

——1981, Etude structurale d'un cisaillement ductile: le charriage ophiolitique corse, dans la région de Bastia: Bulletin de la Société Géologique de France, v. 7, no. 23, no. 4, p. 335–343.

Gibbons, W., Waters, C., and Warburton, J., 1986, The Blueschist Facies Schistes lustrés of Alpine Corsica: a review. This volume.

Giese, P. and Prodehl, C., 1976, Main Features of Crustal Structure in the Alps, *in* Giese, P., Prodehl, C., and Stein, A., eds., Explosion Seismology in Central Europe, Berlin-Heidelberg-New York, Springer-Verlag, p. 347–376.

Graciansky, P. C., Bourbon, M., Lemoine, M., and Sigal, J., 1981, The sedimentary record of Mid-Cretaceous events in the western Tethys and central Atlantic Oceans and their continental margins. Eclogae Géologicae Helvetiae, v. 74, no. 2, p. 353–367.

Hirn, A., and Sapin, M., 1976, La croûte terrestre sous la Corse: données sismiques: Bull. Soc. géol. France, v. 5, p. 1195–1199.

Homewood, P., Gosso, G., Escher, A., and Milnes, A., 1980, Cretaceous and Tertiary evolution along the Besancon-Biella traverse, W. Alps: Eclogae Geologicae Helvetiae, v. 73, no. 2, p. 635–649.

Hussong, D. M., Edwards, P. B., Johnson, S. H., Campbel, J. F., and Sutton, G. H., 1976, Crustal structure of the Peru-Chile trench, 8°12″S Latitude. American Geophysical Union Monograph, no. 19, p. 71–85.

Issacks, B., and Molnar, P., 1971, Distribution of stresses in the descending lithosphere from a global survey of focal-mechanism solutions of mantle earthquakes, Review of Geophysics and Space Physics, v. 9, p. 103–174.

Lahondere, J. C., 1981, Relations du "socle ancien" de la region de Bastia (Corse) avec les Schistes lustrés envionnants: Compte Rendu de l'Academie de Sciences Paris, v. 293, p. 169–172.

Malavieille, J., 1982, Etude tectonique et microtectonique de la deformation ductile dans de grands chevauchements crustaux: Example des Alpes Franco-Italiennes et de la Corse [Thesis]: Montpellier, 117 p.

Maluski, H., 1977, Application de la méthode $^{40}Ar/^{39}Ar$ aux mineraux des roches cristallines perturbées par les évènements thermiques et tectoniques en Corse [Thesis]: Montpellier, 113 p.

Maluski, H., and Schaeffer, O. A., 1982, $^{40}Ar/^{39}Ar$ laser probe dating of terrestrial rocks: Earth and Planetary Science Letters, v. 59, p. 21–27.

Mattauer, M., Faure, M., and Malavieill, J., 1981, Transverse lineation and large-scale structures related to Alpine obduction in Corsica: Journal of Structural Geology, v. 3, p. 401–409.

Matauer, M., and Proust, F., 1976, La Corse alpine: un modèle de genèse du métamorphisme haute pression par subduction de croute continentale sous du matériel oceanique: Compte Rendu de l'Academie de Sciences Paris, v. 282, p. 1249–1252.

Mattauer, M., Proust, F., and Etchecopar, A., 1977, Linéations "a" et méchanisme de cisaillement simple liés au chevauchement de la nappe des Schistes lustrés en Corse: Bulletin de la Société Géologique de France, v. 7, no. 14, p. 841–847.

Mattauer, M., and Tapponier, P., 1978, Tectonique des plaques et tectonique intracontinentale dans les Alpes franco-italiennes: Compte Rendu de l'Academie de Sciences Paris, v. 287, p. 899–902.

Milnes, A. G., Greller, M., and Müller, R., 1981, Sequence and style of major post-nappe structures, Simplon-Pennine Alps: Journal of Structural Geology, v. 3, p. 411–420.

Morelli, C., Giese, P., Carrozzo, M. T., Colombi, B., Guerra, I., Hirn, A., Letz, H., Nicolich, R., Prodehl, C., Reichert, C., Röwer, P., Sapin, M., Scarascia, S., and Wigger, P., 1977, Crustal and upper mantle structure of the northern Apennines, the Ligurian Sea, and Corsica, derived from seismic and gravimetric data: Estratto dal Bolletino di Geophisica, v. 75-76, p. 199–260.

Nardi, R., 1968, Le unità alloctone della Corsica e loro correlazione con le unita della Alpi e dell'Appennino. Memoire della Societa Geologica Italiana, v.7, p. 323–344.

Nicolas, A., and Le Pichon, X., 1980, Thrusting of young lithosphere in subduction zones with special reference to structures in ophiolitic peridotites: Earth and Planetary Science Letters, v. 46, p. 397–406.

Ohnenstetter, d., Ohnenstetter, M., and Rocci, G., 1975, Tholeiitic cumulates in a high pressure metamorphic belt: Petrologie, v. 1, no. 4, p. 291–317.

——1976, Etude des métamorphismes successifs des cumulats ophiolitiques de Corse: Bulletin de la Société Géologique de France, v. 7, no. 18, no. 1, p. 115–134.

Ohnenstetter, M., 1979, La série ophiolitifère de Rospigliani (Corse) est-telle un témoin des phénomènes tectoniques, sédimentaires et magmatiques lies au fonctionnement des zone transformantes?' Compte Rendu de l'Academie de Sciences, Paris, v. 289, p. 1199–1202.

Ollé, J. J., 1981, Etude geologique des unites de la depression central Corse entre Asco et Golo (regions de Ponte-Leccia a Francardo) [Thesis]: Toulouse, 209 pp.

Orsini, J. B., Coulon, C., and Cocozza, T., 1980, Dérive Cénozoique de la Corse et de la Sardaigne et ses marquers géologiques. Geologie en Mijnbouw, v. 59, p. 385–396.

Piccardo, G. B., Messiga, B., and Mazzucotelli, A., 1979, Chemical petrology and geodynamic evolution of the ophiolitic metavolcanites (prasinites) from the Voltri Massif piemontese ophiolite nappe (western Liguria, Italy). Ophioliti, v. 4, no. 3, p. 373–402.

Prince, R. A., and Kulm, L. D., 1975, Crustasl rupture and initiation of imbricate thrusting in the Peru-Chile trench: Geological Society of America Bulletin, v. 86, p. 1639–1653.

Ramsay, J. G., 1963, Structure, stratigraphy and metamorphism in the W. Alps: Proceedings of the Geological Association, v. 74, p. 359–391.

Reutter, K. J., Giese, P., and Closs, H., 1980, Lithospheric split in the descending plate: observations from the Northern Apennines: Tectonophysics, v. 64, p. T1–T9.

Rieuf, M., 1980, Etude Stratigraphique et structurale des unites au nord-est de Corté (Corse) [Thesis]: Toulouse, 234 pp.

Rocci, G., Baroz, F., Bebien, J., Desmet, A., Lapierre, H., Ohnenstetter, D., Ohnenstetter, M., and Parrot, J. F., 1980, The Mediterranean ophiolites and their related Mesozoic volcano-sedimentary sequences, *in* Panayiotou, A., ed., Proceedings of the International Ophiolite Symposium, Nicosia, Cyprus: Geological Survey of Cyprus.

Rocci, G., Ohnenstetter, D., Ohnenstetter, M., 1976, Le log ophiolitique corse. Bulletin de la Société Géologique de France, v. 7, no. 5, p. 1229–1230.

Roeder, D., 1980, Geodynamics of the Alpine-Mediterranean system; a synthesis. Eclogae Géologicae Helvetiae, v. 73, no. 2, p. 353–377.

Saliot, P., 1973, Les principales zones de métamorphisme dans les Alpes françaises. Répartition et significance: Compte Rendus de l'Academie de Sciences, Paris, v. 276, p. 3081–3084.

Saliot, P., and Velde, B., 1982, Phengite compositions and post-nappe high pressure metamorphism in the Pennine zone of the French Alps: Earth and Planetary Science Letters, v. 57, p. 133–138.

Sauvage-Rosenberg, M., 1977, Tectonique et microtectonique des Schistes lustrés et ophiolites de la vallée de la Golo (Corse Alpine) [Thesis]: Montpellier, 91 pp.

Scius, H., 1981, La carte au 50 000ᵉ de Pietra-di-Verde: Etude geologique régionale dans les Schistes lustrés Corses [Thesis]: Strasbourg, 124 pp.

Scholl, D. W., von Heune, R., Vallier, T. L., Howell, D. G., 1980, Sedimentary masses and concepts about tectonic processes at underthrust ocean margins: Geology, v. 8, p. 564–568.

Speed, R. C., and Larue, D. K., 1982, Barbados: Architecture and Implications for accretion. Journal of Geophysical Research, v. 87, no. B5, p. 3633–3643.

Tricart, P., 1974, Les Schistes lustrés du Haute Cristillan (Alpes Cottiennes France): lithostratigraphie, architecture et tectonogenèse: Géologie Alpine, 50, p. 131–132.

Velde, B., 1965, Phengite micas: synthesis, stability, and natural occurrence: American Journal of Science, no. 263, p. 886–913.

Velde, B., and Kienast, J. R., 1973, Zonéographie du métamorphisme de la zone de Sesia-Lanzo (Alpes Piémontaises); études des omphacites et grenats des micaschistes éclogitiques à la microsonde électronique: Compte Rendu de l'Academie de Sciences, Paris, v. 276, p. 1801–1804.

Warburton, J., 1983, The tectonic setting and emplacement of ophiolites: a comparative study of Corsica and the western Alps [Ph.D. thesis]: Swansea, Wales, University College of Swansea, 443 pp.

Welland, M. J., and Mitchell, A. H., 1977, Emplacement of the Oman ophiolite: a mechanism related to subduction and collision: Geological Society of America Bulletin, v. 88, p. 1081–1088.

Westphal, M., Bardon, C., Bossert, A., and Hamzeh, R., 1973, A computer fit of Corsica and Sardinia against Southern France: Earth and Planetary Science Letters, v. 18, p. 137–140.

Westphal, M., Orsini, J., and Vellutini, P., 1976, Le microcontinent Corso-Sarde, sa position initiale: Données palaeomagnetiques et raccords geologiques: Tectonophysics, v. 30, p. 141–157.

MANUSCRIPT ACCEPTED BY THE SOCIETY JULY 29, 1985

Geological Society of America
Memoir 164
1986

High-pressure/low-temperature metamorphic rocks
of Turkey

A. I. Okay
İTÜ, Maden Fakültesi
Jeoloji Bölümü, Teşvikiye
İstanbul, Turkey

ABSTRACT

Five major high-pressure metamorphic complexes are recognised in Turkey. The Tavşanlı Zone in western Turkey is the largest and most important HP/LT metamorphic belt. It consists of a thick basal marble unit overlain by metabasite, metachert, and metashale. HP/LT metamorphism is generally prograde with abundant development of lawsonite, sodic amphibole, and sodic pyroxene in the metabasic rocks. The HP/LT metamorphic rocks are tectonically overlain by a non-metamorphic ophiolite nappe. The metamorphic Alanya nappes, situated on the Mediterranean coast in southern Turkey, consist of three superimposed flat-lying nappes; only the intermediate nappe (Sugözü Nappe), comprising abundant garnet-micaschists and rare bands of metabasite, shows an early eclogite/blueschist facies metamorphism overprinted by a later greenschist facies metamorphism, which also affected the other two nappes and welded them into one unit. Unlike the Tavşanlı Zone, Alanya blueschists are allochthonous and consist mainly of shallow-water sedimentary rocks. The age of metamorphism in the Alanya nappes and in the Tavşanlı Zone is mid-Cretaceous. The Bitlis Massif in eastern Turkey is the third major high-pressure metamorphic complex of the Taurides. Rare lenses and bands of kyanite-eclogites in the Precambrian basement of the Bitlis Massif indicate that the Bitlis Massif has undergone a Precambrian eclogite facies metamorphism, evidence of which was mostly destroyed by the Alpine greenschist facies metamorphism.

The Karakaya Complex of the Pontides has an extensive distribution in northern Turkey, and consists of metabasite, marble, and metagreywacke. It has undergone a high pressure greenschist facies metamorphism during Triassic time with local development of crossite in metabasic rocks. Elekdağ blueschists in north-central Turkey consist of an ophiolite nappe with serpentinite, gabbro and diabase, and an ophiolitic melange metamorphosed in blueschist facies. Glaucophane-bearing eclogites are also described from the Elekdağ area.

Turkish blueschists share two common features with many other Alpine HP/LT metamorphic complexes; the short time span of the HP/LT metamorphism, and blueschist protoliths consisting mainly of sedimentary rocks deposited on continental crust. These features, which differentiate these blueschists from the Circum-Pacific HP/LT metamorphic complexes, are probably a reflection of the peculiar characteristics of the Tethys ocean.

INTRODUCTION

Major blueschist belts are restricted to the Circum-Pacific region and to the Alpine-Himalayan mountain chain. The geology and petrology of the two major blueschist provinces of the Alpine chain, the Western Alps, and the Cyclades belt in Greece, are fairly well known (e.g. Frey and others 1974; Blake and others 1981), whereas knowledge of the widely exposed blueschists in Turkey is only fragmentary. The aim of this paper is to review the stratigraphy, tectonic setting, and petrology of the

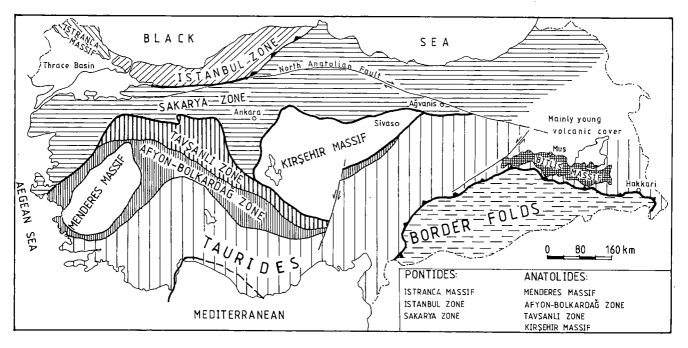

Figure 1. Tectonic map of Turkey showing the tectonic zones in the Pontides and Anatolides. Apart from the Bitlis Massif, Taurides are not subdivided. Heavy lines indicate major sutures.

Turkish blueschists against the background of the tectonics of Turkey.

TECTONIC FRAMEWORK OF TURKEY

Geologically, Turkey was created during the Alpine orogeny caused by the collision of Laurasia and Gondwanaland since the Late Mesozoic. In the classical, four-fold tectonic subdivision of Turkey (Ketin 1966), the Pontides in the north can be assigned to Laurasia, whereas the Anatolides, Taurides, and the Border Folds belong to the Gondwanaland realm with the major Tethyan suture, the İzmir-Erzincan Suture, separating the Pontides from the Anatolides (Figure 1). Characteristically the Pontides show less intense Alpine deformation than the others, with no post-Jurassic regional metamorphism. The Anatolides exhibit both strong Alpine deformation and metamorphism, whereas the Taurides consist of a series of superimposed nappes, which, with some notable exceptions, do not show Alpine regional metamorphism. The Border Folds are the gently folded foreland of the Alpine orogeny, and represent the northern extension of the Arabian Platform. In the terminology of Western Alpine geology, the Pontides can be compared in terms of their tectonic position with the Southern and Austro-Alpine Zone, the Anatolides with the Penninic, the Taurides with the Helvetic zones, and the Border Folds with the Jura Mountains. Unlike the Western Alps, the vergence of the Alpine orogen in Turkey is to the south. Another major difference from the Western Alps is the general absence of Hercynian deformation and metamorphism within the Taurides and Anatolides so that the stratigraphy in the Tauride nappes often spans the whole of the Paleozoic as well as the Mesozoic and commonly the Early Tertiary.

During the Mesozoic and earlier, the Anatolides and Taurides formed a wide and extensive platform, the Anatolide-Tauride platform (Şengör 1979), which was largely contiguous to Gondwanaland but separated from it during the Late Mesozoic by a narrow ocean—the Mesogea of Biju-Duval and others (1977)—whose remnants are the Tekirova and Troodos ophiolites and the present Eastern Mediterranean (Figure 2). The Anatolide-Tauride platform was linked to the Apulian platform in the west, represented today by the major part of the Hellenides and Dinarides, and separated by the Tethys ocean from the Pontides, which were part of Laurasia (cf. Biju-Duval and others 1977). The closure of the Tethys ocean by subduction and the subsequent collision of the two continents led to the Alpine orogeny in Turkey. This simple picture is complicated by the several short-lived small oceans that may have existed during the Mesozoic and earlier in the Pontides and in the Anatolide-Tauride platform. Those oceans caused the zonation of major tectonic units.

The Pontides, Anatolides, and Taurides are subdivided in Figure 1 into several zones commonly separated by minor and often disputed sutures. There are three major zones in the Pontides; in the northwest is the Istranca Massif consisting of Paleozoic and Early Mesozoic sedimentary rocks and granites metamorphosed in the greenschist facies (Aydın 1974). The contact of the Istranca Massif with the İstanbul Zone further east is covered by Tertiary sediments (Figure 2). The İstanbul Zone, equivalent to the İstanbul Nappe of Şengör and others (1984a), has a Precambrian basement of amphibolite, metadiorite, and micaschist overlain by sedimentary rocks of Cambrian to Eocene age with numerous major unconformities, and is separated from

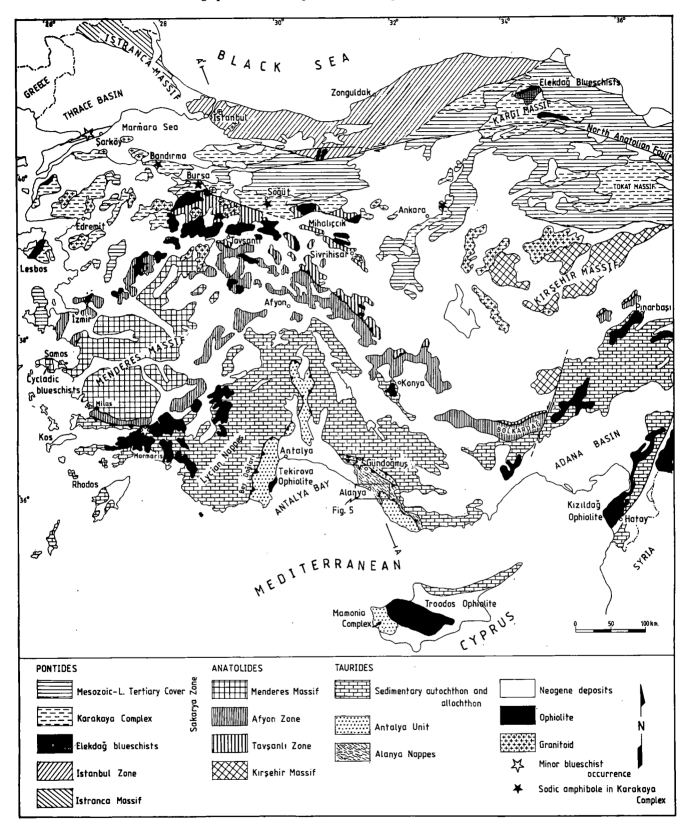

Figure 2. Geotectonic map of western Turkey showing the major blueschist complexes and blueschist localities. Approximate line of section in Figure 3 (A–A′) is also indicated.

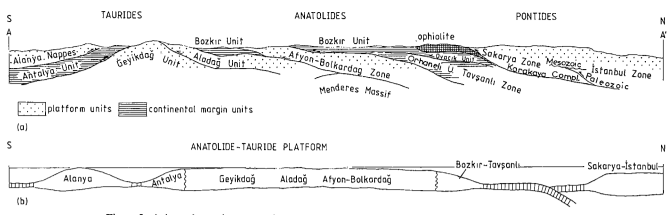

Figure 3. a) A south–north cross section across western Turkey illustrating the gross relations between various tectonic zones. b) A palinspastic reconstruction illustrating the possible position of the various tectonic zones during Early Cretaceous time. Neither section is drawn to scale.

the Sakarya Zone by the Intra-Pontide Suture. The Sakarya Zone (Okay 1984a), comprising the Sakarya Continent of Şengör and Yilmaz (1981) and the eastern Pontides (Figure 1), is characterised by a Permo-Triassic metamorphic basement called the Karakaya Complex (Bingöl and others 1975), which includes HP/LT metamorphic rocks. The Karakaya Complex is overlain by Mesozoic and Tertiary sedimentary and volcanic rocks (Okay 1984a; Figure 2). The juxtaposition of the İstanbul and Sakarya zones across the Intra-Pontide Suture probably occurred during the Early Mesozoic Cimmerian orogeny, which may be related to the closure of the Paleo-Tethys (Şengör and others 1984a). During the Late Mesozoic and Tertiary, the İstanbul and Sakarya zones, and probably the Istranca Massif, formed a single tectonic and paleogeographic unit, the Pontides.

There are four major units in the Anatolides. The Menderes Massif (sensu stricto) in the west consists of gneisses and micaschists representing remobilised Precambrian basement (Dürr and others 1978; Şengör and others 1984b). The Afyon Zone (Okay 1984a) forms the Paleozoic to Lower Tertiary sedimentary cover to the gneisses of the Menderes Massif, and has undergone medium to low grade Barrovian metamorphism along with the Menderes Massif during the Eocene (Dürr and others 1978; Çağlayan and others 1980; Şengör and others 1984b). South of the Menderes Massif, the Afyon Zone consists of a thick basal metaclastic sequence of Paleozoic age overlain by a thick (over a thousand meters) section of Mesozoic marbles with emery horizons. The marbles are succeeded by a slightly metamorphosed thin sequence of red micritic limestone and flysch with serpentinite and limestone olistoliths of Paleocene and Early Eocene age. The switch from neritic to pelagic sedimentation in the Afyon Zone was caused by the emplacement of the Lycian Nappes, which lie tectonically over the Lower Eocene flysch of the Afyon Zone (Dürr and others 1978). The emplacement of the Lycian Nappes was followed by a Barrovian type of regional metamorphism, which has affected the Menderes Massif, the Afyon Zone, and the basal parts of the Lycian Nappes (de Graciansky 1966; Başarır 1970; Dora 1981). In contrast, a well-documented Eo-

cene HP/LT metamorphism exists in the Cyclades immediately to the west of the Menderes Massif (e.g. Altherr and others 1979; Blake and others 1981; Figure 2). The Afyon Zone shows a similar stratigraphy and metamorphism northeast of the Menderes Massif (cf. Okay 1984a).

North of the Menderes Massif, the Northwest Turkish blueschist belt or the Tavşanlı Zone (Okay 1984b), comprised of volcanic and sedimentary rocks metamorphosed in the blueschist facies during the mid-Cretaceous, is partly thrust over the Afyon Zone (Figure 3). The Kırşehir Massif to the east of the Tavşanlı Zone (Figure 2) is made up of high temperature/low pressure metamorphic rocks and abundant granitoids, and represents the deep levels of an off-margin Mesozoic magmatic arc (Seymen 1982). It is surrounded on all sides by sutures; in the north by the İzmir-Erzincan Suture, and on the other sides by the Intra-Tauride Suture (Şengör and others 1982). The Kırşehir Massif and the Tavşanlı Zone probably constitute a paired metamorphic belt of Cretaceous age.

The Taurides consist of a series of superimposed nappes comprising mostly sedimentary rocks of platform and continental margin affinity, and ranging in age from Early Paleozoic up to the latest Cretaceous or Early Tertiary. In the Central Taurides between Konya and Alanya (Figure 2), Özgül (1976, 1984) differentiated a number of tectono-stratigraphic units, which show distinctive stratigraphic, metamorphic, and tectonic features. The Afyon-Bolkardağ Zone in the north constitutes the eastward extension of the Afyon Zone but is here partly allochthonous and is thrust over the Aladağ Unit (Figure 3), a complex of thrust sheets of platform type sediments of Devonian to Late Cretaceous age. Both the Afyon-Bolkardağ Zone and the Aladağ Unit have a local tectonic cover of peridotite and Mesozoic volcano-sedimentary rocks of continental margin affinity (Bozkır Unit, Figure 3). The Aladağ Unit is allochthonous and lies tectonically over the relative autochthon of the Taurides (Geyik Dağı Unit of Özgül 1976) with platform type sediments ranging from Cambrian to Eocene in age. Also tectonically overlying the autochthon in the south is the Antalya Unit (Figure 3), also known as the Antalya Nappes (Lefevre 1967); unlike the Mesozoic platform-type sediments of the Afyon-Bolkardağ, Aladağ, and Geyik Dağı units, the Mesozoic lithologies of the Antalya Unit are of continental margin type and record the rifting and foundering of a Permian carbonate platform leading to the development of a continental margin (Marcoux 1978; Robertson and Woodcock 1982). The Antalya Unit, which is correlated with the Mamonia Complex of Cyprus (Brunn and others 1971), is tectonically overlain in the west of the Antalya bay by the Tekirova ophiolite (Juteau 1970) and in the east by a metamorphic complex made up of three superimposed nappes (Alanya Nappes, Figure 2). The middle of the three crystalline nappes are comprised of eclogites and blueschists (Okay and Özgül 1984). Regional geological arguments indicate that the Antalya Unit was deposited to the south of the Tauride autochthon, in which case a narrow ocean, the Mesogea of Biju-Duval and others (1977), today marked by the Tekirova and Troodos ophiolites isolated the Anatolide-Tauride platform from the bulk of the Gondwanaland during the Late Mesozoic (Figure 3).

Lycian Nappes in the western Taurides, west of Antalya bay, lie tectonically in the west over the slightly metamorphosed Lower Eocene flysch of the Afyon Zone, the autochthonous cover of the Menderes Massif, and in the east over the Miocene flysch of the Tauride autochthon (Geyik Dağı Unit), which outcrops in the Bey Dağları (Figure 2). Lycian Nappes comprise a complex of thrust sheets made mostly of Mesozoic sedimentary and volcanic rocks of continental margin type with a major peridotite nappe (Figure 2; Brunn and others 1971; Bernoulli and others 1974). They can be partly correlated with the Bozkır Unit of Özgül (1976).

During the Mesozoic, the area of the Menderes Massif, Afyon Zone and the major part of the Taurides was a Bahamian-type carbonate platform with a continental crust. It was bordered on the north by the Tethys ocean, and was separated from Gondwanaland by a rifted pelagic basin, where rocks of the Antalya Unit were being deposited and where oceanic crust developed during the Late Mesozoic (Figure 3). During the Cretaceous, the Tethys ocean was consumed by a northward dipping subduction zone, which gave rise to the well-developed magmatic arc of the eastern Pontides (Zankl 1961). Obduction of ophiolite and part of the northward-facing continental margin rocks (Bozkır Unit) on the Anatolide-Tauride platform took place during the Late Cretaceous. Contemporaneously, but from the opposite direction, the Tekirova ophiolite was emplaced on the Antalya Unit. After a lull during the Paleocene, the Anatolide-Tauride platform was internally imbricated and partly metamorphosed during the Eocene with the emplacement of the Afyon-Bolkardağ Zone, Aladağ Unit, and Lycian Nappes from the north, and the Antalya Unit from the south over the Tauride autochthon (Şengör and Yılmaz 1981; Özgül 1984). There was another orogenic phase during the Miocene with the Lycian Nappes moving eastward over the Lower Miocene clastics of the Tauride autochthon in the Bey Dağları (Brunn and others 1971).

In the southeastern Taurides there are several tectonic units not known from the west. One such major tectonic unit is the Bitlis Massif (Figure 1), which is an extensive area of allochthonous metamorphic rocks 300 km long and 40 km wide in southeastern Turkey. Bitlis metamorphic rocks lie with an intervening zone of ophiolite, ophiolitic melange, and flysch over the Paleozoic to Mesozoic shelf sequences of the Arabian platform (Hall 1976). The Bitlis Massif is tectonically sliced and has a strongly imbricated internal structure. However, synthetic stratigraphic sections of the Bitlis Massif differentiate between an old basement composed of gneiss, amphibolite, and rare kyanite-eclogite of pre-Devonian and probable Precambrian age, and a Paleozoic-Mesozoic cover of phyllite, metaquartzite, marble, and metabasite with low-grade Alpine metamorphism (Boray 1975; Yılmaz 1975; Göncüoğlu and Turhan 1983). During the Late Cretaceous, the Bitlis Massif was overthrust by an ophiolite from the north. From the Late Tertiary onward, the Bitlis Massif and the overlying ophiolite cover continued moving southward over the Arabian platform related to the presently continuing convergence of Eurasia and Arabia. For more details on the Turkish geology and for different interpretations of the tectonic evolution of Turkey, the reader is referred to Campbell (1971), Brinkmann (1976), Şengör and Yılmaz (1981), and Ketin (1983).

DISTRIBUTION OF THE HP/LT ROCKS IN TURKEY

Sodic amphibole bearing rocks are reported from a large number of individual localities in Turkey and these were compiled into a map by van der Kaaden (1966). However, when the tectonic setting, age, and regional distribution of the sodic amphibole bearing metamorphic rocks are considered, then these numerous individual localities can be assigned to five major regionally important and genetically distinct HP/LT metamorphic provinces and a small number of geographically and geologically insignificant occurrences. These are listed in Table 1 with data on their areal extent, rock type, age, and characteristic minerals, and will be described in turn.

TAVŞANLI ZONE

This is by far the largest and best studied HP/LT metamorphic complex in Turkey (Table 1, cf. Okay 1984a). It is best exposed in northwest Turkey where it occupies a large triangular area 350 km long and 50 to 100 km wide (Figures 1 and 2). Sodic amphibole bearing metamorphic rocks from north of Konya (Bayiç 1968), from the Bolkardağ region (Blumenthal 1955; Çalapkulu 1980), and crossite-lawsonite schists from the Pinarbaşı area (van der Kaaden 1966; Erkan and others 1978) have a similar tectonic setting as the northwest Turkish blueschists and probably form its attenuated eastward continuation (Figure 2). In the west, the Tavşanlı Zone is situated at the northernmost part of the Anatolides immediately south of the major Tethyan suture; eastward it follows the southern boundary of the Kırşehir Massif (Figure 2). The Tavşanlı Zone apparently

A. I. Okay

TABLE 1. IMPORTANT HIGH-PRESSURE METAMORPHIC COMPLEXES OF TURKEY

Name	Outcrop area	Protolith	Characteristic minerals	Age of deposition	Age of metamorphism
Tavşanlı Zone*	~10,000 km²	Limestone, basic volcanic, chert, shale	Gl., laws., jd., arag., garn., omph.	? Paleozoic- Mesozoic	Mid-Cretaceous
Sugözü Nappe, Alanya*	30-40 km²	Shale (90%) basic volcanic, quartzite	Gl., garn., omph.	?	Mid-Cretaceous
Gündoğmuş blueschists, Alanya	?10-20 km²	Basic volcanic, chert, limestone	Crossite, sodic pyr., garn.	?	? Cretaceous
Basement of the Bitlis Massif†	? 5,000 km²	Shale, basic magmatic rock	Omph., garn., kyanite	?	? Precambrian
Karakaya Complex§	~20,000 km²	Basic volcanic, greywacke, limestone	Crossite	Permo-Triassic	Triassic
Elekdağ blueschists*	600 km²	Basic volcanic greywacke, chert, gabbro, serpentinite	Gl., garn., omph.	?	? Early Mesozoic
Sarköy Blueschists, Thrace	1-3 km²	Basic volcanic chert	Gl., laws., sodic pyr.	?	? Cretaceous

Note: The outcrop areas of the blueschist complexes are highly approximate; The Sugözü Nappe extends for several hundred square kilometers under the overlying nappe cover.
Abbreviations: gl. = glaucophane; laws. = lawsonite; jd. = jadeite; arag. = aragonite; garn. = garnet; omph. = omphacite; sodic pyr. = sodic pyroxene.
*Includes Group C eclogites.
†Includes Group B eclogites.
§Metamorphism is mostly in greenschist facies.

has a parautochthonous position and is tectonically overlain by an ophiolite nappe consisting primarily of peridotite; the mineral assemblage of rare gabbros and dykes in the peridotite indicate that the ophiolite nappe has not undergone blueschist facies metamorphism, and forms a separate tectonic unit (Okay 1984a). In the south, the Tavşanlı Zone is partly thrust onto the Paleozoic-Mesozoic metasediments of the Afyon Zone, which show greenschist facies metamorphism (Figure 3).

Two units are differentiated within the Tavşanlı Zone (Figure 4). The upper Ovacık Unit is an imbricated, strongly tectonised volcano-sedimentary complex consisting of closely intercalated spilite, agglomerate, radiolarian chert, red and green pelagic shale, pelagic limestone, and greywacke (Kaya 1972). The Ovacık Unit generally underlies the ophiolite nappe and includes abundant serpentinite and talc lenses. It has undergone an incipient blueschist metamorphism; although the rocks appear unmetamorphosed in the field, typical blueschist minerals such as lawsonite and aragonite have grown in the amygdales and veins of spilites, and pelagic limestones have been transformed into aragonite marbles. Associated with this incipient blueschist metamorphism there has been a regional metasomatism involving topotactic replacement of augite in spilites by sodic pyroxene resulting in rock with 7-8% Na_2O (Okay 1982).

The lower unit of the Tavşanlı Zone is the Orhaneli Unit, which has unlike the Ovacık Unit a metamorphic field appearance and shows a regular stratigraphy (Figure 4). It forms the lowest tectonic unit, and is thus the relative autochthon; it is tectonically overlain by the Ovacık Unit or by the ophiolite nappe. Graphitic schists with the mineral assemblage of quartz + white mica + chlorite + chloritoid + graphite ± lawsonite make up the lowest observed section of the Orhaneli Unit in the Orhaneli area (Lisenbee 1971). They pass into micaceous phyllites and then into marbles, which are several kilometers thick in the Tavşanlı area (Okay 1980a); the marbles show a distinct lineation defined by the parallel alignment of calcite crystals. The marbles pass upward without a tectonic break into a thick sequence of intercalated metabasites, metacherts, and metashales, which measure several kilometers in the Tavşanlı and Mihaliçcik areas (Figure 4). This metavolcano-sedimentary sequence is lithologically very similar to the Ovacık Unit and is probably its stratigraphic equivalent. Blueschist metamorphism has found its best expression in this metavolcano-sedimentary sequence with the development of lawsonite, glaucophane, jadeite, aragonite, and garnet (Çoğulu 1967; Okay 1980a, b; Servais 1982).

The Orhaneli Unit has undergone a pervasive recrystallisation and ductile deformation with the development of layered

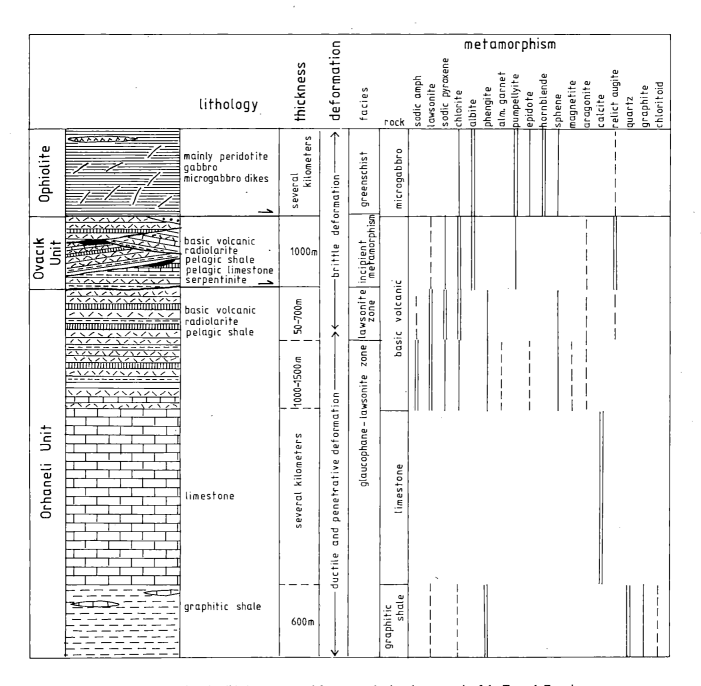

Figure 4. Stratigraphy, lithology, structural features, and mineral paragenesis of the Tavşanlı Zone in northwest Turkey.

parallel schistosity, lineation, and isoclinal folding. In the Tavşanlı area, recrystallisation and penetrative deformation is less intense in the uppermost 500–1000 meters of the Orhaneli Unit directly underlying the ophiolite or Ovacık Unit (Figure 4). In this "lawsonite zone" the rocks generally have a non-metamorphic appearance; the basic volcanic rocks retain their igneous texture but have developed a "lawsonite + sodic pyroxene + chlorite + sphene" assemblage where sodic pyroxene is pseudomorphous after augite. They pass downward along a mapped isograd into

penetratively deformed "glaucophane-lawsonite zone" blueschists (Okay 1980b) with the assemblage of sodic amphibole + lawsonite ± sodic pyroxene + chlorite + sphene in the metabasic rocks; the isograd represents the formation of sodic amphibole at the expense of sodic pyroxene and chlorite. The lawsonite zone assemblages represent an early static stage of metamorphism followed by a sodic amphibole forming event. Sodic amphibole formation is associated with the onset of penetrative deformation, which seems to have had a catalytic effect on the sodic amphibole forming reaction (Okay 1980b). A similar deformation related reaction is described from the blueschists of Syros, Greece (Ridley and Dixon 1984). It is likely that the "lawsonite zone" rocks of the Orhaneli Unit have experienced similar P-T conditions as the underlying glaucophane-lawsonite zone rocks, but could not be penetratively deformed because of the inhibitive effect of the thick and rigid ophiolite lid (Figure 4).

Although prograde textures are prevalent in the Tavşanlı Zone, a greenschist overprint is described in the blueschists of the Sivrihisar area (Gautier 1984), and complex polymetamorphic textures from an area south of Tavşanlı (Kaya 1981). Group C eclogites (cf. Coleman and others 1965) are found as lenticular blocks in the Sivrihisar area (Kulaksız 1978).

There are limited data on the age of deposition of the Ovacık and Orhaneli units. Late Jurassic-Early Cretaceous paleontological ages are obtained from the Ovacık Unit (Akdeniz and Konak 1979; Servais 1981). A speculative Late Paleozoic-Mesozoic depositional age can be assigned to the Orhaneli Unit based on stratigraphic correlation with the Afyon Zone (Okay 1984b). The age of blueschist metamorphism is better known. An upper geological age limit is provided by the abundant blueschist detritus in the Maastrichtian flysch of the Sakarya Zone (e.g. Batman 1978), and by the Middle Paleocene-Middle Eocene sediments that are locally transgressive over the ophiolite and Ovacık Unit. K/Ar dating on phengites from the blueschists (Çoğulu and Krummenacher 1967; Kulaksız 1982 personal communication) indicates a mid-Cretaceous age for the HP/LT metamorphism. The regular stratigraphy of the Tavşanlı Zone suggests that the blueschist metamorphism was largely contemporaneous along the belt. The age data show that immediately after the end of deposition in Early Cretaceous time, the Tavşanlı Zone has undergone blueschist metamorphism during the mid-Cretaceous followed by very fast uplift to the surface. The prograde nature of mineral assemblages, preservation of aragonite, and absence of greenschist overprint in many parts of the Tavşanlı Zone testify to the fast uplift. As the blueschists of the Tavşanlı Zone occur as a coherent terrane, a rigid block several thousand square kilometers in dimension must have been uplifted 30-35 km during the Late Cretaceous to expose blueschists at the surface. As discussed in Okay (1984a), the overload on the blueschists was probably oceanic crust and mantle that was removed by fast erosion giving rise to olistostrome and flysch deposition, and to nappe movements.

The Tavşanlı Zone is the only HP/LT metamorphic belt in Turkey that may have a contemporaneous low pressure/high temperature (LP/HT) metamorphic belt. The Kırşehir Massif to the east of the Tavşanlı Zone (Figure 2) consists of gneiss, micaschist, and marble with LP/HT minerals such as cordierite, wollastonite, and sillimanite (Ayan 1963; Erkan 1976; Seymen 1981; Göncüoğlu 1977), and is intruded by numerous granodiorite and gabbro plutons of Latest Cretaceous-Paleocene age (Ayan 1963; Ataman 1974). A single age determination on the micaschists of the Kırşehir Massif also gives a Late Cretaceous age (Erkan and Ataman 1981). Pending more refined age determinations, it is possible that the Kırşehir Massif forms a paired metamorphic belt with the Tavşanlı Zone, and represents the deep levels of a volcanic island arc. At present, the Kırşehir Massif has a curved internal structure, and is partly separated from the Tavşanlı Zone by the sedimentary sequences of the Sakarya Zone and by the enigmatic "Intra-Tauride suture" (Figure 2). However, the recently demonstrated major rotations in the Sakarya Zone and in the Kırşehir Massif (Evans and others 1982; Sanver and Ponat 1980) indicate that the Cretaceous paleogeography has been strongly modified during the Alpine orogeny.

As discussed in Okay (1984a), the Tavşanlı Zone probably represents the northfacing continental margin of the Anatolide-Tauride platform. The Orhaneli Unit, which includes thick sequences of limestone and shale, was probably deposited on a continental basement while the Ovacık Unit consisting wholly of pelagic sediments and basic volcanics probably represents upper levels of an unusual oceanic crust. This northward facing continental margin was subducted and metamorphosed in blueschist facies during the mid-Cretaceous following the consumption of the Tethys ocean in northwest Turkey (Okay 1984a).

ALANYA BLUESCHISTS

In the very external parts of the central Taurides, east of the Antalya Bay the metamorphic rocks of the Alanya Nappes lie tectonically over the rocks of the Antalya Unit, which in turn are thrust on the autochthon (Geyik Dağı Unit, Figure 2). HP/LT metamorphic rocks are encountered at two different places in this area: in the Alanya Nappes, and in the Gündoğmuş tectonic zone between the Alanya Nappes and Antalya Unit (Figure 2).

Alanya Nappes

The Alanya Nappes (Okay and Özgül 1984) previously known as the Alanya Massif (Blumenthal 1951), lie tectonically over the predominantly Mesozoic continental margin-type rocks of the Antalya Unit, which crop out in a large tectonic window and in a narrow corridor between the Alanya Nappes and the Geyikdağı Unit (Figure 2). Of the three superimposed nappes that make up the "Alanya Massif" (Figure 5), only the intermediate Sugözü Nappe has undergone HP/LT metamorphism. The Sugözü Nappe is only 100 to 800 meters thick but extends for over 40 kms and has a minimum width of 4 km (Figure 5). It is underlain by the Mahmutlar Nappe comprising a heterogeneous series of shale, sandstone, dolomite, limestone, and quartzite, and is overlain by the Yumrudağ Nappe consisting largely of a section of Permian carbonates more than 1000 m thick, with a thin basal shale unit (Figure 5). Unlike the Sugözü Nappe, the Mahmutlar and Yumrudağ nappes have undergone only the late greenschist facies metamorphism, which has also affected the Sugözü Nappe.

The Sugözü Nappe consists predominantly of garnet-micaschists with rare bands and lenses of metabasites. In the western part of the Alanya Nappes, the garnet-micaschists are

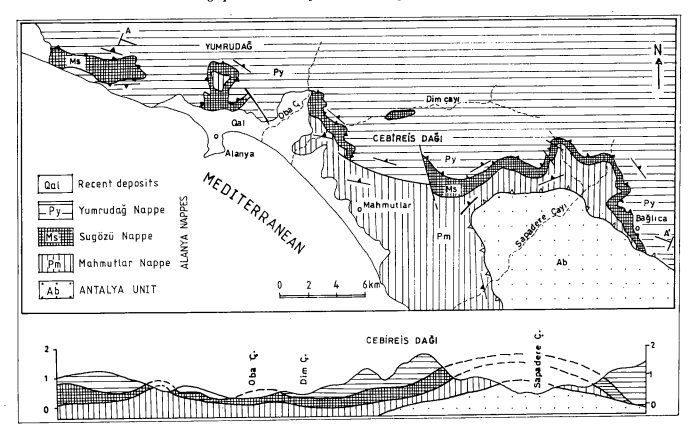

Figure 5. Geological map and cross section of the Alanya area (after Okay and Özgül 1984).

underlain by a thin sequence of metaquartzite, marble, and meta-dolomite, and are intruded by pre-metamorphic aplitic dykes. Garnet-micaschists consist of quartz, albite, phengite, garnet, and chlorite with minor ilmenite, graphite, clinozoisite, biotite, and rutile. The only evidence for HP/LT metamorphism in the garnet-micaschists is rare sodic amphibole inclusions in garnet porphyroblasts (Okay and Özgül 1984). Metabasites collected from different parts of the Sugözü Nappe show polymetamorphic mineral assemblages and textures. The earliest mineral assemblages are preserved in the rare eclogites with a "sodic pyroxene + garnet + sodic amphibole" assemblage. The more common blueschist metabasites have the assemblage of "sodic amphibole + garnet ± sodic pyroxene." Barroisite-amphibolites with the assemblage of "barroisite + garnet + albite + chlorite ± sodic amphibole" represent the last stage in the retrogressive development of the eclogites. In barroisite-amphibolites sodic amphiboles are mostly replaced by barroisite. In most metabasic rocks the three super-imposed mineral assemblages can be petrographically recognised at various stages of preservation. Sugözü blueschists have undergone penetrative and multiphase deformation with the development of schistosity and isoclinal folding. Schistosity planes in the Sugözü Nappe are concordant to those of the Mahmutlar and Yumrudağ nappes.

The early HP/LT metamorphism in the Sugözü Nappe has produced omphacite, garnet, and minor sodic amphibole in met-abasic rocks. This early eclogitic stage was followed by the extensive development of sodic amphibole at the expense of sodic pyroxene. The initial HP/LT metamorphism of the Sugözü Nappe was followed by the tectonic stacking of the Alanya nappes, greenschist facies metamorphism, and deformation, which welded the Alanya nappes into one unit. The greenschist facies metamorphism strongly retrograded the garnet-micaschists and produced albite, chlorite and barroisite in the metabasic rocks. The effect of the greenschist facies overprint and associated deformation was so thoroughgoing that in the field the demarcation of the Sugözü Nappe is made mainly on the presence of conspicuous garnet porphyroblasts in this unit; garnets are lacking in the Yumrudağ Nappe and in the upper parts of the Mahmutlar Nappe.

The protoliths of the Sugözü blueschists must have been a shelf type sequence of shale, quartzite, limestone, and dolomite with rare basic volcanic horizons in the shale unit. Ultramafic rocks do not occur in any of the Alanya nappes. The depositional age of the Sugözü Nappe is not known whereas the HP/LT metamorphism, tectonic stacking, and greenschist facies metamorphism of the Alanya nappes is probably pre-Maastrichtian, as the Antalya Unit underlying the Alanya nappes includes Maastrichtian sediments. Preliminary results of isotopic dating on sodic amphiboles from the Sugözü Nappe indicate a Cretaceous age for the HP/LT metamorphism (M. Satır, personal commun., 1984).

Gündoğmuş Blueschists

Blueschists associated with serpentinite, spilite, and radiolrian chert occur in the tectonic zone between the Alanya Nappes and Antalya Unit in the region of Gündoğmuş (Figure 2, Şengün and others 1978). Very little is known about their tectonic position and areal extent. Basic volcanics, cherts, shales, and limestones are the major rock types; the rocks are completely recrystallised and show a distinct foliation. The mineral assemblage in the metabasites is crossite + garnet + sodic pyroxene + epidote + phengite + sphene (Okay unpub data, 1983). In terms of lithology, associated rocks, and tectonic setting Gündoğmuş blueschists are quite different from the blueschists of the Sugözü Nappe.

The Origin of Alanya Blueschists

The original tectonic setting and the cause of the HP/LT metamorphism of the Alanya nappes are highly problematic. Regional geological arguments are against the derivation of the Alanya nappes and Antalya Unit from north of the Tauride autochthon (Brunn and others 1971; N. Özgül personal commun. 1984). In fact, the Sugözü Nappe and the Tavşanlı Zone exhibit such contrasting features that they have only in common the age and HP/LT nature of the regional metamorphism.

The structural position of the Alanya nappes over the Antalya Unit suggests that the Alanya nappes formed a rifted-off continental fragment, the "Alanya microcontinent," to the south of the Anatolide-Tauride platform; the Alanya microcontinent was separated from the Anatolide-Tauride platform during the Mesozoic by a pelagic basin, the Pamphylian basin of Dumont and others (1972), where rocks of the Antalya Unit were deposited (Figure 3). Minor oceanic crust, represented by the ophiolitic slivers in the Gündoğmus area, may have formed in the Pamphylian basin during the Late Mesozoic. The Alanya microcontinent was also bounded in the south by an oceanic area, the Mesogea of Biju-Duval and others (1977), represented today by the Tekirova, Troodos, and Kizildağ ophiolites and the present Eastern Mediterranean (Figures 2 and 3). HP/LT metamorphism of part of the Alanya microcontinent and Pamphylian basin is probably associated with the partial destruction of the Mesogea during the Late Cretaceous. However, the long and complicated structural and metamorphic history of the Alanya Nappes cannot be related in a simple way to the subduction of Mesogea.

BİTLİS MASSIF

High pressure metamorphic rocks occur in two very different parts of the Bitlis Massif: kyanite-eclogites are found in the old basement of the Bitlis Massif, and locally crossite-bearing metabasites occur in the ophiolite units both overlying and underlying the Bitlis metamorphic rocks.

Bitlis Eclogites

Eclogites were discovered by Türkünal (1980) in the old

basement of the Bitlis Massif south of Muş (Figure 1). In this region eclogites and partially amphibolitised eclogites occur as several centimeter- to several meter-thick lenses and bands in the gneisses and garnet-micaschists. Unaltered eclogites are rare; they consist of omphacite + garnet + kyanite + zoisite + edenitic hornblende + phengite + rutile + quartz + pyrite. Omphacites are commonly altered to a diopside + oligoclase symplectite, and phengites are rimmed by biotite (Okay and others 1985) The field setting, mineral assemblage and garnet compositions with 26–40% pyrope content of the Bitlis eclogites indicate that they are typical Group B eclogites (cf. Coleman and others 1965) generally restricted to the Precambrian gneiss terranes (cf. Bryhni and others 1977). Bitlis eclogites are not related to the Alpine orogeny but have formed during a poorly known Precambrian orogeny, which affected the old basement of the Bitlis Massif. Alpine metamorphism in the greenschist facies in the Bitlis Massif destroyed most of the evidence of the early eclogite facies metamorphism.

Guleman Ophiolite

Local crossite-bearing metabasites are described from an Upper Cretaceous ophiolite complex called the Guleman ophiolite (Perinçek 1979). Because of the strong late tectonic imbrication, the Guleman ophiolite occurs both below and above the Bitlis metamorphic rocks. Hall (1976, 1980) describes closely interlayered crossite and barroisite-bearing metabasites from an ophiolitic melange beneath the Bitlis metamorphic rocks south of Mutki. Similar crossite-bearing metabasites associated with serpentinite, gabbro, radiolarian chert, and pelagic limestone of Late Cretaceous age occur tectonically above the Bitlis Massif (Göncüoğlu and Turhan 1983). Crossite commonly forms small, incipient grains rimming igneous pyroxenes in basic volcanic rocks and has developed only locally in the Guleman ophiolite, which generally shows low-grade greenschist facies metamorphism.

PONTIDE BLUESCHISTS

Unlike the Tauride blueschists, the age of metamorphism of the Pontide blueschists is pre-Jurassic. The two important HP/LT metamorphic complexes of the Pontides are the Karakaya Complex and Elekdağ blueschists (Figure 2).

Karakaya Complex

The Karakaya Complex is a thick, deformed, and variably metamorphosed sequence of basic volcanic rock, limestone, and greywacke of Permo-Triassic age (Bingöl and others 1975; Bingöl 1978). It constitutes the basement for the Mesozoic sedimentary rocks of the Sakarya Zone. The Karakaya Complex has a wide distribution north of the İzmir-Erzincan Suture, and extends for over 1000 kilometers from the island of Lesbos in Greece (Hecht 1972) to the west to the Tokat Massif in the east (Figure 2). In most areas it forms the lowest unit and has an autochthonous position; only in the region of Bursa in northwest Turkey do high-grade gneisses crop out as a tectonic window underneath the Karakaya Complex (Ketin 1984; Okay 1984b). Triassic and Liassic conglomerates are transgressive over the metamorphic rocks of the Karakaya Complex.

In most areas basic pyroclastic rocks, basic lavas, and intercalated limestones constitute 80–90 percent of the Karakaya Complex; the rest is made up of greywacke, shale, acidic volcanic rock, radiolarian chert, limestone olistoliths, and small lenses of serpentinite. Along the whole belt, metamorphism is generally of medium pressure greenschist facies with the common mineral assemblage of epidote + chlorite + actinolite/barroisite + albite + sphene in the metabasites (Yılmaz 1979; Okay 1984c). In several areas, progressive metamorphism of spilites into amphibolites can be observed. For example, Yılmaz (1979) differentiated four metamorphic zones within the Karakaya Complex in northwest Turkey representing the progressive metamorphism of spilites into amphibolites with garnet, barroisite, and albite. Crossite is widely described from the poorly recrystallised basic volcanic rocks and metacherts of the Karakaya Complex; it occurs in the region of Bandırma (van der Kaaden 1966), Bursa (Sağıroğlu and Bürküt 1966), Söğüt (Yılmaz 1979; Ayaroğlu 1979), Ankara (van der Kaaden 1966) and Tokat (Koçyiğit 1979; Figure 2). Foliation and isoclinal folding are well developed in the greenschist facies metabasites of the Karakaya Complex, whereas the lower grade rocks have deformed in a semi-brittle way; the more competent limestone layers within the basic volcanic rocks have broken, giving a melange character locally to the Karakaya Complex.

The limestone beds intercalated with the basic volcanic rocks have locally yielded Late Permian and Early Triassic ages (Okay 1984a). As the metamorphic rocks are unconformably overlain by Middle Triassic to Liassic conglomerates (Bingöl and others 1975; Servais 1981), metamorphism and uplift of the Karakaya Complex must have occurred during the Triassic (Tekeli 1981).

The Triassic events in the Pontides are difficult to decipher because of the interference of later Alpine deformations, so that the Karakaya Complex is variously interpreted as a failed intracontinental rift (Bingöl 1976), an accretionary complex (Tekeli 1981), a narrow oceanic marginal basin (Şengör and Yilmaz 1981), or as a magmatic arc complex (Okay 1984c). Neritic limestones intercalated with voluminous basic volcanic rocks and the presence of dacites and andesites in some parts of the Karakaya Complex (Okay 1984c) suggest that a major part of the Karakaya Complex may represent a Permo-Triassic magmatic arc complex. The metamorphism and deformation of the Karakaya Complex may be related to the collision of the island arc with the İstanbul Zone in the north during the Early Mesozoic.

Elekdağ Blueschists

Although the eclogites from the Elekdağ were described as early as 1907 by Milch, they remain the least known of the major

HP/LT complexes in Turkey. The Elekdağ blueschists occur within a large metamorphic complex in northern Turkey known as the Kargı Massif (Figure 2). The Kargı Massif consists of several northward dipping slices of metabasite, metagabbro, metaserpentinite, metapelite, and marble (Yılmaz and Tüysüz 1984), which can be broadly assigned to the Karakaya Complex. HP/LT metamorphic rocks are mainly described from the northern part of the Kargı Massif from the region of Elekdağ (Figure 2). There, an ophiolite complex has apparently undergone blueschist facies metamorphism; ultramafic rocks consist of antigorite, while metagabbros and metabasites have the mineral assemblage garnet + sodic amphibole + epidote + albite + white mica + sphene (Eren 1979; Yılmaz and Tüysüz 1984). The metaophiolite is thrust southward on a melange unit of tuff, greywacke, radiolarian chert, basic flows, gabbro, and serpentinite, which has similarly undergone blueschist facies metamorphism. Eren (1979) also describes eclogite lenses several meters wide at the contacts of the major metaserpentinite. They are Group C eclogites and consist of omphacite and garnet with minor sodic amphibole, white mica, and sphene. The eclogites are commonly surrounded by antigorite-serpentinite, have chlorite- and actinolite-rich outer shells, and show marginal alteration to sodic amphibole + garnet + epidote metabasite (Eren 1979).

The transgressive Uppermost Cretaceous–Lower Paleocene limestones furnish an upper age limit for the HP/LT metamorphism in the Elekdağ region. Based on regional geological considerations Yılmaz and Tüysüz (1984) favour an Early Mesozoic age for the HP/LT metamorphism. Elekdağ blueschists are probably related to the enigmatic Early Mesozoic Karakaya orogeny.

OTHER OCCURRENCES OF SODIC AMPHIBOLE-BEARING ROCKS IN TURKEY

Apart from the already described regionally important HP/LT metamorphic complexes, there are several minor and local occurrences of sodic amphibole. Tatar (1981) describes fibrous sodic amphibole in marble, spilite, and radiolarian chert from an ophiolitic melange northwest of Sivas (Figure 1). A clear case of metasomatism is observed in the synorogenic Eocene clastics in southeastern Turkey, west of Hakkari (Figure 1). Serpentinite boulders in some conglomerate lenses are partially to completely replaced by interlocking grains of crossite and stilpnomelane. Crossite and stilpnomelane also form large prismatic crystals in some limestone boulders in the same area (Okay, unpublished data, 1982). Two of the more important minor blueschist occurrences are described below.

Şarköy Blueschists

The recently discovered Şarköy blueschists (Şentürk and Okay 1984) occur in a tectonic slice about one kilometer thick uplifted along one of the western offshoots of the North Anatolian Fault in Thrace (Figure 2). Blueschists are associated with tectonic slices of serpentinite, basic volcanic rock, and Maastrichtian pelagic limestone, and are unconformably overlain by Eocene sediments (Kopp 1964). Protoliths of the blueschists are basic volcanic rocks and radiolarian cherts. HP/LT metamorphism has produced crossite, lawsonite, and sodic pyroxene in the basic volcanic rocks, which are commonly completely recrystallised and show a penetrative schistosity. Şarköy blueschists are petrographically and lithologically similar to the blueschists from the Tavşanlı Zone; they are of interest in that they have an unexpected tectonic setting to the north of the Sakarya Zone in the Pontides (Figure 2), and represent slivers of what must be a major regional metamorphic complex.

Marmaris Blueschists

Sodic amphibole bearing metamorphic rocks from Marmaris in southwestern Turkey occur as tectonic slices at the base of a large peridotite nappe (Figure 2). Sodic amphibole is found in frequently mylonitised radiolarian cherts and spilites as incipient fine-grained, often fibrous crystals, or in amphibolites replacing hornblende (van der Kaaden and Metz 1954; Tatar 1968; Okay, unpublished data, 1978). The development of sodic amphibole in the Marmaris region is probably related to the emplacement of the large peridotite nappe and is not the result of a regional metamorphism. Similar very fine grained felty aggregates of sodic amphibole are observed in northwest Turkey at the base of peridotite nappes (Okay 1980c). From the Marmaris region, de Graciansky (1972) also describes blueschist olistoliths from an Upper Cretaceous wildflysch underneath the peridotite nappe.

CONCLUSIONS

A review of the blueschists of Turkey illustrates their diversity in terms of tectonic environment, lithology, and type of metamorphism. In Elekdağ region, the ophiolite itself has undergone HP/LT metamorphism; in the Tavşanlı Zone the protoliths of the blueschists are mostly continental margin type rocks, while the non-metamorphosed ophiolite forms a separate tectonic unit overlying the blueschists. In contrast, Sugözü blueschists in the Alanya area consist of shallow water sediments with no associated ophiolite. Like most major blueschist complexes of the world, the Tavşanlı Zone and Karakaya Complex generally form the lowest tectonic unit, while Sugözü and part of the Elekdağ blueschists occur as nappes. While prograde textures are common in the Tavşanlı Zone and Karakaya Complex, Sugözü blueschists show retrograde and polymetamorphic textures.

Conventional subduction zone models (e.g. Hamilton 1969; Ernst 1970) may be invoked for the genesis of the Elekdağ and Tavşanlı Zone blueschists, but even in the case of Tavşanlı Zone the bulk of blueschist protoliths is not oceanic lithosphere or trench deposits, as expected from such a model, but is continental margin type rocks. A second implication of the model is that the HP/LT metamorphism should be a continuous process not necessarily related to orogenic events generally caused by the collision of continental plates. However, metamorphism of the

Turkish blueschist complexes is apparently restricted in time, and is related to nappe movements and ophiolite obduction (cf. Okay 1984a). These two features, the short time span of the HP/LT metamorphism and blueschist protoliths consisting of sediments deposited on continental crust, are found in most of the Alpine HP/LT metamorphic complexes (e.g. Sesia-Lanzo Zone, Cyclades Belt, External HP/LT Belt of the Hellenides) and differentiates them from the Circum-Pacific blueschist belts. Voluminous trench deposits of Circum-Pacific type were absent in the Alpine Tethys, probably because volcanic arcs were poorly developed and the trenches were bordered by extensive carbonate platforms that restricted the influx of clastic material to the trenches. The preserved blueschists of the Alpine belt were therefore either pieces of ophiolite or more commonly continental margin and carbonate platform sediments, which got subducted during the final phase of the destruction of the Tethys.

A major regional question is the relationship between the Tavşanlı Zone and Cycladic blueschist belt immediately to the west of the Menderes Massif (Figure 2). They have a similar tectonic setting sandwiched between the overlying Laurasian plate and underlying carbonate units of the Apulian-Anatolide-Tauride platform, and a similar stratigraphy with a thick basal marble unit overlain by a metavolcano-sedimentary sequence (Dürr and others 1978; Blake and others 1981, 1984). However, the age of the main HP/LT metamorphism in the Cyclades Belt is Eocene (e.g. Altherr and others 1979) whereas it is mid-Cretaceous in the Tavşanlı Zone. Continental collision, and therefore blueschist metamorphism, could have been earlier in northwest Turkey, probably because it formed a promontory of the Apulian-Anatolide-Tauride platform during the Mesozoic, as suggested by the present geography. Cycladic blueschists extend as far east as the island of Samos (Figure 2; Papanikolaou 1979), which lies a few kilometers west of the Menderes Massif. However, the Eocene metamorphism of the Menderes Massif is in Barrovian type greenschist-amphibolite facies and is probably caused by the deep burial of the Menderes Massif underneath the Lycian Nappes (Şengör and Yılmaz 1981; Okay, unpublished data 1984). Sugözü blueschists of the Alanya area are lithologically and tectonically similar to the blueschists of the external HP/LT belt of the Hellenides (Seidel and others 1982), although metamorphism is again older in the Alanya area.

It is remarkable that in the area of the Aegean and western Turkey, there have been distinct, southward younging HP/LT metamorphic events of Triassic (Karakaya Complex, ?Elekdağ blueschists), mid-Cretaceous (Tavşanlı Zone, Sugözü blueschists), Eocene (Cyclades Belt) and Miocene (external HP/LT belt of the Hellenides) ages, whereas blueschists are scarce in Yugoslavia, western Greece, and eastern Turkey. However, each of these metamorphic events probably marks a critical point in the plate tectonic evolution of the area; they are not a result of the continuous activity of a single subduction zone.

This paper summarizes the features of the presently known HP/LT metamorphic complexes in Turkey. However, there are large metamorphic areas in Anatolia where the petrology and facies type of the rocks are very poorly known; further detailed studies in such areas will certainly reveal new HP/LT metamorphic complexes. It will take several tens of years before the petrology of the Turkish metamorphic rocks are known in satisfactory detail.

ACKNOWLEDGMENTS

Ideas and information on the blueschists of Turkey have been gathered throughout several years by my association with C. Göncüoğlu, N. Özgül, A.M.C. Şengör, O. Tekeli, and Y. Yılmaz in the field and in the office. The author is also indebted to A.M.C. Şengör, who has reviewed this manuscript and provided many helpful suggestions.

REFERENCES CITED

Akdeniz, N., and Konak, N., 1979, Simav, Emet, Tavşanlı, Dursunbey Demirci, Kütahya dolaylarının jeolojisi: Maden Tetkik Arama Enstitüsü, report no. 6547 (unpublished).

Altherr, R., Schliestedt, M., Okrusch, M., Seidel, E., Kreuzer, H., Harre, W., Lenz, H., Wendt, I., and Wagner, G. A., 1979, Geochronology of high-pressure rocks on Sifnos (Cyclades, Greece): Contributions to Mineralogy and Petrology, v. 70, p. 245–255.

Ataman, G., 1974, Revue geochronologique des massifs plutoniques et metamorphiques de l'Anatolie: Hacettepe Bulletin of Natural Sciences and Engineering, v. 3, p. 75–87.

Ayan, M., 1963, Contribution a l'etude petrographique et geologique de la region situee au Nord-Est de Kaman: Maden Tetkik Arama Enstitüsü Publications, no. 115, 332 p.

Ayaroğlu, H., 1979, Bozüyük metamorfitlerinin petrokimyasal özellikleri: Türkiye Jeoloji Kurumu Bülteni, v. 22, p. 101–107.

Aydın, Y., 1974, Etude petrographique et geochimique de la partie centrale du massif d'Istranca (Turquie) [Ph.D. thesis]: University of Nancy, France, 131 p.

Başarır, E., 1970, Bafa Gölü doğusunda kalan Menderes Masifi güney kanadının jeolojisi ve petrografisi: Ege Üniversitesi Fen Fakültesi İlmi Raporlar Serisi, no. 102, 42 p.

Batman, B., 1978, Haymana kuzeyinin jeolojik evrimi ve yöredeki melanjin incelenmesi 1: stratigrafi birimleri: Yerbilimleri, v. 4, p. 95–124.

Bayiç, A., 1968, On metaporphyrites of the Sizma region (Konya): Bulletin of the Mineral Research and Exploration Institute of Turkey, v. 70, p. 142–156.

Bernoulli, D., Graciansky, P. Ch. de, and Monod, O., 1974, The extension of the Lycian nappes (SW Turkey) into the southeastern Aegean islands: Eclogae Geologicae Helvetiae, v. 67, p. 39–90.

Biju-Duval, B., Dercourt, J., and Le Pichon, X., 1977, From the Tethys ocean to the Mediterranean seas: a plate tectonic model of the evolution of the western Alpine System, in Biju-Duval, B., and Montadert, 1., ed., Structural history of the Mediterranean basins: Paris, Editions Technip, p. 143–164.

Bingöl, E., 1976, Evolution geotectonique de l'Anatolie de l'Ouest: Bulletin de la Societe Geologique de France, ser. 7, v. 18, p. 235–254.

—— 1978, Explanatory notes to the metamorphic map of Turkey, in Zwart, H. J., ed., Explanatory text for the metamorphic map of Europe: Leyden, p. 348–354.

Bingöl, E., Akyürek, B., and Korkmazer, B., 1975, Geology of the Biga peninsula and some characteristics of the Karakaya blocky series, in, Congress of earth sciences on the occasion of the fiftieth anniversary of the Turkish Republic: Ankara, Maden Tetkik Arama Enstitüsü, p. 71–77.

Blake, M. C., Jr, Bonneau, M., Geyssant, J., Kienast, J. R., Lepvrier, C., Maluski, H., and Papanikolaou, D., 1981, A geological reconnaissance of the Cycladic blueschist belt, Greece: Geological Society of America Bulletin, v. 92, p. 247–254.

—— 1984, A geological reconnaissance of the Cycladic blueschist belt, Greece: Reply: Geological Society of America Bulletin, v. 95, p. 119–121.

Blumenthal, M., 1951, Recherches geologiques dans le Taurus occidental dans l'arriere-pays d'Alanya: Maden Tetkik Arama Enstitüsü Publications, no. D5, 134 p.

—— 1955, Geologie des hohen Bolkardağ, seiner nördlichen Randgebiete und westlichen Auslaufer: Maden Tetkik Arama Enstitüsü Publications, no. D7, 153 p.

Boray, A., 1975, Bitlis dolayının yapısı ve metamorfizmasi: Türkiye Jeoloji Kurumu Bülteni, v. 18, p. 81–84.

Brinkmann, R., 1976, Geology of Turkey: Stuttgart, Ferdinand Enke Verlag, 158 p.

Brunn, J. H., Dumont, J. F., Graciansky, P. C. de, Gutnic, M., Juteau, T., Marcoux, J., Monod, O., and Poisson, A., 1971, Outline of the geology of the western Taurides, in Campbell, A. S., ed., Geology and history of Turkey: Tripoli, Libya, Petroleum Exploration Society of Libya, p. 225–255.

Bryhni, I., Krogh, E. J., and Griffin, W. L., 1977, Crustal derivation of Norwegian eclogites: a review: Neues Jahrbuch für Mineralogie, Abhandlungen, v. 130, p. 49–68.

Çağlayan, A., Öztürk, E. M., Öztürk, Z., Sav, H., and Akat, U., 1980, Menderes Masifi güneyine ait bulgular ve yapısal yorum: Jeoloji Mühendisliği, v. 10, p. 9–17.

Çalapkulu, F., 1980, Horoz granodiyoritinin jeolojik incelemesi: Türkiye Jeoloji Kurumu Bülteni, v. 23, p. 59–68.

Campbell, A. S., 1971, Geology and history of Turkey: Tripoli, Libya, Petroleum Exploration Society of Libya, 511 p.

Çoğulu, E., 1967, Etude petrographique de la region de Mihaliçcik: Schweizerische Mineralogisch und Petrographische Mitteilungen, v. 47, p. 683–824.

Çoğulu, E., and Krummenacher, D., 1967, Problemes geochronometriques dans la partie N de l'Anatolie Centrale (Turquie): Schweizerische Mineralogisch und Petrographische Mitteilungen, v. 47, p. 825–833.

Coleman, R. G., Lee, D. E., Beatty, L. B., and Brannock, W. W., 1965, Eclogites and eclogites: their differences and similarities: Geological Society of America Bulletin, v. ;76, p. 483–508.

Dora, Ö, 1981, Menderes Masifinde petroloji ve feldispat incelemeleri: Yerbilimleri, v. 7, p. 54–63.

Dumont, J. F., Gutnic, M., Marcoux, J., Monod, O., and Poisson, A., 1972, Le Trias des Taurides occidentales (Turquie). Definition du bassin pamphylien: Un nouveau domaine a ophiolithes a la marge externe de la chaine taurique: Zeitschrift der Deutschen Geologischen Gesellschaft, v. 123, p. 385–409.

Dürr, S., Altherr, R., Keller, J., Okrusch, M., and Seidel, E., 1978, The Median Aegean Crystalline Belt: stratigraphy, structure, metamorphism, magmatism, in Closs, H., Roeder, D., and Schmidt, K., eds., Alps, Appenines and Hellenides: Stuttgart, Schweizerbart, p. 455–476.

Eren, R. H., 1979, Kastamonu–Taşköprü bölgesi metamorfitlerinin jeolojik ve petrografik etüdü [Ph.D. thesis]: Istanbul, Istanbul Technical University, 143 p.

Erkan, E., Özer, S., Sümengen, M., and Terlemez, İ., 1978, Sarız-Şarkışla-Gemerek-Tomarza arasının temel jeolojisi: Maden Tetkik Arama Enstitüsü, Report no. 5646, (unpublished).

Erkan, Y., 1976, Kirşehir çevresindeki rejyonal metamorfik bölgede saptanan isogradlar ve bunların petrolojik yorumlanmalari: Yerbilimleri, v. 2, p. 23–54.

Erkan, Y., and Ataman, G., 1981, Orta Anadolu Masifi (Kirşehir yöresi) metamorfizma yaşı üzerine K-Ar yöntemi ile bir inceleme: Yerbilimleri, v. 8, p. 27–30.

Ernst, W. G., 1970, Tectonic contact between Franciscan Melange and the Great Valley sequence—crustal expression of a Late Mesozoic Benioff zone: Journal of Geophysical Research, v. 75, p. 886–901.

Evans, I., Hall, S. A., Carman, M. F., Şenalp, M., and Çoşkun, S., 1982, A paleomagnetic study of the Bilecik Limestone (Jurassic), northwestern Anatolia: Earth and Planetary Science Letters, v. 61, p. 199–208.

Frey, M., Hunziker, J. C., Frank, W., Bocquet, J., Dal Piaz, G. V., Jäger, E., and Niggli, E., 1974, Alpine metamorphism of the Alps, a review: Schweizerische Mineralogische und Petrographische Mitteilungen, v. 54, p. 247–290.

Gautier, Y., 1984, Deformations et metamorphismes associes a la fermeture Tethysienne en Anatolie Centrale (region de Sivrihisar, Turquie) [Ph.D. thesis]: Centre d'Orsay, Universite de Paris-Sud, 235 p.

Göncüoğlu, C., 1977, Geologie des westlichen Niğde Massivs [Ph.D. thesis]: University of Bonn, 181 p.

Göncüoğlu, C., and Turhan, N., 1983, New ages from the Bitlis Massif: Bulletin of the Mineral Research and Exploration Institute of Turkey, v. 95/96, p. 44–49.

Graciansky, P. C., de, 1966, Le Massif cristallin du Menderes (Taurus occidental, Asie Mineure). Un exemple possible de vieux socle granitique remobilise: Revue de Geographie Physique et de Geologie Dynamique, v. 8, p. 289–306.

——1972, Recherches geologiques dans le Taurus Lycien [Ph.D. thesis]: Centre d'Orsay, Universite de Paris-Sud, 762 p.

Hall, R., 1976, Ophiolite emplacement and the evolution of the Taurus suture zone, southeastern Turkey: Geological Society of America Bulletin, v. 87, p. 1078–1088.

——1980, Unmixing a melange: the petrology and history of a disrupted and metamorphosed ophiolite, southeastern Turkey: Journal of the Geological Society of London, v. 137, p. 195–206.

Hamilton, W., 1969, Mesozoic California and the underflow of Pacific mantle: Geological Society of America Bulletin, v. 80, p. 2409–2430.

Hecht, J., 1972, Zur Geologie von Südost-Lesbos (Griechenland): Zeitschrift der Deutschen Geologischen Gesellschaft, v. 123, p. 423–432.

Juteau, T., 1970, Petrogenese des ophiolites des nappes d'Antalya (Taurus lycien oriental, Turquie): Science de la Terre, v. 15, p. 265–288.

Kaaden, G. van der, 1966, The significance and distribution of glaucophane rocks in Turkey: Bulletin of the Mineral Research and Exploration Institute of Turkey, v. 67, p. 37–67.

Kaaden, G. van der, and Metz, K., 1954, Beiträge zur Geologie des Raumes zwischen Datça-Muğla-Dalaman Çay (SW Anatolien): Türkiye Jeoloji Kurumu Bülteni, v. 5, p. 71–170.

Kaya, O., 1972, Aufbau und Geschichte einer anatolischen Ophiolith Zone: Zeitschrift der Deutschen Geologischen Gesellschaft, v. 123, p. 491–501.

——1981, Preliminary study on the paragenetic relationships in the polymetamorphic blueschist rocks of the Tavşanlı area, West Anatolia: Aegean Earth Sciences, v. 1, p. 27–43.

Ketin, I, 1966, Tectonic units of Anatolia: Bulletin of the Mineral Research and Exploration Society of Turkey, v. 66, p. 23–34.

——1983, Türkiye jeolojisine genel bir bakış: Istanbul Teknik Üniversite Matbaasi, 596 p.

——1984, Türkiye'nin bindirmeli-naplı yapısında yeni gelişmeler ve bir örnek: Uludağ Masifi, in Proceedings, Ketin Symposium, Ankara, February 1984 (19–36).

Koçyiğit, A., 1979, Tekneli bölgesinin (Tokat güneyi) tektonik özelliği: Türkiye Bilimsel Teknik Araştırma Kurumu, Report no. TBAG-262.

Kopp, K., 1964, Geologie Thrakiens 2: Die Inseln und der Chersones: Neues Jahrbuch für Geologie und Paläontologie, Abhandlungen, v. 119, p. 172–214.

Kulaksız, S., 1978, Sivrihisar kuzeybatı yöresi eklojitleri: Yerbilimleri, v. 4, p. 89–94.

Lefevre, R., 1967, Un nouvel element de la geologie du Taurus Lycien: les nappes d'Antalya (Turquie): Comptes Rendus des Seances de l'Academie des Sciences Paris, v. D265, p. 1365–1368.

Lisenbee, A., 1971, The Orhaneli ultramafic-gabbro thrust sheet and its surroundings, in Campbell, A. S., ed., Geology and history of Turkey: Tripoli, Libya, Petroleum Exploration Institute of Libya, p. 349–360.

Marcoux, J., 1978, A scenario for the birth of a new oceanic realm: the Alpine Neotethys: International Congress of Sedimentology, 10th, Jerusalem, Abstracts, v. 2, p. 419–420.

Milch, L., 1907, Über Glaukophan und Glaukophangesteine von Elek Dağ (nördliches Kleinasien) mit Beiträgen zur Kenntnis der chemischen Beziehungen basischer Glaukophangesteine: Neues Jahrbuch für Mineralogie, Geologie und Paläontologie, Festband (1907), p. 348–396.

Okay, A. I., 1980a, Mineralogy, petrology and phase relations of glaucophane-lawsonite zone blueschists from the Tavşanlı region, Northwest Turkey: Contributions to Mineralogy and Petrology, v. 72, p. 243–255.

——1980b, Lawsonite zone blueschists and a sodic amphibole producing reaction in the Tavşanlı region, Northwest Turkey: Contributions to Mineralogy and Petrology, v. 75, p. 179–186.

——1980c, The petrology of blueschists in Northwest Turkey, northeast of Tavşanlı [Ph.D. thesis]: University of Cambridge, 242 p.

——1982, Incipient blueschist metamorphism and metasomatism in the Tavşanlı region, Northwest Turkey: Contributions to Mineralogy and Petrology, v. 79, p. 361–367.

——1984a, Distribution and characteristics of the northwest Turkish blueschists, in Robertson, A.H.F., and Dixon, J. E., eds., The geological evolution of the eastern Mediterranean: Geological Society of London Special Publication 17, p. 455–466.

——1984b, Kuzeybati Anadolu'da yer alan metamorfik kuşaklar, in Proceedings, Ketin Symposium, Ankara, February 1984 (83–92).

——1984c, The geology of the Ağvanis metamorphic rocks and neighbouring areas: Bulletin of the Mineral Research and Exploration Institute of Turkey, v. 99/100 (16–36).

Okay, A. I., and Özgül, N., 1984, HP/LT metamorphism and the structure of the Alanya Massif, Southern Turkey: an allochthonous composite tectonic sheet, in Robertson, A.H.F., and Dixon, J. E., eds., The geological evolution of the eastern Mediterranean: Geological Society of London Special Publication 17, p. 429–439.

Özgül, N., 1976, Toroslarin bazı temel jeoloji özellikleri: Türkiye Jeoloji Kurumu Bülteni, v. 19, p. 65–78.

——1984, Stratigraphy and tectonic evolution of the Central Taurides, in Proceedings, International Symposium on the Geology of the Taurus Belt, Ankara, September 1983 (77–90).

Papanikolaou, D., 1979, Unites tectoniques et phases de deformation dans l'ile de Samos, Mer Egee, Grece: Bulletin de la Societe Geologique de France, ser. 7, v. 21, p. 745–752.

Peringek, D., 1979, Interrelations of the Arab and Anatolian plates (1st Geological Congress of Middle East, guidebook for excursion B): Ankara.

Ridley, J., and Dixon, J. E., 1984, Reaction pathways during the progressive deformation of a blueschist metabasite: the role of chemical disequilibrium and restricted range equilibrium: Journal of Metamorphic Geology, v. 2, p. 115–128.

Robertson, A.H.F., and Woodcock, N. H., 1982, Sedimentary history of the southwestern segment of the Mesozoic-Tertiary Antalya continental margin, southwestern Turkey: Eclogae Geologicae Helvetiae, v. 75, p. 517–562.

Sağıroğlu, G., and Bürküt, Y., 1966, Sur l'age et la petrographie du massif d'Uludağ (Turquie): Compte Rendu des Seances de la Societe de Physique et d'Histoire Naturelle de Geneve, nouvelle serie, v. 1, p. 21–32.

Sanver, M., and Ponat, E., 1980, Palaeomagnetism of the magmatic rocks in Kırşehir and surrounding area: Publication of the Kandilli Observatory, Palaeomagnetism Department, Istanbul, 11 p.

Seidel, E., Kreuzer, H., and Harre, W., 1982, A Late Oligocene/Early Miocene high pressure belt in the External Hellenides: Geologisches Jahrbuch, v. E23, p. 165–206.

Şengör, A.M.C., 1979, The North Anatolian transform fault: its age offset and tectonic significance: Journal of the Geological Society of London, v. 136, p. 269–282.

Şengör, A.M.C., and Yılmaz, Y., 1981, Tethyan evolution of Turkey: a plate tectonic approach: Tectonophysics, v. 75, p. 181–241.

Şengör, A.M.C., Yılmaz, Y., and Ketin, I., 1982, Reply to Comment on "Remnants of a pre-Late Jurassic ocean in northern Turkey: fragments of Permian–Triassic Paleo-Tethys": Geological Society of America Bulletin, v. 93, p. 929–936.

Şengör, A.M.C., Yılmaz, Y., and Sungurlu, O., 1984a, Tectonics of the Mediterranean Cimmerides: Nature and evolution of the western termination of
Palaeo-Tethys, *in* Robertson, A.H.F., and Dixon, J. E., eds., The geological
evolution of the eastern Mediterranean: Geological Society of London Special Publication 17, p. 77–112.

Şengör, A.M.C., Satir, M., and Akkök, R., 1984b, Timing of tectonic events in the
Menderes Massif, western Turkey: evidence of Pan-African basement in
Turkey: Tectonics, 3, p. 693–707.

Şengün, M., Acarlar, M., Çetin, F., Doğan, Z. O., and Gök, A., 1978, Alanya
Masifinin yapısal konumu: Jeoloji Mühendisliği, v. 6, p. 39–45.

Şentürk, K., and Okay, A. I., 1984, Blueschists discovered east of Saros bay in
Thrace: Bulletin of the Mineral Research and Exploration Institute of Turkey, v. 97/98, p. 72–75.

Servais, M., 1981, Donnees preliminaires sur la zone de suture mediotethysienne
dans la region d'Eskişehir (NW Anatolie): Comptes Rendus des Seances de
l'Academie des Sciences Paris, v. 293, p. 83–86.

——1982, Collision et suture tethysienne en Anatolie Centrale, etude structurale
et metamorphique (HP-BT) de la zone nord Kütahya [Ph.D. thesis]: Centre
d'Orsay, Universite de Paris-Sud, 374 p.

Seymen, İ., 1981, Karman (Kırşehir) dolayında Kırşehir Masifi'nin stratigrafisi ve
metamorfizmasi: Türkiye Jeoloji Kurumu Bülteni, v. 24, p. 101–108.

——1982, Kaman dolayında Kırşehir Masifinin jeolojisi (thesis): Istanbul, İstan-
bul Technical University, 164 p.

Tatar, Y., 1968, Geologie und Petrographie des chromitführenden Marmaris
Gebietes (SW Turkei): Maden Tetkik Arama Enstitüsü Publications,
no. 137, 92 p.

——1981, Çamlibel Geçiti (Yıldızeli) yöresindeki ofiyolitik seride metamorfizma:
Karadeniz Teknik Üniversitesi Yer Bilimleri Dergisi, Jeoloji, v. 1, p. 45–65.

Tekeli, O., 1981, Subduction complex of pre-Jurassic age, northern Anatolia,
Turkey: Geology, v. 9, p. 68–72.

Türkünal, S., 1980, Doğu ve güneydoğu Anadolu'nun jeolojisi: TMMOB Jeoloji
Mühendisleri Odası Publications, no. 8, 64 p.

Yılmaz, O., 1975, Cacas bölgesi (Bitlis Masifi) kayaçlarının petrografik ve strati-
grafik incelemesi: Türkiye Jeoloji Kurumu Bülteni, v. 18, p. 33–40.

Yılmaz, Y., 1979, Söğüt-Bilecik dolayındaki polimetamorfizma ve bunlarin jeo-
tektonik anlamı: Türkiye Jeoloji Kurumu Bülteni, v. 22, p. 85–101.

Yılmaz, Y., and Tüyzüz, O., 1984, Kastamonu-Boyabat-Vezirköprü-Tosya ara-
sındaki bölgenin jeolojisi (Ilgaz-Kargı Masiflerinin etüdü): Maden Tetkik
Arama Enstitüsü, report no. 7856.

Zankl, H., 1961, Magmatismus und Bauplan des ostpontischen Gebirges im
Querprofil des Harşit Tales, NE Anatolien: Geologisches Rundschau, v. 51,
p. 218–235.

MANUSCRIPT ACCEPTED BY THE SOCIETY JULY 29, 1985

Printed in U.S.A.

Geological Society of America
Memoir 164
1986

Eclogites from various types of metamorphic complexes in the USSR and the problems of their origin

N . V. Sobolev
Institute of Geology and Geophysics
Siberian Division of the USSR Academy of Sciences
Novosibirsk 630090, USSR

N. L. Dobretsov
Geological Institute of the Buryat Branch
Siberian Division of the USSR Academy of Sciences
Ulan-Ude 670015, USSR

A. B. Bakirov
Institute of Geology
Academy of Sciences of the Kirgiz SSR
Frunze 720000, USSR

V. S. Shatsky
Institute of Geology and Geophysics
Siberian Division of the USSR Academy of Sciences
Novosibirsk 630090, USSR

ABSTRACT

Eclogites in metamorphic complexes of the USSR are of different types with regard to their tectonic conditions and compositions. They are: 1) tectonic inclusions in serpentinite melanges; 2) members of eclogite-glaucophane schist complexes; 3) inclusions in blastomylonite zones; and 4) those within gneiss complexes. Based upon the study of their relationships with country rocks, chemical zoning of coexisting garnets and omphacites, and compositions of gaseous-liquid inclusions in minerals, a polygene character of the eclogites has been shown. P-T equilibrium parameters for eclogites associated with glaucophane schists are commonly close to those of the latter. Eclogites of gneiss complexes tend to be characterized by variable conditions of formation, even within a single complex.

INTRODUCTION

Eclogites of the earth's crust are still mysterious rocks. In spite of intense study in recent years, the problem of their origin is not yet solved. Various hypotheses over the last 50 years are getting unexpectedly strong support. Much evidence has been collected on the mantle origin of eclogites especially from the high-temperature complexes, though the mechanism of their transport from the mantle is still uncertain. On the other hand, eclogites associated with glaucophane schists are more likely formed together with the glaucophane schists derived from crust-al rocks in subduction zones. However, the conditions of rapid upward transport and preservation of eclogites, as well as those factors responsible for sharp boundaries between eclogites and schists, still remain problematic. More and more evidence appears in favor of the fact that eclogites are nearly always not equilibrated with the surrounding host rocks and are constantly subjected to diaphthoresis. These retrograde conditions correspond to the metamorphic grade of the host rock. These facts are usually explained by special "dry" conditions of metamorphism of eclogites or to their magmatic intrusion during high-pressure metamorphism. The latter hypothesis, proposed by the authors about 15 years ago (Dobretsov and Sobolev 1970), seems to explain the mantle origin of eclogites (or primary magma) and conditions of their transport, dryness and non-equilibrium with the country rocks at high pressure being a necessary condition. But this pattern also seems weakly related with the more popular subduction model.

A means of solving these problems generally can be directed toward studying the two most important peculiarities of eclogites: 1) Their structural position and tectonic condition of transport, and 2) peculiarities in their composition and conditions of crystallization (metamorphism) versus those of the surrounding country rocks. These problems are clarified only for inclusions in kimberlites, though there are many unsolved problems with these rocks (Sobolev 1977).

In this paper we make an attempt to discuss these problems as exemplified by the typical eclogite-bearing complexes of the USSR. Unlike another recent review paper (Dobretsov and Sobolev, 1984) devoted to glaucophane schists and associated eclogites, we have also drawn attention to eclogites associated with high-temperature rocks. Many geological schemes and cross sections of the regions studied are included in the previous publication (Dobretsov and Sobolev 1970, 1984; Dobretsov et al. 1974).

ECLOGITE CLASSIFICATION

As has been done earlier (Dobretsov and Sobolev 1984), it is reasonable to use an association of eclogites whose tectonic conditions and compositions are similar.

1. a) Tectonic inclusions in serpentinitic melange, b) in tectonic melange of Franciscan type, or c) olistolites in olistostromes; all these are generally associated with glaucophane schists.

2. a) Eclogite-glaucophane-schist complexes in which eclogites display a regional glaucophane metamorphism or b) are metamorphosed (glaucophanic) olistostromes. Some intermediate cases exist between types 1 and 2.

3. a) Inclusions of eclogites in blastomylonite zones of the low and b) middle temperature facies. In several cases, eclogites of this type contain glaucophane but more often the diaphthoritic minerals correspond to greenschist or kyanite-staurolite facies.

4. Eclogites from gneiss complexes that occur regionally within extended areas. Complexes of this sort are most mysterious since one can hardly invoke tectonic transport within large overthrusting zones. Such tectonic transport seems quite probable for the complexes 1–3.

The complexes of the first type are found to occur most often in the USSR, but they do not form large belts. They are found along the principal tectonic lines in the Pacific area, Koryakia (Markov et al. 1982), the Urals, south Tien-Shan, and the Altai-Sayan region (Dobretsov 1974). Typical examples are the Boruss and Chara belts (see below). The same situation exists for eclogite-glaucophane complexes in the Urals and South Tien-Shan; where typical examples include the Maksyutov Complex in

the South Urals and the Pike River Complex in the Polar Urals. In the latter occurrence there is an uncommon case of transition to higher temperature rock series. The inclusions of eclogites in the blastomylonite zones (type 3) are typical of the Tien-Shan belts, in particular the Atbashy and Makbal complexes. Finally, eclogites from gneiss complexes (type 4) have been found in North and South Kazakhstan (Zerenda and Aktyuz series) as well as in the Mui block in the North Near-Baikalia region.

REGIONAL TECTONIC POSITION AND AGE

With the exception of Koryakia, all the above examples belong to the Urals-Mongolian Paleozoic fold system. According to modern concepts (Zonenshain et al. 1974; Dobretsov et al. 1984), this fold system formed at the site of the Paleo-Asiatic ocean that existed in the Riphean Early Paleozoic and disappeared during the Caledonian epoch (Silurian) and Hercynian time (Carboniferous). The occurrence of the major eclogite and glaucophane schist bearing complexes in the Urals-Mongolian region is shown in Figure 1.

In the Urals, all the eclogite and glaucophane schist localities are confined to the main Urals Overthrust Zone, as determined by the large-scale outcrops of ophiolite rocks of Riphean and Ordovician-Silurian age (Dobretsov et al. 1974). The numbers 1–8 in Figure 1 are as follows:

1. The Eclogite-glaucophane complex, Pike River (Syum-Keu belt).

2. The Voikar belt: zones of serpentine melange with inclusions of glaucophane schists, eclogites, and jadeitites similar to the Boruss belt (see below); also slices/layers of glaucophane schists in the basement of ophiolitic cover.

3. The Kharuta-Pe belt with glaucophane schists formed during metamorphism of Ordovician rocks.

4. The Salatim belts including serpentine melange zone and glaucophane rock series (plate?) situated west of it and structurally below the melange zone.

5. According to our data the eclogite bearing Ufaley complex eclogites are confined to blastomylonite zones along the contact of kyanite-staurolite zone schists with a gneiss core of sillimanite zone schists.

6. The Maksyutov eclogite-glaucophane schist complex.

7. Glaucophane schists and eclogite-like rocks enclosed in the Khabarnin Massif.

8. Eclogites from blastomylonite zones in Mugodzhary.

The age of glaucophane schists and eclogite rocks from many Uralian complexes is considered to be Precambrian (Pike River, Ufaley, Maksyutov, and Mugodzhary complexes), which is supported by some Rb-Sr dating yielding 1100–1200 m.y. ages (Dobretsov 1974; Dobretsov and Sobolev 1984).

In some other cases, the age of metamorphism of the glaucophane schists (from Kharuta-Pe, Salatim, Khabarnin belts) was dated as Late Silurian–Early Devonian.

In South Tien-Shan (Numbers 9–14, Figure 1), most of the complexes with glaucophane schists and eclogite rocks are con-

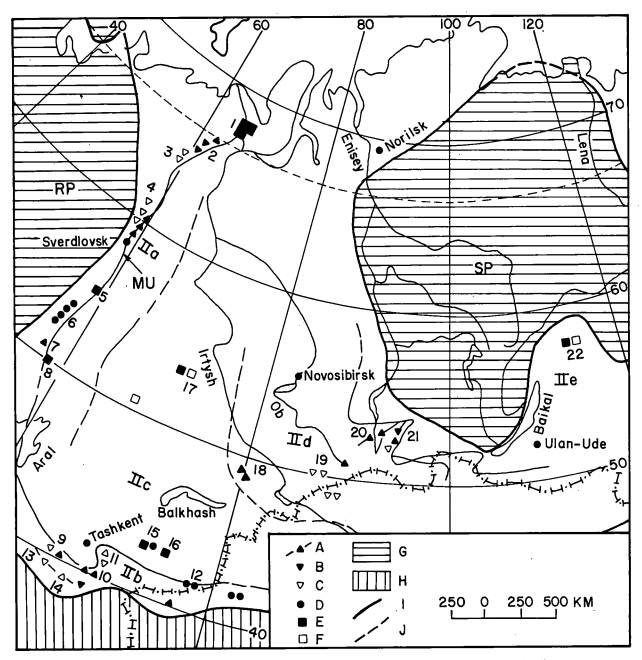

Figure 1. The glaucophane schists and eclogites of the Ural-Mongolia system; see text for locality numbers. A-F: The places of the outcrops of the glaucophane schists and eclogites (A) in a melange; (B) glaucophane schist at the base of an ophiolite nappe; or (C) without close relations to nappes; (D) eclogite-glaucophane schist complexes; (E) eclogites in the blastomylonite zones; (F) eclogites in gneisses (see text); (G) the Russian platform (RP) and the Siberian platform (SP); (H) the Mesozoic and Alpine folded system with blocks of the ancient continental crust; (I) the boundaries of the eugeosyncline zones - the main structural lines (MU - the main Ural line); (J) the approximate boundaries of the Ural (II a), Kazakhstan (II c), and Altai-Sayan (II d) folded systems. IIb - Southern Tien Shan and IIe-Precambrian Baikal folded system.

fined to the border zone with the Middle Tien-Shan which occupies the same tectonic position as the Main Uralian zone of deep-seated overthrusts.

9. The North Nura-Tau crossite-bearing Ittunsai rock series underlain by serpentinitic melange.

10. The Kansk serpentinitic melange with inclusions of glaucophane schists.

11. The Mailisui complex-plate(?) of glaucophane schists associated with ophiolite covers.

12. The Atbashy belt containing zones of tectonic melange

and blastomylonites with inclusions of eclogites and glaucophane rocks.

Also, inside the Zeravshan geosyncline there are tectonic slices of crossite-bearing metamorphic schists in Katarmai and Anzob (13 and 14 in Figure 1). The age problem of these rocks is the same as that of the Urals. Two stages of eclogite-glaucophane metamorphism are suggested: the Riphean (about 1100 m.y.) and Silurian (about 400 m.y.).

In North Tien-Shan and Kazakhstan (Numbers 15–17, Figure 1) all the eclogite localities are confined to the blocks of Precambrian basement.

15. The Makbal complex with blastomylonite zones containing eclogites and glaucophane rocks.

16. The Aktyuz ridge of the Muyunkum block with numerous eclogites from the Aktyuz gneiss series.

17. The Kokchetav block with numerous eclogites in the most ancient Zerenda gneiss series.

In the Altai-Sayan region (Numbers 18–21, Figure 1) all the glaucophane schist and eclogite occurrences are most closely associated with ophiolite belts and zones of serpentine melange.

18. The Chara belt with melange containing eclogites and glaucophane schists.

19. The Uimons belt composed of a glaucophane schist suite and serpentinite bodies.

20. The Boruss belt of melange containing inclusions of eclogites associated with jadeitic rocks and glaucophane schists.

21. The Kurtushibin ophiolitic belt underlain by a zone of glaucophane schists containing sparse eclogite occurrences in the lower melange zones.

22. Mui block in the Near-Baikalia region. Eclogite inclusions in blastomylonite zones and scarce eclogite bodies in gneisses.

The age of most of these eclogites and glaucophane schists is considered Precambrian (Riphean), as fragments are found in the Middle Cambrian olistostrome (in the Boruss and Kurtushibin belts), while the Uimon belt is intruded by 600 m.y. granites.

ECLOGITES AND ASSOCIATED ROCKS FROM SERPENTINITIC MELANGE

The Boruss belt in West Sayan (20 in Figure 1) is a typical example of these rocks and is very important because of the close association of eclogite with monomineralic jadeitic rocks (Dobretsov and Tatarinov 1983).

The Boruss belt represents a series of nappes. The upper slice is composed of ultrabasites and is underlain by melange of the first type with inclusions of the deep-seated rocks. Among the inclusions are monomineralic jadeitites with margins of diopside-jadeitic or garnet-amphibole rocks and eclogites with margins of diopside-jadeitic rocks surrounded by small jadeite pods (Figure 2). This relationship indicates that the jadeitites were formed during Na-metasomatism of the eclogites and were isofacial with them. These inclusions underwent multi-stage diaphthoresis with the formation of albite, cancrinite, analcime, and Cr-jadeite. They

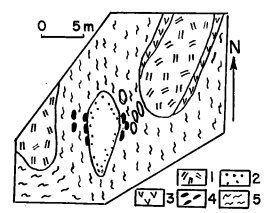

Figure 2. Map showing the inclusions of eclogites and jadeitic rocks in the Boruss melange. 1 - mica and albite-mica rocks; 2 - eclogites, 3 - garnet amphibolites, 4 - jadeite rocks, 5 - serpentinite schists.

are associated with late inclusions of crossite-albite rocks, including rocks with the association albite + jadeite + analcime + crossite (Dobretsov and Tatarinov 1983). The compositions of the pyroxenes (Figure 3, and Table 1) include practically all the fields of the $NaAlSi_2O_6$ - $Na(Cr, Fe^{+3})Si_2O_6$ - $Ca(Mg,Fe)Si_2O_6$ system. For stage 1, the pyroxenes of the series jadeite-diopside are typical in having low contents of Na-Fe and Na-Cr components with some immiscibility between jadeite and omphacite. This immiscibility is due to the different structure of the ordered omphacite (P2/n unlike C/2c in jadeite) and may disappear with increasing temperature and Fe^{+3}-Cr contents (Rossi et al. 1983; Smith et al. 1980). The Cr source came from intermixing with Cr-rich ultrabasites at the diaphthoritic stage 2, when jadeite-kosmochlor and more scarce jadeite-aegirine pyroxenes appear in the margins and veins of the bodies. The aegirine-bearing pyroxenes are more typical of the diaphthoritic stages 3 and 4, whereas Cr enters mostly into the Cr-chlorites or more rarely into Ca-Cr garnets and amphiboles.

According to our temperature and pressure estimates all conditions accompanying the rock formation and transformation in the first type melange zones correspond to the oceanic geotherm. Most likely, this means that the abyssal thrusting occurred in oceanic or intermediate environments, while the continental crust, where the present Boruss ophiolite zone is situated, had developed only during the final stage after crustal thickening. Because of these stages (3 and 4) within the zone of the first type melange we can observe the combination of the low and high-pressure associations.

In the Chara belt (Number 18, Figure 1) the serpentinite melange of the first type, with inclusions of eclogites and glaucophane schists, is placed at the foot of the frontal part of the cover. The cover is composed presumably of the rocks of the upper part of the ophiolitic section (diabases, basalts, siliceous rocks). The latter contain radiolaria of Late Ordovician–Early Silurian age (Ermolov et al. 1981). In the basement of some other slices composed of island arc basalt series, the type 2 melange contains only inclusions of ophiolite rocks (peridotites, gabbros, metaba-

TABLE 1. REPRESENTATIVE MICROPROBE ANALYSES OF MINERALS FROM ECLOGITES AND ASSOCIATED ROCKS IN MELANGE ZONES

| | Boruss belt | | | | | | | | | | Chara belt | | | | |
| | K-57-7 (core) | | | K-57-12a (rim) | | 29/1 | | | 36 (core) | | 1422/1 | | 976/5 | | |
	Ga	Cpx	Hbl	Cpx	Amph	(rim 6) Cpx 6	(rim with Chr.) Cpx 1	Chr	Cpx 1	Cpx 2	Ga	Cpx	Ga	Cpx	Amph
SiO_2	38.45	53.8	54.3	55.9	51.6	54.7	52.8	-	59.4	58.9	39.2	54.5	39.0	52.8	47.0
TiO_2	0.20	0.13	0.36	0.13	0.10	0.10	n.d.	-	0.0	0.34	1.66	0.07	1.14	0.04	0.21
Al_2O_3	21.0	4.97	3.21	9.83	4.16	14.5	6.80	12.7	24.4	21.6	20.8	10.25	19.5	0.49	8.42
Cr_2O_3	0.01	0.0	0.0	0.01	0.03	0.29	19.7	42.4	0.0	0.0	0.02	0.0	0.02	0.0	0.0
Fe_2O_3	0.0	2.77	2.25	3.85	2.52	4.71	1.93	3.0	0.40	1.44	0.34	2.33	3.12	0.0	2.56
FeO	27.1	5.86	10.0	3.47	12.90	0.96	0.35	28.12	0.30	0.45	23.3	4.85	12.4	5.97	10.4
MnO	1.24	0.0	0.10	0.0	0.10	0.0	0.10	1.68	0.0	0.01	0.49	0.01	1.21	0.07	0.23
MgO	3.09	10.0	15.1	6.72	12.7	6.96	2.49	0.85	0.12	1.66	3.27	6.63	4.26	13.25	12.57
CaO	9.01	19.20	10.1	12.65	9.45	7.27	3.08	-	0.40	2.77	10.0	12.97	18.66	24.43	12.21
Na_2O	0.0	4.00	1.92	8.02	2.26	9.70	10.65	-	14.9	12.8	0.1	7.06	0.01	0.20	1.05
K_2O	0.0	0.02	0.12	0.0	0.21	0.0	0.05	-	0.0	0.08	0.01	0.0	0.01	0.01	0.20
H_2O	-	-	2.00	-	2.0	-	-	ZnO 4.98	-	-	-	-	-	-	(2.5)
Total	100.10	100.75	99.46	100.56	98.03	99.19	100.02	100.73	99.92	100.05	99.35	98.61	99.49	97.26	98.35
%Fe/Fe+Mg	88.1	32.0	30.9	36.7	40.0	29.6	32.2	-	66.7	42.9	78.1	36.7	58.0	20.2	36.1
%Jad	-	22.0	-	42.0	-	61.0	30.0	-	97.7	87	-	39.0	-	5.0	-
%Ca-comp.	15.3	76.0	78.2	48.3	75.1	27.1	12.5	-	2.0	10.0	30.2	-	53.0	94.0	91.0

TABLE 1. CONTINUED

| | Voikar belt | | | | | | |
| | | 73/4 | | 344B | | 527 | 265B |
	Ga	Cpx	Hbl	GA	Cpx	Cpx	Cpx
SiO_2	40.0	51.4	44.6	39.3	54.7	59.3	53.8
TiO_2	0.04	0.36	0.9	0.38	0.86	0.06	0.06
Al_2O_3	21.5	4.4	13.0	20.2	11.0	24.15	12.0
Cr_2O_3	0.0	0.0	0.0	0.04	0.05	0.0	0.0
Fe_2O_3	0.0	1.1	1.5	3.81	5.32	0.48	3.44
FeO	20.6	6.4	10.9	22.3	4.50	0.29	0.96
MnO	1.1	0.12	0.08	2.77	0.05	0.36	0.17
MgO	7.1	12.8	13.8	2.44	6.21	1.54	7.97
CaO	9.2	21.3	9.8	7.87	8.96	13.47	14.30
Na_2O	0.03	1.4	1.8	0.0	6.56	0.30	7.28
K_2O	0.0	0.0	0.0	0.0	0.31	-	0.20
H_2O	-	-	2.0	-	-	-	-
Total	99.57	99.08	99.38	99.11	98.52	99.96	100.18
%Fe/Fe+Mg	63.1	24.9	33.3	86.5	46.2	51	50.0
%Jad	-	-	-	-	-	94	-
%Ca-comp.	-	-	-	-	-	5	4

Boruss belt: sample K-57-7 - eclogite, central part of a body; sample K-57-12a - rim of an eclogite body; 29/1 - rim of a jadeitite body with chromite inclusions; 36 - central part of a jadeitite body (Dobretsov and Tatarinov 1983); Chara belt: sample 1422/1 - eclogite; sample 976/5 - eclogite-like rock - Ca-eclogite (Dobretsov et al. 1979). Voikar belt: sample 73/4 - eclogite-like rock from a large lens "Khord-Yus", 344B - glaucophanized eclogite at the foot of ophiolite, Kech-pele river; sample 527 and 265B - from jadeite inclusions (Dobretsov 1974).

Sobolev and Others

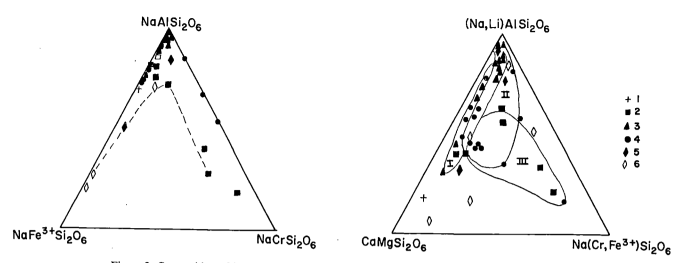

Figure 3. Composition of jadeitic pyroxenes of the Boruss melange rocks: 1 - eclogites, 2 - jadeitic margins, 3 - jadeitites of first stage, 4 - jadeitic rocks and rims of second stage, 5 - albitites and 6 - metasomatic rocks of third stage of diaphthoresis.

salts, plagiogranites, amphibolites, rodingites). Fragments of this melange, including glaucophane schists, eclogites, and minerals such as garnet and omphacite, were found to occur in an olisto-strome of Namurian age underlying the ophiolitic nappes.

A wide range of rock types occurs in the type 1 serpentine melange, which is similar to the Boruss belt. The largest blocks (1 km in diameter) are composed of low temperature schists, glaucophane-bearing greenschists, or more seldom lawsonite-glaucophane schists. Higher temperature rocks such as eclogites, garnet amphibolites, garnet-glaucophane, coarse grained garnet-muscovite-glaucophane rocks, muscovite-crossite and riebeckite quartzites also are found to occur as small, rounded bodies 1–30 m in diameter. The set of these rocks and compositions of the typical minerals (Table 1) are similar to the Uralian complexes from the Pike River and Maksyutov.

In the Voikar belt (Number 2, Figure 1) the first type melange zone occurs at the foot of the large Ray-Iz ultrabasic sheet and inside the ultrabasites of the Voikar Massif (Dobretsov 1974; Kazak 1980). The melange contains inclusions of peridotites, pyrope pyroxenites, eclogites, jadeite rocks, vesuvianite rocks,

glaucophanites and rodingites (Ray-Iz Massif), and jadeitites, albitites, and garnet amphibolites (Voikar Massif). These zones and jadeitites associated with eclogites (Table 1) are very similar to the above-described melange zone of the Boruss belt. In all the above-described cases more deep-seated rocks such as eclogites, jadeitites, and garnet amphibolites were more abundant than the underlying glaucophane schists. Both these rock groups are united by a common oceanic geotherm of different depths (and, in some cases, different formation stages) of the zone of the multistage obduction.

ECLOGITE-GLAUCOPHANE SCHIST COMPLEXES

The Maksyutov belt of the South Urals is a typical example of eclogite-glaucophane schist formation with maximum parameters of metamorphism: P up to 16 kbars at T = 500-550°C (Table 2) (Dobretsov and Lavrentjev 1978). The belt extends in the meridional direction more than 250 km, containing a series of slices inclined to the west. At the foot of the belt a series of sheets of weakly metamorphosed ophiolite rocks of Riphean–Lower

TABLE 2. P-T PARAMETERS OF ECLOGITES EQUILIBRATION

Complexes	P (k bars)	T (°C)
Maksyutov	>15	550-600
Atbashi	>13-14	500-600
Makbal	>12-14	450-600
Kumdy-Kul	>14-16	750-880
Kulet	>14	580-640
Sulu-Tyube	>14	600-650
Enbek-Berlyk	>13	760
Aktyuz	12	480-530

T - estimates after Ellis and Green (1979); P - on jadeite content in omphacite, using data by Kushiro (1969) and Holland (1980).

Paleozoic age are found. The western contact with the younger Suvenyak formation (Pz₁) is of tectonic nature, while in several places between these two formations wedge-shaped bodies of metadiorites with garnet are present.

Inside the Maksyutov complex among the dominating meta-sedimentary schists (metagreywackes, metacherts, meta-arkoses, and metagritstones), lenses and boudins of ultrabasites and eclogites are present. Some sections, in particular, the section near the River Sakmara (see Dobretsov and Sobolev 1970, 1975) reminds one of metamorphosed olistostromes or metamorphosed tectonic melanges of the Franciscan type. Here, among the highly tectonized micaceous schists are intercalations of metacherts including crossite, white mica, jadeite pyroxene, pseudomorphs after lawsonite, boudins, and large lenses of eclogites and metaperidotites. The metaperidotites with coarse-grained radiate-fibrous enstatite include the enstatite + diopside + olivine assemblage, formed at 600°C. In the eclogites, side-by-side with the almandine + omphacite + lawsonite and almandine + omphacite + glaucophane + epidote assemblages are relics of the higher temperature assemblages (the cores of garnet and clinopyroxene with T = 600°C). These temperatures are higher than those that caused metamorphism of the country rock schists (T = 450–500°C with P = 13–14 kbars up to 16 kbars for jadeite + quartz + almandine + mica assemblage). Lennykh (1977) considered apparent high-temperature assemblages as relics of the eclogite facies. The situation here is probably similar to the Franciscan complex in California (Coleman 1967; Hsu 1968; Blake and Jones 1974) but the main mass of schists is more intensely metamorphosed and corresponds to the coarse-grained inclusions in the Franciscan complex. The inclusions in the Maksyutov complex correspond to the most abyssal inclusions of the Franciscan complex discovered on the island of Santa Catalina (Sorensen this volume).

In mineralogical-petrological aspects, as well as structurally, the Maksyutov complex combines the peculiarities of Franciscan and Sesia-Lanzo-Pennine complexes in the Alps. They are the most striking representatives of eclogite-glaucophane schists, and are highly sialic in composition also (non-oceanic).

Many stages of metamorphism and deformation established for the Maksyutov complex (Miller 1977; Moskovchenko 1982) differ sharply from the neighboring formations, suggesting a more ancient age or an allochthonous position. The age of metamorphism is disputable. The U-Th-Pb dating using zircon from the metavolcanics has shown an age of 1200 m.y. (A. A. Krasnobayev personal communication), which is similar to the Rb-Sr isochron of metavolcanites 1200 m.y. (unpublished data by the authors); the K-Ar data correspond to 400–500 m.y. (Dobretsov 1974; Lennykh 1977).

In the Atbashy Range the eclogite-bearing complex is represented by the Choloktor formation (Bakirov 1984), composed mainly of garnet-muscovite-quartz, glaucophane-garnet-muscovite-quartz schists and diaphthorites of these rocks, also by rare marble wedges. The eclogitic boudins and lenses are confined to some definite horizons and occur as individual bodies or aggregations. A continuous series of intermediate varieties from eclogites

to garnet-glaucophane schists were found in the rocks of the Choloktor formation retrogressing to typical greenschists. Structurally there exist two eclogite types: granoblastic and porphyroblastic.

In the Ca-Fe-Mg diagram (Figure 4) are plotted the garnet compositions from eclogite rocks based on full and partial microprobe analyses. The compositions of the garnets as a whole plot in the garnet field of eclogites from glaucophane schist complexes. The considerable spread in the points may be explained by chemical zoning. As shown in the curves of Ca, Mg, Fe, and Mn concentration (Figure 5 and Table 3), there is observed a lowering of Ca and Mn contents and an increase in Fe and Mg contents from the core to the rim of the grains. The Fe/(Fe+Mg) ratio in this case varies from 84 to 64%. Zoning of this sort is characteristic of the garnets from eclogites associated with glaucophane schists (Ghent 1982). Zoning has also been determined in coexisting pyroxenes. Higher Fe-content and jadeite component is observed going from the center to the rim (Table 3). The nature of zoning suggests formation of the eclogites in conditions of progressive metamorphism at rather low temperatures (Figure 6, Table 2).

The Syum-Keu Belt in the Polar Urals is a unique combination of eclogite-gneiss and eclogite-glaucophane complexes of different metamorphic depth. In the southern part of the belt, west of the ophiolite sheet Syum-Keu, the Marun-Keu eclogite-gneiss formation is exposed. It contains inclusions of eclogites of two types: a) pyrope eclogites, of similar chemical composition to troctolites, included in the structure of the metamorphosed banded complex, and possibly meta-ophiolitic; the leucocratic members of this complex such as meta-anorthosites were transformed into garnet-zoisite-kyanite rock with a garnet intermediate in composition between grossular and almandine-pyrope; b) almandine eclogites of diabasic composition predominate in the Marun-Keu complex and are possibly a metamorphosed dike complex (Udovkina 1969; Dobretsov 1974).

North of the Marun-Keu complex the amphibolite-gneiss formation contains garnet amphibolites and eclogite-type rocks; further north there is a garnet-glaucophane formation near the Pike River. This formation is composed of glaucophane metabasites, alternating with garnet-crossite quartzites and metagraywackes. In the central (upper) part of the formation, meta-arkoses, metaconglomerates, and olistoliths(?) are also present. This part corresponds to metamorphosed olistostrome (Dobretsov 1974; Kazak 1980). On the whole, these formations are similar to the above mentioned Maksyutov complex.

These three formations are likely to be three sheets inclined toward the northeast, possibly representing primary metamorphic zoning in the subducted(?) plates. These formations serve as a basement of the Syum-Keu nappe of the ophiolites, composed here mostly of ultrabasites. At the foot of this nappe, the narrow zone of the serpentinite melange is exposed and below it lies the zone of blastomylonites and its micaceous phyllonites with inclusions of the basement rocks (eclogites, eclogite-type, and garnet-glaucophane rocks). In the upper part of the nappe weakly

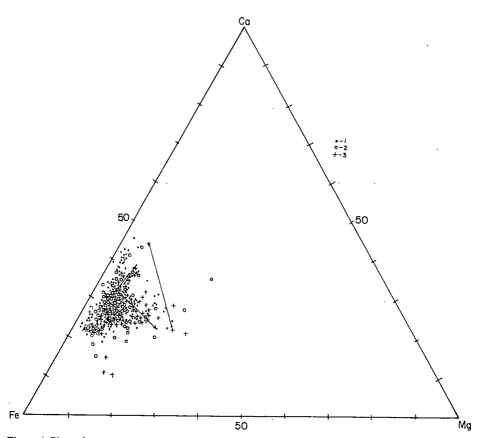

Figure 4. Plots of garnet compositions from eclogites of Tien-Shan. 1 - Atbashi; 2 - Aktyuz; 3 - Makbal.

deformed Ordovician (?) metabasalts and metatuffs contain law-sonite and sometimes glaucophane

ECLOGITES IN BLASTOMYLONITES AND GNEISSES

The Makbal Series forms a dome-shaped structure in the Kirgiz range of the Northern Tien-Shan (no. 15 in Figure 1). All the previous investigators considered the eclogites and the country rocks (micaceous schists, quartzites, marbles) to be formed under the same conditions.

Detailed investigation of this and other eclogite-bearing formations of Tien-Shan (Atbashy, Aktyuz) has shown that the eclogites and their country rocks are primarily allogenic to each . other. As shown in Figure 7, eclogites form boudins that are united into larger lenses with a common diaphthorite shell. Together with eclogites, lenses of the ultrabasites are also present. The country-rock schists are in the form of blastomylonites in the majority of cases. In the Makbal formation, the early associations in metapelites are characterized by the association: garnet + staurolite + biotite + quartz or kyanite + biotite + muscovite + quartz + ore + chloritoid replaced by greenschist associations. The following change of associations is observed in the eclogites: 1) garnet + $omph_1$ + qtz + rut; 2) garnet + cpx_2 + plag + horn + qtz; 3) gar + cpx_2 + ab + czo + glauc + qtz + act; 4) ab + act + czo +

chl; 5) chl + ab + (Fe/Mg) - carb. Association 3 corresponds to the maximum of metamorphism in the country rocks and 4 and 5 correspond to the greenschist diaphthoresis.

Eclogite garnets often contain solid inclusions that are concentrated mainly in the central red-orange colored parts of the grains. Inclusions are represented by amphibole, zoisite, mica, quartz, and rutile. Of the element profiles shown in Figure 5, it may be seen that weakly-zoned cores containing an abundant spessartine component (up to 22%) are sharply separated from the outer zones in which a smoother increase of some and decrease of other components are observed. Like garnets of the eclogite bodies from the Atbashy Range these are also characterized by a decrease in Ca and Mn contents and an increase in Fe^{2+} and Mg from the core to the rim. In this case the Fe/(Fe + Mg) ratio varies from 97 to 80%, while the Ca/(Ca + Mg + Fe + Mn) ratio varies from 0.304 to 0.191. The composition of the central parts of the garnets is very similar to those in glaucophane schists of California.

In coexisting omphacites there is observed an increase of Fe/(Fe + Mg) ratios from the center to the rim with simultaneous increase in jadeitic content. The comparison of the K_D Ga-Cpx/Fe-Mg values of both the central and marginal parts of the grains provides if not a quantitative at least a qualitative idea of the T-variation. Figure 6 illustrates that in the Makbal eclogites, K_D Ga-Cpx/Fe-Mg decreases from the central to the marginal

TABLE 3. REPRESENTATIVE MICROPROBE MINERAL ANALYSES FROM ECLOGITES OF TIEN-SHAN

	Atbashi belt AT-83-28						Makbal 16-192				Ak 81/2		Aktyuz		AK 83/8		
	Cpx_c*	Cpx_r	Ga_c	Gar	Amph	Phen	Cpx	Ga_c	Gar	Amph	Ga	Cpx	Amph	Cpx	Cpx**	Ga_c	Gar
SiO_2	56.2	56.4	36.2	37.9	58.4	51.6	55.7	37.6	38.0	57.4	38.0	53.8	41.2	52.6	56.0	38.5	38.4
TiO_2	0.02	0.03	0.16	0.17	-	0.21	-	0.09	0.15	-	0.19	0.24	0.62	0.15	0.18	0.17	0.11
Al_2O_3	8.90	9.59	19.7	20.3	9.62	23.4	8.27	20.5	20.8	10.8	19.0	8.14	12.6	5.39	3.32	20.2	20.7
FeO	7.68	7.60	27.3	28.0	8.44	2.57	8.93	23.3	27.8	9.32	31.5	9.94	19.8	13.3	12.7	28.1	29.8
MnO	0.10	0.05	1.52	0.40	-	0.05	-	9.30	0.32	-	0.18	-	0.04	0.13	0.13	1.40	0.70
MgO	7.35	6.65	1.43	3.23	12.1	4.53	7.54	0.48	3.92	11.6	1.44	7.75	8.19	7.09	7.98	1.57	2.28
CaO	11.9	11.2	11.2	9.1	1.04	-	12.0	9.45	8.44	1.51	9.52	14.0	10.3	12.6	16.0	10.6	8.72
Na_2O	7.83	8.60	-	-	7.07	0.25	7.87	-	-	6.77	-	6.25	3.06	6.18	3.72	0.03	0.01
K_2O	0.01	0.01	-	-	0.03	10.1	-	0.01	0.01	0.10	1.21	-	1.21	-	0.02	-	-
Total	99.9	100.1	97.5	99.17	96.7	92.7	100.3	99.44	97.5	99.9	100.73	99.9	96.94	97.44	100.0	100.57	100.72
%Fe/Fe+Mg	37.0	39.1	91.9	82.2	-	-	43.3	96.3	79.5	-	92.1	41.0	-	49.7	47.2	90.5	87.7
%Jd	38.5	40					36.6					28.3		23.1	14.6		
%Ca-comp.			32.2	26.1				27.0	23.9		27.3					30.5	26.3

*c - core, r - rim of the same analyzed mineral grain.

**Cpx - clinopyroxene from a clinopyroxene-plagioclase symplectite.

TABLE 4. REPRESENTATIVE MICROPROBE MINERAL ANALYSES FROM ECLOGITES OF KOKCHETAV MASSIF

	Kumdy-Kul 81/10				Sulu-Tyube				Kulet					Enbek-Berlyk E-83-61		
					ST 82/42		ST 81/5		Ku 83/20		K - 21-79			E - 83-61		
	Cpx	Ga_c*	Gar	Bi	Cpx	Ga	Cpx	Ga	Cpx	Ga	Cpx	Ga	Phen	Cpx	Ga_c	Gar
SiO_2	54.6	39.9	39.6	29.5	55.2	37.9	55.5	37.9	54.6	38.7	56.0	37.5	51.0	53.2	38.7	38.7
TiO_2	0.13	0.24	0.25	0.30	0.10	.10	0.15	0.23	0.07	0.14	0.13	0.19	0.19	0.18	0.05	0.03
Al_2O_3	7.47	21.3	21.4	16.2	7.30	20.9	11.2	21.2	7.45	21.4	9.86	21.0	25.8	6.18	21.4	21.1
FeO	5.41	21.5	20.4	22.4	6.94	26.9	3.03	26.8	4.03	24.6	3.09	25.1	1.21	5.92	22.2	22.1
MnO	0.01	0.54	0.61	0.08	-	-	0.08	-	0.03	0.56	-	0.57	-	0.02	0.51	0.55
MgO	10.6	7.19	7.48	18.0	9.30	4.26	8.65	2.11	10.7	6.34	9.64	6.45	4.33	11.0	6.66	5.61
CaO	16.9	11.5	9.9	0.08	15.8	9.7	14.5	10.1	16.8	8.06	15.0	6.73	-	18.3	9.65	10.7
Na_2O	4.23	-	-	0.14	5.34	-	6.23	-	4.39	-	5.65	-	0.44	2.88	-	-
K_2O	0.02	-	-	9.10	-	-	-	0.02	-	-	0.07	-	9.19	-	-	-
Total	99.37	102.17	99.94	95.80	99.98	100.26	100.18	101.35	98.07	99.8	99.44	97.61	92.64	97.66	99.17	98.79
%Fe/Fe+Mg	21.3	62.7	60.4		22.9	77.9	18.9	87.6	17.4	68.0	15.2	68.6		23.2	64.5	68
%Jd	30				30		43		32		41			20		
%Ca-comp.		29.9	27.3			26.5		29.8		22.6		19.1			27.0	30.5

*c - core, r - rim.

Sobolev and Others

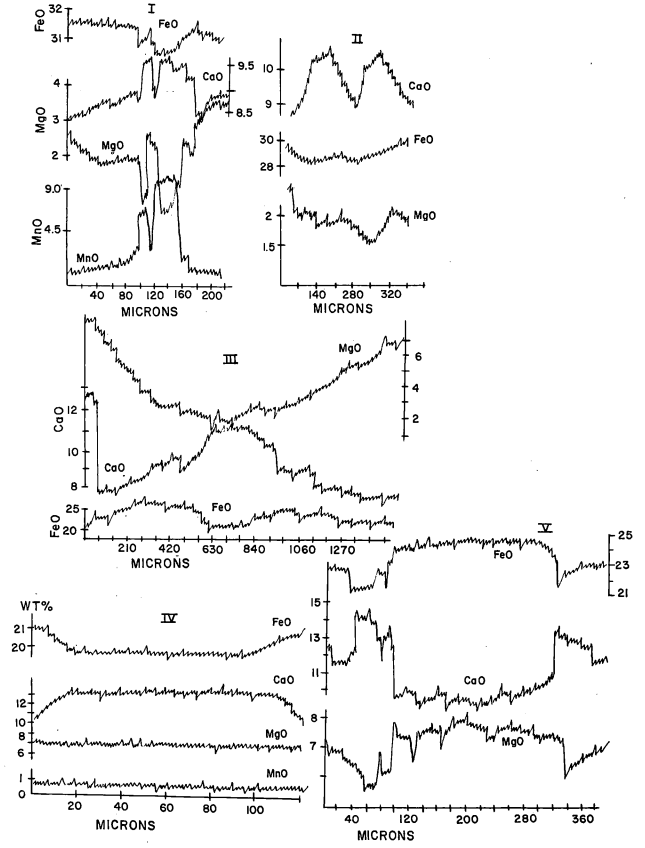

Figure 5. Microprobe traverses of garnet grains: I - Makbal, II - Aktyuz; III - Atbashi; IV - Kumdy-Kul; V - Kulet.

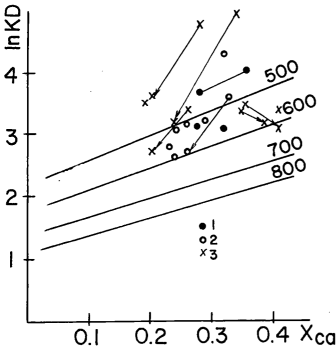

Figure 6. In K_D Ga-Cpx/Fe-Mg - $X_{Ca}V_p$ as a function of T; 1 - Atbashi; 2 - Aktyuz; 3 - Makbal.

parts of the grains. In this case the temperatures of the initial stages of metamorphism did not exceed 500°C.

The Kokchetav Block eclogites occur widely in the lower parts of the Zerendin Precambrian series. The high temperature gneisses, garnet-biotite schists with kyanite and sillimanite, micaceous granite-gneisses, and garnet-kyanite-muscovite and two-mica schists formed as diaphthoresis all contain eclogites. The eclogites from various sites differ both structurally and mineralogically (Figure 8).

The eclogites with the highest equilibrium temperatures occur in the Kumdy-Kul area, where *pyrope serpentinites* are associated with them (Table 2). The absence of zoning in the garnets suggests high equilibrium temperatures for these eclogites. As seen from Figure 5, the garnet grains are homogeneous with the exception of a narrow band (10–15 microns wide) in which zoning is associated with the regressive stage in the eclogite's history. At the same time, eclogites from Sulu-Tyube, Kulet, and Enbek-Berlyk are characterized by lower temperatures of equilibrium. The presence of zoning in the garnets of eclogites from the Kulet area stresses their low-temperature nature.

As mentioned above, when attempting to understand eclogite genesis, one of the most important problems is the comparison of the P-T parameters of eclogite formation as well as that of enclosing rocks. However, the evaluations of the conditions of regional metamorphism of eclogite and enclosing rocks differ

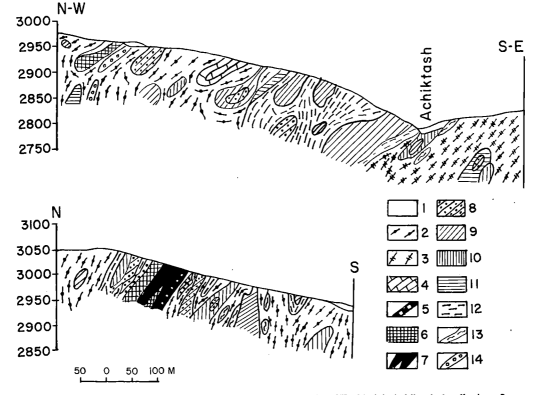

Figure 7. Cross sections of the Makbal Formation in the district of Kyrkbulak-Achiktash; 1 - alluvium; 2 - quartzites and muscovite schists; 3 - garnet-chlorite schists; 4 - marbles; 5 - fine-grained eclogites; 6 - coarse grained eclogites; 7 - the ultrabasites; 8 - the garnet amphibolite; 10 - amphibolite; 11 - garnet-albite-chlorite rocks; 12 - albite-chlorite rocks; 13 - talc rocks (±garnet, chlorite); 14 - carbonate-epidote-chlorite rocks (11-14 - diaphthorites).

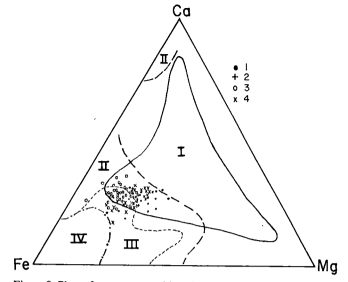

Figure 8. Plots of garnet compositions from eclogites of the Kokchetav massif. 1 - Kumdy-Kul; 2 - Kulet; 3 - Sulu-Tyube; 4 - Enbec-Berlyk; I - compositional field of garnets of eclogite facies (with grospydites); II - of granulite facies; III - of amphibolite facies along with kyanite schists and gneisses facies; IV - of epidote-amphibolite and hornfels facies (Dobretsov et al. 1974).

methods (Sobolev and others 1984). We studied the fluid inclusions of quartz in the two mica schists, and kyanite in garnet-biotite schists and garnet-kyanite-muscovite schists. Quartz inclusions in eclogites were also studied. It has been found that these rocks typically contain metamorphic inclusions of fluid methane, fluid CO_2, water-salt solutions, or their mixtures.

The rocks of the area under study contain samples with quartz of several generations: the earliest quartzes have a predominance of methane + salty water inclusions and the latest contain primarily CO_2 + salty water inclusions. Typical intermediate generations of quartzes have methane + CO_2 + salty water inclusions of variable CO_2 and CH_4 ratios. The variable ratios are probably due to evolution of the fluid composition during the same metamorphic stage. Later stages of metamorphism caused considerable oxidation of the reducing fluid of the earliest state.

As it may be seen from Table 5, maximum pressure values of 9 kbar have been obtained for biotite-garnet-kyanite schists of the Berlyk Suite. For rocks in the same suite, in the area near the Sulu-Tyube hill, highly diaphthorized rocks give a maximum pressure of 7 kbars. Lower values were obtained for the rocks of the Zholdibay suite, 5.5-6 kbars. The same pressure value has been obtained for inclusions in kyanite from the coarse grained garnet-muscovite-kyanite schists from the Kulet area.

The calculated pressure values in eclogites are generally higher than those of the enclosing schists. Eclogites are usually greatly amphibolitized. For the latest eclogite from the Kumdy-Kul area, maximum pressure values of 9 kbars have been obtained. The results suggest that the rocks of the Zholdibay suite

greatly. This was the case in studying eclogites from Kokchetav. In that occurrence we made an attempt to evaluate the pressures of the two rock formations of the Lower suites of the Zerendin series: Berlyk and Zholdibay (Rosen 1971) in whose structure active part was taken by eclogites, using thermobarogeochemical

TABLE 5. ESTIMATION OF FLUID PRESSURE (P_{f1}) ON SPECIFIC VOLUME OF CO_2 AND CH_4 OF INDIVIDUAL METAMORPHOGENIC INCLUSIONS IN QUARTZ FROM THE ROCKS OF KOKCHETAV MASSIF

No.	Sample number	Rock	T_{hom}	s.v.CO_2, CH_4, cm^3/g	P, 10^8 Pa $T°C_{metam}=650°C$	P, 10^8 Pa $T°C_{metam}=700°C$
				Sulu-Tyube		
1	ST-81/7	eclogite	-160 + -152	2.37-2.42	up to 7.5	up to 8.0
2	ST-81/6	eclogite	- 98 + - 90	3.4 -3.85	up to 3.0	up to 3.5
3	ST-82/33	eclogite	-104 + - 83	3.17-5.49	up to 3.5	up to 4.0
4	ST-81/8	eclogite	- 5.0 + 0	1.1 -1.7	up to 4.0	up to 4.5
5	ST-82/14	two-mica-schist	-144 + -100	2.51-3.32	up to 6.5	up to 7.0
6	ST-82/54	gneiss	-114 + - 92	2.83-3.75	up to 5.0	up to 5.0
7	ST-82/57	gneiss	- 89 + - 83	4.04-5.49	up to 2.0	up to 2.5
				Kumdy-Kul		
8	KK-82/62	eclogite	-172 + -148	2.29-2.47	up to 8.5	up to 9.0
9	KK-81/28	eclogite	-100 + - 88	3.32-4.0	up to 3.3	up to 3.5
10	KK-81/10	eclogite	- 50 + - 46	....	....	....
11	KK-81/16	gneiss	-133 + - 96	2.62-3.6	up to 6.0	up to 6.5
12	KK-82/4	Ga-Mu-Ky schist	-130 + -125	2.65-2.74	up to 5.5	up to 6.0
13	KK-83/22	Ga-Mu schist	- 70 + - 68	....	....	....
				Kulet		
14	KU-83/27	amph.-eclogite	-154 + -147	2.42-2.48	up to 7.0	up to 7.5
15	KU-83/26	amph.-eclogite	-151 + -149	2.45-2.46	up to 7.0	up to 7.5
16	KU-83/8	amph.-eclogite	-151 + -149	2.45-2.46	up to 7.0	up to 7.5
17	KU-83/9	Mu-Ga schist	-117 + -113	2.87-2.95	up to 4.5	up to 5.0
18	KU-83/24	schist	- 43 + - 40	....	....	....
19	KU-81/2	Ga-Ky schist	0 + + 10	1.09-1.18	up to 4.0	up to 4.5
20	KU-81/2	Ga-Ky schist	- 15 + 0	0.98-1.09	up to 5.5	up to 6.0
				Enbenk-Berlyk		
21	E-83/72	eclogite	- 21 + - 5.0	1.0-1.1	up to 5.0	up to 5.5
22	E-83/46	Ga-Bi-Ky schist	-172 + -156	2.29-2.40	up to 8.5	up to 9.0

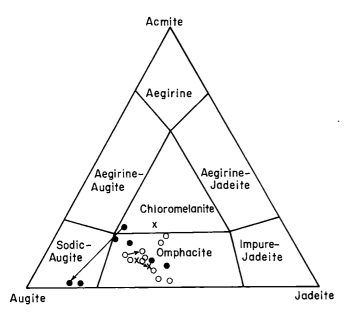

Figure 9. Compositional features of clinopyroxenes in eclogites of Tien-Shan in the system Ac-Jd-Aug (Essene and Fyfe 1967). Solid circles—Aktyuz; open circles—Atbashi; Crosses—Makbal.

enclosed the eclogites of Kumdy-Kul and Kulot areas and were formed by diaphthoresis of the Berlyk suite rocks. This is supported by the presence of highly dense inclusions of methane in the quartz of eclogites of the Zholdibay suite similar to those in quartz of biotite-garnet-kyanite schists of the Berlyk suite.

Based on these data one cannot judge if the eclogites and Berlyk suite rocks are isofacial. The presence of several generations of quartz reflecting various stages of metamorphism favors repeated recrystallization. Therefore we are not sure that the quartz grains of the earliest generation, containing inclusions of high density methane were crystallized under the same conditions . as minerals associated with eclogite. However, the obtained equilibrium P-T values for this eclogite association, for those from Kumdy-Kul area, indicate that they cannot be regarded as products of regional metamorphism of the basic rocks.

In the Aktyuz block of the Muyunkum Massif the eclogite-bearing formation is made up of high-temperature gneisses, plagiogneisses, and diaphthoritic biotite-muscovite and muscovite-chlorite schists. Eclogites are in the form of relics in the inner parts of garnet amphibolite bodies. Amphibole, plagioclase, mica, epidote, calcite, and chlorite are present as alteration products. The associations and relationship between the minerals observed in the enclosing rocks indicate that the rocks of the Aktyuz formation had undergone polyphase metamorphism.

Eclogite garnets are characterized by very high Fe-content and belong to the grossular-almandine series with insignificant admixture of the pyrope component (Figure 4, Table 3). Analysis of both the central and peripheral parts of the grains has shown a lowering in Fe/(Fe + Mg) ratios from the core to the rim as well as in the Ca/(Ca + Mg + Fe + Mn) ratio and Mn contents. However, the concentration curves (Figure 5) show a non-regular

nature of the zoning, which in fact is probably associated with the later processes. The diaphthoresis is displayed by replacement of omphacite by a pyroxene-plagioclase (cpx-pl) symplectite. We made an attempt to analyze omphacite and cpx-pl symplectite evolving around it from the sample Ak 83/8, where the process of omphacite substitution is especially clear. However, extreme chemical inhomogeneity did not permit one to get a reliably complete chemical analysis. The analysis of omphacite listed in Table 3, in addition to the analysis of pyroxene from the symplectite, provides a possibility of fixing the trends in the variations of omphacite composition. The pyroxene composition in this case shifts from the omphacite field to that of Na-augite (Figure 9). Plagioclase in the symplectites is practically pure albite.

It should be stressed that the pyroxenes of eclogites from the Aktyuz block are characterized by high contents of jadeitic and acmitic components, which makes them very similar to pyroxenes from the low temperature eclogites (Dobretsov, 1974) and distinguishes them from pyroxenes of gneiss complexes. The obtained P-T estimate of equilibrium of minerals (Table 2) corresponds to low-temperature conditions.

CONCLUSIONS

From the above review one can conclude that eclogites of various types of metamorphic complexes in the USSR are undoubtedly polygenetic rocks that include most of the variety of eclogites found in the Earth's crust. Thus, eclogites associated with glaucophane schists often have the same P-T parameters as the country rocks. With some exceptions, such as the Maksyutov complex, the garnet zoning usually indicates their growth under conditions of progressive metamorphism. This cannot be the case for eclogites of schist-gneiss complexes. However, eclogite complexes of various types may have similar parameters of equilibrium as, for example, eclogites from Aktyuz. This may indicate that the similar metamorphic stages are inherent to the historical evolution of these complexes. One also has to take into account the fact that eclogites lying within the same complex may be characterized by different conditions of formation, as may be seen from the Kokchetav Massif. Here there are both comparatively low-temperature eclogites (Kulet, Sulu-Tyube) and extremely high temperature ones (Kumdy-Kul), as indicated by a total lack of zoning in the garnets. Also, one cannot discount the uncertainty in determination of the P-T parameters of equilibrium for eclogites and country rocks. This uncertainty is due to imperfection in available geothermometers and geobarometers combined with processes of repeated diaphthoresis of eclogites and their enclosing rocks.

In summary of the above, we may conclude that in a majority of cases the conditions of eclogite formation are characterized by $P_{total} > P_{load}$ (Sobolev 1960; Dobretsov and Sobolev 1975). These conditions are possible in subduction zones and in the zones of deep-seated faults and overthrusts.

The eclogite-glaucophane schist complexes of the Mak-

362   *Sobolev and Others*

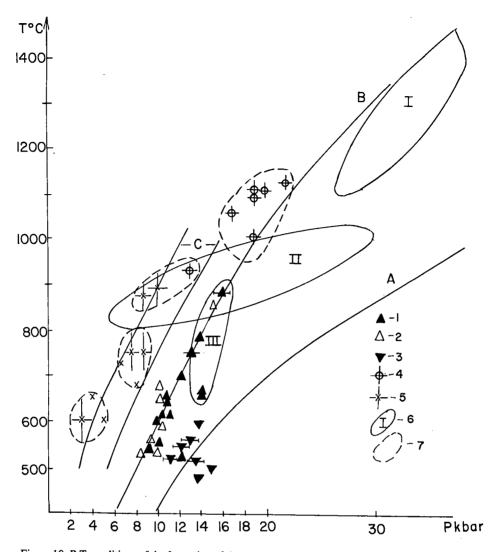

Figure 10. P-T conditions of the formation of the eclogites in the earth's crust: 1-2 - the eclogites in the eclogite-schist-gneiss complexes, cited in this paper (1) and in Dobretsov (1982) (2); 3 - the eclogites in the glaucophane schist complexes; 4 - the mantle inclusions from the gneiss of the Bohemian massif and the Granulitgebirge (Dobretsov, 1982); 5 - the conditions of metamorphism of the country gneiss; 6 - the fields of the origin conditions of pyrope-bearing xenoliths from the kimberlites (I) Sobolev, 1977, spinel-bearing xenoliths in basalts (II) Sobolev, Dobretsov, Sobolev, 1975, and the eclogite field of the Kokchetav massif (III) 7 - the fields of the origin conditions of the definite metamorphic complexes.

syutov-Californian type (line C in Figure 10) may correspond to the model of multistage obduction (Dobretsov 1980; Dobretsov 1981). In our opinion, it combines the model of fluid overpressure and repeated, powerful overthrusting offering the possibility of significant thickening of the crust and an increase in lithostatic pressure. The repetition of obduction leads to the drawing of

olistostrome and graywacke turbidites, formed at the early stages, into the zone of metamorphism.

ACKNOWLEDGMENTS

The authors are indebted to R. G. Coleman and D. C. Smith for useful remarks and a critical reading of the manuscript.

REFERENCES

Bakirov, A. B., 1984, Endogenic geological formation of Kirghizia, vol. 2, Metamorphic formation: Frunze, Ilim, 249 p. (in Russian).
Blake, M. S., and Jones, D. L., 1974, Origin of Franciscan melanges in Northern California: Society of Economic Paleontologists and Mineralogists Special Publication 19, p. 345-357.

Coleman, R. G., 1967, Glaucophane schists from California and New Caledonia: Tectonophysics, v. 4, p. 479-498.
Currie, E. L., and Curtis, L. W., 1976, An application of multicomponent solution theory to jadeite pyroxenes: Journal of Geology, v. 84, p. 179-195.
Dobretsov, N. L., 1974, Glaucophane-schist and eclogite-glaucophane-schist

complexes of the USSR: Novosibirsk, "Nauka", 429 p. (in Russian with English abstr.).

—— 1980, The new overthrusting model for the blueschist metamorphism with reference to the Franciscan-Great Valley problems, California: Ophioliti, v. 4, p. 17–24.

—— 1981, The global petrological processes: Moscow, "Nedra", 236 p. (in Russian).

—— 1982, Origin conditions of pyrope peridotites and eclogites in the crystalline basement of Bohemia and other massives, in Ophiolites—Initiatives 1981; Potsdam, p. 50–54.

Dobretsov, N. L., and Lavrentjev, Yu. G., 1978, The thermobarometry based on white micas, sodium amphiboles and clinopyroxenes: Abstracts of XI General Meeting of IMA, v. II; Novosibirsk, p. 35–36.

Dobretsov, N. L., Sobolev, V. S., Sobolev, N. V., and Khlestov, V. V., 1974, The facies of regional metamorphism at high pressure: (in Russian) Moscow, 328 p. (English translation by D. A. Brown, 1975, ANU Press, Canberra.)

Dobretsov, N. L., and Sobolev, N. V., 1970, Eclogites from metamorphic complexes of the USSR: Physics of Earth and Planetary Interiors, no. 3, p. 462–470.

—— 1984, Glaucophane schists and eclogites in the folded systems of Northern Asia: Ofioliti, v. 9, no. 3, p. 401–424.

—— 1975, Eclogite-glaucophane schist complexes of the USSR, and their bearing on the genesis of blueschist terranes: Geological Society of America Special Paper no. 151, p. 145–155.

Dobretsov, N. L., and Tatarinov, A. V., 1983, Jadeite and nephrite in ophiolites: Novosibirsk, Nauka, 123 p.

Ellis, D. J., and Green, D. H., 1979, An experimental study of the effect of Ca upon garnet-clinopyroxene Fe-Mg exchange equilibria: Contributions to Mineralogy and Petrology, v. 71, p. 13–32.

Ermolov, P. V., et al., 1981, Ophiolites of the Chara zone, in Ophiolites—Alma-Ata; "Ilim". (in Russian).

Essene, E. J., and Fyfe, W. S., 1967, Omphacite in Californian metamorphic rocks: Contributions to Mineralogy and Petrology, v. 15, p. 1–23.

Ghent, E. D., 1982, Chemical zoning in eclogite garnets—a review. Terra Cognita, v. 2, no. 3, p. 301–302.

Holland, T.J.B., 1980, The reaction albite + jadeite + quartz determined experimentally in the range 600–1200°C: American Mineralogist, v. 65, p. 129–134.

Hsu, L. C., 1968, Selected phase relationships in the system Al - Mn - Fe - Si - O - H, a model for garnet equilibria: Journal of Petrology, v. 9, no. 1, p. 40–83.

Kazak, A. P., 1980, The glaucophane schists of the Polar Ural, in Dobretsov, N. L., ed., Petrology of metamorphism formation of Siberia: Novosibirsk,

"Nauka", p. 143–150 (in Russian).

Kushiro, J., 1969, Clinopyroxene solid solution formed by reactions between diopside and plagioclase at high pressures: Mineralogical Society of America Special Paper, v. 2, p. 179–191.

Lennykh, V. I., 1977, Eclogite-glaucophane belt of South Ural: Moscow, "Nauka", 160 p. (in Russian).

Markov, M. S., Nekrasov, G. E., and Palanjan, S. A., 1982, Ophiolites and melanocrate basement of Koryakia Mountains: Reports of tectonics of Koryakia Mountains: Moscow, "Nauka", p. 30–70 (in Russian).

Miller, Yu. V., 1977, Maksyutov complex: Structural evolution of metamorphic complexes: Leningrad, "Nauka", 197 p. (in Russian).

Moskovchenko, N. I., 1982, High-pressure metamorphic complexes of Precambrian in the Phanerozoic folded belts: Leningrad, "Nauka", 161 p. (in Russian).

Rosen, O. M., 1971, Stratigraphy and radiogeochronology of Precambrian Kokchetav massif: Stratigraphy of Precambrian and Tien-Shan: Moscow University, p. 75–84 (in Russian).

Rossi, G., Smith, D. C., Ungaretti, L., and Domeneghetti, M. C., 1983, Crystal-chemistry and cation ordering in the system diopside-jadeite: a detailed study by crystal structure refinement: Contributions to Mineralogy and Petrology, v. 83, p. 247–258.

Smith, D. C., Mottana, A., Rossi, G., 1980, Crystal-chemistry of unique jadeite-rich acmite-poor omphacite from the Nibo eclogite pod, Sorpollen: Lithos, v. 13, p. 227–236.

Sobolev, N. V., 1977, Deep-seated inclusions in kimberlites and the problem of the composition of the upper mantle: American Geophysical Union, 279 p.

—— 1960, Role of high pressure in metamorphism, in Report of the Twenty-First Session Norden, Part XIV: The Granite-gneiss problem: Copenhagen, p. 72–82.

Sobolev, N. V., Tomilenko, A. A., and Shatsky, V. S., 1984, Metamorphic conditions of the rocks of Zerenda series, Kokchetav massif (from studying fluid inclusions), (in press).

Sobolev, V. S., Dobretsov, N. L., and Sobolev, N. V., eds., 1975, Deep-seated xenoliths and Upper Mantle: Novosibirsk, "Nauka", 269 p. (in Russian).

Sorensen, S., 1986, Comparative petrology and geochemistry of blueschist and greenschist facies rocks from the Catalina Schist terrane, southern California: Geological Society of America Memoir 164, this volume.

Udovkina, N. G., 1969, Eclogites of Polar Ural: The peculiarities of their composition and genesis. Moscow, "Nauka," 181 p. (in Russian).

MANUSCRIPT ACCEPTED BY THE SOCIETY JULY 29, 1985

Geological Society of America
Memoir 164
1986

The high-pressure metamorphic belts of Japan:
A review

Shohei Banno
Department of Geology and Mineralogy
Kyoto University
Kyoto 606, Japan

ABSTRACT

High-pressure metamorphic rocks of the Japanese Islands are classified into 9 belts or zones: Tokoro, Kamuikotan, East Abukuma-Motai, Hida Marginal-Joetsu, Sangun, Sanbagawa, Kurosegawa, isolated area of Kyushu, and Yaeyama. The Kamuikotan, East Abukuma-Motai, Hida Marginal-Joetsu, and Kurosegawa zones are associated with abundant serpentinite. All but the last three are paired with low-P, high-T terranes. The Sangun and Sanbagawa belts are paired with the low-P and high-T Hida and Ryoke belts respectively. It is still not clear if the Kamuikotan Zone is paired with any low-pressure terrane in the Japanese Islands. The other terranes are small and of uncertain tectonic significance.

Tectonic position, age, protolith, and metamorphic facies series of the high-pressure metamorphic belts are summarized briefly. Comments on paired belts, eclogites, and the jadeite + quartz assemblage are given in the concluding remarks.

INTRODUCTION

The first systematic attempt to classify the metamorphic belts of the Japanese Islands according to the physical conditions of metamorphism was made by Miyashiro (1961), who developed the concepts of metamorphic facies series and paired metamorphic belts. Twenty years after its publication it is inevitable that Miyashiro's concept of the metamorphic belts of the Japanese Islands needs revision. Since 1961, the geology and petrology of the metamorphic belts have been reviewed intermittently in a number of reports, among which the following are noteworthy (asterisks show literature in Japanese): Hashimoto et al. (1970), Miyashiro (1973b), Hashimoto (1978), Kanmera and Hashimoto (1979*), and Banno (1981). In addition, several memoirs and reports of workshops on Japanese metamorphic rocks have been published by Hide (1977*), Suwa (1979*), Chihara (1980*, 1981*), Hara (1981), and Nishimura (1983*, 1984*).

This paper is a review of research in the metamorphic belts of the Japanese Islands; its purpose is to summarize recent progress toward understanding of the geology and petrology of high pressure metamorphic belts. References pertaining to individual areas listed in Miyashiro (1973b) and those containing the radi-

ometric ages compiled by Nozawa (1977) are not referred to here. As this paper is not particularly concerned with phase relations, metamorphic facies will be used without giving rigorous definitions. The transcription of the Japanese localities follows the system of Miyashiro (1973b), although spellings of the original authors are honored in the references. Thus we use Sanbagawa (Sambagawa) and Bessi (Besshi) etc.

AREAL DIVISION

Based upon the distribution of pre-Tertiary rocks, the Japanese Islands can be divided into five provinces. They are Hokkaido, Northeast Japan, the Inner Zone of Southwest Japan, the Outer Zone of Southwest Japan, and the Nansei Islands (Ryukyu). Among them, the distinction between Hokkaido and the rest of the Japanese Islands is the largest. The Ishikari lowland dividing the central and western parts of Hokkaido is thought to have been the boundary between the Eurasia and Okhotsk or North American Plates possibly until the Pliocene (Den and Hotta 1973; Nakamura 1983*). The western part of Hokkaido is the northern extension of Northeast Japan; thus Hokkaido denotes the area east of Ishikari lowland in this paper. The geological distinctions between the rest of the islands are minor

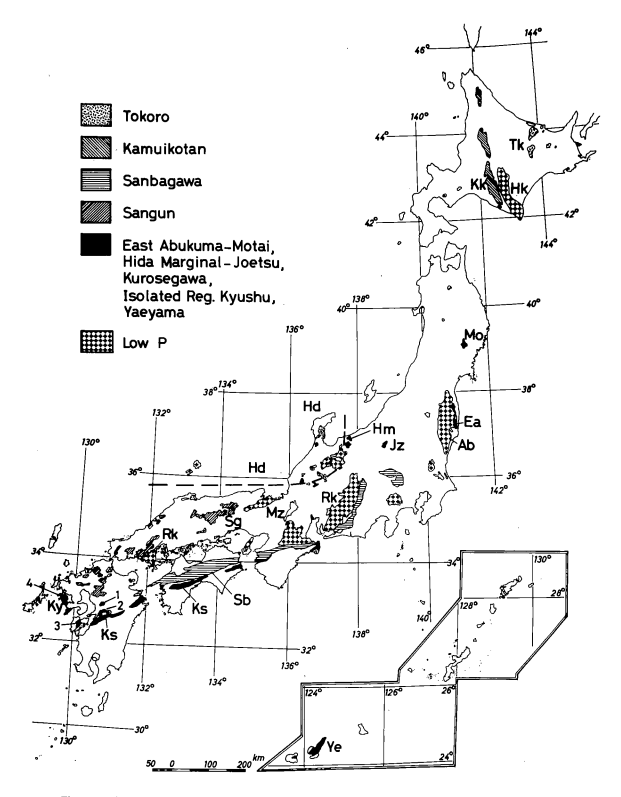

Figure 1. Distribution of high-pressure metamorphic rocks of the Japanese Islands. The size of small areas is exaggerated. Low P stands for low-pressure rock. Abbreviations. Tk = Tokoro Belt, Hk = Hidaka Belt, Kk = Kamuikotan Zone, Mo = Motai, Ea = East Abukuma Zone, Ab = Abukuma, Jz = Joetsu Zone, Hd = Main Hida Belt, Rk = Ryoke Belt, Mz = Maizuru Belt, Sg = Sangun Belt, Sb = Sanbagawa Belt, Ks = Kurosegawa Zone, Ky = isolated areas of Kyushu, 1: Kiyamo, 2: Manotami, 3: Amakusa, 4: Nagasaki, Ye = Yaeyama Belt.

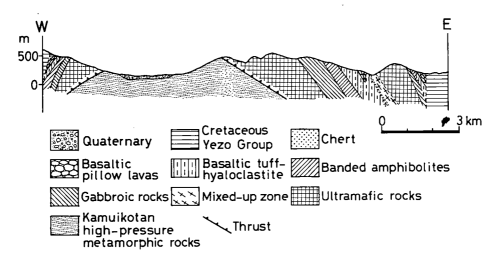

Figure 2. A cross section of the Kamuikotan zone in the Horokanai area. After Ishizuka et al. (1983).

compared to those between Hokkaido and the rest of Japan, but the geology is by no means simple. According to the concept of accretion tectonics, the Japanese Islands are composed of quite a few geologically distinct terranes. A memoir edited by Hashimoto and Uyeda (1983), and the articles of Taira et al. (1982) and Maruyama et al. (1982*) present new ideas on the geological development of the Japanese Islands. These ideas are of minimal importance to this paper.

HOKKAIDO

General Statement

A classical view of Hokkaido's geology is that paired metamorphic belts, the high pressure Kamuikotan and low pressure Hidaka belts, run parallel to each other in a N–S direction that extends to Sakhalin. However, two major finds have caused revisions to be made in regard to this view. One is the finding of another terrane of high pressure metamorphism, the Tokoro Belt (Figure 1) to the northeast of the Hidaka Belt. The other is the finding that the Kamuikotan Belt is not a simple high pressure belt, but a tectonic zone where the high pressure schists and low pressure metamorphics of an ophiolite suite are mixed up. Also, the concept of the Hidaka Belt has been extensively revised by Komatsu and his collaborators (Komatsu et al. 1983) who clarified that the major part of the Hidaka Belt contains a nearly complete section of continental crust ranging from the granulite to the greenschist facies. The Hidaka Belt, which has 17 to 30 Ma radiometric ages, is thrust over the ophiolite sequence from east to west. The ophiolite sequence was also metamorphosed at low pressure, and is considered to be a portion of the low pressure oceanic rock suite of the Hidaka belt (Miyashita 1983*).

Tokoro Belt

According to a preliminary report of Sakakibara (1984*),

the greenstones of the Tokoro belt underwent glaucophanitic metamorphism with resultant formation of aegirinejadeite (up to 40% jadeite content), magnesioriebeckite, winchite, actinolite, epidote, and pumpellyite. The age of the protolith is considered to be late Mesozoic.

Kamuikotan Zone

The presence of low pressure amphibolite in the Kamuikotan Zone has been known for a long time (Igi 1959*) but its significance has largely been ignored. Banno et al. (1978), Asahina and Komatsu (1979), Ishizuka (1980*), and Ishizuka et al. (1983) concluded that the Kamuikotan belt contains an ophiolite suite of rocks that suffered low pressure metamorphism in the zeolite to the granulite facies. Although the high pressure Kamuikotan schists are predominant, the term Kamuikotan Metamorphic Belt is not appropriate because metamorphic rocks of completely different heritages are mixed up, so we prefer to call the belt the Kamuikotan Zone. Figure 2 is a cross section of the Horokanai area and shows the thrust of low pressure ophiolitic rocks over the high pressure Kamuikotan schists.

The high pressure Kamuikotan schists include pelitic, psammitic, siliceous, and basic compositions. Serpentinite is widespread in the schist area, some of which is considered by Maekawa (1983) to be of sedimentary origin. The Kamuikotan schists are separated into several units, ranging from 10 to 100 sq. kilometers that form isolated masses, bordering adjacent units by faults or by sheared serpentinite. Even within one unit, they are lithologically heterogeneous, being much like mélanges. It is worth mentioning that most of the metabasalts and metadolerites have compositions of alkali affinities. It is possible that bulk analyses of the metabasites have nothing to do with the nature of the original basalt, but blastoporphyritic Ti-augite, aegirine, and kaersutite demonstrate the possible identification of alkali basalt (Nakano and Komatsu 1979*; Nakano 1981*). However, Maekawa (1983) has shown that there are tholeiitic metabasalts that contain metamorphic aragonite, as inferred from pyroxene mineralogy.

The Kamuikotan schists contain high pressure minerals such as jadeite + quartz, aragonite, lawsonite, and glaucophane. Although zonal mapping has been attempted, Imaizumi (1984*) insists that the mapping is not complete owing to monotonous mineral assemblages, even though the analysis of sliding equilibrium may make it possible as claimed by Shibakusa (1981). Tectonic blocks, ranging from jadeitized amphibolite to garnet-epidote amphibolite, occur enclosed within serpentinite or pelitic schists. Of these, garnet-epidote amphibolite suffered glaucophane schist facies metamorphism, which resulted in the formation of glaucophane along the grain boundaries of barroisite. A petrologic review of the Kamuikotan schists is given in Ishizuka et al. (1983).

Some K-Ar ages are available for the Kamuikotan schists. The ordinary schists have 70–110 Ma ages, but the tectonic blocks are older, 130–140 Ma (Bickerman et al. 1971; Imaizumi and Ueda 1981*). The facies series, K-Ar ages, tectonic blocks, and the association with ophiolite and covering sediments of the Kamuikotan zone are similar to the Franciscan terrane.

NORTHEAST JAPAN

There are three terranes of high pressure rocks in Northeast Japan; the Kanto Mountains area of the Sanbagawa belt, small high pressure terranes that intermittently continue along the east of the Abukuma Plateau to the southern Kitakami Mountains, and the Joetsu area. The first of these will be treated with the Sanbagawa schists of Southwest Japan. The Joetsu area may well be the extension of the Hida Marginal Zone and will be discussed in the next section. Hence only the last one will be mentioned here.

East Abukuma-Motai Zone

Through the areas of the East Abukuma to Motai (Kitakami), all high pressure terranes are small, fault-bounded areas in between occur Cretaceous granites. The K-Ar ages are around 300 Ma but these dates could represent the youngest terrane.

The largest high pressure terrane, the Motai metamorphics, is situated in the serpentinite zone that borders the south Kitakami terrane, which is probably an exotic terrane of equatorial origin (Saito and Hashimoto 1982) juxtaposed against the Pre-Cretaceous Japan. Serpentinite is widespread in these areas. Maekawa (1981*) suggests that the serpentinite of the Motai area is of sedimentary origin, and the gneiss and amphibolite that were hitherto regarded as the exotic blocks of the fault zone are olistoliths in the Motai Group. The metamorphic facies is mostly epidote-glaucophane, as in the Motai area, but the Yamagami area contains albite-epidote amphibolites (Kuroda and Ogura 1963).

INNER ZONE OF SOUTHWEST JAPAN

There are three terranes of high pressure metamorphic rocks in this province: the Hida Marginal Zone, the Sangun Metamorphic Belt, and the isolated glaucophane schists of Kyushu.

Hida Marginal Zone

This zone intermittently surrounds the complex of granite and gneiss of the Hida metamorphic terrane and is called the Hida Marginal Zone (Figure 3), but the term Circum-Hida (tectonic) zone is also used (Mizutani and Hattori 1983). The Hida terrane probably includes Precambrian rocks but is composed predominantly of 240 Ma gneisses and 180 Ma granitic rocks. According to Hiroi (1981), these continued to the Gyeongii and Okcheon zones of Korea before the opening of the Japan Sea. The Hida Marginal Zone is essentially a terrane of serpentinite mélange composed of high pressure schists and Ordovician to Permian sedimentary formations (Chihara et al. 1979; Igo et al. 1980).

In the Ise area, the high pressure rocks and serpentinites are proposed to be olistoliths in the Late Jurassic Tetori Group (Soma et al. 1981*). In the eastern part of the zone, glaucophanitic schists in the Omi area and greenschists in the Renge area are associated with abundant serpentinite. In the Happo'one area, a huge serpentinite area contains exotic blocks of glaucopohane schist, eclogite, jadeite + quartz rock, and epidote amphibolite (Chihara et al. 1979; Komatsu and Yamasaki 1981*). The radiometric ages of high pressure schists are around 350 Ma, with a few 400 Ma amphibolites (Shibata et al. 1980*; Shibata 1981*). The mélange zone formed before Late Triassic when the Kuruma Group unconformably covered the mélange terrane.

In the Omi area, where the high pressure schists are most coherent, prograde metamorphism from the epidote-glaucophane facies to the epidote amphibolite facies was proposed. Matsumoto (1980*) concluded that the Omi area is a complex of fault-bounded blocks of high pressure schists accompanied by serpentinite. This view explains why the garnet zone without biotite, which should exist between the chlorite and biotite zones of the high pressure facies series, is absent here. His results still fail to interpret a regular distribution of chlorite and biotite zone schists.

The Joetsu Zone is mostly serpentinite mélange with rarely glaucophane schists occurring as exotic blocks (Chihara et al. 1979).

The Hida Marginal Zone, combined with the Joetsu Zone, is a suture zone between the Hida terrane and Mino terranes, the latter being composed of Jurassic sedimentary mélange. Comprehensive reviews on this zone are available in Chihara et al. (1979), and Chihara and Komatsu (1981).

Sangun Belt

The Sangun Belt is a glaucophanitic terrane located southeast of the Hida terrane, which is defined to include the granite

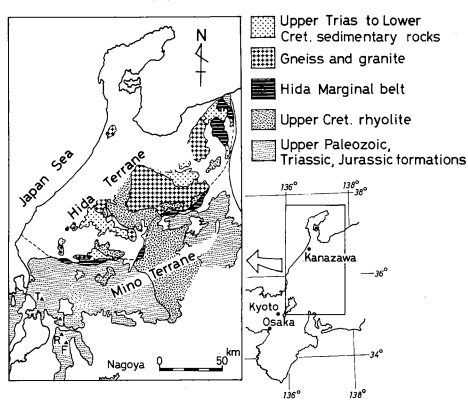

Figure 3. Geological relations of the Hida Marginal Zone (=Circum Hida tectonic zone). Simplified
Figure 1 of Mizutani and Hattori (1983), with permission.

and gneiss of Oki Island off the Sanin coast in the Japan Sea. It is, however, not the continuation of the Hida Marginal Zone. First of all, the cluster of radiometric ages of metamorphic rocks is around 250 and 170 Ma, the latter being definitely post-metamorphic because Middle to Upper Triassic sedimentary rocks unconformably cover the Sangun schists. A preliminary age determination of 300 Ma has been reported from one area (Shibata and Nishimura 1984*).

The Sangun belt is composed of many isolated areas, some of which are intruded by Cretaceous granites that changed the schists to hornfels. Zonal mapping done in the Katsuyama (Hashimoto 1968) and Nishiki-cho (Nishimura 1971) areas has revealed a facies series ranging from the pumpellyite-actinolite facies to the epidote-glaucophane facies, and finally to the albite-epidote amphibolite facies. The last one is characterized by barroisite in basic schists and biotite-free pelitic schists. Lawsonite was recognized in some basic and pelitic schists (Hashimoto and Igi 1970*; Watanabe et al. 1983).

Most of Mesozoic-Paleozoic "eugeosynclinal sediments" of the Japanese Islands are weakly metamorphosed, leading us to regard the sediments as gradational to the lower-grade rocks of adjacent metamorphic belts. For the Sangun belt a gradation to weakly metamorphosed sedimentary rocks has also been envisaged. However, recent progress in biostratigraphy and petrography of Mesozoic-Paleozoic formations in Southwest Japan has revealed that the metamorphism of the Maizuru-Yakuno

zone does not belong to the Sangun metamorphism because of different facies series between them (Ishiwatari 1985), and that the concept of gradual transition from the Sangun to the Tamba terrane should be rejected because of the finding of Jurassic fossils from the Tamba Group (Isozaki and Matsuda 1980). In the western part of Honshu, the Sangun belt shows an apparent synform, having virtually unmetamorphosed rocks at the axis. Here a transition from unmetamorphosed rocks (or pumpellyite-chlorite zone) to the glaucophanitic metamorphic rocks is suggested. However, Hara et al. (1980) have shown that the Sangun schists were thrust from north to south over the Triassic formation. Detailed reviews on the Sangun belt are available in Hashimoto (1972*), Nishimura et al. (1977*) and Nishimura (1981).

Isolated Regions in Kyushu

There are four isolated regions of glaucophanitic metamorphism in Kyushu: Kiyama, Manotani, Amakusa, and Nagasaki. Although small, the Kiyama area has prograde zones from the epidote-glaucophane facies to the albite-epidote amphibolite facies. The radiometric age is around 300 Ma. Glaucophane schist in the Manotani area is a new finding. Minerals of high pressure intermediate type metamorphism, such as albite, epidote, chlorite, actinolite, stilpnomelane, and crossite occur. Rarely lawsonite occurs in pelitic rocks. As this area is said to change gradually to a prograde terrane of low pressure Ryoke type metamorphics, the prevailing opinion is that the high pressure minerals predated the Ryoke metamorphism (Karakida et al. 1984*).

High pressure schists of the Nagasaki and Amakusa areas define another region of glaucophanitic metamorphism. It is composed of schists of the epidote-glaucophane and the albite-epidote amphibolite facies (Nishiyama 1979*). The metagabbros of the Nomo peninsula have a K-Ar age of 280 Ma. The geologic significance of these isolated metamorphics is far from clear. Maruyama et al. (1982*) regards the Kiyama schists as an extension of the Hida Marginal Zone. However, this does not explain the presence of a Permian sea, which is represented by ophiolite of the Maizuru-Yakuno complex, in between the two flanks of high pressure belt. The Manotani, Nagasaki, and Amakusa schists cannot be grouped with the Kiyama schists because of different ages of metamorphism. The classic view of Kobayashi (1941) that the schists belonged to the Sanbagawa belt and attained a N-S trend when the Japanese Islands were separated from the Asian continent may be worthy of re-examination.

OUTER ZONE OF SOUTHWEST JAPAN

The Sanbagawa Belt and the glaucophane schists in the Kurosegawa zone are the high pressure rocks in this province.

The Sanbagawa Belt

During the period of excitement and new ideas in Japanese metamorphic petrology in the 1950s and early 1960s, there was a tendency to overestimate the extent of Sanbagawa metamorphism, and I was to some extent responsible for it. The easternmost extension of the Sanbagawa Belt is the Kanto Mountains. Further eastward, there are reports of phyllitic rocks in the deep drillings of the Kanto Plain, northeast of Tokyo, which may well be Sanbagawa schists, but no exposed Sanbagawa schists exist beyond the Kanto Mountains. The western end of the Sanbagawa belt was once thought to be the lawsonite-glaucophane schists of the Yatsushiro area, but we now think this is part of the Kurosegawa Zone. The transition from the Sanbagawa schists to weakly metamorphosed "Chichibu (Titibu) eugeosynclinal sediments" has been confirmed at many places, but only the northern part of the Chichibu belt is metamorphosed. Thus, for example, a part of Banno's (1964) zone A1 located near serpentinite of the Kurosegawa zone does not belong to the Sanbagawa schists. It belongs to the redefined Kurosegawa zone.

Based on fusilinids in limestone, the original rocks of the Sanbagawa belt were considered to be Upper Paleozoic. During the last fifteen years, however, there have been revolutionary changes in the chronology of the "eugeosynclinal sediments" of the Japanese Islands due to the finding of such microfossils as conodonts in chert and radiolaria in shale. The youngest fossils found in the Sanbagawa schists (in the typical Sanbagawa schist zone) are Upper Triassic conodonts (Suyari et al 1980*). Jurassic radiolaria have been described from the northern part of the Chichibu terrane affected by Sanbagawa metamorphism (Isozaki

et al. 1981*; Sashida, et al. 1982). The radiometric ages of schists are still not abundant and the K-Ar age range from 50–110 Ma still holds good. A whole rock Rb-Sr age of the pelitic schists of 116 Ma was preliminarily reported by Minamishin (1979*).

The original rocks of the Sanbagawa schists are shale, sandstone, chert, limestone, basalts (both sediments and pillow lavas), and tectonic blocks of metagabbro and ultramafics (Kawachi et al. 1982). It is noteworthy that no in situ basaltic rocks are known in the Sanbagawa area, though in the Chichibu terrane there is a report of gabbro and dolerite intrusions (Iwasaki 1978*, 1979).

Sugisaki et al. (1972) stated that the basalt in the Sanbagawa-Chichibu terranes is symmetric. The middle, that is the Mikabu complex, has a tholeiitic composition, and the northern schists terrane and southern unmetamorphosed Chichibu terranes have alkali-basalt affinities. The compositions of basalt, as estimated from the analysis of carefully selected pillow and massive lavas as well as the nature of relic pyroxenes, support Sugisaki et al.'s view. But tholeiitic rocks also occur in the Chichibu terrane (Maruyama and Yamasaki 1978). The nature of exotic blocks of metagabbro and peridotite is discussed by Kunugiza et al. in this volume.

The facies series have been extensively studied in the past 20 years. Metamorphic petrology will be reviewed elsewhere and only briefly commented on here. Figure 4 shows the outline of the facies series established in central Shikoku. The zonal mapping based on the distribution of jadeite, lawsonite, and pumpellyite is not applicable to this belt (Hirajima 1983*). Recently, lawsonite has been found in pelitic schists of the pumpellyite-actinolite zone of Shikoku (Watanabe and Kobayashi 1984).

The stratigraphy and structural geology of the Sanbagawa belt have a long history of research and are known primarily from studies by Kojima, Hara, and their colleagues who are mainly associated with Hiroshima University. Hara et al. (1977*) distinguished two stages in the deformation history, which were named the Ozu-Nagahama and Hijikawa stages. The former resulted in the major structure of south vergency including the Saruta nappe. Recently, a few other groups have joined in this field. Toriumi (1982) has found that deformed radiolaria can be identified even in the schists, and traced the change of strain of the schists from the unmetamorphosed (Chichibu terrane) to the schist areas. The deformation rods of radiolaria described by Toriumi may require uniaxial extension of the Sanbagawa terrane, at least at the earliest stage of deformation. Faure (1982) pointed out that the E-W trending mineral lineation of the Sanbagawa schists in Shikoku, which runs parallel to the trend of the belt, is a stretching lineation due to eastward flow of the Sanbagawa schists. Rotation of albite and garnet porphyroblasts supports his idea. Thus, according to Faure (1982), the first stage of deformation involved eastward flow. Some of the so-called south vergency folds are sheath folds associated with this eastward flow. Faure considers the south vergent folds to be of secondary significance. As garnet rotation with a N-S axis started from an earlier phase of its growth (Sakai personal communication), the author postulates that the N–S compression yielding radiolaria rods took place at

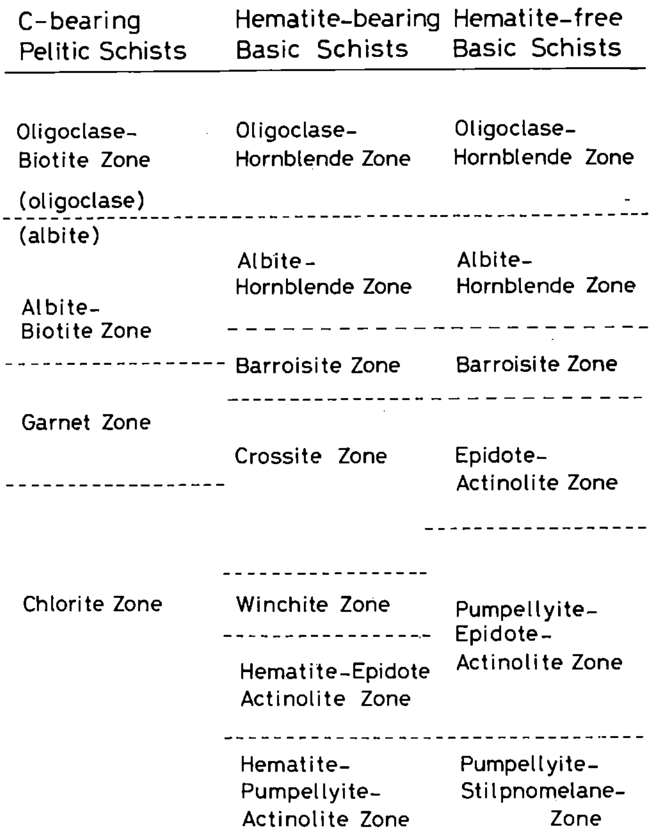

C-bearing Pelitic Schists	Hematite-bearing Basic Schists	Hematite-free Basic Schists
Oligoclase-Biotite Zone (oligoclase)	Oligoclase-Hornblende Zone	Oligoclase-Hornblende Zone
(albite)		
Albite-Biotite Zone	Albite-Hornblende Zone	Albite-Hornblende Zone
	Barroisite Zone	Barroisite Zone
Garnet Zone	Crossite Zone	Epidote-Actinolite Zone
Chlorite Zone	Winchite Zone / Hematite-Epidote Actinolite Zone	Pumpellyite-Epidote-Actinolite Zone
	Hematite-Pumpellyite-Actinolite Zone	Pumpellyite-Stilpnomelane-Zone

Figure 4. Mineral facies series of the Sanbagawa belt in central Shikoku. Hirajima (1983*) has suggested the presence of augite-chlorite zone on the lower-grade side of the pumpellyite-actinolite zone in the Kanto Mountains.

the very beginning of the Sanbagawa metamorphism and was then followed by easterly flow. The south vergency large-scale fold, recumbent or a part of nappe, seems necessary, however, to explain the melange zone where peridotites and eclogitized metagabbros occur (Kunugiza et al. this volume). The tectonic history of the Sanbagawa belt is far from being clear.

From an earlier age of Japanese geology it was thought that the metamorphic grade of the Sanbagawa schists in central Shikoku increased higher in the stratigraphic succession. Petrographic work based on extensive use of microprobe work done in the mid-1970s has elucidated fine aspects of the thermal structure and shown that the highest grade part is located in the middle of the apparent stratigraphic column. Figure 5 is a cross section showing the thermal structure of the Asemi-gawa area of central Shikoku. The mineral zone map of central Shikoku refers to Figure 2 of Kunugiza et al. (this volume). Naturally, many ideas have been presented to explain such a thermal structure. Kawachi (1968) proposed a large scale thrust fault that brought the higher-grade rocks in the midst of the lower-grade ones. Hara et al. (1977*) and Faure (1983) also proposed nappes, but in different position from Kawachi's. Banno et al. (1978) considered the structure to be a recumbent fold because of the absence of a distinct discontinuity of metamorphic grade in the areas of alleged

nappes. It may be absurd to deny explicitly the presence of a thrust that is more or less parallel to lithological boundaries, because a shear zone has been known in the garnet zone and the mineralogy is somewhat anomalous there. However, petrographic evidence leads the author to be skeptical of the concept of the thermal structure having been disturbed by a large scale nappe. The geology and petrology of the Sanbagawa belt may need independent review.

Kurosegawa Zone

The Kurosegawa Zone is a fault zone running more or less E-W in the Chichibu terrain in *sensu lato* or between the Chichibu in *sensu stricto* and Sanbosan terranes. According to Maruyama et al. (1984), it is a serpentinite mélange zone composed of granodiorite and gneiss of 400 Ma, a Siluro-Devonian sedimentary sequence, Permo-Carboniferous and Lower Mesozoic formations, high pressure schists and serpentinite. The serpentinite acts as lubricant to fill their interstices or as media to float them. Two kinds of high pressure rocks were recognized; one has 320–370 Ma K-Ar ages and is characterized by a pumpellyite-glaucophane assemblage in the area studied in detail by Maruyama et al. (1984), who considered the Kurosegawa Zone to be fragments of an island arc or arcs that collided with the main Japanese Islands in Jurassic–Early Cretaceous time.

The eastern limit of the Kurosegawa zone is the eastern tip of Kii peninsula. But Hirajima (1984*) found that the thrust sheet on the Sanbagawa schists of the Kanto Mountains contains jadeite + quartz rock and greenstone mélange, which reminds us of the association of tectonic blocks of the Kurosegawa zone.

NANSEI SHOTO (RYUKYU ISLANDS)

The bedrock geology of the Nansei Shoto (Ryukyu Islands) may be the extension of the Outer Zone of Southwest Japan. The Yaeyama metamorphics occur on Iriomote and Ishigaki islands and associated islets located near the southern end of the Ryukyu Islands. However, they have 180 Ma radiometric ages and therefore cannot be the westernmost part of the Sanbagawa terrane. No high pressure schists of comparable type and age are known in Japan or Taiwan. The assemblages of two metamorphic facies are present in the high pressure rocks, the albite-lawsonite facies and the epidote-glaucophane facies, the latter includes schists containing pumpellyite, crossite, epidote, or actinolite (Kizaki and Watanabe 1977*).

CONCLUDING REMARKS

In concluding the review, a few comments are presented here to emphasize some of the significance and problems related to the high pressure metamorphic rocks of the Japanese Islands.

Paired Metamorphic Belts

Since the concept of paired metamorphic belts was pro-

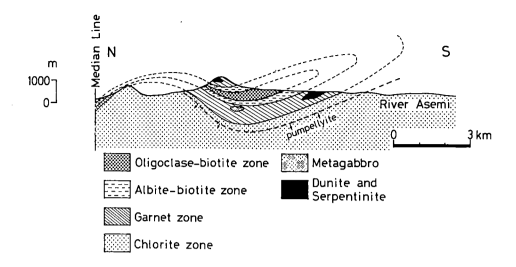

Figure 5. Thermal structure of the Sanbagawa belt in the Asemi-gawa area, central Shikoku. Modified
Figure 5 of Banno et al. (1978), mainly based on personal communication of Sakai.

posed by Miyashiro (1961), it has been largely accepted by Japanese petrologists, if not structural geologists. Quite a few arguments have been made concerning the pairing, age and source regions.

Hashimoto (1978) mentioned that not all of the high pressure terranes are paired with a low pressure metamorphic belt, and pointed out that the East Abukuma-Motai Zone, the Hida Marginal Zone, and the Kamuikotan Zone are not paired with low pressure belts. The high pressure rocks in the Kurosegawa Zone are also not paired with a low pressure metamorphic belt in the present Japanese Islands.

It appears that the high-pressure subducted glaucophanitic metamorphic terranes were originally paired with low P-high T terranes. But the present tectonic positions of some of the terranes are at the sutures of contrasting terranes. They are associated with huge serpentinite bodies, which, in addition to glaucophanites, contain a low pressure ophiolite sequence (Kamuikotan Zone), unmetamorphosed Paleozoic formations (Hida Marginal Zone), or granodiorite and gneiss (Kurosegawa Zone). But the Sangun and Sanbagawa Belts, both large coherent terranes, are not fractured rocks at sutures. Based on age and tectonic positions, they appear to be related to the terrane of granitic intrusion and low pressure metamorphism, the Hida and Ryoke belts, respectively. We may envisage that these two pairs are more or less in situ pairs, genetically related to each other. However, as discussed by Miyashiro (1973a), Faure (1982), and Maruyama et al. (1982*), the present position of the Ryoke and Sanbagawa belts may not be strictly in situ because of their direct contact and the movement of the Median Tectonic Line.

Although the Kamuikotan Zone is associated with abundant serpentinite, there are a few coherent areas such as the Horokanai area. Thus it may not be a serpentinite mélange at the suture of terranes, but is more or less in situ and covered by a thrust sheet of a low pressure terrane. The pairing of the Kamuikotan zone

with the Hidaka belt was once suggested, but the radiometric age of the Hidaka belt is 30–17 Ma, and cannot be combined with the Kamuikotan Zone of 70–110 Ma. A possible candidate to pair with the zone is the granite terrane in southwestern Hokkaido, which is the extension of Northeast Japan.

Jadeite-Quartz Assemblage

So far no regional coherent area in the Japanese Islands contains the jadeite + quartz assemblage. This assemblage has been described and confirmed by microprobe in the following areas.

Kamuikotan Metamorphics. It could be limited to tectonic blocks, but certainly occurs in recrystallized amphibolite and metamorphosed tonalite (or plagiogranite).

Sanbagawa Schists. Jadeite + quartz is not found in Sanbagawa schists. Omphacite occurs in veins. Sodic augite and aegirinaugite commonly occur as matrix minerals containing up to 20 percent of jadeite (Hirajima 1983*). However, the jadeite + quartz assemblage occurs in an exotic tectonic block thrust over the Sanbagawa schists associated with Tertiary non-metamorphic sediments (Hirajima 1984*). Sanbagawa schists in Shikoku contain omphacite + garnet + quartz (eclogite) (Takasu 1984; Kunugiza et al. this volume).

Kurosegawa Zone. Jadeite + quartz occurs in exotic blocks measuring 10 to 20 meters across in serpentinite near Kochi (Maruyama et al. 1978).

Hida Marginal Zone. An exotic block of jadeite + quartz assemblage occurs in serpentinite (Komatsu and Yamasaki 1981*).

Eclogites

The Bessi area of the Sanbagawa metamorphic belt has been the sole locality of eclogitic rocks in the Japanese Islands, but a few new localities of eclogitic rocks are now known. The following is the list of localities of eclogitic rocks in the Japanese Islands.

Sanbagawa Belt. Garnet clinopyroxenite is in Mt. Higashi-Akaishi peridotite (Kunugiza et al. this volume). Eclogites occur in metagabbros near the peridotite body mentioned above (Kunugiza et al. this volume). Garnet-augite rock occurs in the Kawamai body in the Yawatahama area, western Shikoku, which is an olistolith in the original sediments of the Sanbagawa schists. The Kawamai body later suffered the Sanbagawa metamorphism with resultant formation of glaucophane around hornblende of amphibolite (Ishimoto 1974).

Kurosegawa Zone. Garnet-augite rock occurs in garnet amphibolite with Fe-Mg partitioning corresponding to the granulite facies (Yoshikura et al. 1981).

Hida Marginal Zone. Tectonic blocks of eclogite occur in serpentinite (Komatsu and Yamasaki 1981*).

Main Hida Belt. Suzuki (1977) described garnet-augite

REFERENCES CITED

Memoirs and Reports of Workshops

Chihara, K., ed., 1980, 1981, Hida Marginal Belt vols. I and II, Workshop of K. Chihara, Niigata University MARGINALBELT—I & II.

Hara, I., ed., 1981, Tectonics of Paired Metamorphic Belts. Workshop of I. Hara, Hiroshima University PAIREDBELTS.

Hashimoto, M., and Uyeda, S., 1983, Accretion Tectonics in the Circum-Pacific Regions: Tokyo, Terra Science Publishing Company. ACCRETION.

Hide, K., ed., 1977, Sambagawa belt, Commemorate volume to Prof. G. Kojima: Hiroshima, Hiroshima University Press. SANBAGAWA.

Nishimura, Y., ed., 1983, 1984, High Pressure Belt of Inner Zone vols. I-II Workshop of Y. Nishimura, Yamaguchi University. INNERZONE—I & II.

Suwa, K., ed., 1979, *Basement of the Japanese Islands* Commemorate volume to Prof. H. Kano, Akita University. BASEMENT.

The following are the abbreviations for frequently referred to Japanese journals.
JGSJ: Journal of Geological Society of Japan.
JJAMPEG: Journal of Japanese Association of Mineralogists, Petrologists and Economic Geologists.

Papers in English

Asahina, T., and Komatsu, M., 1979, The Horokanai ophiolite complex in the Kamuikotan tectonic belt, Hokkaido, Japan: JGSJ, v. 83, p. 317-330.

Banno, S., 1964, Petrologic studies on Sanbagawa crystalline schists in the Bessi-Ino district, central Sikoku, Japan: Journal of Faculty of Science, University of Tokyo, sec. 2, v. 15, p. 203-319.

Banno, S., 1981, Regional metamorphic rocks: Recent Progress of Natural Science of Japan: Science Council of Japan, v. 5, p. 32-40.

Banno, S., Higashino, T., Otsuki, M., Itaya, T., and Nakajima, T., 1978, Thermal structure of the Sanbagawa metamorphic belt in central Shikoku: Journal of Physics of the Earth, Supplementary, p. 345-356.

Banno, S., Ishizuka, H., Gouchi, N. and Imaizumi, M., 1978, Kamuikotan belt in Hokkaido: the tectonic contact of high pressure metamorphic belt and low pressure ophiolite succession [Abs.]: International Geodynamics Conference, Tokyo, p. 14-15.

Bickerman, M., Minato, M., and Hunahashi, M., 1971, K-Ar ages of the garnet

rock from the amphibolite facies terrane. As granulite occurs locally in the Main Hida Belt, it may belong to the remnant of the granulite facies metamorphism in this terrane.

ACKNOWLEDGMENTS

The author wishes to thank Prof. S. Mizutani for permission to quote Figure 1 of Mizutani and Hattori (1983) with some modification, and to Mr. C. Sakai for drafting some of the figures. Professor E. H. Brown and D. Silverberg helped to improve the manuscript. Professor M. Hashimoto and T. Matsuda, Y. Isozaki, and H. Ishizuka read the manuscript and sent critical comments to the author. These gentlemen are thanked very much, though I did not accept all of their comments. Petrographic work leading to the above discussion was done on the Grants-in-Aid of the Ministry of Education 53321509, 54410904, and 55520105, and a part of the cost of preparing this manuscript was defrayed by another grant, 58540524.

amphibolite of the Mitsuishi district, Hidaka province, Hokkaido Japan: Journal of Association for Geological Collaboration, Japan, v. 25, p. 27-30.

Chihara, K. and Komatsu, M., 1981, Tectonics of Hida Marginal belt and Joetsu belt: PAIREDBELTS, p. 135-142.

Chihara, K., Komatsu, M., Uemura, T., Hasegawa, Y., Shiraishi, S., Yoshimura, T. and Nakamizu, M., 1979, Geology and tectonics of the Omi-Renge and Joetsu tectonic belts: Science Report, Niigata Univ., v. 5, p. 1-61.

Den, N., and Hotta, H., 1973, Seismic refraction and reflection evidence supporting plate tectonics in Hokkaido: Papers in Meteorology and Geophysics, v. 24, p. 31-54.

Faure, M., 1982, Phase précoce Ouest-Est de la zone Sanbagawa dans la partie orientale de Shikoku (Japan SW): Compte Rendus, Académe des Sciences, Paris, v. 295, p. 505-501.

Faure, M., 1983, Eastward ductile shear during the early tectonic phase in the Sanbagawa belt: JGSJ, v. 89, p. 319-329.

Hashimoto, M., 1968, Glaucophanitic metamorphism of the Katsuyama district, Okayama Prefecture, Japan: J. Fac. Sci., Univ. Tokyo, sec. 2, v. 17, p. 99-162.

Hashimoto, M., 1978, Two kinds of glaucophanite terrains in Japan and the environs: Bulletin of the National Science Museum (Tokyo), sec. C, v. 4, p. 157-164.

Hashimoto, M., Igi, S., Seki, Y., Banno, S., and Kojima, G., 1970, Metamorphic facies map of Japan (scale 1:2,000,000) with an explanatory text: Geological Survey of Japan, Tsukuba.

Hiroi, Y., 1981, Subdivision of the Hida metamorphic complex, central Japan, and its bearing on the geology of the Far East in pre-Sea of Japan time: Tectonophysics, v. 76, p. 317-333.

Igo, H., Adachi, S., Furutani, H., and Nishiyama, H., 1980, Ordovician fossils first discovered in Japan: Proceedings of the Japan Academy, v. 56, ser. B, p. 499-503.

Ishimoto, N., 1974, Regional metamorphism of the Yawatahama district in western Shikoku, Japan: Science Report, Tohoku University, Ser. III, v. 12, p. 279-330.

Ishiwatari, A., 1985, Granulite-facies metacumulate of the Yakuno ophiolite, Japan: evidence for unusually thick oceanic crust: Journal of Petrology, v. 26, p. 1-30.

Ishizuka, H., Imaizumi, M., Gouchi, M., and Banno, S., 1983, The Kamuikotan zone in Hokkaido, Japan: tectonic mixing of high-pressure and low-pressure metamorphic rocks: Journal of Metamorphic Geology, v. 1, p. 263-275.

Isozaki, Y. and Matsuda, T., 1980, Age of the Tamba Group along the Hozugawa "anticline," Western Hills of Kyoto, Southwest Japan: Journal of Geoscience, Osaka City University, v. 23, p. 115–134.

Iwasaki, M., 1979, Gabbroic breccia (olistostrome) in the Mikabu greenstone belt of eastern Shikoku: JGSJ, v. 85, p. 481–487.

Kawachi, Y., 1968, Large-scale overturned structure in the Sambagawa metamorphic zone in central Shikoku, Japan: JGSJ, v. 74, p. 607–616.

Kawachi, Y., Watanabe T., and Landis, C. A., 1982, Origin of mafic volcanogenic schists and related rocks in the Sanbagawa belt, Japan: JGSJ, v. 88, p. 797–817.

Kobayashi, T., 1941, The Sakawa orogenic cycle and its bearing on the origin of the Japanese Islands: Journal of Faculty of Science, Tokyo Imperial University, sec. 2, v. 5, p. 219–578.

Komatsu, M., Miyashita, S., Maeda, J., Osanai, Y., and Toyoshima, T., 1983, Disclosing of a deepest section of continental-type crust up-thrust as the final event of collision of arcs in Hokkaido, North Japan: *ACCRETION*, p. 149–165.

Kuroda, Y. and Ogura, Y., 1963, Epidote amphibolite from the western Abukuma Plateau: Science Report, Tokyo Kyoiku Univ., ser. C., v. 8, p. 245–268.

Maekawa, H., 1983, Submarine sliding deposits and their modes of occurrence of the Kamuikotan metamorphic rocks in the Biei area, Hokkaido, Japan: Journal of Faculty of Science, University of Tokyo, sec. II, v. 20, p. 489–507.

Maruyama, S., and Yamasaki, M., 1978, Paleozoic submarine volcanoes in the high-P/T type metamorphosed Chichibu system of eastern Shikoku, Japan: Journal of Volcanology and Geothermal Research, v. 4, p. 199–216.

Maruyama, S., Ueda, Y. and Banno, S., 1978, 208–240 m.y. old jadeiteglaucophane schists in the Kurosegawa tectonic zone near Kochi city, Shikoku: JJAMPEG, v. 73, p. 300–310.

Maruyama, H., Banno, S., Matsuda, T., and Nakajima, T., 1984, Kurosegawa zone and its bearing on the development of the Japanese Islands: Tectonophysics, v. 110, p. 47–80.

Miyashiro, A., 1961, Evolution of metamorphic belts: Journal of Petrology, v. 2, p. 277–311.

Miyashiro, A., 1973a, Paired and unpaired metamorphic belts: Tectonophysics, v. 17, p. 241–254.

Miyashiro, A., 1973b, Metamorphism and metamorphic belts: London, George Allen and Unwin.

Mizutani, S., and Hattori, I., 1983, Hida and Mino: tectonostratigraphic terranes in central Japan: ACCRETION, p. 169–178.

Nishimura, Y., 1971, Regional metamorphism of the Nishikicho district, southwest Japan: Journal of Science, Hiroshima Univ., ser. C, v. 6, p. 203–268.

Nishimura, Y., 1981, Tectonics of the Sangun belt: PAIREDBELTS, p. 143–146.

Nozawa, T., 1977, Radiometric age map of Japan (scale 1:2,000,000): Geological Survey of Japan, Tsukuba.

Saito, Y. and Hashimoto, M., 1982, South Kitakami region: an allochthonous terrane in Japan: Journal of Geophysical Research, v. 87, p. 3691–3696.

Sashida, K., Igo, H., Takizawa, S., Hisada, K., 1982, On the occurrence of Jurassic radiolarians from the Kanto Region and Hida Mountains, central Japan: Annual Report, Institute of Geology, University of Tsukuba, v. 8, p. 74–77.

Shibakusa, H., 1981, Metamorphism of Kamuikotan tectonic belt: PAIREDBELTS, p. 43–48.

Sugisaki, R., Mizutani, S., Hattori, H., Adachi, M., and Tanaka, T., 1972, Late Paleozoic geosynclinal basalt and tectonism in the Japanese Islands: Tectonophysics, v. 14, p. 35–56.

Suzuki, M., 1977, Polymetamorphism in the Hida Metamorphic belt, central Japan: Journal of Science, Hiroshima Univ., Ser. C, v. 7, p. 217–296.

Taira, T., Saito, Y., and Hashimoto, M., 1982, The role of oblique subduction and strike-slip tectonics in the evolution of Japan, *in* Hilde, T.W.C. and Uyeda, S. Geodynamics of the Western Pacific-Indonesian Region: AGU/GSA Geodynamics Series, v. 11, p. 303–316.

Takasu, A., 1984, Prograde and retrograde eclogites in the Sambagawa metamorphic belt, Besshi district, Japan Journal of Petrology, v. 25, p. 619–643.

Toriumi, M., 1982, Strain, stress and uplift: Tectonics, v. 1, p. 57–76.

Watanabe, T., Kobayashi, H., and Sengan, H., 1983, Lawsonite from quartzofeldspathic schist in the Sangun metamorphic belt, Shimane Prefecture: Memoirs of Faculty of Science, Shimane University, v. 17, p. 81–86.

Watanabe, T. and Kobayashi, H., 1984, Occurrence of lawsonite in pelitic schists from the Sanbagawa metamorphic belt, central Shikoku, Japan: Journal of Metamorphic Geology, v. 2, p. 365–369.

Yoshikura, S., Shibata, K., and Maruyama, S., 1981, Garnet-clinopyroxene amphibolite from the Kurosegawa tectonic zone, near Kochi City, and K-Ar ages: JJAMPEG, v. 76, p. 102–109.

Papers in Japanese

Full titles may be requested from the author.

Hara, I., Hide, K., and Nishimura, Y., 1980, Structural Geology Newsletter, no. 25, p. 1–12.

Hara, I., Hide, K., Takeda, K., Tsukuda, E., Tokuda, M., and Shioda, M., 1977, SANBAGAWA, p. 307–390.

Hara, I. Hide, K., and Nishimura, Y. (1980) Structural Geology Newsletter, v. 25, p. 1–10.

Hashimoto, M., 1972, Bulletin of National Science Museum (Tokyo), v. 15, p. 767–775.

Hashimoto, M. and Igi, S., 1970, JGSJ, v. 76, p. 159–160.

Hirajima, T., 1983, JGSJ, v. 89, p. 679–691.

Hirajima, T., 1984, JGSJ, v. 90, p. 629–642.

Igi, S., 1959, JGSJ, v. 65, p. 173–183.

Imaizumi, M., 1984, JJAMPEG, v. 79, p. 1–19.

Imaizumi, M., and Ueda, Y., 1981, JJAMPEG, v. 76, p. 88–92.

Ishizuka, H., 1980, JGSJ, v. 86, p. 119–134.

Isozaki, Y., Maejiama, W., and Maruyama, S., 1981, JGSJ, v. 87, p. 555–558.

Iwasaki, M., 1978, Journal of the Association for Geological Collaboration of Japan (Chikyuka-gaku), v. 32, p. 345–351.

Kanmera, K., and Hashimoto, M., 1979, *in* Kanmera, K., Hashimoto, M., and Matsuda, T., Geology of Japan, Tokyo, Iwanami-Shoten, p. 5–94.

Karakida, Y., Yamamoto, H. and Hayama, Y., 1984, INNERZONE-II, p. 23–30.

Kizaki, K., and Watanabe, T., 1977, SANBAGAWA, p. 283–288.

Komatsu, M. and Yamasaki, T., 1981, MARGINALBELT-II, p. 1–11.

Maekawa, H., 1981, JGSJ, v. 87, p. 543–554.

Maruyama, S., Seno, T., and Engebretson, D., 1982, [Abs.]: 89th Ann. Meeting, Geological Society of Japan.

Matsumoto, K., 1980, MARGINALBELT-I, p. 1–14.

Minamishin, M., 1979, JJAMPEG, v. 74, p. 153.

Miyashita, S., 1983, JGSJ, v. 89, p. 69–86.

Nakamura, K., 1983, Bulletin of the Earthquake Research Institute, v. 58, p. 711–722.

Nakano, N., 1981, JGSJ, v. 87, p. 211–224.

Nakano, N., and Komatsu, M., 1979, JGSJ, v. 85, p. 367–376.

Nishimura, Y., Inoue, T., and Yamamoto, H., 1977, SANBAGAWA, p. 257–282.

Nishiyama, T., 1979, M.S. Thesis, Kanazawa Univ.

Sakakibara, M., 1984, Abstract 91st Annual Meeting, Geological Society of Japan.

Shibata, K., 1981, MARGINALBELT-II, p. 62–63.

Shibata, K. and Nishimura, Y., 1984, INNERZONE-II, p. 31–32.

Shitaba, K., Nozawa, T. and Utsumi, S., 1980, MARGINALBELT-I, p. 110–112.

Soma, T., Matsushima, K., Yamamoto, M., and Maruyama, S., 1981, MARGINALBELT-II, p. 96–104.

Suyari, K., Kuwano, Y. and Ishida, K., 1980, JGSJ, v. 86, p. 827–828.

MANUSCRIPT ACCEPTED BY THE SOCIETY JULY 29, 1985

Printed in U.S.A.

Geological Society of America
Memoir 164
1986

The origin and metamorphic history of the ultramafic and metagabbro bodies in the Sanbagawa metamorphic belt

K. Kunugiza*
Institute of Mineralogy, Petrology, and Economic Geology
Tohoku University
Sendai 980, Japan

A. Takasu
S. Banno
Department of Geology and Mineralogy
Faculty of Science
Kyoto University
Kyoto 606, Japan

ABSTRACT

The higher-grade region of the Sanbagawa metamorphic belt in Shikoku and the Kii peninsula contains metagabbro, peridotite, and serpentinite. These rocks occur exclusively in the garnet, albite-biotite, and oligoclase-biotite zones, and tend to have equilibrium mineral assemblages stable with respect to the mineral zones where they occur. Thermal histories of previous equilibrium stages can be discerned through not only the mineralogy of relics, but also the texture, megascopic structure, and bulk composition of the rocks. The protoliths were layered gabbro, cumulate and residual peridotites, and garnet clinopyroxenite. Some of these rocks were in the eclogite facies before emplacement into the Sanbagawa schists and others were in the granulite facies. They were emplaced into the metamorphic regime by solid intrusion during a later stage of the schistosity formation, that is, emplacement was syntectonic. Some peridotites were serpentinized before attaining equilibrium assemblages of the Sanbagawa metamorphism; their emplacement was during an earlier stage of the metamorphism, but not at the sedimentary stage. The source region of those exotic blocks may have been a lower crust-upper mantle region of an island arc.

INTRODUCTION

The Sanbagawa metamorphic belt is a high-pressure intermediate type (Miyashiro 1961) composed mainly of pelitic and basic schists, but small numbers of ultramafic and metagabbro bodies occur as lenticular and concordant bodies with a maximum size of 7 × 2 km² (Iratsu metagabbro) in the pelitic schists. During the 1970s much effort was devoted to understanding the thermal history of the ultramafic and metagabbro bodies (Mori and Banno 1973; Yokoyama and Mori 1975; Yokoyama 1976,

1980; Banno et al. 1976). Yokoyama and Mori (1975) and Yokoyama (1976, 1980) have indicated that the Nikubuchi peridotite and the Iratsu metagabbro bodies suffered at least two metamorphisms, an earlier granulite facies and a later Sanbagawa event. Banno et al. (1976) have suggested that the Iratsu epidote amphibolite and other coarse-grained epidote amphibolite bodies with gneissose structure are metamorphosed gabbros, based on the conspicuous banding of mafic and felsic lithologies. We call them metagabbros to distinguish them from metamorphosed basalts and associated tuffs, which will be called basic schists hereafter.

*Present address: R & D Laboratory of New Products, Onoda Cement Co., Ltd., Toyosu, Koto-ku, Tokyo 135, Japan.

One of our main concerns about these bodies has been the emplacement tectonics in relation to the Sanbagawa orogeny. It has become clear that the ultramafic and metagabbro bodies of the Sanbagawa metamorphic belt have an origin and thermal history more complex and diverse than hitherto considered (cf. Banno et al. 1976; Banno and Yokoyama 1977). Kunugiza (1984) has indicated that there are at least four types of ultramafic bodies in the Sanbagawa metamorphic belt. Takasu (1984a) has found a relict high-temperature eclogite body, the Sebadani metagabbro, which is accompanied by a peripheral high-pressure contact metamorphic aureole. Since these ultramafic and metagabbro bodies occur only in a well-defined pelitic schist formation, customarily called the Upper Member of the Minawa Formation, Kunugiza (1984) and Takasu (1984a, b) have interpreted them as tectonic blocks in a mélange zone.

In the present paper, after a brief description of the geology, we summarize the origin and the thermal history of the ultramafic and metagabbro bodies, and discuss the emplacement of these bodies in the framework of the Sanbagawa metamorphism, if only to clarify the problems that face us in 1985.

OUTLINE OF GEOLOGY

Figure 1 is an index map showing the location of the study

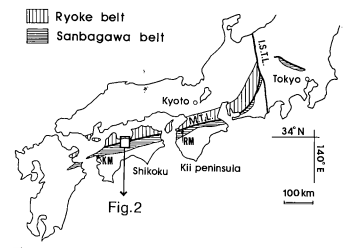

Figure 1. Location of the Sanbagawa metamorphic belt in Southwest Japan. The small area in central Shikoku is enlarged in Figure 2. M.T.L.: Median Tectonic Line; I.S.T.L.: Itoigawa-Shizuoka Tectonic Line; KM: Kawamai peridotite; RM = Ryumon peridotite.

areas. The Sanbagawa metamorphic belt consists of high-pressure schists derived originally from Mesozoic sediments (Suyari et al. 1980) and volcanic rocks with small numbers of ultramafic and metagabbro bodies.

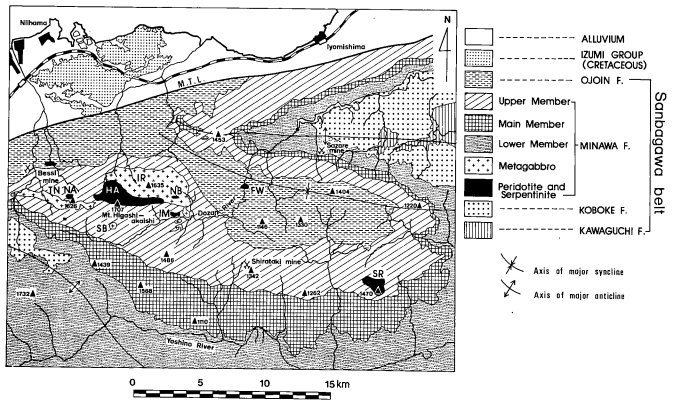

Figure 2. Geological map of central Shikoku (compiled from Geological Section, Besshi Mine, Sumitomo Metal Mining 1966, Takasu and Makino 1980a, and Kenzan Research Group 1984). TN, IR and SB: Tonaru, Iratsu, and Sebadani metagabbro bodies, respectively; NA, HA, NB, IM, FW, and SR: Nishiakaishi, Higashiakaishi, Nikubuchi, Imono, Fujiwara, and Shiragayama ultramafic bodies, respectively; M.T.L.: Median Tectonic Line.

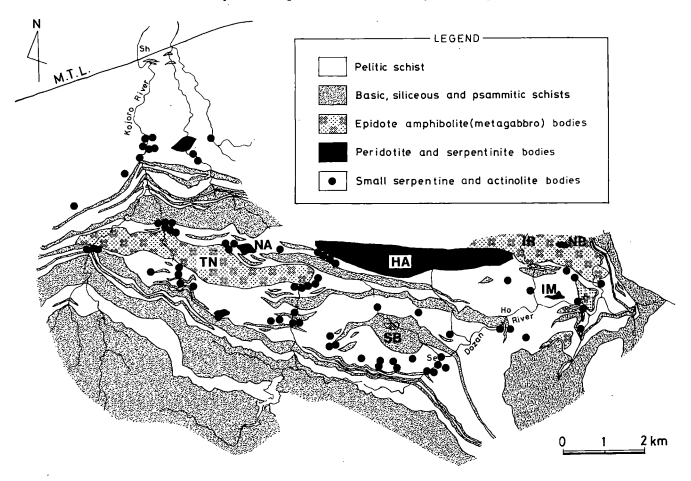

Figure 3. Distribution of actinolitic and small serpentinite bodies in the Bessi district, central Shikoku (after Takasu 1984b). For abbreviations refer to Figure 2.

Figure 2 is a geological map of the Sanbagawa metamorphic belt in central Shikoku, where many ultramafic and metagabbro bodies occur. Ultramafic rocks and metagabbros on the Kii Peninsula occur in an area comparable in lithology and metamorphic grade to that of central Shikoku.

As seen in Figures 2 and 3, ultramafic and metagabbro bodies occur in the Upper Member of the Minawa Formation, but small numbers of serpentinite bodies also occur in the lower part of the Ojoin Formation, which overlies the Minawa Formation. The Upper Member of the Minawa Formation, which surrounds the ultramafic rocks and metagabbros, consists predominantly of pelitic schists with an intercalation of basic, siliceous, and psammitic schists. Some basic and siliceous schists are well traced as marker beds, but in the middle part of the Upper Member there are lenticular and discontinuous layers of basic and siliceous schists adjacent to the ultramafic and metagabbro bodies (Hara et al. 1977; Takasu 1984b). The apparent thickness of the Upper Member varies from 550 to 3300 m in an area about 15×10 km^2 (Takasu and Makino 1980a), and the thicker the Upper Member, the larger the number of bands and volume of the basic and siliceous schists in it.

The ultramafic and metagabbro bodies include prograde serpentinite and retrograde peridotite and metagabbro. Small actinolite bodies are scattered in the pelitic schists in the Upper member (Figure 3), and were derived originally from serpentinite (Takasu 1984b). The ultramafic and metagabbro bodies occur as lenticular or sheet-like bodies, measuring from 7×2 km^2 (Iratsu metagabbros) to several square meters. They are mostly concordant with the pelitic and basic schists.

The compositional banding and gneissose foliation in the metagabbros are generally parallel to the schistosity of the surrounding schists, but are occasionally oblique to it. At the contact between the serpentinite and schists, there are sometimes reaction zones several meters wide, containing chlorite, talc, tremolite, antigorite, diopside, and grossular in various combinations (Kunugiza 1984). No contact metamorphic aureole has been confirmed except in the schists around the Sebadani metagabbro body (Takasu 1984a).

Based on the mapping of marker beds, analysis of structural elements and the distribution of mineral zones, large-scale recumbent anticlines with southward vergence have been proposed by many authors (e.g., Kojima 1951; Hide et al. 1956; Kawachi 1968; Hara et al. 1977; Banno et al. 1978; Faure 1983), but different interpretations are given regarding the location and significance of the axial planes, recumbent synclines, and thrust faults (Takasu and Makino 1980b). The metamorphic grade will be referred to in terms of the mineral zone of the pelitic schists: chlorite, garnet, albite-biotite and oligoclase-biotite zones, in order of increasing grade (Banno et al. 1978; Enami 1982).

ORIGIN AND THERMAL HISTORY

Banno and Yokoyama (1977) have considered that the ultramafic and metagabbro bodies of the Sanbagawa metamorphic belt were materials of the upper mantle and lower crust, and were tectonically emplaced. However, the diversity in origin and thermal history of these rocks has been revealed, and is subsequently far more distinct than earlier supposed. We start our consideration first by summarizing the recent data on the origin and thermal history of six ultramafic and five metagabbro bodies and providing a brief review of small ultramafic and metagabbro bodies. Details for individual bodies may be found in the literature quoted. The thermal history of the ultramafic rocks is summarized in Figure 4, and that of the metagabbros in Figure 6.

Higashiakaishi (HA) Peridotite

The HA peridotite body is lenticular and concordant, measures 5 × 1.5 km², and occurs in the pelitic schists of the biotite zone. It consists predominantly of dunite with small amounts of wehrlite, clinopyroxenite, garnet peridotite, garnet clinopyroxenite, and chromitite. Both compositional and cryptic layerings are developed in the body.

The HA peridotite was formed as a cumulate in the garnet lherzolite facies (Figure 4); the bimineralic nature of the garnet clinopyroxenite and the presence of garnet dunite cannot be attributed to subsolidus reactions (Mori and Banno 1973; Kunugiza 1981). Kunugiza (1981) has suggested that after the crystallization of dunite from a magma, the garnet peridotite, and then garnet clinopyroxenite, which often contain chromite, crystallized as the result of the reaction: a) spinel (chromite in this case) + liquid → olivine + garnet, followed by b) olivine + liquid → garnet + clinopyroxene. The reaction a) is a continuous one, crystallizing olivine + garnet + chromite, when spinel contains Cr as a major component. The crystallization of the HA took place at the pressure range of reactions a) and b); 19–28 kb in the simple system of CMAS of Kushiro and Yoder (1974) or at somewhat higher pressures in the Cr-bearing system (cf. MacGregor 1970).

After its formation as a cumulate body, the HA peridotite suffered metamorphism that is recorded in Fe-Mg partitioning among the cores of garnet, clinopyroxene, orthopyroxene, olivine, and spinel. The stage at which the cores of these minerals were in equilibrium with respect to Fe-Mg partitioning will henceforth be called the HA stage. Mori and Banno (1973) gave 550°C and 7–13 kb for the HA stage, but olivine-spinel (Evans and Frost 1975) and garnet-clinopyroxene (Ellis and Green 1979) geothermometers give about 700°C. Experimental and theoretical P-T grids for the reaction boundary between the

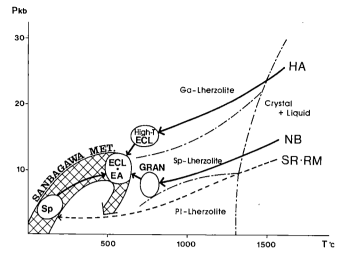

Figure 4. Metamorphic history of the ultramafic bodies. HA, NB, SR, and RM: Higashiakaishi, Nikubuchi, Shiragayama, and Ryumon ultramafic bodies, respectively; ECL: Eclogite; GRAN: Granulite; EA: Epidote amphibolite; Sp: Serpentinite; Ga-Lherzolite: Garnet lherzolite facies; Sp-Lherzolite: Spinel lherzolite facies; Pl-lherzolite: Plagioclase lherzolite facies; SANBAGAWA MET: Prograde and retrograde paths of the Sanbagawa metamorphism.

garnet and spinel lherzolite facies (O'Hara et al. 1971; Obata 1976; Herzberg 1978; Jenkins and Newton 1979) combined with geothermometers, give the equilibrium condition of the HA stage at about 700°C and 15 kb. After the HA stage, the peridotite was subjected to the Sanbagawa metamorphism, which is represented by the rim-rim partitioning of Fe-Mg between garnet and clinopyroxene followed by later transition of garnet-clinopyroxenite to epidote amphibolite accompanied by decrease of temperature, by which chlorite, epidote, hornblende, and almandine garnet were formed. Almandine developed on the rim of the pyrope garnet but also formed euhedrally in the chlorite-epidote-hornblende matrix. For the peridotites, the Sanbagawa metamorphism gave rise to the formation of antigorite and diopside, and ferritchromitization of chromite, although serpentinization only affected small parts of the body.

Nishiakaishi (NA) and Imono (IM) Peridotites

The NA and IM bodies are small peridotites (0.5 × 0.3 km² and 0.5 × 0.2 km²). Both consist of dunites and are accompanied by rare wehrlites. Their mineralogy and texture are similar to those of the HA peridotite except for the absence of garnet. Relict chromite forms layering suggestive of cumulate origin.

Nikubuchi (NB) Peridotite

The NB peridotite, 0.5 × 0.2 km², is completely contained in the Iratsu (IR) metagabbro to be discussed later. Constituent rock types of the igneous stage were primarily dunite, accompanied by spinel wehrlite, spinel pyroxenite, and gabbro. The compositional banding of these rock-types is conspicuous. Green aluminous spinel is a common minor mineral even in dunite.

According to Yokoyama (1980), the thermal history of this body may be summarized as follows (Figure 4). Its layering, both rhythmic and cryptic, suggests a cumulate origin. The presence of olivine + spinel and the absence of olivine + plagioclase in the igneous stage suggest crystallization of 12–18 kb, that is, in the spinel granulite facies. Later, granulite facies recrystallization gave rise to the formation of the following mineral assemblages: olivine + aluminous clinopyroxene and aluminous orthopyroxene, Al-clinopyroxene + Al-orthopyroxene + spinel + plagioclase, and their partial assemblages. The two-pyroxene geothermometer (Mori and Green 1978) and the stability field of the assemblages above suggest 750°C and 5–10 kbars for this stage. A part of the minerals of the granulite stage was recrystallized during the Sanbagawa metamorphism. Major minerals of the Sanbagawa metamorphic stage were garnet, Al-poor clinopyroxene, pargasite, zoisite, kyanite, and quartz. Garnet occurs at the grain boundaries between the Al-pyroxene, spinel, or plagioclase, suggesting the reaction Al-pyroxene + spinel + plagioclase → garnet (Yokoyama and Mori 1975). The two-pyroxene geothermometer (Mori and Green 1978) and the stability field of zoisite-kyanite-quartz (Newton and Kennedy 1963) give about 600°C and 8–13 kb for the Sanbagawa stage (Yokoyama 1980).

Ryumon (RM) Peridotite

Unlike the other ultramafic masses, the RM peridotite is located in the Sanbagawa belt on Kii Peninsula. It measures 1×1 km^2 and consists of dunite with small amounts of wehrlite and clinopyroxenite.

According to Kunugiza (1980), its thermal history may be summarized as follows (Figure 4). The peridotite was originally cumulate judging by the Fe^{3+}- and Ti-rich nature of the relict chromite and the Fe-richness of the relict clinopyroxene. It later suffered serpentinization and then prograde mineralization, which resulted in the assemblages olivine (neoblast) + antigorite + brucite and olivine + antigorite + diopside. The prograde nature of the reaction that produced metamorphic olivine is inferred from the inclusion of serpentinite minerals (antigorite, brucite, and magnetite) in it as well as its zonation with Fe, Mn, and Ni decreasing from the core outward (Kunugiza 1982).

Shiraga (SR) Peridotite

The SR peridotite occupies an area of 1.5×1 km^2. The major constituent minerals are metamorphic olivine, antigorite, and magnetite, and the original peridotite was dunite (Kunugiza 1980). The only relic mineral is the core of chromite. The extensive Al-Cr substitution ($Y_{Cr} = 0.5$–0.85) and low Fe^{3+} and Ti may be comparable to those of chromites of dunite-harzburgite complex at Burro Mountain (Loney et al. 1971), Vulcan Peak (Himmelberg et al. 1973), and in the Seiad Complex (Medaris 1975), suggesting a residual mantle origin (Kunugiza 1984). The thermal history may be postulated as follows: residual mantle peridotite → serpentinization → Sanbagawa metamorphism (Figure 4).

Fujiwara (FW) Peridotite-Metagabbro Complex

The FW body consists mainly of metamorphosed serpentinite derived from dunite and wehrlite interlayered with slightly gneissose metagabbro. Most of the latter rock type has an epidote amphibolite facies mineralogy (Onuki et al. 1978, 1980; Enami 1980). It probably is a cumulate gabbro-wehrlite-dunite complex metamorphosed during the Sanbagawa metamorphism. No trace of granulite or eclogite facies equilibration has been found.

Small Serpentinite Masses

In central Shikoku, where we have worked in greatest detail, there are numerous small metamorphic serpentinite masses in the pelitic schists. Their occurrences are confined to the Upper Member of the Minawa Formation and the lower part of the Ojoin Formation (Figure 3). Kunugiza (1984) has shown that representative mineral assemblages of these serpentinites are, in order of increasing metamorphic grade, antigorite + brucite + diopside (zone I), olivine + antigorite + diopside (zone II), and olivine + antigorite + tremolite (zone III) (Table 1). The grade defined for the ultramafics, both small serpentinites and retrograde peridotite, is consistent with the neighboring Sanbagawa schists, thereby suggesting that the metamorphism is nothing but the Sanbagawa episode.

Because of nearly complete serpentinization, the nature of the source materials of these ultramafics is mostly obscure. However, all the available data suggest a cumulate origin: high X_{Fe} in the clinopyroxene, association with wehrlite, and high Fe^{3+} and Ti in the chromite core (the rim is usually ferritchromite).

The thermal history of these small serpentinite bodies is: cumulate peridotite → serpentinization → Sanbagawa metamorphism.

Actinolite Rock

Small lenses of actinolite rocks, 50 cm or so across, are abundant in the area where small serpentinites are abundant. They also occur at the boundary between the serpentinite and the pelitic schists. They are probably the product of reaction between serpentinite and pelitic schists, which contain epidote and calcite.

Iratsu (IR) Metagabbro

The IR metagabbro, 7×2 km^2, is composed of garnet-clinozoisite amphibolite accompanied by hornblende, eclogite, quartz eclogite, granulite, and marble (Banno et al. 1976). The southwestern border abuts against the HA peridotite. Several serpentinites and the NB peridotite are included in the IR body. A small serpentinite lens, 0.2×0.1 km^2 in size, the Jiyoshiyama (JY) body, separates the main body from the quartz eclogite, which in turn borders the HA peridotite. It is possible that all four units are independent and were tectonically transported to the present position. Another serpentinite, the Higashi-Kojimadani

TABLE 1. REPRESENTATIVE MINERAL ASSEMBLAGES OF
METAMORPHOSED SERPENTINITES (PROGRADE TYPE)

Zone I
YD/TM	antigorite
HD	antigorite + diopside + tremolite
UY	antigorite + carbonate
KO	antigorite + clinohumite + brucite + dolomite
NO	antigorite + brucite
ON/KJ	antigorite + tremolite

Zone II
NT	olivine + antigorite
FW/DO	olivine + antigorite + brucite
MT/OT	olivine + talc + magnesite
TG	antigorite + dolomite
OM	olivine + antigorite + dolomite
TD	olivine + diopside + antigorite + calcite
KK	olivine + antigorite + brucite + magnesite
KG	olivine + diopside + antigorite
KS	antigorite + magnesite

Zone III
IC	olivine + tremolite + antigorite + calcite
TN	olivine + tremolite + antigorite + dolomite + magnesite
JY	olivine + diopside + tremolite + antigorite

Note: YD = Yanagidani; TM = Tomino; HD = Hadeba; UY = Uchiyoke;
KO = Komata; NO = Nakao; ON = Onodani; KJ = Higashi-kojimadani;
NT = Nishi-tanegawa; FW = Fujiwara; DO = Dozangoe; MT = Matsuno;
TO = Oriu-Tonyo; TG = Togubashi; OM = Omorigoe; TD = Tsuchidaira;
KK = Kamikabuto; KG = Kumogahara; KS = Kanseidani; IC =
Ichinomori; TN = Tonaru; JY = Jiyoshiyama.

(KJ) body, may be on the hanging wall of the IR body. However, the NB peridotite is definitely included in the IR body.

The garnet-clinozoisite amphibolite is composed of various combinations of garnet, clinozoisite (often zoisite), hornblende, and plagioclase (albite and oligoclase), with subordinate amounts of omphacite, kyanite, muscovite, paragonite, rutile and sphene. The hornblende eclogite is partly replaced by garnet-clinozoisite amphibolite, sometimes having newly-formed almandine. The end members of the garnet-clinozoisite amphibolite are mono-mineralic hornblende rock, up to 50 cm thick, and zoisite rock with a little hornblende, kyanite and quartz (Figure 5a), 6 m thick maximum.

The banding of melanocratic (hornblende-rich) and leuco-cratic (clinozoisite-rich) layers is conspicuous. Mineralogical grading is observed rarely, and massive pegmatitic metagabbro is not uncommon. These structures demonstrate that layered gabbro was the protolith of the IR body.

Granulite occurs at a few localities. It is composed of Al-clinopyroxene, Al-orthopyroxene, and bytownite, and has con-spicuous banding consisting of melanocratic (pyroxene-rich) and leucocratic (plagioclase-rich) layers. The transition from granulite to garnet-clinozoisite amphibolite has been observed (Yokoyama 1976; Takasu 1984b).

Banno and Yokoyama (1977) postulated the following thermal history, assuming the consanguinity of the NB peridotite and the IR metagabbro (Figure 6): layered gabbro formed at 10 kb or so → granulite facies metamorphism → Sanbagawa meta-morphism. The earlier phase of the Sanbagawa metamorphism formed eclogites that were later replaced by garnet-clinozoisite

amphibolite. Recent detailed study of the eclogites by Takasu (1984a) and Kunugiza (in preparation) suggests a more complex history of crystallization but the essential trend may remain valid.

Tonaru (TN) Metagabbro

The TN body measures 6×1 km^2, and consists mainly of garnet-clinozoisite amphibolite. The banding is more conspicuous (Figure 5b), and the rocks are more foliated than in the IR mass. However, mineralogical grading and gabbro pegmatite structures are sometimes retained (Figure 5c, d). No granulite is known.

The garnet-clinozoisite amphibolite consists of various com-binations of garnet, clinozoisite, zoisite, hornblende, and sodic plagioclase, with minor amounts of omphacite, sodic augite, kya-nite, muscovite, paragonite, quartz, rutile, and sphene. Eclogite has not been found yet, but the presence of a symplectitic aggre-gate of hornblende, sodic augite, and sodic plagioclase after om-phacite, and an aggregate of hornblende and epidote after garnet, suggests that eclogite existed before the garnet-clinozoisite am-phibolite. Peristeritic coexistence of oligoclase and albite is com-mon (Enami 1982). Essentially the same thermal history as for the SB metagabbro discussed below is envisaged for the TN metagabbro.

Sebadani (SB) Metagabbro

This is a rather small body, 0.3×0.2 km^2. It consists of eclogite and garnet-epidote amphibolite, and the banding of hornblende- and epidote-rich layers is also distinct.

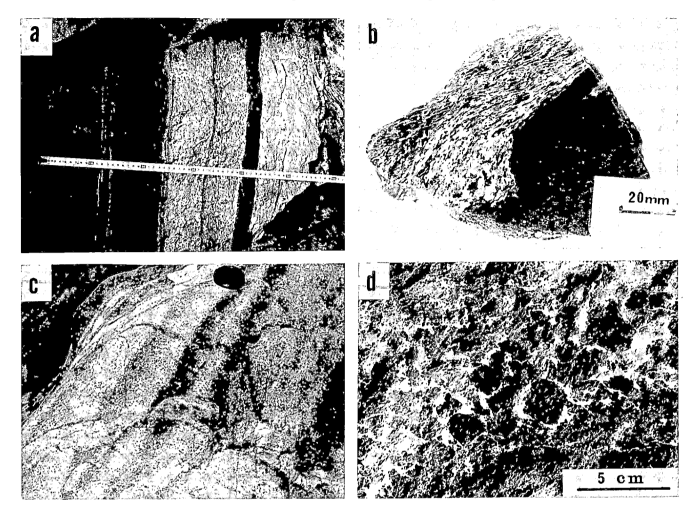

Figure 5. Texture and structure of the metagabbros. (a) An outcrop of the Iratsu (IR) metagabbro, showing a parallel alignment of dark hornblende-rich and light zoisite-rich layers. (b) A hand specimen from the IR metagabbro, showing the contact between two units of the relict igneous grading. (c) Relict igneous grading in the Tonaru (TN) metagabbro body. Diameter of the lens cap as a scale is about 5 cm. (d) Gabbro pegmatitic structure in the TN metagabbro body.

Takasu (1984a) has shown that there are two eclogite parageneses, one being represented by a core-core pair that gives 720–750°C and 12–20 kb, and the other by a rim-rim pair that gives 610–630°C and 10–17 kb (Ellis and Green 1979; Green and Ringwood 1972; Holland 1980). The later eclogite paragenesis develops, replacing the higher temperature eclogite. Notable for the SB body is the fact that within 20 m of the body basic schists underwent high-pressure contact metamorphism that produced omphacite and garnet from albite-epidote amphibolite. The Fe-Mg partitioning between omphacite and garnet in the basic schists also gives 630–650°C by Ellis and Green (1979). Takasu (1984a) concluded that higher-temperature eclogite intruded in solid state into the Sanbagawa schists and contact-metamorphosed the basic schists to form the eclogitic assemblage. Retrograde equilibration of the metagabbro and prograde crystallization of basic schists converged at 610–650°C. Because the epidote amphibolite assemblage, which is the highest grade one in this area, was dehydrated to form eclogite assemblage around the

SB body, this temperature, 610–650°C, must be slightly higher than that of the ordinary Sanbagawa metamorphism. The newly formed eclogite assemblage was also replaced by garnet-epidote amphibolite suggesting that it was an episode during the Sanbagawa metamorphism. The thermal history of the SB metagabbro postulated by Takasu (Figure 6) is gabbro formation → high temperature eclogite → Sanbagawa metamorphism (formation of prograde eclogite and retrograde metamorphism of metagabbro).

Kawamai (KM) Metagabbro

This is exposed in the Sanbagawa belt of western Shikoku, but it occurs in the chlorite zone (Ishimoto 1974). It measures 10×2 km^2, and is composed of amphibolite and peridotite. Rarely, salite-almandine rocks occur in the body. Sometimes there is a leucocratic layer suggestive of derivation from a plagioclase-rich layer. Glaucophane, epidote, chlorite, and albite, that is, the minerals constituting the neighboring Sanbagawa basic

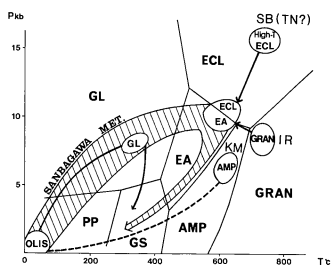

Figure 6. Metamorphic history of the metagabbro bodies. Facies boundaries are slightly modified from Takasu (1984a). SB, TN, IR, and KM: Sebadani, Tonaru, Iratsu, and Kawamai metagabbro bodies, respectively; ECL, GRAN, AMP, EA, GS, GL, and PP: Eclogite, granulite, amphibolite, epidote amphibolite, greenschist, glaucophane schist, and prehnite-pumpellyite facies, respectively. OLIS: Olistolith.

schists, develop in the amphibolite, usually at the hornblende grain boundaries.

The Kawamai metagabbro was originally a layered gabbro. It was subsequently metamorphosed in the amphibolite facies, in which augite-garnet rocks were produced locally. The minerals of the Sanbagawa metamorphism developed along the grain boundaries of the amphibolite minerals.

In the neighborhood of the KM metagabbro there is a greenstone complex of the Sanbagawa metamorphism. The complex contains detrital hornblende that is chemically similar to the hornblende of the KM body. Hence, Nakamura (oral communi-

cation) and Shikano (1981) concluded that the amphibolite was an olistolith emplaced in sediments of largely basaltic origin (Figure 6).

DISCUSSION

Diversity of Thermal Histories

There is quite a variety of thermal histories among the ultramafic and metagabbro bodies of the higher-grade zones in central Shikoku and the Kii Peninsula. Most metagabbros and some peridotites (e.g., HA, NB) have a relict structure of the igneous stage and relict mineralogy of the preceding higher temperature metamorphism. No trace of prograde metamorphism from the greenschist facies is known in them. Therefore we may call them retrograde metagabbros and peridotites. However, many ultramafic bodies cooled down and formed serpentinite before their recrystallization in the Sanbagawa metamorphism, and can be called prograde serpentinite. In central Shikoku, the metagabbro in the FW complex, which consists largely of serpentinite, is the only one without relicts of higher-temperature metamorphism.

In this paper, we are not much concerned with the arguments of geothermometry, but are more concerned with their geological implications. Table 2 summarizes the estimated pressures and temperatures of metagabbro and peridotite bodies referred in the text. To obtain internally consistent values of temperatures, we use the calibration of Fe-Mg partition by Ellis and Green (1979) on garnet-clinopyroxene, and by Mori and Green (1978) on two pyroxenes.

Retrograde metagabbros and peridotites. The thermal histories of the emplacement of metagabbros and some peridotites are rather simple. The SB body was emplaced as a high-temperature eclogite into the Sanbagawa schists, and in its

TABLE 2. ESTIMATED EQUILIBRIUM TEMPERATURE AND PRESSURE OF THE
HIGASHIAKAISHI (HA), NIKUBUCHI (NB), AND SEBADANI (SB) BODIES

Body	Stage	Temperature				Pressure	References
		Opx-Cpx		Ga-Cpx			
		K_D	Mori & Green	K_D	Ellis & Green		
HA	High-T Ecl			6-7	730°C	15 kb	(1) (2)
	Sanbagawa	1.9	600°C			7-13 kb	(1) (2)
NB	Granulite	1.4-1.6	760±40°C			5-10 kb	(3)
	Sanbagawa	1.9-2.0	580-600°C	10	660-670°C	8-13 kb	(3)
SB	High-T Ecl			6-10	720-750°C	12-20 kb	(4)
	Sanbagawa			12-14	610-630°C	10-17 kb	(4)

(1) Mori and Banno (1973); (2) Kunugiza (1981); (3) Yokoyama (1980); (4) Takasu (1984a).

neighborhood partly obliterated the schistosity defined by hornblende and epidote as the result of formation of omphacite and garnet from them. The SB body itself also underwent the later eclogitization of the Sanbagawa metamorphism. The eclogite minerals of the SB body, both of the pre-Sanbagawa and Sanbagawa stages, were then replaced by epidote, barroisitic hornblende, and other minerals. This implies that the emplacement and local contact metamorphism took place while the host schists were in the epidote amphibolite facies. The IR metagabbro, and the NB peridotite included in it, had been basic granulite before its emplacement in the Sanbagawa schists. Even though no contact aureole is recognized around it, no relict of prograde metamorphism of the greenschist or epidote amphibolite facies assemblages is known. The IR metagabbros have eclogites of the Sanbagawa stage, but the major metamorphism is of the epidote amphibolite facies. Lineation and schistosity developed except in the granulitic part of the body. Thus, emplacement of the body while the host schists were in the epidote amphibolite facies is implied. For the HA peridotite, we envisage a similar model to that for the IR body but it belonged to a higher-temperature part of the eclogite facies before emplacement into the schists.

The structural evolution of the Sanbagawa belt is currently in hot dispute (Toriumi 1982; Faure 1983; Hara et al. 1977, 1983). It may be far from being solved, but we postulate the following deformational history (cf. articles in this volume by Banno and by Toriumi and Masui).

1. High angle plate convergence creating south to north compression during an early phase of the metamorphism. Radiolaria are stretched and elongate E–W;
2. Oblique plate convergence creating a transcurrent E–W flow pattern during the peak of metamorphism. E–W mineral lineations formed at this time;
3. Late metamorphic uplift of the highest grade part of the Sanbagawa belt in macroscopic, south verging folds;
4. Undulation of metamorphic complex resulting in E–W crenulation cleavage and kink bands.

The plate interactions during this period are uncertain.

Prograde peridotites and serpentinites. Some peridotites, such as the HA and NB bodies were emplaced into the Sanbagawa schists by processes similar to the metagabbros. However, the SR and RY peridotites and the small serpentinites in central Shikoku that have the texture and mineralogy of prograde metamorphism need further consideration. Two possibilities may explain their thermal histories; prograde metamorphism of the sedimentary serpentinite and metamorphism of the serpentinite that tectonically emplaced into the Sanbagawa schists.

The first model most easily explains the prograde nature of the serpentinites, but fails to explain why they are confined to a certain zone of the metamorphic terrane. These serpentinite bodies are massive and never show sedimentary structures. They are not stratiform but lenticular. To accept the first model, we have to assume the coincidence that the sedimentary formation, containing abundant sedimentary serpentinites, later turned out to be the most intensely folded unit into which deep seated peridotites and metagabbros were emplaced.

We prefer the second model, in which the serpentinite bodies were also emplaced in the Sanbagawa schists in the solid state before the maximum temperature of metamorphism was reached. In other words, unlike the retrograde peridotites, these bodies were already serpentinite at the time of their solid-state emplacement in the Sanbagawa schists. The confinement of serpentinite to certain narrow areas in central Shikoku requires that their emplacement was related to the tectonism that characterized this zone.

Time of peridotite and metagabbro emplacement

The thermal history of the Sanbagawa metamorphism so far deciphered may be summarized as follows. The sedimentary sequence, with intercalations of basaltic volcanic rocks and chert, was metamorphosed in the high pressure intermediate facies series. Figure 6 also illustrates the thermal history of the Sanbagawa schists as revealed by the conventional analysis of mineral assemblages and the analysis of chemical zoning of garnet, amphibole and albite (Banno et al. 1984).

Minerals of the highest grade define the schistosity and lineation. After the maximum temperature was reached, the pressure rather than temperature started to decrease, as revealed by the replacement of garnet and hornblende by biotite, and the increase of anorthite content in the albite porphyroblasts at the very margin. It follows that the metamorphic complex started to rise after the maximum temperatures were reached. The origin of inverted thermal structure is correlated with this uplift of the complex, by which the higher-grade rocks were separated from the heat source and emplaced in the lower-grade rocks with which they cooled down. The metagabbro and peridotite bodies were emplaced contemporaneously or just after the lineation formative stage. The emplacement had to be prior to the commencement of the south-verging folds because the accompanying pressure decrease would not have been favorable for the second stage of eclogite mineralization, a Sanbagawa event. The mineral lineations are interpreted to have formed by eastward flow of the metamorphic complex (Faure 1983). The emplacement of the retrograde bodies may have been related to transcurrent movement of the metamorphic complex.

The prograde serpentinite was emplaced earlier than that of the retrograde bodies into the same area as the prograde bodies, thus suggesting a similar tectonic setting during the emplacement, probably transcurrent movement and associated shear. Even though the thermal histories of the prograde and retrograde bodies appear distinctly different, the essential difference is the temperatures of the bodies at the time of emplacement, which must be ascribed to a difference in the source region rather than that of the temporal or tectonic environments.

Source Region of the Ultramafic and Metagabbro Bodies. One possibility for the source region of the retrograde bodies is the lower crust–upper mantle region of an island arc or continental crust. The characteristic absence of gneiss or acidic to intermediate granulite among the exotic bodies does not favor the mature continental crust (Kay and Kay 1981). Thus, an island arc is suggested, something like the present day lower crust-upper mantle region of Southwestern Japan, whose lower crust is supposed to consist of cumulate olivine gabbro and pyroxene gabbro underlain by cumulate upper mantle, lherzolite, and then partially melted upper mantle (Takahashi 1978). Many of the peridotitic and pyroxenitic cumulates in that region are banded and recrystallized in the granulite facies. The absence of a distinct Moho discontinuity in Southwestern Japan (Hashizume and Matsui 1979) is also consistent with this type of region being the source region of the complex of metagabbro and cumulate peridotite as

seen in the Iratsu and Nikubuchi bodies. The fact that garnet-bearing xenolithic websterite was recently found in the same region (Usui 1983) further strengthens the similarity of Southwest Japan with the source region of the retrograde metagabbro and peridotites of the Sanbagawa metamorphic belt.

ACKNOWLEDGMENTS

We would like to express our sincere thanks to Professor D. A. Brown of Australian National University for his help editing this manuscript. Comments of Professor E. H. Brown and Professor R. N. Brothers helped to improve the manuscript. Kunugiza is deeply indebted to Professor K. Aoki, who supervised his Ph.D. work at Tohoku University. We also thank T. Irino of Kyoto University for drafting the figures.

REFERENCES CITED

Banno, S., Higashino, T., Otsuki, M., Itaya, T., and Nakajima, T., 1978, Thermal structure of the Sanbagawa metamorphic belt in central Shikoku: Journal of Physics of the Earth, v. 26, p. 345–356.

Banno, S., Sakai, C., and Otsuki, M., 1984, Thermal history of the Sanbagawa metamorphic rocks: Journal of Geography, v. 93, p. 515–527 (in Japanese).

Banno, S., Yokoyama, K., Iwata, O., and Terashima, S., 1976, Genesis of epidote amphibolite masses in the Sanbagawa metamorphic belt of central Shikoku: Journal of the Geological Society of Japan, v. 82, p. 199–210 (in Japanese with English abstract).

Banno, S., and Yokoyama, K., 1977, Peridotite-metagabbro complex in central Shikoku; in Hide, K., ed., The Sanbagawa Belt: Hiroshima, Hiroshima University Press, p. 57–68 (in Japanese with English abstract).

Ellis, D. J., and Green, D. H., 1979, An experimental study of the effect of Ca upon garnet-clinopyroxene Fe-Mg exchange equilibria: Contributions to Mineralogy and Petrology, v. 71, p. 13–22.

Enami, M., 1980, Petrology of the Fujiwara mass and the surrounding pelitic schists in the Sanbagawa metamorphic belt, central Shikoku: Journal of the Geological Society of Japan, v. 86, p. 461–473 (in Japanese with English abstract).

——1982, Oligoclase-biotite zone of the Sanbagawa metamorphic terrain in the Bessi district, central Shikoku, Japan: Journal of the Geological Society of Japan, v. 88, p. 887–900 (in Japanese with English abstract).

Evans, B. W., and Frost, B. R., 1975, Chrome-spinel in progressive metamorphism—a preliminary analysis: Geochimica Cosmochimica Acta, v. 39, p. 959–972.

Faure, M., 1983, Eastward ductile shear during the early tectonic phase in the Sanbagawa belt: Journal of the Geological Society of Japan, v. 89, p. 319–329.

Green, D. H., and Ringwood, A. E., 1972, A comparison of recent experimental data on the gabbro-garnet granulite-eclogite transition. Journal of Geology, v. 80, p. 277–288.

Geological Section, Besshi Mine, Sumitomo Metal Mining, 1966, Besshi ore deposits and crystalline schists of the adjacent area: Geological Section, Besshi Mine, Sumitomo Metal Mining.

Hara, I., Hide, K., Takeda, K., Tsukuda, E., Tokuda, M., and Shioda, T., 1977, Tectonic movement in the Sanbagawa belt, in Hide, K., ed., The Sanbagawa Belt: Hiroshima, Hiroshima University Press, p. 307–390 (in Japanese with English abstract).

Hara, I., Shiota, T., Maeda, M., and Miyaoka, H., 1983, Deformation and recrystallization of the Sanbagawa schist with special reference to history of Sanbagawa metamorphism: Journal of Science of the Hiroshima University, ser. C, v. 8, p. 135–148.

Hashizume, M., and Matsui, Y., 1979, Crustal structure of South-Western Honshu, Japan, derived from explosion seismic waves: Geophysical Journal of the Royal Astronomical Society, v. 58, p. 181–200.

Herzberg, C. T., 1978, Pyroxene, geothermometry and geobarometry: experimental and thermodynamic evaluation of some subsolidus phase relations involving pyroxenes in the system $CaO-MgO-Al_2O_3-SiO_2$: Geochimica Cosmochimica Acta, v. 42, p. 945–957.

Hide, K., Yoshino, G., and Kojima, G., 1956, Preliminary report on the geologic structure of the Besshi spotted schist zone: Journal of the Geological Society of Japan, v. 62, p. 574–584 (in Japanese with English abstract).

Himmelberg, G. R., and Loney, R. A., 1973, Petrology of the Vulcan Peak alpine-type peridotite, southwestern Oregon. Geological Society of America Bulletin, v. 84, p. 1583–1600.

Holland, T.J.B., 1980, The reaction albite = jadeite + quartz determined experimentally in the range 600–1200°C. American Mineralogist, v. 65, p. 129–134.

Ishimoto, N., 1974, Regional metamorphism of the Yawatahama district, in western Shikoku, Japan: Science Reports of Tohoku University, Ser. III, v. 12, p. 279–330.

Jenkins, D. M., and Newton, R. C., 1979, Experimental determination of the spinel peridotite to garnet peridotite inversion at 900°C and 1000°C in the system $CaO-MgO-Al_2O_3-SiO_2$, and at 900°C with natural garnet and olivine: Contributions to Mineralogy and Petrology, v. 68, p. 407–419.

Kay, R. W., and Kay, S. M., 1981, The nature of the lower continental crust: Inferences from geophysics, surface geology, and crustal xenoliths: Reviews of Geophysics and Space Physics, v. 19, p. 271–297.

Kawachi, Y., 1968, Large-scale overturned structure in the Sanbagawa metamorphic zone in central Shikoku, Japan: Journal of the Geological Society of Japan, v. 74, p. 607–616.

Kenzan Research Group, 1984, Stratigraphy and geologic structure of the Sanbagawa metamorphic belt in the Oboke area, central Shikoku, Japan: Earth Science, v. 38, p. 53–63 (in Japanese with English abstract).

Kojima, G., 1951, Stratigraphy and geological structure of the crystalline schist region in central Sikoku: Journal of the Geological Society of Japan, v. 57, p. 177–190.

Kunugiza, K., 1980, Dunite and serpentinite in the Sanbagawa metamorphic belt, central Shikoku and Kii peninsula, Japan: Journal of the Japanese Association of Mineralogists, Petrologists, and Economic Geologists, v. 75, p. 14–20.

——1981, Petrogenesis of ultramafic bodies of the Sanbagawa metamorphic belt [Ph.D. Thesis]: Tohoku University.

——1982, Formation of zoning of olivine with progressive metamorphism of serpentinization—an example from the Ryumon peridotite body of the Sanbagawa metamorphic belt, Kii peninsula: Journal of the Japanese Association of Mineralogists, Petrologists and Economic Geologists, v. 77, p. 157–170.

——1984, Metamorphism and origin of ultramafic bodies of the Sanbagawa metamorphic belt in central Shikoku, Japan: Journal of the Japanese Association of Mineralogists, Petrologists, and Economic Geologists, v. 79, p. 20–32 (in Japanese with English abstract).

Kushiro, I., and Yoder, H. S., 1974, Formation of eclogite from garnet peridotite: Liquidus relations in the portion of the system $MgSiO_3$-$CaSiO_3$-Al_2O_3 at high pressure: Carnegie Institute Year Book, v. 73, p. 266–269.

Loney, R. A., Himmelberg, G. R., and Coleman, R. G., 1971, Structure and petrology of the alpine-type peridotite at Burro Mountain, California, USA: Journal of Petrology, v. 12, p. 245–309.

MacGregor, I. D., 1970, The effect of CaO, Cr_2O_3, Fe_2O_3 and Al_2O_3 on the stability of spinel and garnet peridotites: Physics of the Earth and Planetary Interiors, v. 3, p. 372–377.

Medaris, L. G., Jr., 1975, Coexisting spinel and silicates in alpine peridotite of the granulite facies. Geochimica Cosmochimica Acta, v. 39, p. 947–958.

Miyashiro, A., 1961, Evolution of metamorphic belts: Journal of Petrology, v. 2, p. 277–311.

Mori, T., and Banno, S., 1973, Petrology of peridotite and garnet clinopyroxenite of the Mt. Higashi-Akaishi mass, central Sikoku, Japan—Subsolidus relation of anhydrous phases: Contributions to Mineralogy and Petrology, v. 41, p. 301–323.

Mori, T., and Green, D. H., 1978, Pyroxenes in the system $Mg_2Si_2O_6$ at high pressure: Earth and Planetary Science Letters, v. 26, p. 277–286.

Newton, R. C., and Kennedy, G. C., 1963, Some equilibrium reactions in the join $CaAl_2Si_2O_8$-H_2O: Journal of Geophysical Research, v. 68, p. 2967–2983.

Obata, M., 1976, The solubility of Al_2O_3 in orthopyroxene in spinel and plagioclase peridotites and spinel pyroxenite: American Mineralogist, v. 61, p. 804–816.

O'Hara, M. J., Richardson, S. W., and Wilson, G., 1971, Garnet-peridotite stability and occurrence in crustal and mantle: Contributions to Mineralogy and Petrology, v. 32, p. 48–68.

Onuki, H., Yoshida, T., and Suzuki, T., 1978, The Fujiwara mafic-ultramafic complex in the Sanbagawa metamorphic belt of central Shikoku. 1. Petrochemistry and rock-forming mineralogy: Journal of Japanese Association of Mineralogists, Petrologists, and Economic Geologists, v. 73, p. 311–322 (in Japanese with English abstract).

Onuki, H., Yoshida, T., and Nedachi, M., 1980, The Fujiwara mafic-ultramafic complex in the Sanbagawa metamorphic belt of Shikoku. 2. Zoned chromite and its paragenesis: Journal of Japanese Association of Mineralogists,

Petrologists and Economic Geologists, v. 75, p. 14–20 (in Japanese with English abstract).

Shikano, K., 1981, Geology and petrology of the Mikabu greenstones and associated rocks at the Yawatahama area in the western Shikoku [Master's thesis]: Kanazawa University, Japan, p. 1–155 (in Japanese with English abstract).

Suyari, K., Kuwano, Y., and Ishida, K., 1980, Discovery of the Late Triassic conodonts from the Sanbagawa Metamorphic Belt proper in western Shikoku: Journal of the Geological Society of Japan, v. 86, p. 827–828 (in Japanese).

Takahashi, E., 1978, Petrologic model of the crust and upper mantle of the Japanese Island Arcs: Bulletin of Volcanology, v. 41, p. 529–547.

Takasu, A., 1984a, Prograde and retrograde eclogites in the Sanbagawa metamorphic belt, Besshi district, Japan: Journal of Petrology, v. 25, p. 619–643.

——1984b, Geology and petrology of the Sanbagawa metamorphic belt in the Besshi district, central Shikoku, Japan [Ph.D. Thesis]: Kyoto University, 1–174 (MS).

Takasu, A., and Makino, K., 1980a, Stratigraphy and geologic structure of the Sanbagawa metamorphic belt in the Besshi district, Shikoku, Japan—Reexamination of the recumbent structures: Earth Science, v. 34, p. 16–26 (in Japanese with English abstract).

——1980b, Author's reply to the discussion by Y. Kawachi of "Stratigraphy and geologic structure of the Sanbagawa metamorphic belt in the Besshi district, Shikoku, Japan—Reexamination of the recumbent fold structures: Discussion": Earth Science, v. 34, p. 240–244 (in Japanese with English abstract).

Toriumi, M., 1982, Strain, stress and uplift: Tectonics, v. 1, p. 57–76.

Usui, M., 1983, Garnet-bearing xenolith in alkaline basalt from Mt. Onyama, Okayama Prefecture [abs.]: Joint Meeting of the Japanese Association of Mineralogists, Petrologists, and Economic Geologists, the Mineralogical Society of Japan, and the Society of Mining Geologists of Japan in 1983, p. 42 (in Japanese).

Yokoyama, K., 1976, Finding of plagioclase-bearing granulite from the Iratsu epidote amphibolite mass in central Shikoku: Journal of the Geological Society of Japan, v. 82, p. 549–551.

——1980, Nikubuchi peridotite body in the Sanbagawa metamorphic belt; Thermal history of the "Al-pyroxene-rich suite" peridotite body in high pressure metamorphic terrain: Contributions to Mineralogy and Petrology, 9. 73, p. 1–13.

Yokoyama, K., and Mori, T., 1975, Spinel-garnet-two pyroxene rock from the Iratsu epidote amphibolite mass, central Japan: Journal of the Geological Society of Japan, v. 81, p. 29–37.

MANUSCRIPT ACCEPTED BY THE SOCIETY JULY 28, 1985

Geological Society of America
Memoir 164
1986

Strain patterns in the Sanbagawa and Ryoke paired metamorphic belts, Japan

Mitsuhiro Toriumi*
Megumi Masui
Department of Earth Sciences
Faculty of Science
Ehime University
Matsuyama 790, Japan

ABSTRACT

Although the Sanbagawa and Ryoke paired metamorphic belts of Japan show great metamorphic contrast, their deformation histories are similar. Consequently, it is inferred that the tectonic configurations of the two metamorphic belts were the same and that they have undergone a single process of deformation and metamorphism during Cretaceous time.

The ductile deformation giving rise to the unusual uniaxial extension type of strain throughout the paired metamorphic belts is interpreted to be a result of two stages of deformation during the metamorphism: initial north-south simple shortening, followed by strong east-west tangential ductile shear.

INTRODUCTION

In Japan, the configuration of a high P/T type regional metamorphic belt (Sanbagawa belt) juxtaposed against a low P/T type metamorphic belt (Ryoke belt), has been called a paired metamorphic belt by Miyashiro (1961). The metamorphic ages of the two belts are coeval, but there are major contrasts in physical conditions and facies series between the high P/T and low P/T type metamorphism (Table 1). As first noted by Miyashiro (1961), the contrast in physical conditions is the direct manifestation of the thermal structure in arc-trench systems; subduction is the mechanism responsible. Sedimentation, metamorphism, and uplift processes of both belts occurred between 150 Ma to 80 Ma. Miyashiro (1972) considered that the high P/T type metamorphic belt formed in the subduction zone and was widely separated from the low P/T type metamorphic belt that was formed in the arc. The belts were interpreted to have been later juxtaposed by movement on a large-scale transform fault, marked by the present day Median Tectonic Line. Alternatively, Ernst (1971) suggested that nappe movements caused later juxtaposition of the two metamorphic belts. A very different interpretation was offered by Dickinson (1971) who suggested that the

Ryoke metamorphic belt was coupled with the Shimanto subduction complex.

Recent studies of strained radiolaria document the patterns of strain in radiolarian-bearing cherts. Toriumi and Noda mapped regional strain geometry and magnitudes in the Sanbagawa metamorphic belt of Shikoku. They confirmed the results of Toriumi (1982) that deformation occurred during metamorphism. In the previous studies, the regional strain ellipsoid, as

TABLE 1. PHYSICAL AND MECHANICAL CONDITIONS
OF THE RYOKE AND SAMBAGAWA METAMORPHISM

	Ryoke metamorphism	Sambagawa metamorphism
Temperature	300-800°C	200-600°C
Pressure	2-5 kb	7-15 kb
Shape of strain ellipse	prolate	prolate
Direction of maximum axis of strain ellipse	N70W-EW	N80W-EW
Strain path	prolate	prolate
Strain magnitude	1-120%	1-200%
Age	60-90 Ma	80-110 Ma

*Present address: Geological Institute, University of Tokyo, Hongo, Bunkyo-ku, Tokyo 113, Japan.

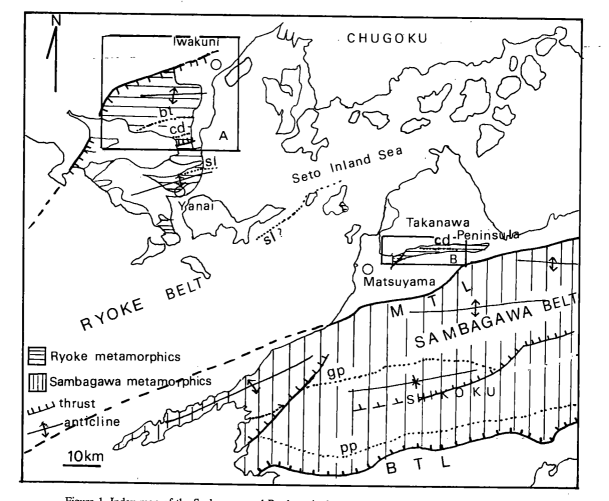

Figure 1. Index map of the Sanbagawa and Ryoke paired metamorphic belts in southwestern Japan. The map shows the biotite isograd (bt), the cordierite isograd (cd) and the sillimanite isograd (sl) in the Ryoke belt, and boundaries between the prehnite-pumpellyite zone and the pumpellyite-actinolite zone (pp), and the boundary between the pumpellyite-actinolite zone and the epidote-glaucophane zone (gp) in the Sanbagawa belt. The abbreviations are as follows: MTL: Median Tectonic Line; BTL: Butsuzo Tectonic Line; A: Iwakuni-Yanai district; and B: Takanawa Peninsula of the Ryoke belt.

determined by the radiolarian strain meter (Toriumi 1982), indicates a uniaxial extension type with the magnitude of strain increasing with metamorphic temperature.

In this paper we compare the magnitude, geometry, and strain paths, both in the Ryoke and Sanbagawa metamorphic belts from the region of the Iwakuni-Yanai district to the Takanawa Peninsula and to western Shikoku (Figure 1) in an attempt to relate the deformation patterns and metamorphic history of the two belts.

OUTLINE OF THE RYOKE AND SANBAGAWA METAMORPHIC BELTS

Ryoke Metamorphic Belt

The geological structure of the Ryoke metamorphic belt in the Yanai-Iwakuni district is characterizead by large scale south

verging thrusts in the north, N-verging thrusts in the cordierite zone, and large scale flow folds with mushroom sections with axial traces trending east-west (Figure 1). There are also large scale gentle upright folds. Nureki (1960) and Shimizu (1984) showed that the schistosity planes (S_1) and the mineral lineation (L_1) are well-developed and often refolded by the later upright folds in the lower grade metamorphic zones. Intrafolial isoclinal folds having east-west axial traces are common in alternating layers of metamorphosed chert and mudstone (Shimizu 1984). In the higher grade zones, the irregular shaped intrafolial and flow folds are abundant in the core region of the large scale flow folds mentioned above. They are coaxial and their axes trend east-west parallel to the mineral lineation.

The mineral lineation can be recognized by parallel alignment of elongated cordierite, plagioclase, biotite, sillimanite, and quartz. Pressure shadows around quartz aggregates are common in pelitic schists (Shimizu 1984). Garnet and cordierite with pull-apart structures also are common. Their extension orientation is nearly east-west. Therefore, it is concluded that the first phase deformation showing the east-west extension orientation is synmetamorphic, as suggested by Nureki (1960). The c-axis preferred orientations of quartz in metaquartzites were recognized by Nureki (1960) and their patterns display the weak cross-girdle II type (Miller and Christie 1981) with some asymmetry in the c-axis concentration. This fabric may be related to the east-west shear deformation of the Ryoke metamorphic rocks (Behrmann and Platt 1981). Biotite, cordierite, and sillimanite isograds run subparallel to the general trend of the Ryoke belt (Oho 1969, 1977; Shimizu 1984) and the extension direction of the metamorphic belt as determined from strained radiolaria.

Physical conditions of Ryoke metamorphism are inferred from equilibrium relations among K-feldspar, cordierite, biotite, andalusite, sillimanite, and muscovite. Shimizu (1984) concluded that metamorphic conditions of the Yanai district were about 500°C and 2–4 kbar at the cordierite isograd and about 700–800°C and 3–5 kbar at the sillimanite isograd.

Sanbagawa Metamorphic Belt

The geological structure of the Sanbagawa metamorphic belt in Shikoku is characterized by large-scale thrusts, nappes, and gentle upright folds (Kawachi 1968; Ernst et al. 1970; Hara et al. 1977; Faure 1983). In the higher grade zones, large recumbent folds dominate, with nearly horizontal to north dipping axial planes. In the lower grade zones, large thrusts having southward transport directions are common. Gentle upright folds occur in the whole region of the Sanbagawa belt (Hara et al. 1977).

Hara et al. (1977) and Faure (1983) recognized three deformation stages, resulting in: schistosity planes (S_1), axial cleavage planes (S_2), and the crenulation cleavage planes (S_3); a mineral lineation (L_1), a lineation (L_2) formed by the intersection of S_1 and S_2, and a lineation (L_3) formed by the intersection of S_2 and S_3. The orientations of these lineations are close to each other and trend east-west. Since S_1 and L_1 are refolded by the large scale nappes mentioned above, and since S_2 is also refolded by the gentle upright folds and cut by S_3, Hara et al. (1977) and Faure (1983) concluded that S_1 and L_1 were formed during the first phase of deformation, before the nappe and S-vergent thrust movements and the late upright folding.

Metamorphic minerals such as amphiboles (glaucophane-crossite, hornblende, actinolite), epidote, biotite, albite, and garnet are elongated in an east-west direction and their alignments form the mineral lineation (L_1). Hara et al. (1977) and Faure (1983) concluded that the first phase deformation is therefore synmetamorphic. In contrast, since the metamorphic isograds are cut by the later thrusts and nappes as shown in Figure 1, the second phase deformation forming S_2 and L_2 is postmetamorphic (Hara et al. 1977). Recently, Faure (1983) found that folds belonging to the first phase of deformation commonly accompany sheath folds and intrafolial isoclinal folds with mushroom sections elongated in an east-west direction. He concluded

that the synmetamorphic deformation was produced by ductile shear with an east-west transport direction. This type of ductile deformation is consistent with asymmetric pressure shadows of quartz aggregates, porphyroblastic albite, and garnet that have north-south rotational microstructure axes, and with an asymmetric crossed girdle quartz c-axis fabrics in metaquartzites (Faure 1983). Toriumi (1982) also suggested that strain patterns inferred from deformed shapes of radiolarians in the metamorphic chert in the Kanto Mountains resulted from east-west ductile shear.

Physical conditions of the Sanbagawa metamorphism are inferred from phase petrology to be T = 200–300°C in the low grade zone (prehnite-pumpellyite zone) and 500–600° in the highest grade zone (oligoclase-biotite zone), with a range of 7–15 kbar of pressure (Seki 1958; Banno 1964; Higashino 1975; Toriumi 1975; Banno et al. 1978; Enami 1983).

The metamorphic zonation of the Sanbagawa belt is simple in that temperature increases northward from the chlorite zone, through the garnet zone and biotite zone, to the oligoclase-biotite zone (Banno et al. 1978; Enami 1983). The garnet and biotite isograds run east-west. However, the highest grade zone forms the core of a nappe transported into lower grade zones. Thus the oligoclase-biotite zone has a cigar-like east-west elongated shape within the biotite zone (Enami 1983). In the oligoclase-biotite zone and the biotite zone, metamorphosed gabbro and peridotite, which recrystallized in the granulite and eclogitic facies, predate Sanbagawa metamorphism (Kunugiza et al. this volume). They were tectonically emplaced into Sanbagawa schists during metamorphism. Consequently, upper mantle eclogites and subducting Sanbagawa metamorphics were mechanically mixed during metamorphism. Similar relations have been suggested for the Shuksan area in the Cascades of Washington by Brown et al. (1983).

Recently, the metamorphic P-T trajectory was investigated by means of detailed analyses of zoned garnet coexisting with chlorite (Sakai et al. in press) and actinolite and glaucophane (Toriumi 1975). The findings suggest that cooling began in the typical high P-T conditions and ended in an intermediate P-T condition. The metamorphic ages of the Sanbagawa belt range from 80 to 110 Ma (Banno and Miller 1965; Yamaguchi and Yanagi, 1970). This range overlaps the period of the Ryoke metamorphism (60-90 Ma; e.g., Nozawa 1977).

STRAIN PATTERNS IN THE RYOKE AND SANBAGAWA METAMORPHIC BELTS

Method

The shape and orientation of the strain ellipsoid and the strain magnitude can be determined from deformed radiolarians in metacherts and metapelites in the regional metamorphic belts (Toriumi 1982). This method (called hereafter radiolarian strain meter) is readily applied to deformation of the low P/T type metamorphic rocks of the Ryoke belt, because radiolarian-bearing metacherts and metapelites are abundant. As described in a previous paper (Toriumi 1982), the maximum elongation axis of the strain ellipsoid is parallel to the mineral lineation, and the maximum shortening axis is normal to the foliation plane (Figure 2).

Metamorphic rocks containing deformed radiolarians were oriented in the field, and three thin sections were prepared to measure the aspect ratios. The aspect ratios of the strain ellipse on each plane are calculated from the mean values of 30 to 50 measurements. The longest axis of the strain ellipsoid is a_1, the shortest axis a_3, and the intermediate is a_2.

Strain Patterns in the Ryoke Metamorphic Rocks

The variation in the shape of the strain ellipsoid determined by the radiolarian strain meter is shown in a Flinn diagram (Figure 3). Most of the strain ellipsoids are of the uniaxial extension type. The a_1/a_2 ratio increases with increasing metamorphic temperature while a_2/a_3 ratio remains nearly constant (Table 2). This trend has been defined as a strain path by Toriumi (1985).

The orientations of a_1 of strain ellipsoids are almost always parallel to the mineral lineation defined by parallel alignment of elongated amphiboles, chlorite, and biotite; to the direction of pull-apart of pyrite and magnetite; and to the orientation of the pressure shadows of quartz aggregates in the plane of schistosity. Consequently, we have measured the a_1-orientation of the metamorphic rocks in the field from the mineral lineation. The orientation of the lineations is shown in Figure 4, and indicates that in lower grade zones, that is the lower strained zones, the a_1-orientation trends N80°W to east-west with horizontal plunge, and in higher grade zones it trends N70°W to east-west.

The strain magnitudes (ϵ (%) = 2/3 ln (a_1/a_3) × 100) of metacherts were determined in the chlorite zone to the biotite zone, but the radiolarian shape could not be used above the cordierite isograd. A contour map of the strain magnitudes is shown in Figure 5. From the map of strain zones, it is clear that the contours of strain magnitude run nearly parallel to the biotite and cordierite isograds shown on Figure 1. This indicates that the deformation occurred during metamorphism. Similar results are obtained from the Sanbagawa metamorphic belt described below.

The map of strain zones, together with a_1-orientation data, indicates directly the strain pattern during metamorphism. The map of strain zones is cut into two areas by large faults in the northern zone (the Iwakuni-Yanai district): the western area has a complicated pattern of concentric contours but the eastern area has straight contours of strain magnitude.

In the southern border zone (Takanawa Peninsula) of the Ryoke metamorphic belt, the map of strain zones is not so clear because we examined only three radiolarian-bearing metamorphic rocks (Figure 5B). However, there is no doubt that the strain magnitude is much lower (less than 20%) than in the Yanai-Iwakuni district (over 100%), although the metamorphic grade belongs to the biotite and cordierite zones.

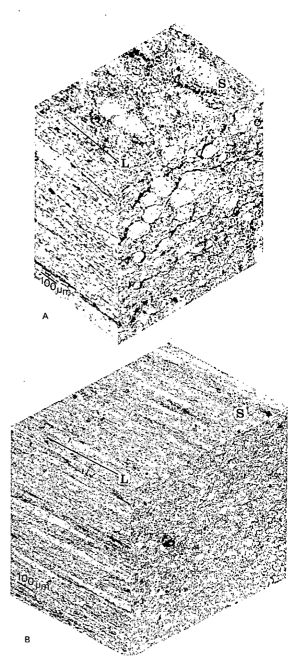

Figure 2. Representative shapes of deformed radiolarians in metacherts of the (A) Ryoke and the (B) Sanbagawa belts. L is the mineral lineation and S is the schistosity plane.

TABLE 2. VARIATION OF AXIAL RATIOS OF THE STRAIN ELLIPSE WITH METAMORPHIC GRADE IN THE RYOKE METAMORPHIC BELT

	Metamorphic zone		
	Chlorite	Chlorite/ biotite	Biotite and lower cordierite
a_1/a_2	1.25-1.6	1.6-2	2-6
a_2/a_3	1-1.6	1-1.6	1-1.25

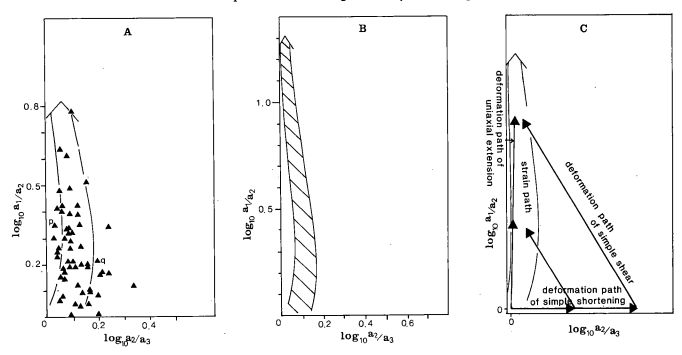

Figure 3. Flinn diagram showing strain ellipsoids determined by deformed radiolarians of metaquartzites in (A) the Ryoke metamorphic belt; and (B) the Sanbagawa metamorphic belt. (C) Hypothesized deformation paths (X, Y) producing the unusual prolate strain path. In (A), the p and q represent the data from axial and limb parts of a 10-meter flow fold from the northern part of the Ryoke belt. The arrow indicates the strain path that shows the trend of strain patterns with increasing metamorphic temperature. Most specimens show strong uniaxial extension.

Strain Patterns in the Sanbagawa Metamorphic Belt

Toriumi (1982) and Toriumi and Noda (in preparation) have established that the strain ellipsoids determined by the radiolarian strain meter of the Sanbagawa metamorphic belt are close to the uniaxial extension type. According to Toriumi and Noda (in preparation), the ratio of a_1/a_2 increases from 1.25 to 20 with nearly constant a_2/a_3 about 1.0 to 1.4 (Figure 3B). In addition, the a_2/a_3 ratios decrease slightly with increasing a_1/a_2 ratio. Therefore, the deformation trend with increasing metamorphic temperature displays the uniaxial extension type strain path in the sense of Toriumi (1985), similar to that of the Ryoke metamorphic rocks (Figure 3A). This fact is very important for tectonic models of paired metamorphism.

The a_1-orientations measured by mineral lineation are plotted in lower hemisphere of a Schmidt net as shown in Figure 4. It is significant that the a_1-orientation of the Sanbagawa belt is regionally parallel to that of the Ryoke metamorphic belt. This is a key point for clarifying the tectonic regime of the Sanbagawa belt. The a_1-orientation distribution of the lower strained zones is slightly oblique to that of higher strained zones as shown in Figure 4.

Maps of strain zones of western Shikoku have been constructed by Toriumi and Noda (in preparation) as shown in Figure 5C. The contours of strain magnitude run east-west, parallel to the metamorphic zonation defined by Nakajima et al. (1977),

and Aiba (1983), and are also consistent with the results of Toriumi (1982) in the Kanto Mountains. It is significant that the contours of strain magnitude are sharply cut by the large scale thrusts (second phase deformation) (Figure 5C), thereby suggesting that the distribution of strain magnitude and strain patterns inferred by deformed radiolarians were formed before the second stage deformation.

The strain magnitude increases from 10-20% in the prehnite-pumpellyite facies zone to 140-200% in the epidote-glaucophane schist zone. In the Kanto Mountains of the Sanbagawa belt, one of the authors found that strain magnitude may be as much as 200-300% in the epidote-glaucophane schist zone. Judging from the presence of more elongated porphyroclastic quartz grains, we consider that the strain magnitude is much larger than that of the Ryoke metamorphic belt.

DISCUSSION

Our data indicate the similarity of strain patterns in the two metamorphic belts (Table 1). Since the strain magnitude increases regularly with increasing metamorphic temperature, and the contours of strain magnitude are cut sharply by post-metamorphic thrusts and nappes (Figure 5), it is concluded that the plastic deformation of initially spherical radiolarians took place during the first phase of deformation in both the Ryoke and Sanbagawa metamorphic belts. The similarity in strain patterns suggests that

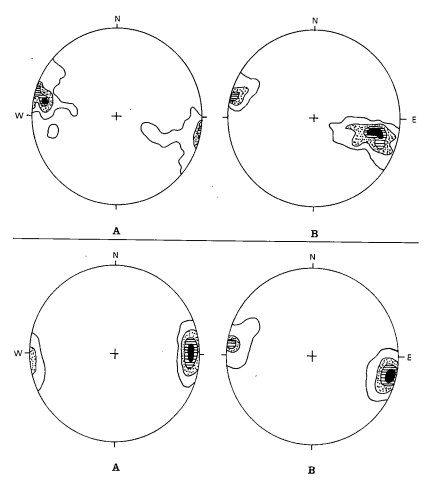

Figure 4. Projections of the maximum elongation orientation a_1 of the Ryoke belt (upper) and of the Sanbagawa belt (bottom) in the lower hemisphere of Schmidt (equal-area) net. Ryoke belt (upper): A—lower strain zones, B—higher strain zones. Sanbagawa belt, A—(bottom): lower strain zones, B—higher strain zones. The contours show 4-8-12-16%.

both metamorphisms took place in a single tectonic framework. Further, the parallel alignment of the lineations in both metamorphic belts suggests that the later displacements during uplift and faulting (e.g., the Median Tectonic Line shown in Figure 1) did not significantly change their original arrangement. It follows that the paired metamorphic belts are cogenetic.

Since the metamorphic minerals in the basic rocks of the Sanbagawa belt defining the mineral lineations include minerals of both prograde and retrograde stages of the metamorphism (Higashino 1975; Toriumi 1975; Otsuki 1980), it is strongly suggested that the first phase deformation occurred at the main stage of the metamorphism and continued into the uplift stage. On the other hand, the metamorphic minerals are often bent and cut by the S_2 and S_3 planes. Commonly, calcite and quartz fill cracks formed by S_2 and S_3.

As shown in Figure 3, the shape change of strain ellipsoids with increasing metamorphic temperature does not indicate the deformation path. It shows a trend of settled points of deformation paths along which the metamorphic rocks of each grade underwent the synmetamorphic deformation. Thus, in order to

discuss the tectonic history of the Sanbagawa and Ryoke metamorphic belts, we must determine the deformation path with increasing metamorphic temperature (this is called the "strain path" here).

Two alternative models of deformation path (Hobbs et al. 1976) are discussed by Toriumi and Noda (in preparation) and Toriumi (1985). The first model is that the strain path, as shown in Figures 3A and 3B, directly represents the deformation path (X in Figure 3C), and the second is that the strain path is the result of a combination of two different deformation paths (Y in Figure 3C). The former case is a single stage deformation. To explain the latter case, two synmetamorphic events are needed, because the strain magnitude increases regularly with increasing metamorphic temperature, but the shape of strain ellipsoids is everywhere prolate.

Any reasonable model of the strain path must reconcile the strong east-west directed ductile shear with the near uniaxial prolate strain ellipsoids and the flow folds parallel to the axis of maximum extension. Simple shear alone should produce ellipsoids with $a_1/a_2 = a_2/a_3$, and would not produce the style of

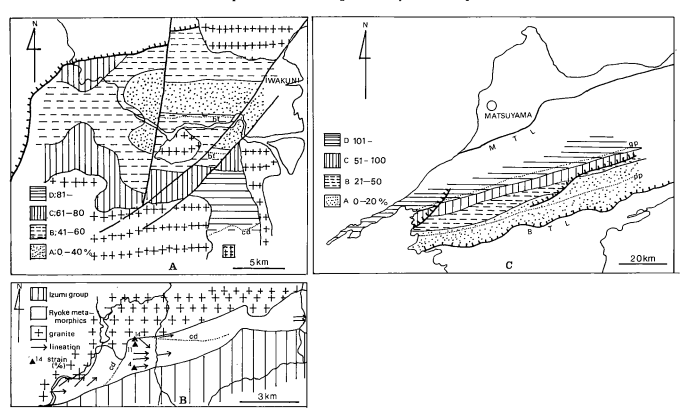

Figure 5. Maps of strain zones of (A) the Yanai-Iwakuni district; (B) the Takanawa Peninsula in the Ryoke metamorphic belt; and (C) the western Shikoku in the Sambagawa metamorphic belt. Abbreviations are the same as those in Figure 1.

folds observed. On the contrary, the prolate ellipsoids require considerable shortening in the a_2 direction. A complex strain history is therefore indicated.

A two stage deformation model (Figure 3C) is proposed by Toriumi (1985), in which initial north-south simple shortening, resulting in large folds, is followed by east-west tangential shear deformation producing the strong elongation of radiolaria and shearing of the folds. By this model we are able to interpret the unusual trend of the change of strain ellipsoids with increasing metamorphic temperature. There is, however, no unique deformation path, but the prolate ellipsoids are seen to be the result of shear roughly parallel to a_1 superimposed on a first phase shortening approximately parallel to the present direction of a_2.

The uplift of the Sanbagawa metamorphic belt was considered to correlate with the second deformation phase, forming S_2 and L_2 by Ernst et al. (1970) and Hara et al. (1977). Alternatively, Faure (1983) and Toriumi (1982) suggested that the uplift accompanied the east-west tangential shear deformation. Judging from the occurrence of large scale post-metamorphic thrusts and nappes in the Sanbagawa and the Ryoke belts, some uplift of the paired metamorphic belts occurred during the second, post-metamorphic deformation. On the other hand, as noticed in the previous sections, since the mineral lineations are defined in part by retrograde minerals that trend east-west, significant amounts

of the uplift seem to be associated with the first tangential shear.

Recently, a synmetamorphic tangential ductile shear deformation having the same direction of transport as the extension of the metamorphic belt has been reported in other regions of high P/T metamorphism (Corsica; Faure and Malavieille 1981; Calabria; Faure 1980). On the other hand, many nappes in the high P/T Alpine metamorphic belt have transport directions oblique to the extension direction of the root zone (Siddans 1983). As the shear deformation should be closely related to the uplift process of deeply subducted high P/T type metamorphic belts, the tectonic relations between the tangential and oblique directions of ductile shear deformation are significant. This is a future problem to be answered.

CONCLUSION

Studies on strain geometry and strain path using strain markers of initially spherical radiolarians of the Cretaceous paired metamorphic belts of Japan have produced the following results:

1. The strain ellipsoid of both the Sangabawa and Ryoke metamorphic belts is the uniaxial extension type (prolate) in every metamorphic grade.

2. The strain magnitude increases with increasing meta-

morphic temperatures in the Sanbagawa metamorphic belt, but in the Ryoke belt it increases with increasing temperature in the lower metamorphic zones but remains constant in the higher grade zones.

3. The orientation of maximum elongation direction of strain ellipsoids is aligned in an east-west direction in both the Sanbagawa and Ryoke belts.

Consequently, it appears that the paired metamorphic belts were formed by a single orogenic process during subduction, despite their differences in facies series.

ACKNOWLEDGMENTS

We thank E. H. Brown, J. Talbot, D. S. Cowan, and S. Banno for many suggestions and critical reading of this manuscript.

REFERENCES CITED

Aiba, T., 1983, Sanbagawa metamorphism of the Nakatsu-Nanokawa district in the northern Chichibu belt of the middle east of Shikoku, Journal of the Geological Society of Japan, v. 88, p. 875–885 (J).

Banno, S., 1964, Petrologic studies on Sanbagawa crystalline schists in Bessi-Ino district, central Sikoku, Japan, Journal of Faculty of Science, University of Tokyo, sec. II, v. 15, p. 203–319.

Banno, S., and Miller, J. A., 1965, Additional data on age of metamorphism of the Ryoke-Abukuma and Sanbagawa metamorphic terranes, Japanese Journal of Geology and Geography, v. 36, p. 17–22.

Banno, S., Higashino, T., Otsuki, M., Itaya, T., and Nakajima, T., 1978, Thermal structure of the Sanbagawa metamorphic belt in central Shikoku. Journal of Physics of the Earth, v. 26, suppl., p. 345–356.

Behrmann, J. H., and Platt, J. P., 1981, Sense of nappe emplacement from quartz c-axis fabrics; an example from Betic Cordilleras (Spain); Earth and Planetary Science Letters, v. 59, p. 298–315.

Brown, E. H., Wilson, D. L., Armstrong, R. L., and Harakal, J. E., 1983, Petrologic, structural, and age relations of serpentinite, amphibolite, and blueschist in the Shuksan Suite of the Iron Mountain-Gee Point area, North Cascades, Washington, Geological Society of America Bulletin, v. 93, p. 1087–1098.

Dickinson, W. R., 1971, Clastic sedimentary sequences deposited in shelf, slope, and trough settings between magmatic arcs and associated trenches, Minato, M. ed., "Pacific Geology," v. 3, p. 15–30: Tokyo, Tsukiji Shokan.

Enami, M., 1983, Petrology of pelitic schists in the oligoclase-biotite zone of the Sanbagawa metamorphic terrain, Japan: phase equilibria in the highest grade zone of a high-pressure intermediate type of metamorphic belt, Journal of Metamorphic Geology, v. 1, p. 141–161.

Ernst, W. G., 1971, Metamorphic zonations on presumably subducted lithospheric plates from Japan, California and the Alps: Contributions to Mineralogy and Petrology, v. 34, p. 43–59.

Ernst, W. G., Seki, Y., Onuki, H., and Gilbert, M. C., 1970, Comparative study of low-grade metamorphism in the California Coast Ranges and the outer metamorphic belt of Japan: Geological Society of America Memoir 124, 276 p.

Faure, M., 1980, Microtectonique et charriage Est-Ouest des nappes alpines profondes de Sila (Calabre, Italie mérdionale): de Géologie Dynamique et de Géographie Physique, v. 22, p. 135–146.

——1983, Eastward ductile shear during the early tectonic phase in the Sambagawa belt, Journal of the Geological Society of Japan, v. 89, p. 319–329.

Faure, M., and Malvieille, J., 1980, Etude structurale d'un cisaillement ductile: le charriage ophiolitique corse dans la région de Bastia: Bulletin de la societe Géologie de France, v. 23, no. 4, p. 335–343.

Hara, I., Hide, K., Takeda, K., Tsukada, E., Tokuda, M., and Shiota, T., 1977, Tectonic movement in the Sambagawa belt, Hide, K. ed., "The Sanbagawa Belt:" Hiroshima, Hiroshima University Press, p. 307–390.

Higashino, T., 1975, Biotite zone of the Sanbagawa metamorphic terrain in the Shiragayama area, central Shikoku, Japan, Journal of the Geological Society of Japan, v. 81, p. 653–670 (J).

Hobbs, B. E., Means, W. D., and Williams, P. F., 1976, An Outline of Structural Geology: New York, John Wiley & Sons, 571 p.

Kawachi, Y., 1968, Large scale overturned structure in the Sanbagawa metamorphic zone in central Shikoku: Journal of the Geological Society of Japan, v. 74, p. 607–616 (J).

Kunigiza, K., Takasu, A., and Banno, S., The origin and metamorphic history of the ultramafic and metagabbro bodies in the Sanbagawa metamorphic belt, this volume.

Miller, D. M., and Christie, J. M., 1981, Comparison of quartz microfabric with strain in recrystallized quartzite: Journal of Structural Geology, v. 3, p. 129–142.

Miyashiro, A., 1961, Evolution of metamorphic belts: Journal of Petrology, v. 2, p. 277–311.

——1972, Metamorphism and related magmatism in plate tectonics: American Journal of Science, v. 272, p. 629–656.

Nakajima, T., Banno, S., and Suzuki, T., 1977, Reactions leading to the disappearance of pumpellyite in low-grade metamorphic rocks of the Sanbagawa metamorphic belt in central Shikoku, Japan: Journal of Petrology, v. 18, p. 263–284.

Nozawa, T., 1977, "Radiometric age map of Japan, 1:2,000,000": Geological Survey of Japan, Tokyo.

Nureki, T., 1960, Structural investigation of the Ryoke metamorphic rocks of the area between Iwakuni and Yanai, southwestern Japan: Journal of Science of Hiroshima University, ser. C, v. 3, p. 69–141.

Ono, A., 1969, Zoning of the metamorphic rocks in the Takato-Shioziri area, Nagano Prefecture: Journal of the Geological Society of Japan, v. 75, p. 521–536.

——1977, Petrologic study of the Ryoke metamorphic rocks in the Takato-Shioziri area, central Japan: Japanese Journal of Association of Mineralogy, Petrology, and Economic Geology, v. 32, p. 453–468.

Otsuki, M., 1980, Petrological study of the basic Sambagawa metamorphic rocks in central Shikoku, Japan [Ph.D. thesis]: University of Tokyo.

Sakai, C., Banno, S., Toriumi, M., and Higashino, T., in press, Growth history of garnet in pelitic schists of the Sanbagawa metamorphic terrain in central Shikoku: Lithos.

Seki, Y., 1958, Glaucophanitic regional metamorphism in the Kanto Mountains, central Japan: Japanese Journal of Geology and Geography, v. 29, p. 233–258.

Shimizu, Y., 1984, Petrologic studies of Ryoke metamorphic rocks in the Yanai area, western Japan [M.Sc. thesis]: Ehime Univ (J).

Siddans, A.W.B., 1983, Finite strain patterns in some Alpine nappes: Journal of Structural Geology, v. 5, p. 441–448.

Toriumi, M., 1975, Petrological study of the Sanbagawa metamorphic rocks, Kanto Mountains, central Japan: University Museum, University of Tokyo Bulletin, no. 9.

——1982, Strain, stress and uplift: Tectonics, v. 1, p. 57–76.

——1985, Two types of ductile deformation/metamorhic belts, Tectonophysics, v. 113, p. 307–326.

Toriumi, M., and Noda, H., Spatial distribution of plastic strains of Sanbagawa metamorphic belt, in preparation.

Yamaguchi, M., and Yanagi, T., 1970, Geochronology of some metamorphic rocks in Japan: Eclogae Geologicae Helvetiae, v. 63, p. 371–388.

Manuscript Accepted by the Society July 29, 1985

Geological Society of America
Memoir 164
1986

A low P/T metamorphic episode in the Biei area, Kamuikotan blueschist terrane, Japan

*Hirokazu Maekawa**
Geological Institute
Faculty of Science
University of Tokyo
Hongo Bunkyo-ku
Tokyo 113, Japan

ABSTRACT

The Kamuikotan metamorphic rocks in the Biei area, central Hokkaido, Japan, consist of submarine slide blocks of basalt, amphibolite, and ultramafic protoliths in a matrix of pelitic-psammitic and mafic sediments. Two stages of metamorphism are recognized: The first was a low P/T event (referred to as the initial-stage metamorphism) that can be detected by the common presence of relict calcic amphiboles. These occur in the amphibolite of the Oichanunpe Formation as detrital fragments in mafic sediments and as relicts in metabasalt. They have high Ca contents (Ca = 1.9–2.0 for O = 23) that are analogous to those in amphibolites from other low-pressure terranes. The second metamorphism was a high P/T event (referred to as the main-stage metamorphism) characterized by the widespread occurrence of low calcium sodic amphibole, sodic pyroxene, and aragonite. All rocks, including ultramafites, have been affected by the high P/T metamorphism. Rock types and primary mineral parageneses of basites from the Biei area are similar to those from present intraplate oceanic islands; the probable origin of these rocks is ancient oceanic islands. The initial-stage metamorphism may have taken place at the base of the islands, as a type of sea floor metamorphism, where relatively high temperatures prevailed prior to subduction metamorphism.

INTRODUCTION

The Kamuikotan tectonic belt extends more than 330 km along the N-S trending axial zone of Hokkaido, Japan. The belt is characterized by large amounts of Jurassic–Early Cretaceous high P/T metamorphic rocks and associated ultramafic rocks. The Biei area is situated at the central part of the belt, southwest of Asahigawa, central Hokkaido (Figure 1). High P/T minerals such as sodic amphibole (magnesioriebeckite–crossite), sodic pyroxene (aegirine-augite) and aragonite, occur widely in rocks from this area. In addition, relict calcic amphiboles (actinolite–hornblende) are common constituents of these rocks. The chemical composition of the amphiboles suggests that they were formed under relatively low P/T conditions. Recently, a low P/T metamorphic event has been recognized in another high P/T meta-

morphic terrane in Japan (e.g. Mikabu greenrock complex: Nakamura 1984). These low P/T metamorphic events are thought to be oceanic. The purpose of the present paper is to describe the modes of occurrence and chemical composition of these relict calcic amphiboles and to discuss their significance. The distribution of calcic amphiboles in the Biei area aids in the refinement of tectonic models of Hokkaido. A geologic map of the Biei area was published by Suzuki et al. (1964), and petrological studies in the Biei area were reported by Tazaki (1964) and Hervé (1975). Details of the Kamuikotan belt were recently summarized in Ishizuka et al. (1983) and Watanabe and Maekawa (in press).

OUTLINE OF GEOLOGY

The Kamuikotan metamorphic rocks in the Biei area are divided into three N-S trending units. From west to east these

Present address: Department of Earth Sciences, Faculty of Science, Kobe University, Nada 657, Japan.

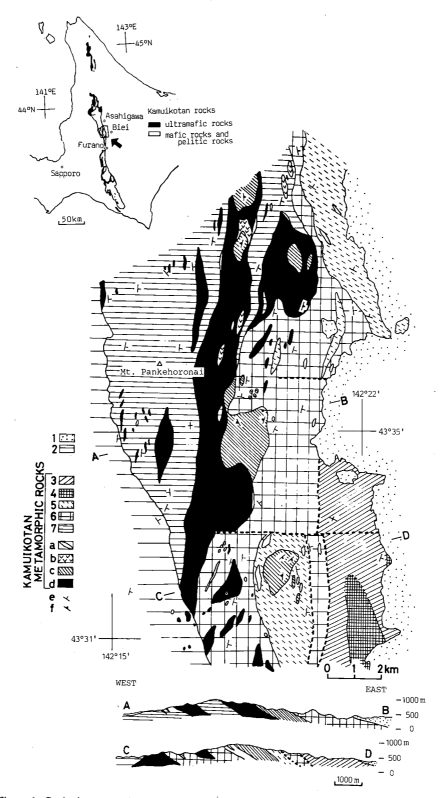

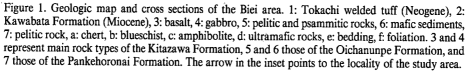

Figure 1. Geologic map and cross sections of the Biei area. 1: Tokachi welded tuff (Neogene), 2: Kawabata Formation (Miocene), 3: basalt, 4: gabbro, 5: pelitic and psammitic rocks, 6: mafic sediments, 7: pelitic rock, a: chert, b: blueschist, c: amphibolite, d: ultramafic rocks, e: bedding, f: foliation. 3 and 4 represent main rock types of the Kitazawa Formation, 5 and 6 those of the Oichanunpe Formation, and 7 those of the Pankehoronai Formation. The arrow in the inset points to the locality of the study area.

three units (Figure 1) include the Pankehoronai Formation (pelitic rock unit), the Oichanunpe Formation (mafic rock unit), and the Kitazawa Formation (basalt unit). The planar features of each unit, such as bedding planes or shear planes trend north-south and dip eastward at 30°–40°, suggesting a simple homoclinal structure (Tazaki 1964; Maekawa 1983). A serpentinite-rich zone lies between the two areas occupied by the Pankehoronai and the Oichanunpe Formations. Along the western boundary of the belt, the Kamuikotan rocks are unconformably overlain by the Miocene Kawabata Formation (alternating beds of sandstone and mudstone), and are covered by the Tokachi welded tuff of probable Neogene age along their eastern boundary.

The calcic amphiboles, that are related to low P/T metamorphism occur only in the Oichanunpe Formation. The geology of Oichanunpe Formation will be explained here exclusively. Detailed geology of the Biei area was given in Maekawa (1983).

Oichanunpe Formation

The Oichanunpe Formation extends for more than 50 km north to south, from approximately 10 km northwest of Asahigawa to Furano, south of the Biei. It consists of picrite, basalt, mafic sediments, psammitic and pelitic rocks, red chert, ultramafic rocks, and amphibolite. For simplicity, rock names of the protoliths are used in this paper. Although these rocks have undergone regional high P/T metamorphism, in most cases the original rock types can be easily inferred due to their weak recrystallization and deformation. The maximum thickness of the formation is about 2000 m.

Metamorphic Protoliths. Basalt consists of clinopyroxene as a relict igneous mineral with sodic–calcic amphibole, sodic pyroxene, epidote, chlorite, pumpellyite, albite, aragonite, and sphene as metamorphic minerals. Lawsonite occurs rarely, and prehnite and zeolites are absent. Chlorite pseudomorphs after olivine (occasionally including relict spinel) and albitized plagioclase laths are commonly observed. Ophitic, subophitic, and variolitic textures are well preserved. Determination of basaltic composition is based on the relict clinopyroxene compositions and is divided into two types: 1. an alkali rock suite and 2. a tholeiitic suite (Maekawa 1982). The alkali basalt commonly contains titanaugite and kaersutite. Tholeiites consist primarily of olivine and clinopyroxene as phenocrysts or microphenocrysts, with clinopyroxene and plagioclase comprising the groundmass.

Picrite consists of more than 15 volume percent chloritized olivine with a dendritic clinopyroxene matrix that contains subordinate amounts of spinel and plagioclase (Figure 2A, B).

Mafic sediments contain clinopyroxene (augite and titanaugite), dendritic clinopyroxene aggregates, detrital glass grains (replaced by chlorite or clay minerals), and a dusty matrix (Figure 2C, D). The detrital fragments are partly fractured by deformation at the post-depositional stage, which probably occurred during the high P/T metamorphism, and the rocks show a weak foliation. Aggregates of glass fragments ("hyaloclastite") are frequently observed as 0.5- to 2-mm lenses in the mafic sediments

(Figure 2E). Mafic sediments commonly contain well-developed laminations parallel to the mafic sediments boundary. Mafic sediments commonly intercalate with 0.5- to 5-cm mud layers and frequently contain pebbles or cobbles of basalt (showing ophitic or subophitic texture), picrite, and red chert (Figure 2F).

Pelitic and psammitic rocks consist of quartz, albite, white mica, chlorite, pumpellyite, stilpnomelane, aragonite, and carbonaceous matter (disordered graphite). The clastic grains are commonly quartz and albite. Their original shapes are well preserved, and are angular or subangular.

Red chert is predominant. The chert consists of cryptocrystalline quartz with small amounts of white mica, limonite, and chlorite. It commonly contains radiolarian fossils. The shapes of radiolaria are slightly elongated by deformation during the low-grade high P/T metamorphism.

Ultramafic rocks are partly serpentinized harzburgite, dunite, and lherzolite. Chlorite rocks and magnesite-quartz rocks derived from ultramafites occur locally as minor amounts. Rodingites are frequently found in ultramafite masses.

Amphibolite consists of brown-green hornblende, albite, epidote, rutile, sphene, and clinopyroxene. Epidote (clinozoisite) pseudomorphs after plagioclase are common. Incipient crystallization of actinolite and sodic amphibole occurs at the rim or cleavage trace of brown-green hornblende. Pumpellyite, aragonite, and saussurite (mainly composed of fine-grained epidote) also occur as secondary products. Well-developed 1- to 2-mm bands comprised of alternative hornblende-rich and plagioclase-rich layers commonly occur in the amphibolite. Olivine (Fo_{76-77}) porphyroblast-bearing amphibolite and tremolite-chlorite-talc rocks are rarely observed in the large amphibolite mass.

Field Occurrence. **Picrite, alkali basalt, tholeiite, red chert, and psammitic rocks** commonly occur in the mafic or pelitic sediments as blocks of varying shapes and sizes (commonly up to 10 m). The contacts between these blocks and surrounding sediments appear to be concordant with bedding surfaces. Abundant 0.5- to 2-cm fragments of the larger blocks are frequently observed in the sediments. The blocks of various rock types are thought to have been mixed during the probable premetamorphic depositional stage while the matrix was still unconsolidated. These field relationships may be explained as large-scale slides that included ophiolitic blocks within the trench where sediments were tectonically accumulated and squeezed down in the subduction zone (Maekawa 1982). As such, the Oichanunpe Formation may be regarded as an olistostrome.

Ultramafic rocks and amphibolite have similar modes of occurrence in relation to the rocks mentioned above. The boundary between ultramafic rocks and sediments is not a fault contact, but is concordant with the bedding surface of the surrounding sediments. Chromian spinel and serpentinite detrital fragments commonly occur in the sediments near the ultramafic rocks (Maekawa 1983). Small amphibolite blocks in the mafic sediments form an S_0 plane that is roughly concordant with the pelite and psammite laminations. The structure is analogous to that of a conglomerate of a non-deformed terrane. The occurrence of ul-

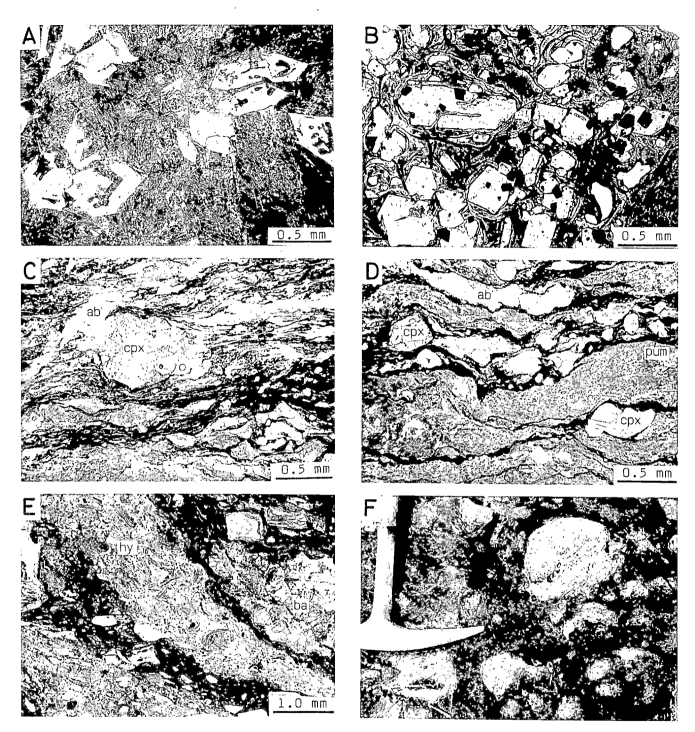

Figure 2. A, B: Dendritic and skeletal olivine crystals in picrite. Olivine is completely replaced by chlorite (white part). Dark grayish matrix are aggregates of fine-grained clinopyroxene. Opaque mineral in B is spinel. plane light. C, D: Mafic sediments consisting of clinopyroxene fragments and dusty matrix. ab: albite, cpx: clinopyroxene, and pum: pumpellyite. plane light. E: Mafic sediments consisting of hyaloclastite (hy), basalt (ba), and dusty matrix. plane light. F: Mafic conglomerate containing rounded pebbles and cobbles of picrite and basalt with ophitic to subophitic texture.

TABLE 1. REPRESENTATIVE BULK ROCK COMPOSITIONS

	Tholeiite		Alkali Rock		Picrite	
	79057	79067	80515	80675	80526	80627
SiO_2	48.91	50.26	55.14	48.16	42.08	44.82
TiO_2	1.24	1.23	3.62	3.00	0.45	0.74
Al_2O_3	15.40	13.98	15.40	15.54	5.09	9.38
FeO^*	13.76	12.00	10.33	14.14	12.64	11.07
MnO	0.20	0.21	0.13	0.23	0.20	0.19
MgO	9.44	8.88	3.54	7.49	32.63	22.80
CaO	7.47	9.35	7.06	7.32	5.04	9.77
Na_2O	2.85	3.10	3.90	3.45	0.27	0.48
K_2O	0.48	0.15	0.20	0.02	0.03	0.00
P_2O_5	0.01	0.02	0.40	0.13	0.00	0.00
Total	99.76	99.18	99.72	99.48	98.43	99.25
Fe/Mg	1.46	1.35	2.92	1.89	0.39	0.49

*total Fe as FeO

tramafite and amphibolite blocks suggests a sedimentary origin for the blocks. Mafic sediments that enclose amphibolite blocks contain many fragments of brown hornblende grains from the blocks. Sodic amphibole (magnesioriebeckite) commonly occurs at the rim of such sedimentary hornblende grains. Sodic amphibole is also formed as a secondary mineral in amphibolite blocks. This clearly suggests that both blocks and surrounding rocks underwent the final stage of high P/T metamorphism.

TWO STAGES OF METAMORPHISM

Two stages of metamorphism can be recognized in the Biei area. The first occurred prior to the depositional stage. It can be detected by the common existence of relict calcic amphiboles in low-grade mafic rocks and amphibolite. Calcic amphibole occurs as the main constituent of the amphibolite, as detrital fragments in the mafic sediments, and as relict metamorphic minerals in the basalt. The second took place as a post-depositional metamorphic stage and is characterized by the widespread occurrence of sodic amphibole, sodic pyroxene, and aragonite. For convenience, metamorphism at the predepositional stage is referred to as the initial-stage metamorphism, and that of the post-depositional stage as the main-stage metamorphism.

Analytical Procedure

The constituent minerals of the rocks were analyzed with the JEOL electron probe microanalyzer Model JXA-5. Detailed descriptions of the analytical procedure and the correction factors have been given by Nakamura and Kushiro (1970). Bulk rock compositions were obtained for 10 major elements using the X-ray fluorescence spectrometer, Rigaku IKF 3064 (Table 1). Detailed analytical methods are described in Matsumoto and Urabe (1980).

MAIN-STAGE METAMORPHISM

Kamuikotan metamorphic rocks in the Biei area belong to the pumpellyite-actinolite facies based on the definition of Naka-

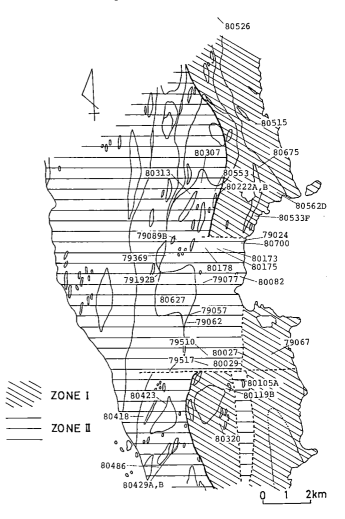

Figure 3. Distribution of Zone I and Zone II. Sample localities are also shown in this figure.

jima et al. (1977). The metamorphic grade increases westward from lower-grade pumpellyite + hematite subfacies (Zone I) to higher-grade epidote-actinolite subfacies (Zone II) (Figure 3).

The eastern half of the Biei area (Zone I) is characterized by the pumpellyite + hematite assemblage, and commonly contains pumpellyite, sodic pyroxene (aegirine-augite), sodic amphibole (magnesioriebeckite), aragonite, albite, and rarely lawsonite. On the other hand, the western half (Zone II) is characterized by the epidote + actinolite + chlorite assemblage, and contains ubiquitous pumpellyite, epidote, actinolite, chlorite, and albite (Figure 4). All rock types underwent this main-stage metamorphism. The ultramafic rocks contain the serpentine minerals chrysotile and/or lizardite in Zone I and antigorite in Zone II. Amphibolites in Zone I commonly include pumpellyite, sodic amphibole, aragonite, and fine-grained sphene as secondary minerals, whereas the main-stage actinolite is frequently formed along the rim or cleavage trace of the initial-stage hornblende in amphibolites from Zone II.

In the southern part of the mapped area, a narrow area of

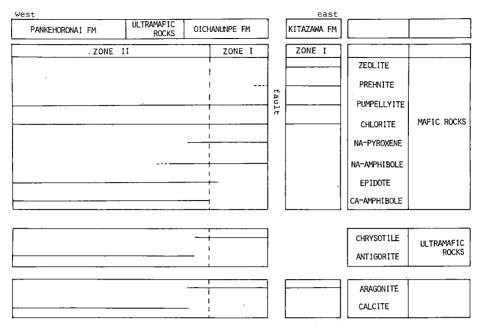

Figure 4. Range of occurrence of metamorphic minerals.

Zone II metamorphism is fault-bounded by the rocks typical of Zone I (Figure 3). This may suggest a tectonic displacement at the later uplift stage.

Chemistry of Main-Stage Amphiboles

Picrite, mafic sediments, and basalts in the Biei area are characterized by Mg-rich and Si-poor composition. This characteristic has a strong effect on amphibole chemistry.

Chlorite. Chlorite is a ubiquitous mineral occurring in varying amounts in the mafic rocks. Chemical compositions of chlorite are given in Table 2. The chlorites show a small range in $Al/(Fe+Mg)$, but show an appreciable range in Fe/Mg. The range in each sample, however, is relatively small, less than 0.07. The ratio of $Fe/(Fe+Mg)$ in chlorite depends strongly on the amount in the host rock.

Amphiboles. Amphiboles formed during the main-stage metamorphism are actinolite ($1.5 < Ca < 2.0$ for $O = 23$), subcalcic actinolite ($1.0 < Ca < 1.5$) and sodic amphibole ($Ca < 1.0$ $< Na$) (see Table 3). These amphiboles form complete solid solution between actinolite and magnesioriebeckite (Figure 5). Actinolite is colorless and subcalcic actinolite and sodic amphibole are pale blue to violet. These amphiboles are commonly found as fine acicular crystallites. They occur along the rim or cleavage trace of relict clinopyroxene and hornblende, and as individual grains in the matrix. Subcalcic-sodic amphiboles are not uncommonly mantled by actinolite.

The amounts of ferric iron were estimated as maximum possible Fe_2O_3 on the basis of stoichiometry of amphibole, using the same constraints (after Brown 1977). Plots of element partitioning among coexisting phases are useful to evaluate the attainment of chemical equilibrium in the rocks (Ernst et al. 1970;

Kawachi 1975; Coombs et al. 1976). Kawachi (1975) and Coombs et al. (1976) have shown a strong correlation of Fe-Mg distribution between actinolite and chlorite pairs in the pumpellyite-actinolite facies rocks. The coexisting actinolite and chlorite pairs in the present area are plotted on Figure 6. In order to exclude the initial-stage calcic amphibole and actinolite with a sodic amphibole component, only actinolite with high Ca and

TABLE 2. CHEMICAL COMPOSITIONS OF CHLORITE

	79089B	79510	80178	80222A
SiO_2	30.16	29.33	29.19	27.52
Al_2O_3	16.76	17.93	18.72	16.55
TiO_2	0.00	0.00	0.00	0.01
FeO^*	12.14	12.57	10.47	22.47
MnO	0.16	0.14	0.16	0.34
MgO	27.38	26.15	27.40	18.34
CaO	0.10	0.04	0.04	0.05
Na_2O	0.06	0.03	0.00	0.00
K_2O	0.01	0.00	0.00	0.00
Cr_2O_3	0.73	0.00	0.00	0.00
Total	87.50	86.19	85.98	85.28
O=28				
Si	5.94	5.87	5.79	5.89
Al iv	2.06	2.14	2.21	2.11
Al vi	1.83	2.08	2.16	1.97
Ti	0.00	0.00	0.00	0.00
Fe^{2+}	2.00	2.10	1.74	4.02
Mn	0.03	0.02	0.03	0.06
Mg	8.04	7.79	8.10	5.85
Ca	0.02	0.01	0.01	0.01
Na	0.02	0.01	0.00	0.00
K	0.00	0.00	0.00	0.00
Cr	0.11	0.00	0.00	0.00
$Fe/(Fe+Mg)$	0.20	0.21	0.18	0.41

*total Fe as FeO

TABLE 3. CHEMICAL COMPOSITIONS OF AMPHIBOLES

	Actinolite		Sodic-Subcalcic Amphibole		Edenite or Edenitic Hornblende			
	79077	79510	79062	80119B	79192B	79192B	79369	79369
SiO_2	57.17	56.42	56.21	53.44	49.44	48.17	44.44	45.10
Al_2O_3	0.72	1.72	1.30	3.03	6.92	7.68	10.21	9.96
TiO_2	0.01	0.00	0.01	0.06	0.12	0.15	0.10	0.14
Fe_2O_3	1.83	1.85	7.68	13.28	1.30	4.06	3.18	1.82
FeO	5.30	7.77	3.50	3.72	9.09	6.35	8.48	9.30
MnO	0.18	0.16	0.00	0.00	0.00	0.00	0.00	0.00
MgO	20.72	17.82	17.75	13.83	16.18	16.61	14.45	14.65
CaO	12.47	10.80	7.35	4.27	11.58	11.53	11.64	11.98
Na_2O	0.29	1.81	3.83	5.13	2.98	2.87	3.13	3.03
K_2O	0.01	0.17	0.04	0.03	0.21	0.19	0.19	0.17
Total	98.70	98.52	97.67	96.79	97.82	97.61	95.82	96.15
O=23								
Si	7.88	7.88	7.86	7.65	7.11	6.92	6.60	6.67
Al^{iv}	0.12	0.12	0.14	0.35	0.89	1.08	1.40	1.33
sum	8.00	8.00	8.00	8.00	8.00	8.00	8.00	8.00
Al^{vi}	0.00	0.17	0.08	0.16	0.29	0.22	0.38	0.40
Ti	0.00	0.00	0.00	0.01	0.01	0.02	0.01	0.02
Fe^{3+}	0.19	0.19	0.81	1.43	0.14	0.44	0.36	0.20
Fe^{2+}	0.61	0.91	0.41	0.45	1.09	0.76	1.05	1.15
Mn	0.02	0.02	0.00	0.00	0.00	0.00	0.00	0.00
Mg	4.26	3.71	3.70	2.95	3.47	3.56	3.20	3.23
sum	5.08	5.00	5.00	5.00	5.00	5.00	5.00	5.00
XM_{1-3}	0.08	0.00	0.00	0.00	0.00	0.00	0.00	0.00
Ca	1.84	1.62	1.10	0.65	1.78	1.78	1.85	1.90
Na	0.08	0.49	1.04	1.42	0.83	0.80	0.90	0.87
sum	2.00	2.11	2.14	2.07	2.61	2.58	2.75	2.77
Na	0.00	0.11	0.14	0.07	0.61	0.58	0.75	0.77
K	0.00	0.03	0.01	0.01	0.04	0.03	0.04	0.03
$Fe^{2+}/(Fe^{2+} + Mg)$	0.13	0.20	0.10	0.13	0.24	0.18	0.25	0.26

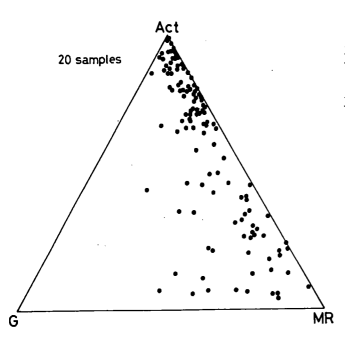

Figure 5. Amphiboles plotted in the Act [actinolite, Ca_2 (Mg, Fe^{2+})$_5Si_8O_{22}$ $(OH)_2$]-G [glaucophane, $Na_2Mg_3Al_2Si_8O_{22}(OH)_2$]-MR [magnesioriebeckite, $Na_2Mg_3Fe_2^{3+}Si_8O_{22}(OH)_2$] diagram.

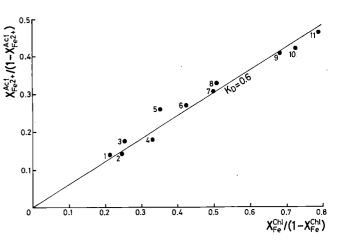

Figure 6. Distribution of Fe and Mg between actinolite (Si >7.8, Al <0.2, and Ca >1.7 for O = 23) and coexisting chlorite. 1: 80178, 2: 79077, 3: 80175, 4: 80307, 5: 80429B, 6: 79517, 7: 80700, 8: 80429A, 9: 80029, 10: 79024, 11: 80173.

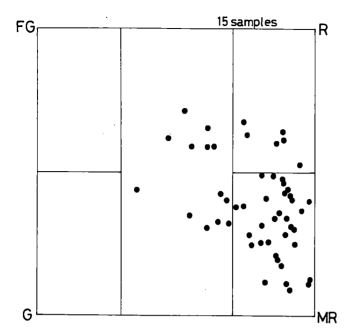

Figure 7. Compositions of sodic amphiboles. G: glaucophane [Na₂ Mg₃Al₂Si₈O₂₂(OH)₂], FG: ferroglaucophane [Na₂Fe₃²⁺Al₂Si₈O₂₂ (OH)₂], R: riebeckite [Na₂Fe₃²⁺Fe₂³⁺Si₈O₂₂(OH)₂], MR: magnesiorie-beckite [Na₂Mg₃Fe₂³⁺Si₈O₂₂(OH)₂]

low Al contents (Ca >1.7, Si >7.8 and Al <0.2 for O = 23) is used in Figure 6. The figure shows that the Fe-Mg distribution between actinolite and chlorite is relatively consistent (K_D = 0.6), which suggests the attainment of equilibrium.

Subcalcic-sodic amphibole is higher in Fe/Mg than actinolite. Sodic amphiboles are plotted on the Miyashiro diagram (1958) in Figure 7. They are mostly magnesioriebeckite, and

partly riebeckite and crossite. The sodic amphiboles in the Biei area have a characteristically low Fe/Mg ratio and a high Ca content. Figure 8 depicts Ca versus Fe/(Fe+Mg) for subcalcic and sodic amphiboles. Ca contents of these amphiboles tend to decrease with increasing Fe/(Fe+Mg). Each sample enclosed by a dotted line is labeled from "1" to "18" in increasing Fe/ (Fe+Mg) and decreasing Ca order. As the Fe/(Fe+Mg) content in chlorite corresponds well to that in host rock, low Fe/(Fe+Mg) ratio and high Ca contents in subcalcic to sodic amphibole in mafic rocks is considered to depend on low Fe/(Fe+Mg) ratio in host rock.

MODES OF OCCURRENCE OF INITIAL-STAGE AMPHIBOLES

Aluminous brown, green to pale green amphiboles are commonly found as clastic grains in mafic sediments, and as relict (metamorphic) minerals in the basalt and the amphibolite. These amphiboles are commonly mantled by the main-stage colorless actinolite and bluish subcalcic–sodic amphibole.

Amphibole sandstone composed mainly of amphibolite fragments and detrital amphibole grains is widespread in the Oichanunpe Formation. Layers or lenses of amphibole sandstone tend to be abundant near the large amphibolite blocks. The rocks can easily be recognized by their peculiar texture in Zone I (Figure 9A). On the other hand, recognition by texture is difficult in Zone II because relict amphibole grains are more or less replaced by actinolite formed during the main-stage high P/T metamorphism (Figure 10A). Relict amphibole phases (except for Al-poor amphibole), however, can be distinguished from the later stage actinolite by chemical composition, as discussed later.

Basalts (tholeiite and alkali basalt) and picrite commonly

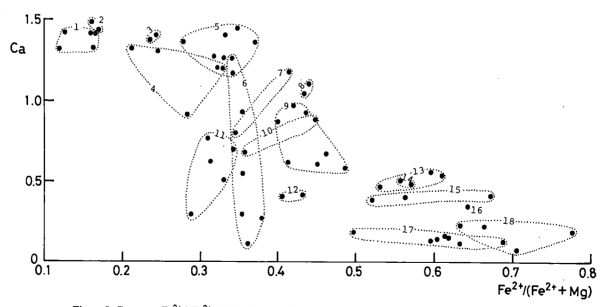

Figure 8. Ca versus Fe²⁺/(Fe²⁺ + Mg) diagram of subcalcic to sodic amphiboles. 1: 80178, 2: 79517, 3: 80429A, 4: 79062, 5: 80027, 6: 80423, 7:79024, 8: 80533F, 9: 80222A, 10: 80418, 11: 80105A, 12: 80562D, 13: 80082, 14: 80313, 15: 80320, 16: 80553, 17: 80486, 18: 80222E.

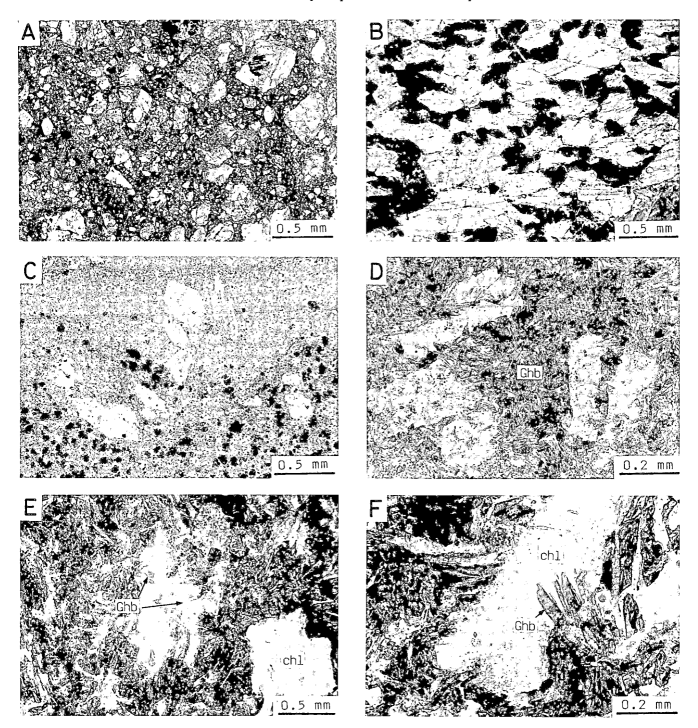

Figure 9. A: Mafic sediments (amphibole sandstone). Detrital fragments consist of green hornblende and saussuritized plagioclase and the matrix is filled with albite, chlorite, sphene, white mica, and dusty materials. plane light. B: Amphibolite composed of pale brown hornblende and saussurite. plane light. C, D: Picrite. Fine-grained groundmass clinopyroxene is wholly replaced by pale green hornblende (Ghb). Fine-acicular hornblende also occurs within chloritized olivine pseudomorphs (white part). plane light. E, F: Pale green hornblende (Ghb) in basalt. chl: chlorite. plane light. It occurs within cavities filled with chlorite. plane light.

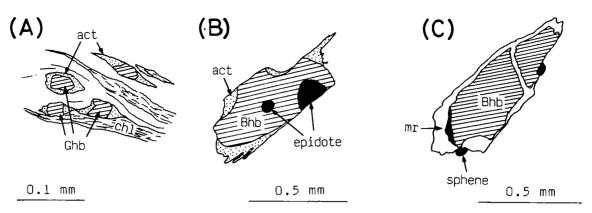

Figure 10. Mutual relations between the initial-stage amphibole and main-stage amphibole in mafic sediments (A) and amphibolite (B, C). Ghb: green hornblende (initial stage), Bhb: brown hornblende (initial stage), act: actinolite (main stage), mr: magnesioriebeckite (main stage), chl: chlorite.

contain various amounts of relict metamorphic pale green amphibole phases (Figure 9C, D, E, F). Some of the rocks consist largely of pale green amphibole and albite (Figure 9C, D). In such rocks, the pale green amphibole is more or less replaced by the main-stage actinolite and subcalcic-sodic amphibole assemblage. Although the primary clinopyroxenes were obviously replaced by the pale green amphiboles, the rocks have always preserved their original ophitic, subophitic, and variolitic textures.

Brown and green amphiboles of the initial-stage metamorphism are the main constituent minerals in the amphibolite (Figure 9B). They are commonly mantled by actinolite and subcalcic–sodic amphibole (Figure 10B, C).

CHEMICAL COMPOSITION AND P-T CONDITION OF INITIAL-STAGE AMPHIBOLES

In general, Al_2O_3 content in calcic amphibole tends to increase with increasing metamorphic grade (Miyashiro 1958; Banno 1964). The relict (initial-stage) calcic amphiboles show a wide variety in Al_2O_3 contents, up to 11 weight percent, and they

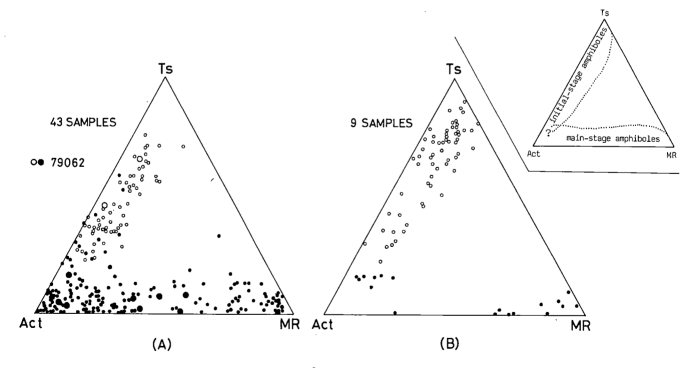

Figure 11. Ts: tschermakite $[Ca_2(Mg, Fe^{2+})_3Al_2Si_6Al_2O_{22}(OH)_2]$-Act: actinolite $[Ca_2(Mg, Fe^{2+})_5 Si_8O_{22}(OH)_2]$-MR: magnesioriebeckite $[Na_2Mg_3Fe_2^{3+}Si_8O_{22}(OH)_2]$ diagrams of amphiboles. A: basalt, picrite, and mafic sediments, B: amphibolite. Large symbols in (A) show amphiboles from one specimen 79062.

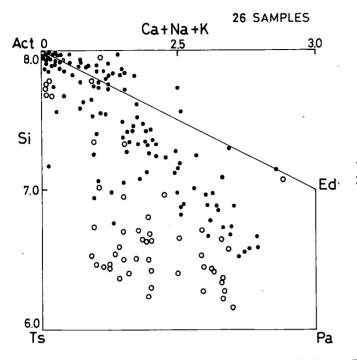

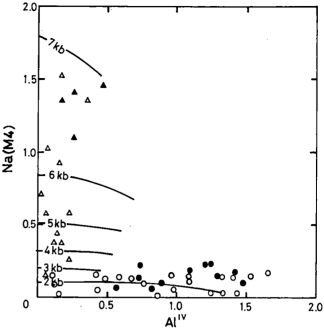

Figure 12. Si versus Ca + Na + K plot of calcic amphiboles. Solid circle: basalt, picrite, and mafic sediments in the Oichanunpe Formation; Open circle: amphibolite in the Oichanunpe Formation.

Figure 13. Na (M4) versus Al^{IV} plot of the initial-stage and main-stage amphiboles from basalt, picrite, mafic sediments, and amphibolite in the Oichanunpe Formation. Open circle: initial-stage amphibole from oxide-free specimen; Solid circle: initial-stage amphibole from oxide-bearing specimen; Open triangle: main-stage amphibole from oxide-free specimen; Solid triangle: main-stage amphibole from oxide-bearing specimen. Isobars after Brown (1977).

appear to represent various grades of a metamorphic complex that once had a low P/T metamorphic sequence. The content does not correlate to Fe/Mg in amphibole in the present area. Amphibole from Zone I contains a minimum Al_2O_3 content, about 4–5 weight percent. The Al_2O_3 content tends to be less in Zone II than Zone I. Where Al_2O_3 content is less than 4–5 weight percent, such amphiboles are compositionally indistinguishable from actinolite formed during the high P/T (main-stage) metamorphism.

Amphiboles of both stages in basalt, picrite, mafic sediments, and amphibolite are plotted on tschermakite [Al^{IV}]-actinolite [$Ca-Al^{IV}$]-magnesioriebeckite [Na(M4)] triangular diagrams (Figure 11). Open circles in Figure 11 show relict (initial-stage) amphiboles, so far detected by the modes of occurrence. The main-stage amphiboles plot on the tschermakite-poor side, whereas the initial-stage amphibole shows a tschermakite-rich trend. Two trends for amphiboles can also be recognized in amphibolites (Figure 11B).

Calcic amphiboles are plotted on the Si versus Ca+Na+K diagram of Miyashiro (1973) in Figure 12. Calcic amphiboles from the amphibolite are richer in tschermakite and poorer in edenite component than that in other mafic rocks. Calcic amphiboles in mafic rocks are characterized by their high edenite component, and some of them are edenite or edenitic hornblende of Leake (1978). Representative chemical compositions are given in Table 3 (specimens 79192B and 79369). Calcic amphiboles of both stages in mafic rocks characteristically contain high Na_2O contents, the component occurring mainly as an edenite compo-

nent. As proposed by Shido (1958) and Seki et al. (1959), the possible reaction of an edenite component entering into calcic amphibole is as follows:

$$Ca_2Mg_5Si_8O_{22}(OH)_2 + NaAlSi_3O_8 = NaCa_2Mg_5AlSi_7O_{22}(OH)_2 + 4SiO_2$$

Tremolite   Albite   Edenite   Quartz

Shido (1958) suggested that amphibole from quartz-free rocks tends to contain a higher edenite component than those from quartz-bearing rocks. The mafic rocks in the present area are characterized by the SiO_2-poor composition, and lack quartz in most cases, so a high edenite component in the amphiboles from the present area reflects mainly a SiO_2 deficient bulk composition.

Figure 13 is Na(M4) versus Al^{IV} plot of Brown (1977) for initial-stage and main-stage amphiboles in basalt, picrite, mafic sediments, and amphibolite. The diagram suggests a high-pressure condition for the main-stage of metamorphism, and a low-pressure condition of the initial-stage.

CONCLUDING REMARKS

As discussed above, modes of occurrence and chemical compositions of relict calcic amphiboles suggest a low P/T metamorphic episode prior to high P/T metamorphism in the Biei area. Rock types and igneous mineral parageneses of basalts and picrites from the area are similar to those from the present intraplate oceanic islands, such as Hawaii and Tahiti, suggesting an oceanic island origin of these rocks (Maekawa 1982). The initial-stage metamorphism may have taken place below the islands, where relatively high temperatures prevailed, and thereafter the islands may have been transported to a trench and collapsed due to down-bending of the underlying oceanic slab. Many blocks or fragments supplied from the oceanic islands were rapidly deposited by gravity slides. They have undergone the main-stage of metamorphism in the subduction zone.

In the Horokanai area, 30 km northwest of the Asahigawa, the low P/T metamorphic rocks within an ophiolite succession (Horokanai ophiolite) overlie high P/T Kamuikotan metamorphic rocks (Asahina and Komatsu 1979; Ishizuka 1980 and 1985). The Kamuikotan belt is considered to be situated along an ancient consuming plate boundary and the belt, in its premetamorphic stage, is regarded as a basin accumulating various rocks derived from oceanic settings. The following possibility is worth considering: Many blocks and fragments of basalt, amphibolite, and ultramafic rocks scattered in the Biei area were supplied from a once-exposed ophiolitic complex such as the Horokanai ophiolite, which may have been emplaced by local obduction and thus escaped subduction related metamorphism. However, further study is necessary to evaluate the hypothesis more comprehensively.

ACKNOWLEDGMENTS

I would like to express my appreciation to professors Y. Nakamura and I. Kushiro (University of Tokyo) for instructive suggestions and valuable discussions. I am much indebted to professors S. Banno (Kyoto University) and E. H. Brown (Western Washington University) for useful suggestions, and to professors C. A. Landis (University of Otago), M. Toriumi (University of Tokyo), and C. Ziegler (Western Washington University) for critical reading of the manuscript. I sincerely thank professors D. S. Coombs (University of Otago), P. Schiffman (University of California), and H. Ishizuka (Kochi University) for reviewing the manuscript.

REFERENCES CITED

Asahina, T. and Komatsu, M., 1979, The Horokanai ophiolitic complex in the Kamuikotan tectonic belt, Hokkaido, Japan: Journal of the Geological Society of Japan, v. 85, p. 317–330.

Banno, S., 1964, Petrologic studies on Sanbagawa crystalline schists in the Bessi-Ino district, central Shikoku, Japan: Journal of the Faculty of Science, University of Tokyo, Section II, v. 15, p. 203–319.

Brown, E. H., 1977, The crossite content of Ca-amphibole as a guide to pressure of metamorphism: Journal of Petrology, v. 18, p. 53–72.

Coombs, D. S., Nakamura, Y., and Vuagnat, M., 1976, Pumpellyite-actinolite facies schists of the Taveyanne Formation near Loéche, Valais, Switzerland: Journal of Petrology, v. 17, p. 440–471.

Ernst, W. G., Seki, Y., Onuki, H., and Gilbert, M. C., 1970, Comparative study of low-grade metamorphism in the California Coast Ranges and the Outer Metamorphic Belt of Japan: Geological Society of America, Memoir 124, 276 p.

Hervé, F., 1975, Petrography of the Kamuikotan metamorphic belt at the Ubun-Oroen cross section, central Hokkaido, Japan: Journal of the Faculty of Science, Hokkaido University, Series IV, v. 16, p. 453–470.

Ishizuka, H., 1980, Geology of the Horokanai ophiolite in the Kamuikotan tectonic belt, Hokkaido, Japan: Journal of the Geological Society of Japan, v. 86, p. 119–134 (in Japanese with English abstract).

——— 1985, Prograde metamorphism of the Horokanai ophiolite in the Kamuikotan zone, Hokkaido, Japan: Journal of Petrology, v. 26, p. 391–417.

Ishizuka, H., Imaizumi, M., Gouchi, N., and Banno, S., 1983, The Kamuikotan zone in Hokkaido, Japan: Tectonic mixing of high-pressure and low-pressure metamorphic rocks: Journal of Metamorphic Geology, v. 1, p. 263–275.

Kawachi, Y., 1975, Pumpellyite-actinolite and contiguous facies metamorphism in part of upper Wakatipu district, South Island, New Zealand: New Zealand Journal of Geology and Geophysics, v. 18, p. 401–441.

Leake, B. E., 1978, Nomenclature of amphiboles: American Mineralogist, v. 63, p. 1023–1052.

Maekawa, H., 1982, Sedimentation, metamorphism and tectonics of high pressure and low temperature metamorphic terranes: central part of the Kamuikotan belt, and Motai Group [Ph.D. thesis]: University of Tokyo, 334 p.

——— 1983, Submarine sliding deposits and their modes of occurrence of the Kamuikotan metamorphic rocks in the Biei area, Hokkaido, Japan: Journal of the Faculty of Science, University of Tokyo, Section II, v. 20, p. 489–507.

Matsumoto, R., and Urabe, T., 1980, An automatic analysis of major elements in silicate rocks with X-ray fluorescence spectrometer using fused disc samples: Journal of Japanese Association for Mineralogists, Petrologists, and Economic Geologists, v. 75, p. 272–278 (in Japanese with English abstract).

Miyashiro, A., 1958, Regional metamorphism of the Gosaisyo-Takanuki district in the central Abukuma Plateau: Journal of the Faculty of Science, University of Tokyo, Section II, v. 11, p. 219–272.

——— 1973, Metamorphism and metamorphic belts: London, George Allen & Unwin, 492 pp.

Nakajima, T., Banno, S., and Suzuki, T., 1977, Reactions leading to the disappearance of pumpellyite in low-grade metamorphic rocks of the Sanbagawa metamorphic belt in central Shikoku, Japan: Journal of Petrology, v. 18, p. 263–284.

Nakamura, Y., 1984, Chemical characteristics of relict amphiboles from the Mikabu greenrock complex, Toba district [abs.]: Annual Meeting, Geological Society of Japan, Tokyo, p. 445 (in Japanese).

Nakamura, Y., and Kushiro, I., 1970, Compositional relations of coexisting orthopyroxene, pigeonite and augite in a tholeiitic andesite from Hakone volcano: Contributions to Mineralogy and Petrology, v. 26, p. 265–275.

Seki, Y., Aiba, M., and Kato, C., 1959, Edenite in Sanbagawa crystalline schists of the Sibukawa district, central Japan: Japan Journal of Geology and Geography, v. 230, p. 233–243.

Shido, F., 1958, Plutonic and metamorphic rocks of the Nakoso and Iritono districts in the central Abukuma Plateau: Journal of the Faculty of Science, University of Tokyo, Section II, v. 11, p. 132–217.

Suzuki, M., Watanabe, J., and Kasugai, A., 1964, Geologic map of Japan, 1:50,000 "Biei" sheet and its explanatory text: Hokkaido Development Agency (in Japanese with English abstract).

Tazaki, K., 1964, Alkali amphibole-bearing metamorphic rocks in the Kamuikotan belt of the south-western part of Asahikawa, central Hokkaido: Earth Science, no. 71, p. 8–17 (in Japanese with English abstract).

Watanabe, T., and Maekawa, H., Early Cretaceous dual subduction system in and around the Kamuikotan tectonic belt, Hokkaido, Japan: in N. Nasu, K. Kobayashi, S. Uyeda, I. Kushiro, and H. Kagami, Editors, Formation of Active Ocean Margins: Terra Scientific Publishing Company, Tokyo, (in press).

MANUSCRIPT ACCEPTED BY THE SOCIETY JULY 29, 1985

Geological Society of America
Memoir 164
1986

Regional eclogite facies in the high-pressure metamorphic belt of New Caledonia

K. Yokoyama*
R. N. Brothers
P. M. Black
Department of Geology
Auckland University
Auckland, New Zealand

ABSTRACT

In the New Caledonian mid-Tertiary metamorphic belt, a continuous progression is shown by isograds in metasedimentary rocks (with intercalated metavolcanic rocks) from a lawsonite zone (lawsonite-albite-chlorite and blueschist facies) through an epidote zone (blueschist facies) into an omphacite zone (eclogite facies). Other isograds within the omphacite zone are compositions of almandine and phengite as defined by sliding equilibria. Isogradic surfaces have dips 10° to SW; the epidote zone thickness is 300–500 m and the omphacite zone 500 m. Metamorphic P-T conditions were: epidote isograd = 390°C, 9.5 kb; omphacite isograd = 430°C, 11 kb; and the deepest part of the omphacite zone equilibrated finally at 550°, 10.5 kb. The T gradient of 7–10°C/km in the lawsonite zone increased to 50–60°C/km across the omphacite isograd.

In epidote zone metasedimentary rocks, the blueschist total assemblage is quartz - glaucophane - phengite - paragonite - chlorite - clinozoisite - almandine - rutile - sphene plus jadeite now preserved only as relics within retrograde albite porphyroblasts; however, interbedded metabasites are "low-grade" or "transitional" eclogites with omphacite ± almandine - clinozoisite - glaucophane - actinolite - chlorite - sphene ± rutile. In the omphacite zone (eclogite facies) the total assemblage is similar in metasedimentary rocks and metabasites: omphacite - almandine - clinozoisite - glaucophane - barroisite - phengite - paragonite - rutile - sphene - quartz with differences in specific associations controlled by bulk chemical composition. In both zones, hydrous phases coexist with omphacite-almandine in equilibrium associations.

INTRODUCTION

The occurrence of eclogites or eclogitic metamorphic assemblages within the New Caledonia Tertiary high-pressure schists has been reported by numerous authors (Brière 1919; Lacroix 1941; Coleman and others 1965; Brothers 1974; Black 1977; Brothers and Yokoyama 1982) but no comprehensive description of these rocks has been given yet. This region contains an unusually well-developed low-temperature high-pressure sequence that has been preserved throughout a variety of sedimentary and igneous parent lithologies. The brief (20 m.y.) Tertiary event produced, on a regional scale and in a single facies series, a direct progression from lawsonite - albite - chlorite schist through blueschist facies to the eclogite facies in a wet environment of metamorphic recrystallization.

The present paper describes omphacite - almandine associations that have developed on a regional scale in a protolith com-

*Present address: National Science Museum, Shinjuku-ku, Tokyo 160, Japan.

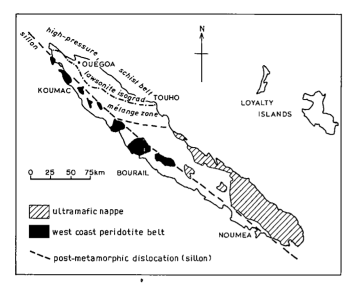

Figure 1. Regional setting for the high-pressure metamorphic belt on the main island of New Caledonia.

posed originally of sediments (tuffs, sandstones, siltstones) with intercalated basic and acid lava flows and tuffs. Metamorphic rocks derived from all of these parent lithologies carry the eclogite index assemblage (omphacite - almandine) and they form a narrow coastal zone of gneisses, 3–5 km wide, along the highest-grade (deepest) margin of the high-pressure belt (Figure 1). Detailed mapping and mineralogy of this eclogite zone show that regional isograds for omphacite and other phases can be recognised within the *metasediments*; the common, but more sporadic, basic metavolcanics within the same sequence contain eclogite assemblages that are similar to those of the enclosing metasedimentary rocks.

From the field and mineralogical data there seems to be no doubt that the metasedimentary and metavolcanic rocks are cofacial within the eclogite facies, and the metasedimentary members are estimated to constitute at least 80% by volume of the eclogite terrane. For the petrological discussion that follows, some definition of terms is necessary. A review of terminology (e.g. Coleman and others 1965; Banno 1970; Miyashiro 1973; Winkler 1974; Dobretsov and others 1975; Turner 1981) indicates that, at present, the name *eclogite (sensu stricto)* is being restricted to rocks that contain the assemblage omphacite + almandine and are basaltic in bulk composition; this usage will be maintained in this paper. For the associated cofacial metasediments that also contain omphacite and almandine, the terms *eclogitic gneisses* (or paragneisses or metasediments) will be applied exclusively, thus following the sense of the nomenclature already used, for example, by Velde and Kienast (1973) for comparable rocks in the Sesia-Lanzo area of the European Alps.

GEOLOGICAL SETTING

The regional high-pressure schists and gneisses in New

Caledonia form a NE–SW trending metamorphic belt, measuring 175 km by 25 km, at the northern end of the main island (Figure 1), with the highest grade rocks located along the northeastern Pacific coastline. The southern and western boundary of the belt is a regional mélange zone (Brothers and Blake 1973) about 30 km wide containing large schuppen of basement Permian-Jurassic metagreywackes with overlying Cretaceous-Eocene sediments that were thrust northeastwards to form a capping over the high-pressure schists during metamorphism (Brothers 1974). These thrust-slices vary in maximum metamorphic grade from prehnite-pumpellyite to pumpellyite-actinolite and low-grade greenschist assemblages. They are bounded by major shear surfaces with occasional lawsonite-albite schists developed on the soles. Many of the shear systems have been penetrated by tectonically-injected slivers of ocean crust derived from beneath the basement metagreywackes; these meta-ophiolites (Black and Brothers 1977) have a different high-pressure mineralogy and an older K-Ar age (41 m.y.) than the rocks of the regional metamorphic belt (39–21 m.y.).

There is no field evidence to suggest that the regional high-pressure schists and the mélange zone were once part of a subduction system with a convergence zone dipping either eastward into the Pacific Ocean or westward into the Coral Sea. Tectonism associated with the metamorphism seems to have consisted dominantly of low-angle oceanward thrusting of the continental margin during the Late Eocene when the basement metagreywackes were emplaced as flat-stacked schuppen over an adjacent Cretaceous-Eocene sedimentary-volcanic marine basin that was effectively immobile (Brothers and Yokoyama 1982). During a coeval event, the southern end of the island was covered by a large nappe of oceanic crust and upper mantle (Massif du Sud, 6000 km^2) that was obducted in the opposite direction, from northeast to southwest (Guillon and Routhier 1971; Guillon 1975; Paris 1981; Podvin and others in press). Although positive evidence is lacking, it is possible that this ultramafic sheet originally extended as far north as the regional high-pressure belt, and has since been eroded. In this case, the metamorphic history of the blueschist and eclogite facies could have contained two immediately successive phases of rapid tectonic burial: at first beneath the basement metagreywackes, with this whole complex then covered by the ultramafic nappe. The locus for these tectonic movements has been interpreted (Briggs and others 1978; Paris 1981) as a mid-Tertiary intra-plate suture, or plate-plate transform boundary, the fossil trace of which is now marked by the mélange zone (Figure 1). On the western side of the island, both the mélange zone and the regional high-pressure belt terminate against a major fault zone, the *sillon* of Routhier (1953).

THE HIGH-PRESSURE BELT

The most detailed studies of lithology and sequence have been made in the northern part of the belt (Espirat and Millon 1965, 1967; Brothers 1970; Briggs and others 1978), especially in the vicinity of Pam-Ouégoa (Figure 2) where Cretaceous-Eocene

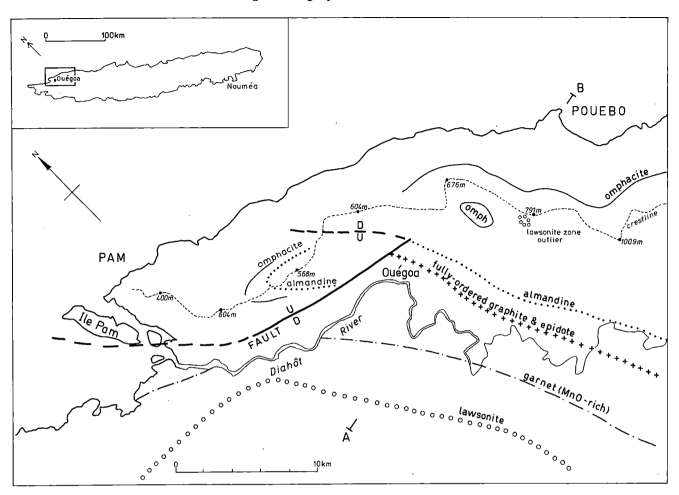

Figure 2. Isograd map for metasedimentary rocks across the full width of the high-pressure belt in the Pam-Ouégoa-Pouébo region (see Figure 5), with the coastal mountain range topography indicated by the crestline and Diahot River. Cross section AB is given in Figure 4.

sedimentary and igneous parent rocks can be recognised within the metamorphites. In general, the stratigraphic succession appears to young from east to west, so that metamorphosed Cretaceous clastic lithologies dominate in the northeastern coastal area. However, all major fold movements pre-dated the metamorphic climax of recrystallisation and there must have been some structural closure toward the east because Paleocene foraminifera have been identified in a limestone in high-grade gneiss (omphacite zone) near the coastline at Tao, south of Pouébo (Arnould 1958; Paris 1981). The parent Cretaceous sequence included carbonaceous claystones, siltstones, and sandstones, with intercalated dolerite sills and rhyolite tuffs and flows; minor lithologies are calcarenites, basic tuffs, intermediate igneous rocks, and rare conglomerate and concretionary layers. *Inoceramus* fossils within the lawsonite zone date the stratigraphically upper part as Senonian. In the metasediments, metamorphic textures (schistose to gneissic) and grade (lawsonite zone to omphacite zone) increase from southwest to northeast across the belt, but the persistent main foliation is parallel to original bedding. The general absence of penetrative deformation is evident in the

omphacite zone where ribboned relict bedding is common in eclogitic paragneisses that have normal contacts with intercalated, unsheared eclogite metabasite pillow lavas and breccias. The Cretaceous sequence, probably of shallow water origin, is overlain conformably to the southwest by fine-grained, deeper-water Early Eocene metasedimentary rocks consisting of thinly-bedded siliceous argillites, massive cherts and limestones, along with rare and weakly metamorphosed basalts and dolerites.

During Cretaceous sedimentation, rhyolite volcanism produced sedimentary exhalative stratiform Cu-Pb-Zn (-Ag-Au) mineralisation associated with black carbonaceous shales (Briggs and others 1977). Acid volcanics are present in the northern part of the metamorphic belt, but are absent in the southern part. In these two sectors there are comparable differences in the composition of the metasediments, which become less siliceous and more calcareous toward the south. The eclogitic paragneisses described in this paper were crystallised from the more siliceous northern sediments.

Within the high-grade epidote and omphacite zones, particularly near Pam-Ouégoa, major fault zones contain large tectonic

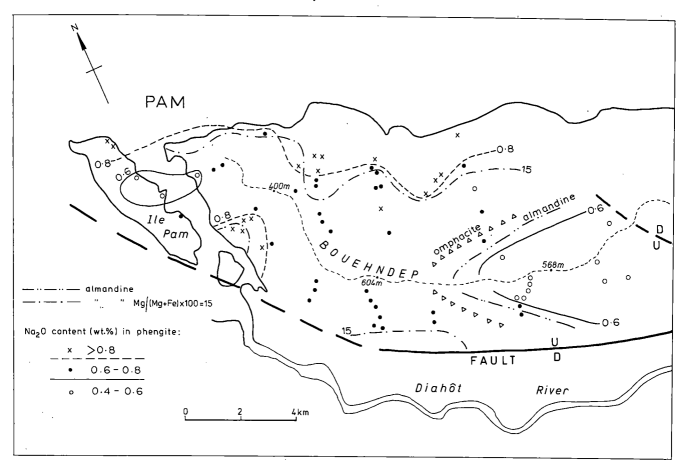

Figure 3. Metasedimentary rock isograds in Pam Peninsula for sliding equilibria rim compositions in almandine, and in phengite coexisting with paragonite.

blocks of serpentinite, cherts, and metabasites presumably derived from the ocean floor substrate of the Cretaceous marine basin which lies immediately beneath the metamorphic belt. Their grade of metamorphism (Paterson 1980) is from blueschist to eclogite facies, but they will not be considered in this paper.

Based on K-Ar dates on glaucophane and white mica (Coleman 1967; Blake and others 1977), the main period of recrystallisation in the high-pressure belt occurred between 39–36 m.y. The whole metamorphic event, including prograde and retrograde phases, occupied a span of about 20 m.y. from 39–21 m.y. (Brothers and Yokoyama 1982).

ISOGRADS, ISOGRADIC SURFACES, AND ZONES

In the Pam-Ouégoa-Pouébo area (Figures 2, 3 and 4) metamorphic grade increases northeastward across the high-pressure belt from the continental to the oceanic side. In field traverses the progressive changes in mineral assemblages can be followed and sampled regularly from metasedimentary and metaigneous lithologies. The description of index minerals in this area by previous workers (Brothers 1974; Black 1977; Briggs and others 1978; Diessel and others 1978) has identified the following progressive

sequence of isograds in metasediments, up to the epidote zone (Figure 5): lawsonite, Na-amphibole, spessartine (rim), and fully-ordered graphite that coincided with the epidote (clinozoisite) isograd. A narrow transitional subzone where lawsonite coexists with epidote (LET in Figure 5) is defined on the high-grade side by a lawsonite-out isograd that also coincides with the appearance of almandine compositions in garnet rims.

On the high-grade side of this almandine isograd, and at a deep level in the schist body, the earlier descriptions had also recorded a zone for barroisitic hornblende that was well-substantiated by mapping first appearances of the blue-green amphibole in metasediments. Subsequent studies have revealed variations in the composition of this amphibole from actinolite to barroisite, and replacement textures indicate a time-span of crystallisation that pre-dated and post-dated glaucophane; for these reasons the location of an appropriate isograd needs to be defined by detailed correlation between composition and texture.

Brothers and Yokoyama (1982; Figure 2) described an omphacite isograd in the deeper paragneisses, and from additional work this has been accurately located along with two further isograds that have been constructed by connecting similar rim compositions in minerals generated by sliding equilibria (Banno

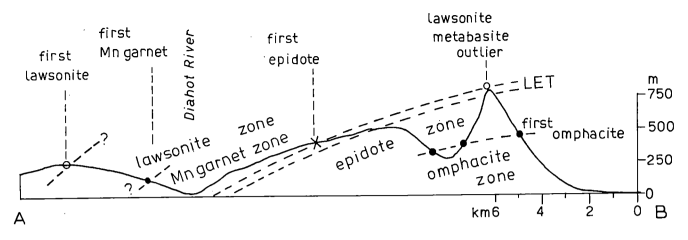

Figure 4. Cross section AB of Figure 2 with the crestline lawsonite metabasite outlier projected onto the line of section.

and others 1978): for almandine with Mg/(Mg+Fe) ratio = 0.15, and for phengite with Na_2O content = 0.8 wt% coexisting with paragonite (Figure 3).

The isograd map of Figure 2 includes a narrow mountain range that extends from SE to NW through the region, from behind Pouébo to the end of Pam Peninsula; the crestline (as indicated in Figure 2) lies at altitudes of 400–1000 m, with flanking ridges that descend steeply northeastward to the coastline and southwestward into the valley of the Diahot River. The isograds for omphacite and for almandine and phengite compositions appear on both flanks of the range, thus allowing calculation of the attitude of the isogradic surfaces. These are only gently inclined, with dips from 0° to 10°W or SW (Figure 4). Lawsonite-bearing metabasites form a small residual on the range crest to the west of Pouébo where the topographic upwards sequence on the southern flank of the mountains is omphacite zone (thickness concealed at depth), epidote zone (460 m vertical ≈ actual thickness), and lawsonite zone (outlier); the sequence is well exposed and no tectonic breaks were found. From completed structural maps for the Pam-Pouébo district (Bell and Spörli in preparation), it is clear that the isogradic surfaces for epidote and omphacite post-date and cut across a major antiform, and were not flexed by regional folding.

The metasedimentary rocks of the epidote zone (Figure 5) have assemblages of blueschist type (Turner 1981) which, with passage into the omphacite zone, become conspicuously coarse (grain size up to several cm). The character of their total assemblage is changed to eclogitic paragneisses with the regional appearance of omphacite accompanied by zoisite and epidote as minor phases.

The metabasites of the epidote zone contain chlorite, sodic and calcic amphibole, clinozoisite, and the eclogite mineral pair omphacite-almandine. At or near the omphacite isograd, as defined from the associated metasediments, the metabasites (eclogites) show some notable changes in mineralogy; chlorite and actinolite disappear and zoisite, barroisite, and paragonite appear

as common phases, but omphacite and almandine continue through the omphacite zone without significant variation in their chemical compositions.

In the omphacite zone there are only minor differences, controlled by bulk chemical composition, between the total assemblage for the eclogites and the eclogitic paragneisses. However, there is considerable variation in the percentage modal occurrences of minerals within these eclogitic assemblages: for example, in metabasites omphacite + almandine may form up to 95% of an assemblage; epidote group minerals and amphiboles (glaucophane and/or barroisite) may dominate in some lithologies while paragonite is less than 10%. In metasediments quartz + mica may form up to 70%, paragonite 20%, and up to 10% for each of omphacite, almandine, glaucophane, and the epidote group.

Acid igneous rocks (meta-acidites) form massive bodies or bedded tuff layers intercalated with paragneisses in the epidote and omphacite zones of Pam Peninsula, but they are not known in the Pouébo area. Common prograde minerals are quartz, sodic pyroxene (chloromelanite and omphacite) and phengite; albite is present, but only as a retrograde phase. Some assemblages contain garnet (in skeletal crystals embayed by quartz), ferroglaucophane, K-feldspar, and stilpnomelane; the amphibole and stilpnomelane are idioblastic and show no sign of replacement. The massive rhyolitic body (leptynite) of Bouehndep, lying across the omphacite isograd of Figure 3, contains jadeite (Jd_{87-98}) that was almost entirely replaced by albite + aegirine-jadeite or chloromelanite (Black and others in press).

ZONAL MINERAL SEQUENCES

Within the epidote and omphacite zones it is possible to recognise those phases that were P-T prograde and those that originated from later retrograde recrystallisation. The onset of retrogression was marked by rapid decompression (Bell and

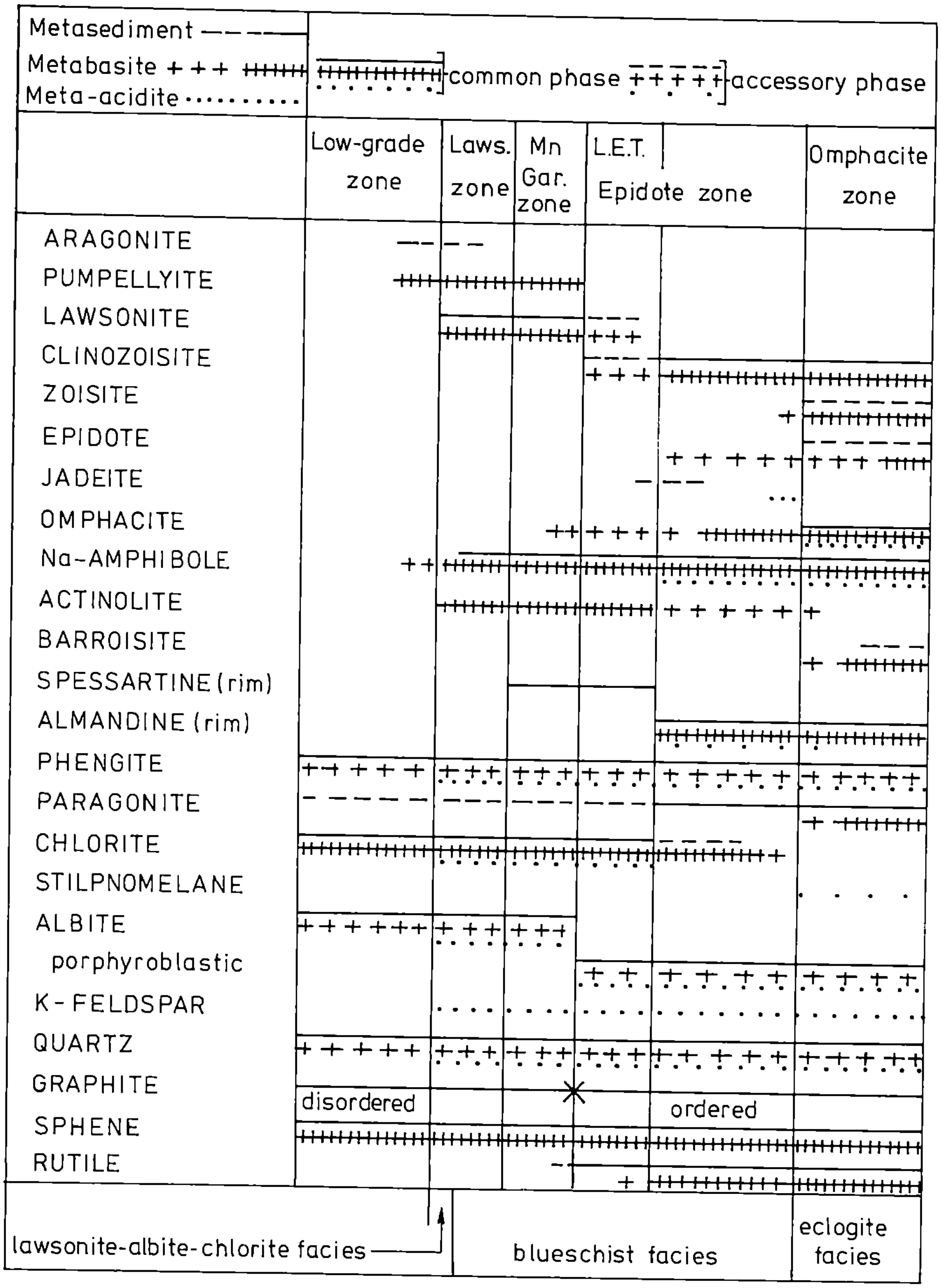

Metasediment ————
Metabasite + + + ++++++   ++++++++++ common phase + + + + accessory phase
Meta-acidite ··········

	Low-grade zone	Laws. zone	Mn Gar. zone	L.E.T. Epidote zone	Omphacite zone
ARAGONITE	—	— —			
PUMPELLYITE	++++	++++++	++++++		
LAWSONITE		++++++	++++++	+ + +	
CLINOZOISITE				+ + +	
ZOISITE				+ + +	++++++++++ ++++++++++
EPIDOTE				+	++++++++++
JADEITE				+ + + + + +	+ + + ++++
OMPHACITE			— —		
Na-AMPHIBOLE	+ +	++	+ + +	+ +++++++	++++++++++
ACTINOLITE	+ +	++++++	++++++	++++++	++++++++++
BARROISITE		++++++	++++++	+ + + + +	+
SPESSARTINE (rim)					+ — — + ++++++
ALMANDINE (rim)				++++++++++	++++++++++
PHENGITE	+-+ + + +	+.+.+.	+ + .+	+. + +	+ + + +.+. +	+. +.+.+ +
PARAGONITE	— — — — —	— — — —	— — —	— — —	+ ++++++
CHLORITE	++++++++++	++++++++++	++++++++++	++++++++++ +	
STILPNOMELANE					· · · ·
ALBITE porphyroblastic	+ + + +++	+ + +	+ ++	.+.+	+. +. +. +	+. +.+ +
K-FELDSPAR		· · · ·	· · · ·	· · · ·	· · · · · ·
QUARTZ	+ + +++	+.+.+.	+ ++	.+-+ +	+.+. + .+.+.	+-+.+ +.+
GRAPHITE	disordered		✕	ordered	
SPHENE	++++++++++	++++++	++++++	++++++	++++++++++++	+++++++++++++
RUTILE				—	+ ++++++++++	+++++++++++

lawsonite-albite-chlorite facies ——— | blueschist facies | eclogite facies

Figure 5. Zonal distribution of minerals in the metamorphic sequence across the full width of the high-pressure belt in Ouégoa district. Zones are defined from isograds in metasedimentary rocks. The distribution of rutile and sphene is taken from Itaya and others (in press).

Brothers 1985) and replacement textures identify secondary mineral associations of distinctly lower pressure character.

In the high-grade schists of the Mn garnet zone (Figure 5), the critical prograde metasedimentary rock paragenesis is quartz - albite - lawsonite - ferroglaucophane - Mn garnet - chlorite - phengite. Within the lawsonite-epidote transition subzone several changes occur: (a) clinozoisite appears and lawsonite gradually disappears; (b) chlorite and Na-amphibole become more magnesian; (c) garnet rims are Fe-enriched towards almandine; and (d) jadeitic pyroxenes (Jd_{84}) are present as relics within albite porphyroblasts near the lawsonite-out isograd. Beyond the lawsonite-epidote transition subzone, and throughout the epidote zone, albite forms large poikiloblasts that appear to be retrograde after Na-pyroxene, the original persistence of which across the epidote zone is indicated by the presence of relict jadeite, enclosed by retrograde albite, in meta-acidite close to the omphacite isograd. The latter occurrence is most important for definition of the progressive PT crystallisation path at the higher grades of metamorphism. At this level in the metamorphic sequence across the omphacite isograd, the three interbedded lithologies with contrasted bulk compositions contain quite different mineralogies. These are listed below; retrograde minerals, identified from replacement textures, are italicised for ready recognition.

Metasedimentary Rocks

(i) High-grade epidote zone assemblage: quartz - paragonite - phengite - chlorite - glaucophane/crossite - clinozoisite - almandine - rutile - sphene plus *albite* (after jadeite).

(ii) Omphacite zone assemblage: quartz - omphacite - almandine - paragonite - phengite - glaucophane/crossite - clinozoisite - rutile - sphene ± barroisite plus *albite - chlorite - epidote - glaucophane.*

The metasedimentary reaction that defines the omphacite isograd is:

$$4\,(Mg, Fe)_5Al_2Si_3O_{10}(OH)_8 + 39\,NaAlSi_2O_6 + 10Ca_2(Al)_3Si_3O_{12}(OH) + 17SiO_2$$
$$\text{chlorite} \quad + \quad \text{jadeite} \quad + \quad \text{clinozoisite} \quad +$$

$$\rightleftharpoons 40\,(Na_{0.5}Ca_{0.5})\,(Mg, Fe)_{0.5}Al_{0.5}\,Si_2O_6 + 19NaAl_2AlSi_3O_{10}(OH)_2 + 2H_2O$$
$$\text{omphacite} \quad + \quad \text{paragonite}$$

The occurrence in metasedimentary rocks of porphyroblastic albite enclosing and in apparent equilibrium with quartz and omphacite grains also suggests the more simple reaction:

$$\text{impure jadeite + quartz} \rightleftharpoons \text{omphacite + albite}$$

Metabasite

(i) High-grade epidote zone assemblage: quartz - omphacite - almandine - phengite - chlorite - glaucophane/crossite - actinolite - clinozoisite - rutile - sphene plus *albite - epidote - chlorite.*

(ii) Omphacite zone assemblage: quartz - omphacite - almandine - phengite - paragonite - barroisite - rutile - sphene - clinozoisite plus *albite - epidote - chlorite - actinolite - glaucophane.*

Meta-acidite

(i) High-grade epidote zone assemblage: quartz - jadeite (Jd_{87-98}) - K feldspar - phengite ± almandine ± ferroglaucophane/crossite ± allanite ± sphene plus *albite-epidote.*

(ii) Omphacite zone assemblage: quartz-omphacite or aegirine-jadeite and/or chloromelanite - orthoclase - phengite ± ferroglaucophane/crossite ± stilpnomelane plus *albite-epidote.*

PROGRADE MINERALOGY

Compositional variation with metamorphic grade has been reported for garnet, pyroxenes, amphiboles, micas, chlorites

(Black 1973a, b, 1974a, 1975), sulphides, and oxides (Itaya and others in press). However, most of the data refers to the lawsonite and epidote zones and, with the exception of pyroxenes, little detail has been given for epidote zone metabasites or for all rock types from the omphacite zone. Compositions of the major minerals, together with a considerable amount of new data, are summarised below.

Pyroxenes

Apart from the minor occurrence of jadeite, common sodic pyroxenes in the paragneisses and metabasites all plot in the omphacite compositional field with their diopside + hedenbergite mole fraction mainly 40-50% in metasediments and mainly 50-60% in basic rocks (Figure 6). The hedenbergite mole fraction is usually less than 15% in both lithologies. Some pyroxenes coexisting with epidote are chloromelanite. Fe-rich pyroxenes, such as aegirine-jadeite and Fe-omphacite, are found in meta-acidite along with jadeite which, when coexisting with quartz, contains up to 98% jadeite mole fraction (Black and others in press).

Garnets

Garnets have pronounced compositional zoning from spessartine-rich cores to pyrope-rich rims, particularly within metasedimentary rocks. Although their composition is affected by mineral assemblage and metamorphic grade, rims of garnets from sedimentary and basic parents show a similar compositional range (Figure 7). MnO contents in most garnet rims in the omphacite zone are less than 2 wt%, thus allowing variation in composition to be represented by almandine, pyrope, and grossular components. The grossular mole fraction ranges from 20-35% and Mg/(Mg+Fe) ratios of the rims are from 4-29%. The 100 Mg/(Mg+Fe) ratios of garnet rims show systematic variations within omphacite zone metasediments and represent sliding equilibria. One arbitrarily selected value (15%) has been mapped as an isograd (Figure 3) and it closely parallels sliding equilibria isograds for other minerals.

Amphiboles

In the epidote zone, basic rocks contain glaucophane and actinolite, whereas the associated metasediments carry only glaucophane, but in the omphacite zone glaucophane and barroisite occur in both lithologies (Figure 5). Sodic amphiboles show solid solution toward calcic amphiboles, although there is a compositional gap between these two varieties (Black 1973b). Continuous solid solution exists from actinolite to barroisite in these New Caledonia assemblages. Two amphibole analyses plot close to the winchite endmember (Figure 8) with unusually high K_2O contents (0.63 and 1.04 wt%) compared with other omphacite zone amphiboles $(K_2O \approx 0.3$ wt%). Crossite is confined to epidote-bearing, Fe_2O_3-rich assemblages.

Yokoyama and Others

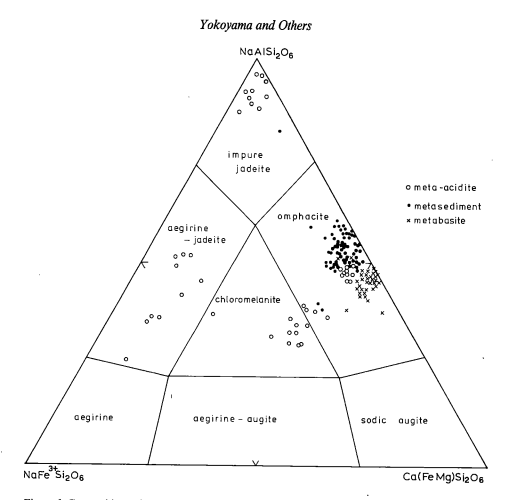

Figure 6. Compositions of metamorphic pyroxenes in the epidote and omphacite zones, Pam-Ouégoa-Pouébo region. End-member compositions were calculated according to Carpenter (1979).

Epidote Group

Three members of the epidote group occur in these regional eclogites: zoisite, clinozoisite, and epidote. Clinozoisite is the most common epidote-group mineral in metabasites and metasediments. Zoisite is present in about 15% of all samples from Pam Peninsula and usually coexists with clinozoisite. Prograde epidote occurs mostly along the eastern and western coast of Pam Peninsula, but in the central range it is a retrogressive phase with chlorite and albite as a replacement of glaucophane and paragonite. The compositional variation of epidote-group minerals is shown in Figure 9.

Micas

Paragonite and phengite coexist in most paragneisses of the epidote zone and in both ortho- and paragneisses of the omphacite zone, but the Na-mica is chemically variable on a microscale because it is frequently altered to albite + chlorite. All phengites have a high celadonite content (35–40 mole %). Paragonite contains essentially negligible amounts of CaO, MgO, and FeO; where coexisting with phengite in the omphacite zone, K_2O values vary from 0.6 to 1.3 wt%. The Na_2O content of phengite

coexisting with paragonite is constant within each sample and it increases with metamorphic grade: <0.4 wt% in the lawsonite zone, 0.4–0.6 wt% in the lawsonite-epidote transitional subzone and the epidote zone, and 0.6–1.2 wt% in the omphacite zone. Two levels of Na_2O in phengite, 0.6 wt% and a value in the middle of the omphacite zone (0.8 wt%), have been plotted as isograds in Figure 3 and are parallel to other mineral isograds including the sliding equilibrium composition for $Mg/(Mg+Fe)$ in garnet.

Feldspar

In all zones, and as a prograde or retrograde phase, albite has compositions in the range $An_{0-1.5}$. In the epidote and omphacite zones, K-feldspar has not been observed in the metasediments, but it is common in the meta-acidites and has been seen surrounding paragonite in one metabasite.

RETROGRADE MINERALOGY

The isograds or isogradic surfaces for omphacite and other prograde phases are essentially parallel, thus indicating a regular progression in metamorphic grade throughout the whole region.

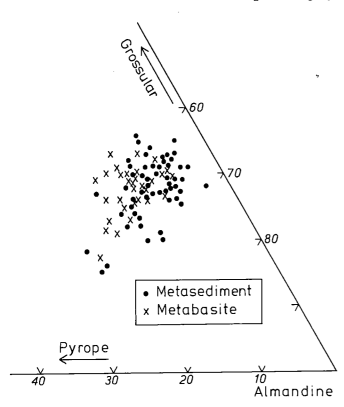

Figure 7. Compositions of garnet rims in the epidote and omphacite zones, Pam-Ouégoa-Pouébo region: MnO <2wt% in all rims.

However, there are notable differences in the nature and degree of retrogressive metamorphism between the Pam Peninsula and Pouébo areas, which are separated by a major NW–SE fault zone containing many serpentinite bodies and tectonic blocks (Figure 2). Gneisses in the northern fault block at Pam Peninsula show some retrogressive alteration, but prograde minerals are well preserved. In the Pouébo area, the prograde assemblages have been extensively altered by retrogression and in many localities only relics remain of the earlier eclogite mineralogy. As a result, within the 600 samples examined from the omphacite zone, detailed mineral and phase equilibrium analyses are numerically biased towards Pam Peninsula assemblages. The most widespread retrogressive changes were replacement of rutile by sphene, jadeite and omphacite by albite, and the breakdown of glaucophane + paragonite to give albite and chlorite.

Pam Peninsula

Retrogressive recrystallisation of omphacite to albite occurred mainly where the pyroxene was jadeite-rich (Jd_{50-60}) and associated with quartz-bearing assemblages. Consequently, albite replacement of omphacite is most prevalent in metasediments and is rare in the quartz-poor or quartz-free basic rocks containing pyroxenes that are less jadeitic ($<Jd_{50}$). Omphacite pseudomorphs consisting of albite + glaucophane are present in some

metasediments. Breakdown of the assemblage glaucophane + paragonite to albite + chlorite is common in epidote- and clinozoisite-bearing rocks. In contrast, zoisite assemblages rarely contain retrogressive albite and decomposition textures, and prograde minerals such as omphacite, glaucophane and paragonite are abundant. Prograde barroisite rims glaucophane in clinozoisite-bearing metasediments and in zoisite-bearing metabasites on the high-temperature side of the omphacite zone (Na_2O in phengite >0.8 wt%). The reverse relationship, with retrograde glaucophane (crossite) replacing barroisite, appears only in metabasites containing epidote-group minerals that are strongly zoned from Al-rich cores to Fe-rich rims; this suggests that later crystallisation of sodic amphibole was the result of a late increase in fO_2.

Pouébo Area

The eclogitic metasediments are intensively altered to lower grade assemblages. Omphacite and garnet are replaced mainly by albite and epidote-chlorite respectively, so that the pyroxenes usually are present in small relic grains within feldspar porphyroblasts. Glaucophane and paragonite are enclosed in albite, and glaucophane is extensively chloritised so the amphibole-mica pair is rare. The occurrence of unaltered glaucophane is restricted to paragonite-free rocks in the epidote and omphacite zones. The metasediments are dominated by retrogressive phases and the most frequent assemblage is quartz - albite - phengite - chlorite - epidote group-relict garnet, plus the two amphiboles that have remained stable under retrograde conditions: barroisite in the omphacite zone, and glaucophane in the epidote zone.

ECLOGITE MINERAL ASSEMBLAGES

The essential characteristic of prograde mineral assemblages is the association omphacite - almandine - paragonite - zoisite - barroisite - rutile in ortho- and paragneisses. The typical retrograde assemblages contain albite as porphyroblasts and parageneses such as glaucophane - albite - chlorite and albite - epidote group-glaucophane. Using mineral compositions that approximate those of natural minerals in the Pam-Ouégoa-Pouébo area, prograde assemblages for omphacite zone ortho- and paragneisses are shown in Figure 10 and fields for retrogressive mineral assemblages in Figure 11. In the Pam area, however, few retrogressive reactions have gone to completion so that the prograde and retrograde phases are seen in frozen reaction relationships that could be represented by combining Figures 10 and 11.

Assemblage Stability Fields

The stability relations of the various mineral assemblages can be discussed and illustrated by means of Schreinemakers' analysis (Figure 12). The characteristic minerals in metasedimentary and metabasic parageneses across the omphacite zone are omphacite, garnet, barroisite, glaucophane, paragonite, phengite,

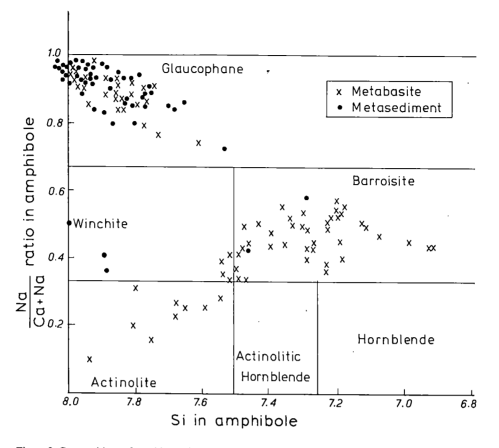

Figure 8. Compositions of amphiboles in the omphacite zone of the Pam-Pouébo region. Compositional boundaries according to the nomenclature of Leake (1978) assuming no A-site occupancy.

epidote-group minerals, chlorite, rutile, sphene, albite, and quartz. These minerals can be described compositionally by 11 components: SiO_2, Al_2O_3, Fe_2O_3, FeO, MgO, CaO, Na_2O, K_2O, TiO_2, MnO, and H_2O.

In order to simplify the system for chemographic analysis, the following assumptions have been made:

1. K_2O and TiO_2 are each present in only one phase (phengite, and rutile or sphene) and can be classed as accessory.

2. SiO_2 is present as free quartz and is thus an excess component.

3. MnO is a minor constituent in all ferromagnesian phases

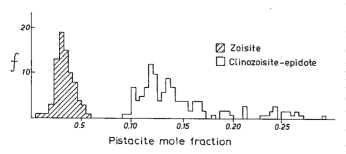

Figure 9. Frequency distribution of the pistacite mole fraction in epidote-group minerals in the epidote and omphacite zones, Pam-Ouégoa-Pouébo region.

except garnet cores, which are not in equilibrium with other minerals; it is therefore regarded as interchangeable with FeO.

4. Fe_2O_3 is regarded as exchangeable with octahedral Al_2O_3 in amphiboles, chlorites, pyroxenes, and epidote, provided there is no change in oxidation state of the system.

5. Oxygen isotope data (Black 1974b) indicate that the minerals have exchanged with an oxygen-bearing fluid phase. In addition, hydrous fluid inclusions are numerous in quartz-rutile veins in the omphacite zone and there is an abundance of hydrous prograde and retrograde phases in both ortho- and paragneisses. Accordingly, it is assumed that H_2O was present in the system as an excess phase.

6. Mineral composition data suggest that FeO and MgO are not interchangeable. Although the Fe/(Mg+Fe) ratio of garnet changes with metamorphic grade, garnet is always an FeO-rich phase, although other mafic minerals such as omphacite, glaucophane, and barroisite have lower Fe/(Mg+Fe) ratios and thus consistently project to the MgO side in any graphical representation. Therefore garnet is regarded as an FeO-rich phase and all other mafic minerals as MgO-rich members, following Holland (1979a) and Laird and Albee (1981).

If the above assumptions and approximations are accepted, then the natural mineral assemblages can be reduced to, and described by, a four-component system (Mg, Fe^{2+})O, (Al,

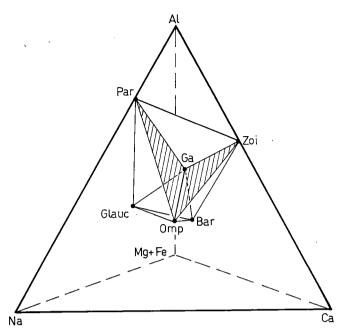

Figure 10. Prograde mineral associations as defined within the tetrahedron Al - Na - Ca - (Mg + Fe). Mineral compositions used approximate to those of natural minerals in Pam-Ouégoa-Pouébo region.

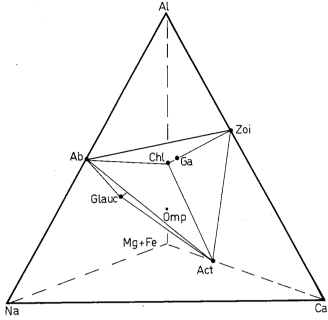

Figure 11. Retrograde mineral associations as defined within the tetrahedron Al - Na - Ca - (Mg + Fe). Mineral compositions used approximate to those of natural minerals in Pam-Ouégoa-Pouébo region; prograde omphacite and garnet are shown for reference to Figure 10.

$Fe^{3+})_2O_3$, CaO, and Na_2O as shown in Figures 10 and 11. Most of the entropy and volume data for the minerals are from Helgeson and others (1978). The entropy and volume data for glaucophane and barroisite are calculated by structural algorism and equations (57) and (62) of Helgeson and others (1978). Although the mineral compositions are idealised, compositional variations in the natural phases do not affect the topology of the system or the discussion.

In a Schreinemakers' analysis of assemblages in high-grade schists of the Ouégoa district, Black (1977) did not distinguish between the prograde and retrograde mineral phases generated by two distinct metamorphic episodes. Schreinemakers' bundles, projected on the kyanite-diopside-albite plane, are shown in Figure 12, which contains only representative stable (critical) assemblages from the PT field enclosed by reaction lines; H_2O quartz, and epidote-group minerals are excess phases and can be regarded as present in all assemblages. Slopes of reactions were calculated using thermodynamic data (see Appendix) for phases at 500°C and 10 kb. The effect of substituting Fe^{3+} for Al on the position of the reaction lines (Figure 12) has been estimated using partition coefficients from natural minerals analysed by wet-chemical methods (Yokoyama and Black, in preparation). Introduction of Fe_2O_3 moved the omphacite-glaucophane-paragonite field towards the high-pressure side, and the barroisite-omphacite-paragonite field towards the high-temperature side.

A critical assemblage in ortho- and paragneisses is omphacite - glaucophane - paragonite and in basic rocks omphacite - barroisite - paragonite. Neither of these assemblages is associated with those textural relations which suggest retrogressive re-

equilibration so that their stability fields can be safely regarded as part of the prograde sequence. Since both assemblages are often observed with zoisite, the barroisite association is considered to have equilibrated at higher temperatures than the association carrying glaucophane (Figure 12). This conclusion is supported by the occurrence of omphacite - barroisite - paragonite only along the high-temperature side of the omphacite zone ($\geqslant 0.8$ wt%, Na_2O in phengite).

In the Pam area, the omphacite - paragonite - glaucophane assemblage is often found interstratified in the same outcrop with albite - paragonite - glaucophane paragneiss, the omphacite - paragonite - glaucophane association containing zoisite in Fe_2O_3-poor rocks and the albite - paragonite - glaucophane assemblage including epidote in Fe_2O_3-rich rocks. Hence, it is reasonable to assume that stability of the assemblages is related to Fe_2O_3 content rather than to pressure differences. The omphacite - paragonite - glaucophane and albite - paragonite - glaucophane assemblages can "coexist" in the low-pressure side (Figure 12, shaded area) of the omphacite - paragonite - glaucophane field. The absence of retrograde albite in the Fe_2O_3-poor omphacite - paragonite - glaucophane - zoisite assemblages suggests that, during the retrogressive stage, the pyroxene in the Fe_2O_3-rich (epidote-bearing) omphacite - paragonite - glaucophane associations became unstable and recrystallised almost completely to albite. This is shown in Figure 12 by the shift in Fe_2O_3 that enlarged the stability field for albite-paragonite-glaucophane.

In the omphacite zone metasediments of the Pouébo area, prograde pyroxene-bearing assemblages have been replaced ex-

Yokoyama and Others

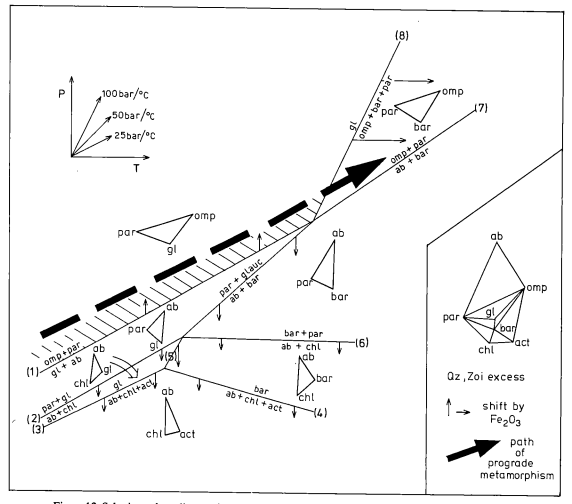

Figure 12. Schreinemakers diagram for critical assemblages in the hydrous Ca - Al - Na - Mg compositional field with excess quartz-zoisite-H_2O and accessory almandine. Projection on kyanite-diopside-albite plane. Thermodynamic data for slopes of the numbered reactions are given in Table 1.

tensively by retrograde minerals, including albite. Along the high-temperature side of the omphacite zone, barroisite is present as a stable prograde amphibole that has persisted in association with retrogressive albite + chlorite; on the lower temperature side of the zone, glaucophane similarly coexists with albite + chlorite. These two assemblages are located on the low-pressure side of the prograde fields for omphacite - paragonite - barroisite and omphacite - paragonite - glaucophane in Figure 12. Thus, the distribution of these assemblages within Figure 12 and across the metamorphic field suggests that retrogressive recrystallization in the Pouébo area was largely a function of depressurisation at essentially constant temperature, as indicated by curve B in Figure 13.

THE ROLE OF FLUIDS

Anhydrous conditions for crystallisation of eclogites have been accepted from experimental results (Yoder and Tilley 1962; Green and Ringwood 1967; Bryhni and others 1970; Lambert

and Wyllie 1972) and from bulk-composition equivalence with hydrous amphibole-rich rocks that are also of deep-seated origin (e.g. Ernst and Dal Piaz 1978; Maresch and Abraham 1981). Additional support for a water-free paragenesis has come from other studies (e.g. Morgan 1970; Feininger 1980). Nevertheless, petrographic descriptions (e.g. Coleman and others 1965; Reinsch 1977, 1979) show that eclogite and hydrous blueschist assemblages are associated on the scale of a thin-section or hand-specimen as well as in regional metamorphic belts (Velde and Kienast 1973; Okrusch and others 1978). "Hydrous eclogite facies" parageneses are very common in the Selje district of Norway (Smith 1971, 1976, 1980; Lappin and Smith 1978, 1981). From occurrences in veins and vugs in metamorphosed basic and sedimentary rocks, Essene and Fyfe (1967) believed that omphacite crystallised in the presence of a fluid phase; this deduction can be extended to eclogites in which garnet behaves as an iron-rich phase and the stability of the assemblage is tied to that of the pyroxene (Holland 1979a; Laird and Albee 1981, this study). Holland (1979b) calculated and measured a fluid phase in equili-

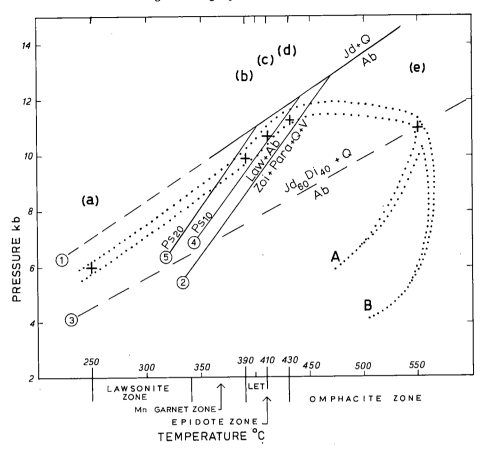

Figure 13. Inferred P-T metamorphic path (stippled) from lawsonite zone to omphacite zone in the Ouégoa-Pam-Pouébo region. Crosses identify reference points (a)-(e) on the P-T path as discussed in the text, and retrograde stages are shown separately for the Pam area (A) and the Pouébo area (B). Curve 1 from Holland (1980); curve 2 from Holland (1979a): for curves 3-5 see Appendix. LET = lawsonite-epidote transition subzone.

brium with eclogite where the activity of H_2O was high and near to unity (0.98). Lappin and Smith (1981) deduced very low X_{CO_2} and hence possibly very high X_{H_2O} in the Norwegian carbonate eclogites. From phase relations and thermodynamics, Brown and Bradshaw (1979) and Laird and Albee (1981) concluded that some eclogites are products of a metamorphic environment containing water-rich fluids under high pressure.

In the New Caledonian gneisses, crystallisation of omphacite-garnet was prevalent in stable equilibrium with hydrous minerals and hydrous fluid, in both metabasites and metasediments. In meta-acidites the assemblage jadeite-almandine-quartz is stable. From reaction 1 in Table 1 it is noteworthy that prograde omphacite-paragonite assemblages are more hydrous than retrograde albite - epidote group - glaucophane, suggesting that with increased fluid pressures omphacite - paragonite ± garnet may be the more stable association. A main conclusion from the present study of eclogites regionally interstratified with eclogitic paragneisses is that eclogites can crystallise and are stable in hydrous conditions, and that eclogite mineralogy is controlled by bulk chemical composition and the prevailing combination of pressure and temperature variables.

TEMPERATURE AND PRESSURE

Temperatures of 410°C for the almandine (lawsonite-out) isograd, 430°C for the omphacite isograd, and 550°C for the deepest part of the omphacite zone (Figure 13) are indicated by oxygen isotope data for coexisting quartz and phengite (Black 1974b). Garnet-omphacite geothermometry for the deepest part of the omphacite zone also indicates temperatures in the 500-600°C range (Paterson 1980; Ghent and others in preparation). The quartz and phengite are commonly present in both prograde and retrograde assemblages; thus it is uncertain which stage of recrystallisation they represent, but the temperatures can safely be regarded as minimal for the respective prograde zones. Identification of the flat-lying isogradic surfaces along the mountain range allows a reasonable estimate of metamorphic thickness within the high-grade gneisses. As measured west of Pouébo and in Pam Peninsula (Figures 2 and 3) the epidote zone has a thickness between 0.3 and 0.5 km. Consequently, a temperature gradient of at least 50-60°C/km is apparent from the oxygen isotope data. The presence of such a steep gradient in the high-grade zones is a significant petrogenetic factor. In view of the

Yokoyama and Others

TABLE 1. THERMODYNAMIC DATA FOR BALANCED REACTIONS 1-9 GIVEN IN FIGURE 12

500°C 10 kb	V cm³ Mol⁻¹	S cal Mol⁻¹(°K)⁻¹	1	2	3	4	5	6	7	8	9
Omphacite $NaCaAlMgSi_4O_{12}$	126.5	158.24	12						3	-7.5	
Albite (High) $NaAlSi_3O_8$	100.43	110.51	-13	-13	-50	-50	2	-27	-4.5		2.5
Zoisite $Ca_2Al_3Si_3O_{12}(OH)$	135.90	170.85	-6		6	-18	-6	-10	-1	7	-1
Quartz (α) SiO_2	22.688	23.17	14	-1	7	-21	-1	-14	4	-5.5	-1
Glaucophane $Na_2Mg_3Al_2Si_8O_{22}(OH)_2$	262.3	313.72	-4	5	25		-7			9	-2
Barroisite $NaCaMg_3Al_3Si_7O_{22}(OH)_2$	161.4	319.54				50	12	20	-1	-6.5	2
Chlorite (14A) $Mg_5Al_2Si_3O_{10}(OH)_8$	207.11	271.92		-3	-9	-23	-3	-12			
Paragonite $NaAl_3Si_3O_{10}(OH)_2$	132.53	158.22	9	3				7	2.5	-4	-0.5
Tremolite $Ca_2Mg_5Si_8O_{22}(OH)_2$	272.92	316.31			-6	-7					
H_2O*	17.86	30.10	-2	4	14	58	10	26	-1	-2	1
slope bar/°C			30.5	27.7	23.6	-13.6	77.8	-0.2	35.0	102.4	44.4

*Data from Delany and Helgeson (1978).
Data for omphacite, glaucophane, and barroisite calculated according to description in text.
All other data from Helgeson et al. (1978).

short time span for the metamorphic event as recorded by geological data (Late Eocene) and K-Ar dates (39–21 m.y.) and the low potential of the rock mass for heat conduction, the gradient was probably a feature inherited from a heatflow profile that was already established within the sedimentary pile during the prograde metamorphic phase.

From field mapping, assemblage sequence and the oxygen isotope data of Black (1974b), previous estimates of temperature gradients in the lawsonite zone by Brothers (1974) and Diessel and others (1978) had indicated T values between 7 and 12°C/km along a recurved P-T path that steepened rapidly to 30°C/km into the epidote zone. The accelerated increment in temperature relative to depth, now recognized within the epidote zone, is part of the total P-T gradient from the lawsonite zone to the omphacite zone, as shown in Figure 13. Within this diagram, important constraints on P-T relationships are imposed by the experimental or computed curves, relative to which the reference points (a)–(e) determine the shape of a P-T path containing estimated P values subject to errors of about 1 kb.

a. Lawsonite isograd: with aragonite present (Brothers 1970, 1974) and oxygen isotope T = 250°C (Black 1974b), P≥6 kb from the aragonite-calcite data of Johannes and Puhan (1971), but jadeite is absent.

b. Passage from the Mn garnet zone to the lawsonite-epidote transition (LET) subzone: first appearance of clinozoisite containing 0.10-0.19 mole fractions of pistacite (Figure 9) as defined by reaction lines 4 and 5 (see Appendix); with no jadeite present and oxygen isotope T = 390°C, P ≥ 9.5 kb.

c. Passage from the lawsonite-epidote transition subzone into the epidote zone, across a lawsonite-out and almandine-in isograd (Figure 5): with relict jadeite (Jd₈₄) in metasediment blueschist on both sides of the isograd that has oxygen isotope T = 410°C, P ≥ 10.5 kb.

d. Transition from the epidote zone to the omphacite zone: on the high-grade side of the epidote zone and near the omphacite isograd, meta-acidite contains jadeite (Jd₈₇₋₉₈) and blueschist metasediment has oxygen isotope T = 430°C, thus indicating P ≥ 11 kb.

e. Omphacite zone: from among 600 metasediment samples distributed throughout the zone, the jadeite component in omphacite coexisting with quartz + albite lies always in the range Jd₅₀₋₆₀ and is therefore regarded as the prograde composition for the pyroxene in equilibrium with a slightly calcic parent sediment; oxygen isotope T = 550°C in the deepest part of the zone.

The P-T path from (d) to (e) is thus confined between curves 1 and 3, and (as drawn) it must include a temperature-prograde phase of climax crystallisation that prevailed during a decline in pressure in the order of 2 kb. Meta-acidites in the omphacite zone also contain assemblages indicating depressurisation where jadeite + quartz has been replaced by omphacite + albite (Black and others in press). Below the stability field of metasediment omphacite with composition Jd_{50}, depressurisation of the metamorphic system was accompanied by decrease in temperature, accelerating retrograde re-equilibrium marked by the appearance of albite (replacing omphacite) and other retrogressive phases. Crystallisation during pressure decline but without significant loss of temperature is the "pressure retrogression" of Brothers and Yokoyama (1982) or "decompression recrystallisation" of Ernst and Dal Piaz (1978) and it is a feature of many high-pressure metamorphic belts (e.g. Miller 1977; Banno and others 1978; Lappin and Smith 1978; Reinsch 1979; Maresch and Abraham 1981).

As pointed out by Brothers (1974) and Diessel and others (1978), the absolute values and increments of pressure across the New Caledonian metamorphic belt cannot be explained as due to deep burial, since well-documented geological evidence for the area does not support such a concept. From present-day exposures that lie structurally above the lawsonite isograd, a tectonic-stratigraphic thickness on the order of 12 km (≈ 3.5 kb) can be shown for the apparent overburden on the lawsonite zone (Briggs and others 1978). However, the association lawsonite-aragonite near the lawsonite isograd, with T = 250°C, would require experimental data pressures on the order of 6 kb (≈ 21 km of overburden) that cannot be proved from the regional geology. Similarly, pressure increase to at least 10.5 kb (≈ 37 km by burial) at the lawsonite-out and almandine isograd with relict Jd_{84} (T = 410°C) is required by the experimental curves of Figure 13 in order to follow the demonstrated continuous progression to assemblages with Jd_{87-98} + quartz in the high-grade part of the epidote zone near the omphacite isograd (T = 430°C); according to the diagram, a pressure increment of about 0.5 kb ($\approx$ 1.8 km) is required for transition across the epidote zone, which is known to be no more than 0.5 km in thickness.

From the evidence of field data, mineral associations, isogradic surfaces, and temperature gradient it appears that the metamorphic environment did not necessarily evolve within a pile of sediments and volcanics that had been buried to extreme depths in the crust. More likely, the high-pressure system was developed at a shallow level in a continental margin setting where oceanward-directed low-angle thrust tectonics had transmitted a large quantum of tangential compression across the main thrust interface (the regional lawsonite isogradic surface) to enhance deformation and subsequent recrystallisation within the immobile substrate; this protolith contained an unusually high heatflow, which during metamorphism was cooled upwards to low thermal gradients in the lawsonite-albite-chlorite and blueschist zones immediately beneath the cold capping of the over-riding continental slab.

The eclogites in the deepest part of the regional high-pressure sequence entered a final phase of depressurisation and cooling controlled partly by blockfaulting movements that eventually terminated the metamorphic event. The blocks were differentially elevated and in the high-standing Pam Peninsula (Figure 2) the flat-lying omphacite isogradic surface is at an altitude at least 500 m above its counterpart in the Pouébo area. Thus, with the cessation of thrust compression plus accelerated uplift and erosion, the Pam eclogites were subjected to rapid decompression and heat loss occupying a short time-span (curve A, Figure 13). As a result, the period available for retrograde recrystallisation was brief, as shown by the general preservation of eclogite assemblages and the limited development of retrogressive phases. In the relatively less-elevated Pouébo omphacite zone, which still carries the full sequence of the epidote zone surmounted by an outlier of the lawsonite zone, a slower cooling path during decompression (curve B, Figure 13) allowed extensive retrogressive recrystallisation within the prograde assemblage. From microstructural evidence, Bell and Brothers (1985) have identified a deformation-crystallisation sequence indicating that retrogressive recrystallisation commenced along the P-retrograde T-prograde section of the metamorphic path, between (d) and (e) in Figure 13.

ACKNOWLEDGMENTS

This work is part of a regional project being carried out in collaboration with the Bureau des Recherches Géologiques et Minières in New Caledonia, and grateful acknowledgment is made of the support and hospitality received especially from MM Chiron, Gerard, Paris, Eberlé, and Maurizot. Fieldwork in New Caledonia, equipment and a Postdoctoral Fellowship (KY) were funded by the Research Committee of Auckland University. Reviews of the manuscript by E. H. Brown, R. G. Coleman, E. D. Ghent, and D. C. Smith were valuable for formulation of the final version.

APPENDIX

The equilibrium line (3) in Figure 13 for Na-pyroxene (Jd_{60}) + quartz + albite has been calculated (following Essene and Fyfe 1967; Wood and others 1980; Turner 1981) from the Jd_{100}-quartz-albite curve by the equation:

$$(P_{Jd_{100}}) - P_{(Na-px)} = \frac{RT}{\Delta V} \ln a_{jd}^{px}$$

where ΔV is the volume change in the Jd_{100} + quartz = albite reaction.

There are practical difficulties in estimating a a_{jd}^{px} for pyroxenes in the jadeite-diopside system because two different cations must be mixed in each of two different sites; however, in the jadeite-acmite system two cations are mixed in one site, so that simple ideal mixing is assumed and a $a_{jd}^{px} = X a_{jd}^{px}$. An acidic rock containing aegirine-jadeite (Jd_{30-40}) outcrops close to the omphacite isograd and 50 m above a metasediment with omphacite ($Jd_{60}Di_{40}$) + quartz + albite and therefore reaction line (3) has been calculated using ($Jd_{36}Ac_{64}$) + qz + ab. Reaction lines (4) and (5) in Figure 13 have been constructed by the same method as reaction line (3). The activity of $Ca_2Al_3Si_3O_{12}(OH)$ is calculated as $(X_{Al}^{M1}) \times (X_{Al}^{M2}) \times (X_{Al}^{M3})$; intracrystalline exchanges of Al and Fe among

M1, M2, M3 sites of the epidote group are those of Helgeson and others (1978).

REFERENCES CITED

Arnould, A., 1958, Etude géologique de la partie nord-est de la Nouvelle-Calédonie [Thesis]: Paris, University of Paris, 436 p.

Banno, S., 1970, Classification of eclogites in terms of physical conditions of their origin: Physics of the Earth and Planetary Interiors, v. 3, p. 405–421.

Banno, S., Higashino, T., Otsuki, M., Itaya, T., and Nakajima, T., 1978, Thermal structure of the Sanbagawa metamorphic belt in central Shikoku: Journal of Physics of the Earth, v. 26 supplement, p. 345–356.

Bell, T. H., and Brothers, R. N., 1985, Development of P-T prograde and P-retrograde, T-prograde isogradic surfaces during blueschist to eclogite regional deformation metamorphism in New Caledonia as indicated by progressively developed porphyroblast microstructures: Journal of Metamorphic Geology, v. 3, p. 59–78.

Black, P. M., 1973a, Mineralogy of New Caledonia metamorphic rocks. I. Garnet from the Ouégoa district: Contributions to Mineralogy and Petrology, v. 38, p. 221–235.

——1973b, Mineralogy of New Caledonia metamorphic rocks. II. Amphiboles from the Ouégoa district: Contributions to Mineralogy and Petrology, v. 39, p. 55–64.

——1974a, Mineralogy of New Caledonia metamorphic rocks. III. Pyroxenes, and major element partitioning between pyroxenes, amphiboles and garnets from the Ouégoa district: Contributions to Mineralogy and Petrology, v. 45, p. 281–288.

——1974b, Oxygen isotope study of metamorphic rocks from the Ouégoa district, New Caledonia: Contributions to Mineralogy and Petrology, v. 47, p. 197–206.

——1975, Mineralogy of New Caledonia metamorphic rocks. IV. Sheet silicates from the Ouégoa district: Contributions to Mineralogy and Petrology, v. 49, p. 269–284.

——1977, Regional high-pressure metamorphism in New Caledonia: phase equilibria in the Ouégoa district: Tectonophysics, v. 43, p. 89–107.

Black, P. M., and Brothers, R. N., 1977, Blueschist ophiolites in the mélange zone, northern New Caledonia: Contributions to Mineralogy and Petrology, v. 65, p. 69–78.

Black, P. M., Brothers, R. N., and Yokoyama, K., in press, Mineral parageneses in eclogite-facies meta-acidites in northern New Caledonia, *in* Smith, D. C., ed., Developments in petrology: eclogites and eclogite-facies rocks: New York, Elsevier.

Blake, M. C., Brothers, R. N., and Lanphere, M. A., 1977, Radiometric ages of blueschists in New Caledonia, *in* International Symposium on Geodynamics in South-west Pacific, Nouméa, August-September 1976: Paris, Editions Technip, p. 279–281.

Brière, Y., 1919, Les éclogites françaises—Leur composition minéralogique et chimique; leur origine: Bulletin de la Société Française de Minéralogie, t. 41, p. 72–222.

Briggs, R. M., Kobe, H. W., and Black, P. M., 1977, High-pressure metamorphism of stratiform sulphide deposits from the Diahot region, New Caledonia: Mineralium Deposita, v. 12, p. 263–279.

Briggs, R. M., Lillie, A. R., and Brothers, R. N., 1978, Structure and high-pressure metamorphism in the Diahot region, northern New Caledonia: Bulletin du Bureau de Recherches Géologiques et Minières, Sect. 4, p.171–189.

Brothers, R. N., 1970, Lawsonite-albite schists from northernmost New Caledonia: Contributions to Mineralogy and Petrology, v. 25, p. 185–202.

——1974, High-pressure schists in northern New Caledonia: Contributions to Mineralogy and Petrology, v. 46, p. 109–127.

Brothers, R. N., and Blake, M. C., 1973, Tertiary plate tectonics and high-pressure metamorphism in New Caledonia: Tectonophysics, v. 17, p. 337–358.

Brothers, R. N., and Yokoyama, K., 1982, Comparison of the high-pressure schist belts of New Caledonia and Sanbagawa, Japan: Contributions to Mineral-

ogy and Petrology, v. 79, p. 219–229.

Brown, E. H., and Bradshaw, J. Y., 1979, Phase relations of pyroxene and amphibole in greenstone, blueschist and eclogite of the Franciscan complex, California: Contributions to Mineralogy and Petrology, v. 71, p. 67–83.

Bryhni, I., Green, D. H., Heier, K. S., and Fyfe, W. S., 1970, On the occurrence of eclogite in western Norway: Contributions to Mineralogy and Petrology, v. 26, p. 12–29.

Carpenter, M. A., 1979, Omphacite from Greece, Turkey and Guatemala: compositional limits of cation ordering: American Mineralogist, v. 64, p. 102–108.

Coleman, R. G., 1967, Glaucophane schists from California and New Caledonia: Tectonophysics, v. 4, p. 479–498.

Coleman, R. G., Lee, D. E., Beatty, L. B., and Brannock, W. W., 1965, Eclogites and eclogites: their differences and similarities: Geological Society of America Bulletin, v. 76, p. 483–508.

Delany, J. M., and Helgeson, H. C., 1978, Calculation of the thermodynamic consequences of dehydration in subducting oceanic crust to 100 kb and 800°C: American Journal of Science, v. 278, p. 638–686.

Diessel, C.F.K., Brothers, R. N., and Black, P. M., 1978, Coalification and graphitization in high-pressure schists in New Caledonia: Contributions to Mineralogy and Petrology, v. 68, p. 63–78.

Dobretsov, N. L., Sobolev, V. S., Sobolev, N. V., and Khlestov, V. V., 1975, The facies of regional metamorphism at high pressures: Australian National University Department of Geology Publication, no. 266, 363 p.

Ernst, W. G., and Dal Piaz, G. V., 1978, Mineral parageneses of eclogitic and related mafic schists of the Piemonte ophiolite nappe, Breuil-St. Jacques area, Italian Western Alps: American Mineralogist, v. 63, p. 621–640.

Espirat, J. J., and Millon, R., 1965, Carte géologique à l'échelle du 1/50000, Pam-Ouégoa, avec notice explicative: Bureau de Recherches Géologiques et Minières, Paris.

——1967, Carte géologique à l'échelle du 1/50000, Paagoumêne, avec notice explicative: Bureau de Recherches Géologiques et Minières, Paris.

Essene, E. J., and Fyfe, W. S., 1967, Omphacite in California metamorphic rocks: Contributions to Mineralogy and Petrology, v. 15, p. 1–23.

Feininger, T., 1980, Eclogite and related high-pressure regional metamorphic rocks from the Andes of Ecuador: Journal of Petrology, v. 21, p. 107–140.

Green, D. H., and Ringwood, A. E., 1967, An experimental investigation of the gabbro to eclogite transformation and its petrological implications: Geochimica et Cosmochimica Acta, v. 31, p. 767–833.

Guillon, J. H., 1975, Les massifs péridotitiques de Nouvelle-Calédonie. Type d'appareil ultrabasique stratiforme de chaîne récente: Mémoire de l'Office de la Recherche Scientifique et Technique Outre-mer, no. 76, p. 1–120.

Guillon, J. H., and Routhier, P., 1971, Les stades d'évolution et de mise en place des massifs ultramafiques de Nouvelle-Calédonie: Bulletin du Bureau de Recherches Géologiques et Minières, Section IV, no. 2, p. 5–38.

Helgeson, H. C., Delany, J. M., Nesbitt, H. W., and Bird, D. K., 1978, Summary and critique of the thermodynamic properties of rock-forming minerals: American Journal of Science, v. 278, p. 1–229.

Holland, T.J.B., 1979a, Experimental determination of the reaction paragonite = jadeite + kyanite + H_2O, and internally consistent thermodynamic data for part of the system $Na_2O-Al_2O_3-SiO_2$ -H_2O, with applications to eclogites and blueschists: Contributions to Mineralogy and Petrology, v. 68, p. 293–301.

——1979b, High water activities in the generation of high pressure kyanite eclogites of the Tauern Window, Austria: Journal of Geology, v. 87, p. 1–27.

——1980, The reaction albite = jadeite + quartz determined experimentally in the range 600–1200°C: American Mineralogist, v. 65, p. 129–134.

Itaya, T., Brothers, R. N., and Black, P. M., in press, Sulfides, oxides and sphene in high-pressure schists from New Caledonia: Contributions to Mineralogy and Petrology.

Johannes, W., and Puhan, D., 1971, The calcite-aragonite transition , reinvestigated: Contributions to Mineralogy and Petrology, v. 31, p. 28–38.

Lacroix, A., 1941, Les glaucophanites de la Nouvelle-Calédonie et les roches qui les accompagnent, leur composition et leur genèse: Mémoires de l'Académie des Sciences de l'Institut de France, t. 65, p. 1–103.

Laird, J., and Albee, A. L., 1981, High-pressure metamorphism in mafic schist from northern Vermont: American Journal of Science, v. 281, p. 97–126.

Lambert, I. B., and Wyllie, P. J., 1972, Melting of gabbro (quartz eclogite) with excess water to 35 kbars, with geological applications: Journal of Geology, v. 80, p. 693–708.

Lappin, M. A., and Smith, D. C., 1978, Mantle-equilibrated orthopyroxene eclogite pods from the Basal Gneisses in the Selje District, western Norway: Journal of Petrology, v. 19, p. 530–584.

——1981, Carbonate, silicate and fluid relationships in some Norwegian eclogites: Transactions of the Royal Society of Edinburgh: Earth Sciences, v. 72, p. 171–193.

Leake, B. E., 1978, Nomenclature of amphiboles: American Mineralogist, v. 63, p. 1023–1052.

Maresch, W. V., and Abraham, K., 1981, Petrography, mineralogy, and metamorphic evolution of an eclogite from the island of Margarita, Venezuela: Journal of Petrology, v. 22, p. 338–362.

Miller, C., 1977, Mineral parageneses recording the P-T history of Alpine eclogites in the Tauern Window, Australia: Neues Jahrbuch für Mineralogie Abhandlungen, B. 130, p. 69–77.

Miyashiro, A., 1973, Metamorphism and metamorphic belts: London, George Allen and Unwin Limited, 492 p.

Morgan, B. A., 1970, Petrology and mineralogy of eclogite and garnet amphibolite from Puerto Cabello, Venezuela: Journal of Petrology, v. 11, p. 101–145.

Okrusch, M., Seidel, E., and Davis, E. N., 1978, The assemblage jadeite-quartz in the glaucophane rocks of Sifnos (Cyclades archipelago, Greece): Neues Jahrbuch für Mineralogie Abhandlungen, B. 132, p. 284–308.

Paris, J. P., 1981, La géologie de la Nouvelle-Calédonie, un essai de synthèse: Bureau de Recherches Géologiques et Minières Mémoire 113, 240 p.

Paterson, L. A., 1980, Petrology of high-pressure metamorphic tectonic blocks from northern New Caledonia [BSc Honors thesis]: Auckland, University of Auckland, 62 p.

Podvin, P., Brothers, R. N., and Paris, J. P., in press, Ophiolites of New Caledonia, in Bogdanov, N. A., ed., Ophiolites of continents and comparable oceanic floor rocks: Report of International Geological Correlation Programme Project 39.

Reinsch, D., 1977, High-pressure rocks from Val Chiusella (Sesia-Lanzo Zone, Italian Alps): Neues Jahrbuch für Mineralogie Abhandlungen, B. 130, p. 89–102.

——1979, Glaucophanites and eclogites from Val Chiusella, Sesia-Lanzo Zone (Italian Alps): Contributions to Mineralogy and Petrology, v. 70, p. 257–266.

Routhier, P., 1953, Etude géologique du versant occidental de la Nouvelle-Calédonie entre le Col de Boghen et la pointe d'Arama: Mémoires de la Société Géologique de France, no. 67, p. 1–271.

Smith, D. C., 1971, A tourmaline-bearing eclogite from Sunnmöre, Norway: Norsk Geologisk Tidsskrift, v. 51, p. 141–147.

——1976, The geology of the Vartdal area, Sunnmöre, Norway, and the petrochemistry of the Sunnmöre eclogite suite [Ph.D. thesis]: Aberdeen, University of Aberdeen.

——1980, Highly aluminous sphene (titanite) in natural high pressure hydrous-eclogite-facies rocks from Norway and Italy, and in experimental runs at high pressures. 26ᵉ Congrès Géologique International, Paris, Résumés 02.3.1, p. 145.

Turner, F. J., 1981, Metamorphic petrology (second edition): Washington, New York, London, Hemisphere Publishing Corporation, 524 p.

Velde, B., and Kienast, J. R., 1973, Zonéographie du métamorphisme de la zone de Sezia-Lanzo (Alpes piémontaises): étude des omphacites et grenats des micaschistes éclogitiques à la microsonde électronique: Compte Rendu de l'Académie des Sciences Paris, Série D, p. 1801–1804.

Winkler, H.G.F., 1974, Petrogenesis of metamorphic rocks (third edition): New York, Springer-Verlag Incorporated, 320 p.

Wood, B. J., Holland, T.J.B., Newton, R. C., and Kleppa, O. J., 1980, Thermochemistry of jadeite-diopside pyroxenes: Geochimica et Cosmochimica Acta, v. 44, p. 1363–1371.

Yoder, H. S., and Tilley, C. E., 1962, Origin of basaltic magmas: an experimental study of natural and synthetic rock systems: Journal of Petrology, v. 3, p. 342–532.

MANUSCRIPT ACCEPTED BY THE SOCIETY JULY 29, 1985